Process Operations Safety

Process Operations Safety

The What, Why, and How Behind Safe Petrochemical Plant Operations

M. Darryl Yoes
Safety Consulting International, LLC
Jackson, LA, USA

Published by John Wiley & Sons, Inc., Hoboken, New Jersey.
Published simultaneously in Canada.

Library of Congress Cataloging-in-Publication Data

Names: Yoes, M. Darryl, author.
Title: Process operations safety : the what, why, and how behind safe petrochemical plant operations / M. Darryl Yoes.
Description: First edition. | Hoboken, New Jersey : Wiley, [2025] | Includes index.
Identifiers: LCCN 2024018000 (print) | LCCN 2024018001 (ebook) | ISBN 9781394271931 (hardback) | ISBN 9781394271955 (adobe pdf) | ISBN 9781394271948 (epub)
Subjects: LCSH: Chemical plants–Safety measures. | Petroleum chemicals industry.
Classification: LCC TP155.5 .Y64 2025 (print) | LCC TP155.5 (ebook) | DDC 660/.2804–dc23/eng/20240515
LC record available at https://lccn.loc.gov/2024018000
LC ebook record available at https://lccn.loc.gov/2024018001

Cover Design: Wiley
Cover Image: © kimly/Adobe Stock Photos

Set in 9.5/12.5pt STIXTwoText by Straive, Chennai, India

Contents

About the Author

M. Darryl Yoes is a practical experience refinery supervisor and manager, having spent over 50 years in the refining industry. He began his career as a process apprentice at one of the largest US Gulf Coast refineries. After quickly advancing through several supervisory and management assignments, including first- and second-line supervisor positions in process, mechanical, and technical, he advanced to the position of risk management advisor. He then became the safety, health, and environmental manager of one of the largest petroleum refineries in the world.

He has either led or participated in numerous process safety incident investigations, including those involving the tragic loss of life. He served as emergency coordinator and incident manager in the role of refinery superintendent and has extensive experience in managing other emergency operations, including oil spill response operations. He has also led or participated in numerous process hazards analysis studies and led or participated in operations integrity and safety risk assessments at major US and European petrochemical sites. He has also served as the safety manager for major construction projects at US refineries.

Mr. Yoes continues to provide process safety management consulting for refining and petrochemical plants and for construction safety management for major construction projects across the United States. He is a course leader and facilitator for petrochemical plant operations safety management training designed for refinery and petrochemical plant management, supervision, engineering, and plant operators. He has led comprehensive process safe operations training courses in essentially all regions of the world, including North and South America, Asia, and Europe, sharing lessons learned from industry process safety incidents involving the loss of containment of flammable or toxic materials leading to fire, explosions, and injuries or fatalities. The safe operations training presents practical methods or work practices to help avoid the reoccurrence of these types of incidents. To a large extent, the content of the safe operations training course is reflected in this book.

He is the founder and president of Safety Consulting International, LLC, the developer of this helpful guide to preventing process safety incidents. He is currently employed by EcoScience Resource Group, LLC, in Baton Rouge, a firm specializing in developing process operations training manuals, process operating and environmental procedures, and operator training guides for the petroleum refining and petrochemical industries.

Foreword

There have been, and continue to be, far too many incidents at refining and chemical plant sites worldwide, resulting in fires, explosions, and loss of life and property. Fortunately, catastrophic process incidents resulting in loss of life and the destruction of process equipment generally don't occur regularly at a single site. However, this contributes to the lack of what is commonly called "organizational memory," the memory of what has happened before, and the lack of understanding of the root causes of process incidents. Refining and petrochemical organizations should continually ask, "Do we know what we don't know?"

This book is designed as a reference and training guide for refinery and petrochemical plant operators and front-line supervisors. It presents "what has happened" and "what can happen," discussing how similar incidents can be prevented. It is not designed as a technical or operations manual for site management. This book is also designed as a textbook for universities, community colleges, vocational schools, and other educational facilities to help educate and expand the role of managers, operators, and safety professionals in our industry.

While this book explicitly targets the petroleum and petrochemical industries, other process industries can also benefit greatly from the material presented. For example, if this material is applied, the power generation industry, chemical blenders and processors, and others involved in operating complex operations in the presence of chemical, electrical, or mechanical hazards will experience a safer operating environment and a reduction in the number of near misses and process-related accidents.

Understanding the principles of process safety management and applying lessons learned from previous catastrophic incidents to current operations is critical to ensuring that the incident will not be repeated. What has happened before can happen again unless we learn from the past and from others who have experienced these catastrophic events. We must understand what has occurred and apply the lessons learned to help prevent the same incident from occurring at our facility.

Process safety incidents are not high-frequency events; however, due to the potential for very high consequences, including the loss of life, the destruction of equipment and property, and the impact on the surrounding communities in which we operate, we cannot wait until one of these events occurs to determine if our safety systems are functioning.

Our workers must fully understand the causes of these incidents and what they can do to prevent the incident from occurring at our facility. This leads to another question we should continually ask ourselves: how do we know? How do we know if our safety management systems are working to prevent a significant loss of containment of flammable or toxic material or a major fire or explosion at our facility? Are our safety management systems functioning to ensure the integrity of our equipment, the status and compliance with our operating procedures, the quality and availability of the process safety information, and that process hazards are adequately identified and mitigated during the process hazards analysis?

The tragic accident at the BP Texas City refinery in March of 2005 presented a wealth of process safety "learnings" that should be examined as opportunities for continued improvement implemented at petrochemical sites worldwide. However, as documented on both the US Chemical Safety Board (CSB) and the US Occupational Safety and Health (OSHA) websites, in the relatively brief period since the BP explosion, there has been a continuing series of fires and explosions in our industry. This is also true for the related industries, such as coal mines, natural gas-powered electrical power sites, and offshore oil drilling rigs, which have resulted in significant loss of life and destruction of plants and equipment. Much was learned from the investigation and legal inquiries into the tragedy in the Gulf of Mexico due to the Deepwater Horizon oil rig fire, the resulting 11 fatalities, and the unprecedented environmental impact.

These incidents must be a wake-up call for our industry to place additional focus on process safety. Our industry must apply the same dogged determination to process safety that we have placed on personnel safety over the past several decades. We have made great strides in the improvement of personnel safety. We can have the same results in process safety if emphasis is applied to continuous improvement, to the lessons learned from past incidents, and to the identification and mitigation of process safety hazards.

This book focuses on the operational safety of petroleum refining and petrochemical plants, although many of the "learnings" can be easily adapted to other industries. It is a practical guide written in straightforward and easy-to-understand language so anyone can understand the principles of process safety management. Highly technical terms and complicated formulas are not used, and graphs and other illustrations are designed to be simple and easy to understand. We will explore the causes of incidents with the potential for the "loss of containment" of flammable or toxic materials that can lead to a significant fire, explosion, or release of toxic materials into the environment.

Case studies of actual incidents are presented to help illustrate the key lessons learned. Most importantly, they are reviewed as an opportunity to learn from what has occurred, to determine how a reoccurrence may be prevented, and how the lessons can be incorporated into everyday operations systems and standards. This book is unique from other similar Process Safety books in that it outlines practical solutions to prevent a reoccurrence of the event, including steps that operators, supervisors, and managers can take.

Most chapters include a brief quiz at the end of the chapter. I encourage the reader to read through these questions as a quick review of the key points made in the chapter. The questions are duplicated in Appendix A, along with the answers.

The author has over five decades of experience in the petroleum industry, with operational experience in nearly all aspects of refining operations. Formally a Safety and Environmental Manager at a major US Gulf Coast refinery, he has led process safety training worldwide for petroleum refineries and petrochemical plants for the past 16 years.

It is believed that the information provided in the book will help lead to improved safety performance in the petroleum and petrochemical industry. However, neither Safety Consulting International LLC, the author, M. Darryl Yoes, nor the company producing the documents contained herein warrants or represents, expressly or by implication, the correctness or accuracy of the content or the information presented here. This material is presented to improve operations safety worldwide and is not intended to be used as operating guidance or procedures. Any use of the material contained herein is done with the user accepting legal liability or responsibility for the consequence of its use or misuse.

Safety Consulting International, LLC, the author, or any other company or person as outlined above, make any claim, representation, or warranty, or express or implied, that acting in accordance with this book or its contents will produce any particular results with regard to the subject matter contained herein, or satisfy the requirements of any applicable federal, state or local laws and regulations; and nothing in this document constitutes technical advice. If such advice is required, it should be sought from an attorney, a qualified company, or a qualified individual.

No material known to be copyrighted or proprietary has been intentionally used in this document without the owner's express approval or permission. The author has made extraordinary efforts to contact all owners of images used in this book to secure consent for the use of the photographs. All information presented is either readily available in the public sector or has been approved by the copyright owner. The author has over 50 years of collecting process safety information and those images without attribution were taken by the author. There are several images, where after an extensive search, the original could not be found; these are noted as such.

1

The Guiding Principles of Process Safety Management

If petroleum refining and petrochemical facility management instills the following six Guiding Principles of Process Safety Management in their people as core values, many process safety incidents will be prevented or significantly mitigated. However, I also realize that these six principles sound too easy and that implementation is far more complex. This requires continuous attention to detail, comprehensive procedures, safe work practices, rigorous compliance with those practices, and constant management commitment and support.

The Six Guiding Principles of Process Safety Management:

- Prevent Loss of Containment (of a Flammable or Toxic).
- Minimize Fuel and Air Mixtures.
- Minimize Sources of Ignition.
- Prevent Uncontrolled Exothermic Reactions.
- Ensure Rigorous Field Verification and Audits.
- Maintain a Strong Organizational Culture of Process Safety.

When implemented through the process safety management system and embedded into the company or site culture, the Guiding Principles of Process Safety Management will help prevent the reoccurrence of process safety incidents.

The guiding principles of process safety are considered "Layers of Protection" to help prevent significant site incidents such as explosions or fire. Each barrier should be applied and rigorously enforced as individual critical protection barriers. In the event of the unexpected failure of one barrier, for example, loss of containment of a flammable due to a pump seal failure, the remaining barriers are still available to help prevent the incident from escalating to a significant fire or explosion.

Overview of the Principles of Process Safety

Prevent Loss of Containment (of a Flammable or Toxic)

Loss of containment of flammable or toxic material can lead to an explosion, fire, and harm to people and the environment. These consequences are significantly reduced or eliminated without the accidental or intentional release of flammable or toxic materials. Subsequent chapters of this book will focus on the "how-to" to prevent loss of containment, including the importance of:

- Ensuring safe design for processing or handling flammable or toxic materials, including Inherently Safer Technology where appropriate.
- Establishing and enforcing Safe Operating Limits.
- Ensuring equipment inspection and Follow-up.
- Maintaining Operating Procedures up-to-date and Procedure Compliance.
- Properly managed changes to equipment, procedures, chemicals, roles, responsibilities, etc.
- Ensure effective control of the disablement of critical devices.
- Closely controlled response to "minor" drips/leaks to prevent the occurrence of a catastrophic release.
- Ensure the timely operator response to Alarms.
- Verify the proper process line-up of piping and valves to prevent vessel overfills.

Process Operations Safety: The What, Why, and How Behind Safe Petrochemical Plant Operations, First Edition. M. Darryl Yoes.

Minimize Fuel and Air Mixtures

Allowing fuel (hydrocarbons) to mix with air can result in a flammable mixture leading to a vapor cloud explosion (VCE) and an explosion in a confined space such as a tank or other process vessel. This consequence can be reduced or eliminated by ensuring compliance with equipment preparation procedures, such as rigorously enforced procedures to eliminate air from process equipment before admitting hydrocarbons into the process vessel. Subsequent chapters of this book will focus on procedures and tactics to minimize fuel and air mixtures that have the potential for creating a flammable mixture or atmosphere, such as:

- An improved understanding of the properties and characteristics of hydrocarbons and other related chemicals by Operators, including Flash Temperature, flammable or explosive range, Autoignition Temperature, and Vapor Pressure (Reid and True Vapor Pressure). This includes guidance on where this data can be found for materials associated with their units and operations.
- An improved understanding of the causes, potential, and consequences of VCE or boiling liquid expanding vapor explosion (BLEVE) and guidance for the prevention of these catastrophic events.
- Effective air freeing of process equipment before admitting hydrocarbons during startup or when preparing to return equipment to service.
- Ensuring adequate preparation of process equipment when breaking containment for equipment opening for maintenance or repairs.
- Ensuring safe rundown to atmospheric tanks by strictly applying operating limits or envelopes.

Minimize Sources of Ignition

Uncontrolled sources of ignition in process areas where hydrocarbons are present, although normally contained, can result in ignition, explosion, and fire, should a leak or other loss of containment occur. Ignition sources should be minimized or controlled to only those needed for the process to operate, such as process furnaces or heaters. Subsequent chapters of this book will provide emphasis on the minimization of Sources of Ignition with the discussion of such issues as:

- Rigorous control of open flame or other hot work with the potential for ignition of flammable atmospheres.
- Minimizing access to process areas by motorized equipment or other similar potential ignition sources through the site work permit process.
- Ensuring Operator knowledge of Static Electricity hazards and strict compliance with site static electricity control procedures.
- Application and enforcement of Electrical Hazardous Area Classification.

Prevent Uncontrolled Exothermic Reactions

Undetected and uncontrolled reactions have occurred in the process industry resulting in loss of life, severe injuries, and damage to plants and equipment. Uncontrolled reactions can occur in process equipment such as reactors and other process vessels where hydrocarbons are present in the presence of catalyst, temperature, and hydrogen. For example, an uncontrolled exothermic reaction in a Hydrocracker reactor can quickly exceed the reactor vessel's design temperature and pressure, resulting in a potential explosion. Subsequent chapters of this book will focus on the "how-to" to identify an uncontrolled exothermic reaction and the immediate actions required to stop the reaction, such as:

- Hazards of undetected or uncontrolled exothermic reactions (Chapter 8.28).
- Hazards of chemical incompatibility or reactivity (Chapter 8.31) covers the hazard of inadvertently mixing two or more incompatible chemicals or exposing chemicals to high pressures or temperatures; for example, mixing concentrated sulfuric acid with a strong base such as sodium hydroxide (results in extreme heat and liquid explosion).
- The characteristics and hazards of uncontrolled Pyrophoric materials such as iron sulfide scale are discussed in Chapter 8.21, including the hazards of an exothermic reaction, fire, and severe equipment damage when opening equipment to the atmosphere.

Ensure Rigorous Field Verification and Audits

There must also be a periodic (scheduled) and comprehensive process of field verification and audits to ensure these principles are consistently applied, which former US President Ronald Reagan referred to as "Trust but Verify." Without a comprehensive verification process to ensure ongoing compliance, people can become complacent and lose focus or entirely fall out of compliance with the critical application of these critical work processes. The verification process must be detailed and written, with individual system audits regularly scheduled and carried out, and "findings" should be reported to site management for follow-up and actions.

Chapter 6, "Process Safety Management Systems," includes more detail on the field verification process.

Organizational Culture

This book will review a variety of process safety incidents and the severity of those incidents at refining and petrochemical sites in all regions of the world. We will also cover a significant number of case studies of actual incidents to understand the following:

1. What happened?
2. Why or how did it happen?
3. Most importantly, how do we prevent it from happening again?

As you read through the incident case studies, please consider the organization's culture at the time of the incident.

In most organizations, the mission, values, and objectives are set by senior management and communicated throughout the organization, thereby helping to establish the organizational culture of the organization. A combination of written policies, procedures, rules, and guidelines reinforces this culture. However, organizational culture also includes unwritten rules, practices, and expectations that are well understood as the "way things are done here." Organizational culture reflects what is important to the people who make up the organization and plays an essential role in how employees make decisions.

As you review the upcoming case studies, the following represents some questions to ask related to the organizational culture of the organization:

- What was important to the organization at the time of the event?
- Where is the organization's focus?
- Where are they directing the company's resources?
- What are the employees being rewarded for?
- What is the first question being asked in company meetings or the first topic of discussion?

More on "Organizational Culture"

Organizational Culture is the values and behaviors that contribute to an organization's unique social and psychological environment. They are the written and unwritten rules that drive the organization. The organization's culture includes its expectations and experiences, philosophy, and values that hold it together and is always expressed in its self-image, inner workings, and interactions with the outside world.

The culture of the organization also drives the expectations for performance. It is based on shared and common beliefs, customs, and written and unwritten rules that are considered valid over time, sometimes called corporate culture. The corporate culture is demonstrated by:

- How the organization conducts business, and how the employees and customers are treated.
- The extent to which employees can participate in making decisions or expressing new ideas.
- The healthy flow of communication between the workers and their supervisors and managers. Do the workers feel empowered to discuss or bring up concerns or issues that affect them?
- The healthy flow of authority and information through the organization and its leadership.
- The employee's commitment to the organization's overall objectives and goals.

The culture of the organization affects the organization's overall performance and the attention to things like customer care and service, product quality, safety, attendance and punctuality, and the overall concern for the environment and community.

I believe you will note that in most of the case studies presented in this book, the organizational culture was lacking regarding safety, especially process safety. Our industry says, "You get what you inspect, not what you expect." We can all set high expectations for process safety; however, employees respond when they see the managers setting those expectations and then following up by inspecting (or auditing) for full-field compliance with programs and practices to meet those expectations. For example, the managers expected full compliance with the work permit system on the Piper Alpha offshore oil platform. However, following the devastating explosion and fire on 6 July 1988, which destroyed the platform and resulted in 167 fatalities, the investigators described the work permit system as a "formality" or "check-the-box" exercise. The tragic Piper Alpha incident occurred due to a loss of containment of LPG, resulting in the enormous loss of life and the destruction of the entire platform.

How Does This Happen?

How does a core work process like a work permit system become a formality? Does it happen overnight? It does not happen quickly – it happens over a long period. With a process like work permits, if the managers and supervisors are not looking at the system, if they are not inspecting or auditing the process, if they are not making improvements as deficiencies are found, if they are not continuously striving for continuous improvement, the workers get the impression that this is not all that important after all. The process becomes a "formality" or a "check-the-box exercise," and the system's quality degrades over time. This happened on Piper Alpha and contributed to the tragic accident.

Organizational culture is critical to the organization. You want to ensure you contribute to a sound organizational culture in your company, including a strong focus on process safety.

Additional References

3rd International Conference on Multidisciplinary Research and Practice Page I "A Study on Diverse Parameters of Organizational Culture and Its Impact on Employee's Performance" Dr. Chetna Parmar, Prof. Kalagi Shah*

Organizational Culture vs. Business Results Grzegorz Starybrat Operations and Supply Chain Leader | Strategic Business Development | Organizational Transformation | Change Management | Process and Performance https://www.linkedin.com/pulse/organizational-culture-vs-business-results-grzegorz-starybrat/

Elements of Process Safety Management © Safety Services 2018–2019 A Christian Run and Operated Website https://www.safetyinfo.com/process-safety-management-elements-psm-free-index/

Chapter 1. Guiding Principles of Process Safety Management

End of Chapter Quiz

1 What are the six guiding principles of process safety management?

2 What is meant by "Organizational Culture"?

3 How do we ensure that the process safety management principles are continuously applied?

4 How does a core safety management system degrade to a "formality"?

5 What was one of the contributing factors in the tragic Piper Alpha fire?

6 How can we prevent the reoccurrence of a major process safety incident?

7 Guiding Principles of Process Safety are considered __________ ____ __________ to help prevent significant site incidents such as explosions or fire.

8 This chapter discusses nine practices that can help prevent the loss of a flammable or toxic containment. Can you name at least five?

9 Allowing fuel (hydrocarbons) to mix with air can result in a ___________ ___________ leading to a Vapor Cloud Explosion and an explosion in a confined space such as a tank or other process vessel.

10 Organizational Culture is the set of ________ ____ _________ that contribute to an organization's unique social and psychological environment. They are the written and unwritten rules that drive the organization.

2

Process Safety Background and Federal Regulations

Today, it would be easy to take process safety management for granted, given the standards and laws governing refining and petrochemical plant operations. However, we should keep in mind how we got here. The refining and petrochemical industry experienced numerous process safety failures during the 1970s, 1980s, and 1990s, resulting in the tragic loss of life and significant equipment damage losses. Although there has been a reduction in the number of incidents and lives lost, process safety incidents still occur today. Lessons must be learned and applied to help prevent future catastrophic incidents to avoid further loss of life and human suffering. We must learn from our past and apply the lessons learned the hard way to help prevent a similar event at our facility.

So, what is process safety? Process safety management is concerned with managing safety hazards arising from process operations and is distinct from the management of conventional personnel safety (such as slips, trips, and falls). In a typical personnel safety incident, one person usually has an injury; a process safety incident, such as a catastrophic fire or explosion, may result in many personnel injuries, fatalities, and significant equipment damage or destruction. A significant process safety incident can also impact the community where we live and work. Additionally, process safety incidents impact the company's reputation and affect the ability of the company to continue operations.

Process safety management focuses on preventing "loss of containment" (releases) of flammable or toxic materials into the atmosphere or process vessels or spaces not intended to contain these materials. Proper process safety management includes procedures, facilities, and work processes that minimize the potential for mixtures of flammables and air and control ignition sources near process equipment.

The generally accepted Principles of Process Safety are (discussed in more detail in Chapter 1):

- Prevent the loss of containment of flammable and toxic materials.
- Minimize the potential for uncontrolled mixtures of fuels and air.
- Minimize the potential sources of ignition in process areas.
- Prevent uncontrolled exothermic reactions (typically described as a catalyst temperature runaway).
- Ensure rigorous field verification and audits.
- Maintain a robust organizational culture of process safety.

Good process safety management requires detailed knowledge of the chemical and process hazards associated with the operation of the plant. It also requires an ongoing organizational emphasis or "culture" to ensure that the correct focus and resources are applied to identify and control these hazards. This book will examine several case studies of actual refining and petrochemical incidents; we will explore what has happened or what can happen and the critical lessons learned from incidents.

What Has Happened (Process Safety Incidents – A Partial Listing)

In the 1970s, 1980s, and 1990s, numerous catastrophic process safety incidents occurred in the refining and petrochemical industry in the United States and internationally. Most incidents involved the loss of containment of flammable or toxic materials. Many resulted in the tragic loss of industry workers and citizens in the communities where the incident occurred. In most cases, the facilities were either heavily damaged or destroyed. Thousands of lives have been changed forever due to these catastrophic events.

Appendix B lists process safety incidents that have occurred worldwide and provides examples of some of the industry's catastrophic process safety incidents and the number of lives lost during this period in our history (not intended to be

Process Operations Safety: The What, Why, and How Behind Safe Petrochemical Plant Operations, First Edition. M. Darryl Yoes.

a complete list). Many of these incidents resulted in catastrophic equipment damage, lost production, and significantly impacted the facility's communities. However, the real tragedy is the loss of life and injury to personnel and the number of lives that were changed forever by the event.

US Federal Regulations Governing Process Safety Management

The CAA of 1990 required OSHA and EPA to issue regulations governing process safety.

As a direct result of the CAA and the number of incidents and the resulting impact on people, communities, and facilities, the OSHA developed the process management rule for facilities that handle or process highly hazardous chemicals. This rule was promulgated in 1992 (29 CFR 1910.119 – "Process Safety Management of Highly Hazardous Chemicals").

In what appears to be an almost parallel path, and as required by Section 112(r) of the CAA Amendments, the United States EPA enacted similar regulations when the EPA Risk Management Program (RMP) regulations (Title 40 CFR Part 68) were promulgated into law. Requirements for the EPA RMP very closely mirror those of OSHA PSM. The EPA must publish rules and guidance for chemical accident prevention at facilities that use certain hazardous substances.

OSHA PSM (29CFR 1910.119)

On 24 February 1992, the federal OSHA promulgated the final rule for process safety management (PSM) (Title 29 of CFR Section 1910.119). This was a milestone in managing process safety in the United States. The PSM regulation contains requirements for managing hazards associated with processes using "highly hazardous chemicals." It mainly applies to manufacturing industries, particularly those pertaining to chemicals, transportation equipment, and fabricated metal products. The standard includes special provisions for contractors working in "covered" facilities. In each industry, PSM applies to those companies that manufacture or store more than 130 specific toxic and reactive chemicals in listed quantities (the quantity threshold is listed in the standard). The standard applies to inventories of flammable liquids and gases in amounts of 10,000 pounds (4,535.9 kilogram) or more.

This OSHA PSM standard contains requirements for preventing or minimizing the consequences of catastrophic releases of chemicals that are:

- Toxic
- Reactive
- Flammable
- Explosive

The PSM rule provides specific and detailed regulations for the following 14 process safety elements:

1. Employee Participation	8. Mechanical Integrity
2. Process Safety Information	9. Hot Work Permit
3. Process Hazard Analysis	10. Management of Change
4. Operating Procedures	11. Incident Investigation
5. Training	12. Emergency Planning and Response
6. Contractors	13. Compliance Audits
7. Pre-startup safety review	14. Trade Secrets

OSHA PSM elements are interdependent and supportive of the other PSM elements. All elements are required, essential components of the process safety management of a site or facility and related to other elements.

Requirements and guidance for requirements of each element are spelled out in the regulation.

OSHA bulletin 3132 provides additional details on the development and requirements of the PSM standard and helps with the interpretation of the standard. State plans approved under section 18(b) of the Occupational Safety and Health Act of 1970 must adopt standards and enforce requirements that are as effective as the federal requirements.

Environmental Protection Agency Risk Management Plan (40CFR, Part 68)

Similarly, the EPA enacted regulations on 24 May 1996, with the EPA RMP under the CAA regulation (CAA Section 112(r), Title 40 CFR Part 68). The requirements for the EPA RMP closely mirror those of OSHA PSM but include a more focused approach to the community's "right to know" and the facility's communication and interaction with the local community and emergency response groups. For example, the 14 OSHA PSM elements were expanded to address an incident or release's off-site effects and require the facility to develop plans for a "worst-case" incident and an alternate release scenario. The RMP requires the facility to develop and communicate information about the hazardous materials stored or processed at the site and how those hazards are managed, including the response plans for the two incident scenarios.

The following is directly from the EPA website (epa.gov):

Section 112(r) of the CAA Amendments requires EPA to publish regulations and guidance for chemical accident prevention at facilities that use certain hazardous substances. The RMP rule contains these regulations and guidance. The RMP rule requires facilities that use extremely hazardous substances to develop an RMP, which:

- Identifies the potential effects of a chemical accident,
- Identifies steps the facility is taking to prevent an accident, and
- Spells out emergency response procedures, should an accident occur.

These plans provide valuable information to local fire, police, and emergency response personnel to prepare for and respond to chemical emergencies in their community. Making RMPs available to the public also fosters communication and awareness to improve accident prevention and emergency response practices at the local level.

The RMP rule was built upon existing industry codes and standards. It requires facilities that use listed regulated Toxic or Flammable Substances for Accidental Release Prevention to develop an RMP and submit that plan to the EPA.

Facilities holding more than a threshold quantity of a regulated substance in a process are required to comply with EPA's Risk Management Program regulations.

The regulations require owners or operators of covered facilities to implement a risk management program and submit an RMP to the EPA.

Each facility's program should address three areas:

- Hazard assessment details the potential effects of an accidental release, an accident history of the last five years, and an evaluation of worst-case and alternative accidental releases.
- A prevention program that includes safety precautions and maintenance, monitoring, and employee training measures, and
- An emergency response program spells out emergency health care, employee training measures, and procedures for informing the public and response agencies (e.g., the fire department), should an accident occur.

The plans are revised and resubmitted to EPA every five years.

Additional assistance and guidance are available on the EPA website (www.epa.gov).

US Chemical Safety and Hazard Investigation Board

A significant step forward in addressing process safety occurred in the United States in 1990 when the US Chemical Safety and Hazard Investigation Board was authorized by the CAA Amendments of 1990 as an independent investigation agency for process safety incidents.

The following was extracted directly from the CSB website (csb.gov):

"The U.S. Chemical Safety and Hazard Investigation Board (CSB) is authorized by the Clean Air Act Amendments of 1990 and became operational in January 1998." The Senate legislative history states: "The principal role of the new CSB is to investigate accidents to determine the conditions and circumstances which led up to the event and to identify the cause or causes so that similar events might be prevented." Congress gave the CSB a unique statutory mission and provided in law that no other agency or executive branch official may direct the activities of the Board. Following the successful model of the National Transportation Safety Board and the Department of Transportation, congress directed that the CSB's investigative function be completely independent of the rulemaking, inspection, and enforcement authorities of EPA and OSHA. Congress recognized that Board investigations would identify chemical hazards that those agencies did not address. The legislative history states:

> "The investigations conducted by agencies with dual responsibilities tend to focus on violations of existing rules as the cause of the accident almost to the exclusion of other contributing factors for which no enforcement or compliance actions can be taken. The purpose of an accident investigation (as authorized here) is to determine the cause or causes of an accident and whether those causes violated any current and enforceable requirement".

"Although the Board was created to function independently, it collaborates with EPA, OSHA, and other agencies in important ways. The Board has entered into several memorandums of understanding (MOUs) that define the terms of collaboration. For example, in cases where several agencies conduct investigations of an accident, the MOUs outline mechanisms for coordination in the field. The goal of the MOUs is to allow each agency to carry out its statutory mission efficiently and without unnecessary duplication of effort."

The following are excerpts from the Chemical Safety and Hazard Investigation Board website (csb.gov) and explain the role/mission of the Chemical Safety Board:

> "The CSB is an independent federal agency charged with investigating industrial chemical accidents. Headquartered in Washington, DC, the agency's board members are appointed by the President and confirmed by the Senate. . . ."
> "The CSB conducts root cause investigations of chemical accidents at fixed industrial facilities..."
>
> "The agency does not issue fines or citations but does make recommendations to plants, regulatory agencies such as the Occupational Safety and Health Administration (OSHA) and the Environmental Protection Agency (EPA), industry organizations, and labor groups. Congress designed the CSB to be non-regulatory and independent of other agencies so that its investigations might, where appropriate, review the effectiveness of regulations and regulatory enforcement".
>
> "Both accident and hazard investigations lead to new safety recommendations, which are the Board's principal tool for achieving positive Change. Recommendations are issued to government agencies, companies, trade associations, labor unions, and other groups. Implementation of each safety recommendation is tracked and monitored by CSB staff. When recommended actions have been completed satisfactorily, the recommendation may be closed by a Board vote".

The CSB website is a ready resource for outstanding process safety training materials with ready access to process safety incident information, including:

- Periodic updates on ongoing incident investigations.
- Comprehensive incident reports for completed investigations include photos, schematics, technical data (when appropriate), and recommendations to prevent reoccurrence.
- Outstanding process safety videos or clips typically include animations to help explain what happened or contributed to the incident. In many incidents, the videos also have CCTV footage of the incident.

The CSB videos are of excellent quality and explain what happened, how it happened, and recommendations to help prevent it from happening again. These are great training tools for shift teams and others for process safety. I frequently use the CSB videos in the Safe Operations Training courses for Refining and Chemical Plants worldwide. They are always very well received.

Additional References

OSHA Bulletin 3132 "Process Safety Management" https://www.osha.gov/sites/default/files/publications/osha3132.pdf

OSHA Booklet 3918 Process Safety Management for Petroleum Refineries https://www.osha.gov/sites/default/files/publications/OSHA3918.pdf

OSHA Process Safety Management Standard (29CFR 1910.119) PSM Regulation https://www.osha.gov/laws-regs/regulations/standardnumber/1910/1910.119

Environmental Protection Agency (40CFR Part 68) "Risk Management Plan" https://www.epa.gov/rmp

US Chemical Safety Board web site https://www.csb.gov/

Quick Summary of Refining and Petrochemical Plant Incidents Article "A Deadly Industry" by EHS Today https://www.ehstoday.com/safety/article/21916835/a-deadly-industry

Chapter 2. Process Safety Background and Federal Regulations

End of Chapter Quiz

1 Name the guiding regulation for Process Safety Management.

2 Which federal agency is charged with investigating significant process safety incidents?

3 What is the most common cause of process safety events?

4 Besides the employees of a petrochemical facility, who may be affected by a large process safety incident at a facility?

5 What valuable training resources does the US Chemical Safety Board make available free of charge?

6 What other federal regulation is very similar to the Process Safety Management regulation?

7 Process safety management focuses on preventing "______ __ _________" (releases) of flammable or toxic materials into the atmosphere or process vessels or spaces not intended to contain these materials.

8 Good process safety management requires detailed knowledge of the _________ ___ __________ hazards associated with the operation of the plant.

9 The Clean Air Act (CAA) of 1990 required OSHA and EPA to issue regulations governing process safety. Note: Select the correct answer:
a. EPA and NFPA
b. OSHA and EPA
c. DOT and USCSB
d. NTSB and OSHA

10 The United States Occupational Safety and Health Administration (OSHA) developed the Process Management Rule for facilities that handle or process ________ ________ chemicals. This rule was titled 29 CFR ______.____.

11 Where can the quantity threshold for a specific toxic chemical be found?

12 How many different process safety elements are covered in the OSHA PSM rule?

13 A significant step forward in addressing Process Safety occurred in the US in 1990 when the US Chemical Safety and Hazard Investigation Board was authorized by the Clean Air Act Amendments of 1990 to function as an _________ _________ agency for process safety incidents.

14 The CSB website is a ready resource for outstanding process safety training materials with ready access to ________ _________ safety incident information.

15 Both _________ ____ _______ investigations lead to new safety recommendations, which are the Board's principal tool for achieving positive change.

3

Properties of Hydrocarbons (Fire and Explosion Hazards)

When working around or with hydrocarbons, whether on an oil platform, oil refinery, or petrochemical plant, it is essential to be familiar with the characteristics and hazards of the hydrocarbons and other chemicals with which we may be working. Knowledge of the hydrocarbon characteristics can be very helpful daily, especially when confronted with an incident such as a loss of containment or fire. Properties such as flash point, vapor pressure, vapor density, flammable range, autoignition temperature, and boiling point are critical to understanding the materials or products' hazards and safe handling requirements.

We are familiar with the conventional fire triangle shown in Figure 3.1. There must be fuel, air (or oxygen), heat (source of ignition), and the resulting chain reaction to start a fire. To extinguish the fire, we must remove one side of the triangle by removing the fuel source, the air or oxygen, or cooling the fire to eliminate the heat of combustion, breaking the chain reaction.

Flammable Range

Where hydrocarbons are the fuel source, the fire triangle becomes a bit more complicated due to the properties of hydrocarbons. This is illustrated in Figure 3.2. The hydrocarbons must be in the proper fuel-to-air ratio for ignition to occur. If hydrocarbons are present as a fuel source, but the concentration or mixture of hydrocarbons in the air is too low (the fuel-to-air mixture contains too few hydrocarbons), a fire will not occur. In this case, the mixture is too lean to support combustion or start a fire. Likewise, ignition will not occur if the concentration or mixture of hydrocarbon in the air is too high or too rich (the fuel-to-air mixture contains too much hydrocarbon). The mixture must be within the flammable range for ignition to occur. This chapter discusses the flammable range and other properties of hydrocarbons relative to fires and explosions.

If the fuel source is a hydrocarbon, the fuel and air mixture must be within the flammable range for a fire to occur.

The mixture of fuel and air must be within the flammable range, or fire will not occur. As illustrated here, for butane, this would be between 1.8% and 8.4% butane vapor in the air.

Some hydrocarbons have a wide flammable range, increasing the probability of a fire or explosion when released into the atmosphere where ignition sources may be present. For example, the flammable range for hydrogen is very wide, from 4% LEL to 75% LEL; acetylene has an even wider flammable range, from about 2% LEL to essentially 100% LEL (as per most Material Safety Data Sheets or MSDS).

As indicated above, the Explosive Range can be found on the MSDS for most hydrocarbons. You will note in Figure 3.5 that the flammable range for several products and intermediates found in a typical refinery or chemical plant is illustrated. The wider the flammable range, the higher the probability of the material being in the flammable range when released into the environment where ignition sources may be present. Therefore, the more hazardous the material becomes from a fire or explosion perspective.

Flash Point

Flash point is the lowest temperature at which a liquid gives off enough vapor to form a flammable mixture in air. The National Fire Protection Association (NFPA) has established NFPA 30 ("Flammable and Combustible Liquids Code") as the standard for determining if a material is flammable or combustible. The code specifies that material with a flash point

Process Operations Safety: The What, Why, and How Behind Safe Petrochemical Plant Operations, First Edition. M. Darryl Yoes.

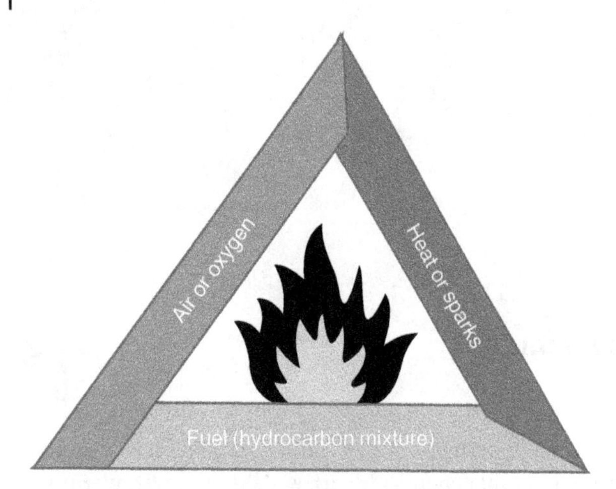

Figure 3.1 The Fire Triangle.

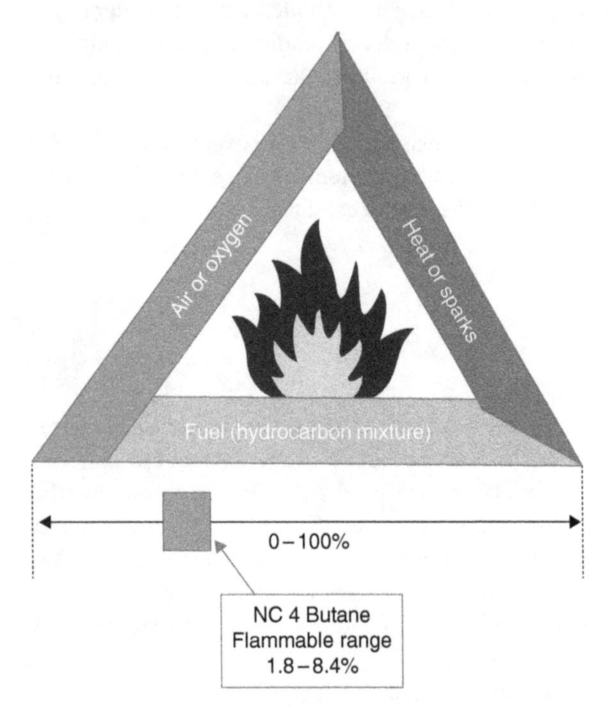

Figure 3.2 The significance of flammable range (% lower explosive limit [LEL]). The example below indicates the significance of the flammable range. An example is N-butane, with a 1.8% lower explosive limit (LEL) to 8.4% upper explosive limit (UEL). When between the LEL and UEL, the mixture is in the flammable range. Notice how this relatively narrow flammable range affects the flammability.

below 100 °F (37.8 °C) is flammable, while a flash point of 100 °F or above is considered combustible. This standard has been adopted in most other countries.

Appendix B to OSHA CFR 1910.1200 defines flash point as four distinct classifications, from Category 1, the most hazardous, to Category 4. This is illustrated in the following table extracted from the regulation. Note that Categories 1 and 2 use a combination of flash point and initial boiling point (IBT) to categorize flammable liquids. However, most safety professionals use the Standard established by the NFPA as the definition for flash point. OSHA also enforces the NFPA standard. (See Chapter 8.31: for additional discussion.)

From OSHA CFR 1910.1200 Appendix B: "A flammable liquid shall be classified in one of four categories in accordance with Table 3.1."

Flash point is an indication of the volatility of the product and one of the main characteristics when considering the storage of the product. Products with a low flash point will be more volatile and susceptible to generating flammable vapors. Therefore, these products are significantly more hazardous when fire or explosion concerns are considered during handling and storage. For example, gasoline has a flash point of around −36 °F and is highly flammable. However, the typical flash point for diesel fuel is around 125 °F (52 °C) and is relatively stable at ambient temperatures. However, if the temperature of the product is increased to above its flash point, the material will readily ignite in the presence of a flame or other arcing or sparking device. The flash point of a specific material can be found on the product's Material Safety Data Sheet (MSDS).

Table 3.1 Criteria for Flammable Liquids

Category	Criteria
1	Flash point < 23 °C (73.4 °F) and initial boiling point ≤ 35 °C (95 °F).
2	Flash point < 23 °C (73.4 °F) and initial boiling point > 35 °C (95 °F).
3	Flash point ≥ 23 °C (73.4 °F) and ≤ 60 °C (140 °F).
4	Flash point > 60 °C (140 °F) and ≤ 93 °C (199.4 °F).

The machines are enclosed when the flash point test is performed in the laboratory. The result is reported even though the operator cannot see the test being run. However, most laboratories can still run a "manual" flash where they slowly heat a small beaker of the product while observing an electric arc in the space just above the liquid. At the same time, the lab technician closely monitors the liquid's temperature. When they see a "pop" flash (not necessarily a continuous flame) in the vapor space, the liquid temperature is recorded, which is the flash point of the product being tested.

In the field, it is important to note that the ignition source need not be an open flame but could also be an electrical arc or the surface of a steam pipe or other hot process equipment. Lower flash flammable liquids can easily be ignited when released into the atmosphere when ignition sources are present. Liquids with a higher flash point (combustibles) are less susceptible to ignition but will ignite if heated above the product's flash point in the presence of an ignition source, for example, by coming into contact with hot equipment where the product is heated above its flash point or due to another ignition source such as hot work being done nearby.

Significant "Learning" Related to Flash Points

- It is important to remember that a product will generate flammable vapors and can readily ignite when heated above its flash point. For example, even though diesel is relatively stable at ambient temperatures, heating diesel above its flash point of around 125 °F (52 °C) can result in a fire or explosion.
- A gas test for LEL may not indicate the presence of trapped liquid hydrocarbons (liquids at a temperature below their flash point). When gas testing equipment is used for opening or hot work, the gas test instrument tests the percentage of the LEL. A zero reading does not necessarily indicate that the vessel or equipment is hydrocarbon-free.
  For example, a vessel containing a heavier hydrocarbon (a hydrocarbon with a flash point above 100 °F such as diesel or gas oil) at a temperature below its flash point may not be vaporizing. It, therefore, will not read on the gas test instrument. Should this product be heated above its flash point, for example, by a hot stream going to the tank or by hot work being performed on or near the equipment, it may readily ignite or explode.
- Hydrocarbons released in fine mists or aerosols act as a gaseous material and may ignite at temperatures significantly above their flash points. For example, lube oils with very high flash points may behave like a low-flash material when released into the atmosphere in an aerosol or mist.
- Remember that an ignition source need not be an open flame but could also be, for example, an electrical arc or the surface of a steam pipe or other hot process line or equipment.

Figure 3.3 illustrates the relative hazards when handling or storing hydrocarbons with a given flash point.

Typical flash points for common hydrocarbon products are provided in Figure 3.4.

Hazard	Flash point	
Very low hazard	Flash point > 200°F	Combustible
Moderate to low hazard	Flash point 150–200°F	
High to moderate hazard	Flash point 100–150°F	Flammable
Extreme to high hazard	Flash point 0–100°F	
Extreme hazard	Flash point < 0°F	

Figure 3.3 Relative hazards of hydrocarbons by flash point. Source: EPA Cameo Chemicals and Product MSD Sheets.

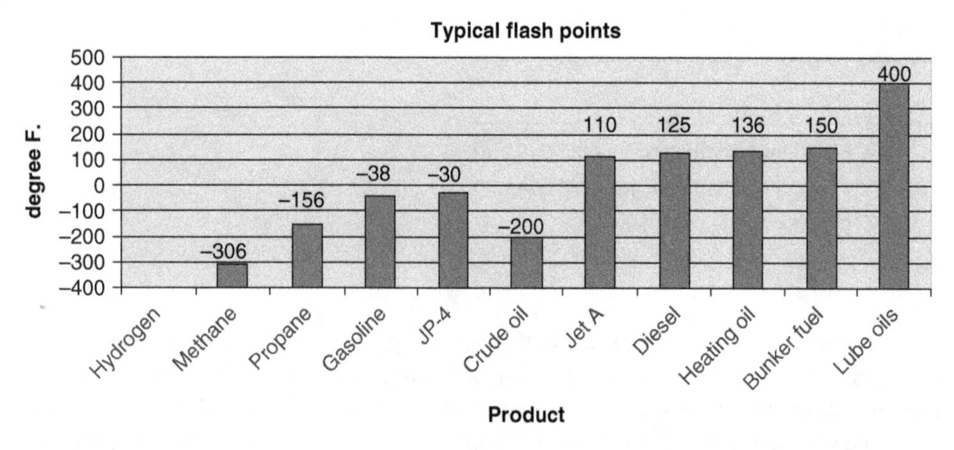

Figure 3.4 Typical flash points (°F). Source: EPA Cameo Chemicals and Product MSD Sheets.

More on the Flammable Range or Flammable Limits (LEL or UEL) (Sometimes Referred to as the Explosive Range)

Flammable limits (or flammable range) are the range of hydrocarbon vapors in a mixture with air (by percent volume) that will ignite when in the presence of an ignition source (such as a spark, static electricity, flame, or hot surface). To support combustion, the mixture of fuel-in-air must be within the fuel-in-air range for that hydrocarbon or mixture of hydrocarbons. If the mixture contains too few hydrocarbons to support ignition, it is below the Lower Explosive Limit (LEL) and is too lean to ignite. Should the concentration or mixture of hydrocarbon vapors contain too much hydrocarbon to burn, it will be above the Upper Explosive Limit (UEL), and the mixture will be too rich to ignite.

The LEL is a key indicator when determining the hazards of hydrocarbons. For a hydrocarbon, the lower the LEL, the less vapor is needed in the air before igniting. When performing gas testing before conducting hot work, we are testing for the presence of hydrocarbons by testing the percentage of LEL in the work area.

Most hydrocarbons have a flammability range of about 1–10% (ratio of vapor in the air). For example, the flammability range of gasoline is typically about 1.4–7.4%. The mixture is too lean to burn for typical gasoline blends with a ratio of less than 1.4% gasoline vapor in the air. Above 7.4%, the mixture will be too rich to burn. Remember these are typical; always verify data on the product's Material Safety Data Sheet (MSDS).

It is important to note that some gases have a wide range of flammable concentrations, such as hydrogen, ether, ethylene, acetylene, carbon monoxide, and hydrogen sulfide. Since a very small amount of vapor is needed for ignition, these materials may find an ignition source a significant distance from the release point. The flammable limits or flammable range (LEL and UEL) can also be found on the product's MSDS.

Typical flammable ranges for some common hydrocarbon streams/products are provided in Figure 3.5. Note the wide flammable range for hydrogen, ethylene (C2″), hydrogen sulfide, carbon monoxide, and acetylene. This is important to ensure the equipment is completely free of these materials when preparing equipment for entry or hot work when the equipment previously contained these materials.

This is equally important when air-freeing equipment containing materials with a wide flammable range before returning to service. Operating procedures should require equipment purging to remove air, and oxygen levels should be confirmed by gas testing for O_2 before introducing hydrocarbons. Most sites require less than 1% oxygen before introducing hydrocarbons, even lower where hydrogen may be present.

It should also be recognized that using acetylene for conducting hot work in confined spaces requires very rigorous control due to the wide flammable range. Due to these hazards, some sites have replaced acetylene with other blended gases for welding.

For example, Figure 3.5 indicates the flammable range for typical hydrocarbons found in a refinery or petrochemical plant.

Note: The number in the bar represents the "width" of the flammable range. For example, H_2 has a UEL of 75% and an LEL of 4% (75% – 4% = 71%). H_2 has a total flammable range of 71%; propane has a total flammable range of 8% (10% UEL – 2% LEL = 8%).

The wider the flammable range, the more likely the vapor cloud will be within the flammable range during a release.

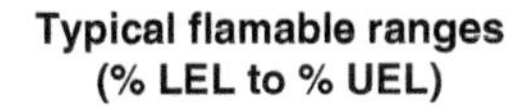

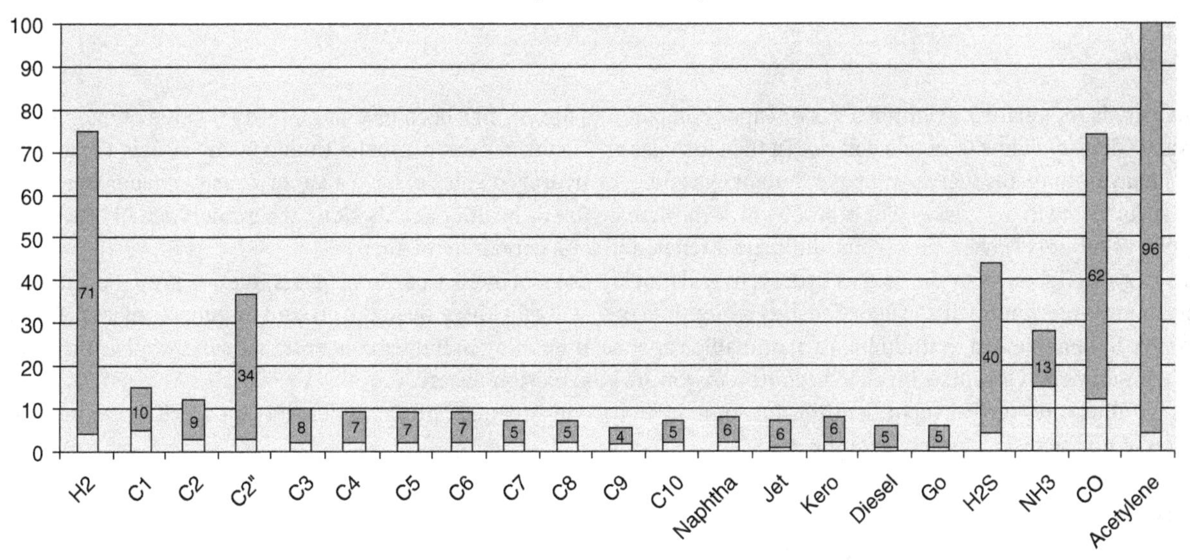

Figure 3.5 Flammable range for typical hydrocarbons. Source: EPA Cameo Chemicals and Product MSD Sheets.

Autoignition Temperature

The autoignition temperature is when hydrocarbon vapors spontaneously ignite without an external ignition source. Many streams on operating units operate at temperatures well above their autoignition temperature. This means that these streams are hot enough that in the event of a release or loss of containment, these hydrocarbons will ignite immediately in the presence of air and without a source of ignition ("autoignition").

For most hydrocarbons, this temperature is approximately 480 °F (248 °C) to approximately 600 °F (315 °C). Figure 3.6 illustrates typical autoignition temperatures for several refinery hydrocarbon streams. The autoignition temperature for a specific product or stream can also be found on the product MSDS.

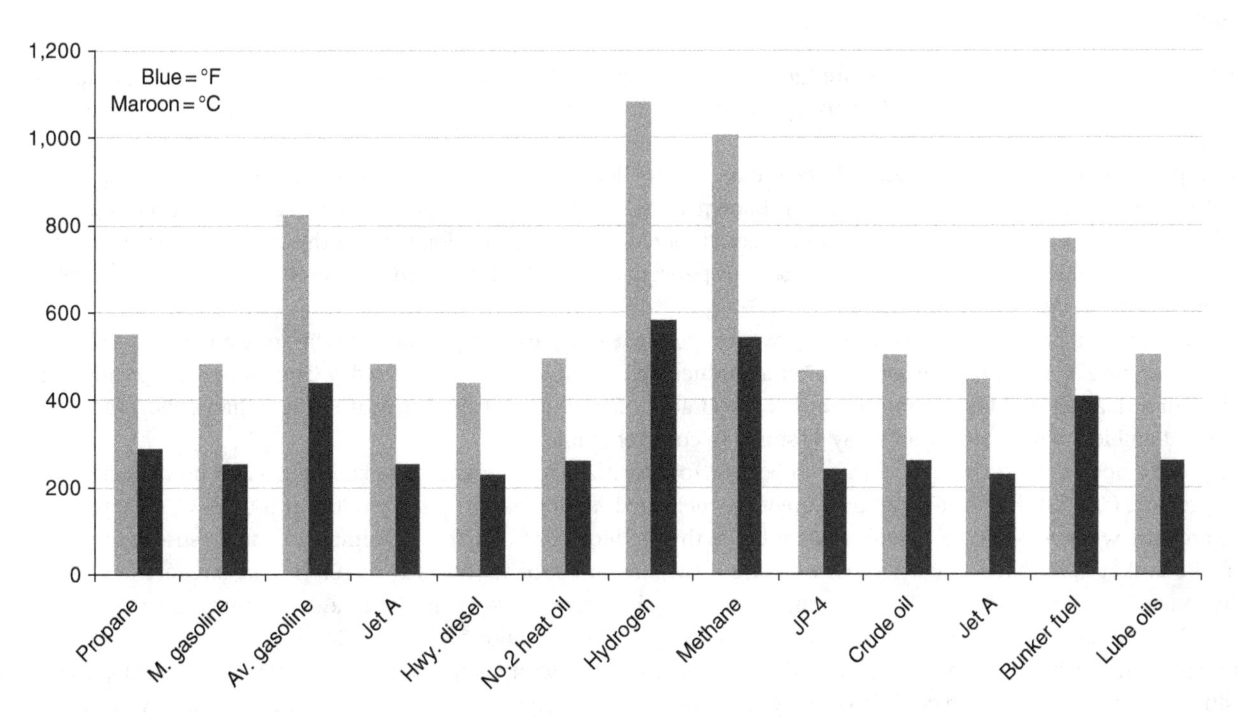

Figure 3.6 Typical product autoignition temperatures (°F). Source: EPA Cameo Chemicals and Product MSD Sheets.

Figures 3.7 and 3.8 indicate the typical Flash and Autoignition data for specific petroleum refinery products or streams and similar data for a typical petrochemical plant.

Vapor Density

Vapor density is the relative weight of a gas or vapor compared to air. Air has been given an arbitrary value of one. A gas with a vapor density of less than one will rise in the air. A gas with a vapor density greater than one will sink in the air.

Vapors heavier than air, such as propane, butane, gasoline, or hydrogen sulfide, have a vapor density greater than one and can accumulate in low places. For example, they may create fire or health hazards along the ground and in other low places such as refinery sewers, pump pits, drainage ditches, and tank containment areas.

Vapors or gases lighter than air, such as hydrogen, will quickly rise when released. Some gases, such as propane, butane, gasoline, or any other gases with a vapor density greater than one, will quickly settle to the ground or into low places. These are also very flammable and result in highly flammable vapor accumulating in higher concentrations in low places during a loss of containment. Of course, this will become a major fire or explosion hazard.

Figure 3.9 illustrates typical vapor densities for some hydrocarbon streams. The vapor density of a specific product can also be found on the product safety data sheet.

Boiling Point

The boiling point is the temperature at which a liquid will boil (measured in degrees F at sea level). For example, water at sea level and atmospheric pressure will boil at 212 °F (100 °C). Increasing the pressure will increase the boiling point; decreasing the pressure (i.e., in a vacuum) will reduce the boiling point. Propane has a boiling point of −44 °F (−42 °C); therefore, propane is a gas at ambient temperature and atmospheric pressure. However, propane will become liquid under pressure, such as during storage in a sphere, a tank car, or another pressurized container. Compounds or blended products generally have a boiling point specified as a range reflecting the characteristics of the blended components. For example, gasoline may have a higher percentage of lighter components such as propane (C3s) and butane (C4s) during winter and, therefore, a lower boiling point.

A low boiling point means the substance will be in gas form at room temperature unless it is pressurized. Flammable materials with a low boiling point generally present a more significant fire hazard. The boiling point can also be found on the product MSDS.

Vapor Pressure

Vapor pressure is a measure of the pressure the vapor exerts on the atmosphere or the container, such as a floating roof tank or the gasoline tank in your vehicle. Vapor pressure indicates how easily a liquid will evaporate, releasing hydrocarbon vapors.

Vapor pressure is typically measured with the product at 100 °F (37.7 °C), with the result expressed in pounds per square inch atmospheric (PSIA) (or BAR or KPA). This is known as the Reid vapor pressure (RVP) (vapor pressure at 100 °F). Some MSDS may express the vapor pressure in millimeters of mercury (mm Hg). Understanding the vapor pressure relative to the normal atmospheric pressure of about 14.7 pounds per square inch absolute (PSIA), or about 760 mm Hg (at sea level, lower for higher elevations), is important.

Liquids with a vapor pressure above 14.7 PSIA will evaporate readily, releasing vapors into the atmosphere. For example, gasoline may have a vapor pressure above 14.7 and produce significant vapors on a warm day. Therefore, we typically store gasoline in a floating roof tank to minimize this evaporation. Diesel has a vapor pressure of less than 1 PSIA, does not vaporize at ambient temperatures, and may be stored in cone roof tanks.

For process operations personnel, it is also important to recognize the difference between Reid vapor pressure and true vapor pressure (TVP). While Reid vapor pressure is the measured vapor pressure of the product at 100 °F (37.7 °C), the TVP is the pressure when tested at the actual temperature of the product. For example, a gasoline component stream produced to off-site tankage will have a higher vapor pressure if the stream temperature is above 100 °F. The stream's amount of "light" material can also affect TVP. For example, allowing the operating temperature of a debutanizer tower to go low can result in lighter material in the tower bottoms stream to tankage by allowing C3s or C4s to go out with the bottoms, significantly increasing the vapor pressure of the stream. A product with a vapor pressure greater than 13 PSIA TVP is classified as liquefied petroleum gas (LPG) and requires pressurized storage such as spheres, horizontal storage vessels, or other pressurized storage.

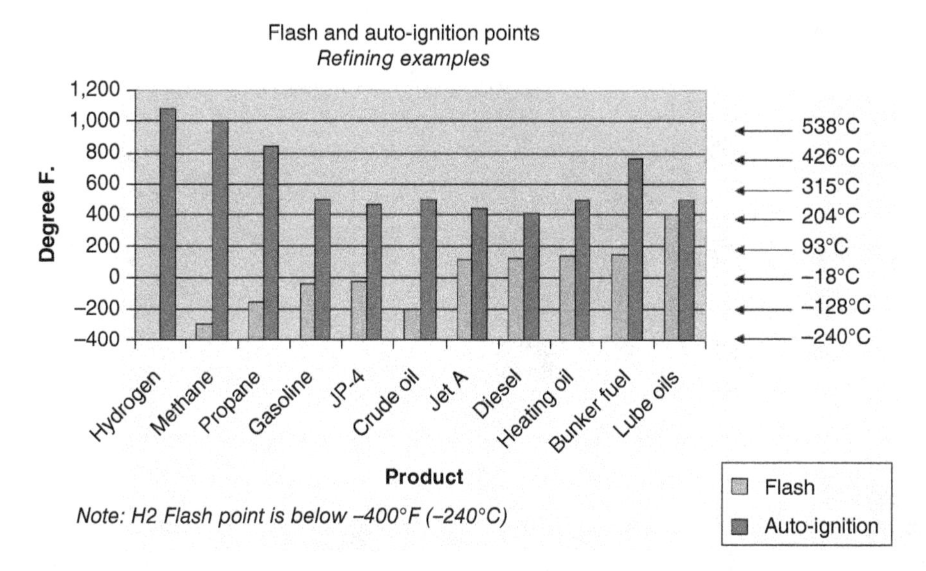

Figure 3.7 Typical flash point and autoignition temperature data for petroleum refinery products or streams. Source: CAMEO & Product MSDS.

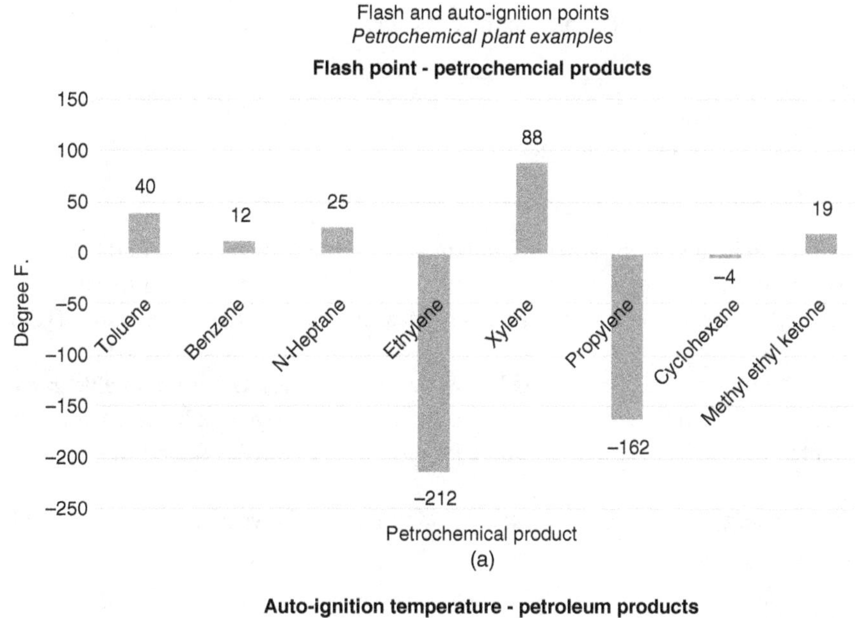

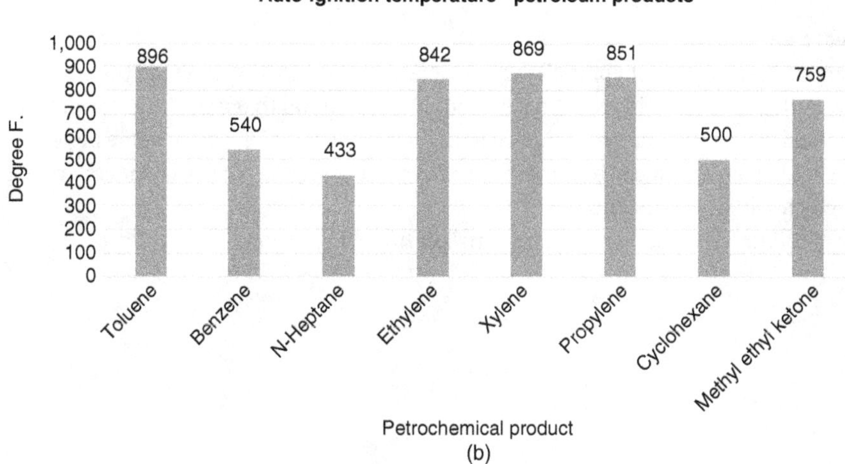

Figure 3.8 Figure (a) illustrates the typical flash point data, Figure (b) illustrates typical autoignition data, both for typical petrochemical plant streams or products. The data source was CAMEO and the Product MSDS.

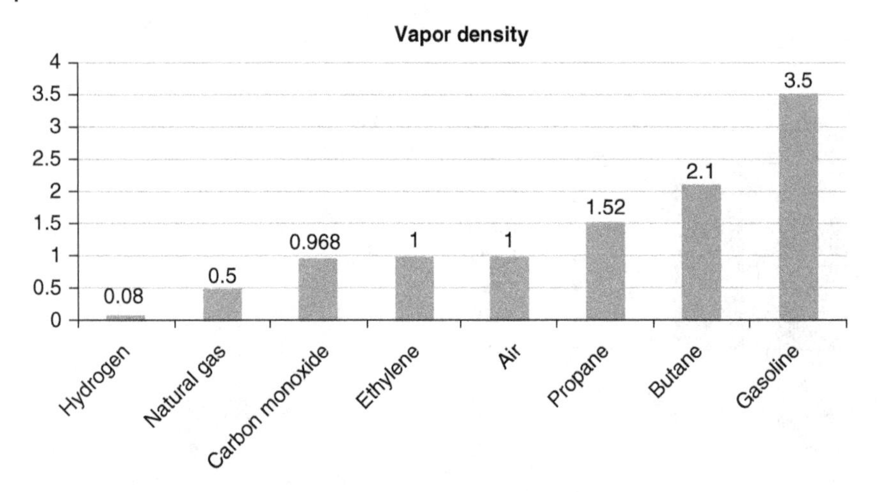

Figure 3.9 Vapor density. Source: Product MSD Sheets.

Minimum Ignition Energy

The minimum amount of energy required for the combustion of a hydrocarbon is the minimum ignition energy (MIE). Different fuel/air mixtures will have different MIE requirements. Some of the critical factors that affect the MIE are:

- The temperature of the hydrocarbon mixture.
- The total amount of energy that is available for ignition.
- The rate at which energy is applied or the period over which it is delivered.
- The surface area over which the energy is delivered.

The MIE values are usually the energy required to ignite the most reactive mixture of fuel and air, generally in the middle of the flammable range. A flammable mixture close to the upper or lower limit may require more energy than specified for the product. The electrostatic discharge energy necessary for ignition varies significantly over the flammable range (LEL to UEL).

A hydrogen–air mixture's MIE is only 0.019 millijoule (and may be as low as 0.004 millijoule). Other flammable gases such as gasoline, methane, ethane, propane, butane, and benzene are usually 0.1 millijoule (as reported by Lewis and von Elbe – see References at the end of this chapter) and as high as 0.48 millijoule for propane. The MIE is sometimes (but not always) found on the MSDS.

Figure 3.10 illustrates the minimum ignition energy to ignite a flammable fuel/air mixture relative to the ratio of fuel and air in the mix.

Properties of Flammable Liquids that Increase the Risk

In a product spill or release, the properties outlined in this chapter all come together and determine how the product will react to the environment. Lighter, more volatile materials will result in a spreading vapor cloud that will "hug" the ground and can quickly result in a massive vapor cloud explosion when reaching an ignition source. Heavier, combustible materials will be less susceptible to forming a vapor cloud and may instead become more of an environmental concern. Understanding the characteristics of the materials with which we work is essential.

In most cases, the following properties of flammable liquids increase the risk of fire or explosion:

- Lower explosive limit (lel).
- Wider flammable range.
- Lower flash point (flammable.)
- Lower autoignition temperature.
- Lower ignition energy.
- Higher Vapor Pressure (RVP or TVP).

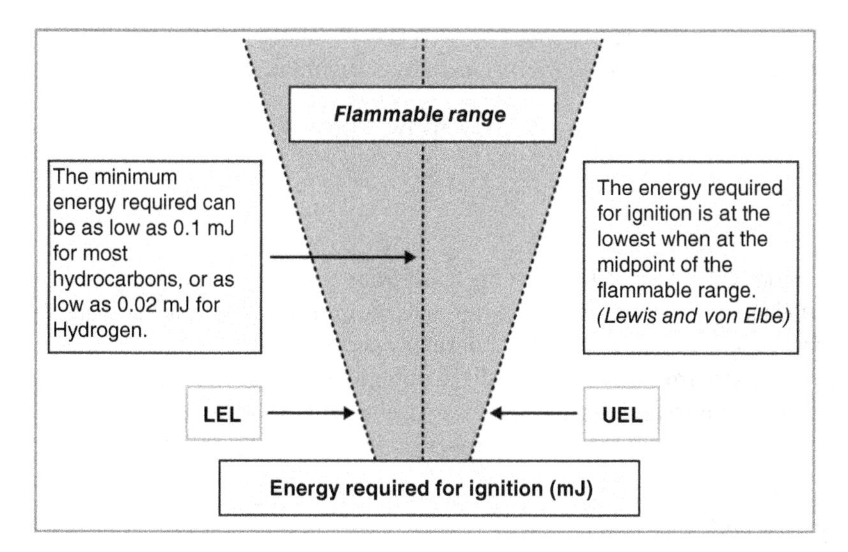

Figure 3.10 Minimum ignition energy. Ignition energy required relative to ratio of fuel/air mixture. Source: Image courtesy of Ecosciene Resource Group Visual Artists.

Safety Data Sheets

OSHA requires safety data sheets to communicate hazards to those affected when working with or near the chemicals. They can help determine appropriate personnel protective equipment (PPE) and other hazard prevention or mitigation and safely perform work with hazardous chemicals.

The US Environmental Protection Agency (EPA) has also developed and published the CAMEO Chemical Database, an excellent reference for MSDS and other emergency response information that can be easily downloaded to your computer.

The product safety data outlined in this chapter for products and other intermediate hydrocarbon process streams are readily available in the product Safety Data Sheets and CAMEO. When opening process equipment or performing maintenance on systems either currently containing flammable or toxic materials or previously containing these materials, the product hazards should always be communicated to those doing the work.

Types of Explosions

An explosion is a rapid transformation of physical or chemical energy into mechanical energy and involves gas expanding rapidly. The following describes the types of explosions:

Physical Explosion

A physical explosion is defined as an explosion arising from the sudden release of stored energy, such as from the failure of a pressure vessel. For example, a pressure vessel is pressurized above the design pressure or fails due to a brittle fracture.

Boiling Liquid Expanding Vapor Explosion (BLEVE)

A boiling liquid expanding vapor explosion (BLEVE) is another example of a physical explosion. A BLEVE occurs when the liquid is heated above its atmospheric boiling point and rapidly released into the atmosphere (e.g., as the result of a vessel failure). A BLEVE almost always results in rapid overpressure, vessel rupture, flying shrapnel, and when involving a flammable liquid, a massive fireball, and extreme radiant heat. A BLEVE is covered in more detail in Chapter 5.

Chemical Explosion

A chemical explosion is caused by the very rapid conversion of solid or liquid explosive material into a gas with a greater volume than the substances from which they were generated. This process requires only a fraction of a second, produces extremely high temperatures (several thousand degrees), and is accompanied by overpressure (shock wave) and loud noise. An explosion involving a mixture of natural gas (methane) and air (oxygen) or a mixture of ammonium nitrate and hydrocarbon are examples of a chemical explosion.

Deflagration or Detonation?

The difference between deflagration and detonation is the speed (velocity) of the flame front moving through the unreacted fuel. Deflagration describes a combustion event where the flame front velocity through the unreacted fuel is lower than the speed of sound. In a deflagration event, shock waves or overpressure does not normally occur.

A detonation is an event where the velocity of the flame front moving through the unburned fuel is greater than the speed of sound. The resulting shock waves or overpressure can result in catastrophic damage to people, plants, and equipment. A detonation can travel in any direction, and when this occurs in a pipeline or other similar enclosure, it can travel with or against the direction of gas flow.

Additional References

Cameo Chemical Database (Properties of hydrocarbons – searchable in the database), US Environmental Protection Agency.

Measurement of minimum ignition energy in hydrogen-oxygen-nitrogen premixed gas by spark discharge, Journal of Physics: Conference Series - Ayumi Kumamoto et al. 2011 J. Phys.: Conf. Ser. 301 012039.

Combustion, Flames, and Explosion of Gases (Third Edition), Combustion and Explosives Research, Inc. Pittsburgh, Pennsylvania, Bernard Lewis Ph.D., Sc.D. (Cantab.), Guenther von Elbe Ph.D. (Berlin).

Lower and Upper Explosive Limits for Flammable Gases and Vapors (LEL/UEL). www.mathesontrigas.com

Static Electricity Guidance for Plant Engineers, Graham Hearn, Wolfson Electrostatics Limited.

Chapter 3. Properties of Hydrocarbons (Fire and Explosion Hazards)

End of Chapter Quiz

1 A hydrocarbon mixture in air must have the correct ratio of hydrocarbons to air for ignition to occur when exposed to a flame or other ignition source. What do we refer to as this mixture?

2 Where can most of the hydrocarbon hazards be found?

3 Which hydrocarbon characteristic is referenced when considering the volatility of the product and especially when considering storage of the product?

4 What hazard may go undetected when gas testing a vessel for LEL?

5 What property of acetylene makes it particularly hazardous in welding operations, especially in or near confined spaces?

6 It is important to remember that when heated above this characteristic, a product will generate flammable vapors and ignite readily.

7 How is the autoignition temperature of a hydrocarbon defined?

8 Most hydrocarbons will spontaneously ignite at what temperature? Where can a more exact temperature be found?

9 What is Vapor Density?

10 What are the hazards of gases such as propane, butane, and gasoline vapors which have a vapor density making them heavier than air?

11 What is meant by the term vapor pressure?

12 What is the difference between Reid Vapor Pressure and true vapor pressure?

13 What is the difference between deflagration and a detonation explosion?

14 There must be _______, ________, ________, and the resulting chain reaction to start a fire.

15 When gas testing for trapped liquids in a process system, the gas test for LEL may ______ indicate the presence of trapped liquid hydrocarbons if the liquid is at a temperature below the flash point.

16 The boiling point is the temperature at which a liquid will ______ (measured in degrees F at sea level).

17 Increasing the pressure will ________ the boiling point; decreasing the pressure will ________ the boiling point, for example, in a vacuum.

18 How many of the properties of flammable liquids that can increase the risk of fire or explosion can you name?

19 A BLEVE occurs when the liquid is heated above its atmospheric _________ ________ and rapidly released into the atmosphere (e.g., as the result of a _________ failure).

20 An explosion is a rapid transformation of _________ or _________ energy into mechanical energy and involves gas expanding rapidly.

4

Vapor Cloud Explosions (VCEs)

In this chapter, we will discuss what can happen in the event of loss of containment of a light hydrocarbon such as light naphtha, gasoline, or LPG materials like propane, butane, or ethylene. When mixed in the right proportions with air (flammable range), such products can result in a devastating explosion referred to as a VCE or vapor cloud explosion. VCEs can be very destructive and have destroyed numerous manufacturing facilities. VCEs have also resulted in many lives lost and many very seriously injured people. We will discuss what causes a VCE, some examples of VCE incidents, and actions we can take to help minimize the potential for VCE.

A vapor cloud explosion can occur when a flammable mixture of hydrocarbon vapor and air vapor is ignited and is more severe in a partially confined area. For example, an explosion occurs when a loss of containment of a flammable occurs in the open area of a process unit, where the mixture is partially confined by obstructions such as pipe racks, fractionating columns, and unit structures. These "obstructions" create turbulence and contribute to thoroughly mixing the hydrocarbon vapor and air, producing a much more uniform mixture of fuel and air. If this highly volatile mixture "finds" an ignition source, the rapidly expanding flame front and hot gases, in the presence of the obstruction, result in a high velocity of the hot gases. The high velocity results in "overpressure," or a pressure wave. The overpressure can result in catastrophic equipment damage and, in many cases, a secondary loss of containment, which contributes to additional fuel released into the fire area and a rapidly spreading event.

In some cases, primary fractionation columns have been blown down from their foundation, even while in-service. VCE damage can result in other equipment, such as process drums and other process vessels or equipment, being completely blown over or from their foundations, causing further loss of containment of flammable or toxic materials. The overpressure may also cause damage to process piping systems, adding more fuel to the expanding fire. A VCE can potentially destroy buildings not designed for explosion overpressure. For example, wood frame trailers and even concrete structures without adequate internal reinforcement have been destroyed, resulting in severe injuries and fatalities to the occupants. The resulting VCE overpressure has resulted in catastrophic damage to the unit involved, but in some cases, shrapnel and debris have been blown miles from the explosion site.

In a vapor cloud explosion event, most personnel injuries and fatalities occur when people are struck by flying debris or building collapse. Generally, most fatalities occur when people are indoors and the buildings are severely damaged or collapse due to the explosion pressure wave. For example, all fifteen deaths occurred in March 2005 at the BP Texas City refinery explosions were in portable buildings (wood frame trailers) located relatively close to the explosion. These trailers were not designed for the overpressure which resulted from the blast. Many mobile trailers were destroyed, resulting in the significant loss of life and traumatic injuries to the occupants.

So, how do we prevent a vapor cloud explosion or minimize the effect of one? The most important way to avoid the VCE is to prevent the loss of containment of flammable or toxic materials in the first place, and many causes of loss of containment and their prevention are presented in subsequent chapters of this book. Preventing a loss of containment requires continuous management focus on process safety, ongoing training, the diligence of operating personnel, and the rigorous inspection and maintenance of process equipment. Also, sites should continually assess their procedures and practices to help eliminate opportunities for uncontrolled fuel and air mixtures and ensure potential ignition sources are controlled or minimized. Examples of how this can be achieved are strict energy isolation and work permit procedures, accurate and current operating procedures, including detailed procedures for startup and shutdown and for other "transient" operations such as a temporary operating mode or plant tests, strict adherence to electrical classification for arcing or sparking equipment, controls for access of motorized vehicles onto process areas or units, and rigorous inspection procedures for process piping and vessels.

Process Operations Safety: The What, Why, and How Behind Safe Petrochemical Plant Operations, First Edition. M. Darryl Yoes.

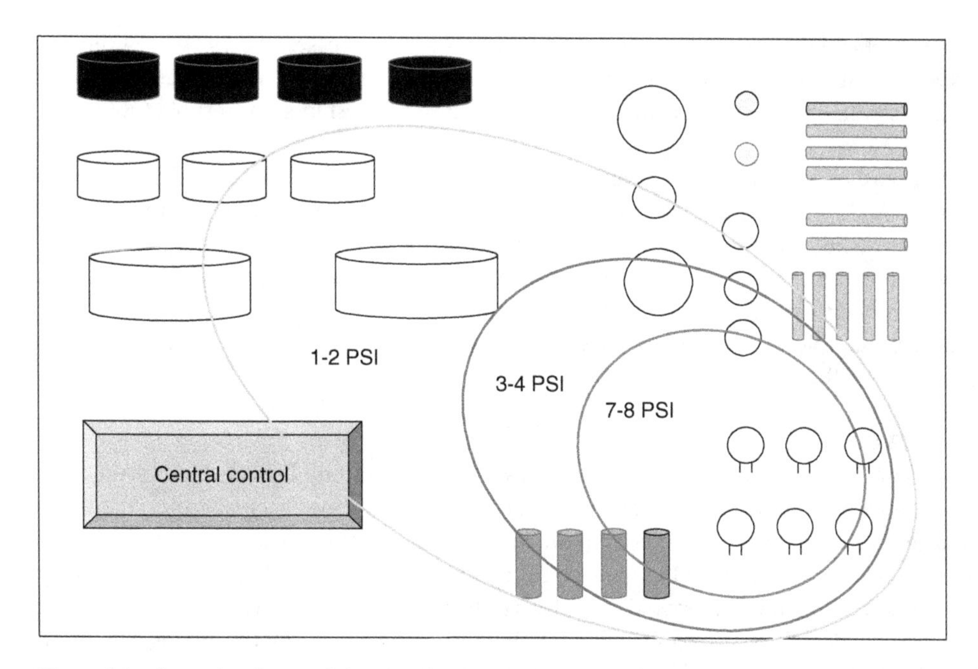

Figure 4.1 Example of potential vapor cloud explosion overpressure.

All loss of containment or near loss of containment events should be reported and investigated for potential lessons learned and include recommendations implemented to help guard against a reoccurrence of the event. Even minor flammable or toxic vapor release events should be reported and investigated. This means that site management must foster an environment where it is acceptable, even encouraged, to report these incidents. Every loss of containment should be viewed as a "learning opportunity," with lessons learned turned into actions to prevent the next one. The more significant incidents, even catastrophic events, can be prevented by thorough and accurate reporting of minor incidents (minor releases). A detailed investigation of the minor incidents with corrective actions implemented can prevent the more serious incidents, even those that could have had catastrophic consequences.

Safety engineers and risk management specialists can use specialized computer programs to model the overpressures resulting from a potential vapor cloud explosion. The computer model considers the specific hydrocarbons or other chemicals involved and their characteristics, the inventories of these materials contained in the respective processing equipment (towers, drums, reactors), the operating conditions (temperature, pressure, levels), and the degree of congestion in the area. The model develops a series of overpressure curves starting with the highest overpressure in the zone immediately surrounding the target vessel. The pressures cascade lower as the distance increases from the target vessel.

This overpressure data is extremely useful to site management in helping to understand the potential blast hazards for the unit and developing site emergency response plans and procedures. The potential overpressure data is also used to assess the location and type of construction for buildings that will be occupied in or near the process area. For example, this data is essential when considering the placement of occupied temporary buildings for use during unit turnarounds or construction projects. An example of a typical overpressure study plot plan is shown in Figure 4.1. Additional building siting details, including temporary structures, can be found in the American Petroleum Institute Standard API-752 "Management of Hazards Associated with Location of Process Plant Permanent Buildings."

Figure 4.1 an overpressure study was used to evaluate the siting of occupied buildings in process areas. The building design takes into consideration the potential overpressure should an explosion occur.

The example shown in Figure 4.1 is hypothetical but represents a severe overpressure event. An actual overpressure study with overlapping overpressures will be much more complicated than this simple example. Also, we have seen overpressures that were calculated to be much higher than in this example, mainly where hot or light hydrocarbons, such as LPG, are involved in the release. An actual event can result in significant equipment damage and threaten human life.

What Has Happened/What Can Happen

Unfortunately, there are many VCEs to select from as examples. In their listing of the Largest 100 Losses (1972–2001), Marsh Risk Consulting Services lists 18 of the most significant losses experienced by the refining and petrochemical industry (incidents with greater than $150M in property damage). Six of these incidents were catastrophic VCEs, with the single largest property loss at nearly $1B ($839M). The most current version of their Largest 100 Losses, the 1974–2019 version, indicates that "of these new losses, a remarkable eight were among the 50-largest industry losses of all time, and four of these were among the 20-largest losses ever. Several more occurred just below the 100-largest losses threshold (now at US $175 million)."

Of course, this does not include the cost of human life and suffering, the cost of business interruption, and the resulting impact on the company's reputation. A VCE event can devastate the site, resulting in loss of life and severe personnel injuries, massive damage or destruction to the processing facilities, and a significant impact on the surrounding community. These incidents have also resulted in the loss of the company's "license to operate." A license to operate is an unwritten consent by the community and regulatory agencies to operate in the area.

The incidents below were selected as examples of What Has Happened and What Can Happen Again. This is partly due to the event's nature and the importance of the lessons learned from the incident.

Pasadena, Texas; Polyethylene Plant (23 October 1989)

A large flammable vapor release occurred during maintenance operations to unplug settling legs on the polyethylene reactor. Routine maintenance operations were underway to clear a high-density polyethylene reactor settling the leg of accumulated polymer. The settling leg piping had been isolated from the in-service reactor with a single 8-inch ball valve equipped with a provision for a lock-out device to prevent the inadvertent operation of the valve. Further, the air supply lines to the pneumatic valve actuator had been removed to avoid the valve's unintentional opening. The settling leg piping was disassembled during maintenance to allow the unplugging operation. Following the explosion, this piping was found with portions still disconnected, indicating the job was still in progress when the incident occurred.

Most importantly, the ball valve was found in the fully opened position even though the maintenance work had not been completed. The pneumatic actuator lines to the ball valve actuator were connected but, unfortunately, had been reversed, allowing the valve to travel to the fully opened position (with the settling leg piping not fully connected). This allowed the product to be released directly into the atmosphere, resulting in the massive loss of containment of the highly flammable product.

At the time of the incident, the reactor was operating normally at a pressure of approximately 700 psig and contained a highly flammable mixture of ethylene and isobutane. The released vapor quickly ignited, resulting in the VCE. This incident resulted in 23 fatalities and 314 injuries and destroyed two of the site's high-density polyethylene manufacturing plants. The explosion was felt miles from the facility and registered between 3 and 4 on the Richter scale on the Rice University seismographs, about 15 miles from the site. There were also several secondary explosions, including the BLEVE of other on-site pressurized storage vessels.

Figure 4.2 Phillips polyethylene plant. An aerial photo provided by the Federal Emergency Management Agency taken before the 1989 explosion. Source: Photo courtesy of the U.S. Emergency Management Agency (Ind).

Lessons from this incident include needing more rigorous energy isolation before opening process equipment, such as double block and bleed isolation valves, and using blinds and blanks to ensure isolation from hazardous energy. The incident also highlights the requirement for strict worker adherence to energy isolation procedures. Other causal factors may have contributed to the incident, including timely process hazard analysis and a more rigorous work permit process for opening process equipment. Other vital learnings include the location of occupied buildings relative to process units, improved fire protection equipment, and improved personnel training in work permits, maintenance procedures, and emergency/evacuation response. See Figures 4.2–4.4 for photos of this catastrophic event.

(a)

(b)

Figure 4.3 Phillips polyethylene plant explosion. Aerial photos taken by the Federal Emergency Management Agency following the explosion (23 October 1989). Source: Photo courtesy of the U.S. Emergency Management Agency (FEMA). Additional information may be found at https://www.fema.gov/photo-video-audio-use-guidelines.

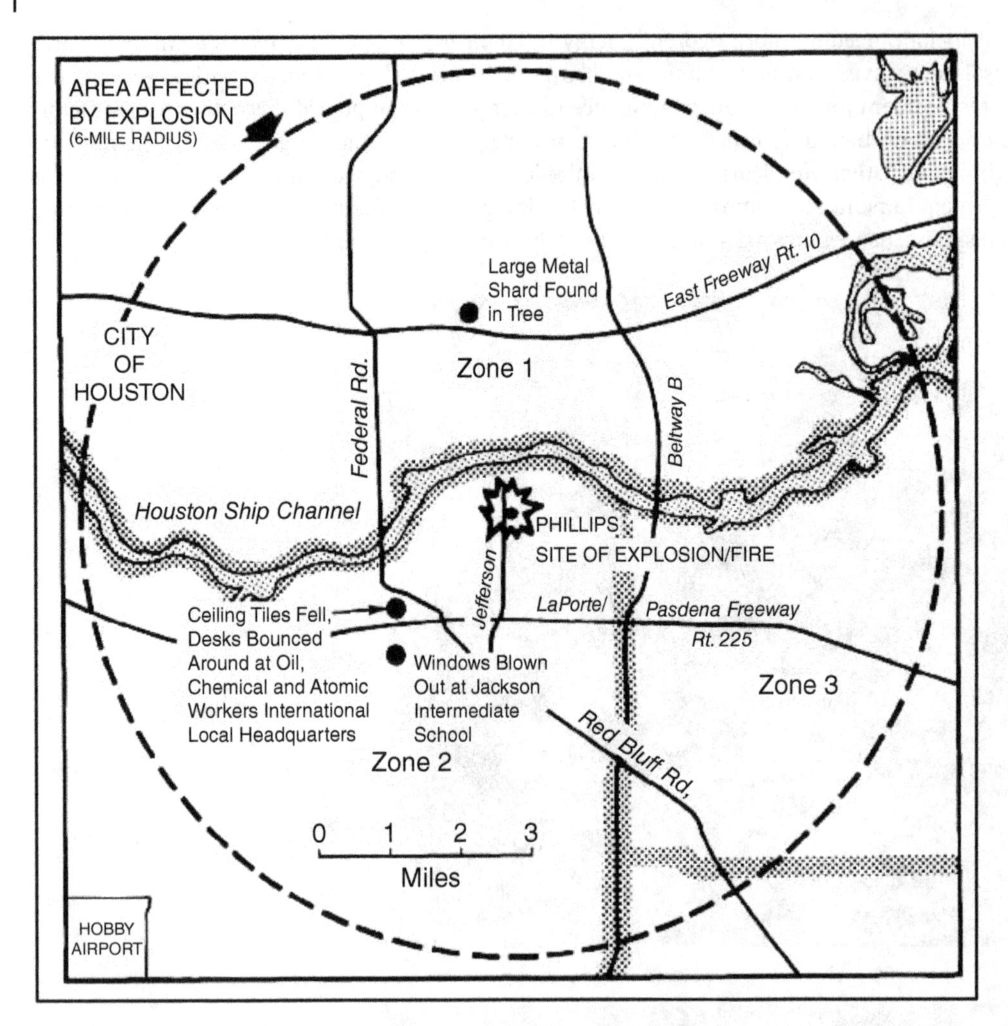

Figure 4.4 Area impacted by the phillips explosion (23 October 1989). Source: Photo use courtesy of the US Federal Emergency Management Agency (FEMA). Additional information may be found at https://www.fema.gov/photo-video-audio-use-guidelines.

ExxonMobil Refinery Baton Rouge, Louisiana; Petroleum Refinery (24 December 1989)

A large vapor release occurred at this large US Gulf Coast refinery following the rupture of an 8-inch pipeline within the refinery containing a mixture of ethane and propane. The resulting large vapor cloud was ignited a few minutes following the release, resulting in a vapor cloud explosion, causing significant damage to the pipe rack and surrounding process equipment and multiple secondary fires, including several tank fires, including two large diesel tanks. The incident resulted in two fatalities and a total shutdown of the refinery and adjacent chemical plant.

The incident occurred immediately following a severe freeze where the ambient temperature was below freezing for several days and reached a low ambient temperature of about 10 °F. It is believed that the piping failure resulted from the thermal expansion of the trapped liquids. The pipeline was blocked during the freezing weather and ruptured the following afternoon, 24 December. The ambient temperature had climbed to just above freezing before the rupture occurred.

This incident highlights the importance of recognizing and mitigating the significant pressures from thermal expansion, especially in light hydrocarbon lines such as naphtha and gasoline and liquefied petroleum gases such as ethane, propane, and butane. Thermal expansion of trapped liquids in pipelines, pressure vessels, and heat exchangers, which are blocked in and then subjected to temperature increases, can reach pressures significantly exceeding the maximum design pressure of the equipment.

This can result in gasket failure or rupture of the piping or vessels and catastrophic loss of containment, leading to toxic release, fire, and explosion. Equipment that is required to be blocked in should be protected from the hazard of thermal expansion pressure increase by either overpressure protection devices such as thermal expansion relief valves or detailed operating procedures such as venting the equipment to a tank, flare, or other safe disposition to relieve any accumulated pressure due to thermal expansion. See Figures 4.5 and 4.6 for photos of the fire that followed this explosion.

Figure 4.5 Exxon Baton Rouge Refinery following rupture of 8″ LPG line due to thermal expansion (24 December 1989). The above image is courtesy of WAFB TV, Channel 9, Baton Rouge, LA, 24 December 1989.

Figure 4.6 Exxon Baton Rouge Refinery following the 1989 Christmas Eve Explosion due to thermal expansion in an isolated LPG pipeline. Sixteen petroleum storage tanks were burning as a result of the explosion. The ambient temperature reached 8 ° Fahrenheit overnight. In addition to the two fatalities, the incident resulted in severe damage to a major crude oil and product transfer pipe rack, 16 storage tanks, and a long-duration outage of most of the refinery. The Louisiana State Capitol building, the tallest state capital in the United States, is in the foreground. It stands 460 feet tall with 34 floors and appears dwarfed by this incident. The above image is courtesy of WAFB TV, Channel 9, Baton Rouge, LA.

Geismar, Louisiana; Olefins Plant (13 June 2013)

Unfortunately, another tragic incident involving overpressure from thermal expansion occurred at an Olefins Plant facility in Geismar, Louisiana, on 13 June 2013. This incident resulted in 2 fatalities and 167 injuries.

The US Chemical Safety and Hazard Investigation Board investigated this incident. They reported that the exchanger shell ruptured due to starting flow through the tube side or "hot" side of an exchanger, a Propylene Fractionator reboiler, with the cold side blocked in and blocked away from the exchanger's overpressure protection. This resulted in a rapid pressure increase on the shell side of the exchanger, rupture of the exchanger shell, release of propylene, and the vapor cloud explosion.

A photo of the Geismar Olefins Plant explosion is available in Figure 4.7 (Courtesy of the US Chemical Safety and Hazard Investigation Board). The Chemical Safety Board completed a full incident investigation, which can be found on its website (CSB.Gov).

Figure 4.7 Williams Olefins Plant Explosion, Geismar, Louisiana, 13 June 2013. Source: Photo courtesy of the U.S. Chemical Safety and Hazard Investigation Board. This photo is also used as the cover for this book courtesy of the CSB.

Killingholme, England; Petroleum Refinery (16 April 2001)

This refinery experienced a major vapor cloud explosion following the major release of ethane, propane, and butane vapor due to the catastrophic failure of a 6-inch elbow on the overhead line of the Deethanizer at the Saturated Gas Plant. The failure occurred suddenly at the first 90° piping ell located 26 inches (670 mm) downstream of a water injection point between the deethanizer column and the overhead condensers. It was reported that the water injection had changed from intermittent to continuous use about six months before the failure, resulting in severe internal corrosion and erosion.

The U.K. Health and Safety Executive (HSE) investigated this incident and reported considerable equipment damage. The resulting fire caused other secondary losses of containment and severely escalated the event. The saturated gas plant and an adjacent unit were extensively damaged. At least one process vessel was dislodged from its foundation on the adjoining unit. Buildings some distance away were also significantly damaged by the overpressure. Fortunately, there were no fatalities or serious injuries. The incident occurred during a holiday, which limited personnel exposure. Also, the incident happened near the shift change time, and two process crews were available to respond and help with equipment isolation. See Figures 4.8 and 4.9 courtesy of the UK Health and Safety Executive. These two photos help to illustrate the force of the VCE and the resulting equipment damage.

The Killingholme incident was remarkably similar to another vapor cloud explosion in 1988 at a refinery near New Orleans, LA. Unfortunately, this VCE resulted in fatal injuries to seven process operators. Both VCE incidents were initiated by the failure of piping elbows downstream of water injection points on the overhead piping on light ends units. In the New Orleans refinery incident, the water injection mix point was in a similar location on the overhead line of an FCC Depropanizer column. In this case, the FCC main column was blown completely down, and the FCC unit was destroyed.

These incidents illustrate the awesome force that can occur during a VCE. They also highlight the importance of controlling corrosion/erosion around mixing and injection points and the criticality of rigorous equipment inspections. Piping downstream of a mixing point should be considered a potential high corrosion area and routinely inspected. As noted above, failures can and have occurred in relatively short periods.

However, a VCE can occur in many other ways. Any loss of containment of a light product such as a liquefied petroleum gas (LPG), gasoline, naphtha, or similar products at a temperature above its boiling point (the boiling point at atmospheric pressure) can result in an uncontrolled vapor release and a VCE.

Figure 4.8 Killingholme, England, damage to unsaturated gas plant. Contains public sector information published by the Health and Safety Executive and licensed under the https://www.nationalarchives.gov.uk/doc/open-government-licence/version/3/. Source: Reproduced by the kind permission of HSE. HSE would like to make it clear that it has not reviewed this product and does not endorse the business activity of Safety Consulting International, LLC.

Figure 4.9 Killingholme, England, damage to the adjacent unit. Contains public sector information published by the Health and Safety Executive and licensed under the https://www.nationalarchives.gov.uk/doc/open-government-licence/version/3/. Source: Reproduced by the kind permission of HSE. HSE would like to make it clear that it has not reviewed this product and does not endorse the business activity of Safety Consulting International, LLC.

Strict adherence to procedures is required during higher hazardous operations, such as:

- Drawing water from pressurized storage tanks or process vessels.
- Opening equipment for maintenance when the equipment has been in light hydrocarbon service.
- Returning equipment in light hydrocarbon service to service.
- Routing light hydrocarbon streams to tankage (avoid sending light streams to atmospheric tankage).
- Responding to "minor" leaks on light or hot hydrocarbon piping or equipment.
- Closely monitoring tank transfers to ensure tanks are not overfilled.
- Ensure considerations for thermal expansion when isolating light hydrocarbon equipment, for example, blocking long runs of light hydrocarbon piping and exposing them to the hazard of thermal expansion.

A very similar incident occurred at the Shell Refinery located near New Orleans on 5 May 1988. In that incident, a catastrophic vapor cloud explosion occurred following a piping failure immediately downstream of a water injection point in the 8″ Depropanizer overhead piping system. Seven Operators were killed in this incident. See Chapter 8.9 for more information on injection and mixing point failures.

In the Shell incident, the explosion was so powerful that the main FCC fractionator, typically the largest tower in a refinery, was blown completely down in the explosion. The main fractionation column was in service at the time of the explosion. I was unsuccessful in locating the owner of the image; therefore, it is not shown here.

Additional References

"Management of Hazards Associated with Location of Process Plant Permanent Buildings." American Petroleum Institute Standard API-752

Petrochem History "The Phillips Explosion" Pasadena, TX: 23 October 1989 https://www.pophistorydig.com/topics/phillips-petroleum-explosion-1989/

Federal Emergency Management Agency (FEMA) Investigation Report on the 1989 Phillips Chemical Plant Explosion and Fire https://ncsp.tamu.edu/reports/USFA/pasadena.pdf

U.S. Chemical Safety and Hazard Investigation Board B.P. Texas City Refinery Explosion and Fire final report and training video Report No. 2005-04-I-TX https://www.csb.gov/bp-america-refinery-explosion/

U.K. Safety and Health Executive final report Killingholme, England Refinery Explosion and Fire https://www.hse.gov.uk/comah/conocophillips.pdf

U.S. Chemical Safety and Hazard Investigation Board Williams Olefins Plant Explosion and Fire final report and training video Report No. 2013-03-I-LA https://www.csb.gov/williams-olefins-plant-explosion-and-fire-/

Process Plant Layout IChemE – Process Plant Layout, Second Edition Sea'n Moran https://vbook.pub/documents/process-plant-layout-4wlgd1095y26

100 Largest Losses in the Hydrocarbon Industry, 1972–2001, 26th edition of Marsh JLT Specialty's 100 Largest Losses in the Hydrocarbon Industry report, 1974–2019. 27th edition of Marsh's 100 Largest Losses in the Hydrocarbon Industry – incidents between 1974–2021 https://www.marsh.com/us/industries/energy-and-power/insights/100-largest-losses-in-the-hydrocarbon-industry.html

OSHA Investigation of the Shell Refinery Explosion, which occurred in Norco, LA, on 5 May 1988. https://www.osha.gov/pls/imis/establishment.inspection_detail?id=100478866

Chapter 4. Vapor Cloud Explosions (VCE)

End of Chapter Quiz

1 What effects can increase the force and contribute to the destructive nature of a vapor cloud explosion?

2 How does confinement increase the force of a vapor cloud explosion?

3 In a vapor cloud explosion, most personnel injuries occur outdoors or indoors?

4 What is the most effective way to prevent a vapor cloud explosion?

5 How can a facility prevent loss of containment incidents which may lead to a vapor cloud explosion? What are several examples of management systems? Name three examples of a management system.

6 How can a site ensure that personnel can be protected in the event of a loss of containment incident?

7 All loss of containment or near loss of containment events should be ________ and ___________ for potential lessons learned and include recommendations ______________ to help guard against a reoccurrence of the event. Even _______ flammable or toxic vapor release events should be reported and investigated.

8 What is the purpose of the site overpressure study?

9 What are several causal factors that can result in a loss of containment and lead to VCEs? Name all you can.

10 Name two references to learn more about vapor cloud explosions.

11 What causes extreme damage that can occur in a VCE?

12 Why is it important to investigate all loss of containment incidents involving flammable or toxic releases, even the small ones that have no impact?

13 How do safety engineers and risk management specialists model the predicted overpressures from a potential VCE?

14 How can the risk management overpressure model help prevent incidents during unit turnarounds when many people are present on the unit?

5

Boiling Liquid Expanding Vapor Explosion (BLEVE)

A Boiling Liquid Expanding Vapor Explosion (BLEVE) is an explosion that can occur when a pressurized vessel containing a liquid stored at a temperature at or above its boiling point suddenly fails (a liquid above its boiling point if it were at atmospheric pressure). Liquid petroleum gas (LPG) such as propane or butane are present in their vapor phase at atmospheric pressure and normal ambient temperatures but are liquefied under pressure or very cold conditions. When this pressure is released, such as a sudden vessel failure, the liquid LPG is instantly and violently released into the atmosphere as large quantities of highly flammable vapor. When this loss of containment occurs, it most often results in a major explosion, a massive fireball, and a sudden huge radiant heat release. In almost all cases, this results in an extremely high overpressure (or pressure wave) and debris or shrapnel being hurled a long distance from the incident.

When the LPG is released from the pressurized container, it expands from a liquid to a vapor. The expansion ratio from liquid to vapor for Propane is about 1–270; for every 1 volume of Propane liquid released into the atmosphere, it will expand to about 270 volumes of highly flammable propane vapor, creating a massive vapor release. The expansion rates for other LPGs are similar. This rapid and massive expansion is one of the characteristics of LPG that makes it one of the most hazardous materials in the industry.

Damage from a BLEVE can be catastrophic and may result in numerous secondary fires or explosions due to the impact on adjacent facilities. The explosion results in a significant pressure wave. Large sections of the tank or other containment vessel may become projectiles and "rocket" thousands of yards, causing further damage to adjacent equipment.

LPG storage tanks or transportation vessels are susceptible to BLEVE should the containment vessel (such as a tank truck, tank car, storage sphere, or process unit vessel) be compromised by physical damage such as corrosion, damage from the derailment, or by fire impingement from an adjacent fire. The most common cause of BLEVE is a fire on or near the storage vessel, resulting in flame impingement on the vessel, causing failure of the vessel shell, and the resulting catastrophic loss of containment. A BLEVE has been known to occur within minutes, hours, or even days after the initial fire. The real hazard of a BLEVE is that it is instantaneous, with no warning when one does occur. BLEVEs have caught many responders off guard, resulting in tragic consequences, including large loss of life.

Most BLEVEs occur when an adjacent fire is impinging directly onto the vessel shell in an area above the liquid level, causing the steel to weaken and fail. Generally, when a fire is impinging on the vessel below the liquid level, the liquid absorbs the heat and helps prevent the vessel shell from failing. However, as the fire continues, the contents are heated, resulting in a pressure increase on the tank to the point where the relief valves on the vessel will open to protect the vessel from overpressure. Over time, this lowers the liquid level and may expose the vessel to direct fire impingement to the vessel shell above the liquid level, causing the vessel to BLEVE. The massive explosion results in a huge fireball as the LPG flashes and expands 270 times its volume (expansion for Propane), resulting in overpressure (or blast wave), radiant heat release, and shrapnel propelled for long distances. See Figure 5.1a,b.

BLEVE is not limited to LPG or even to flammable materials. Vessels containing other liquids above the boiling point may also experience a BLEVE should the containment vessel be compromised. For example, a refinery steam drum can BLEVE should the vessel fail due to accidental damage, overpressure, or corrosion. This steam explosion and the resulting blast and release of hot water/steam can also be devastating. BLEVEs have also occurred on pipelines containing gasoline or naphtha when the pipeline is exposed to fire, especially if the pipeline is isolated (blocked in). The illustrations in Figure 5.1a,b below help to explain how a BLEVE can occur.

Shrapnel and projectiles from the storage tank can be propelled for long distances, resulting in secondary loss of containment, leading to additional fires and explosions. A massive fireball or flying debris can severely injure or kill people.

Process Operations Safety: The What, Why, and How Behind Safe Petrochemical Plant Operations, First Edition. M. Darryl Yoes.

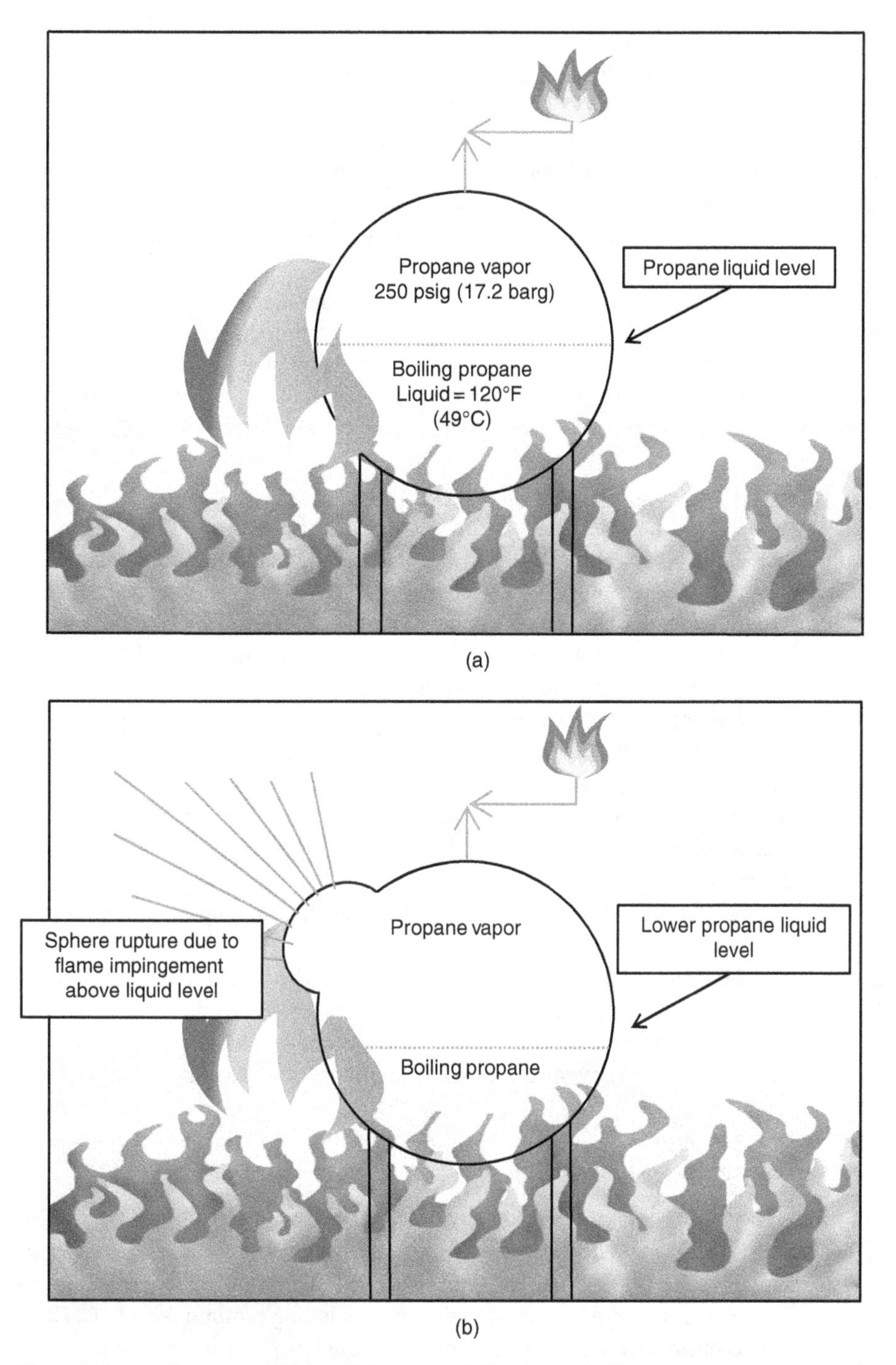

Figure 5.1 (a) Illustrates a fire impinging on an LPG sphere. (b) Illustrates the sphere rupturing releasing its contents in a massive fireball or BLEVE (Boiling Liquid Expanding Vapor Explosion).

Preventing BLEVEs means avoiding failure of the storage vessel. Particular attention should be given to facilities, procedures, and operating practices for equipment containing pressurized liquids stored at temperatures above their boiling point, especially liquefied petroleum gas such as Propane or Butane. This means continuous diligence by site management and guarding against complacency by operations and maintenance personnel handling LPG or LPG equipment. LPG equipment and procedures should undergo periodic audits and assessments, including a review of operating practices and routine walk-down of facilities. Some considerations for regular reviews:

- Robust work permit procedures should be in place and enforced to open all equipment containing LPG to prevent the accidental release of liquefied LPG.

- Before opening, permit procedures must ensure that equipment is completely depressurized and free of liquid hydrocarbons. Verify that vents and drains are not plugged and that the equipment is at zero pressure.
- Flange bolts should be loosened when breaking the flange seal, allowing the flange to be quickly reclosed if a liquid is present (never remove all flange bolts before breaking the flange seal).

- Pay special attention to water draw operations; Operators should be trained on the proper procedures to draw (drain) water from a light end's vessel. Please see Chapter 8.20, "Sampling and Drawing Water from Process Vessels" for details.
- All water draw valves should remain closed, and valves for sample facilities should also be closed, plugged, and opened only when used.
- Small piping in LPG service should be minimized. Small piping should be seal welded, braced, and supported to prevent failure. Small piping should also be protected against accidental contact with vehicles or motorized equipment.
  - Gauge glasses should be avoided on LPG storage vessels due to the potential for small piping or glass failure.
- All bleeders and vents in LPG service should be closed and plugged with a bar stock (solid) pipe plug.
- Dead-ended piping should be minimized due to the potential of internal corrosion or freezing of trapped water – either can result in a major loss of containment.
  - All dead-ended piping should be protected from freezing during periods of cold weather.
  - They should also be routinely inspected for internal corrosion.
- Are pipelines in LPG service insulated? Consider the potential for corrosion under insulation and/or fireproofing, which may result in loss of containment. Ensure these areas are routinely inspected for potential corrosion under insulation (CUI). Please refer to Chapter 8.13, Corrosion under insulation and/or Under Fireproofing for details.
- Ensure that vents and drains from pressure relief valves, sample points, or similar devices are not oriented toward the vessel. This can result in fire impingement on the vessel should a leak or release occur.
- Are controls in place to prevent a truck from accidentally driving away with a connected tank truck or a freight engine pulling connected LPG railcars at loading racks? For example, the track derailer should remain in place at railcar loading racks until the railcars are completely disconnected and all rack and car connections secure.
- Is vessel firefighting equipment in good condition and routinely tested? The vessel should be equipped with firewater sprays and deluge capability, which should be tested periodically. A good practice is to test the deluge system annually, checking for plugged spray nozzles or dry spots on the vessel.
  - The water deluge activation valves should be located remotely from the vessel and accessible by Operators during emergencies.
- Most LPG storage vessels are equipped with emergency water flood connections to allow the bottom filling of the LPG vessel with firewater. These should also be remotely located and tested periodically to ensure readiness.
- Does the site have the capability to deliver an uninterrupted supply of firewater in the event of flame impingement on the vessel? The National Fire Protection Association (NFPA) recommends a minimum of 500 GPM of firewater for each point of flame impingement.
- And finally, does the site have an emergency response plan that considers the potential for BLEVE, including evacuation of personnel, and has the plan been practiced during exercises and drills?

BLEVE emergency response preplanning is critical and should be periodically tested in drills and exercises. Should a fire occur on or near a pressurized storage tank, starting the water deluge system supplemented by direct water streams from available hydrants is critical. The firewater systems should be started during the first few minutes of fire exposure to help minimize the risk of a vessel BLEVE. Large amounts of firewater are required, including a minimum of 500 GPM at each point of flame impingement (see NFPA recommendation). Special attention should be given to protecting potentially exposed personnel, including evacuation from the surrounding areas. If the vessel is leaking near the bottom or from the connected piping, consideration should also be given to internally flooding the vessel with firewater. Firewater can be introduced into the vessel using the quick connections on the vessel fill lines. The vessel should not be emptied while exposed to an impinging fire, as this will quickly expose more of the vessel shell to flame impingement. Emergency preplans should also recognize that even with our best efforts, we may not successfully prevent a BLEVE, and therefore all personnel must be protected.

- Included in the design of these new spheres:
  - Fire-Water Deluge Spray System.
  - Overpressure Protection is designed to avoid impingement (relief valve outlets directed away from the vessels).
  - Fireproofing of Structural Supports.
  - Fire-Water Emergency Flooding (to fill the sphere with firewater during a loss of containment or fire impingement).

Figure 5.2 Photo of new spheres in propane service. Source: Photo printed with permission; Lian Rag Co., Consulting and Technical Inspection Engineers (http://www.rag.ir//11411/).

- Drainage sloping away from the vessel to prevent the pooling of LPG spills under or near the vessels.
- The impounding of potential spills in the enclosed firewall (in some designs).
- Water and Sample facilities are designed to prevent the auto-refrigeration and potential freezing of LPG valves in the open position (double block valves, including spring-loaded to close).

Some of these safety facilities are visible in Figure 5.2 above.

Key Lessons Learned

Vessels containing pressurized liquids above their boiling points have the potential for BLEVE should they be exposed to a fire. These types of incidents have the potential for large loss of life and catastrophic damage to equipment and facilities. Operators should be well informed on the hazards of LPG and the operations procedures and practices to prevent loss of containment. Pressurized storage vessels and associated equipment should receive regular inspections, including periodic walk-downs by supervisors and managers, to help ensure the integrity of the equipment and strict adherence to safe operating procedures and practices.

Emergency response preplanning for a potential BLEVE situation is critical. It should be documented in the site's emergency plans manual and include response integration with local municipalities and their emergency response personnel. Emergency procedures should include evacuating all personnel exposed to the effects of a potential BLEVE, including personnel outside the facility.

What Has Happened/What Can Happen

There have been numerous BLEVEs in the transportation and petrochemical industry. BLEVEs are most common in transportation incidents (railcars, tank trucks) but have occurred in refineries and chemical plants. Significant industrial BLEVEs include the accidents at Feyzin, France (1966), Texas City, Texas (1978), Romeoville, IL (1984), and San Juan Ixhuatepec, Mexico (1984). All these incidents resulted in catastrophic loss of life and equipment damage. There were also catastrophic BLEVEs at the Chiba, Japan refinery following the Japanese earthquake of 2011; however, no fatalities occurred at that site due to extensive evacuations of personnel.

Figure 5.3 The three images shown in (a, b, and c) show the devastation from multiple BLEVEs which occurred at Feyzin, France in 1966.

Feyzin Refinery – Feyzin, France (4 January 1966)

The Feyzin disaster occurred in a refinery near the town of Feyzin, 6.2 miles (10 kilometers) south of Lyon, in Southeast France, on 4 January 1966. An LPG spill occurred when the water draw (drain) valves froze in the open position while operators attempted to collect a sample from the water draw (drain) from a 7,500 barrel (1,200 cubic meter^3) pressurized Propane sphere.

Unfortunately, the proper procedures for draining water were not followed, and propane vaporized, creating auto-refrigeration and the valves freezing in the open position. The vapors were subsequently ignited by a passing freight train, resulting in an impinging fire leading to vessel failure and a BLEVE. The initial BLEVE damaged adjacent storage spheres and resulted in a catastrophic fire. Four other spheres experienced BLEVE following the initial fire, resulting in 18 fatalities (including 12 firefighters). See Figure 5.3a–c for images of this incident provided with kind courtesy of the archives at le Progrès (France). Several additional images of the Feyzin event are available in Chapter 8.20 (Sampling and Drawing [Draining] Water from Process Vessels).

This incident highlights the requirement for Operator training in LPG handling procedures, including water draw operations. Proper sequencing of the water draw valves is critical to prevent LPG auto-refrigeration, which can freeze the valves in the open position. The valves should be operated with the upstream (vessel side) valve in the fully open position and the downstream (drain side valve) being throttled to control flow. Operators conducting water draw operations must never leave the water draw open and unattended, even for minutes, as a massive vapor cloud can form should LPG be released from the open water draw.

Texas City, Texas (30 May 1978)

On 30 May 1978, a series of catastrophic BLEVEs ripped through the Texas City Refining Company at about 2:00 a.m., killing 5 workers and injuring 10 others. Industry reports indicate that the incident started with an Isobutane transfer and the receiving tank being overfilled, resulting in a rupture. Reports indicate the sphere was nearly "liquid full" and failed when exposed to 300 psi. The Marsh report titled "Large Property Damage Losses in the Hydrocarbon-Industries" details the Texas City Refining Company incident in May of 1978.

The following is summarized from the Marsh "The 100 Largest Losses, 1972–2001":

The cause of this release and the massive explosions that started in the alkylation unit tank farm are unknown. What is known is that an unidentified failure led to the release of light hydrocarbons (LPG), quickly finding an ignition source. In less than five minutes, one of the 5,000-barrel spheres failed, causing a tremendous fireball and sending large pieces of shrapnel (sections of the sphere) throughout the plant. Within 30 minutes, five 1,000-barrel horizontal vessels, four 1,000-barrel vertical vessels, and one more 5,000-barrel sphere failed from shrapnel damage or BLEVEs. Large sections of the storage tanks were propelled in all directions and onto the operating units and tank farms, resulting in additional major fires. Several pieces of shrapnel also structure the firewater storage tank and electrically driven fire pumps, leaving only the two diesel firewater pumps operational.

This Marsh report portrays the significant impact of a BLEVE explosion on a refinery or petrochemical plant. A more recent version of the Marsh report is now available.

Romeoville, Illinois (23 July 1984)

A disastrous explosion and fire occurred at this refinery when an amine absorber pressure vessel ruptured, resulting in a BLEVE. Seventeen people were killed in this incident, including operating crew members and emergency responders, and an additional 17 people were hospitalized. Before the event, a leak was reported on the amine absorber vessel, and operations personnel and emergency responders responded to the incident before the vessel failed catastrophically. It was later discovered that amine absorbers with high concentrations of hydrogen sulfide had experienced hydrogen attack, hydrogen-induced cracking, and internal corrosion.

The Occupational Safety and Health Administration (OSHA) investigated the Romeoville incident and recommended subsequent metallurgical inspections at other US sites. Those inspections revealed significant cracking in amine processing equipment, primarily in or near welds, and where modifications or repairs had been made to the vessel. The OSHA study found that field welding procedures can significantly contribute to brittle failures of these vessels (vessels with greater susceptibility to hydrogen embrittlement and localized corrosion).

It was noted in the OSHA report that the failed vessel had been field repaired, and the repair welds had not been post-weld heat-treated (PWHT) to relieve residual stresses in or near the welds. Industry codes provide additional details on preventing this metallurgical failure and inspection techniques to detect a pending failure (API-510, ASME boiler and pressure vessel codes, ANSI codes). OSHA also developed a technical manual highlighting the "recent" cracking experience and general pressure vessel guidelines, including inspection techniques and mitigation steps. This information is available on the OSHA website (www.osha.gov).

This incident also highlights the need to quickly evacuate personnel from the immediate process area when an abnormal situation exists, such as a leak. This is also important during a unit upset or during periods of transient operations such as startup and shutdown.

Mexico City, Mexico (San Juanico 19 November 1984)

A major fire, quickly followed by a series of catastrophic explosions, occurred at approximately 05:35 hours on 19 November 1984 at the government-owned and operated LPG storage and distribution Terminal in San Juanico, Mexico. The entire storage terminal was destroyed. This was one of the most devastating process safety events in terms of loss of life and equipment damage (except for the Bhopal, India methyl isocyanate [MIC] release, which also occurred during 1984 [3 December 1984], which resulted in an estimated 2,700 fatalities. However, other estimates place the number of fatalities in the Bhopal incident much higher).

In Mexico City, the LPG pressurized tanks included six large spherical tanks and 48 horizontal storage vessels ("propane bullets") with a combined capacity of over 4.2 million gallons. At the time of the incident, the terminal was operating at 90% of capacity and receiving propane by pipeline. The two largest spheres and the 48 bullets were full, and propane was being received into the remaining four spheres when an 8-inch pipeline to one of the spheres failed. The failure resulted in a sudden drop in pressure on the propane pipeline supplying product to the terminal, which was detected in the facility control room. The operators could not contact the pipeline terminal to stop the LPG delivery, and the LPG was released for 5–10 minutes before an adjacent ground pit flare ignited the resulting large vapor cloud. The resulting overpressure,

Figure 5.4 Pemex LPG Terminal located at San Juan Ixhuatepec, Mexico, before the series of BLEVEs. Several additional spheres were added to the terminal after this photograph was taken. Source: Image courtesy Mr. Dick Hawrelak (Presented to ES-317Y at UWO in 1999. Dick Hawrelak).

shrapnel from the damaged tanks, and radiant heat caused a secondary loss of containment and a number of other ground fires.

San Juan Ixhuatepec, home to about 40,000 people, was essentially destroyed by the explosions at the adjacent LPG terminal, killing many inhabitants. An estimated 500–600 people were killed due to the incident, and 5,000–7,000 suffered severe burns. The heat generated by the devastating inferno left a tiny percentage of the remains recognizable.

The first BLEVE occurred about fifteen minutes after the initial release, followed by a series of BLEVEs as other LPG storage vessels violently exploded. The four smaller spheres and 44 of the bullets BLEVE'ed. Reports indicated that LPG was raining down, causing other fires. The explosions were so severe that they were detected on a University of Mexico seismograph. The terminal's firewater system was disabled in the initial blast, and water for firefighting and cooling the remaining two large spheres was delivered by 100 tank cars. Several photographs of this devastating series of BLEVEs follow in Figures 5.4, 5.5a–c, and 5.6.

Lessons learned from this incident include the following:

- Emergency isolation equipment should be available to quickly isolate a leaking or failed system. These isolation valves should be readily accessible and tested during periodic emergency drills and exercises.
- Adequate overpressure protection for vessels and pipeline systems, including periodic reviews to consider changes in process capacity.
- Consider equipment spacing during plant design.
- Emergency response equipment for LPG facilities should include an adequate firewater supply, including water deluge systems.
- Effective gas detection can help with the early detection of a release.
- Availability of emergency services equipment (arrival of emergency services was hindered by area traffic due to residents attempting to evacuate the area).
- Well-documented and practiced emergency response plans; practiced on-site emergency drills and exercises.
- Never allow complacency when working on or near equipment containing hazardous or toxic materials. Never take anything for granted.

(a)

(b)

(c)

Figure 5.5 Mexico City, 19 November 1984, following a series of devastating BLEVE explosions. Source: (a) The Center for Chemical Process Safety (CCPS) (An AIChE Technology Alliance), and Engro Polymer and Chemicals Limited. From the Safety Briefing for the First Regional CCPS Meeting in Pakistan (November 2018 – 34 years after this catastrophic event). (b and c) Image with the kind courtesy of the American Institute of Chemical Engineers (AICHE), The Center for Chemical Process Safety (CCPS), and the Beacon. http://www.aiche.org/CCPS/Publications/Beacon/index.aspx.

Figure 5.6 LPG cylindrical storage tank propelled 3,940 feet (1,200 meters) into a housing area. Source: Image courtesy Dick Hawrelak: Presented to ES-317Y at UWO in 1999 by Dick Hawrelak.

The Cosmo Refinery Series of BLEVEs – Ichihara City (Chiba), Japan (11 March 2011)

The Cosmo Refinery series of BLEVEs occurred following the 11 March 2011, Japanese earthquake and resulting Tsunami. When the earthquake occurred, one of the spheres at this site was undergoing a hydrostatic pressure test and was filled with water, 1.8 times as heavy as LPG and the heaviest weight the vessel will see in its lifetime.

When the earthquake occurred, the supporting structure for the sphere failed, and the vessel toppled from the supports, landing on some LPG piping inside the sphere firewall. The sphere piping was also undergoing maintenance, and the

Figure 5.7 Cosmo Refinery (Image before LPG explosions). Source: Photos courtesy of Google Earth Pro.

Figure 5.8 Cosmo Refinery, mage after LPG explosions (multiple BLEVEs. Source: Photos courtesy of Google Earth Pro. Note that some tanks were propelled into the Sea of Japan by the explosions.

isolation valves for the piping were locked in the open position and could not be closed. This resulted in a very large pool fire directly below the other spheres containing various grades of LPG.

The impinging fires resulted in failures of several of the spheres and catastrophic BLEVEs of the vessels, causing the vessels to be propelled into the Sea of Japan. The resulting fires spread to an adjacent asphalt storage tank farm. Fortunately, the site had evacuated the area, no one was killed, and the injuries were not serious.

Following the Japanese earthquakes, another series of catastrophic BLEVE explosions occurred at the Cosmo Refining complex on 11 March 2011. Images of the Cosmo Refining complex before and after the series of BLEVE explosions are available in Figures 5.7 and 5.8 Cosmo Refinery (Japan).

Additional References

"The 100 Largest Losses in the Hydrocarbon Industry" Marsh and McLennan Companies The 100 Largest Losses in the Hydrocarbon Industry (marsh.com)

"Union Oil Company of California Explosion Report" Romeoville, Illinois (23 July 1984) Occupational Safety and Health Administration (OSHA) https://www.osha.gov/laws-regs/standardinterpretations/1986-04-11

"Mexico City Explosion and Resulting Fire" Marsh and McLennan, 'Large Property Damage Losses in the Hydrocarbon-Chemical Industries a thirty-year Review,' 16th Edition, Marsh and McLennan Protection Consultants, 1995.

Mexico City Explosion and Resulting Fire UK Health and Safety Executive (hse.gov.uk) https://www.hse.gov.uk/comah/sragtech/casepemex84.htm

National Fire Protection Association (NFPA) Training Video "BLEVE Update" (This video is dated but still excellent from a training perspective)

European Commission, Joint Research Centre eNatech – Natural hazard-triggered technological accidents database Report on explosions and fires at Cosmo Oil refinery's LPG storage area and adjacent asphalt tank farm https://enatech.jrc.ec.europa.eu/Natech/14

French Ministry of the Environment – DPPR/SEI/BARPI – CFBP BLEVE in an LPG storage Facility at a refinery 4 January 1966 Feyzin (Rhône), France https://www.aria.developpement-durable.gouv.fr/wp-content/files_mf/FD_1_feyzin_GC_ang.pdf

British Health and Safety Executive Refinery fire at Feyzin. 4 January 1966 https://www.hse.gov.uk/comah/sragtech/casefeyzin66.htm

Chapter 5. Boiling Liquid Expanding Vapor Explosion (BLEVE)

End of Chapter Quiz

1 What is the mechanism or cause of a Boiling Liquid Expanding Vapor Explosion (BLEVE)?

2 What is by far the most common cause of a BLEVE occurring?

3 How long after the initial fire and vessel impingement must we prepare for the potential BLEVE?

4 What is meant by LPG at a temperature "above its boiling point"?

5 What is the hazard of a BLEVE?

6 What are the warnings, and how do we know the vessel is about to BLEVE?

7 What precautions do we take to help prevent or mitigate a potential BLEVE?

8 How can a water draw operation on an LPG sphere result in a BLEVE of the vessel? Describe the proper procedure for draining water on an LPG sphere.

9 Can you name several sites where a BLEVE has occurred? Try to name at least three.

10 When Propane flashes from a liquid to a vapor, as when a vessel initially BLEVEs, how much does it expand when it vaporizes?

11 Is it possible for a vessel that is in a service other than hydrocarbon service to BLEVE?

12 Can metal debris be propelled long distances from a vessel during a BLEVE?

13 The BLEVE potential is considered during the initial vessel design. Name at least four special considerations incorporated into the LPG vessel design.

6

An Introduction to Process Safety Management Systems

Process Safety Management Systems (PSMS) are the guiding documents that transcribe the company's corporate Safety, Health, Environmental, and Security values and culture into documented and actionable management plans. The PSMS forms a Framework by which the corporate values are documented and communicated to all employees, contractors, and anyone who works for the company.

With the framework in place, the PSMS system adds additional detail and objectives and ensures the goals are translated into actionable plans and procedures aligned with the corporate values. The Objectives, Plans, and Procedures are communicated to all corporation members (or site, as the case may be) for implementation. Responsibilities are assigned to individuals, generally members of management with corresponding roles in the organization, or to subject matter experts for PSMS implementation. Those members Responsible are also Accountable for implementation and the ongoing verification and measurement to ensure complete execution and continuous improvement.

The importance of Safety Management Systems is emphasized in the Piper Alpha training video "Spiral to Disaster." There were 167 lives lost in the Piper Alpha explosion and fire, which occurred in the North Sea in 1988. In the training video, critical work processes such as Permit to Work, Shift Handover, and Critical Device Management were characterized as "*sloppy, ill-organized, and unsystematic.*" The video and the documented investigation reports for this tragic incident clearly show the gaps in these critical work processes on Piper Alpha and how this contributed to the tragic loss of life and the platform's destruction. Safety management systems like PSMS, when properly implemented, maintained, current, and enforced, help ensure that these systems are organized and systematic.

The PSMS is documented and, at a minimum, includes the following written components:

• The "What"	• Defines the Scope/Objectives of the system.
• The "How"	• Provides the Procedures/Practices to make it happen
• The "Who"	• Who is Responsible and Accountable for implementation?
• The "How Do We Know?"	• Are we getting the expected results?
• The "How Can We Do Better"	• Ensures continuous improvement.

The United States Occupational Safety and Health Administration (OSHA) prepared a sample Safety Management System named the "Safety and Health Program Management Guidelines." The OSHA guideline outlines the following four elements as the guiding principles for the program:

- Management Leadership and Employee Involvement
- Worksite Analysis
- Hazard Prevention and Control
- Training

While this is a helpful guideline for implementing personnel safety, one designed to manage petrochemical process safety must be more comprehensive and contain elements specific to process operations and equipment integrity. Process safety management systems should meet the requirements outlined above and all provisions of the OSHA Process Safety Management standard (29 CFR 1910.119, "Process safety management of highly hazardous chemicals").

Process Operations Safety: The What, Why, and How Behind Safe Petrochemical Plant Operations, First Edition. M. Darryl Yoes.

Several refining and petrochemical companies have developed management systems covering Safety, Health, Environmental, and Operations Integrity, including the additional elements protecting process safety. This chapter provides examples of other PSM Management System elements and a general discussion of how the process safety management system (PSMS) is used to help establish a sound process safety culture and ensure ongoing compliance verification and continuous improvement.

A key element to any sound PSMS is the support and commitment from senior management. The direction for a sound PSM culture must come from the top of the organization. Due to the nature of these systems and the resources required for initial implementation and ongoing support, the commitment from senior-level managers at the site and the corporate level is critical and essential.

PSMS sets the minimum standard/expectation for key PSM performance parameters. These are supported by management and reinforced by ongoing stewardship and management reviews to help ensure compliance and corrective actions when required. For example, vessel inspection schedule: the PSMS will define the expectations for vessel inspections (internal and external) and track actual inspection performance through stewardship and management reviews. In this example, routine reports will track vessel inspection compliance using predefined PSM "leading indicators." These reports are routinely reviewed by the site's senior management levels.

The PSMS must be designed with a defined process for self-improvement, including periodic audits to evaluate the PSMS program documentation and field compliance to meet the program's deliverables. Generally, these audits are conducted by peers with PSMS management system experience and experience in the specific program, procedure, or assessed equipment. Audits should be designed to provide feedback to the PSMS stewards and include any identified gaps and opportunities for continued improvement.

Most PSMS management systems use a tiered approach for system documentation and requirements, with higher-level documents providing a broad overview of management expectations for the specific system and the details provided in the lower-level supporting documents. For example, the Management System for "Training" will typically have broad training requirements in the higher-level system documents, requiring training to be conducted periodically and comply with OSHA PSM and other regulatory requirements. The training required by specific organization members is then detailed in the supporting documents, including training requirements for initial operator training, operator refresher training, and training for maintenance employees.

The overall Training system should also utilize leading indicators to track compliance with the specific system requirements. Finally, to ensure compliance, the program documentation should also define the ongoing audit process, including audit frequency, qualifications/experience requirements of those conducting the audit, and documentation of the audit results.

The following list provides examples of typical process safety management systems:

I. Management Leadership and Employee Involvement[1]
II. Worksite Analysis[1]
III. Hazard Prevention and Control[1]
IV. Training[1]
V. Emergency Plans and Preparedness
VI. Equipment Design and Construction
VII. Equipment Integrity and Inspection
VIII. Managing Change
IX. Operating Procedures, maintenance procedures, and drawings
X. Contractor Safety Management
XI. Learning from Incidents (incident investigation and sharing of results)
XII. Community and Right to Know
XIII. Continuous Improvement Process

1 Included in OSHAs "Safety and Health Program Management Guidelines."

Example Management System (Overview)

PSM System IV	"Training"
Sub-Systems:	I. Initial Operator Training II. Operator Refresher Training III. Console Operator Initial Training IV. Console Operator Refresher Training V. Maintenance Employee Training VI. First-Line Supervisor Initial Training
Sub-System II	
Operator Refresher Training:	
Site Requirement 1a:	All Operators are to receive annual refresher training on operations procedures and process safety information, including hazardous materials used in the process.
Measurement:	Review training records annually for compliance.
	Metric = % Operators who have completed operator refresher training within the planned schedule (expected to be 100%).
Reported to:	Site operations manager
Responsibility	
For follow-up:	Operator's Supervisor
Continuous Improvement:	The internal team reviews results annually and reports recommendations for improvement to the site operations manager.
	Ensure Corporate assessment reviews periodically (typically threeto five years), and audit results and recommendations for improvement are reported to site and corporate management.

Process Safety Management Systems are structured to help prevent and minimize the consequences of the accidental release of highly hazardous chemicals. The PSMS will be reviewed regularly and updated to reflect new or modified tasks and/or procedures. A copy of the program shall be made available to everyone working for the company and others to help ensure the applicable OSHA regulations are complied with.

Example Process Safety Management System

Note: This is an example and not a representation of any company, facility, or site management system. This is written as an example of a system based on the requirements of the Occupational Safety and Health Administration's regulation covering process safety (29 CFR 1910.119). Some provisions closely mirror the OSHA standard requirements, and sections of the OSHA Standard are included in the following text. This should not be used as a Process Safety Management System; however, it is intended as a guide in developing such a standard.

SCOPE

The site PSMS program covers the following:

- Each activity involving a chemical listed on the OSHA PSM Table I at concentrations at or above the specified threshold quantity.
- Each activity involving a flammable liquid or gas as defined in 29 CFR 1910.1200 that is stored or used on-site in one location in a quantity of 10,000 pounds or more. The exception is hydrocarbon fuels used solely for workplace consumption and flammable liquids stored in atmospheric tanks.
- It does not apply to oil or gas drilling or servicing operations (as outlined in OSHA PSM (29CFR 1910.119)).

Process Safety Information

The process safety information shall also be provided to those individuals working at the site and includes the following:

1. A compilation of written process safety information before conducting any process hazard analysis required by the PSM Standard. This includes the following information related to the hazards of the highly hazardous chemicals used in the activities at the site:

- Toxicity
- Permissible exposure limits (PEL)
- Physical data
- Reactivity data
- Corrosivity data
- Thermal and chemical stability data
- The hazardous effects of inadvertent mixing of different materials could foreseeably occur.

2. Information pertaining to the technology of the process shall include at least the following:
   - A block flow diagram or simplified process flow diagram
   - Process chemistry
   - Maximum intended inventory
   - Safe upper and lower limits for such items as temperatures, pressures, flows, or compositions
   - Evaluation of the consequences of deviations, including those affecting the safety and health of employees.
3. Where the technical information no longer exists, such information may be developed in sufficient detail in conjunction with the process hazard analysis to support the analysis.
4. Information pertaining to the equipment in the process:
   - Materials of construction (e.g., metallurgy)
   - Electrical classification
   - Relief system designs and design basis
   - Piping and instrument diagrams (P&IDs or PFDs)
   - Ventilation system design
   - Design codes and standards employed.
   - Material and energy balances for processes built after 26 May 1992.
   - Safety systems (such as interlocks, detections, and suppression systems)

Process Hazard Analysis

The process hazard analysis (PHA) shall be performed by a team with expertise in engineering and process operations. The team shall include at least one employee with experience and knowledge specific to the process being evaluated. Also, one team member must be knowledgeable in the process hazard analysis methodology.

The team will conduct the process hazard analysis. The analysis will identify, evaluate, and control the hazards involved in the process. The Committee will determine and document the priority order for conducting the study based on the following:

- The extent of the process hazards
- Number of potentially affected employees
- Age of the process
- Operating history of the process

The process hazard analysis shall be updated and revalidated at least every five years by the same or equally competent Committee.

One or more of the following methodologies are appropriate to determine and evaluate the hazards of the process being analyzed:

- What-If or What-If/Checklist
- Checklist
- Hazard and Operability Study (HAZOP)
- Failure Mode and Effects Analysis (FMEA)
- Fault Tree Analysis
- An appropriate equivalent methodology

The process hazard analysis will address the following:

- The hazards of the process
- The identification of any previous incident with likely potential for catastrophic consequences in the workplace

- Engineering and administrative controls applicable to the hazards and their interrelationships, such as applying detection to provide early warning of releases.
- Consequences of failure of engineering and administrative controls
- Facility sitting
- Human factors
- A qualitative evaluation of a range of possible safety and health effects of failure of controls on employees in the workplace.

A system will be established to promptly address the findings and recommendations; ensure that the recommendations are resolved in a timely manner and the resolution is documented; document what actions are to be taken; complete actions as soon as possible; develop a written schedule of when these actions are to be completed; communicate the actions to operating, maintenance, and other employees whose work assignments are in the process and who may be affected by the recommendations or actions:

- Assure that the recommendations are resolved in a timely manner and that resolutions are documented.
- Document what actions are to be taken.
- Complete actions as soon as possible.
- Develop a written schedule of when these actions are to be completed.
- Communicate the actions to site management and those that could be affected by the recommendations or actions.
- Retain the process hazard analysis, all updates, and revalidation of each process for the life of the process.

Operating Procedures

Written operating procedures shall be developed and implemented that provide clear instructions for safely conducting activities involved in each covered process consistent with the process safety information and shall address at least the following elements.

The procedures will provide clear instructions for safely conducting activities and will include the following:

1. The steps for each operating phase
   - Initial startup.
   - Normal operation.
   - Temporary operations.
   - Emergency shutdown, including the conditions under which emergency shutdown is required and the assignment of shutdown responsibility to qualified operators to ensure that emergency shutdown is executed safely and promptly.
   - Emergency operations.
   - Normal shutdown.
   - Startup following a turnaround or after an emergency shutdown.
2. Operating limits
   - Consequences of deviation.
   - Steps required to correct and avoid deviation.
   - Safety systems and their functions.
3. Safety and health considerations
   - Properties of and hazards presented by the chemicals used in the process.
   - Precautions necessary to prevent exposure include engineering controls, administrative controls, and protective personnel equipment (PPE).
   - Control measures are to be taken if physical contact or airborne exposure occurs.
   - Safety procedures for opening process equipment, such as pipe breaking.
   - Quality control for raw materials and management of hazardous chemical inventory levels.
   - Any special or unique hazards.
4. Operating procedures shall be readily accessible to employees who work in or maintain a process.
5. Safe work practices to provide for the control of hazards during operations shall be developed and implemented, such as the following:
   - Lockout/tag-out (energy isolation before mechanical work).
   - Confined space entry.

- Opening process equipment or piping.
- Control over the entrance to a facility by maintenance, contractor, laboratory, or other support personnel.
- Chemical Hygiene Plan.
- These safe work practices shall apply to employees and contractor employees.

Training

A. Initial Training
   Each employee presently involved in operating a process, and each employee, before they operate a newly assigned process, shall be trained in an overview of the process and the operating procedures. The training shall emphasize the specific safety and health hazards, emergency operations including shutdown, and safe work practices applicable to the employee's job tasks.
B. Refresher Training
   Refresher training shall be provided at least every three years and more often, if necessary, to each employee involved in operating a process to ensure that the employee understands and adheres to the current operating procedures of the process. In consultation with the employees involved in operating the process, the employer shall determine the appropriate frequency of refresher training.
C. Training Documentation
   The site shall ascertain that each employee involved in operating a process has received and understood the training required by this paragraph. Records shall be implemented containing the employee's identity, the training date, and the means used to verify that the employee understood the training.

Contractors

A. This paragraph applies to contractors performing maintenance or repair, turnaround, major renovation, or specialty work on or adjacent to a covered process.
B. It does not apply to contractors providing incidental services that do not influence process safety, such as janitorial work, food and drink services, laundry, delivery, or other supply services.
C. When selecting a contractor, the site shall obtain and evaluate the contract employer's safety performance and programs:
   - The employer shall inform the contract employees of the known potential fire, explosion, or toxic release hazards related to the contractor's work and the process.
   - The employer shall explain to contract employers the applicable provisions of the emergency action plan.
   - The employer shall develop and implement safe work practices to control the entrance, presence, and exit of contract employers and contract employees in covered process areas.
   - The employer shall periodically evaluate the performance of contract employers in fulfilling their obligations as specified in the OSHA standard, §1910.119(h)(3), and in this section; and
   - The employer shall maintain a contract employee injury and illness log related to the contractor's work in process areas.
D. The contract employer shall ensure that the contract employee is trained in the work practices necessary to safely perform their job:
   - The contract employer shall ensure that each contract employee is trained in the work practices necessary to perform their job safely.
   - The contract employer shall ensure that each contract employee is instructed in the known potential fire, explosion, or toxic release hazards related to their job and the process and the applicable provisions of the emergency action plan.
   - The contract employer shall document that each contract employee has received and understood the training required by the OSHA PSM standard and that the contract employer will prepare a record that contains the identity of the contract employee, the date of training, and the means used to verify that the employee understood the training.
   - The contract employer shall ensure that each contract employee follows the safety rules of the facility, including the safe work practices required by the OSHA PSM standard (lockout/tagout, confined space entry, etc.), and
   - Advise the employer of any unique hazards presented by the contract employer's work or any hazards found by the contract employee's work.

Pre-startup Safety Review

A. A pre-startup safety review shall be conducted for new or modified facilities when the modification is significant enough to require a change in the process safety information.

B. The pre-startup safety review shall confirm that before the introduction of highly hazardous chemicals to a process:
   - Construction and equipment follow design specifications.
   - Safety, operating, maintenance, and emergency procedures are in place and are adequate.
   - For new facilities, a process hazard analysis has been performed, recommendations have been resolved or implemented before startup, and modified facilities meet the requirements of management of change.
   - Safety, operating, maintenance, and emergency procedures are in place and are adequate.
   - Training of each employee involved in operating a process has been completed.

Mechanical Integrity

A. The site shall establish and implement written procedures to maintain the ongoing integrity of the following items of process equipment:
   - Pressure vessels and storage tanks.
   - Piping systems (including piping components such as valves).
   - Relief and vent systems and devices.
   - Emergency shutdown systems.
   - Controls (including monitoring devices and sensors, alarms, and interlocks).
   - Pumps.

B. Each employee involved in maintaining the ongoing integrity of process equipment shall be trained in an overview of that process and its hazards and in the procedures applicable to the employee's job tasks to assure that the employee can safely perform the job tasks.

C. An inspection and testing program will be established, including the following:
   - Inspections and tests shall be performed on that process equipment.
   - Inspection and testing procedures shall follow applicable codes and standards and generally accepted good engineering practices (e.g., the American Society of Mechanical Engineers, the American Petroleum Institute, the American Institute of Chemical Engineers, the American National Standards Institute, the American Society of Testing and Materials, and the National Fire Protection Association).
   - The frequency of inspections and tests of process equipment shall be consistent with applicable manufacturers' recommendations and good engineering practices.
   - They will be conducted more frequently if determined necessary by prior operating experience.
   - Each inspection and test shall be documented to identify the following:
     (a) The date of the inspection or test,
     (b) The name of the person who performed the inspection or test,
     (c) The serial number or another equipment identification method for the Inspection,
     (d) A description of the inspection or test performed, and
     (e) The results of the inspection or test.

D. Deficiencies in equipment that are outside acceptable limits (defined by the process safety information mentioned in Part IV of this Program and §1910.119(d) of the OSHA standard) must be corrected either before further use or in a safe and timely manner provided necessary means are taken to assure safe operation.

E. Whenever a new plant or new equipment is constructed, the site will ensure that:
   1. The equipment, as built, is suitable for the process application for which it is to be used.
   2. Equipment, as fabricated, meets design specifications.
   3. Appropriate checks and inspections are performed as necessary to ensure that equipment is installed correctly, consistent with design specifications, and the manufacturer's instructions; and
   4. The maintenance equipment and spare parts meet design specifications.

F. The employer shall correct deficiencies in equipment that are outside acceptable limits as defined in the OSHA PSM Standard before further use or in a safe and timely manner when necessary, and means are taken to assure safe operation.

Hot Work Permit

A. The site shall issue a hot work permit for hot work operations conducted on or near a covered process (welding, burning, and similar work).
B. The hot work permit shall ensure that the fire prevention and protection requirements in 29 C.F.R. §1910.252(a) are documented (the OSHA standard that specifies the precautions for fire prevention in welding or cutting work have been implemented before starting any hot work).
C. Each permit shall indicate the date(s) authorized for hot work and identify the object on which hot work is to be performed.
D. The permit shall be kept on file until the hot work operations are completed.

Management of Change

A. The site shall establish and implement written procedures to manage changes (except for "replacements in kind") to process chemicals, technology, equipment, procedures, and changes to facilities that affect a covered process.
B. The procedures shall ensure that the following considerations are addressed before any change:
   - The technical basis for the change.
   - Impact of the change on employee safety and health.
   - Modifications to operating procedures.
   - The necessary time period for the change; and,
   - Authorization requirements for the proposed change.
C. All site employees involved in operating a process and maintenance and contract employees whose job tasks will be affected by a change in the process shall be informed of and trained in the change before the startup or any affected part of the process.
D. If such a change results in a change to the process safety information covered by this Program and 29 C.F.R. §1910.119(d), that information will be appropriately updated.
E. If a change covered by the PSM Standard results in a change to the process safety information required by OSHA PSM, such information will be updated accordingly.

Incident Investigation

A. The site shall investigate each incident that resulted in or could reasonably have resulted in, a catastrophic release of highly hazardous chemicals in the workplace.
B. An incident investigation shall be initiated as promptly as possible but not later than 48 hours following the incident.
C. An incident team shall be established and consist of at least one person knowledgeable in the process involved, including a contract employee if the incident involves the contractor's work and other persons with appropriate knowledge and experience to thoroughly investigate and analyze the incident.
D. The investigation team must prepare a report at the conclusion of the investigation, which includes, at a minimum:
   - The date of the incident,
   - The date that the investigation started,
   - A description of the incident,
   - The factors that contributed to the incident, and
   - Any recommendations resulting from the investigation.
E. The site shall establish a process to promptly address and resolve the incident report's findings and recommendations. All resolutions and corrective actions shall be documented.
F. The report shall be reviewed with all affected personnel whose job tasks are relevant to the incident findings, including contract employees, where applicable.
G. Incident investigation reports shall be retained for five years.

Emergency Planning and Response

The site shall establish and implement an Emergency Action Plan for the entire facility in accordance with the OSHA PSM Standard. In addition, the emergency action plan shall include procedures for handling small releases. Employers covered under this standard may also be subject to the hazardous waste and emergency response provisions contained in the OSHA regulation.

Compliance Audits

A. The site shall certify that they have evaluated compliance with the provisions of OSHA PSM at least every three years to verify that the procedures and practices developed under the standard are adequate and are being followed.
B. The compliance audit shall be conducted by at least one person knowledgeable in the process.
C. A report of the findings of the audit will be developed.
D. The site shall promptly determine and document an appropriate response to each of the compliance audit findings and document deficiencies that have been corrected.
E. The site shall retain the two (2) most recent compliance audit reports.

Trade Secrets

A. Regardless of the trade secret status of such information, all information needed to comply with the provisions of the OSHA PSM standard will be made available to those responsible for compliance.
B. It is recognized that a confidentiality agreement may be required for those persons provided access to information that can be considered a trade secret as provided in the OSHA Hazard Communication standard, 29 C.F.R. 1910.1200.
C. Subject to the rules and procedures contained in that standard (see 1910.1200(I)(1) through 1910.1200(I)(12)), employees and their designated representatives shall be permitted access to trade secret information contained within our process hazard analysis and other documents that are required to be developed by the OSHA Process Safety Management standard, 29 C.F.R. 1910.119.

End of the example Process Safety Management System

US OSHA Sample Safety Management System

The United States Occupational Safety and Health Administration (OSHA) prepared a sample Safety Management System named the "Safety and Health Program Management Guidelines." The OSHA guideline outlines the following four elements as the guiding principles for the example program:

- Management Commitment and Employee Involvement.
- Worksite Analysis.
- Hazard Prevention and Control.
- Safety and Health Training.

Note: The OSHA Guideline is available at the following website: https://www.osha.gov/shpmguidelines/SHPM_guidelines.pdf.

While this is a sound guideline for implementing personnel safety, process safety management systems should meet the requirements outlined above and all OSHA Process Safety Management standards (29 CFR 1910.119, "Process safety management of highly hazardous chemicals"). A process safety management system must be more comprehensive and contain elements specific to process operations involving highly hazardous materials, including those that address equipment integrity (preventing loss of containment).

The OSHA Process Safety Management Regulation mentioned above is made up of fourteen elements, which are:

1. Employee Participation	8. Mechanical Integrity
2. Process Safety Information	9. Hot Work Permit
3. Process Hazard Analysis	10. Management of Change
4. Operating Procedures	11. Incident Investigation
5. Training	12. Emergency Planning and Response
6. Contractors	13. Compliance Audits
7. Pre-startup Safety Review	14. Trade Secrets

For efficiency purposes, the following list consolidates the two OSHA lists, the four elements from the OSHA example Safety Management System plus the fourteen elements from the OSHA PSM Regulation to make up the critical Elements of a Process Safety Management System (PSMS).

1. Management Commitment, Accountability, and Employee Involvement *(Renamed Management Commitment, Accountability, and Employee Involvement, but includes Employee Participation)*	8. Mechanical Integrity
2. Process Safety Information	9. Permit to Work *(Renamed from Hot Work Permit to include other types of work permits)*
3. Process Hazard Analysis and Risk Assessment *(Renamed from Process Hazard Analysis to include Worksite Analysis and other types of risk assessment)*	10. Management of Change
4. Procedures *(Includes Operating and Maintenance Procedures)*	11. Incident Investigation
5. Training *(Includes Safety and Health Training)*	12. Emergency Preparedness *(Renamed from Emergency Planning and Response to cover all aspects of emergency readiness and response)*
6. Contractors	13. Compliance Audits
7. Pre-startup Safety Review	14. Trade Secrets

With this consolidation, we now have the elements to make up an example Process Safety Management System, made up of the PSM required 14 elements plus the four elements recommended by OSHA for personnel safety.

As a recap, the example PSMS is made up of the following "Elements":

1. Management Commitment, Accountability, and Employee Involvement	8. Mechanical Integrity
2. Process Safety Information	9. Permit to Work
3. Process Hazard Analysis and Risk Assessment	10. Management of Change
4. Procedures	11. Incident Investigation
5. Training	12. Emergency Preparedness
6. Contractors	13. Compliance Audits
7. Pre-startup Safety Review	14. Trade Secrets

This is not a management system, as it only defines elements. We must develop the expected results for each management system, specific action items for each expected outcome, and the process to measure the results or verify that the desired result is being achieved.

As illustrated below, consider the Management System as several management work items or processes in a series flow.

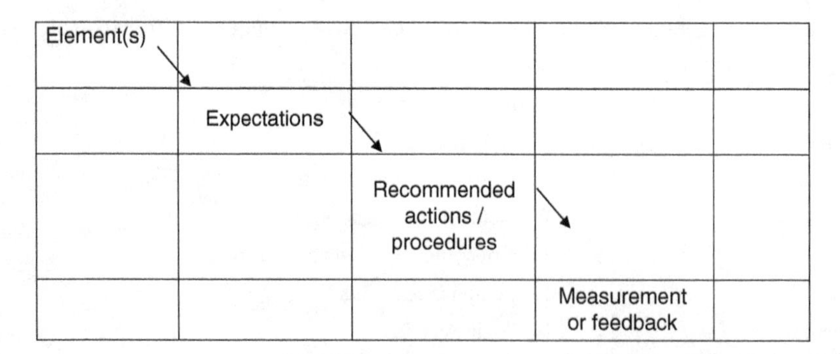

For example, the OSHA example Safety Management System has "Elements" or main safety areas of concern as listed below:

- Management Commitment and Employee Involvement (8)
- Worksite Analysis (5)
- Hazard Prevention and Control (4)
- Safety and Health Training (3)

In the OSHA example, the four Elements have several "Expectations" or expected results identified by management, and each Expectation has several "Recommended Actions" to help ensure the system is delivering the desired results. The Expected Results (3) and associated Recommended Actions (8) for Element 1 in the OSHA example safety management system are listed below. The "online" approach did not include measurements or prescribe a process to ensure the system performs as expected and delivers the desired results. This would be expected in most Process Safety Management Systems and will be recommended in other methods listed in this chapter.

Element 1 "Management Commitment and Employee Involvement" (3 Expectations, 8 Recommended Actions):

Expectations:

1.1 Management commitment and employee involvement complement other company systems and standards.

1.2 Management commitment provides the motivating force and resources for organizing and controlling activities within an organization.

1.3 Employee involvement provides how workers develop and express their commitment to safety and health protection.

Recommended Actions:

1.1a. State a worksite safety and health policy

1.1b. Establish and communicate a clear goal and objective for the safety and health program

1.2a. Provide visible top management involvement in implementing the program

1.2b. Encourage employee involvement in the program and in decisions that affect their safety and health (e.g., inspection or hazard analysis teams; developing or revising safe work rules; training new hires or co-workers; assisting in accident investigations)

1.2c. Assign and communicate responsibility for all aspects of the program

1.2d. Provide adequate authority and resources to responsible parties

1.2e. Hold managers, supervisors, and employees accountable for meeting their responsibilities

1.3a. Review program operations at least annually to evaluate, identify deficiencies, and revise, as needed

Element 2 "Worksite Analysis" (2 Expectations, 5 Recommended Actions):

Element 3 "Hazard Prevention and Control" (4 Expectations, 4 Recommended Actions):

Element 4 "Safety and Health Training" (3 Expectations, 4 Recommended Actions):

Note: The additional Expectations and Recommended Actions for Elements 2–4 in the OSHA example are not listed here but are available on the OSHA website.

Several refining and petrochemical companies have developed management systems covering Safety, Health, Environmental, and Operations Integrity, including the additional elements protecting process safety. This chapter provides examples of other PSM Management System elements and a general discussion of how the Process Safety Management System (PSMS) is used to help establish a sound process safety culture and ensure ongoing compliance verification and continuous improvement.

To recap:

The site or local PSMS, once completed, should be made up of the following data or sections:

- The overarching Corporate Guiding Documents provide the corporate expectations (these documents may be titled appropriately to their content to illustrate the corporate involvement).
- Specific Action Plans or Procedures (linked to the Corporate Guiding documents).
- Responsibilities for Implementation.
- Stewardship (tracking the status of Action Plans vs. documented in the PSMS).
- Continuous Improvement Plans.

Most PSMS management systems use a tiered approach for system documentation and requirements, with higher-level documents providing a broad overview of management expectations for the specific system and the details provided in the lower-level supporting documents. For example, the Management System for "Training" will typically have broad training requirements in the higher-level system documents, such as a requirement that training is conducted periodically and complies with OSHA PSM and other regulatory requirements. The detailed training required by specific members of the organization is then detailed in the supporting documents to include training requirements for initial Operator training, Operator refresher training, and training for maintenance employees.

The overall Training system should also utilize leading indicators to track compliance with the specific system requirements. Finally, to ensure compliance, the program documentation should also define the ongoing audit process, including audit frequency, qualifications/experience requirements of those conducting the audits, and documentation of the audit results.

The following list provides examples of typical Process Safety Management Systems:

1. Management Leadership and Employee Involvement[2]
2. Worksite Analysis[2]
3. Hazard Prevention and Control[2]
4. Training[2]
5. Emergency Plans and Preparedness
6. Equipment Design and Construction
7. Equipment Integrity and Inspection
8. Managing Change
9. Operating Procedures, Maintenance Procedures, and Drawings
10. Contractor Safety Management
11. Learning from Incidents (Incident Reporting, Investigation, and Sharing of Results)
12. Community and Right to Know
13. Continuous Improvement Process

Chapter 6. An Introduction to Process Safety Management Systems

End of Chapter Quiz

1. What is the purpose of the Process Safety Management System?

2. What are the five key written components of each Process Safety Management System?

3. Can you name the four key elements the OSHA example Safety Management system requires?

4. Why is not the OSHA sample Safety Management System unsuitable for a petroleum refining or petrochemical plant management system?

5. Can you name the 14 key elements required by the OSHA Process Safety Management regulation (29 CFR 1910.119)?

6. What is the first Management Element in each of the referenced Management Systems?

7. The Process Safety Management Systems (PSMS) are the guiding documents that transcribe which of the following values and culture into documented and actionable management plans?
   a. Safety
   b. Health
   c. Environmental
   d. Security
   e. All of the above

8. The PSMS forms a Framework by which the corporate values are documented and communicated to all _________, _________, and _________ who works for the company.

2 Included in OSHAs "Safety and Health Program Management Guidelines."

9 Match the components of the management system with the proper description.

A. The "What"	A. Who is Responsible and Accountable for implementation?
B. The "How"	B. Are we getting the expected results?
C. The "Who"	C. Defines the Scope/Objectives of the system.
D. The "How Do We Know?"	D. Ensures continuous improvement.
E. The "How Can We Do Better"	E. Provides the Procedures/Practices to make it happen

10 The importance of Safety Management Systems is emphasized in what training video, where 167 lives were lost, and the critical work processes were characterized as "sloppy, ill-organized, and unsystematic."

11 PSMS sets the minimum standard/expectation for key ________ performance parameters. These are supported by management and reinforced by ongoing stewardship and management reviews to help ensure compliance and corrective actions when required.

12 The PSMS must be designed with a defined process for self-improvement, including periodic _______ to evaluate the PSMS program documentation and field compliance to meet the program's deliverables. Generally, these audits are conducted by peers with PSMS management system experience and experience in the ________ ________, __________, or __________ _____________.

13 The training required by specific organization members is detailed in the supporting documents to include training requirements for ________ Operator training, Operator __________ training, and training for _____________ employees.

14 The overall management system should also utilize _____________ indicators to track compliance with the specific system requirements.

a. leading
b. lagging
c. leading and lagging
d. critical

7

Importance of Operating Procedures

The OSHA PSM Standard prescribes detailed requirements for operating procedures and ongoing owner verification for accuracy and relevance. Operating procedures are critical to safe operations and must be maintained current, and managers/supervisors should enforce procedure compliance in the field and on the console.

A significant number of catastrophic incidents, some with loss of life, have occurred due to deviation from established procedures and have contributed to major process safety events, including those with catastrophic loss of life, severe damage to site facilities, and resulting in a major impact on the community. For example, the US Chemical Safety and Hazard Investigation Board found that Operators at BP Texas City (March 2005) were routinely deviating from written procedures during the startup of the Isomerization Unit by operating with the fractionation column level above the range of the tower level control instrument, and that these deviations were tolerated by site management. This contributed to operating the tower "blindly," which resulted in overfilling the column and resulting in the vapor cloud explosion and fire. Unfortunately, several other significant and noteworthy events have occurred.

An Introduction to Process Safety Management Systems

Process Safety Management Systems (PSMS) are the guiding documents that transcribe the company's corporate Safety, Health, Environmental, and Security values and culture into documented and actionable management plans. The PSMS forms a Framework by which the corporate values are documented and communicated to all employees, contractors, and anyone who works for the company.

With the Framework in place, the PSMS system adds additional detail and objectives, where inadequate compliance with existing procedures plays an important role. Among these are the 1988 Piper Alpha platform (North Sea) explosion and fire (167 fatalities), the 1998 Esso Longford (Australia) explosion and fire (2 fatalities), the 2005 Buncefield Terminal (United Kingdom) explosion and fire, and the 2010 Deepwater Horizon (Gulf of Mexico) explosion and fire (11 fatalities).

The OSHA PSM Standard recognizes the importance of procedures and requires PSM-covered sites to maintain detailed written operating procedures. These procedures should provide clear instructions for safely conducting routine and non-routine tasks for each covered process.

The following operations should be covered:

(Note: Sections of the OSHA PSM Standard are included below as examples and should be included in the site standard for procedures.)

- Steps for each operating phase,
- Initial startup,
- Normal operations,
- Temporary operations,
- Emergency shutdown includes the conditions under which emergency shutdown is required and the assignment of shutdown responsibility to qualified operators to ensure that emergency shutdown is executed in a safe and timely manner,
- Emergency Operations,
- Normal shutdown and startup following a turnaround or after an emergency shutdown,
- Operating limits, consequences of deviation, and steps required to correct or avoid deviation,
- Safety and health considerations,
- Properties of, and hazards presented by, the chemicals used in the process,

Process Operations Safety: The What, Why, and How Behind Safe Petrochemical Plant Operations, First Edition. M. Darryl Yoes.

- Precautions necessary to prevent exposure, including engineering controls, administrative controls, and personal protective equipment,
- Control measures to be taken if physical contact or airborne exposure occurs,
- Quality control for raw materials and control of hazardous chemical inventory levels and any special or unique hazards,
- Safety systems and their functions, and
- Operating procedures shall be readily accessible to employees who work in or maintain a process.

The OSHA PSM Standard also requires that "operating procedures be reviewed as often as necessary to ensure that they reflect current operating practice, including changes resulting from changes in process chemicals, technology, and equipment, and changes to facilities. The employer shall certify annually that these operating procedures are current and accurate" (OSHA 1910.119) (f)(3).

The following pages and bullet points in this chapter provide additional thoughts on how procedures can be enhanced and how some of the more serious process safety hazards can be addressed. They are not necessarily intended to replace existing procedures but are additional considerations as procedures are being reviewed or updated.

Procedures should be readily available to Operators, and Operators should be trained in the procedures and receive periodic refresher training. OSHA requires that refresher training shall be provided at least every three years and, more often, if necessary, to each employee involved in operating a process to ensure that the employee understands and adheres to the current operating procedures of the process (OSHA 1910.119) (g)(2). The employer shall determine the appropriate frequency of refresher training in consultation with the employees operating the process. A good practice is to track the Operator training as a process safety leading indicator (more on leading indicators in Chapter 11).

Many of the operations procedures are complex, detailed, and long. Some procedures contain critical operations such as startup, shutdown, and other transient operations. Procedures of this type may potentially result in a catastrophic incident if the procedure is not implemented correctly or is not followed in the correct sequence. A good practice is for procedures of this type to be designated so that they receive special attention. For example, these procedures could be designated "SSHE Special" (Safety Security, Health, or Environmental) or a similar designation. Procedures so designated would allow special attention to their structure or layout during auditing and, most importantly, during their use by the Operators. For example, procedures designated "SSHE Special" should be formatted in a checklist-style format with requirements that the Operators sign or initial each step in the procedure as it is implemented.

Of course, this would also require supervisors to enforce these requirements once implemented. This helps ensure that procedures are executed correctly, in the correct sequence, and that critical steps are not omitted or missed.

In some higher-hazard procedures, inserting a "hold point" may be appropriate immediately before the procedure. This is important where special verification is warranted directly before carrying out the subsequent step, such as verifying valve position. Higher-hazard procedures requiring special attention by the Operators should also be immediately preceded on the same page by the appropriate signal word to notify the Operator of the particular hazards present during this portion of the procedure.

Note:

The signal words below are defined by ANSI (American National Standards Institute) to designate a degree or level of safety alerts. Additional information is available on the ANSI website at http://www.ansi.org (standard ANSI Z535.2).

- **Danger:**
  Danger indicates a hazardous situation that, if not avoided, will result in death or severe injury. This signal word is to be limited to the most extreme situations. (Generally, the signal word "Danger" should not be used for property damage hazards unless personal injury risk appropriate to the level is also involved.)
- **Warning:**
  Warning indicates a hazardous situation that could result in death or serious injury if not avoided. (The signal word "Warning" should not be used for property damage hazards unless personal injury risk appropriate to the level is also involved). See the example in Figure 7.1 for an example of a typical "Warning."
- **Caution:**
  Caution indicates a hazardous situation that could result in minor or moderate injury if not avoided. (The signal word "Caution" may be used to alert against unsafe practices that may cause property damage).
- **Notice:**
  "Notice" is the preferred signal word to address practices unrelated to personal injury. (The signal word may be used to indicate a statement of company policy).

Warning:
Failure to monitor the tower level during level control valve commissioning can send vapor or light material to the naphtha tank and the potential for tank roof damage and fire or explosion.

Figure 7.1 Example of a typical warning.

Also recommended is the use of a color code for the signal word or a colored border for the signal word to help with the hazard identification; for example, Red for "Danger," Orange for "Warning," Yellow for "Caution," and Black for "Notice."

Emergency procedures should include all foreseeable scenarios such as flammable or toxic vapor release, fire, loss of plant utilities such as steam or instrument air, and major weather events such as a hurricane or severe freeze. Operators should be trained in emergency procedures and periodically practice procedure execution through tabletop discussions and drills/exercises.

All procedure changes, including updates and revisions, should be controlled using the site Management of Change (MOC) process. The MOC process ensures that the changes are carefully reviewed and approved before the change is commissioned in the field. Operators should be informed or trained on the changes before operating the equipment. Additionally, it is a good practice to periodically audit the operating procedures using a process similar to the Process Hazard Analysis (PHA) process, where the procedure is closely analyzed to ensure that the procedure clearly details the proper steps in the correct sequence and that potentially hazardous steps are appropriately identified.

Furnace Operations

Special emphasis should be placed on furnace procedures due to the nature of furnace operations, that is, the intentional burning of fuel and air in a confined space (the furnace) and the heating of petroleum flowing through the furnace tubes to very high temperatures and pressures. The significant number of furnace explosions in the refining and petrochemical industry reinforces this requirement. I have witnessed or been involved in numerous furnace events during my career, as either an investigation team member or the team leader following the incident.

The following are potential furnace procedure enhancements that should be considered for inclusion in furnace procedures:

Purging of unburned fuel from the firebox before attempting to light the pilots and burners:

- During furnace startup, there is the potential for an accumulation of unburned fuel and air present in the firebox, which may lead to an explosion during an attempt to light or relight the furnace pilots or burners. Procedures should include detailed guidance for purging furnaces to ensure unburned hydrocarbon removal before attempting to light furnace pilots or burners during unit startup. This is especially important for relighting pilots and burners immediately after a hot furnace trip.
  - For furnace relight of pilots and burners following a furnace trip, the procedure should have a WARNING about attempting to light pilots and burners without a complete air purge of the furnace if the pilots and burners are not operating. The warning should indicate that this could result in a furnace explosion and fire, personnel injury, and the potential loss of life.
  - The hot trip procedure may indicate that the purge is not required if the pilots are still operating following the trip.
- Typically, the purge should include the operation of induced and forced draft fans for furnaces equipped with combustion air preheat equipment. Special procedures for natural draft furnaces (without combustion air preheat fans) should be included, including purging with steam or a stack air educator (if available). Otherwise, a minimum of 30 minutes purge with the natural draft with the furnace damper in the fully open position. In either case, there should be a consideration for gas testing the furnace for LEL at the burners and peep doors before attempting to light pilots or burners (LEL should be zero). Gas testing is critically important after purging with a natural draft (no blowers or steam purge).

Furnace procedures related to furnace flooding:

- Many furnace explosions have occurred due to the furnace being in a fuel-rich or "flooding" situation, where the ratio of fuel to air is out of balance (where the firebox fuel is too rich). Some sites refer to this as "bogging."

- Furnace procedures should address this potential for furnace flooding, including symptoms of furnace flood and response to a flooding scenario.
  - Typical flooding symptoms include:
    - The firebox O_2 is very low (<0.5%) – One of the first indications of a flooded firebox.
    - The firebox flue gas has high combustibles (>1,000 ppm), which is one of the first indications of a flooded firebox.
    - These indications may be followed by a rapid drop in coil outlet temperatures or firebox bridge wall temperature.
  - Other signs of flooding that may be noticeable to the outside operator include:
    - Loud rhythmic "whoosh" sound or vibration.
    - Pulsating flames.
    - Local draft gauge swinging widely.
    - Hazy appearance in the firebox (but do **not** look in the firebox if flooding is suspected – opening the peep doors would allow air into the firebox and could result in an explosion).
    - Smoke from the stack (severe flooding or bogging).
- The procedure should include the steps the Operator should take to respond quickly to indications of a furnace flood event:
  - Immediately remove personnel away from a fired heater that is suspected of flooding.
  - Immediately reduce fuel intake by 25%.
  - Place fuel gas controls on manual control and open the instrument cascade to prevent the DCS from increasing fuel.
  - Place the damper on manual control to prevent additional air ingress to the firebox (otherwise, the O_2 analyzer will cause the damper to open, admitting more air).
  - Do not automatically trip the fired heater. Suddenly removing all fuel from a rich firebox may result in a flammable mixture that could autoignite in the firebox, causing an explosion. The natural draft will admit air into the firebox, resulting in an explosion.
  - Do not take any steps to increase air into the firebox, for example, damper adjustments, burner adjustments, or the opening of the observation doors.
- One of the more common causes of furnace flooding is caused when the furnace is operating in a turndown mode with one or more burners turned off, with the air registers for the out-of-service burner(s) left in the open position. This results in excess air being educated into the furnace firebox, where this significant amount of excess air bypasses the operating burners. The furnace draft pulls this excess air up to the furnace breeching area, which affects the O_2 analyzer, resulting in the O_2 controls reducing the damper. This causes the in-service burners to operate without adequate air, flooding the firebox with fuel.
  - Procedures should include the requirement that the air registers be fully closed anytime the burners are turned off. This prevents excess air from entering the furnace.
- Procedures should also provide guidance or proactive steps that can be taken to help prevent a furnace flood, especially for furnaces with a history or propensity for flooding.
  - A proactive response during periods of transient operation can reduce the potential for flooding. Consideration should be given to temporarily increasing the heater excess O_2 before significantly increasing feed rate, feed quality, or changes in fuel gas heating values that will increase the heater firing rate. This gives the heater flexibility in O_2 to absorb the additional firing and minimize the potential for flooding.
  - The O_2 should be optimized after the changes are made and the heater has lined out and is stable.

Other General Furnace Procedures

Furnace procedures should also caution lighting burners from other burners or a hot furnace wall due to the potential for delayed light off and potential explosion due to unburned fuel in the firebox. Procedures should specify that pilots and burners should be lit individually and only with an approved torch or other approved lighting device. Other General Furnace Procedures for consideration include the following:

- Install pilot and fuel gas blinds promptly after a furnace shutdown to prevent unburned fuel from entering the firebox (potential for explosion).
- Include caution about operating above maximum allowable Tube Metal Temperatures due to significantly reduced tube life and potential coking or ruptured tube.

  - Timely problem-solving is needed when a sudden change in TMT occurs.
- Never ignore alarms indicating loss of pass flow as severe tube damage may occur with the potential for ruptured tube and explosion or fire.
  - Prompt significant reduction in firing should accompany the loss of pass flow.

Unit Operating Limits and Critical Alarms

- As indicated earlier, OSHA PSM requires each site to establish a list or table of unit operating limits (OLs), consequences of deviation, and the expected Operator actions to take. These would include references to critical alarms to ensure that safe upper and lower OLs are adhered to for process variables such as Temperature, Pressure, and Flow.
- A table listing this data should be provided in the appropriate data section of the procedure and referenced in the appropriate procedure sections (startup, shutdown, abnormal operations).
- The procedure should include guidance/procedure for prompt site notification in the event of deviation from the OL limits.

Air-freeing Equipment Before Introducing Hydrocarbons (During Startup or Bringing Equipment Back into Service Following Repairs and for New Equipment)

- Ensure all procedures for commissioning equipment, such as startup or returning equipment to service, contain caution/warnings about proper purging of air from equipment before introducing hydrocarbons.
- Purging may be done with steam or Nitrogen if the unit cannot tolerate moisture. For example, nitrogen may be required due to the presence of a catalyst, higher metallurgy, or hygroscopic chemicals (which absorb moisture from the air).
- The procedure should contain caution/warning about the risk of inadvertently leaving air trapped and the potential for internal explosion or fire.
- The procedure should also contain guidance for testing for the presence of O_2 before introducing hydrocarbons. A good practice is to use the "concertina" purging process (pressure and depressurization) a minimum of three times before testing for O_2 concentration.
- The Procedure should specify the maximum allowable O_2 level following the O_2 purge.

Startup Procedures

- Ensure cautions/warnings about losing the fractionator tower bottom level during unit startup or other transient operations.
  - Loss of the tower's bottom level results in the potential to blow light/hot vapor from the tower to downstream equipment, which is not designed for it. For example, blowing light material into an atmospheric storage tank can sink the floating roof and result in a full surface tank fire.
- Ensure startup procedures include checklists for draining free water from unit equipment and piping before introducing hot hydrocarbons or raising unit temperatures. Include Cautions/Warnings that hot hydrocarbons contacting free water inside process equipment can overpressure equipment, opening atmospheric relief valves and may overwhelm their capacity, resulting in an explosion. Flashing of free water can also damage equipment internals, including fractionation tower trays.
  Note: Based on industry experience, it may be an excellent reminder to have the Operator verify there is no free water standing above the discharge check valve on the unit pump-around pumps. They should also operate pumps during flushing procedures to remove all free water.
- Ensure startup procedures include guidance for "Tightness Testing" the unit before introducing hydrocarbons. The procedure should include guidance for the following:
  - Specify the test medium to be used (Steam, Nitrogen, Other)
  - Specify the test pressure (typically around operating pressure)
  - Specify the leak detection method (unit walk down with bubble solution, acoustics, FLIR camera, etc.)
  - Criteria for acceptance (how do we know we have a pass or fail?)

  - o Note: Pressure testing at pressures exceeding normal operating pressures may be considered pneumatic pressure testing and require a detailed review by management and the appropriate site risk management group or committee.
- Ensure all instruments, alarms, and other devices identified as SHE Critical are function-verified before introducing hydrocarbons into the unit.
  - A checklist of the devices identified as SHE Critical should be available in the manual.
- The procedure manual should list the previously identified Operating Envelopes (OEs) and Operating Limits (OLs).
  - The startup procedure should reference these OEs or OLs and prescribe reporting requirements for deviations during unit startup.
  - Reporting should include alarm flood issues or other similar issues with alarm systems.
- Start-up procedures should ensure that the site "clears the unit" in preparation for a startup before introducing hydrocarbons. This includes the following considerations:
  - Clear all nonessential personnel from the unit (not directly involved with the startup).
  - Disconnect, remove, and store all temporary hoses and similar connections from process equipment (those not directly used during the startup) in a safe location.
  - Remove scaffolding and other equipment that impede process personnel's access to critical operating equipment.

Procedures for Prevention of Brittle Fracture Failures

Note: Brittle Fracture (BF) is an instantaneous failure of process piping or vessels and occurs when the pipe or vessel is pressurized when the metal temperature is below the minimum design metal temperature (MDMT). This can occur during hydrostatic testing when the temperature of the hydrotest water is below the MDMT for the pipe or vessel. It can also happen when the equipment is in service due to flash vaporization of LPG. Brittle fracture has the potential to release flammable or toxic materials and has resulted in sudden, unexpected explosions, and fires. When in service, a brittle fracture or a process vessel occurs most often during transient operations such as startup, shutdown, process upsets due primarily to flash vaporization of LPG, and periods of cold weather when the equipment is pressurized but with no heat input. Thicker wall carbon steel equipment is especially vulnerable to brittle fracture.

- Common brittle fracture definitions:
  - Minimum Design Metal Temperature (MDMT); The lowest safe temperature that the vessel should see when it is pressurized to the equipment design pressure. The MDMT is determined by the material(s) of construction and equipment design (metallurgy, wall thickness, etc.).
  - Minimum Safe Operating Temperature (MSOT); Minimum Safe Operating Temperature at a corresponding pressure (a sliding scale).
  - Critical Exposure Temperature (CET); The CET is the lowest temp that the equipment can experience. The CET is derived from normal and abnormal operating conditions (including unanticipated excursions).
  - General guidelines to help prevent brittle fracture: During unit startup, do not pressurize equipment above 25% of design pressure until the metal temperature is above the MDMT.
- Unusual situations can result in temperatures lower than MDMT:
  - Leaking equipment or exchanger internals with C2 or C3 (cold spots due to auto-refrigeration).
  - Safe parking or squatting a unit at pressure for a longer period, particularly over a cold period.
  - Hydrostatically testing equipment using cold water (water at a temperature below the vessel MDMT).
  - Pressurizing vessels with LPG (e.g., C2 or C3) after they are already pressurized with nitrogen (due to the partial pressure difference, the LPG flash vaporizes, creating auto-refrigeration and cold temperatures).
- Procedures should reference the potential for brittle fracture in the relevant sections (startup, shutdown, and other transient operations) with appropriate warnings and references to the vessel MDMT data table.
- The first line of defense against possible Auto-Refrigeration and Brittle Fracture excursions:
  - Understand what causes auto-refrigeration (flash vaporization of LPG) (e.g., Ethane, Propane, Butane, and other LPGs.).
  - Know the limits of your equipment (minimum design metal temperature) (MDMT).
  - Avoid the excursion (operating with the vessel pressurized when below the (MDMT).
- A table listing the MDMT for all unit process vessels should be included in the appropriate section of the procedures and referenced in each pertinent section (startup, shutdown, other transient operations, and the special section for cold weather operations).

Procedures should also include a response to a BF excursion to aid Operations in recovering from the excursion (operating below the MDMT). A generic response procedure is provided below:

- Recognize that BF is possible and clear personnel from the area.
- Stabilize the unit, and do not make any sudden moves.
- If possible, drain liquid LPG from the vessel remotely (do not send anyone to the vessel to drain liquid).
- If LPG is still feeding the vessel, stop the LPG.
- Do not increase vessel pressure or add a dry gas such as Nitrogen to the vessel.
- After the unit is stabilized, begin slowly reducing the vessel pressure until the pressure is below 25% of the vessel design pressure.
- During depressurization, closely monitor the vessel temperature and control the depressurization rate to prevent the vessel from becoming colder.
- As a precaution, the Materials Engineer and the site Metallurgist should be consulted before returning the vessel to service.

Managing Hydrocarbon Thermal Expansion Hazards in Process Piping and Vessels

Note: Thermal expansion results from applying heat to piping and vessels when liquid-filled (little or no vapor space above the liquid). Thermal expansion results from the volume expansion due to the temperature increase in "trapped" liquids. Even small temperature changes, such as daytime heating, have resulted in catastrophic loss of containment and explosion due to relatively small temperature changes. For example, inadvertent isolation of a gasoline or naphtha pipeline without providing thermal expansion can result in a nearly 100 psi increase for each 1 °F temperature rise. This can potentially exceed recommended piping hydrostatic testing pressures on an average day.

- The procedures should include warnings and procedures for protecting process equipment from the hazards of thermal expansion in the appropriate sections. Examples of protection against thermal expansion include:
  - The best case is providing thermal expansion relief valves, especially in long pipelines in the gasoline boiling range and lighter materials, including LPG.
  - Alternatively, provide the provision for venting one end of long pipelines to an operating system or storage tank to relieve the thermal expansion of trapped liquids.
- The procedures should specifically address the hazard of overpressure if the equipment is blocked in without a positive relief path. Some examples are listed below:
  - Piping in intermittent service is blocked at both ends (especially light materials, as outlined above).
  - Vessels that are blocked in when removed from service (or returning).
  - Heat exchangers are blocked in when preparing for cleaning or inadvertently blocking the "cold" side while the "hot" side of the exchanger is still flowing.
  - Exchanger commissioning or decommissioning procedures should provide warnings about avoiding commissioning the hot side before the cold side due to the potential for thermal expansion on the cold side.
  - Heat-traced piping or vessels (if liquid-filled).
  - Thermal relief valves were removed for servicing.
- Procedures should provide special guidance for preventing thermal expansion in double-seated valves when in liquid service due to the potential for trapped liquids in the cavities above and below the internal valve gates. These valves are designed with external pressure relief for the gate cavities (thermal relief valves and pressure vent tubing). Owners should verify with the valve manufacturer if and how the double-seated block valves are protected from this hazard.

Identification and Management of Imported Hazardous Chemicals

- The procedure should include a listing of imported hazardous chemicals, considering the following:
  - Regulatory requirements (e.g., OSHA PHS listings, API751 – Specific to HF Acid) (if applicable).
  - Potential for a credible Consequence 1 scenario.
  - Need for specific procedural mitigations to be performed by a Process Technician during off-loadings.
- The procedure should reference the following as appropriate to the chemical involved:

  - IH procedures/MSDS.
  - Loading, unloading, handling, and storage procedures.
  - Emergency response guides.
  - Drills on emergency response.
  - Training personnel on proper handling of the imported chemical.
  - Site checklist for each imported chemical.
- The procedure should also indicate the MOC process is in place and required for all new-to-site chemicals.

Pressure and Vacuum Hazards

Note: Some process vessels are not designed as pressure vessels. For example, atmospheric storage tanks such as fixed-roof tanks are typically only designed for very little pressure (typically measured in inches of water). Also, most process vessels are not designed for vacuum conditions (typically measured in inches of water vacuum). Subjecting these vessels to either pressure or vacuum can result in tank or vessel failure with the potential for loss of containment.

Tank pressure and vacuum requirements are found in API-620, API-650, and API-2000. For example, fixed-roof storage tanks are typically designed for a maximum vacuum of about −4″ water (−0.14 psi) with the PV Vent set to open at around −2″ of water vacuum (−0.07 psi).

The same low-pressure tanks have a typical overpressure design of about 22″ water (about +0.8 psi) with the PV Vent set to open at around 10″ water pressure (or about +0.4 psi). With these very low limitations, one can see the potential for either overpressure or vacuum damage (Note: These data are typical – always verify the design).

Vessels at risk include fixed-roof atmospheric storage tanks, process pressure vessels (primarily vacuum hazards), railcars, tank trucks, and other similar process vessels:

- Procedures should include appropriate cautions and warnings about the potential for overpressure and damage to storage tanks and process vessels.
- Storage tank PV Vents should be classified as SHE Critical and included in the regular inspections and maintenance intervals.
- PV Vents should not be defeated, even during tank maintenance operations, unless alternate protection is provided.
- Include procedures for managing the rate of fill and rate of drain limitations (e.g., when filling in preparation for a hydrostatic test or draining after the test is completed).
- Procedures should include cautions and warnings to prevent the tank from being blocked in without vacuum protection after steaming or hot purging operations.
- Procedures should ensure Operators make monthly API 653 tank walkaround inspections of tank vents to help ensure the vents are not obstructed. Observations include the following potential obstructions:
  - Scale.
  - Product collecting/condensing on a screen, if installed.
  - Ice.
  - Birds/Wasp nests.
  - Plugging/blanking off.
  - Painting/grit or sandblasting.
  - Temporary protective covering.

LPG Handling and Prevention of BLEVE (Boiling Liquid Expanding Vapor) – Includes Procedures for Draining Water from LPG Vessels (Water Drawing, Decanting Operations)

Note: A pending BLEVE is one of the worst-case scenarios for a refinery or chemical plant due to the magnitude of the explosion, the amount of secondary damage, the shrapnel released, and the impact on the community. Vessels primarily at risk contain material stored at temperatures above their boiling point (boiling point at atmospheric pressure, for example, Butane or Propane). BLEVEs are most often caused by loss of containment of LPG or other flammable material, resulting in fire and resulting flame impingement on process vessels or pressurized storage tanks. The vessel can fail within minutes when the flame impinges on the vessel shell above the liquid level.

Other LPG hazards include potential auto-refrigeration in piping and vessels when pressure is reduced or when taking a pressure drop across valves or other equipment. The pipe or vessel can be subjected to temperatures well below the design temperature (MDMT), resulting in the potential for freezing valves in the open position and/or the catastrophic brittle fracture of equipment.

When released into the atmosphere, LPG expands nearly 300 times when flashing from a liquid to vapor, creating a major vapor cloud with potential explosion and fire.

Procedures should include procedures for LPG handling with special emphasis on preventing loss of containment during routine operations such as water draw operations, sampling of LPG, opening equipment that previously contained LPG, emergency response, and general safety observations in LPG-containing equipment/areas.

- **LPG water draw operations procedure:**
  - Water draw procedures should include appropriate warnings (proper operation of valves during water draw operations (upstream fully open/throttle on downstream), never leaving the water draw unattended, ensuring both valves are fully closed when the task is complete, and never defeat a spring-loaded valve).
  - LPG process and storage vessels should include two block valves for water draw operations.
  - Both valves remain fully closed when not in use.
  - When drawing water, the upstream valve (valve closest to the vessel) should be in the fully open position (to prevent auto-refrigeration and freezing of the valve).
  - The flow should be controlled by throttling the downstream valve.
  - Close both upstream and downstream valves when the task is complete.
    - Note: In cold areas, there may be a third drain valve from the vessel to displace water from the water draw piping before closing the upstream valve (refer to site procedure).
- **Procedure for the sampling of LPG liquids:**
  - LPG sampling procedures should include appropriate warnings (proper operation of valves during LPG sampling operations (upstream fully open/throttle on downstream), never leaving the sample valves unattended, ensuring both valves are fully closed when the task is complete, never defeating a spring-loaded valve).
  - LPG process and storage vessels should include two block valves for sampling operations.
  - Both valves remain fully closed when not in use.
  - When sampling, the upstream valve (valve closest to the vessel) should be in the fully open position (to prevent auto-refrigeration and freezing of the valve).
  - The flow should be controlled by throttling the downstream valve.
  - Close both upstream and downstream valves when the task is complete.

LPG Emergency Response Procedures (Include in Emergency Response Section of Procedures)

- **Fire near or on LPG pressure tank (sphere, horizontal tank, or process vessel):**
  - Immediately activate the emergency response team.
  - Turn on the water deluge system and ensure complete coverage of the vessel.
  - Supplement the water deluge system with fixed fire monitors to cover any "dry spots" and locations where there is a direct flame impingement on the vessel.
  - Supplement as needed with additional portable fire monitors.
  - Water flooding connections, if provided, can also be used to displace leaking products from a bottom connection.
  - If the initiating event has ruptured a fire water main, and there is no firewater available or any other means of getting water onto the exposed vessels within 10 minutes, the area should be evacuated!

Unplugging Process Bleeders (Process Vents and Drains)

Personnel has been severely burned or otherwise injured attempting to clear or unplug process vents and drains. Most incidents occur because of the use of improper tools or equipment or failure to follow the manufacturer's recommendations for their use. See Figure 7.2 below.

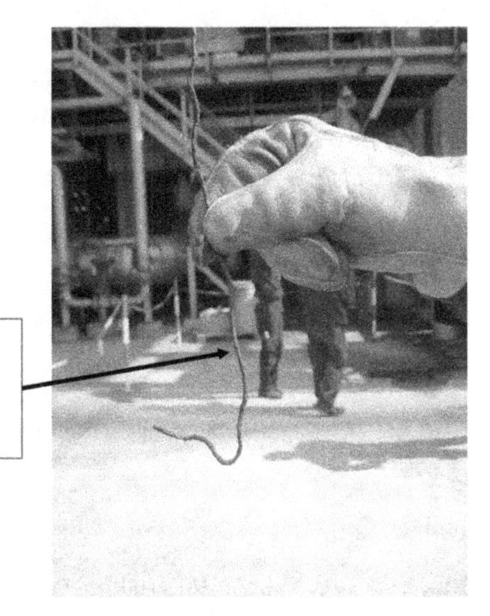

Figure 7.2 Example of an unauthorized bleeder unplugging device; a welding rod. This was found on the 3rd floor of a crude unit structure near a control valve station.

Figure 7.3 Enclosed bleeder unplugging device (available in multiple configurations from Lawton industries). Source: Photo courtesy of Lawton Industries (https://www.lawtonindustries.com/).

- The procedure should include warnings about the potential loss of containment and personnel exposure to flammable, toxic, or hot materials if appropriate equipment is not used.
- Include specific caution about using welding rods or wire (or similar) devices as bleeder unplugging tools.
- An approved bleeder unplugging tool should be used for the unplugging operation (currently available from Lawton Industries).
- Before being used, the equipment should be verified as rated for the system pressure, temperature, construction materials, and product compatibility.
- The manufacturer's instructions should be followed closely (available from the Lawton Industries website).

Images of improper and proper bleeder unplugging tools are available in Figures 7.2 and 7.3.

Proper bleeder unplugging tool (process safety device):

- The bleeder unplugging tool is a fully enclosed device.
- It is both pressure and temperature-rated.

- It is also equipped with a pressure gauge and venting capability.
- The bleeder unplugging tool facilitates a connection to a pressurized bleeder and the drilling through an obstruction. After pressure is shown on the gauge, the tool can be safely withdrawn, the bleeder can be isolated, and the tool safely vented to relieve the pressure.
- The tool is completely self-contained.

Additional References

OSHA Process Safety Management Standard (29CFR 1910.119).

ANSI standard ANSI Z535.2 (Guidewords to designate a degree or level of safety alerting), American National Standards Institute (www.ansi.org).

Lawton Industries, Information and procedures for a bleeder unplugging device, (http://www.lawtonindustries.com).

Investigation Report for the 1988 Piper Alpha platform (North Sea) explosion and fire (167 fatalities). "A Post-Mortem Analysis of the Piper Alpha Accident: Technical and Organizational Factors", M Elisabeth Pate'-Cornell, Department of Industrial Engineering and Engineering Management, Stanford University.

Investigation Report for the 1998 Esso Longford (Australia) explosion and fire (2 fatalities). "The Esso Longford Gas Plant Accident", Report of the Longford Royal Commission, June 1999. By Authority: Government Printer for the State of Victoria (Australia).

Investigation Report for the 2005 Buncefield Terminal (United Kingdom) explosion and fire., COMAH Final Report (Control of Major Accident Hazards) (02/11), British Health and Safety Executive – The Competent Authority.

Investigation Report for the 2010 Deepwater Horizon (Gulf of Mexico) explosion and fire (11 fatalities). US Chemical Safety and Hazard Investigation Board Final Report, Deepwater Horizon Rig Mississippi Canyon 252, Gulf of Mexico 20 April 2010, REPORT NO. 2010-10-I-OS 04/12/2016 (and CSB Final Reports Volumes 1–4).

Chapter 7. Importance of Safe Operating Procedures

End of Chapter Quiz

1. Which regulation mandates required procedures for covered processes at petrochemical facilities?

2. Why is it so important that this regulation be followed so closely?

3. Operating procedures are critical to safe operations and must be maintained ________, and managers/supervisors should __________ procedure compliance in the field and on the console.

4. The US Chemical Safety and Hazard Investigation Board found that Operators at BP Texas City (March 2005) were routinely __________ from written procedures during the startup of the Isomerization Unit.

5. The OSHA PSM Standard recognizes the importance of procedures and requires PSM-covered sites to maintain detailed __________ operating procedures.

6. What does the PSM Regulation say about the frequency of procedure reviews to ensure accuracy?

7. Does the OSHA PSM Regulation require periodic training on the procedures?

8. Can you name a good practice for tracking the required Operator refresher training?

9 What is a good practice for the designation of procedures for higher-risk tasks that have already resulted in or could result in a catastrophic incident?

10 Can you name the four key signal words defined by ANSI and used in procedures to designate a degree or level of safety alerting and tie each to their associated colors?

11 Specific procedures require special emphasis due to the nature of risk and industry experience.

12 What are the requirements laid out by the PSM regulation relative to process procedures for sites covered by the PSM regulation?

13 Emergency procedures should include all foreseeable scenarios, such as:

a. flammable or toxic vapor release
b. fire
c. loss of plant utilities such as steam or instrument air
d. major weather events such as a hurricane or severe freeze
e. none of the above
f. all the above

14 Operators should be trained in emergency procedures and periodically practice procedure execution through __________ discussions and __________/__________.

15 How should all changes or modifications to procedures be approved and controlled?

16 Which class of process equipment merits special attention to detail due to the nature of the hazards present?

17 What special tools are available specifically for unplugging bleeders?

8

Significant Causes of Process Safety Incidents "What Has Happened, What Can Happen, and Key Lessons Learned"

This section will deviate from talking about process safety in general terms. We will now focus on the more specific causes of process safety incidents. We will discuss What Has Happened, What Can Happen, and Key Lessons Learned from specific incidents. These include some of the more common causes of loss-of-containment events that have resulted in or could have resulted in major process safety process incidents. Many cases covered in this chapter have occurred repeatedly in our industry. When discussed in an open forum, these almost always lead to a discussion of "Organizational Memory" or "what has happened before can happen again." Therefore, in most of the causal factors covered here, we will also discuss examples of these in our industry. We will discuss the causal factor briefly and then follow up with "What Has Happened" or "What Can Happen" and discuss the "Key Lessons Learned." Our objective here is to learn from others and apply the "lessons learned" at our site. If we do this, it will help ensure the incident is not repeated at our facility.

Examples of causal factors to be discussed in this chapter include the following:

- Hazards of Startup, Shutdown, or other similar transient operations.
- Safe Permit to Work – Controlling Work.
- Importance of Managing Change.
- Controlling Disablement or Impairment of Critical Devices.
- Managing Critical Alarms.
- Development of and strict adherence to Safe Operating Limits.
- Product Rundown and Storage Tank Safety.
- Managing Mixing Points and Injections Points.
- Managing Process Piping Dead Legs.
- Managing potential for Brittle Fracture Failures.
- Hazards of Furnaces and Heaters.
- Managing Corrosion Under Insulation or Corrosion Under Fireproofing.
- Managing the hazards of Thermal Expansion of trapped liquids.
- Managing the hazards of Pressure Testing with a Pneumatic (Pneumatic Pressure Testing).
- The hazards of Piping and Vessels with Internal Refractory ("Cold Wall Design").
- The hazards of incorrect Metallurgy in alloy piping circuits (Positive Materials Identification when working with alloy piping or vessels).
- The hazards of Nitrogen (Asphyxiation properties of Nitrogen and other inert gases).
- The hazards of Hydrogen Sulfide (H_2S).
- Safe Product Sampling and Draining of Water from LPG and other process streams.
- Pyrophoric Ignition Hazards, including Fractioning Columns with Internal Packing.
- Controlling Static Electricity hazards.
- Hazards of Hydrofluoric Acid.
- Safe Use of Utilities, including Connecting Utilities to Process Equipment.
- The hazards associated with Steam and Hot Condensate.
- Electrical Safety Hazards for Operations Personnel (Including Hazardous Area Classification System).
- Hazards of Combustible Dust.
- Hazards of Undetected or Uncontrolled Exothermic Reactions.
- Response to "Minor" Leaks and "Temporary" Repairs.
- Hazards of Chemical Incompatibility.

Process Operations Safety: The What, Why, and How Behind Safe Petrochemical Plant Operations, First Edition. M. Darryl Yoes.

- Hazards of "Normalized Deviation."
- Hazards of Unintended Process Flow (The Mixing of Hydrocarbons and Air) in Fluid Catalytic Cracking Units (FCCU).
- Car Seal Valve Management Plan.
- Small Piping Guidelines for Operations Safety.

We are trying to ensure an awareness of the types of incidents that have happened and can happen again in process areas. We can learn from what has happened before and ensure training and mitigation are in place to help ensure that it never happens in our unit or work area.

8.1

Returning Equipment to Service

It's probably not a surprise that many process safety events, some with catastrophic consequences, have occurred during unit startup or periods of transient operations. After all, most units are designed for continuous and steady-state operations, which is what Operators and other support personnel are accustomed to. However, the units must come down periodically for turnaround, maintenance, or project tie-ins, and at other times, sometimes controllably, due to economic reasons. This means units will be in periods of start-up operations and other types of transient operations. I have seen and participated in many unit startups during my career, most often units returning from a turnaround and occasionally brand new or highly modified units following a lengthy construction process.

In each startup, it is always the same concern: will we have a leak or severe loss of containment? Have we got all the equipment back together, correct? Are all the gaskets in? Are all the bolts tight? Are they the right bolts and gaskets? Are they the correct metallurgy? Have we left any vents or drains open? Any of these issues may lead to a loss of containment. They can result in an explosion or fire where people could be injured or severely damage the equipment or experience extended downtime.

They may also threaten the community and the company's reputation. A significant event of this type can even have a disastrous impact on the company's survivability. These can be severe events.

In this section, we'll discuss all of these concerns and others and provide methods and practices to help ensure that we don't have a loss of containment incident and other ways to help ensure a smooth and uneventful return to service.

What Are the Most Obvious Hazards?

The start-up procedures and the start-up team must recognize and address many hazards. I'll list a few of those here, and then we'll address each of these and others in the following paragraphs.

Potential hazards during unit startup:

Potential hazard	Potential consequence
Leak or release or other loss of containment	Explosion, fire, or personnel injury
Air left inside process equipment	Vessel internal explosion or fire
Water left inside process equipment	Safety relief valve release to the atmosphere, internal vessel damage, and potential loss of containment
Overfilling process vessels	Release to the atmosphere, explosion, or fire
Loss of tower liquid level (tower bottoms level)	Blow through of hot vapor into downstream equipment, potential explosion or fire, or sunken tank roof, also with the potential for explosion or full surface tank fire
Instrumentation failures	Loss of process control leading to loss of containment, explosion, or fire
Backing hydrocarbons or other process chemicals into utility systems (through temporary hose connections)	Contamination of the utility leading to potential explosion or unexpected fire
Inadvertently leaving a blind in a process line or system	Unscheduled downtime for removal or inadvertent overpressure of process equipment
Operator fatigue leads to operating errors	Loss of containment, explosion, or fire
Inadequate information exchange at shift change or poor communication between shift and day organizations	This can lead to various process operational issues, including sending the product to the wrong disposition, loss of containment, explosion, or fire.

Process Operations Safety: The What, Why, and How Behind Safe Petrochemical Plant Operations, First Edition. M. Darryl Yoes.

Preventing a Leak or Release or Other Loss of Containment

To prevent a leak of release during startup, we conduct a rigorous test of all equipment, including piping, pressure vessels, and flanges. This pressure test is generally called a "tightness test" and is done with steam, Nitrogen, or Helium in rare cases. Even though this test uses a pneumatic to check for leaks, it is generally not considered a pneumatic pressure test since we only pressure to about normal operating pressure. Chapter 8.15 provides additional details on the hazards of pneumatic pressure testing at higher pressures exceeding design pressure. The tightness test verifies the pressure envelope is leak-free on equipment following mechanical handover. It includes checking for leaks from flanges, valve packing, vents, drains that may be left open, and instruments and fittings. The tightness test tests the entire pressure envelope for leaks.

Leak Testing Techniques (Attribution to Lian Rag Co. for Providing This Section)

Bubble Leak Testing

As the name implies, Bubble Leak Testing relies on the visual detection of a gas (usually air) leaking from a pressurized system. Small parts can be pressurized and immersed in a tank of liquid, and larger vessels can be pressurized and inspected by spraying a soap solution that creates fine bubbles onto the tested area. For flat surfaces, the soap solution can be applied to the surface, and a vacuum box can be used to create a negative pressure from the inspection side. If there are through leaks, bubbles will form, showing the location of the leak.

Pressure Change Testing

Pressure Change Testing can be performed on closed systems only. Detection of a leak is done by either pressurizing the system or pulling a vacuum and then monitoring the pressure. Loss of pressure or vacuum over a set period indicates a leak in the system. Changes in temperature within the system can cause changes in pressure, so readings may have to be adjusted accordingly.

Halogen Diode Testing

Halogen Diode Testing is done by pressurizing a system with a mixture of air and a halogen-based tracer gas. After a set period, a halogen diode detection unit, or "sniffer," is used to locate leaks.

Mass Spectrometer Testing

Mass Spectrometer Testing can be done by pressurizing the test part with helium or a helium/air mixture within a test chamber and then surveying the surfaces using a sniffer, which sends an air sample back to the spectrometer. Another technique creates a vacuum within the test chamber so that the gas within the pressurized system is drawn into the chamber through any leaks. The mass spectrometer is then used to sample the vacuum chamber, and any helium present will be ionized, making very small amounts of helium readily detectable.

More on Leak Testing

The leak testing methods vary from plant to plant; however, the procedures should specify the test medium used, usually steam or Nitrogen (N_2), when process circuits must be kept dry. For example, when a catalyst is present or for cryogenic processes, Nitrogen is used to avoid water contaminating the catalyst or freezing in cold equipment. The procedure should also specify the test pressure, low vs. operating pressure, and leak detection method. The leak detection process could involve taping flanges, using a soap solution and line walking to identify leaks, and using ultrasonic leak detectors or special leak detection analyzers in the case of Helium.

A standard leak detection method is to pressure the system with Nitrogen, apply masking tape to all flanges, punch a small hole in the tape at the top of the flange, and then walk the systems one by one, spraying small holes with a soap solution to detect any flange leakage. While holding pressure on the system, the pressure is monitored. If there is a loss of pressure of typically more than 10% of the test pressure over 30 minutes, this is interpreted as an undetected leak. In this case, the leak testing process is continued until the leak is found.

We must protect the process equipment from overpressure during the leak test. The test is, therefore, done with the relief valves in service. If the relief valves are not in service, a pressure regulator and a pressure relief valve (PRV) may be installed on the N_2 supply with the PRV set to protect the equipment from any potential overpressure. Also, since the equipment is being pressurized, the vessel temperatures must be maintained above the minimum design metal temperatures (MDMT) to guard against a potential brittle fracture.

This is especially important when vaporizing N_2 from a truck as the N_2 will auto-refrigerate when vaporized, and the cold temperatures can carry into the process equipment. The leak testing procedures should be documented as an integral part of the start-up procedure and include the test medium, the test pressure, the method of detection, and the criteria for acceptance.

Unit Walk Down and Preparation of Unit for Startup (Before Bringing in Hydrocarbons)

The Operators must do a comprehensive walk down of the unit before hydrocarbons are brought into the unit. They should be trained to look for full thread engagement on all flange studs (no short bolting). The ASME Standard B-31-3, 335.2.3 requires a minimum of full engagement, meaning, ("bolts shall extend through their nuts such that there is complete thread engagement for the full depth of the nut"). There should be no studs missing or loose. See Figure 8.1.1a,b for examples of

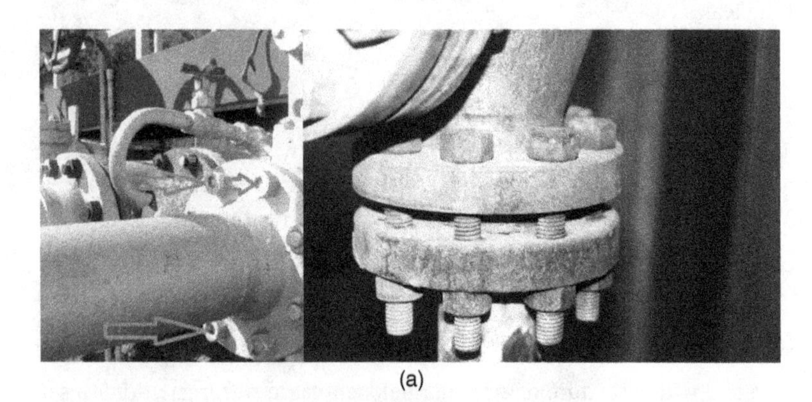

(a)

(b)

Figure 8.1.1 Examples of "Short bolted" flanges and missing studs.Note in Figure (a) that several of the nuts are "short bolted" with no threads showing outside the nut. Figure (b) is a flange where two of the studs have been left out entirely.

short-bolted flanges and flanges with missing studs. Short bolted, loose, or missing studs should never be accepted and should be repaired before hydrocarbons are brought into the unit.

Also, the Operators should verify that no vents or drains are left open and that all utility hoses are disconnected, rolled up, and safely stored off walkways and platforms. All scaffolds that interfere with the Operator's access to pumps and other process equipment should be dismantled or removed from the unit. All sewer covers should be removed and safely stored to ensure good drainage and prevent spilled material accumulation should a leak or spill occur during startup.

Finally, and very importantly, all nonessential personnel should be off the unit and removed from the unit and surrounding areas before bringing in hydrocarbons. I like to call this "clearing the unit;" all personnel not directly supporting the startup should be removed from the unit and the surrounding area. This is a critical process to help ensure the unit is ready for startup and that personnel in the area are out of harm's way during the startup.

During the walk down, the Operators should observe control valves and block valves for valve packing concerns. Figure 8.1.2 illustrates a control valve that needs attention. The packing nuts are not tightened evenly, indicating that the packing gland retainer is probably cocked. This packing may fail under pressure and should be addressed before startup. Figure 8.1.3 highlights a block valve with the packing nuts tightened almost down. This shows that not much packing is left in the valve, which could also fail after startup. Both are examples of the types of checks the Operators should be trained to observe before bringing hydrocarbons into the unit. These should be addressed before startup.

Figure 8.1.2 Example of control valve with packing gland uneven or cocked.

Figure 8.1.3 Example of block valve with packing nuts tightened all the way down (this means there is very little packing left in the valve).

Figure 8.1.4 Example of a Vent or Drain valve missing the bar stock plug.

Also, during the unit walk down, the Operators should verify that all vents and drains are closed, and bar stock plugs should be installed and tightened with a wrench. Bar stock plugs should have the same metallurgy as the coupling or valve it is threaded into. Don't forget about the battery limits; sometimes, we get so busy with the main unit that we forget about the battery limits. Loss of containment incidents have occurred at the battery limits due to open vents or drains. The best way to ensure that none are missed is by using a checklist, typically built into the startup procedure, and signing each off as closed and plugged. See Figure 8.1.4 for a photo of a vent or drain valve that is missing the plug. Ensure seal flush lines and oil mist piping are installed correctly and tight on pumps and compressors.

Verify that all instruments are installed and connected. All instruments should be verified as functional from the control room. Control valves should be stroked to ensure they are fully functional before bringing in hydrocarbons. All devices classified as Safety Critical should receive complete operational checks to verify the functionality before bringing hydrocarbons to the unit, including critical alarms such as low and high levels, temperature, low or high flow, and pressure alarms. For

Figure 8.1.5 Verify that all instruments are connected before startup.

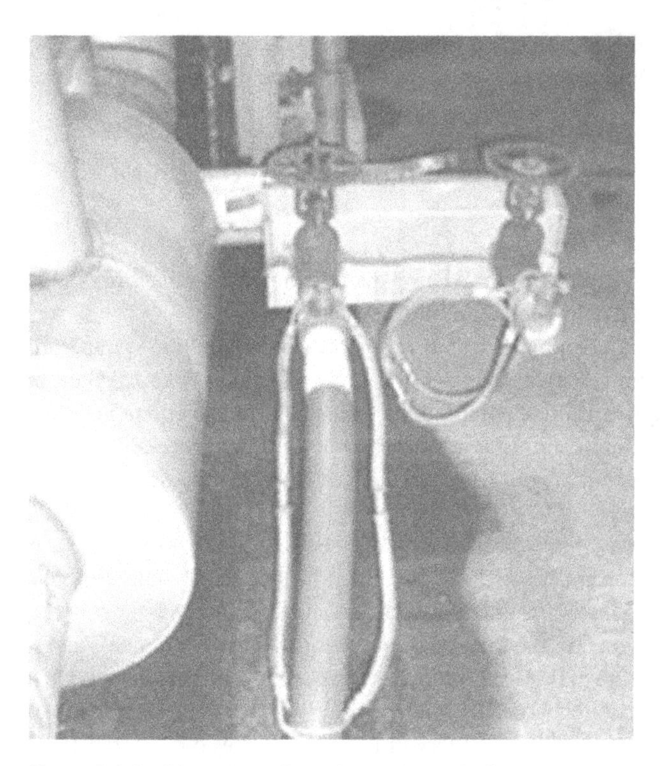

Figure 8.1.6 Disconnect all temporary hoses before startup.

example, all unit trip devices, such as furnace and reactor trips, should be thoroughly tested from the initiating element to the final one before startup commences. The start-up procedure should have a critical device checklist listing all the critical alarms and devices, and it should be followed to ensure none are missed. Figure 8.1.5 illustrates an incomplete installation of the instrumentation.

All temporary hoses should be disconnected to prevent the backflow of hydrocarbons and safely stored away from pipe racks, work platforms, and personnel walkways where they can become tripping hazards. Temporary hoses should be equipped with check or non-return valves to prevent backflow from the process into the utility connection. Figure 8.1.6 shows a utility hose connection that remains in place, a hazard during start-up operations.

The Air-freeing Process: Ensuring That Air Is Reduced to a Safe Level (Before Bringing in Hydrocarbons)

The procedures for air-freeing will vary from process to process and company to company. What's provided here is a general discussion of air-freeing, which should help understand the key concepts.

The objective of air-freeing the equipment is to essentially eliminate all the air from the unit before allowing hydrocarbons into the unit. Air and hydrocarbons don't mix in confined spaces, and explosions have occurred due to the failure to properly air-free the unit before bringing in hydrocarbons. Our objective is to eliminate air down to a concentration of about 1% or less Oxygen by volume. Some plants where Hydrogen is involved may have targets much less than the 1% mentioned here; some have Oxygen targets in the parts per million (ppm) range.

The start-up procedure should describe the air-freeing process and be followed. Generally, the equipment is purged into the atmosphere using steam or Nitrogen as the purge medium. Remember, if using Nitrogen, the vents should be barricaded, and people should be made aware of the Nitrogen purging operations due to the asphyxiation properties of Nitrogen. More on this in Chapter 8.18, "Hazards of Nitrogen." If using Nitrogen, the most effective way to remove the air is with concertina purging, sometimes called huff-and-puff or simply pressure and depressure.

Verify the normal operating pressure and increase the pressure to a pressure less than the normal operating pressure. Then quickly release the Nitrogen into the atmosphere. Follow this process three times and then test the vent at the top of the equipment with an Oxygen analyzer (recently calibrated). Continue the process until the target of 1% O_2 is achieved. Don't forget about blocked equipment, dead legs, and spare equipment when purging. Ensure that all equipment is purged before allowing hydrocarbon into the unit. The best way to do this is with a checklist to help ensure nothing is missed. The list should include all the equipment that must be freed of air and require a sign-off as the equipment is verified air-free.

The Water Removal Process: Ensuring That All Free Water Is Removed from the Unit (Before Bringing in Hydrocarbons)

Likewise, the procedures for water-freeing will vary from process to process and company to company. Again, what's provided here is a general discussion of water-freeing, which should help understand the key concepts.

We work very hard for some units to keep free water out of the equipment. For example, we exclude free water from units like Reformers, Hydrocrackers, and Hydrotreaters, where water can damage the catalyst, or units like Alkylation, where free water can react with the acid, resulting in extreme corrosion of equipment and the cold ends section of Olefins plants (steam crackers) and other cryogenic processes where water can interfere with the refrigeration process. However, we typically use steam for purging or tightness testing for process units like crude and delayed Cokers, and a small amount of free water will accumulate in the equipment.

This water must be removed during startup as hydrocarbons are introduced and before heat is added to the process. If water is circulating in the process, in the pumps, piping, and/or towers, it will flash to steam as heat is added. Water expands 1,600 times its volume when it flashes to steam. This rapid expansion results in a pressure surge, severely damaging the vessel's internals. The pressure surge can damage or destroy the trays in the fractionation columns, it can open relief valves, and if the relief valves vent to the atmosphere, the release can result in loss of containment, explosion, and fire. Vessels have been known to explode catastrophically during startup when water contacts the hot hydrocarbons inside the vessel due to the extreme pressures generated. For example, this occurred at the CITGO Refinery in Port Arthur, LA, on 3 March 1991. During the startup of an FCCU unit, the slurry settling drum exploded, resulting in the deaths of six Operators. OSHA determined the cause to be a failure to adequately drain water from the Slurry Settling Drum during startup, resulting in a catastrophic explosion when the free water flashed to steam after contacting hot hydrocarbons.

For most units, the process we follow to remove water from equipment before bringing it back into service involves a cold flush with a heavier hydrocarbon, such as a light gas oil-type material. The free water is drained at low points as the unit is being cold flushed. It is important to ensure all pumps are operated to flush water from the pumps and discharge piping. For example, pump-around pumps on crude units or fluid catalyst cracking units (FCCU) must be operated on gas oil; otherwise, water can be trapped in the discharge piping above the check valve. In this case, during startup, when the pump is started to control the temperatures in the tower, the flow will push the free water directly into the hot tower, where it will flash, resulting in a large pressure surge in the column and over pressuring the column.

For this reason, it is a good idea to change over the pumps and ensure all pumps are operated during the flushing operation. Running the pumps on gas oil before heat is added will push the water back to the tower, where it can be safely drained away before heat is added. Also, don't forget control valve bypasses and exchanger bypasses. They must be opened to flush any standing water while the unit is being flushed. The same is true for any blocked-in spare equipment; ensure that the spare equipment is also operated to remove free water trapped in the pump or the discharge piping.

Ensuring That Instrumentation Is Active and Functional

Due to unit startup and equipment commissioning, instrumentation must be active and fully functional. All instrumentation must work correctly, including level, flow, pressure, temperature indicators, and controllers. All critical instrumentation, such as low-level and high-level alarms and emergency shutdown systems, must be verified as fully functional before the start-up commences. A critical alarms and instrumentation checklist should be included in the start-up procedures and completed and signed off before bringing in hydrocarbons.

Verification of instrument functionality is a joint activity between the Instrument Technician and Operator working together to perform the checks. Each critical instrument or critical shutdown system should have a specific procedure for the device that details exactly how the instrument should be checked. The checks should be performed from the initiating element through the final element. For example, suppose the initial element is a flow indicator designed to trip a furnace by closing the fuel gas trip valves in the event of low process flow. In that case, the device should be checked from the process flow indicator through the trip valve. This would include the logic solver, pneumatic relays, or a triple modular redundant programmable logic controller such as a Triconex-type system.

A unit startup is not a passive activity, and the field Operators have an extremely important role in supporting the start-up operation. The field Operators are the eyes and ears of the Console Operator. The console Operator has no windows to look out and see what is happening on the unit; they can't hear or smell what's happening. They rely on the field Operators to inform them of the unit status as the start-up progresses. Regular and frequent communication between the field Operator and the Console Operator should exist.

When establishing liquid levels and during warm-up operations, the field Operators should periodically check the levels in the sight glasses and relay this information to the Console Operator. The Operators should not rely solely on the control room instrumentation. For example, if a level gauge is full, look for other indicators, like pressure gauges, to indicate where the level is. A simple pressure gauge on the bottom of the tower can tell an Operator a lot about what's going on inside the tower. This is especially true if the gauge is ranged correctly and marked with visual cues when the tower is operating normally.

The console Operator should periodically verify that the bottoms-level controllers are functioning correctly. Remember, if the tower bottoms control valve is stuck closed, it can result in a tower overfill; if stuck open, it can blow through the hot gases inside the tower to downstream equipment that is not designed to see the hot materials. Either case can result in a catastrophic incident.

Importantly, if there is a conflict with expected instrumentation readings during the startup, do **not** assume the instruments have failed or are not working. There have been numerous incidents where the instruments displayed what the Operators interpreted as unreasonable results. In many of those cases, it turned out in hindsight that the instruments were correct, and the Operators were unaware that something else was going on.

Ensuring Compliance with Start-up Procedures

Chapter 7 covers "The Importance of Operating Procedures" and provides a comprehensive overview of procedures and expectations for their use in process operations. This section will not attempt to repeat this guidance but instead provide more specific information on the importance of procedures to start-up operations.

Procedures must be maintained and current to add value to start-up operations. Each site should have a process to periodically review the procedures for accuracy and ensure they are current. If the team discovers that a specific procedure is not current during a startup operation, they are responsible for taking time out and updating the procedure by redlining the changes. Typically, these redlined changes must be reviewed for accuracy by a supervisor, and the procedural changes must comply with the site's MOC process.

Most sites have a process where procedures are reviewed by a group of experienced Operators and supervisors to identify the most critical procedures or those critical to the process and where non-compliance may have the potential for loss of containment or explosion and fire. Most sites identify these as "Critical Procedures" or sometimes "SSHE Critical Procedures" and emphasize ensuring they are maintained current and are followed in the field. Generally, Critical procedures are developed in a checklist format and require the Operator (Console and Field) to sign as each step is completed during the startup. They must sign with the date and time when the step is completed. This is to help ensure that all steps are completed in the correct sequence and that steps in the procedure are not overlooked or missed.

Furnace operations are just one example of critical procedures where having detailed and current procedures in place and following those procedures is critical. An example of this is illustrated in a tragic incident on 13 November 2007 at the Chevron Pascagoula, Mississippi, Refinery. In that incident, a furnace tripped offline, and an Operator attempted to quickly return the furnace to service by relighting the burners. Unfortunately, the Operator failed to purge the firebox and was killed in the resulting explosion. Following that incident, OSHA cited Chevron for failing to provide operators with clear instructions for proper furnace startup after an emergency shutdown.

In the 23 March 2005, BP Texas City, Texas explosion and fire, the US Chemical Safety and Hazard Investigation Board found that the Operators had developed a practice of routinely operating the ISOM Raffinate Splitter Tower with the level above the top tap of the level instrument. This led to operating the tower "blind" with no way to monitor the level or know just how high the level was. This eventually led to grossly overfilling the tower, resulting in the loss of containment and the explosion that quickly followed. Fifteen people died in that incident, and more than 180 others sustained injuries, many of which were life changing.

Unfortunately, many similar examples could be provided involving inadequate procedures or procedures that were not followed. Remember, deviations to written procedures require approved Management of Change documents before the deviation is implemented in the field.

Importance of Development and Compliance with Operating Limits

OSHAs Process Safety Management standard (29CFR 1910.119) requires sites with covered processes to develop information pertaining to the technology of the process, including safe upper and lower limits for such items as temperatures, pressures, flows, or compositions and an evaluation of the consequences of deviations, including those affecting the safety and health of employees.

Safe upper and lower limits can be extremely valuable during transient operations such as unit startup and shutdown. As indicated in the regulation, safe limits apply to a wide range of process parameters, including those mentioned in the regulation and others such as tower levels, PH, vapor pressure, water or moisture content, etc.

At BP Texas City, the Operators had developed a habit of routinely deviating from written procedures and filling the fractionator tower above the top tap on the level instrument. Had safe upper-level limits for the fractionator bottoms level been established, and if those limits had been enforced, it is possible that the incident that occurred there would not have occurred.

Operating limits should be established for all key process parameters, and those limits should be enforced. Expectations should be in place for prompt reporting of all deviations to established limits, including an investigation to determine why the deviation occurred and corrective actions to help prevent a reoccurrence of the deviation.

Loss of Fractionator Tower Bottoms Level

The liquid level in the fractionator bottoms provides an effective seal between the hot vapor in the fractionator and the equipment downstream of the fractionator that is not designed to see the hot vapor. For this reason, the fractionator tower bottoms-level control and valve should be considered a safety-critical device, and its functionality should be thoroughly verified before the startup begins. A significant incident can occur if the tower bottoms control valve fails in the open or closed position.

If the control system or the control valve fails in the open position, it will quickly dump the liquid level from the tower. This allows the hot vapors from the tower to be vented directly into downstream equipment. For example, if the tower bottoms stream is directed to a floating roof tank, this hot vapor will flow directly to the tank, resulting in a sunken roof and a full surface tank fire or tank explosion.

Should the control system fail and close the control valve, or if the control valve fails to close, the feed will back up into the fractionator above the top tap of the level instrument. Should the level reach the flash zone level, where the feed is entering the tower, the liquid can be picked up by the feed velocity and carried into the overhead condensers, resulting in a "liquid lock" in the overhead condensers. This can result in rapid over-pressurization of the fractionator and the opening of the fractionator overhead relief valves. If the relief valves vent to the atmosphere, this can result in a loss of containment, and a significant fire or explosion can occur.

These are just a couple of scenarios that may occur should the tower's bottoms level control system or control valve fail to open or close. Therefore, this system should be classified as safety-critical and subjected to rigorous testing and verification before unit startup and periodic testing during operation.

Fatigue Management During Startup

Fatigue of Operating personnel also played a significant role in the BP Texas City explosion and fire. Workers at BP had completed 29 days of working twelve-hour shifts when the catastrophic incident occurred. It has been demonstrated that workers' fatigue can significantly negatively impact workers' situational awareness and job performance.

Studies indicate that acute sleep loss and cumulative sleep debt are two main underlying issues directly related to fatigue. Both appeared to play a key role in the BP Operators' ability to recognize abnormal situations and respond to return the operations to a safe level.

Fatigue can result in the inability to concentrate or focus and can significantly affect the decision-making process. A symptom of situational awareness is failure to recognize that something was wrong after feeding a fractionation tower for hours while not taking anything out of the tower, and the level instrument indicates a falling level. The CSB discovered this in their investigation of the incident on 23 March 2005.

As a direct result of the recommendations from the Chemical Safety and Hazard Recognition Board to the BP incident, the American Petroleum Institute has issued a revised edition of API RP-755, "Fatigue Risk Management Systems for Personnel in the Refining and Petrochemical Industries."

The revised version (2nd edition) guides employees, managers, and supervisors in understanding, recognizing, and managing fatigue in the workplace and expands on the previous version to include guidance in the following areas:

- Limits to hours of service (including guidance on how to manage call-outs for covered positions).
- New guidance on the work environment (including requirements for lighting and fatigue assessment).
- Individual risk assessment and mitigation (provides new information on the availability of technology that can help detect fatigue).
- "Shoulds" vs. "Shalls": Several sections of RP 755 were changed from "should perform" to "shall perform" to clarify that all components of Fatigue Risk Management Systems are needed.

Significant progress has been made in this critical area since the BP incident in 2005. In addition to the revised API standard, many companies have implemented versions of a revised fatigue management plan. Of course, there is still much work to do, and workers should feel free to discuss issues at their workplace with their supervisors. We must work together to ensure another incident like the one discussed throughout this section does not happen again.

Ensuring a Thorough Shift-to-Shift and Shift-to-Day Turnover

During unit startup, the conditions in the unit will change very rapidly. The Operators will conduct tightness testing on piping and vessels, air-freeing and water-freeing equipment, and establish levels in process vessels. They will also verify the functionality of critical instruments and other safety-critical devices. Occasionally, there may be safety-critical equipment undergoing final or preventative maintenance (PMs) and may be in a defeated state, with active Controls of Defeat.

This information must be clearly communicated between the outgoing process shift team and the incoming shift team (Shift Superintendents, Shift Team Leaders, Console Operators, and Field Operators). Likewise, clear and comprehensive communication between the shift organization and the day organization is equally essential. For example, the day organization is typically responsible for developing stream disposition during startup; there must be ongoing communication between the day and shift organization on this and other pertinent issues.

More information on the Shift Turnover process is available in Chapter 10; however, the minimum required includes the following:

- A face-to-face exchange of information between the outgoing process shift team member and the incoming shift team member before the outgoing team member leaves their post. This may result in a visit to the field to discuss the more complicated issues.
- A comprehensive and legible shift turnover log detailing the events on the prior shift and any ongoing or anticipated issues for the coming shift.
- The shift log should contain the following type of information (this is not intended to be a complete list):
  - Any changes made during the shift (i.e., instruments verified, levels established, blinds installed or removed, etc.).
  - The status of safety or critical environmental devices that are disabled or unavailable for any reason. Include the contingency plan for monitoring the process and protecting the equipment during the outage.
  - Any flawless operations items, including safety, health, and environmental incidents, near misses, or any issues with safety-critical equipment.
  - For example, any environmental issues, flaring, or discharge to the sewer.
  - Alarms – especially any alarms that are inhibited or disabled.
  - The status of fire protection equipment and other emergency response systems.
  - Any emergency management of change documents generated during the shift.
  - Equipment preparation for maintenance – to be verified as prepared by the incoming shift (opening process equipment, lock-out/tag-out isolation, equipment that is drained and opened).
  - Any interfaces with Mechanical, Technical, or Operations, and/or with upstream or downstream units, etc.

Figure 8.1.7 Example of an installed blind.

Ensuring Blinds Installed for Maintenance or Turnaround Are Removed and Ensuring "Running" Blinds Are Reinstalled

Blinds and blanks are extremely important in preparing process equipment for turnaround and maintenance operations. They are how we ensure positive isolation of process energy from equipment open for mechanical work. The placement of blinds and blanks requires a well-thought-out plan before the unit or equipment is taken out of service. The blinds and blanks must be installed in the right places, typically as close to the vessel or equipment as possible and in the right sequence. The sequence is important to avoid installing blinds that cannot be safely removed later due to process pressure against the blind, creating a "trapped" blind. Each blind should also have a visible tag to ensure it is identified in the field. Typically, the process organization develops the blinding strategy using the P&ID drawings to guide blind placement and the blind installation and removal sequence.

The equipment Blind Isolation List, sometimes referred to simply as the blind list, is extremely important. Every blind that is installed must be accounted for on the blind isolation list. This ensures that installed blinds are not forgotten and left in the unit, only to be discovered during startup or later when attempting to switch a stream to its final disposition. Every blind that is installed must be listed on the isolation list. Likewise, when a blind is removed from the unit, that blind must be signed off the blind isolation list. The list must always reflect the number of installed blinds and list each blind separately.

Most units also have what are called "running blinds." A running blind refers to blinds installed during normal unit operations but removed during maintenance or turnaround operations. For example, running blinds are installed to prevent accidentally sending a light stream like LPG to an atmospheric storage tank or sending a hot stream to a cold tank. The running blinds removed from the unit must also be reflected on the blind isolation list as removed. This helps ensure the running blinds are reinstalled before the unit is returned to service. See Appendix J for an example of a Blind Isolation List and Simplified Blind Location Drawing. Figure 8.1.7 illustrates a blind installed for equipment isolation. Note the clearly visible tag indicating when the blind was installed and the equipment being protected by the blind. The tag also identifies a contact person who can provide more information relative to the blind.

Siting of Temporary Building or Trailers to Support the Downtime

It is not uncommon for the maintenance or mechanical organization to request that temporary trailers or portable buildings be placed on or near the unit to house their people during the downtime. This occurred at BP Texas City and has also happened at other sites.

However, having occupied buildings, including temporary buildings like trailers, which are not designed for the potential overpressure that may occur in the event of a loss of containment, explosion, and fire, presents a real hazard. The siting and other requirements for occupied buildings in or near process areas are covered in two API Recommended Practices:

- API-752 "Management of Hazards Associated with Location of Process Plant **Permanent** Buildings"
- API-753 "Management of Hazards Associated with Location of Process Plant **Portable** Buildings"

As the name implies, API-752 covers the siting and design requirements for occupied permanent buildings in process areas. This covers control rooms, warehouses, offices, laboratories, etc. These buildings typically have personnel assigned or are used for recurring group functions such as meetings. API-752 helps managers ensure that personnel are protected by having occupied temporary or permanent buildings designed for the potential overpressure in the event of a loss of containment.

API-753 covers the siting of temporary buildings, such as turnaround trailers or portable buildings, that are placed on the unit for a specific purpose and a specific period. These buildings may also be occupied and may not be designed for the potential overpressure that would occur in an explosion. API-753 provides managers with guidance to help ensure portable buildings not designed for overpressure are not placed in vulnerable areas or ensure buildings in those areas are not occupied. API-753 only allows temporary buildings designed to withstand the potential overpressure that could occur in the process areas.

The guiding principles in both documents ensure that:

- Personnel should normally be located away from process areas if possible.
- Occupied buildings should be minimized within process areas.
- The occupancy of buildings located near process areas should be minimized.
- Occupied buildings (permanent or temporary) should be designed for the potential overpressure that may occur.
- Occupied buildings should be managed as an integral part of the facility's design, construction, and maintenance work processes.

Site Building Study

A building siting study can also provide valuable data by identifying the potential hazards and overpressure that buildings may be exposed to in the event of a loss of containment. The building study will identify potential sources of loss of containment, the types of materials or products that may be released, the potential ignition sources in the area, the degree of congestion in the surrounding area, and calculate the potential overpressure from an uncontrolled explosion. The most common type of explosion is a Vapor Cloud Explosion or VCE, which occurs when a release of flammable vapor like gasoline or naphtha forms a vapor cloud. When this vapor cloud comes in contact with an ignition source, such as an arcing or sparking electrical component, it results in an explosion. When this explosion occurs in the presence of obstacles or confinement, it results in a VCE. The VCE results in a large pressure wave or "overpressure" that can damage process areas (process units and storage areas). The building study will calculate the potential overpressure resulting from a VCE, and this data can be used when assessing the types and designs of occupied buildings placed in or near process areas.

Before placing an occupied temporary building in or near the process, please consult with the building study to understand the potential overpressures that could result from the release of flammable materials and refer to the API Recommended Practices (API-752 and API-753) for guidance. Also, ensure the Management of Change (MOC) process is completed for each temporary building installed.

Ensure the Pre-startup Safety Review (PSSR) Is Completed Prior to Startup

The PSSR is a special safety review conducted on a new unit or modified equipment prior to the startup or commissioning of the equipment. The PSSR is intended to help ensure that the new or modified facilities meet the design criteria and operating intent and to detect any potential issue or hazard that may have been introduced during the equipment design and construction. Simply stated, the PSSR is essentially the last audit to help ensure that the unit or equipment, the procedures, and the Operators are ready for startup.

A Pre-startup Safety Review is a broad-based review that covers the process equipment and ensures the operating procedures are complete and up-to-date and Operator training is current, especially for any new or modified equipment. The PSSR procedure and checklist are typically prepared as an integral part of the pre-commissioning and commissioning documents.

The Pre-startup Safety Review is required by the US Occupational Safety and Health Administration (OSHA) (OSHA 29CFR 1910.119(i). The following is directly from the regulation:

> The employer shall perform a pre-startup safety review for new facilities and modified facilities when the modification is significant enough to require a change in the process safety information. The pre-startup safety review shall confirm that before introducing highly hazardous chemicals to a process, Construction, and equipment are following design specifications; Safety, operating, maintenance, and emergency procedures are in place and are adequate. For new facilities, a process hazard analysis has been performed, recommendations have been resolved or implemented before startup, and modified facilities meet the requirements contained in management of change, paragraph (l). Training of each employee involved in operating a process has been completed.

It is noteworthy that OSHA requires a PSSR to be completed for all new or modified facilities (when the modification requires a change in the process safety information). This regulation also requires a Process Hazard Analysis (PHA or HAZOP) to be completed on new facilities before commissioning, and the HAZOP recommendations must be resolved or implemented.

Finally, Remember the Following

We think of a startup as almost routine and consider it "just another startup." This is certainly not true. Every startup is unique, and every startup is different. For example, modifications may have been made to the unit since the last time it was started, or other interfacing process units or off-site facilities may have different situations. It could be that a pump normally used for the startup is out of service, or a different tank lineup is being used for this startup.

For this reason, it may be necessary to update the start-up procedure to address the conditions that will exist at the time. Remember, always follow the MOC process when doing this type of work. It is a good idea to get the startup team together in advance, discuss the procedure that will be used, and discuss what differs from previous startups. Develop the start-up plan and then implement the plan.

It is not uncommon to experience a small bump in the road while starting up a large process unit. A pump typically used in the process may fail, a feed tank that is usually used may not be available as expected, or an instrument or analyzer may fail. Should this happen, don't overreact to the event. Take a time out, get the team together, discuss what has happened, develop a response plan, and implement the plan. So many times, the unit has been very close to achieving complete startup, and the team hits a small bump and overreacts, and suddenly, the unit is back down, or it results in an incident. This can almost always be avoided by implementing a well-developed plan.

What Has Happened/What Can Happen

BP Refinery
Texas City, Texas
23 March 2005
15 Fatalities, 180 Injured

The US Chemical Safety and Hazard Investigation Board investigated this incident and prepared a detailed investigation report and an excellent training video.

Much has been said about this incident in various industry reports, the media, and this book. However, there were so many lessons learned here I would be remiss if they were not covered in more detail in this section. After all, the incident occurred during a start-up operation and resulted in a catastrophic explosion, which resulted in the most significant industry event in several decades in the United States. The incident cost BP billions in property damage, victim's compensation, lost production, and damage to the company's reputation. It ultimately resulted in BP selling the Texas City Refinery to one of its competitors.

By now, almost everyone understands what happened at BP; the Isomerization Raffinate Splitter Tower was overfilled, releasing hot naphtha that quickly ignited, causing a catastrophic explosion. The fifteen fatalities and many injuries were contract employees who occupied wood frame trailers near the Isom unit, working for a project unrelated to the Isom unit.

Figure 8.1.8 BP Texas City explosion damage, including the area containing the occupied trailers. Source: Photo courtesy of the Chemical Safety and Hazard Investigation Board (CSB). The photo is from the CSB report "BP America (Texas City) Refinery Explosion." The arrow is an annotation by the CSB to show the location of the Isom Unit blowdown drum.

The US Chemical Safety and Hazard Investigation Board conducted a very comprehensive investigation of this incident. It published a detailed report indicating a long list of causal factors contributing to the incident. The CSB also developed an excellent training video on this incident titled "Anatomy of a Disaster," which I frequently use in my Safe Operations Training course. I'll not replay the CSB report here, but instead, I'll provide a list of the causal factors that were identified (not intended to be a complete list). Please refer to the CSB's final report for more details on this incident. Figure 8.1.8 provides a good view of some of the explosion damage, including the portable trailers and occupied trailers, that were destroyed or heavily damaged by the explosion.

BP Texas City incident causal factors:

- Significant budget cuts and failure to invest.
- Location of occupied trailers near operating process equipment handling highly hazardous materials (13 trailers were destroyed, 27 were damaged).
- Inadequate shift turnover (also referred to as shift handover):
- Operating procedures were not current, not maintained, and, most importantly, not followed.
- Operating limits were not enforced.
- Inadequate instrumentation design.
- Critical instruments are not being maintained and not checked for functionality before startup.
- Operator Fatigue.
- Frequent management turnover.
- The Blowdown system design appeared inadequate (vented to the atmosphere).
- Control panel layout.
- Inadequate process staffing.
- Inadequate Operator training – No process simulators.
- Lack of incident reporting, investigation, and development of corrective actions to prevent a recurrence of the incident.
- Personnel not involved in the startup were not removed from the surrounding area during hazardous operations (unit startup).
- Ignition sources were allowed near the unit while it was starting up.
- Lack of process safety leading and lagging indicators.
- Process Safety functions are not at a high level in the organization.

Quoted by the US Chemical Safety and Hazard Investigation Board following their investigation (not a complete list):

- "Widespread tolerance of non-compliance with HSE rules."
- "Poor implementation of safety management programs."
- "Lack of leadership competence and understanding."
- "Managers job to know."
- "Check the box mentality."
- "A learning disability."
- "Blind to Process Safety."

What Has Happened/What Can Happen (Trapped Air During Startup)

Citgo Refinery
Lake Charles, LA
24 September 2001
Three injuries (none reported to be serious)

An explosion and resulting fire occurred at the Citgo Refinery during the startup of equipment on a Unicracker (Hydrocracker) unit. When the explosion occurred, it ruptured a 6-inch gasoline line that served as the primary fuel source for the extended fire. The fire burned for more than 18 hours; three persons were treated for minor burns at local hospitals.

An investigation determined the accident occurred due to an internal detonation in the Unicracker Hydrogen supply filter system as that system was being brought back into service following a routine filter change.

OSHA investigated this incident. Theresa Schmidt, reporting for KPLC News, a local media outlet, said that a company safety manager explained what led to the explosion: She is paraphrased here; "There was a vessel in which there was an accidental mixing of air and Hydrogen that resulted in an explosion. The explosion sent a piece of metal out of the vessel, striking a support column and a 6-inch gasoline line. The leaking gasoline led to a huge explosion, which most people heard."

This incident emphasizes why procedures and field compliance should adhere to equipment air-freeing procedures. This is especially important when Hydrogen may be present due to the extremely wide flammable range of Hydrogen.

Key Lessons Learned

Remember, air and hydrocarbons, especially air and Hydrogen, do not mix in confined spaces and quickly become a recipe for disaster. Like the one in Lake Charles, LA, numerous explosions have occurred by failing to complete a comprehensive air-freeing process.

This incident highlights the importance of a thorough air-freeing process before admitting hydrocarbons to the process equipment. It matters less as to which medium is used for the air-freeing, Nitrogen or Steam. When carrying out the air-freeing operation, it is more important to ensure that vessels are not missed and that gas testing is done for Oxygen content before allowing hydrocarbons to be brought in.

The best way to do this is to develop a comprehensive checklist of all equipment and check off each piece after the air-freeing step is completed using the checklist as a guide. It is equally important to ensure that all spare equipment, dead legs, and blocked-in equipment are included in the air-freeing process.

What Has Happened/What Can Happen (Trapped Water During Startup)

United Kingdom Refinery
Rapid overpressure in atmospheric fractionator on a Crude Distillation Unit during startup
March 1977

The Crude Distillation Unit had just completed a turnaround and was in start-up mode when the safety relief valves opened on the atmospheric fractionation tower to the atmosphere. The relief valves remained open for about 30 minutes, discharging crude oil over the refinery and into the surrounding neighborhood onto homes, cars, and vegetation, including trees. This incident resulted in a significant cleanup effort in the surrounding community.

Free water was trapped in one of the pump-around circuits, causing rapid overpressure in the crude atmospheric fractionating tower. The water was determined to be above the check valve on the pump around the pump discharge. As the unit was circulating and the temperature was coming up on the tower, the operating crew started the pump-around pump, which pushed the water directly into the hot tower, where it contacted the hot oil. The water quickly flashed to steam, resulting in a pressure surge inside the tower. All trays in the fractionating column were damaged and required repairs, and some required replacement.

A very similar event occurred at a refinery in South Africa in 2004. The South Africa incident also occurred during the startup of the APS when the bottom pump-around pump was commissioned. Shortly afterward, the main APS tower safety valves lifted for three minutes, with 30% of the release falling onto cars and homes outside the refinery fence line. It was determined that an approximately 10-foot (3-meter) pipe above the check valve on a pump around pump discharge had not been drained of water. This resulted in ~80 gallons (~300 liters) of water being introduced into the hot tower.

It is unsettling to see a relief valve on an atmospheric crude fractionator or any other process unit releasing crude into the atmosphere due to tower overpressure. This has happened often in our industry due to free water trapped inside the vessel during startup. The water quickly flashes to steam and results in the tower overpressure.

What Has Happened/What Can Happen (Trapped Water During Startup)

Citgo Refinery
Lake Charles, LA
3 March 1991
Six Fatalities

A major explosion and fire occurred at the Citgo Refinery in Lake Charles while the FCC Unit Citgo Petroleum Refinery was starting up. The resulting explosion and fire resulted in six Operator Fatalities. OSHA determined the cause to be the failure to adequately drain water from the Slurry Settling Drum during startup. The FCCU Slurry Settling Drum was over-pressured and failed catastrophically when the hot Slurry came into contact with the accumulated water in the vessel.

OSHA issued a total of 409 citations to Citgo Petroleum. The following is directly from the OSHA Investigation Report on this incident:

> "On or about March 3, 1991, the main block valve at the bottom of the F-7 was closed, which prevented water from draining out of the CUU "A" during the start-up procedure." "Consequently, hot oil was mixed with water in the F-7 Slurry Settling Drum for the CUU "A," thereby causing a catastrophic failure of the slurry settling drum, releasing flammable and combustible liquids, gases, and vapors." (Note: F-7 was the Slurry Settling Drum, CUU "A" was the FCC main column).

OSHA went on to recommend a method to help prevent an incident of this type from happening again as follows: "A feasible and useful method among others of correcting this hazard is to provide and strictly enforce the use of a detailed written start-up procedure with signed and dated logs for the specific task(s)."

See Figure 8.1.9 for a representation of a typical slurry settling drum (Not all FCCUs have a slurry settling drum, and this drawing is not intended to represent the unit where this incident occurred).

Key Lessons Learned

The incidents described here emphasize the importance of a thorough dewatering process before and during the time the temperature is being increased during unit startup. Again, the best way to do this is to develop and follow a detailed equipment checklist where free water can accumulate. All low point drains must be checked before the unit's temperature increases to ensure the unit is completely water-free and continued during the period heat is added. Be especially aware of the water that can become trapped above the check valves in the pump discharge piping. This is particularly true for pump-around pumps where the water can be pushed directly into the tower. If the tower is at operating temperature (hot), the water will flash to steam, resulting in an extreme pressure surge inside the tower. This pressure surge can open relief valves to the atmosphere and can, in some cases, overwhelm the relief valves, resulting in the potential for vessel rupture.

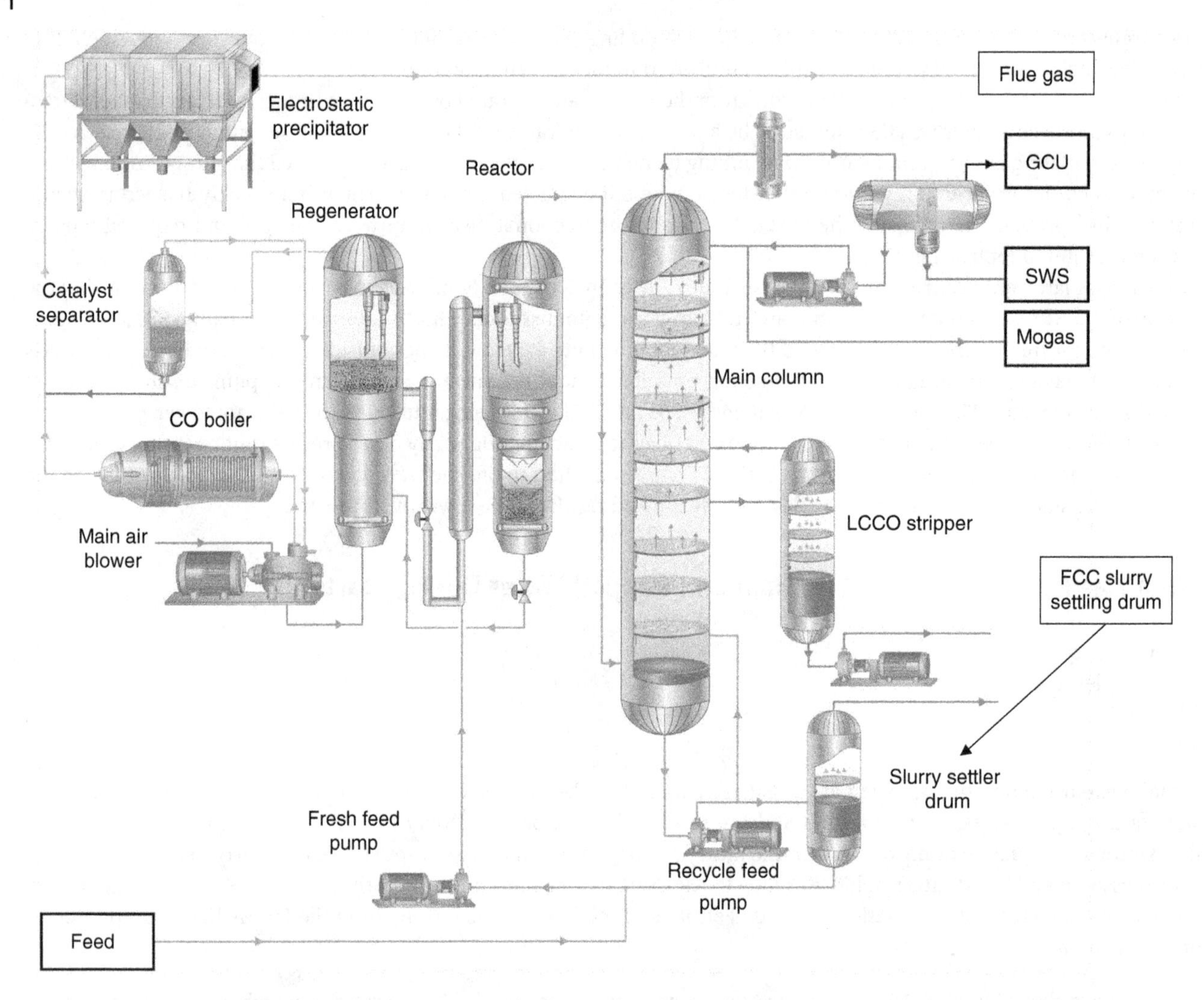

Figure 8.1.9 Example of an FCC slurry settling drum (Not the actual unit described here). This image is courtesy of Ecoscience Resource Group Visual Artists. It was created to illustrate the location of the slurry settling drum (if the unit is so equipped).

What Has Happened/What Can Happen

Ethylene Release and Fire
Kuraray America, Inc. EVAL Plant
Pasadena, Texas
19 May 2018

The US Chemical Safety and Hazard Identification Board investigated and issued a detailed report of this incident.

A release of ethylene from a relief valve occurred on 19 May 2018, resulting in a fire that injured 23 workers, including two who were life-flighted to a local burn center. One of the workers remained in the hospital for several days due to life-threatening injuries.

The ethylene and vinyl alcohol plant was in start-up mode, returning equipment to service following a maintenance turnaround when the release and fire occurred. At the time of the event, 266 employees and contractors were onsite, some of whom were performing work on the unit unrelated to the startup.

During the startup, a high pressure developed in the reactor and caused the emergency activation of the overpressure relief system, which is designed to vent the pressure from piping oriented in a horizontal open-ended vent pipe. The release was directed toward a pickup truck located at grade where contract employees were welding on a section of unrelated piping, resulting in the immediate ignition of the hydrocarbon release. Some contractors were directly impacted by the fire,

Figure 8.1.10 Kuraray America EVAL Unit safety valve vent piping. Source: Photo courtesy of the Chemical Safety and Hazard Investigation Board (CSB). The photo is from the CSB report "Ethylene Release and Fire, Kuraray America, Inc., EVAL Plant Pasadena, Texas, May 19, 2018." The CSB added annotation highlighting the reactor overpressure relief piping installed in the horizontal orientation.

Figure 8.1.11 Kuraray America EVAL Unit fire from the safety valve vent piping and the truck where the welding was in progress. Figure (a) illustrates the fire which occurred due to the relief valve opening. Figure (b) is a photo of the location of the work crew where welding was ongoing and their equipment, including their vehicle.

resulting in injuries. Other workers sustained slips and falls as they attempted to run or jump to escape the area. The reactor pressure returned to normal after about three minutes, allowing the relief valve to close and extinguish the fire.

Figure 8.1.10 illustrates the process vent location where the release to the atmosphere occurred. Figure 8.1.11a,b show the significant fire and damage to the adjacent pickup truck.

Key Lessons Learned

As typical in an event like this one, several lessons are learned to help prevent a reoccurrence of the incident. In this case, the unit was in a transient mode of operation, a startup. However, there were 266 workers onsite during the startup. Many were working on support tasks that were unrelated to the startup. A good practice is to ensure that all nonessential personnel is removed from the unit anytime the unit is in a transient mode of operation, such as a startup, shutdown, emergency operation, or any other form of transient operation.

The emergency release vent pipe was designed with a horizontal relief pipe, and unfortunately, it vented directly in the area where hot work was being performed. The CSB felt this was most likely the source of ignition following the release. Following this incident, the site redesigned this pipe to a vertical release with the emergency release point high above the unit structure. Another alternative would have been to route the safety valve release to a closed system, such as a flare. In hindsight, this should have been identified in a unit Hazard and Operability Analysis (HAZOP) or other similar types of risk assessment.

Strict enforcement of operating envelopes may have prevented the PRV release by operator action to prevent the reactor overpressure by earlier detection and rapid action following approved procedures to reverse the causal factors for the overpressure.

These and others can become part of the organizational culture through a functioning operations integrity management system.

These photos are courtesy of the Chemical Safety and Hazard Investigation Board (CSB). Both images are from the CSB report "Ethylene Release and Fire, Kuraray America, Inc., EVAL Plant Pasadena, Texas, May 19, 2018." The left image shows the fire, which occurred immediately after the release. The right image is of the pickup truck where the welding was in progress when the release occurred.

Additional References

US Chemical Safety and Hazard Investigation Board: BP Texas City explosion and fire – 23 March 2005, Report No. 2005-04-I-TX. https://www.csb.gov/bp-america-refinery-explosion/

US Occupational Safety and Health Administration (OSHA): Process Safety Management standard (29CFR 1910.119)

- Includes the Pre-startup Safety Review (PSSR) (29CFR 1910.119(i)). https://www.osha.gov/laws-regs/regulations/standardnumber/1910/1910.119

Elements of Process Safety Management; Safety Services 2018–2019, Safety Info; A Christian Run & Operated Website, PO BOX 477 Guntersville, AL 35976. https://www.safetyinfo.com/process-safety-management-elements-psm-free-index/

American Petroleum Institute:

- RP 752: "Management of Hazards Associated with Location of Process Plant Permanent Buildings"
- RP 753: "Management of Hazards Associated with Location of Process Plant Portable Buildings"
- API Press Release; RP-755 2nd Edition; 2 May 2019
- RP-755: "Fatigue Risk Management Systems for Personnel in the Refining and Petrochemical Industries"

ASME B-31-3: "Process Piping", Oil and Gas Journal On-Line Staff Report; Report on Citgo Explosion (Incident update), OSHA Incident Database:

- OSHA Investigation on Citgo Lake Charles, LA Explosion and Fire (3 March 1991). https://www.osha.gov/pls/imis/accidentsearch.accident_detail?id=14391726
- OSHA Investigation on the Chevron Pascagoula, MS, Furnace Explosion, and Fire (13 November 2007). https://www.osha.gov/pls/imis/establishment.inspection_detail?id=954037.015

US Chemical Safety and Hazard Investigation Board: Ethylene Release and Fire at Kuraray America, Inc. EVAL Plant, CSB Report No. 2018-03-I-TX, Pasadena, Texas, 19 May 2018 | No. 2018-03-I-TX. https://www.csb.gov/kuraray-pasadena-release-and-fire/

https://www.wlox.com/story/25333553/osha-releases-findings-penalties-for-deadly-chevron-explosion/ (news article)

Report by local Lake Charles media (KPLC News) on the Citgo Lake Charles explosion and fire which occurred on 24 September 2001: https://www.kplctv.com/story/511333/citgo-reports-on-cause-of-fire/

US Chemical Safety and Hazard Investigation Board, Safety Digest: CSB Investigations of Incidents during Startups and Shutdown. https://www.csb.gov/assets/1/17/csb_start_shut_02.pdf?16301

Chapter 8.1. Returning Equipment to Service – Safe Startups and Other Transient Operations

End of Chapter Quiz

1 The start-up procedures and the start-up team must recognize and address many hazards. See how many you can list below – name as many hazards as possible.

2 What procedure is completed during preparation for a startup to help ensure that a leak or major loss of containment does not occur?

3 Explain how the start-up procedure should be used during the unit startup.

4 Prior to unit startup, the Operators do a comprehensive unit walk down to ensure the unit is ready for startup. What are examples of what the Operators are looking for during this walk down?

5 How are instruments and other critical devices prepared and checked before startup?

6 Explain the process for air-freeing process equipment before bringing in hydrocarbons.

7 Are Operating Limits important during startup?

8 What is the purpose of the Pre-startup Safety Review (PSSR)?

9 Is there a process for tracking and accounting for blinds, and is this referenced during startup?

10 Do startups ever become routine and become "just another startup"?

11 What is the consequence of water left trapped in a fractionation column during startup?

12 Describe the potential consequences of losing the bottoms level in a fractionation column when the tower is operating.

13 Why is a detailed shift log and shift turnover discussion important during a unit startup or similar unit transition?

14 List at least five topics that should be discussed during shift turnover during a unit startup.

15 Why is it critical to maintain the unit blind list up-to-date?

16 Operating limits should be __________ for all key process parameters, and those limits should be __________. Expectations should be in place for prompt __________of all deviations to established limits, including an investigation to determine why the deviation occurred and corrective actions to help prevent a reoccurrence of the deviation.

17 At BP Texas City, the Operators had developed a habit of routinely __________ from written procedures and filling the fractionator tower above the top tap on the level instrument.

18 What is meant by "running" blind?

8.2

Safe Permit-to-Work – Breaking Containment and Controlling Work

A permit-to-work system is a formal written system that controls certain potentially hazardous work. A competent, authorized person authorizes the permit for a workgroup to proceed with the specified job under the stated conditions. It is a "legal" binding document, and when the work permit is approved and issued to a workgroup, it is a "license" for work to start. The key objective of the permit-to-work is to ensure that personnel are informed of hazards and the work is executed safely, protecting personnel safety, health, and the environment.

The plant should have clear expectations for full compliance (zero tolerance) with applying work permit processes and procedures. These expectations should be extended to all management and Supervision, who should be expected to routinely verify that the activities associated with the work requirements are correctly executed in the field and ensure that the hazards are properly controlled through a system of regular effectiveness reviews. Fieldwork should fully comply with the work permit requirements, including the procedures for energy isolation, opening process equipment, and confined space entry.

As the equipment owner, the Process organization is accountable and responsible for ensuring that equipment or area is handed over to the workgroup and that the equipment is safe to work on. Each designated individual or position in the work permit system has specific roles and responsibilities with qualifications and training requirements specific to those roles.

The Work Permit process ensures that jobs are planned, equipment is prepared, precautions are taken, and effective communications are maintained between the permit issuer and the permit acceptor. Other impacted individuals/groups have occurred. An effective work permit process helps ensure that all personnel have been informed of hazards and proper procedures and that work is executed safely, protecting the health and the environment.

The work is clearly and legibly specified on the work permit form and includes the precautions during execution. Permit-to-work is an essential part of safe systems for many maintenance activities. Work is authorized to start only after defined safety procedures are implemented and appropriate protection, including equipment isolation and personnel protective equipment, is in place. All personnel involved in the work permit process have defined roles and must be trained to carry out their responsibilities. The work permit system also helps ensure that jobs are planned and the equipment is properly prepared to carry out the work.

By ensuring that effective communications have taken place between the permit issuer, permit acceptor, and other impacted individuals/groups, including other work near the work being performed, the permit-to-work is one of the key communication tools between those preparing the equipment and authorizing the work and those who will be executing the work in the field and others that the task may impact.

Types of Work Permits

It is common for a site to have several types of work permits. For example, work involving an open flame may be classified as hot work and require a "Hot Work Permit." Hot work without an open flame, such as grinding or soldering, may require a "Non-flame Hot Work Permit". Work that does not involve heat or sparks may require a "Cold Work Permit." Any entry into a confined space typically requires a "Confined Space Entry Permit," and the opening of process equipment may require an "OPE" or "Opening Process Equipment Permit." Each permit is typically a different color to help identify the type of permit being authorized and has the specific precautions and types of gas testing required noted on the permit form for the type of work authorized. Of course, there may be variations of this at different manufacturing sites across the United States

Process Operations Safety: The What, Why, and How Behind Safe Petrochemical Plant Operations, First Edition. M. Darryl Yoes.

and worldwide. Some sites combine the two kinds of hot work into one Hot Work Permit; some combine the requirements of opening process equipment onto the existing Cold Work Permit.

For sites that use the Hot Work permit form, this permit generally authorizes work involving an open flame, such as welding, burning, arc gouging, etc. This permit will ensure proper gas testing for flammables, that the sewers are properly covered to prevent gas from entering the area, and that sparks do not enter the sewer system. The permit will also ensure that fire protection equipment is readily available and may require the continuous wetting of resulting sparks in the area.

A permit for nonflame hot work may authorize the operation of motorized equipment, such as air compressors, motor vehicles, and electrically powered tools, in process areas. Likewise, this permit will ensure gas testing for flammables and proper sewer sealing before work is authorized. This permit will also address the fire protection required in the nonflame hot work area.

The Confined Space Entry permit authorizes entry into confined spaces such as storage tanks, process vessels, vessel skirts, or other spaces not designed for occupancy. This permit will ensure proper gas testing of the vessel for O^2, lower explosive limit (LEL), H_2S, and other chemicals known to exist in the process. The Entry Permit will also ensure that a hole watch is available to track each person who enters the vessel, record the time in and out of the vessel, and ensure that communication is in place with the entrant and that a rescue plan is available if needed.

The Cold Work permit is typically used to authorize routine maintenance not involving hot work, confined space entry, or opening process equipment. A typical cold work permit may involve scaffold erection or maintenance on a pump or motor after the equipment has been isolated. This permit will specify the precautions for the work being performed, the appropriate PPE for the workers, and any other special precautions that should be in place to perform the work safely.

Energy Isolation (Lock-out/Tag-out, or LOTO)

Controlling hazardous energy is an extremely important part of the permit-to-work. Hazardous energy may consist of process pressure, electrical, mechanical (such as springs), temperature, etc. Failure to recognize energy sources and adequately control hazardous energy has resulted in many process safety incidents, some with tragic outcomes.

To achieve energy isolation, the operations group, typically Operators, will ensure the equipment is depressurized and drained; valves for isolation are identified, closed, and secured in the closed position. Most sites use a system of process isolation tags and locks (with a chain if required) to secure the valves in the isolated position. The objective should always be "zero energy," which should be confirmed with open (and unobstructed) process vents and drains.

A higher-level review and formal risk assessment to identify mitigation will be required if zero energy cannot be achieved; in some cases, for example, when flammable or toxic materials are involved, the equipment will be shut down to achieve proper isolation before work can be performed. Those performing the work must also install their individual lock to protect against the unauthorized release of energy while work is in progress.

Electrical energy isolation is normally accomplished by opening the electrical breaker and installing the isolation tags and a process lock to secure the breaker in the open position. While work is in progress, the personnel performing work must also install their individual locks on the opened breaker(s). An attempt to start the equipment at each pushbutton location should be made to ensure the electrical energy has been isolated BEFORE the work can begin.

Gas Testing of Equipment to Ensure Readiness for Work to Begin

When opening process equipment, verifying that hazardous materials such as flammables or toxins have been properly purged from the equipment before work begins is important. For example, the gas test can help identify the need for additional PPE (Personal Protective Equipment), such as self-contained breathing apparatus (SCBA), while installing blinds in the piping systems. The limits for work to proceed should be clearly defined and readily available, typically defined on the work permit form. Gas testing should only be done by people trained on the specific gas testing equipment they will be using. When testing for LEL, it is always good to gas test for O^2 first, as some gas testing equipment will not provide an accurate LEL reading if the O^2 is low.

Remember, it is important to recognize that a gas test for flammables will not detect combustibles trapped in piping or other process equipment due to the flashpoint of the combustibles, e.g.: flashpoints above 100 °F (38 °C). This material will not be vaporizing at most ambient temperatures; therefore, vapor will not be present for the gas test apparatus to read. Therefore, if combustibles are present, the gas test device may provide a false indication indicating hydrocarbon free, even though combustibles are present.

Safety Hazard Analysis

A Jobsite Hazard Analysis should be conducted before the work begins on jobs with a higher potential for hazards to be present or deemed to have the potential for injury. This process identifies the individual steps to be undertaken during the job, the hazards associated with each step, and a plan for mitigating or controlling the identified hazards. The Jobsite leader and supervisors associated with the task should document and review this formal hazard identification process. For example, a task to install blinds in a naphtha line may include hazards relative to trapped energy or the potential for leaking valves, which could release the product when the flanges are opened. Mitigation could include reviewing the equipment isolation process and verifying vents and drains to ensure they are properly opened and free from internal obstructions or plugging.

Joint Field Walk down and Verification of Equipment Readiness for Work

Before work begins, the permit issuer should conduct a joint field visit with the person accepting the work permit to review the work, the precautions and mitigation in place, and verify equipment isolation for mechanical work. The exact work to be done should be discussed in the field; for example, when opening process equipment, the visit should highlight which flanges to be opened and, if applicable, the proper sequence to be followed. The flanges or points where the equipment is to be opened should be clearly identified with tags, painting, or another method selected by the site. The isolation methods, including secured valves, open vents, drains, etc., should be explained to those performing the work.

The joint field walk down is an extremely important requirement and should be a key element of the site's permit-to-work process. It should be periodically audited to ensure compliance and follow up with corrective actions if noncompliance is observed.

And finally, before the work begins, the work crew should participate in a nonformal risk management process. This process has various names, but the crew should always ask, "What is the worst thing that could happen?" and then develop a plan to ensure that mitigation or precautions are in place to prevent such incidents from occurring during the work.

Permit Audits and Feedback

Due to the importance of work permits, each site should have a work permit audit process to help ensure that work permits do not become a "formality" like what happened at Piper Alpha. Work permits, both paper audits and fieldwork in progress, should be audited periodically to ensure full compliance with the site work permit standards and requirements.

I like to say that each tick box on the work permit form is there for a reason, and many are "built on blood." They are there because of incidents that have already occurred in our company or our industry. They are also there to help ensure it does not happen again. Suppose we are not looking at permits regularly, not providing feedback on our observations, and not constantly seeking continuous improvement. In that case, the guys and gals at the field level will conclude that they must not be all that important.

Motorized Equipment as Potential Ignition Sources

Frequently, a running engine, especially a diesel-powered engine, may become the ignition source when hydrocarbon vapors are ingested into the engine. In the event of a release of hydrocarbon vapor in the area where a diesel engine is operating, the engine will ingest the vapor, causing the engine to overspeed. The engine will also begin misfiring and back-firing, resulting in the ignition of the hydrocarbon vapor. When a diesel engine goes into overspeed, it cannot be shut down by simply turning the key as in a gasoline engine. The diesel vapor and air mixture compression in the cylinder fires a diesel engine. When a diesel engine ingests hydrocarbon and is overspeeding, this will continue until either the engine overrevs and explodes, the hydrocarbon release is stopped, or the air intake to the engine is shut off. Diesel engines can be equipped with an automatic shutoff of the air intake in the event of overspeed, which is required by regulation for engines operating in petroleum refineries and petrochemical plants in some US-regulated areas.

Please refer to the OSHA Safety Fact Sheet in Appendix M for more details. Also, Appendix N lists known cases where overspeeding diesel engines have been identified as the source of ignition in petroleum refining and petrochemical plant fires and explosions.

What has Happened/What can Happen

Shell Singapore Refinery
Singapore Pulau Bukom
21 August 2015 (six burn injuries, three reported to be critical)

According to the Ministry of Manpower (MOM) in Singapore, the Shell Pulau Bukom Refinery was conducting maintenance on a crude distillation unit when the incident occurred.

On 21 August 2015, two groups of workers were simultaneously conducting maintenance and project activities on two different jobs. One group of workers was executing hot work from a scaffold using an acetylene torch to cut and dismantle piping. The other workers were dismantling a solvent pipeline at grade and near the hot work activities. The second group was removing a joint connection to a valve and connecting a hose to the valve to drain residual flammable hydrocarbon from inside the piping to a nearby pit.

"When one of the workers opened the valve to start the draining process, flammable vapors from the draining of hydrocarbons came into contact with the sparks from the hot works," said MOM. When the worker draining the flammable material was alerted about the hot work, he immediately closed the valve. However, the flammable liquid ignited almost immediately, injuring six workers.

According to the MOM, the fire was contained within about 30 minutes by the Bukom Emergency Response Team. The MOM investigation found that "there was a systemic failure in Shell's oversight to check for compatibility of different work activities carried out within the same vicinity at the same time." According to the MOM, the hot work by one group was not coordinated with the cold work, resulting in the flammable vapor from the cold work being ignited by sparks from the nearby hot work.

Key Lessons Learned

The cause of this incident appears to be very simple. However, most would be surprised at how often incidents like this happen in our industry. Workers should be trained and continuously reminded to be constantly aware of what is happening around them. They should ask, "will my work impact others in the area?" They should also ask, "will work being carried out by others in the area impact the work I am about to do?" As I said, this sounds simple, but it does happen often.

The permit-to-work process should also cover this for each of the two jobs. The work permit for hot work should require the workers to verify that the area is hydrocarbon free before any hot work is started. This should normally require a walk down of the area and a gas test for flammable vapors before any hot work is authorized. Hot work jobs also typically require a fire watch who is continuously aware of work near the hot work.

The workers about to disjoint piping should also be checking for ignition sources in the area and continuously hosing down sparks to minimize the potential for ignition. The opening process equipment (OPE) job should also require a walk down of the area, checking for potential sources of ignition in the area. OPE tasks also typically require the erection of barriers or barricades set to keep others out of the area where the equipment is being opened.

We depend on the detailed permit-to-work process to see that these minimum requirements are in place to ensure this type of incident does not reoccur.

What has Happened/What can Happen

Tosco Avon, CA.; Loss of Containment and Fire (23 February 1999)
Fire During Opening process equipment for maintenance
Four fatalities and one serious injury

The Tosco Avon Refinery experienced a devastating fire on 23 February 1999 while removing the naphtha draw-off piping on the operating atmospheric crude tower. During the fire, there were four fatalities, and one other contract person was critically injured when he jumped from the scaffold where the workers were attempting to drain naphtha from the naphtha draw-off piping with the tower operating. An image of the fire location is illustrated in Figure 8.2.1 below. The US Chemical Safety Board investigated this incident and developed a detailed report.

The naphtha draw-off piping leak was discovered on 10 February 1999 (13 days before the fire occurred). Despite attempts by the operations personnel to isolate the leak, it reoccurred at least three times before the fire. On the day of the incident,

Figure 8.2.1 A devastating fire occurred at the Tosco refinery, martinez, California, on 23 February 1999 while workers were replacing the naphtha draw-off piping on the crude unit atmospheric column. The unit was in service when the loss of containment and fire occurred. The fire resulted in four fatalities and one serious injury. Source: The image is from the CSB Final Investigation Report of this incident.

the piping contained a significant amount of naphtha, pressurized by a leaking bypass isolation valve. A work permit had been prepared by operations personnel, and maintenance employees were authorized to "drain and remove" the naphtha draw-off piping. Due to continued valve leakage and several unsuccessful attempts to drain the line, a Tosco maintenance supervisor directed workers to cut out the piping with a pneumatic saw. The first cut was made successfully at the 104′ 6″ elevation. When the second cut was started at the 78′ 7″ elevation, the leak reoccurred at the cut. Work was stopped, and a Tosco Maintenance Supervisor instructed the crew to open a flange at the 38′ 1″ elevation. As the line drained, naphtha was suddenly released from the open end of the piping at the previously cut 104′ 6″ elevation. The naphtha ignited almost immediately, most likely from contacting the nearby hot surfaces of the bottom's exchangers, and quickly engulfed the tower structure and personnel. Refer to the unit schematic in Figure 8.2.2 below, indicating the pipework and location of the pipe cut.

One contractor jumped from the scaffold and sustained significant injuries due to the fall. The other four contractors died because of their burn injuries.

The US Chemical Safety Board investigated the incident, and a detailed report was issued, along with a two-page summary of the incident indicating the identified causal factors and a separate news release. The causal factors identified by the US CSB included the following (source; extracted from the CSB report):

- Tosco Avon refinery's maintenance management system did not recognize or control serious hazards posed by performing nonroutine repair work while the crude processing unit remained in operation.
  - Tosco Avon management did not recognize the hazards presented by sources of ignition, valve leakage, line plugging, and inability to drain the naphtha piping. Management did not conduct a hazard evaluation of the piping repair during the job planning stage. This allowed the execution of the job without proper control of hazards.

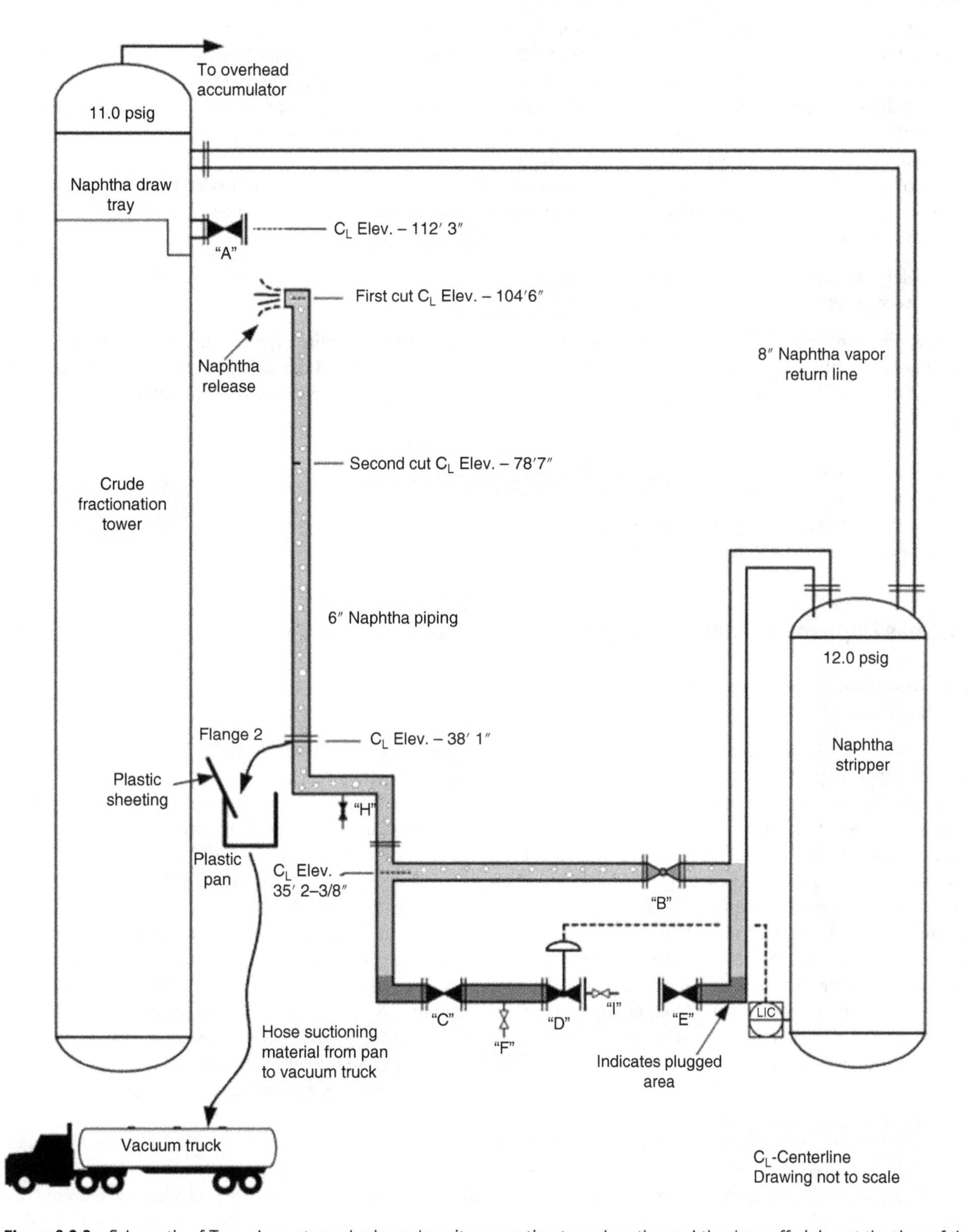

Figure 8.2.2 Schematic of Tosco Avon atmospheric crude unit preparation to replace the naphtha draw-off piping at the time of the release and fire. The workers were draining naphtha from the opened flange into a plastic pan and to a vacuum truck when the release from the cut section of the piping (at the 104′6″ elevation). Source: Image courtesy of the Chemical Safety and Hazard Investigation Board (CSB). The image is with kind courtesy of the US Chemical Safety and Hazard Investigation Board from the CSB Final Investigation Report of the incident. See the report for additional details.

  - Management did not have a planning and authorization process to ensure that the job received appropriate management and safety personnel review and approval. The involvement of a multidisciplinary team in job planning and execution, along with the participation of higher-level management, would have likely ensured that the crude unit was shut down to safely make repairs once it was known that the naphtha piping could not be drained or isolated.
  - Tosco did not ensure that supervisory and safety personnel maintained a sufficient presence in the unit during the execution of this job. Tosco's reliance on individual workers to detect and stop unsafe work was an ineffective substitute for management oversight of hazardous work activities.
  - Tosco's procedures and work permit program did not require that ignition sources be controlled before opening equipment that might contain flammables, nor did it specify what actions should be taken when safety requirements such as draining could not be accomplished.

Tosco's safety management oversight system did not detect or correct serious deficiencies in the execution of maintenance and review of process changes at its Avon refinery.

- Neither the parent Tosco Corporation nor the Avon facility management audited the refinery's line breaking, lock-out/Tag-out, or blinding procedures three years before the incident. Periodic audits would have likely detected and corrected the pattern of serious deviations from safe work practices governing repair work and operational changes in process units. These deviations included practices such as:
  - Opening of piping containing flammable liquids prior to draining.
  - Transferring flammable liquids to open containers.
  - Inconsistent use of blind lists.
  - Lack of supervisory oversight of hazardous work activities.
  - Inconsistent use of MOC reviews for process changes.

What has Happened/What can Happen

Piper Alpha Oil and Gas Platform Fire (167 Fatalities)
North Sea (6 July 1988)

The Cullen Inquiry was set up in November 1988 to investigate the incident and determine its cause. The chair of the inquiry was the Scottish judge, William Cullen. Judge Cullen released the public inquiry report in November of 1990.

The Piper Alpha oil platform was in the North Sea on the Continental Shelf of the United Kingdom sector. On 6 July 1988, the platform experienced a catastrophic explosion and fire, which destroyed the complex platform in about two hours. There were 226 souls on board the platform when the incident occurred, 165 lives were lost on the platform, and two more were lost during the rescue efforts. There were total of 167 lives were lost on the platform, making this one of the most significant process safety incidents of all time.

At about 9:45 p.m. on the evening of the incident, one of the condensate injection pumps tripped. When the pump tripped, the other injection pump was out of service for maintenance (although the maintenance had not started). The safety relief valve for the other pump had also been removed, and a temporary blank was installed in its place, with the blanks held in place by only four bolts in each blank. When the condensate pump tripped, the on-duty crew reopened the valves on the opposite condensate pump and returned it to service. A large condensate (primarily liquid propane) release occurred immediately after the condensate pump was started from the area where the safety valve had been removed. It is believed that the safety valve blank was not adequately rated for the pressure and failed when it was pressurized after the pump was restarted, resulting in the release.

The failure in communications regarding the safety valve and the condensate pump status was attributed to an inadequate Shift Handover procedure (See Chapter 11) and a less-than-adequate Work Permit process.

The fire was further exacerbated by the initial explosion, causing serious equipment damage and a large loss of containment of crude oil and secondary crude fires. These fires impinged onto the high-pressure gas lines, previously identified as high risk due to limited isolation and backflow potential.

The fire persisted for a longer time due to delays of the other two platforms in shutting down their crude flow, which flowed through the Piper Alpha platform en route to onshore. The two high-pressure gas lines ultimately failed, resulting in the total destruction of the massive exploration, drilling, and production platform.

This event was investigated by a British Public Inquiry led by the Honorable Lord Cullen. There were many lessons learned from the Piper Alpha incident, including an ineffective shift handover process, lack of communication and coordination between the platforms, lack of a process to control critical device disablement, lack of risk assessment regarding the platform modifications when adding the natural gas and LPG modules, emergency preparedness, and of course the work permit process. The investigation described the work permit process as a "formality." We must constantly seek continuous improvement in our critical systems and processes to ensure that emphasis is maintained.

Energy isolation is a critical part of the work permit process, and the blanks on the pressure relief valve were not rated for the pressure and were installed with only four bolts in place of the normal number. The blanks are more of a dust cover than isolation blinds. This did not meet the requirements for energy isolation. When the condensate pump was started, the release of LPG was almost instantaneous, resulting in the initial explosion and fire that ultimately destroyed the platform and led to the tremendous loss of life.

See the photographs in Figures 8.2.3 and 8.2.4a before and after images of the Piper Alpha oil platform's devasting fire and the enormous loss of life that occurred on that fateful evening.

This devasting fire and the massive loss of life, coupled with several other tragic process safety events during this same period, resulted in some very significant changes and improvements in our industry, many of which are reflected in this book. I believe that sometimes it is good to reflect on our industry and recognize these kinds of incidents are exactly why we do some of the things we do today. Improvements like improved permit-to-work and equipment isolation procedures, systematic management of change process, equipment lock-out and tag-out, and numerous improvements are designed to help prevent a reoccurrence of these kinds of events now and in the future. However, our industry still heavily depends on our people, and we are still subject to complacency and other similar human factors. We must have robust and approved procedures that must be followed. Otherwise, these kinds of events can still happen today.

Figure 8.2.4b is a photograph of a memorial to the one-hundred-sixty-seven lives lost in this tragic event. One-hundred sixty-five were lost on the platform, and two additional lives were lost on a support vessel that was attempting to rescue others from the burning platform. This monument is placed in a memorial park in Hazlehead, Scotland to help remember all who were killed on that tragic event on 6 July 1988. The names of those killed in this tragic event are engraved on all four sides of the monument.

Figure 8.2.3 Photograph of the Piper Alpha Oil platform in the North Sea before the devastating fire on 6 July 1988. Note the on-board housing and control room, the drilling deck, processing facilities, the flare system, and the helicopter deck. Source: PA Images/Alamy Stock Photo.

(a)

(b)

Figure 8.2.4 (a) Photograph of the Piper Alpha Oil Platform in the North Sea during the deadly and devastating fire. At this stage, the fire is nearly out, and major sections of the platform are gone, including the accommodations module, where many of the fatalities occurred and (b) Photograph of the memorial to the Piper Alpha Platform in Hazlehead Park in Aberdeen, Scotland. It was placed there to commemorate the one hundred sixty-seven men who died in this disaster. *The names of the lost are engraved on all four sides of the memorial.* Source: (a) PA Images/Alamy Stock Photo. (b) Anastasia Yakovleva/Alamy Stock Photo.

What has Happened/What can Happen

Phillips Petroleum Chemical Plant (Pasadena, Texas) (23 Fatalities)
23 October 1989

At about 1:05 p.m. on 23 October 1989, a massive explosion occurred with devastating results at a manufacturing site near Houston, Texas. The explosion and fire destroyed a large portion of the Phillips 66 Company's Houston Chemical Complex (HCC). Twenty-three plant workers were killed in the explosion, and three hundred and fourteen were injured. The two polyethylene production plants at the source of the explosion were destroyed. The overpressure from the explosion had a force estimated to be equal to the force of 2.4 tons of TNT and an earthquake that registered 3.5 on the Richter Scale about 20 miles away. It was about 10 hours before the massive fire was brought under control.

The explosion resulted from inadequate energy isolation while maintenance was being performed on the reactor settling legs. Energy isolation consisted of closing the single isolation valve and disconnecting its associated pneumatic actuator lines. There was no secondary physical isolation such as a lock-out device, a blind, or other "hard" isolation.

The investigation team found the pneumatically operated isolation valve in the fully open position following the explosion, which released ethylene at 700 psig into the atmosphere. The associated instrument air lines, which powered the pneumatic-operated isolation valve, were found reconnected to the isolation valve. However, they were found to be cross-connected, which resulted in the valve traveling to the fully open position.

The explosion was investigated by the US Occupational Safety and Health Administration (OSHA), by the US Department of Homeland Security, and the Federal Emergency Management Agency (FEMA). See those reports for additional details on this tragic event.

Lack of positive energy isolation was a key causal factor in this incident. Positive energy isolation, such as a lock, would not have allowed the valve to be operated until the piping was reconnected.

See the section on Vapor Cloud Explosions for more information and photographs of this tragic incident.

Please refer to the images in Chapter 4 (Vapor Cloud Explosions) of this site before and after this series of devastating explosions.

What has Happened/What can Happen

Bethune Municipal Wastewater Tank
Daytona Beach, FL.
11 January 2006
(Two Fatalities, One Critical Injury)

The US Chemical Safety Board investigated this incident and prepared a report with recommendations.

Figure 8.2.5 Images of the Bethune Municipal Wastewater Worksite. A man lift, and a crane were used to perform a welding operation above a methanol storage tank when a major fire occurred, directing the flame directly onto the man lift basket and the crane cab. Source: Courtesy of US Chemical Safety and Hazard Investigation Board. Photos are from the CSB report on this incident. Annotations are by the CSB.

The explosion occurred at the Bethune Point Wastewater Treatment Plant, owned and operated by the City of Daytona Beach, Florida, on 11 January 2006. Two municipal workers were fatally burned, and a third suffered grave injuries. The explosion occurred inside a methanol tank when the lead mechanic and another worker were cutting the metal roof directly above the tank vent. Sparks showered down from the cutting torch and ignited methanol vapors coming from the vent, creating a fireball on top of the tank. The CSB report recommended improving procedures to control hot work, such as cutting torches. Figure 8.2.5a,b show the work location and location of the crane used by the workers.

Key Lessons Learned

- The work permit is a legally binding document, and when being prepared, each block and tick mark should be considered carefully. When the permit is issued, it is a "license" for work to begin.
- As the equipment owner, the process organization must ensure that equipment is prepared correctly and the site is ready for the work to begin before the permit is issued.
- The process should ensure that a joint field walk down is required for all OPE, confined space entry, hot work tasks, and other similar tasks that represent higher hazards. The walk down should include the permit issuer and permit acceptor and verify that the work area has been properly prepared and inspected before issuing the work permit.
- The permit acceptor is responsible for ensuring that all workers know the hazards and specifics of how the equipment has been isolated and prepared for work.
- Each manufacturing site should have a documented and rigorous work permit process with special attention to opening process equipment (OPE), confined space entry, and hot work tasks, especially hot work on or near vessels with confinement.
- Opening Process Equipment (OPE) should always be considered "High Hazard Work" and requires attention to detail and close monitoring of work activities during field execution.
- When preparing process equipment for opening, the site process should verify energy isolation to as near zero energy as possible. Good practice requires a risk assessment and a higher authorization level when zero energy is not attainable, for example, leaking valves. Equipment isolation typically consists of blinds in process lines with process tags, open electrical breakers with process locks and tags, chains with process locks and tags on manual wheels on motor-operated valves, etc.
- The work permit process should require an accurate description of the work to be done, including the type of work, the location, and the equipment where the work is to be performed.

- Hot work permits should consider gas testing, including what gas tests are to be done and the acceptable ranges for each type of gas test (O^2, LEL, H_2S, and other gases specific to the process). If work is on a vessel, the gas test should include the vessel interior and the surrounding area.
- People responsible for gas testing should be trained on the specific gas test equipment or analyzers they will be using before conducting field gas tests.
- The work permit should specify the appropriate fire protection equipment readily available at the job site. For example, a pressurized fire water hose, and a properly rated fire extinguisher (Class A, B, C, or D) should be readily available at the job site.
- The site process should ensure that job site hazards are eliminated or mitigated before issuing a work permit, including the potential for flammables in or near the hot work area. Examples are process sewers or drains, stored flammables, sample points, water draws from tanks or other process vessels, open pits, etc.
- The site should have a process for auditing work permits and providing feedback to management regularly. Audits should include paper and fieldwork in progress to ensure full compliance with the site requirements and standards. Feedback should lead to continuous improvement actions as appropriate.
- Break-in work should be avoided to the extent possible. Experience has shown that break-in work is typically not as well planned as routine work and, therefore, has a higher incident rate.

And always – If the job is not going right, or if safety issues are identified:

- STOP THE JOB!

Additional References

U.S. Chemical Safety and Hazard Investigation Board CSB Investigation Report (Tosco, Avon refinery crude unit fire) Martinez, CA – 23 February 1999. https://www.csb.gov/tosco-avon-refinery-petroleum-naphtha-fire/

U.S. Chemical Safety and Hazard Investigation Board CSB Investigation Report (Bethune Municipal Wastewater Tank explosion), Daytona Beach, FL. – 11 January 2006. https://www.csb.gov/csb-releases-findings-from-fatal-daytona-beach-wastewater-plant-explosion-investigation-at-public-meeting-cites-inadequate-engineering-lack-of-public-worker-safety-coverage/

Coastal Safety "Spiral to Disaster" Video (Piper Alpha Explosion and Fire) 6 July 1988, North Sea Oil Platform.

Reports on the Phillips Polyethylene Plant Explosion and Fire US Occupational Safety and Health Administration (OSHA), and by the US Department of Homeland Security, Federal Emergency Management Agency (FEMA).

U.S. Chemical Safety and Hazard Investigation Board CSB Videos: "Hot Work: Hidden Hazards" and "Dangers of Hot Work". https://www.csb.gov/videos/

The Public Inquiry into the Piper Alpha Disaster (the Lord Cullen report).

Center for Chemical Process Safety of the American Institute of Chemical Engineers "Building Process Safety Culture: Tools to Enhance Process Safety Performance".

U.S. Chemical Safety and Hazard Investigation Board Work permit safety bulletin "Seven Key Lessons to Prevent Worker Deaths During Hot Work In and Around Tank". https://www.csb.gov/assets/1/17/csb_hot_work_safety_bulletin_embargoed_until_10_a_m__3_4_101.pdf?14

U.S. Chemical Safety and Hazard Investigation Board Investigation Report: Motiva Enterprises LLC Delaware City Refinery Explosion and Fire 17 July 2001. https://www.csb.gov/assets/1/20/motiva_final_report.pdf?13758

Singapore Ministry of Manpower "Company fined $400,000 for fire at Petroleum Refinery in Pulau Bukom". https://www.mom.gov.sg/newsroom/press-releases/2019/0108-company-fined-for-fire-at-petroleum-refinery-in-pulau-bukom

Chapter 8.2. Safe Permit-to-Work, Breaking Containment, and Controlling Work

End of Chapter Quiz

1. What is a Work Permit?
2. What is the expectation for compliance with the permit-to-work process in the facility?
3. Is this also true for work being done in the field?
4. Who is responsible for preparation of the equipment to be worked on?
5. What does the permit-to-work process do to help ensure the job is performed safely?
6. Is it common for a work site to have more than one work permit form? Please expand on your answer.
7. Energy isolation is an important part of the work permit. What kind of energy is controlled?
8. When process equipment is involved, how is hazardous energy controlled?
9. What is this energy isolation process called?
10. Describe the process for ensuring lock-out tag-out for electrical equipment.
11. A higher-level review and formal ______ __________ to identify mitigation will be required if zero energy cannot be achieved; in some cases, for example, when flammable or toxic materials are involved, this will require the equipment to be shut down to achieve proper isolation before work can be performed.
12. What is the main purpose of gas testing before opening process equipment?
13. If a gas test instrument indicates 0% flammables, does this always indicate that the equipment is "hydrocarbon free"?
14. When a task represents a potential for a more serious hazard, what can be done to identify and control hazards in addition to the traditional permit-to-work process?
15. Just before the permit is issued, what is the final process to help ensure good communication between the permit issuer and permit acceptor and that proper precautions are in place?
16. How do we ensure that the work permit process is applied properly at our site and that it does not become a "formality"?
17. If you are about to issue a permit to burn or weld on a process vessel, is it enough to have an acceptable gas test before issuing the permit?
18. Why should "break-in" work be avoided to the extent possible?
19. The work permit should specify the appropriate _____ ________equipment that should be readily available at the job site. For example, a pressurized fire water hose and a properly rated fire extinguisher (Class A, B, C, or D) should be available at the job site.
20. Opening Process Equipment (OPE) should always be considered "_____ ______" work and requires attention to detail and close monitoring of work activities during field execution.

8.3

Managing Change

The US OSHA Process Safety Management Standard (29 CFR 1910.119) ("PSM") requires the employer to establish and implement written procedures to manage changes (except for "replacements in kind") to process chemicals, technology, equipment, and procedures; and changes to facilities that affect a covered process.

One thing that is certain about operating an asset-intensive operation like a petroleum refinery or petrochemical plant is change. There is a saying, "Nothing is consistent but change." We know that change is inevitable if we continue to be efficient, competitive, and exist. However, change must be properly reviewed to ensure unidentified risks or hazards are not allowed to enter the process and create havoc, such as loss of containment, fires, or explosions.

We must look no further than BP Texas City to understand how that can happen. During the US Chemical Safety and Hazard Investigation Board (CSB) investigation, the CSB determined that operators were allowed to change procedures without formal reviews or approvals. These unmanaged changes allowed the ISOM unit fractionator column to become liquid-filled, releasing hot naphtha directly into the atmosphere, leading to a catastrophic explosion and fire, which killed fifteen and injured 180 others.

This is only one example where unmanaged change has resulted in catastrophic consequences. There have been many industry events where failure to effectively manage change has directly led to the loss of containment and has resulted in fires or explosions or has been a significant causal factor in the loss of containment of a flammable or toxic.

Another significant and more recent incident with similar lessons learned occurred at the Williams Olefins Plant in Geismar, Louisiana. This incident occurred on 13 June 2013 and resulted in 2 fatalities and 167 injuries. The CSB investigated this event and prepared a detailed report and an excellent training video covering the incident and the lessons learned.

The Williams Olefins Plant had previously retrofitted two reboiler exchangers on a propane fractionator with new valves to facilitate the removal of either of the exchangers for maintenance while the unit continued in operation. Williams did not complete the Management of Change review and documentation until after the project was completed and in service, not beforehand as required by the standard. During the construction and startup of the new facilities, one of the new valves was left in a closed position, resulting in one of the exchangers being isolated from the overpressure protection. This resulted in a rupture of the exchanger shell and a large explosion and fire when a flow of hot water started through the exchanger. Suppose the MOC had been done as required by the standard, along with a rigorous Process Hazard Analysis (PHA) and a thorough Pre-startup Safety Review (PSSR). In that case, it is possible that the hazard would have been identified and this incident would not have happened.

An effective Process Safety Management System focuses on proactively identifying risks, assessing them, and taking action to control them. An effective process to identify changes, review the changes, and implement controls and approvals almost always includes the Management of Change as a key component.

We generally think of physical changes to the equipment as requiring a "Management of Change" (MOC). Physical changes such as upgrading equipment metallurgy, replacing a pump impeller with a larger one, and installing a new pipe connection are covered under the OSHA Process Safety Management standard. Other changes that are also straightforward Management of Change include revisions to operating procedures, the introduction of a new or different chemical into the process, and the introduction of new processing technology.

Many other changes may be more subtle but never-the-less should be considered requiring an MOC. Examples include unit feed quality changes, catalyst technology and/or activity, and program changes in the distributed control system (DCS) should all trigger the MOC process.

Process Operations Safety: The What, Why, and How Behind Safe Petrochemical Plant Operations, First Edition. M. Darryl Yoes.

Other examples of changes that should trigger the MOC process include the following:

- Any equipment changes or modifications other than “replacement in kind.”
- Any change in materials used in the process, such as a new chemical, catalyst, crude, or other new feedstock.
- A shift in operation to conditions outside the established process or mechanical design envelope.
- A change in operations or maintenance procedure.
- A change in the setpoint of critical instrumentation outside of the approved range.
- A change in the use of equipment not covered by existing and approved procedures.
- This change could increase the security risk to personnel and the plant facilities.
- A change in the organizational structure, such as adding additional staff or removing existing positions (typically not including personnel moves within the existing organizational structure).

A change applies to both temporary and permanent changes, and the type of change should be noted on the MOC form. Temporary changes should be assigned a duration; the MOC documentation should specify the life of the change, and if the change runs beyond the specified due date, the change must be reauthorized.

A formal Management of Change document or form should be completed once a proposed change has been identified (and before it is implemented). This process is designed to ensure that the change is managed correctly, that the change is reviewed/approved by authorized company representatives, and that proper engineering and management oversight/approval is established before the change is implemented.

The MOC documentation should specify the technical basis for the change and any safety and health effects resulting from the change. Additionally, the MOC process ensures that unit Operators and other affected personnel are made aware of the change and properly trained before operating the unit or equipment after the change has been implemented. Management of Change documentation is legal and retained with other PSM documentation associated with the covered process.

When a change is proposed, it must undergo a rigorous PHA to ensure that the proposed change does not introduce hazards into the process or equipment. This review ensures that safety, health, and environmental risks arising from these changes remain acceptable. The PHA will consider the following during the review:

- Hazards of the process.
- Engineering and administrative controls to ensure the hazards are controlled or mitigated.
- The potential consequences of failure of controls.
- Facility siting, such as the addition of occupied structures or temporary buildings.
- Any human factor issues being introduced by the change.
- The potential effects on on-site personnel from the failure of controls are also evaluated.
- A review of the incident history and a review of previous incidents.

For most changes, this review may be done by a group of experienced personnel and should normally be a line-by-line review, considering the proposed change and the potential impact it may have. At most sites, following this review, the proposed change is given the “approval to progress,” meaning that the change is approved for work to begin in the field on the construction or modification. Approval-to-progress does not mean the change has been approved to operate or commission.

Operations, maintenance, and technical personnel are trained in the change following the PHA. Training is an important step in ensuring the personnel are ready to assume responsibility for the change and are aware of the operating procedures and the impact the change will have on the operations. Training is a routine part of the MOC process, and a record of training should be completed for review.

After field construction is completed, the change should be subjected to a PSSR. The PSSR helps confirm that the field construction conforms with the original design and that procedures have been developed and are adequate and available to the Operators. The PSSR also verifies that the PHA recommendations have been resolved or implemented and that workers have been trained in the new or modified process or equipment.

The “approval to commission” is given once the PSSR has been completed, all follow-up items have been resolved, and personnel have been trained in the change. This means the change can be placed in service.

In-Kind, Not In-Kind and Emergency Changes

- In-kind change is a replacement identical to the original and satisfies all relevant requirements, standards, and specifications. An in-kind change does not require the MOC process.
- Not-in-kind change is when there is any difference between the original and the replacement. A not-in-kind change requires the complete MOC process.
- Emergency Change is an unplanned change needed immediately to reduce the risk of an SH&E incident or respond to an unforeseen event or situation that affects continued efficient operations. Generally completed on-shift by a pre-designated and trained on-site emergency MOC contact leading the evaluation.
  - Emergency MOC Contact: a supervisory employee assigned during "off-hours" to oversee the handling of Emergency Changes, should the need arise, and ensure they are processed through all the essential elements of the MOC Procedure.

All operations changes, such as product assignments of storage tanks, piping, and associated facilities, require a detailed and approved MOC, as do alarm and interlock set points, such as interlock between an automatic shut-off valve and a high-level tank alarm. Chances to operations and maintenance procedures, including modifications to any step of any operational procedure or emergency procedure and the unloading procedures required to unload a vessel when there are no standard manifold/connections, all require an MOC. Other changes that need a detailed and approved MOC change package include the following:

- Changes to site Standards, including relocation of equipment and changing key vendors.
- The introduction or testing of new equipment at the loading rack.
- Changing the location of fire extinguishers or other emergency response equipment.
- Replacement of an electric motor for another with a different rotation, horsepower, frame size, voltage requirement, or any other similar change.
- Reducing the number of personnel in a process area or the consolidation of posts or units on a process console.
- A change to reuse the same facilities for different products or changes in additives percentage.
- New or different software for a computerized control system, such as a computer-controlled truck loading system.

Types of Changes Requiring MOC

Changes to process equipment tend to be the focus of most MOC systems. Most MOC-related failures are often associated with small items that appeared insignificant and, in most cases, did not get a thorough, comprehensive review. MOC is required for some changes that may appear to be administrative, such as changes to routine or emergency operating procedures or training programs. Changes to process chemistry also require a detailed change package and approvals, such as changes to process chemistry, new crude oils, and new catalysts or chemicals. This includes any changes to operating conditions that may affect the operating envelopes or operating limits or changes to the process flow scheme, such as using a system other than originally designed and approved. Also, changes to the process control system, such as the DCS, computer control programs, safety systems such as interlocks or emergency shutdown systems, and alarm systems, all require a detailed MOC change request form.

More Examples of Equipment Changes Requiring MOC

- Relocation of equipment.
- Installing new bleeders in existing lines.
- Installation of new vents on lines or tanks.
- Temporary Lines involving Process Streams.
- New Piping Tie-ins or hot taps and stopples.

- Changes in overpressure relief valve size or type (e.g., spring to pilot).
- Changes to storehouse stock and/or spare parts inventory.
- Change in metallurgy, strength, temperature, or pressure rating of bolts and studs.
- Changes to gaskets or packing materials.

Other Types of Changes that Require a MOC

- Changing information on a unit drawing (P&ID, electrical one-line) due to a field check by the draftsman.
- Changes affecting facility Siting (turnaround trailers, modification of existing structures, new buildings that will be occupied, including temporary occupied buildings, etc.).
- Changes that affect electrical hazardous area classification.
- The addition of new sources of flammable or toxic materials.
- Changes that affect Environmental permits.
- New water draws, adding/deleting streams to the flare or sewer, permitting flares or units, sewer system modifications, and new wastewater generation.

Change Package Outline

Whether a paper-based system or a computerized process, the MOC Change Package should be developed in advance with appropriate approvals, generally from the Process Business Team Leader or Basic Equipment Owner.

The change package document should include the following documents:

- The completed and approved MOC Change Request form.
- Documentation detailing the scope of the change.
- Details of the impact of Safety, Health, and Environmental considerations or concerns.
- Any impact on other systems and operations.
- Design considerations for facilities and equipment changes.
- Marked-up drawings or sketches detailing the planned changes.
- Documentation of any personnel effects due to the change.

Temporary and Permanent Changes

Permanent Changes are the types of change anticipated to be in place for the long term and are expected to be durable. On the other hand, temporary changes are expected to be relatively shorter-term. They are required to accomplish a special purpose, address a short-term but immediate risk, improve efficiency, or permit continued operations until a formal permanent change can be implemented.

In temporary changes, the scope, purpose, and time limitations should be documented and communicated to all involved personnel. The approved time period should be specified for the temporary change, and the temporary change should be rereviewed and reapproved before the initially approved period expires. A higher level of authority should be required to reapprove MOC extensions of temporary MOCs to prevent arbitrary extensions. Responsibility for supervision and follow-up of temporary changes are assigned and documented.

The changes in scope, purpose, and time limitations are subjected to the same review and approval procedures as the initial approval. Examples of temporary changes that must be carefully considered to prevent the temporary changes from becoming permanent changes include the following: temporary pipe clamps as a short-term repair, short-term changes to operating limits to address equipment issues, such as deviating from a crude unit desalter capacity limit, or changes to other operating envelopes or limits for any reason.

The following are examples of post-accident investigations, which found that ineffective management of change led to the loss of containment and resulting fire, explosions, and loss of life.

What Has Happened/What Can Happen

Nypro Explosion and Fire,
Flixborough, United Kingdom
June 1974 (28 Fatalities, 89 Injuries)

On 1 June 1974, a catastrophic explosion occurred at the Nypro chemical plant near Flixborough, England, in North Lincolnshire. The report by the British Health and Safety Executive (HSE) documented the incident. That report is the basis for this summary below.

The Flixborough chemical plant, located close to the village of Flixborough (near Scunthorpe), North Lincolnshire, England, and owned by Nypro (UK) (a joint venture between Dutch State Mines and the British National Coal Board), had been in operation since 1967, producing caprolactam, a chemical used to produce nylon. The process involved the oxidation of cyclohexane in six reactors, which were configured in series flow, producing a mixture of cyclohexanol and cyclohexanone.

The plant was shut down for investigation and repairs due to a crack and resulting cyclohexane leak in reactor No.5. A decision was taken to remove the failed reactor and install a temporary 20″ bypass pipe to connect reactors No. 4 and No. 6 to enable the plant to continue operations.

On 1 June 1974, after the unit had been returned to service, the new 20-inch bypass system ruptured, resulting in the loss of containment of a large quantity of cyclohexane, quickly forming a very large vapor cloud. At about 4:53 p.m., a massive explosion destroyed the plant and caused significant damage to the surrounding businesses and homes.

Twenty-eight people died, and more than 100 were injured, with around 100 homes in the adjacent village being destroyed or badly damaged. Eighteen fatalities occurred in the control room due to the windows shattering and the collapse of the roof. No one escaped from the control room. The fires burned for several days, and after ten days, those still raging were hampering the rescue work.

The British Health and Safety Executive (HSE or the close equivalent to US OSHA) made an official inquiry to determine the cause of the incident. The HSE determined that the failure was due to plant modifications made without a full assessment of the potential consequences. The bypass had been designed by engineers who were not experienced in high-pressure pipework; no plans or calculations were produced, the pipe was not pressure-tested, and it was mounted on temporary scaffolding poles that allowed the pipe to twist under pressure and thermal stress. Only limited calculations were undertaken on the integrity of the bypass line. No calculations were undertaken for the dog-legged shaped line or bellows, and no drawing of the proposed modification was produced. The HSE inquiry indicated that the failure may have occurred due to a smaller fire on a nearby 8-inch pipe, resulting in stresses in the 20-inch bypass pipe and the associated bellows assemblies.

This investigation implied that the piping bypass was not adequately engineered to consider the piping stresses associated with the high-temperature operations. As indicated above, temporary pipe supports were used to support the temporary piping, which was not reviewed for adequacy. Failure of the temporary piping and/or the associated expansion bellows resulted in a catastrophic release of hot hydrocarbons, which quickly found an ignition source. The resulting large explosion and fire caused the deaths of 28 people and 89 injuries. Damage to the surrounding community was also extensive, with 1,800 homes and businesses severely damaged.

Key Lessons Learned

The Flixborough incident directly resulted from a poorly managed or engineered change. The replacement pipe was sketched out with chalk on the shop floor without consideration for the thermal expansion it would experience when placed in service. The pipe was installed with bellows on each end, and when the pipe was placed in service, the high temperature and resulting thermal growth and pipe stress resulted in one of the bellows buckling and rupturing. Figure 8.3.1 illustrates the installation of the reactor's temporary bypass piping. Note that it is supported by scaffolding and includes bellows to help control the thermal expansion as the pipe is heated during operations. Figure 8.3.2 shows the devastating series of explosions and fires that resulted when the pipe failed. Figure 8.3.3 illustrates the failed and damaged bellows that resulted in the release and explosion.

Of course, this incident occurred well before the advent of formal management of change as we know it today. A proper MOC would have ensured that the pipe was subjected to a stress analysis with provisions for thermal growth and adequate support.

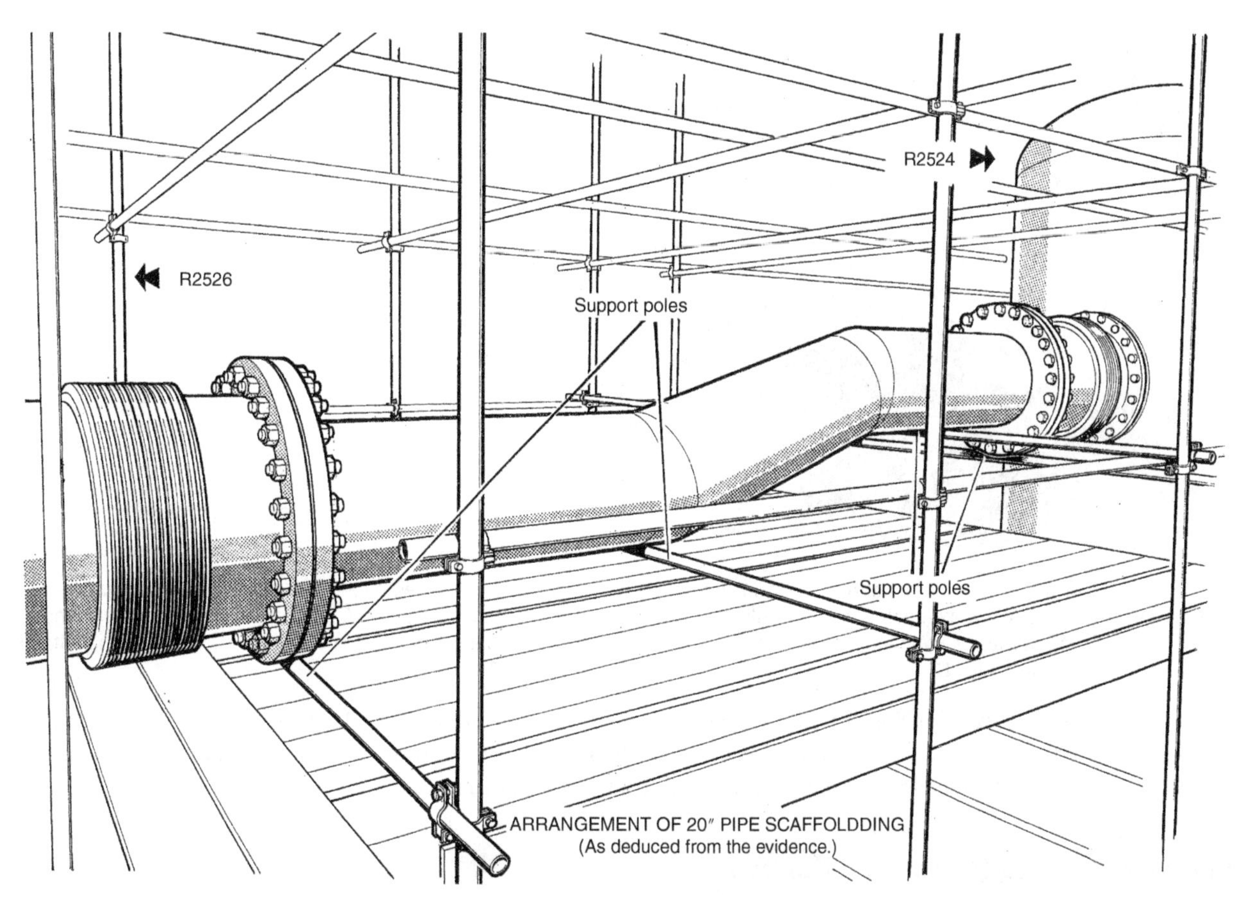

Figure 8.3.1 Flixborough temporary 20-inch reactor bypass pipe as installed. Contains public sector information published by the health and safety executive and licensed under the https://www.nationalarchives.gov.uk/doc/open-government-licence/version/3/. Source: Reproduced by the kind permission of HSE. HSE would like to make it clear that it has not reviewed this product and does not endorse the business activity of Safety Consulting International, LLC.

Figure 8.3.2 Photo of flixborough explosion and fire. Contains public sector information published by the health and safety executive and licensed under the https://www.nationalarchives.gov.uk/doc/open-government-licence/version/3/. Source: Reproduced by the kind permission of HSE. HSE would like to make it clear that it has not reviewed this product and does not endorse the business activity of Safety Consulting International, LLC.

Figure 8.3.3 Flixborough – the failed bellows. Contains public sector information published by the health and safety executive and licensed under the https://www.nationalarchives.gov.uk/doc/open-government-licence/version/3/. Source: Reproduced by the kind permission of HSE. HSE would like to make it clear that it has not reviewed this product and does not endorse the business activity of Safety Consulting International, LLC.

It is our role to ensure this is done properly for similar operations on our facilities today. Lack of adequate MOC can result in an almost immediate loss of containment, as occurred at Flixborough, or can result in a failure many years later. I am convinced that if we use the management systems as intended, these incidents will not happen.

What Has Happened/What Can Happen

Tosco Avon Refinery
Martinez, CA
23 February 1999 (4 Fatalities)

A major refinery flash fire occurred at the Tosco Avon Refinery in February 1999 when maintenance workers attempted to replace a leaking section of naphtha draw-off piping between the atmospheric fractionating column and the sidestream stripper on a crude distillation unit. Four contract maintenance workers were killed due to the resulting flash fire.

The US Chemical Safety and Hazard Investigation Board (CSB) investigation determined that the piping and associated block valves were severely corroded due to operating the crude desalting equipment beyond its capability. The CSB investigation noted that operational changes included changes in crude composition, which contributed to overloading the desalting equipment and operating the desalter beyond its capacity. Changes in crude oil composition resulted in water and corrosive materials, including ammonium chloride, being carried over into the fractionator, which deteriorated the piping and valves. The CSB investigation identified that failure to manage the change in feedstock composition resulted in the piping and valve corrosion and contributed to the fire.

The CSB investigation also determined other causal factors resulting in the loss of containment and resulting fire, including procedures related to safe work practices. These are covered separately in Chapter 8.2 of this book.

Key Lessons Learned

This is a good example of how the management of change applies to more than just physical change. With our current MOC systems, we would be required to do a detailed review to identify the hazards introduced by operating the unit desalter beyond normal limits. Not only would we identify the hazards, but we would also be required to develop mitigation and monitoring procedures to ensure damage is not done by the planned operational changes.

Had the hazards of operating the desalter beyond the normal limits been recognized at Tosco Avon, and had those hazards been properly addressed, the corrosion would have been mitigated, and the naphtha leaks would not have occurred in the first place. Management of Change is designed to do this when it is properly implemented.

What Has Happened/What Can Happen

Giant Industries Refinery Explosion and Fire
Jamestown, NM
8 April 2004 (6 employee injuries, (4 serious burns and trauma-related injuries)

This incident was investigated by the US Chemical Safety and Hazard Investigation Safety Board.

This loss of containment and large fire incident occurred on 8 April 2004 at the Giant Industries refinery in Jamestown, New Mexico, at the refinery's HG Alkylation Unit. The release occurred while routine maintenance was done on the Alkylation Unit Iso-Stripper Alkylate Recirculation Pump during the opening to replace the pump seal.

The accident seriously injured four workers with burns and trauma-related injuries. Two other plant personnel suffered minor injuries fleeing the area of the explosions. All nonessential plant personnel were temporarily evacuated from the refinery, and personnel were also evacuated from adjacent businesses. Fortunately, no HF Acid was involved or released in this incident.

The Alkylate pump was being prepared for maintenance for a leaking pump seal (Alkylate is a volatile high-octane light naphtha gasoline blend component). The mechanics planned to remove the rotating element for a shop replacement of the pump seal. Unit Operators prepared the pump for removal by the mechanics by closing the suction and discharge ¼ turn valves and securing both valves with tags and locks. The valve position was verified by using the position of the valve handles (not the valve steam indicator placed on the valve by the valve manufacturer).

As the Operators prepared the pump for turnover to the maintenance personnel, they depressurized the pump by connecting a temporary hose to a bleeder on the pump discharge piping. This was done instead of using the pump case drain for depressurizing the pump because the case drain was not equipped with a drain valve. Following the closure of the pump isolation valves and bleeding the pressure with the hose connection, the Operators assumed the pump was isolated and depressurized, and the pump was turned over to the mechanics for seal replacement.

As the mechanics removed the bolts between the pump case and the rotating element, a sudden release of hot alkylate at 350 °F (177 °C) and 150 psig (10.3 bar) was released, resulting in the major unit fire and injuries to personnel. After the incident, the pump suction valve was found to be in the fully open position, and the vent hose was plugged internally, preventing the draining or depressurizing of the pump.

The Alkylate pumps had originally been equipped with a gear-operated operator; however, before this incident, the valve had been modified by removing the gearbox and replacing it with a manual operator; basically, a two-foot-long metal bar

Figure 8.3.4 Photo of a 2-foot bar handle used as the valve operator for the Alkylate pump. Note the original valve position indicator placed on the valve by the valve manufacturer. Source: Photo courtesy of the US Chemical Safety and Hazard Investigation Board (Annotations are by the CSB).

Figure 8.3.5 Photo of the temporary hose connection as an attempt to depressurize the pump before turning it over to maintenance personnel for seal replacement. Note the case drain (low point bleeder) located at the bottom of the pump and the absence of a block valve. Source: Photo courtesy of the US Chemical Safety and Hazard Investigation Board (Annotations are by the CSB).

was loosely fit over the square shaft on the pump valve stem. This bar-style handle could easily be repositioned in different directions.

Some Operators quickly developed a practice of verifying the valve position by the position of the manual operator and not the original valve indicator placed on the valve by the valve manufacturer. The original indicator is very small and more difficult to see than the large metal bar currently used as the valve handle.

As the mechanics were opening the flanges, the suction valve was fully open, and the pump remained completely pressurized, resulting in the loss of containment of the hot and light product, which ignited immediately following the release, most likely due to autoignition.

See Figure 8.3.4 for photos of the valve handle and original valve position indicator. Figure 8.3.5 indicates the temporary hose used in an attempt to depressurize the pump before turning it over to the maintenance personnel for seal replacement.

Key Lessons Learned

Several key lessons are to be learned and shared from this loss of containment incident. This incident clearly illustrates the importance of verifying energy isolation by the Process personnel before turning equipment over to the mechanics. In hindsight, the hose used by the Operators to depressurize the pump was plugged internally, and the pump remained pressurized.

Vents and Drains should be verified as open and unobstructed to ensure the equipment is prepared and ready to turn over for maintenance. A thorough field walk down between the permit issuer and the workers must be done before work begins to ensure a clear understanding of the work to be done and to communicate how energy isolation is achieved. The equipment owner (Process) owns and is accountable for energy isolation. Isolation must be field verified including verification of all vents and drains and other potential energy sources. The equipment must be verified free of energy (including but not limited to pressure, temperature, and flammable or toxic materials).

The Operators verified the isolation valve position by observing the position of the loosely fitted metal bar that had previously replaced the gearbox operator. This bar could easily be removed, replaced on the valve stem, and placed in different orientations. The valve handle was found to be placed in the opposite position, indicating visually that the valve was fully closed while the valve was in the fully open position.

As highlighted in this section, the OSHA Process Safety Management standard (29 CFR 1910.119) requires that any change that may affect a process covered by that standard, except a "replacement in kind," requires detailed management of change (MOC) evaluation. When the valve was modified to remove the gearbox actuator and replace it with a manual bar, this should have triggered a MOC to determine if hazards were being introduced. The metal bar should have been permanently attached so that it could not be easily removed and replaced in a different orientation. Management of Change should be implemented for all modifications to equipment, including valve actuators.

Additional References

Occupational Safety and Health Administration (OSHA) Process Safety Management Regulation 29CFR 1910.119, specifically 1910.119(l). https://www.osha.gov/laws-regs/regulations/standardnumber/1910/1910.119

Environmental Protection Agency Risk Management Plan Appendix A (the regulation Appendix A) 40 CFR part 68 (specifically Para 68.75). https://www.epa.gov/rmp/risk-management-plan-rmp-rule-overview

U.S. Chemical Safety and Hazard Investigation Board Final Investigation Report TOSCO Avon Refinery Martinez, California 23 February 1999 Report No. 99-014-I-CA Issue Date: March 2001. https://www.csb.gov/assets/1/20/tosco_final_report.pdf?13752

U.S. Chemical Safety and Hazard Investigation Board Final Investigation Report Giant Industries Refinery Explosions and Fire 8 April 2004 Report No. 2004-08-I-NM October 2005. https://www.csb.gov/giant-industries-refinery-explosions-and-fire/

British Health and Safety Executive Flixborough (Nypro UK) Explosion 1 June 1974 Health and Safety Executive, "The Flixborough Disaster: Report of the Court of Inquiry," HMSO, ISBN 0113610750, 1975. https://www.hse.gov.uk/comah/sragtech/caseflixboroug74.htm

The Wikipedia – The Free Encyclopedia Flixborough Disaster. https://en.wikipedia.org/wiki/Flixborough_disaster.

Chapter 8.3. Managing Change

End of Chapter Quiz

1. Which US Regulation requires management of change?

2. Why is the MOC process such an important part of process safety management?

3. What kinds of changes require strict adherence to the MOC process?

4. Are their other changes that are not necessarily so easily recognized but still require compliance with MOC?

5. What is the Pre-startup Safety Review, and when is it required?

6. As defined by the MOC system, what are the main types of changes?

7. One thing that is certain about operating an asset-intensive operation like a petroleum refinery or petrochemical plant is ________. There is a saying, "Nothing is consistent but change."

8. At the Williams Olefins plant, if the MOC had been done as required by the standard, along with a rigorous Process Hazard Analysis (PHA) and a thorough Pre-startup Safety Review (PSSR), it is possible that the hazard would have been __________ and this incident would not have happened.

9. Which of the following does NOT require a MOC to be developed?
   a. Replacement of a pump impeller with a larger one.
   b. Replacement of a seal on a floating roof tank with the same type seal.
   c. Modification of a turbine overspeed trip to make it more reliable.

d. Replacing the catalyst in a reactor with catalyst that is thought to be more active.
e. Updating a start-up procedure with better drawings.
f. Changing the unit feedstock to one with a higher yield.

10 The MOC documentation should specify the ________ basis for the change and any ______ ___ ______ effects resulting from the change.

11 What is the purpose of the Process Hazard Analysis (PHA), and why is it done?

12 Does "Approval to Progress" mean that the site has the approval to operate the equipment that the engineers have designed?

13 What is required by the MOC process to commission the equipment?

14 Describe what constitutes an emergency change and how this can be implemented.

15 The approved time period should be specified for the temporary change, and the temporary change should be rereviewed and reapproved before the initially approved time period expires.

- True
- False

16 The OSHA Process Safety Management standard (29 CFR 1910.119) requires that any change that may affect a process covered by that standard, except a "________ ___ ______," requires detailed management of change (MOC) evaluation.

17 What other US regulatory standard has this same MOC requirement?

8.4

Ensuring Availability of SSHE Critical Devices (Controlling the Unauthorized and Extended Disarmament of Critical Devices)

Oil refineries and petrochemical plants have safety equipment and systems designed to protect personnel, the community, the environment, and equipment from process upsets, abnormal process conditions, and security intrusions. These devices are Safety, Security, Health, and Environmental (SSHE) critical devices or, most often, just SCD.

These critical devices are designed to be the last line of defense to protect people, equipment, and the environment without the benefit of Operator intervention. For example, pressure relief devices (PRVs) protect pressure vessels and/or piping from overpressure, low-flow trip valves in fuel gas systems protect heaters and furnaces by cutting the flow of fuel in the event of low process flow, reactor trip systems trip and depressure the reactor in the event of a catalyst temperature excursion.

Safety-critical equipment includes equipment, instruments, analyzers, or other devices, fixed, mobile, or portable, whose failure to operate or function could result in one or more of the following:

- Serious personal injury to a plant worker.
- Loss of containment can result in a significant fire or explosion.
- Serious disruption to the off-site community.
- Loss of containment is likely to have a serious environmental impact.
- Inability to mitigate or carry out emergency response to an adverse event.
- Unauthorized access to plant facilities or computer systems.

The following are examples of devices that are generally defined as SSHE critical devices (this is not intended to be a complete list):

- Pressure/vacuum relief valves
- Car seal valves
- Steam turbine Overspeed trip devices
- Protective systems for fired equipment.
- Compressor protective systems
- Process unit shutdown systems
- Fixed and mobile firefighting facilities
- Firewater pumps
- Breathing apparatus sets
- Safety showers
- Flame/detonation arrestors
- High/low temperature and pressure protective systems
- Storage tank high-level alarms
- Level alarms/cut-outs on flare and blowdown drums
- Protective systems for heaters on tanks
- Hydrocarbon gas detectors
- Smoke detectors, fire detectors, and fire suppression systems
- Power-operated emergency isolation valves

In the past, pneumatic valves functioned as safety-critical devices (SCD). These were relatively simple air-to-operate valves with air pressure acting against a spring to maintain the valve in operating mode. The valve would move to the fail-safe position upon instrument activation or loss of instrument air pressure. Today, many critical devices are electronic, utilizing triple modular redundant (TMR) computer-based technology (such as Triconex, Modicon, or Mark VI systems). The electronic systems have onboard diagnostics, voting logic, and triple redundancy. However, the challenges are the same in that these systems must be in service to protect the equipment and people in case of a unit upset or emergency. Their defeat and the duration of the defeat must be controlled, regardless of the reason and duration of the defeat.

Process Operations Safety: The What, Why, and How Behind Safe Petrochemical Plant Operations, First Edition. M. Darryl Yoes.

Each manufacturing site needs to identify and develop a list of all SSHE critical devices. This list should be developed as a controlled document and periodically verified as current (up to date), with copies readily available to field and console Operators. The SSHE devices should be identifiable in the field and on the Piping and Instrumentation Diagrams (P&ID)s A common practice is to paint each SSHE device a distinctive color in the field, with no other devices painted the same color.

Most SSHE devices lie dormant, although they continuously monitor the process against the trip target; if the process remains stable and on the trip point's safe side, the device does not function. This means that it must be periodically tested to ensure that the device is functional and will work when expected. The test interval is directly related to the expected service factor or device availability target. Therefore, the manufacturing site should develop the expected service factor or availability target for each SSHE device. A typical availability target for most SSHE devices will fall between 98% and 99%, which means that a device with an availability target of 98% will function as expected 98 times out of 100 times it is called upon to function. It is generally accepted that no devices will have 100% availability. One way of addressing this deficiency is to provide multiple devices to help ensure the availability of critical services.

Each site should have a well-documented and communicated process for protecting the equipment when one of the SSHE devices is out of service ("defeated"). This means that each defeat must be controlled, have proper authorization, and must be communicated to all involved. Safety mitigation plans are developed and implemented to provide alternate protection while the critical device is unavailable. It is also essential that the device is properly returned to full online service after the PM or maintenance is completed.

The Car Seal Valve Management Plan is related to protecting critical devices. This requires a protective car seal placed on the isolation valves upstream or downstream of each critical device. The Car Seal Valve Management Plan is described in Chapter 8.34. Any breaking of a car seal on a designated car seal-protected valve should require an SSHE Device Disablement Authorization Form, sometimes also called the Control of Defeat form, before the car seal is broken.

A list of "defeated" SCD should be posted in clear view of the Unit Operator and the Console Operator, and the Operators should be aware of the alternate mitigation in place. An example of mitigation to provide alternate protection is the car sealing of valves to ensure an adequate relief path during periods when a pressure relief valve (PRV) is "defeated." If Safety Critical Equipment is taken out of service, deactivated, or bypassed for any reason, this information must be communicated as a routine part of the Shift Turnover Process (more information on the Shift Turnover Process is available in Chapter 10).

Another good practice is to require a "defeat tracking board" located in the control center in clear view of managers and supervisors, listing all active SSHE device defeats. This can help by emphasizing the number of defeats and the duration of the defeats.

Whether the SSHE critical device is the modern safety instrumented system, including the stand-alone PLC controller, or the conventional air-to-actuate trip device, the defeat must be controlled if the system is bypassed or defeated for any period or any reason. The device disablement process should apply to either system to ensure appropriate approvals and alternate mitigation to provide the safety function while the device is defeated.

Unfortunately, one practice that is all too common is for the Operator to "bypass" SCD or other safety-critical instrumentation (such as a furnace or compressor trip) when the processing unit is in an upset condition or experiencing abnormal operations. Generally, this is done with the best intentions to help prevent a unit trip caused by the process upset. However, experience has shown that this is exactly when we need the devices to function. Management expectations should be clear that the critical devices always remain in service, except when the device disablement process specifically authorizes a defeat and then only when alternate protection is in place.

The device disablement authorization should be for a specific period, for example, one process shift. Control of Defeat (COD) extensions for longer periods should require escalating levels of management approval. The longer the duration of the defeat, the higher the level of approval that is required. Defeat extensions should also require consideration for additional layers of protection.

Steam turbine overspeed trips are considered SCDs since they protect the turbine from speeding to destruction. However, a general-purpose steam turbine should never be operated without a functioning overspeed device. Since the overspeed can occur almost instantaneously, there is no alternate protection for a nonfunctioning overspeed trip. This should be the one exception to the device disablement process for safety-critical equipment. A steam turbine with a nonfunctioning overspeed trip must be immediately removed from service until the overspeed trip device is repaired.

Many Process Safety Management Systems referenced in Chapter 6 include detailed requirements for managing and controlling defeats or disablement of SSHE critical equipment. These requirements lead to developing and implementing detailed procedures at each petroleum refining or petrochemical manufacturing site.

An example of a sample SSHE Device Disablement Authorization form is available in Appendix F.

What Has Happened/What Can Happen

Piper Alpha Oil Platform (6 July 1988)
Explosion and Fire
167 Fatalities and the oil production platform was destroyed.

As I think of the consequences of the failure of device disablement systems, I recall several significant events, including some in the company I worked for all these years. However, although some were undoubtedly tragic for the people involved, the Piper Alpha disaster stands out above all. This is due to the extreme loss of life on Piper Alpha and the number of lives that were changed forever by this event. One hundred and sixty-seven lives were lost on Piper Alpha, and those who survived did so by jumping into the North Sea, where life expectancy is measured in minutes due to the cold waters of the North Atlantic, even in July.

Piper Alpha was designed as a crude oil exploration and drilling platform located in the tumultuous and cold waters of the North Sea, about 120 miles (193 kilometer) N/E of Aberdeen, Scotland, in waters about 474 feet (145 meter) deep. The platform was commissioned in 1976 and was operated as a joint venture between Occidental Petroleum (Caledonia) Ltd and Texaco. The platform was later modified to handle natural gas and natural gas liquids, primarily ethane and propane, from other platforms in the North Sea. Piper Alpha was connected by undersea crude oil pipelines and high-pressure gas lines to two other producing rigs (Tartan and Claymore). The two 36-inch diameter natural gas pipelines ran for miles on the seafloor and operated at 1,750 psig (124 BARG).

At about 10:30 p.m. on the evening of 6 July 1988, and with a typical staffing of 240 people on board the platform, there was a loss of containment of Liquidfied Petroleum Gas (LPG), which resulted in a significant fire. The fire impinged onto a crude line, which resulted in the failure of the two high-pressure gas lines, resulting in a catastrophic explosion and fire that destroyed the entire platform. Due to a lack of emergency preparedness, 165 lives were lost on the platform, and two additional lives were lost on one of the support vessels that attempted to rescue people from the platform. In the final analysis, essentially, no one was rescued from the platform; those saved were rescued after jumping over 100 feet into the cold waters of the North Sea below the platform.

Piper Alpha received the natural gas from the 36″ pipelines into a receiver drum. A separate line from the top of this receiver drum supplied gas as fuel to the gas turbines, powering the electrical generators and supplying electrical power to the platform. This drum had two pumps to maintain the level in the drum by pumping the accumulated liquid, primarily LPG (ethane and propane), injecting the LPG into the crude line, and going onshore for further processing. These pumps were high-head discharge pumps operating at about 2,000 psig (138 bar).

There were two different but related tasks ongoing on Piper Alpha on the evening of 6 July 1988. One of the LPG pumps had just been removed from service and was in the early stages of overall. It had been electrically disconnected but otherwise was still connected to the system. The second task involved servicing the relief valve associated with this LPG pump. The relief valve had been removed, and the flange was covered with a temporary but non-rated blank.

At about 9:45 p.m., the in-service LPG pump suddenly and unexpectedly stopped pumping. The crew worked with the pump for a few minutes to return it to service. However, due to the quickly rising level in the separator drum, a decision was made to return to service the pump, which had been scheduled for an overhaul. The Operators took this action due to concerns about the high separator level tripping the gas turbine generators, which would trip off the electrical power on the platform. The electrical connections to the alternate pump driver were reconnected, and the pump was quickly returned to service.

When the pump started, a large release of LPG occurred at the temporary blank installed in place of the relief valve. This resulted in an almost instantaneous explosion and a large fire. This large fire ruptured an adjacent crude line, and the flames from the resulting fire impinged onto the incoming high-pressure gas lines, which failed within several minutes. The explosion and massive fire from the high-pressure gas lines resulted in the total loss of the platform and the catastrophic loss of life.

The Operators were using the work permit process to communicate the status of ongoing maintenance tasks and critical equipment status between shifts. However, it was discovered after this incident that the work permit process had degraded over time to a formality. When the engineer for the PRV job reported to the control room at the end of the shift, he found the Operators busy with other discussions. He signed off on the permit himself and left the work permit on the desk without communicating the status of the relief valve to the Operators. The work permit was not filed correctly, and this was reportedly the last time it was seen. The operators on the following shift did not know the relief valve had been removed and a blank had been installed. The information was lost from the work permit, and unfortunately, the valve's location was out of sight. The valve was some distance from the pump, and other machinery obstructed the view. The LPG pump was started with the relief valve removed and a non-rated blank left in place of the relief valve; the LPG release occurred from the non-rated blank.

Many lessons came out of this industry-changing event, including emergency preparedness, effective drills and exercises, effective work permit systems, lock-out and tag-out procedures, well-documented operations procedures, field compliance with policies and procedures, effective employee training, and coordination between platforms.

Another critical learning came during the emergency response phase of this incident. As it turns out, the firewater pumps serving the firewater deluge system had been placed in the manual position and could only be started by hand. Operators developed a habit of placing the pumps in manual mode to prevent the pumps from auto-starting when divers were working near the pump inlets. Unfortunately, on Piper Alpha, this quickly evolved to where the pumps were placed in a manual mode almost every day, regardless of where the divers were working. During the fire, two workers attempted to access the pump controls to start the pumps but never made it, and the firewater pumps were never started. This is another example where effective control of the disablement process may have provided an alternate means of activating the firewater pumps had it been provided.

A valuable lesson that applies here is the importance of a well-documented and rigorously implemented process to control all defeats of SSHE critical equipment. Had such a process been in place on Piper Alpha, this incident could have been avoided, and those 167 lives would never have been at risk.

Key Lessons Learned

- The Device Disablement process applies to all SSHE critical equipment functioning as the "last line of defense" (except turbine overspeed trips).
- Effective Device Disablement requires the following:
  - Management approval for the defeat before the device is defeated.
  - A well-thought-out and documented alternate safety mitigation plan replaces the device's functionality that is being defeated.
  - Communication to all potentially affected personnel.
  - Verification that the defeated device has been returned to full online service after the maintenance or PM work is completed.
- Defeat extension requires a higher level of management and consideration for additional layers of protection.
- Each manufacturing site has the responsibility to do the following:
  - Identify and develop a list of all site SSHE critical devices and make a list available to all field and console Operators.
  - Develop the expected service factor for each SSHE critical device and the ongoing PM schedule for the device based on the service factor.
  - Ensure the site has an effective Device Disablement process that is documented and enforced and subjected to periodic verification.
  - Use the "layered" approval with higher approval authority required for longer periods of defeat to help ensure the number and duration of defeats are minimized.
  - Ensure that a process is in place to communicate the status of critical devices that are defeated for longer periods or for situations where large numbers of devices are defeated and are appropriately communicated to the site leadership. Resources should be available to prevent long periods of non-availability of critical devices, even if additional mitigation exists.
- Site Management is responsible for creating and supporting a strong culture of compliance with the process for preventing and controlling unauthorized or extended defeat of critical devices.

Additional References

Article by Lean Compliance: "Critical Defeats - Managing the Last Line of Defense" Lean Compliance™ Ontario, Canada. https://www.leancompliance.ca/post/critical-defeats-managing-the-last-line-of-defense

Article: "A Practical Approach to Managing Safety Critical Equipment and Systems in Process Plants" Tahir Rafique – Lead Electrical and Instruments Engineer Douglas Lloyd – Senior Electrical and Instruments Engineer Ken Evans – Senior Electrical and Instruments Engineer. https://dvikan.no/ntnu-studentserver/reports/Practical%20Approach%20to%20Managing%20Safety%20Critical%20Equipment%20and%20Systems%20in%20Process%20Plants.pdf

Article by Control Engineering: "When should you bypass your safety system?" This article includes some good schematics to explain their logic. https://www.controleng.com/articles/when-should-you-bypass-your-safety-system/

Flight Safety; "Fire and Fury: The destruction of Piper Alpha". https://www.flightsafetyaustralia.com/2018/07/fire-and-fury-the-destruction-of-piper-alpha/

Video; "Spiral to Disaster" (the Piper Alpha story) Coastal Safety and Environmental. www.coastal.com email: sales@coastal.com

Chapter 8.4. Ensuring Availability of Safety, Security, Health, and Environmental (SSHE) Critical Devices

End of Chapter Quiz

1. What does the term Safety, Security, Health, and Environmental Critical Devices mean?
2. Which is by far the most common of these devices?
3. What is the potential consequence if a SSHE Critical Device fails to function as expected?
4. Can you name other examples of SSHE Critical Devices? Name all that you can.
5. What is the best practice for each manufacturing site relative to identifying and tracking safety-critical devices?
6. Is it important that the SSHE Critical Devices are periodically tested in the field to ensure full functionality? Why is this important?
7. How is the test interval established for each Critical Device?
8. How are the people and equipment protected when the critical device is out of service for testing or maintenance?
9. What are these systems or procedures called?
10. Can you name the key features of a typical Critical Device Impairment plan or a Control of Defeat plan? Name all the features that you can.
11. In years past, these critical systems were simple pneumatic-controlled devices. In today's world, many of these systems are now ______________-_______________ and are minicomputers with triple redundant technology.
12. In addition to being triple redundant, these new electronic systems are designed with ________ __________ and onboard diagnostics. For example, they may require two sensors out of three for the device to act. This dramatically improves their reliability and eliminates false trips.
13. What is a good practice for the identification of SSHE critical devices?
14. The ______ ______ ______ management plan is a related process for protecting critical devices. This requires a protective car seal placed on the isolation valves upstream or downstream of each critical device.
15. What is a good practice for keeping the site management and Operators aware of the number and types of critical devices that are bypassed or otherwise disabled?

16 When do we most need the SSHE devices to be fully functional? Mark all that apply.

a. When the unit is in an upset condition.
b. During a unit startup or shutdown.
c. When a control variable is out of range (above or below limits).
d. When a relief valve is relieving.
e. During a utility failure.

17 What is the only SSHE device where a disablement process like Control of Defeat does not apply? Why?

8.5

Managing Critical Alarms

In the past, most process units had few alarms to alert the Operator of abnormal unit conditions or operations. These were typically "hard-wired" alarms connected to a display panel in the unit control room. However, with the advent of computerized process control and today's sophisticated distributed control systems (DCS), it is now relatively easy to install alarms on all process variables. Therefore, it is essential that in today's process units, a management system is in place to oversee the process of installing alarms into the computer software and that those alarms being installed provide the essential information required by the Operator in the event of a unit upset or process emergency. This is often a case where less is better, providing they are well-thought-out and properly prioritized.

The alarm management system should oversee the number of alarms available to each operator and ensure that the alarms are correctly prioritized, comply with proper human factor considerations, are adequately checked for operability, and are maintained in good working order. During steady-state operations, a realistic target for a sound-functioning alarm system should be a maximum alarm rate of about 10–20 alarms per shift per Console Operator (including low-priority alarms).

This topic is where a lot of good information is readily available but has not been applied to many DCS systems. A good example is the Abnormal Situation Management Consortium (ASM); a significant amount of alarm management information is readily available to ASM Consortium members and can be purchased from Amazon by members of the general public. Another source for information on alarm management is the Engineering Equipment and Materials Users Association (EEMUA) Publication 191, "Alarm Systems, a guide to design, management, and procurement." This guide is available for purchase on the EEMUA website; https://www.eemua.org/Products/Publications/Digital/EEMUA-Publication-191.aspx.

A relatively new American National Standards Institute (ANSI) standard also contains detailed information on refining and petrochemical alarm management systems' design, operation, and maintenance. ANSI/ISA-18.2, which is available at https://www.isa.org/standards-and-publications/isa-standards/list-of-all-isa-standards. The ISA-18.2 index is listed below:

- ANSI/ISA-18.2-2016, Management of Alarm Systems for the Process Industries
- ISA-TR18.2.1-2018, Alarm Philosophy
- ISA-TR18.2.2-2016, Alarm Identification, and Rationalization
- ISA-TR18.2.3-2015, Basic Alarm Design
- ISA-TR18.2.4-2012, Enhanced, and Advanced Alarm Methods
- ISA-TR18.2.5-2022, Alarm System Monitoring, Assessment, and Auditing
- ISA-TR18.2.6-2012, Alarm Systems for Batch and Discrete Processes
- ISA-TR18.2.7-2017, Alarm Management When Utilizing Packaged Systems
- ISA-18.1-1979 (R2004), Annunciator Sequences and Specifications

Alarm management is one of the key focus areas since alarm issues can undoubtedly lead to loss of containment events. Before, there was an emphasis on alarm management. When I was the Risk Management Advisor at a large refinery on the US Gulf Coast, I had the opportunity to lead a small team as we developed the alarm management philosophy and the alarm management procedures for the site. I learned a lot about alarms in that assignment, and most of the following came from that opportunity. This was a long time ago, and I am sure a lot has changed. I urge each site to carefully consider the resources listed below and apply the lessons learned to your alarm system.

Process Operations Safety: The What, Why, and How Behind Safe Petrochemical Plant Operations, First Edition. M. Darryl Yoes.

Critical Elements of Alarm Management

The most valuable basis for an effective alarm management system is a well-thought-out and documented Alarm Management Philosophy, which forms the basis for the alarm management system. This can be developed using the Alarm Management Guideline from the ISA Standard, the ASM Consortium, or The EEMUA publication as a basis.

The Alarm Management Philosophy should address the overall philosophy for the alarm systems used at the site and typically includes the following critical information at a minimum:

- The alarm definitions are used at the site.
- The alarm priorities are based on required operator response times (typically based on an alarm priority matrix).
- The alarm performance indicators.
- Alarm maintenance procedures.
- Alarm management of change procedures.
- The procedure for disabling alarms, including risk mitigation.
- Alarm interface design (interface to console and field operators).
- Alarm system training for Console Operators.
- A description of the Alarm Database (including the designated owner).

The following is intended to be only a starting point in describing the issues regarding alarm management and the key elements that may be considerations in developing alarm management guidelines.

Alarm Flood

An alarm flood is a condition where alarms come into the Operator faster than the Operator's ability to comprehend them. Without diligent oversight of the alarm system, the number of alarms can grow to the point where even a mild process upset or abnormal condition can create a condition known as "alarm flood." I have also heard this called an "avalanche," with which I agree. This can be catastrophic to the Console Operator in severe unit upsets.

In the alarm flood condition, the Operator cannot easily distinguish between alarms needing immediate attention and those with relatively lower priority. This rapid rate of alarms prevents the operator from determining the cause of the process upset or emergency. The alarming rate can inhibit the operator's ability to identify the cause of the incident and respond in the time needed to address it.

Nuisance Alarms

Another factor that often contributes to alarm issues is nuisance alarms. Alarms continuously in and out of the alarm state distract the Operator and lead to complacency. Generally, it is usually a few individual alarms that are causing a high alarm rate. These can easily cause the Operator to miss a critical alarm due to being distracted by these nuisance alarms. To stop a nuisance continuously in and out of the alarm state, some operators have even resorted to disabling or inhibiting the alarm.

One cause of nuisance alarms is setting the alarm limit near operating conditions. An effective alarm management process should continuously evaluate the overall alarm rate and the rate of individual alarms to identify and address nuisance alarms.

Human Factors and Alarm Priority

An effective alarm management process should have an approval process for adding new alarms, including an engineering review and management approval. The alarm display should be in proximity to and visible to the Operator. Alarm priority should be established, including considerations such as alarm panel display, alarm indicator color, and the auditable alarm tone.

Alarm color and tone should be consistent with alarm priority and expected Operator response time. The guidance documents mentioned above have developed guidelines to improve operator effectiveness during "abnormal situations," including alarm display and management recommendations.

Generally, alarms prioritized as "high priority" should not be DCS-based. Due to potential reliability issues, all high-priority alarms should bypass the DCS system and go directly to a "hard-wired" alarm, fully independent of the DCS system or subsystems. These alarms should also have a distinct sound and color to easily distinguish high-priority alarms from other lower-priority alarms.

Rationalization

Consideration should be given to established units for conducting a thorough review of existing alarms and a "rationalization" process for those found to be noncritical. Our refinery eliminated several thousand alarms and reset priorities in the remaining alarms to reduce the number of "high" priority alarms significantly. We eliminated many alarms, such as alarms on valve position. Other alarms designed to help ensure product quality were moved to print on a desktop printer and off the DCS process alarm screens. These were still available to the Operator but were no longer competing with the higher-priority process alarms.

The alarm rationalization process is typically done by a team including representation from process operations, generally an experienced console Operator, someone from Technical, usually an experienced contact engineer, someone from computer engineering, and an experienced instrumentation engineer or supervisor. This team will typically go through the entire alarm list and reset priorities on existing alarms based on required console and field operator expected response times, eliminating or moving other lower priority alarms to other systems such as a printer or onto a system other than the high- priority alarm system.

Alarm Management Database

Each console should have an Alarm Management Database with a designated owner whose responsibility is to ensure data integrity. The site should determine what data is to be entered into the database; generally, this is the data for all unit or area alarms, including alarms associated with the DCS and those that bypass the DCS and go directly to the hard-wired alarms. The site should also determine the process for adding new alarms and manage the change process for deleting or modifying alarms and resetting the alarm priorities. The Alarm Enforcer (described below) is generally run directly from the alarm management database.

Alarm Enforcement

The Alarm Management Enforcer (The Enforcer) is an option that can be built into the Alarm Management Database. The Enforcer is a software program that can automatically reset all alarms that have been disabled, inhibited, or changed from their original state back to their original state or value as described in the Alarm Management Database at the beginning of the following process shift. This is to help ensure that a changed or altered state is not overlooked and an alarm priority becomes permanently changed or permanently disabled without proper authorization and management of change documents.

Typical Alarm Performance Indicators

Alarm Performance Indicators must be developed and rigorously tracked to understand any performance issues with the site alarm management system. The following are typical alarm performance indicators that can be used to monitor and determine the overall performance of the alarm system. This list is only an example; each site should develop a list of performance indicators that work for that site and provide the information to help determine the effectiveness of the alarm management system.

Definitions should be created for each of the site alarm performance indicators. The alarm indicators should be an integral part of the site's Operational Integrity Management System, and the report of alarm performance tracking should be reported to site management regularly.

- **Average number of alarms on:**
  Alarms in a continuous state of alarm are not a good indication, are a nuisance, and represent a condition that has become "a normalized deviation."
- **Alarm rate:**
  The number of new alarms generated each hour should be monitored. This number should normally be low to prevent overloading the Operator.
- **Alarm repeats:**
  An alarm with excessive repeats can quickly become a "nuisance" and be either ignored by the Operator or inhibited and become unavailable.
- **Alarm Acknowledgment time:**
  Tracking alarm acknowledgment time can determine other alarm or operator issues, especially if the acknowledgment times are in minutes rather than seconds.
- **Chattering alarms:**
  Alarms constantly going in and out of alarm significantly add to alarm overload and may represent a mechanical or alarm setpoint issue.
- **Nuisance alarms:**
  A Nuisance alarm is alarming by a condition other than that intended when the alarm was installed and activated; the risk is that nuisance alarms may be ignored, and the actual event will not be noticed.
- **Redundant alarms:**
  An alarm that warns the Operator of an event that is already being alarmed by a separate alarm.

What Has Happened/What Can Happen

Texaco Refinery (Pembroke Cracking Company [PCC])
Milford Haven, England
24 July 1994

Following an investigation of a large refinery fire in July 1994 at the Texaco refinery in Milford Haven (UK), the British Health and Safety Executive (HSE) concluded that the Operators were attempting to deal with an excessive number of alarms at the time of the incident causing them to miss critical alarms. The incident resulted in a large explosion and fire, causing injuries to 26 people, significant equipment damage, and extended unit downtime. The HSE listed alarm system design and specifically "alarm flood" as a causal factor in the incident investigation report.

The events that led to the accident started on Sunday, 24 July 1994, when a severe electrical storm passed through the area. An electrical power interruption resulted in upsets and the shutdown of most of the refinery. The power failure affected the crude vacuum distillation, alkylation, butamer units (butane isomerization), and the fluidized catalytic cracking unit (FCCU) units. The crude distillation unit was shut down because of a fire that a lightning strike had started. After a short period, all refinery units except the FCCU were shut down. The FCCU was still operating, although in an upset state.

After about five hours of FCCU operations in an upset state, a flare distribution line located directly adjacent to the FCCU structure ruptured, releasing a large volume of light hydrocarbons, resulting in a large explosion and fire. The prevailing wind pushed the fire into the FCCU structure, causing severe damage to the FCCU and adjacent infrastructure.

The flare line rupture and loss of containment occurred due to light flammable hydrocarbon liquid being continuously pumped into a process vessel with the outlet valve on this process vessel closed due to a valve malfunction. The vessel overfilled, and the relief valve opened, releasing approximately 140 barrels (20 meter tons) of light liquid hydrocarbons into the flare header, causing the flare line to become overloaded, leading to the collapse and rupture. Due to the release of light materials and the rupture of the flare piping, the fire was allowed to burn out and was finally extinguished the following Tuesday, 26 July 1994.

As cited in the British Health and Safety report, a significant contributor was the alarm flood that occurred. Two Operators reportedly received 275 different alarms in the 11 minutes before the explosion. Due to this flood of alarms, the console Operator missed a critical alarm on the process vessel, indicating the high drum level. It was noted in the HSE report that a considerable percentage of alarms at this site were classified as critical alarms, and the Operator was facing many alarms all at once, and most were critical alarms.

Key Lessons Learned

Alarms are critical process equipment and must be carefully and rigorously maintained. It is important that each site be aware of any issues or concerns with alarms and that a process is in place that always ensures the availability of alarms. If alarms are defeated or disabled for any reason, they should be subject to the control of defeat type process to ensure that the proper authority authorizes the disablement. The defeat process should also ensure that an alternate protection plan is in place that replaces the functionality of the defeated device and that the alarm is properly returned to service as soon as possible.

A sound alarm management process typically includes all the following characteristics:

- Each new alarm is carefully reviewed to ensure the need before adding new alarms.
- Existing alarms are periodically reviewed to ensure they are properly prioritized.
- Reviews ensure that alarms use sound human factors considerations (location, color, an audible tone, etc.).
- Nuisance alarms are identified and addressed.
- The alarm management process should support a periodic rationalization of alarms, including eliminating, reprioritizing, or placing noncritical alarms onto other systems.
- The Alarm Management Performance Tracking should be routinely reported to the site management as an integral part of the site's Operations Integrity Management Process.
- Changes to the alarm database or individual alarms must be managed through the site's MOC process.

Additional References

UK Health and Safety Executive Report Texaco Milford Haven Explosion and Fire – 24 July 1994.

UK Health and Safety Executive HSE Information Sheet Better Alarm Handling – Chemicals Sheet No. 6.

Intelligent Alarming Effective alarm management improves safety, fault diagnosis, and quality control Martin Hollender Carsten Beuthel (https://library.e.abb.com/public/0d024150cfb0dfd0c125728b0036f2be/20-23%201M703_ENG72dpi.pdf).

Engineering Equipment and Materials Users Association EEMUA Publication 191 Alarm systems - a guide to design, management, and procurement. https://www.eemua.org/Products/Publications/Digital/EEMUA-Publication-191.aspx

ANSI/ISA-18.2-2009 "Management of Alarm Systems for the Process Industries." Abstract available at https://www.isa.org/getmedia/55b4210e-6cb2-4de4-89f8-2b5b6b46d954/PAS-Understanding-ISA-18-2.pdf

IEC 61508-1999 "Functional safety of electrical/electronic/programmable electronic safety-related systems."

IEC 61511-2004 "Functional safety - Safety instrumented systems for the process industry sector".

Chapter 8.5. Managing Critical Alarms

End of Chapter Quiz

1 What has changed that makes the Alarm Management System so very important?

2 Why is Alarm Management one of the key focus areas?

3 What are the key elements of a site Alarm Philosophy document?

4 Alarm color and tone should be consistent with alarm ________ and expected Operator ________ time.

5 What are some of the issues that can be experienced with the alarm system that the Alarm Management System should address? Name all the issues that you can.

6 Can these be utilized as Leading Indicators for process safety and improving the alarm management system?

7 Can you name the key human factors issues that should be part of the alarm management system?

8 What is meant by alarm rationalization?

9 What is meant by alarm "enforcement"?

10 Generally, alarms prioritized as "high priority" should not be DCS-based. Due to potential reliability issues, all high-priority alarms should bypass the ________ system and go directly to a "_______-_______" alarm, fully independent of the DCS system or subsystems. These alarms should also have a distinct sound and color to easily distinguish high-priority alarms from other lower-priority alarms.

11 To whom should the routine Alarm Management Performance Tracking be routinely reported? Should this be an integral part of the site's Integrity Management Process?

12 All changes to the alarm database or individual alarms must be managed through the site's MOC process.
- True
- False

13 If alarms are defeated or disabled for any reason, they should be subject to the control of defeat type process to ensure that the proper authority ____________ the disablement. The defeat process should also ensure that an alternate protection plan is in place that replaces the ______________ of the defeated device and that the alarm is properly __________ ___ __________ as soon as possible.

8.6

Managing Safe Operating Limits or Operating Envelopes

In the United States, the federal OSHA Process Safety Regulation (OSHA 29 CFR 1910.119) (d) requires each OSHA Process Safety Management covered process to define the safe upper and lower limits for such items as temperatures, pressures, flows, or compositions; and an evaluation of the consequences of deviations, including those affecting the safety and health of employees.

Using the BP Texas City March 2005 explosion as an example, the operators routinely deviated from the safe upper limits of the fractionator level by running the tower level above the range of the level instrument. This eventually resulted in the overfill of the tower and loss of containment of hot light naphtha to the atmosphere, resulting in a devastating explosion and fire. The safe upper and lower limits should be defined and recorded in the process data section of the Operating procedures for access and compliance by all Operators. Alarms should also be set to the safe side of the operating limit to provide the operators with time to respond and prevent exceeding the safe limit.

Establishing a safe operating envelope or limit requires input from the key stakeholders in the process, technical (engineering) and mechanical. The expectation is that all personnel will ensure full compliance with the operating envelope once it is established. The objective is to ensure the safe operating envelope is as close to the minimum and maximum integrity values as possible to allow operations flexibility and optimization. However, a competing objective is to ensure time for the Operators and instrumentation to respond to avoid tripping over the limits, resulting in excursions and reportable deviations.

Equipment safe upper and lower mechanical integrity values are determined by the equipment design and selection of equipment metallurgy, fabrication methods, and process dynamics. For example, stainless steel has a very low minimum design metal temperature. It, therefore, would be expected to have mechanical integrity values much lower than carbon steel from a brittle fracture perspective and thus can operate at much lower temperatures without sustaining damage to the vessel. Once the equipment's maximum and minimum mechanical integrity values are established, this data is used to develop the maximum and minimum safe operating limits.

The minimum and maximum safe operating limits are typically established for all critical parameters, including flows, temperatures, levels, pressures, and other key operating parameters such as PH and maximum or minimum velocities. Generally, the safe operating limits are established by technical support with involvement and concurrence by Operations. The Center for Process Safety has published an excellent reference article on setting safe operating limits on its website (www.aiche.org).

When establishing safe operating limits or operating envelopes, it is recommended that the following be considered:

- Determine the design limits for all parameters (upper and lower). Design limits may include design minimum and maximum limits, mechanical integrity limits, equipment and instrumentation systems, relief valve settings, and other similar parameters.
- Obtain the planned or acceptable operating limits for all modes of operation. This becomes a minimum for the safe operating limits.
- Ensure the safe operating limits are set within the design limits. For a starting point, 10–15% below the maximum design limit and 10–15% above the minimum design limit are good guides to ensure that the operators and instrumentation are available to respond.
- The normal operating limit must be within the safe operating limit.
- When setting limits, consider the potential for abnormal operations, process upsets, alarms, high-integrity protection systems, and safety instrumented systems. In some cases, these values may need to be derived from Operator response times and instrument or trip system response times.

Process Operations Safety: The What, Why, and How Behind Safe Petrochemical Plant Operations, First Edition. M. Darryl Yoes.

- Once the safe operating limits are established, the final step is to evaluate and document the consequences of the deviations beyond that limit as required for OSHA PSM (for US facilities covered by OSHA PSM 29 CFR 1910.119 (d)).

This effort to develop the Safe Operating Limits generally falls to technical support but is strongly supported by operations. Once established, Operators must strive to maintain operations within the operating envelope or operating limits. All deviations from the operating limits should be reported and tracked as a process safety-leading indicator. Follow-up for each deviation should be developed to understand what caused the deviation, and recommendations should be developed to prevent a reoccurrence.

If no safe operating limits currently exist, a good practice is to establish minimum and maximum safe operating limits (or envelopes) and the associated alarm limits about 10–15% to the safe side of the upper and lower defined limits (10–15% below the high design limit and/or 10–15% above the low design limit). This should allow the Operator time to react and prevent exceeding the actual design limit. As time is available and engineering support is available, these values can be adjusted considering the dynamics of the process and Operator or instrument response time.

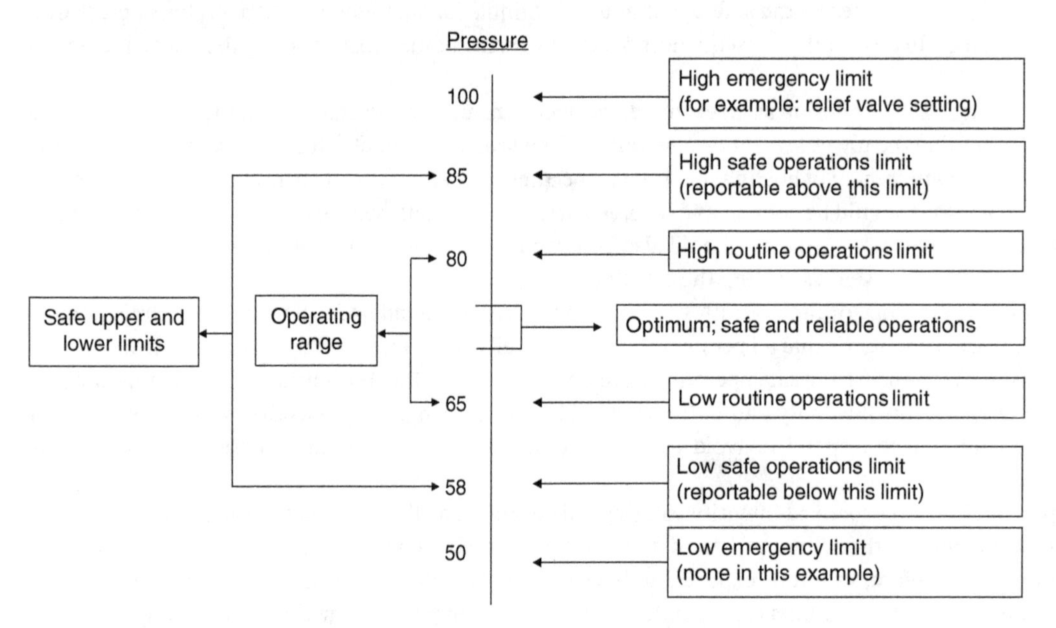

Figure 8.6.1 Safe upper and lower operating limits using operating pressure as the example.

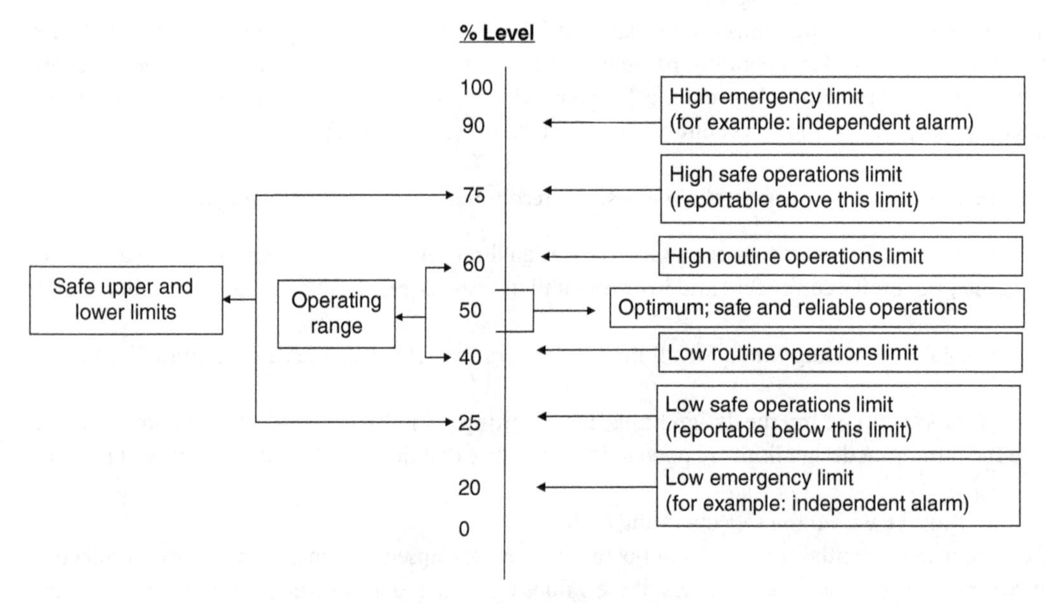

Figure 8.6.2 Safe operating limits using tower level as the example.

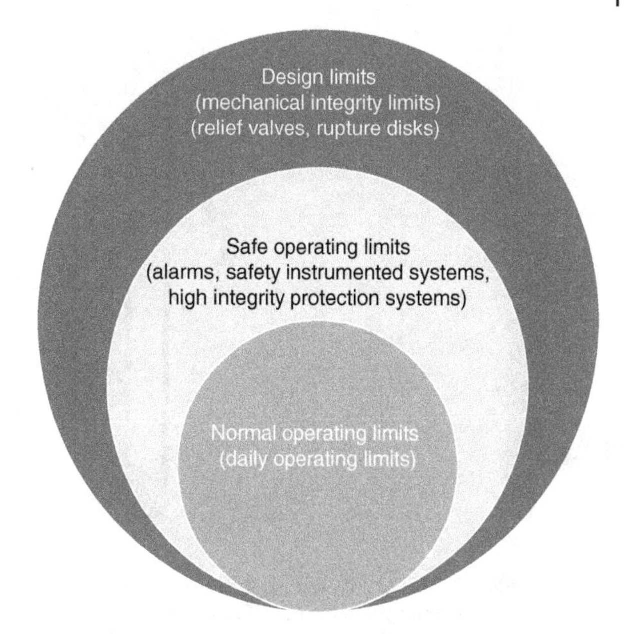

Figure 8.6.3 Relationship of safe operating limits to design limits and normal or routine operating limits.

Once the minimum and maximum safe operating limits are established, site management should set expectations that operations shall report and investigate any deviations from the safe operating limit and track each deviation as a process safety-leading indicator.

One can readily see that if BP Texas City had established a maximum safe operating limit for the level in the fractionator column below the top tap on the level gauge and below the independent high-level alarm, and if the Operators were accountable for tracking the number of excursions above this level, it is quite possible that the March 2005 incident would not have happened.

See Figures 8.6.1–8.6.3 for examples of Safe Upper and Lower Operating limits.

What Has Happened/What Can Happen

BP Texas City explosion and fire
Texas City, Texas
23 March 2005

Going back to the earlier comments on BP Texas City. The US Chemical Safety and Hazard Investigation Board issued a comprehensive report and a great video on the 23 March 2005 explosion and fire in which fifteen lives were lost and over 180 people were injured, significantly damaging equipment and the company's reputation. The CSB report indicated that the Isom Raffinate splitter tower was equipped with a typical displacer-type level indicator to monitor and transmit the tower level to the Operators in the control center. This device uses a float-type displacer to monitor the level between the top and bottom level taps on the device.

The CSB report indicated that the Operators had developed a practice of routinely filling the tower above the top tap on the level instrument during unit startups, which is above nine feet from the bottom of the tower. This deviated from the written procedure to operate the tower at 50% during start-up operations. This deviation was done to prevent potentially damaging the furnace during unit startup, which could occur by loss of flow through the furnace. However, operating with the level above the top tap on the level instrument left the Operators blind to the actual level in the tower. The displacer level instrument can only monitor the level when the level is between the device's lower and upper taps. With the level above the top tap or below the bottom tap, there is no other way of knowing the actual level of the tower.

The displacer level instrument is also equipped with a high-level alarm designed to alert the panel Operator that the level is high in the tower. However, the Operators developed a practice of operating the tower with this alarm activated; the CSB report indicated that the tower's high-level alarm set point was exceeded 65 times during the last 19 startups before the explosion. This indicates that the Operators were unaware of the potentially serious consequences of operating with the level above the top tap of the level instrument and with the high-level alarm activated. The CSB report also indicated that

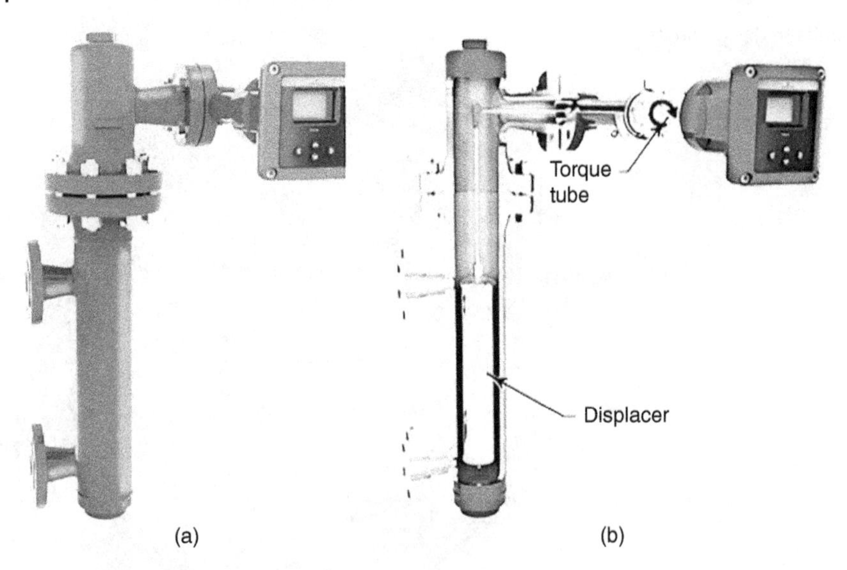

Figure 8.6.4 Photos of a typical displacer-type level indicator. Figure (a) illustrates the complete level displacer indicator unit, including the tube for the level float. Figure (b) is a close up image of the control box for the level displacer unit. Source: Images provided by courtesy of Emerson Electric Co. *Fisher Displacer | 249 Caged* (emerson.com) https://www.emerson.com/documents/automation/product-bulletin-fisher-dlc3100-dlc3100-sis-digital-level-controller-en-1278940.pdf.

the level instrument was calibrated for a different service and had not been recalibrated for several years. See Figure 8.6.4a,b for photos of a typical Displacer-type level indicator device.

Fisher™ FIELDVUE™ DLC3100 Digital Level Controller on Type 249 Series Sensor (image number X1458 attached).

For these devices to work properly, they must be operated within the operating range of the instrument, and the instrument must be properly maintained and calibrated.

At BP Texas City, the Raffinate splitter tower was also equipped with a separate high-level alarm completely independent of the displacer level indicator. A device like this should typically be classified as safety-critical and receive periodic checks to ensure it is fully functional. Safety-critical devices should also be checked to ensure functionality before unit startup. These checks should be performed from the initiating element to the final element. However, this device was not checked prior to startup and failed to function even after the tower was overfilled.

The tower was also equipped with a sight glass as an independent way for the outside Operator to verify the level in the tower. These devices are also crucial to the facility's operations and must be maintained (cleaned) periodically to ensure they are available to the Operators. Outside Operators should regularly check the level in the tower using the sight glass and radio the reading to the console Operator to verify the gauge glass observations with the actual reading from the displacer level device. In this case, the sight glass was reported to be dirty and unreadable. It was obvious that the sight glass was not being maintained.

In the final analysis, the Chemical Safety Board reported several issues related to the operation of the raffinate splitter tower and its level monitoring equipment at BP Texas City. These issues contributed to the loss of containment and the resulting explosion and fire that occurred there on 23 March 2005. Our focus here is that the Operators accepted this, and they were deviating from the written procedures and had previously, and were at the time, operating the tower with a level above the top tap on the level instrument. This means that they were operating the tower blind to the actual level and could not know where the actual level was in the tower.

Had management implemented the concept of safe operating limits for both high and low tower levels, and if these limits had been enforced, requiring reporting of each exceedance. Had an investigation been required to determine the reason for the exceedance and recommendations developed and implemented to prevent a reoccurrence, it is possible that this incident may not have occurred.

Key Lessons Learned

Developing Critical Operating Limits or Envelopes and strictly enforcing these parameters by site management can dramatically affect process safety. Incidents have occurred when deviations from procedures and operating limits are allowed to become routine practices.

Strict Operating Limits and the requirement for reporting and investigation, when excursions exceed these limits, result in more rigorous operating discipline and lead to safer operations.

Of course, there were many additional lessons learned in this incident, and those will be discussed throughout Chapter 8.

Additional References

Occupational Safety and Health Administration (OSHA).
Process Safety Management standard: 29 CFR 1910.119 (d).

US Chemical Safety and Hazard Investigation Board, BP Texas City Refinery Explosion and Fire (Report No. 2005-04-I-TX). file:///C:/Users/Owner/AppData/Local/Packages/Microsoft.MicrosoftEdge_8wekyb3d8bbwe/TempState/Downloads/CSBFinalReportBP%20(1).pdf

Inglenook Engineering – Fireside Chat Series. http://www.inglenookeng.com/_blog/fireside_chats/post/safe-operating-limits-part-1

Center for Chemical Process Safety, Safe Operating Limits. https://www.aiche.org/ccps/resources/glossary/process-safety-glossary/safe-operating-limits

Chapter 8.6. Managing Safe Operating Limits or Operating Envelopes

End of Chapter Quiz

1. What are the OSHA Process Safety Regulation requirements respective to safe operating limits?
2. Generally, the safe operating limits are established by ______________ ______________ with involvement and concurrence by ______________.
3. The normal operating limit must be within the ____________________.
4. Once the safe operating limits are established, what is the final step in the process?
5. Following the development of the safe operating limits, what should the expectation be if there is an exceedance of these limits?
6. In the BP Texas City explosion, the Operators routinely ___________ from the safe upper limits of the fractionator level by running the tower level above the range of the level instrument. This eventually resulted in the overfill of the tower and loss of containment of hot light naphtha to the atmosphere resulting in a devastating explosion and fire.
7. Establishing a safe operating envelope or limit requires input from the key stakeholders in the ________, ________, and __________. The expectation is that all personnel will ensure full ____________ with the operating envelope once it is established.
8. What are some examples of operating parameters for which operating envelopes should be developed?
9. How should all deviations from the operating limits be reported and tracked?
10. Follow-up for each deviation should be developed to understand what caused the deviation and ________________ should be developed to prevent a reoccurrence.

8.7

Maintaining Safe Product Rundown Control

Many product streams flow from the process units to downstream tanks during refinery and chemical plant operations. Many product types and conditions are involved, including several types of tanks, tank operating conditions, and tank limitations. For example, these range from low volatility, heavy gas oil streams at cold temperatures to high-pressure propane or butane going to pressurized storage and everything in between. The operating conditions, primarily the temperature and vapor pressure of the product rundown, must be aligned with the receiving tank, or the results can be catastrophic.

Loss of control of a product stream that is streaming to offsite tankage or as feedstock to downstream units can lead to a severe loss of containment incident and loss of life. This is especially true during unit startups, shutdowns, and upset when the product may be off-specification or unit conditions are not well established. Examples include the inadequate cooling of the product stream, resulting in sending hot products to a cold tank, especially a tank with a water bottom, which can result in a tank boil over or froth over, or sending light materials to an atmospheric tank such as LPG to a floating roof tank which can result in sinking of the floating roof and a full surface tank fire.

Product specifications are established for normal process operations based on safety and environmental considerations for the downstream tank or process unit. For example, an atmospheric storage tank with a floating roof is typically designed to contain a product with a true vapor pressure (TVP) of 13 psia (TVP is vapor pressure in PSIA at actual product temperature). Normal product rundown parameters are set to ensure the product compositions (% of C3, C4, and C5) and the product temperature do not exceed the 13 psia requirement. A product with a TVP greater than 13 psia is considered LPG and must be handled/stored in a pressurized tank such as a sphere or horizontal storage drum.

During a processing unit upset, if the percentage of light materials increases, or should the rundown temperature increase, the TVP may exceed the design maximum of 13 psia TVP. Vaporizing the offsite tank contents can result in a large uncontrolled vapor release at the tank, leading to an explosion or a major fire. Vaporization of tank contents can also result in loss of roof buoyancy and the sinking of a floating tank roof, adding to the incident scope.

Fixed roof tanks (cone roofs) have additional limitations or restrictions. A fixed roof tank is limited to a maximum temperature rundown of 200 °F (93 °C) due to the water bottoms. If the tank temperature exceeds the boiling point of water, it can result in a boilover of the tank contents and a catastrophic fire (more on boilover later). The temperature of the product in fixed roof tanks must also be limited to no more than 15 °F (8 °C) below the product's flash point. This is due to the tank being vented into the atmosphere. Heating the product above the flash point will result in the vaporization of the product, resulting in an explosive mixture from the tank vent and accumulating around the base of the tank.

Most refineries also operate some tanks as "hot tanks." These tanks operate well above the boiling point of water and must operate completely free of water. Even small amounts of water entering these tanks with the product can result in a tank "boilover." A tank boilover is a catastrophic event that occurs when the tank has a layer of water on the tank's bottom and is heated above the boiling point of water. When the water reaches the boiling point, the free water instantaneously flashes to steam, expanding 1,700 times, and the resulting pressure surge pushes all the tank contents over the top of the tank. In most boilovers, the roof is blown clear, and the resulting release results in a catastrophic fire, raining down burning liquid over a wide area.

Fixed roof tanks in "hot service" must be operated water-free, with no free water in the tank or sent to the tank. The tank temperature is typically maintained above 265 °F (130 °C) and should not be operated at temperatures between 200 °F (93 °C) and 265 °F (130 °C). This ensures that trace amounts of water are quickly flashed off and that water does not accumulate in the tank. Should the tank level exceed 34 feet (10.36 meters), the 265 °F should be increased by 1 °F (0.56 °C) per foot above 34 feet.

Operating limits must be established for product rundown streams, and those limits must be enforced. The limits must consider the product involved and the design and limitations of the rundown tanks. Alarms should be set to identify

Process Operations Safety: The What, Why, and How Behind Safe Petrochemical Plant Operations, First Edition. M. Darryl Yoes.

escalating conditions BEFORE the limitation is reached. For example, an atmospheric floating roof tank containing light naphtha may have a maximum temperature limitation of 100 °F to prevent the naphtha from vaporizing. In this example, the unit rundown alarm may be set to alarm at 90 °F to allow the Operator time to act before reaching the tank limit. Any deviation from the operating limit should result in an investigation and corrective actions to prevent the incident from occurring again. These alarms should also be classified as Safety Critical and receive periodic function tests to ensure their functionality.

Blinds should normally be installed on operating units to prevent the inadvertent sending of light streams to atmospheric tanks, hot streams to cold tanks, wet streams to hot tanks, etc. We generally refer to these blinds as "running blinds," blinds that are installed to prevent sending streams to tanks where they can do harm. These blinds are typically removed during unit turnarounds, and it is essential to account for these blinds on turnaround blind lists and ensure that the running blinds are reinstalled before the unit is returned to service.

Unit rundown constraints are not limited to product temperatures, vapor pressure, and free water. Numerous other limitations must be considered when establishing operating limits, such as pH (acid or caustic content), solids and particulates such as catalyst content, etc. It is also important to consider all these different parameters and other similar issues present during unit upsets or other abnormal operating conditions.

What Has Happened/What Can Happen

REPSOL Refinery, Puertollano, Spain
Product Rundown Control Results in Refinery Explosion and Fire
14 August 2003 (Nine Fatalities)

On 13 August 2003, the Repsol Refinery near Puertollano, Spain, experienced an electrical power interruption. The following day, the operating crew restarted the FCC unit. The startup was not going well, as catalyst circulation issues were experienced. These issues diverted the Operators' attention from the Debutanizer tower, which was also in start-up mode. It has been reported that alarms were activated, indicating the low level and low temperatures in the Debutanizer tower, indicating light materials (unstabilized naphtha) were being sent from the Debutanizer tower bottoms into a floating roof tank. It was also reported that the low-temperature alarms were either not noticed or possibly ignored by the Operators. At the same time, start-up attention was directed to problems with the catalyst circulation on the FCC.

When unstabilized naphtha (naphtha containing significant amounts of LPG) was sent from the debutanizer to the floating roof tank, a vapor cloud quickly formed around the base of the tank and in the surrounding tank farm area. A major explosion occurred when the tank's floating roof sank, releasing an even larger cloud of LPG into the atmosphere, where it was ignited by a contractor's truck. The vapor pressure of the unstabilized naphtha was calculated at about 30 psia (2 kg/cm^2) or about twice atmospheric pressure.

The explosion caused major damage to the atmospheric crude unit and resulted in 6 other tank fires. This incident resulted in the loss of seven storage tanks and significant damage to the refinery infrastructure and other process units. There were nine fatalities due to the explosion and resulting tank fires.

Please see the photographs from this incident in the Chapter on Storage Tanks

What Has Happened/What Can Happen

Hot Coker Feed Tank Boil Over during Crude Unit Startup
Baton Rouge, LA
1966

I remember this incident almost like it was yesterday. This incident happened just as I started my career in the industry in 1966. The refinery was in the process of starting up a Crude Distillation Unit following a unit turnaround. After the unit was up and online, one of the last steps was to switch the streams to their typical dispositions. Operators were switching the vacuum tower bottoms to the hot Coker Feed disposition (a cone roof tank in the adjacent tank farm). There was a small amount of free water in the piping or pumps. This water entered one of the hot Coker Feed tanks, a cone roof storage tank that operates at temperatures well above 400 °F (204 °C). The resulting pressure surge from the rapidly expanding water in the hot cone roof tank blew the roof from the tank, causing it to land in an adjacent tank firewall. The hot Coker

Feed material, essentially asphalt, was sprayed throughout the refinery, coating many other process units and equipment. Fortunately, this spray was contained inside the refinery, and there was no outside contamination.

No one was injured in this event, but it severely damaged the storage tank and resulted in a significant clean-up operation in the refinery that continued for some time. Indeed, there was the potential for serious injury, highlighting the attention to detail required relating to unit rundown streams. This is particularly true when preparing to restart refining equipment or when in startup mode.

Key Lessons Learned

Both incidents highlight the importance of establishing unit rundown operating limits or envelopes and strict adherence to those limits. The organization's culture should be such that deviations, even minor deviations, are always reported and investigated, with corrective actions developed and implemented to help prevent a recurrence of the deviation. This culture should be continuously enforced by refinery management.

Once the operating limits are established, alarms should be set with ample time for the Operators to respond before exceeding the limits. Unit rundown alarms, such as temperature alarms, should also be classified as safety-critical devices and periodically checked to ensure functionality.

Additional References

"Incidents That Define Process Safety" Repsol Refinery Explosion, Puertollano, Spain (Page 111), John Atherton, Frederic Gil. Wiley Inter-science.

Article: "Repsol blames human error for Puertollano refinery blast". https://www.icis.com/explore/resources/news/2003/09/18/519437/repsol-blames-human-error-for-puertollano-refinery-blast/

"Guidelines for Engineering Design for Process Safety". This book contains guidance for rundown limits and rundown alarms and general guidance for the engineering design of product rundown systems. Available from Wiley Publishing (Amazon and your local bookstore).

Chapter 8.7. Maintaining Safe Product Rundown Control

End of Chapter Quiz

1 What is the vapor pressure limit for a rundown naphtha stream to product storage?

2 What are the potential consequences of inadequate cooling of rundown streams to storage tanks by the process units?

3 What are the considerations for establishing the operating limits for temperature and vapor pressure for product rundown streams?

4 What temperature limitations are in place to protect against the flashing of lower flash products and creating a hazardous vapor near fixed roof tanks?

5 What are the typical guidelines for operation of fixed roof tanks in "hot service" such as asphalt and hot residuum (tanks that normally operate at temperatures above 200 °F (93 °C)?

6 The operating conditions, primarily the __________ and ________ ________ of the product rundown, must be aligned with the receiving tank, or the results can be catastrophic.

7 Loss of control of a product stream, for example, inadequate cooling of the product stream resulting in sending hot products to a cold tank, especially a tank with a ______ bottom which can result in a tank ______ ________ or froth

over, or sending light materials to an atmospheric tank such as LPG to a _______ _______ roof tank which can result in sinking of the floating roof and a full surface tank fire.

8 Why is it important to operate a fixed roof tank in hot service, for example, asphalt andresiduum , at a temperature of 265 °F (130 °C) or above, and they should not be operated at temperatures between 200 °F (93 °C) and 265 °F (130 °C)?

9 Tank alarms designed to prevent exceeding operating limits should be classified as critical ________ devices and be subjected to periodic function ________ to ensure operability.

10 Running blinds should normally be installed on operating units to prevent the inadvertent sending of_______ streams to atmospheric tanks, _______ streams to cold tanks, ________ streams to hot tanks.

11 What types of operating conditions are more likely to lead to periods where refineries and petrochemical plants are most susceptible to deviations from operating limits?

8.8

Storage Tank Safety

This chapter covers the various types of storage tanks (cone roof tanks, floating roof tanks, spheres, and refrigerated LPG tanks) and the causes of fires and explosions in storage tanks. We will discuss the kinds of products each type of tank is designed to handle and the limitations of each type of tank. We will then review the hazards associated with storage tanks, followed by a review of several catastrophic incidents with storage tanks and the critical lessons learned to help prevent a reoccurrence.

One of the more frequent causes of tank fires and other tank-related incidents is inadequate control of product rundown from the producing units (see Product Rundown Control, Chapter 8.7 for more details). Sending products to atmospheric tankage too hot, too light (a product containing light hydrocarbons such as C3, C4s, or C5s), or sending water to hot tanks has resulted in a significant loss of containment as a vapor release and resulting fires and explosions. For this reason, tank safety is not just for off-site personnel. Unit personnel and good unit operations can have a significant positive impact on storage tank safety.

Types of Storage Tanks and Limitations of Each Type

External Floating Roof Tanks

These tanks are designed to store products with a flashpoint of less than 100 °F (38 °C), such as naphtha, gasoline, crude oils, and chemical products such as benzene or toluene. Floating roof tanks at sea level are limited to products with a True Vapor Pressure (TVP) less than 13 psia. 13 psia TVP defines atmospheric storage, and products with greater than 13 psia (89 kPa) require pressurized storage, including products such as LPG (propane, butane, and pentanes are stored in spheres or bullets due to their vapor pressure). Because of atmospheric pressure, the allowable TVP is reduced by 0.5 psi for each additional 1,000 feet in elevation for tanks located above sea level. External floating roof tanks can also store products at a temperature within 15 °F (8 °C) of their flashpoint or above their flashpoint, whereas products stored in cone roof tanks must not be within 15 °F of the flashpoint.

For purposes of this discussion, atmospheric tanks with floating roofs are designed to operate with a maximum of 13 psia TVP (vapor pressure at the product temperature), when located at or near sea level (less when located at higher elevations). However, it is not uncommon for local or state air quality regulations to establish lower vapor pressure limits for environmental purposes.

The floating roof design allows the roof to float on top of the product, thereby preventing evaporation or flashing of the product into the atmosphere. Floatation is accomplished by the roof pontoons, which are individual liquid-free compartments built into the roof for buoyancy. See Figures 8.8.1–8.8.4 to illustrate the floating roof design (single deck design) and note the roof pontoons are arranged around the outer periphery of the roof. These are separate compartments and are sealed to prevent liquid from entering the compartment. Each pontoon has an inspection hatch to allow Operators to inspect for a leakage into the pontoon periodically.

The illustration also shows the roof support legs, designed to support the roof when the tank is at a low level ("landing" the roof). The illustration also shows a roof vent designed to break the vacuum under the roof when it lands on the support legs. For example, when taking the tank out of service for inspection or maintenance. The drawing also illustrates the gauging hatch and the center roof drain. The central roof drain allows rainwater to drain from the roof through the center sump and exit through an internal articulated pipe or drain hose out the tank's sidewall. The center roof drain is equipped with a check valve to prevent the backflow of product onto the roof in the event of failure of the internal drainpipe or drain hose.

Process Operations Safety: The What, Why, and How Behind Safe Petrochemical Plant Operations, First Edition. M. Darryl Yoes.

Figure 8.8.1 Typical floating roof storage tank.

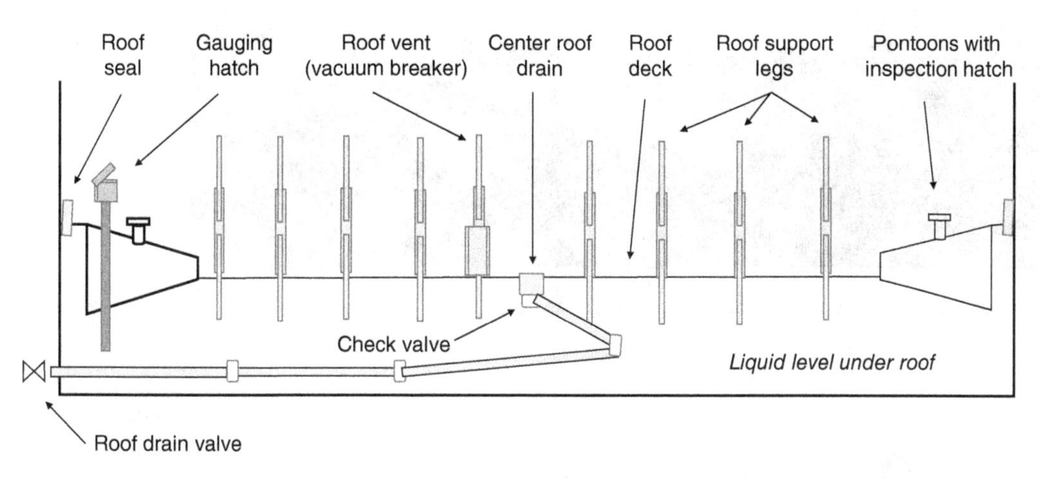

Figure 8.8.2 Single deck pontoon roof.

Roof Drain and Drain Hose (or Articulated Pipe) and other Features of a Floating Roof Tank. Note that the rolling ladder is available to access the top of the floating roof. This design is a "single" deck where the pontons are positioned around the outer perimeter of the roof. Some tanks have the "double" deck design, where the entire roof serves as a large, partitioned pontoon.

Internal Floating Roof Storage Tanks

An internal floating roof storage tank may be a retrofit of an existing external floating roof design or a new tank built with an aluminum internal floating roof with the geodesic dome cover. Geodesic dome covers are frequently used to help reduce emissions or to eliminate rainwater from entering the traditional floating roof design tanks.

These tanks may also be retrofitted from a cone roof storage tank to an internal floating roof to change the tank service to a product with a lower flashpoint. For example, a product can be changed from a combustible, such as diesel, to a flammable, like naphtha or gasoline. The internal floating roof is less likely to have environmental emissions and less likely to leak rainwater into the product and accumulate water.

The roof in an internal floating roof tank may be supported by cables suspended from the external roof and will not be equipped with roof support legs like a conventional floating roof tank. In this case, the roof cannot be accessed for maintenance with the tank in service due to the potential for vapor accumulation between the fixed and internal floating roofs. See Figure 8.8.5a,b for conversions of floating roof tanks to an internal floating roof with a geodesic dome cover.

Figure 8.8.3 Overhead view of a single deck pontoon floating roof storage tank.

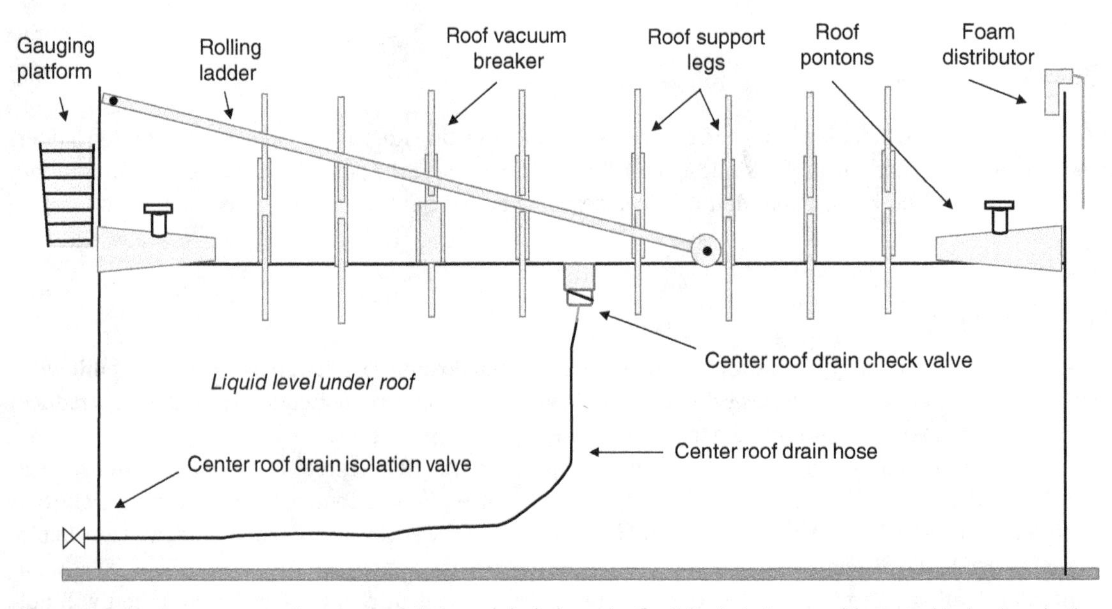

Figure 8.8.4 Arrangement of center roof drain and drain hose (or flexible drain hose as illustrated below) and other features of a floating roof tank.

(a)

(b)

Figure 8.8.5 Typical retrofit of an existing floating roof storage tank to an aluminum geodesic dome cover. Converting the tank to an internal floating roof. Source: (a) Photo courtesy of United Industries Group, Inc. (UIG Tanks). (https://unitedind.com/uig-everdome-geodesic-domes-and-aluminum-reservoir-covers/); (b) Photo courtesy of Deep South Crane and Rigging (https://deepsouthcrane.com/projects/lifting-the-dome).

Cone Roof Storage Tanks

Cone roof storage tanks are used for services where floating roofs are not required to suppress the vapors. They are typically used to store diesel, fuel oils, lubricants, waxes, and similar nonvolatile products. They operate below 200 °F (93 °C) due to the likely presence of free water on the tank bottom and the risk of flashing water if operated at higher temperatures. Cone roof tanks must also operate at temperatures 15 °F (8 °C) below the flashpoint of the stored material to prevent vapor release to the atmosphere from the tank vents. Due to the atmospheric vents, cone roof storage tanks cannot be used to store products at temperatures above the flashpoint of the product stored, even if the tank is inerted with nitrogen or vapor enriched with natural gas. See Figure 8.8.6 for a photo of typical cone roof storage tanks.

Cone roof tanks are also frequently used in areas of high seismic activity or where heavy snow load conditions may exist. They are occasionally equipped with vapor recovery systems in environmentally sensitive areas.

Figure 8.8.6 Typical cone roof or fixed roof storage tanks.

Figure 8.8.7 Cone roof storage tanks in asphalt (hot) service.

Cone Roof Storage Tanks in Hot Service

Cone roof or fixed roof tanks may occasionally be used in "hot Service" for asphalt or other heavy, intermediate unit feed products such as hot Coker feed or hot FCC feed. These tanks should be operated at temperatures above 265 °F (130 °C) to prevent free water accumulation on the tank bottom (note add 0.7 °F per foot for levels above 35 feet). At these temperatures, any accumulation of water can result in a boilover, a devastating event. At these temperatures, trace amounts of water entering the tank are quickly flashed off, and water cannot accumulate. Cone roof tanks in hot service also operate at 15 °F (8 °C) below flash of material stored to prevent vaporization to the atmosphere from the roof vents. See Figure 8.8.7 for an example of a cone roof storage tank in asphalt service.

Note that asphalt or heavy oil service tanks are often painted black for energy conservation.

Pressurized Storage Tanks

Pressurized storage tanks such as spheres or horizontal storage vessels ("bullets") are used for storage of liquefied petroleum gases (LPG) such as propane, butane, pentane, or light chemical products such as propylene, or butadiene and light slop (containing LPG, etc.). These products typically are gases at atmospheric pressure and are stored at more than 13 psia TVP. See Figure 8.8.8 for photos of typical pressure storage tanks (spheres and horizontal storage vessels).

Pressurized storage tanks typically have a range of safety features in the event of exposure to fire or loss of containment. For example, the terrain under the sphere should slope away from the sphere to avoid low spots where a pool fire could impinge on the vessel. For this reason, there are typically no catch basins directly below the spheres.

Spheres and horizontal storage vessels ("bullets") typically have a water deluge system designed to completely blanket the vessel with firewater in case of fire exposure or another emergency. The water deluge system must be checked periodically to ensure nozzles are not plugged and there is even water spray distribution. When the water deluge system is working properly, there should be no dry spots on the vessel.

The support legs are fireproofed to prevent failure during a fire event. In some rare cases, the entire sphere may be fireproofed, especially when the sphere is near other LPG-containing equipment. Most spheres and bullets also have fire water injection facilities to facilitate "water flooding" of the vessel in case of a leak or release near the bottom. Water flood is normally done with a hose or other break-away type connection to inject fire water directly into the vessel to float the LPG up in the vessel. Water flood can also be used to add cooling to the vessel in the event of fire or exposure to fire.

Figure 8.8.8 Pressurized storage tanks (sphere and horizontal storage vessels).

The firewall or containment dykes for LPG-containing storage tanks are typically equipped with firewall drain valves. These are normally maintained closed in the event of a spill or release of LPG into the tank firewall and opened only when necessary to drain rainwater.

Refer to the Chapter on BLEVEs (Chapter 5) for more details on the equipment and features associated with LPG storage vessels.

Refrigerated Storage Tanks

Refrigerated storage tanks store liquefied petroleum gas, such as propane, butane, propylene, etc., at atmospheric pressure. These refrigerated tanks are double-walled, insulated vessels designed to store cold liquids. The internal wall is typically a metal alloy suitable for the low temperature and pressure of the materials stored.

The LPG is maintained in a liquid state at atmospheric pressure by auto-refrigeration of the boil-off gas (auto-refrigeration). Then, it is by the reliquefication by compression of the vapors. Loss or outage of the refrigeration system or overheating of contents may result in loss of containment and vapor release; therefore, the refrigeration systems should be treated as safety-critical systems. Facilities typically include compressors for reliquefication, including standby compressors and backup electrical power and instrumentation. These systems are also typically equipped with hydrocarbon detection and water deluge systems.

The temperature of products stored in refrigerated tanks must be maintained below the boiling point of the material stored to prevent product vaporization and loss of containment. The tank elevation must be considered due to lower atmospheric pressure at higher altitudes and lower temperatures required to maintain LPG in a liquid state at higher elevations.

Typical boiling points for LPG (at sea level), which must be maintained in refrigerated storage tanks:

- *n*-Butane = 31 °F (−0.5 °C)
- Propane = −43 °F (−45 °C)

Figure 8.8.9 A photo of a double-walled and insulated refrigerated storage tank for LPG (propane, butane, propylene, butylene, etc.).

- Propylene = −48 ° (−54 °C)
- Ethylene = −153 °F (−103 °C)

A photo of a refrigerated LPG storage tank is included in Figure 8.8.9.

Storage Tank Hazards

Storage Tank Hazards include the following potential scenarios. Each of these will be discussed in the following sections:

- Tank overfills.
- Vapor release.
- Boil over or froth over.
- Sinking a floating roof.
- Explosion or fire due to hot work (welding, etc.).
- Tank exposed to vacuum or overpressure.
- Pyrophoric Materials (covered in Chapter 8.21, “Pyrophoric Ignition Hazards”).

Tank Overfill

Overfilling a storage tank, especially one containing flammables or toxins, can be one of the most severe incidents facing a refinery or chemical plant due to the potential consequences. Numerous catastrophic incidents in the industry's recent past involving tank overfills, including the Buncefield Storage Terminal in Hemel Hempstead, Hertfordshire, on 11 December 2005, when 20 storage tanks were involved in the initial explosion and fire.

Other similar incidents highlighting the significance of tank overfills include the 23 October 2009 Caribbean Petroleum Tank Terminal explosion and fire involving 17 of the 48 petroleum storage tanks and other equipment onsite and in neighborhoods and businesses off-site. The Caribbean Petroleum fires burned for almost 60 hours. In the same month, another devasting fire occurred at the IOC Product Depot in Jaipur, India, and involved 12 storage tanks (and reportedly resulted in 11 fatalities). More recently, a major fire occurred at the Intercontinental Tank Terminal in Deer Park, Texas, on 17 March

2019, resulting in damage to 11 storage tanks and resulted in forced shelter-in-place orders and school closures, while toxic runoff from the site closed a seven-mile stretch of the Houston Ship Channel for three days.

One can see from these examples that a tank overfill can involve additional tanks and impact operations and the community for miles around the site. These incidents can also affect the site's "license to operate." For example, the Buncefield terminal has not been rebuilt and will most likely not be rebuilt due to the significant community impact of that incident.

Tanks in volatile service (containing flammable or toxic materials) should typically be equipped with independent high-level alarms to help prevent tank overfill. The high-level alarms should be classified as safety-critical devices and subjected to periodic testing to ensure their functionality, generally on a predetermined inspection interval. Current technology is evolving to include storage tanks used for these volatile materials and to be equipped with automatic tank shutoff devices to stop or divert flow from the tank in the event of a high level to prevent a potential overfill.

Possible causes of tank overfill:

- Gauge reading incorrectly or stuck.
- The operator is not aware of tank levels and/or product movements (Operator awareness).
- Valve misalignment in the field – product going to the wrong tank.
- Lack of complete and thorough Operator rounds.

One of the most common causes of tank overfill is a faulty level indication, such as a stuck gauge or failure of the high-level alarm. Following the devastating fire at the Caribbean Petroleum Company (CAPECO), the US Chemical Safety and Hazard Investigation Board found that malfunctioning of the tank side gauge (the float and tape apparatus) during filling operations led to the recording of inaccurate tank levels. The CSB found that CAPECO and similar storage tank terminals were exempted from the OSHA and EPA Process Safety Management regulations, and the tank-level monitoring system at CAPECO was not properly maintained and had no redundancy. The CSB recommended that US OSHA and EPA enact regulations to ensure that terminals operating with gasoline, jet fuel, and other materials meeting the NFPA flammability rating 3 or higher be equipped with an independent high-level cutout system (totally independent from the tank-level gauge), see the CSB Investigation Report for more details).

There are other things that the operators can do to help avoid a tank overfill. Operators should be continuously aware of "moving" tanks and the approximate levels in those tanks. They should periodically calculate a material balance and double-check the calculated volumes with the actual tank gauges; this is a good way to determine if the gauge is stuck or hung, and this process has detected tank line-up issues in the past.

Operators should make complete and thorough operator rounds near the beginning of the shift and several additional times during the shift to verify tank levels and manifold status. When switching rundown or receiving tanks, it is standard practice to confirm that the new designated tank level is moving as expected and that the tank switched out has stopped moving. Be aware of the potential for gravitating from a tank located at an upper elevation or a taller tank. A good practice is always to keep the main tank valve closed when the tank is not in use; however, ensure that pipelines are protected against thermal expansion.

Vapor Release

Sending "unstabilized" naphtha (naphtha containing light ends or products that vaporize at atmospheric temperature) to atmospheric tankage can exceed the maximum design vapor pressure and cause a vapor release at the tank roof vent or floating roof seal.

Sending a light product such as naphtha to an atmospheric tank too hot can also result in high vapor pressure ("true vapor pressure") and a loss of containment to the atmosphere from the tank roof seal.

Process units should strictly adhere to maximum operating temperature limits when sending products to storage. These limits should be a part of the organization's culture, and compliance should be enforced. See Chapter 8.7 for more details on Product Rundown control.

Tank Boil Over or Froth Over

Sending a hot product to tankage is another significant safety concern. Most atmospheric tanks have standing or "free" water on the bottom of the tank (below the product level). This is a normal condition and is to be expected. Sending a hot product to the tank (product above 200 °F) (93 °C) can heat the tank contents, including the water bottoms. Water boils at 212 °F (100 °C) at atmospheric pressure; heating the water to its boiling point can result in flash vaporization of the water to steam and result in a tank "boil over." A tank boil over releases large amounts of product, which may be volatilized by the heat of steam, resulting in a fire or explosion. At a minimum, boilover results in a large product spill and the potential for a major explosion or fire, severe personnel injuries, and significant tank damage.

Tanks in "hot" service should operate at temperatures above 265 °F to allow the flashing of trace amounts of moisture from the product. If operated at colder temperatures, the moisture can accumulate on the tank bottom as free water, eventually heating it to the boiling point and causing the tank to boil over. Atmospheric storage tanks should not operate between 200 °F (93 °C) and 265 °F (130 °C).

A significant boilover of a crude oil storage tank occurred at a storage terminal at Three Rivers, Texas, on 25 August 1990, while the tank was burning with a full surface tank fire. The released hot crude sprayed over a large area and caught responders and many sightseers' completely off guard. A video on this event that includes a good description of what leads to a boilover is available on YouTube at the following link: https://www.youtube.com/watch?v=qxL37WWynqw.

An excellent example of a catastrophic storage tank boilover occurred at the Milford Haven Amoco oil refinery in Pembrokeshire, Wales, on 30 August 1983. There is a good description of this incident online at https://www.ife.org.uk/firefighter-safety-incidents/milford-haven/33996.

According to reports, the storage tank (Tank T011) was 256 feet (78 meter) in diameter and 65 feet (20 meter) high. The fire started around 10:53 a.m. when the tank contained approximately 322,000 barrels of crude oil (46,000 tonnes), and the response began almost immediately. The fire quickly evolved and spread to the entire storage tank. At approximately midnight, the first of several boilovers occurred. See Figures 8.8.10 and 8.8.11 for images associated with this catastrophic event.

During this fire event, the heat spread through the crude oil and heated the free water on the bottom of the tank, flashing the water to steam. When the water flashes to steam, it expands 1,600 times, resulting in a massive pressure wave inside the tank. The pressure wave lifted the crude oil, causing it to erupt from the storage tank, releasing burning liquid onto the surrounding area. A fireball with a radius of 300 feet (90 meter) and a flame attaining almost 500 feet (150 meter) in height resulted from the boil over. A large quantity of burning crude oil was released, spreading the fire into the containment dike and destroying most of the firefighting equipment. This forced the response team to cease operations. Six firefighters sustained mild injuries during their retreat.

It is not unusual for the burning liquid to blow the floating roof from the tank, causing it to land on response equipment or personnel. It is not clear if this happened in this incident.

Figure 8.8.10 Photos during the initial stages of the Milford Haven Tank T011 fire. Source: Images courtesy of Mid and West Wales Fire and Rescue Service.

Figure 8.8.11 Photos during the boilover of the Milford Haven Tank T011 fire. Source: Images courtesy of Mid and West Wales Fire and Rescue Service.

Cause of This Incident

The investigation team concluded that hydrocarbon vapors were present on the floating roof due to cracks in the roof. There were previous reports of leakage of crude oil onto the roof due to these cracks.

Ignition was found to be caused by embers from heavy flaring of refinery gases due to an unplanned compressor outage on the FCC Unit.

Sinking a Floating Roof

The sinking of the roof on floating roof tanks is another cause of major tank incidents. Floating roofs are designed with pontoons, which are product-free compartments within the roof to provide adequate buoyancy. If rainwater or product enters these pontoons, the roof can sink. Tank roofs are also equipped with roof drains to remove rainwater from the tank roof. These drains are usually piped through the tank's internal area with either hose connections or piping with flexible joints. If the center roof drain is obstructed with debris, such as rags or other trash, rainwater flow can be restricted, resulting in water entering the pontoons. In the event of a drain hose or piping failure, the roof drain normally has a check valve at the roof to prevent product backflow and a valve at the drain outlet (located at the base of the tank shell), which may be closed to prevent loss of containment. This check valve can fail in the closed position and restrict rainwater flow, resulting in water accumulation on the floating roof and entering the pontoons.

Causes of sunken roofs include:

- Leaks in roof pontoons result in products entering the pontoons.
- Severe rainfall events such as hurricanes or tropical storms overwhelm the capacity of the roof drain.
- The roof drain is blocked or plugged with debris.
- The roof drain check valve is stuck in the closed position, resulting in rainwater accumulation on the roof.
- Unstabilized naphtha was sent to the tank, causing a loss of roof buoyancy.
- Mechanical failure of roof or roof guide pole.

Figure 8.8.12 Photo of an open-top floating roof tank with a sunken roof. Note the foam used to suppress the vaporization and potential ignition.

In the event of a sunken roof on a tank containing a volatile product (such as crude oil, naphtha, or gasoline), firefighting foam should be applied to the product surface area as soon as possible to suppress vapors and help prevent ignition.

Figure 8.8.12 illustrates an open-top floating roof tank with a sunken roof. In this example, the incident occurred at a Gulf Coast refinery during a tropical storm with deluge amounts of rain. There was incomplete water removal from the roof due to the check valve in the center roof drain being stuck in the partially closed position. This allowed rainwater to back into the roof pontoons, sinking the floating roof. Note that the refinery fire team promptly applied firefighting foam to the naphtha surface to help prevent the ignition of the vapors.

Explosion or Fire Due to Hot Work (Welding, Etc.)

Many explosions and fires occurred when the tank was out of service, opened for maintenance, and supposedly "gas-free." One example occurred at the TEPPCO Gasoline Terminal in Garner, Arkansas, on 12 May 2009. This was a gasoline storage tank that was opened to the atmosphere and had been cleaned for maintenance. The hot work permit indicated that gas testing occurred at 7:00 a.m., the start of the work shift. However, the first arc was not struck until workers returned from lunch at approximately 2:30 p.m. when the incident occurred. No documentation indicated that gas testing was conducted after the workers returned from lunch or when they started the hot work activities just before the explosion. Vapors were released from the tank during the heat of the day, resulting in the explosion.

This is an excellent example of where continuous gas monitoring or procedures that ensured that hot work began immediately following the gas testing could have saved three lives. Work atmospheres can change rapidly; gas monitoring needs to be conducted immediately before and during hot work activities to ensure that workers are constantly aware of the potential development of an explosive atmosphere. I am a fan of continuous gas testing as it is real-time testing with audible and visual alarms in the event gas is detected. Today's continuous gas testers are available with a wide range of sensors, including most toxics, and continuously test for oxygen concentration and the lower explosive limit.

Many tank fires are directly attributable to hot work during maintenance. Therefore, ensure that rigorous procedures are in place and are followed before work begins, including:

- Risk assessment; a job safety analysis (JSA) or Job Loss Analysis (JLA) to identify the hazards and mitigation steps before any work is done.
- Verify the training, experience, and competence of personnel.
- Ensure that the flammable atmosphere has been eliminated, including equipment isolation, purging of residual gas, gas testing, and the elimination and control of ignition sources.

Figure 8.8.13 TEPPCO gasoline storage tank explosion, Garner, Arkansas, 12 May 2009. Source: Photo courtesy of the US Chemical Safety and Hazard Investigation Safety Board.

- Ensure adequate spacing between the vessel, potential ignition sources, other process equipment, and buildings where ignition sources may be present.
- Be aware of hot work in other vessels adjacent to the vessel where you are working (Figure 8.8.13).

Tank Exposed to Vacuum or Pressure

Many storage tanks in our industry have been severely damaged or destroyed by vacuum or overpressure exposure. Storage tanks are not designed to see more than just a few inches of water pressure or vacuum and are easily destroyed if they are exposed to slightly more than this. The pressure/vacuum vents, the protection against overpressure or vacuum, are designed to open at very little pressure or vacuum. The formula for calculating the force that a storage tank can be exposed to is:

- Pressure × Area = Force; or $P \times A = F$

Storage tanks have an immense surface area; therefore, it takes very little pressure or vacuum to put a significant amount of force against the sidewall of a storage tank. For example, at a Gulf Coast site, they emphasized the complete blinding of all connections associated with the preparation of equipment for opening. The work crew did precisely that, and they blinded all connections in a storage tank, including the atmospheric vent. There was a small change in atmospheric pressure overnight, and someone noticed a large deflection in the tank sidewall the following morning. The tank had been damaged by forces induced by the change in atmospheric pressure overnight. This illustrates how little pressure is required to damage the tank shell and why a pressure/vacuum vent must protect the tank.

Storage tanks have been damaged in numerous other ways. The most common has been the intentional or inadvertent blocking of the pressure/vacuum (P/V) vent. For example, painters have the company's interest at heart when they "protect" the P/V vent by wrapping it in plastic or by a hoard of insects that plug a P/V vent due to the sweet odors emanating from the vent.

I recall an incident at a refinery near the Gulf Coast where a work crew was charged to resolve hydrocarbon staining to the side of a freshly painted tank due to drippage from the P/V vent. They added a drainpipe to the outlet of the P/V vent. The new vent pipe was extended down the side of the tank to just above grade. This worked well until heavy rains hit the area and covered the new vent line, effectively sealing the new vent. This occurred while the tank was emptied, which pulled a vacuum on the tank, collapsing and destroying the freshly painted tank.

Steaming equipment or hot purging to storage tanks is another vacuum hazard. If equipment is steamed to a storage tank, and if the tank is blocked following the steaming without a working vent, the tank can be subjected to significant vacuum,

Figure 8.8.14 Storage tank damaged by overpressure or vacuum. Source: European Industrial Gases Association (EIGA/AISBL) Safety Newsletter SAG NL 82/05/E "Tanks damaged by vacuum". Photo courtesy of the European Industrial Gases Association (EIGA).

causing the tank to collapse. Condensing steam or hot vapors can create a powerful vacuum, and if the P/V vent is blocked in or blinded, the tank can collapse.

Please see Figure 8.8.14, an image of a collapsed cone roof tank. This storage tank was in the process of being painted. In preparation for sandblasting and painting, the workers erected scaffolding and placed plastic covers over the attachments to prevent sand and paint from getting into the equipment. As the workers placed the plastic, they covered the pressure/vacuum vent valve on the top of the tank.

Unfortunately, the painters failed to remove the plastic from the vent valve, and when the Operators started pumping liquid from the tank, it collapsed due to a vacuum inside the tank. The plastic cover was found sucked into the vent piping blocking the airflow into the tank and creating a vacuum.

Again, this is pressure × area = force. There was very little vacuum; however, when multiplied by the storage tank's surface area, this created enormous force, causing the tank to collapse.

See Chapter 8.25 for Figure 8.25.6 for an example of the hazard of a vacuum formed by condensing steam in a tank vessel. This image is of a railcar that was steamed and then blocked in without providing a vent to break the vacuum created. A tank or railcar can collapse due to a vacuum from the condensing steam.

Tank Inspection

API-653 details generally accepted engineering practices for storage tank internal and external inspections, including daily walk-around external inspections and periodic internal inspections. Also, periodic inspections of tank roofs should be completed to verify roof integrity, including checks for pontoon leaks and to ensure the roof drains are not clogged with debris. Any entry onto the tank roof should be done under controlled conditions, including atmospheric gas testing for the presence of hydrocarbons and toxic material such as Hydrogen Sulfide (H_2S).

The tank bottom-to-shell joint (sometimes called the chime) is a high-stress joint and warrants rigorous inspection and follow-up. This joint should never be covered by soil, gravel, or sand. Any indication of corrosion or a leak occurring at the chime may indicate a serious condition and warrant follow-up by management and engineering.

Sampling and Water Drawing

Many incidents have occurred from the relatively simple task of sampling or drawing water from the bottom of a hydrocarbon storage tank. A cardinal rule for sampling and drawing water is to ensure that the operation is continuously monitored to prevent the accidental release of the product. The sample or water valve should never be left unattended when sampling or drawing water from a tank or other vessel containing hydrocarbons. Also, when sampling from the top of the tank (from the gauging hatch), precautions should be taken to ensure proper bonding of the sampling apparatus to the tank. This should be done before the sampling or gauging apparatus is lowered into the vapor space. This helps to prevent an arc from a static electrical charge into the tank vapor space. See Chapter 8.22 for more details on the hazards of static electricity.

Monitoring Tank Levels

Not too many years ago, tank levels were monitored by the Operator going to the tanks and "pulling the bob." This was a manual tank-level float with a direct reading gauge on the side of the tank. Some facilities still use this method for high-flash products such as lubricants.

DCS control and today's technology allow tank monitoring remotely and automatically. The process computer can continuously monitor tank levels and rate of change alarms set for any unanticipated changes or movements. For example, a failed roof drain was detected remotely when the change alarm on the process computer prevented a significant product spill. The rate of change alarms has also identified tank manifold line-up errors, preventing product contamination or tank overfills due to product gravitation.

Even with the latest technology, the tank levels for "moving" tanks must be continuously monitored. This can be done by the Operator or Controller printing out the gauge sheets hourly and reviewing the tank gauges to ensure the tanks are moving as expected. This approach can easily detect a "hung" gauge that has stopped moving as expected.

The typical liquid level indicator includes an independent alarm limit switch (ALS) with at least two normally closed switch outputs and provides a local High-High alarm (audible and visible). The design should ensure that the level indicator measures the tank's liquid level and is not attached to the roof. If attached to the roof, a false indication may be present if the roof either hangs up or sinks.

Some newer designs may use radar-based technology but should ensure the independent level measurement for tank level and High-High alarms, that is, the inputs for alarms should be separate from inputs for measuring the tank level.

Landing a Floating Roof onto Its Support Legs

To prevent damage to storage tanks with the floating roof design, they are equipped with support legs designed to support the roof when the roof is at a low level. These roof vents have two positions (normal and maintenance positions). The maintenance position allows for a working room under the roof during tank maintenance when the tank is out of service.

The floating roof is also equipped with a roof vent that opens to break the vacuum on the roof to prevent roof damage when the roof is "landed" on the support legs. When the roof is landed, this vent opens and allows air to enter, breaking the vacuum. Landing the roof should normally be avoided to the extent possible as this allows air and hydrocarbons to mix under the roof, creating the possibility of explosion and fire. Landing the roof also releases hydrocarbon vapors into the atmosphere through the open roof vent. In many states, it is a violation of the state or federal air permit to land the roofs of floating roof tanks for this reason.

Switch Loading and Static Charge

Switch loading is filling a storage tank with a high-flash product after a low-flash product has been pumped from the tank, for example, filling a tank with diesel over a small gasoline heel. In switch loading, during the first couple of minutes of the transfer, the fuel/air mixture is too rich to burn due to the gasoline vapors; however, within the first minutes, some of the flammable vapor is purged from the tank, and air enters with the fill. The tank vapor space goes into the flammable range within the first minutes of the filling operation and stays in the flammable range for the entire product transfer. For this reason, switch loading is extremely hazardous and should be avoided. It has been estimated, for example, that 75% of fires that have occurred at loading racks are the direct result of switch loading.

A static electrical charge can be induced in storage tank operations by the product flowing into the tank, especially during periods of high velocity or turbulence, especially with products with low electrical conductivity, such as jet fuel, gasoline, benzene, toluene, and a wide variety of similar products. This can also occur if the product has entrained water or particulates in the product. This is generally an issue during a tank-filling operation, where the tank level is low, and the floating roof is not floating (the tank has a vapor space under the roof).

API RP 2003 contains protective measures to control the generation of electric charge in a storage tank, including limiting the velocity of the liquid in the fill line until the fill line inside the tank is covered to a specific depth, generally about 2 feet. Procedures should be in place and rigorously enforced to minimize the initial fill rates until the fill line is covered by 2 feet, or in the case of floating roof tanks, until the roof is floating.

This same procedure should be in place for tanks with an internal floating roof; the recommended practice recommends that the velocity of liquid in the fill piping and the discharge velocity of the incoming liquid stream be limited until the internal floating roof is floating and the vapor space under the roof has been eliminated.

The RP also establishes guidelines for a relaxation period before allowing sampling and gauging storage tanks. Generally, the tank must be still for 30 minutes to allow for the static charge to dissipate before any sampling or gauging is allowed. See Chapter 8.22 on "Hazards of Static Electricity" for additional information.

Tank Containment Valve

Each tank should have an operable tank containment valve and an isolation valve located at the tank on each product line. This valve should isolate the tank in case of a pipe or manifold leak. Some sites operate with the tank containment valves normally closed except when the tank is used. This practice aims to prevent accidental flow into or out of the tank and help avoid product contamination. However, always ensure protection against thermal expansion when isolating with tank containment valves closed.

Tank berms should be designed to contain the tank's contents in the event of a spill and maintained in good condition. Berms should also be kept clear of debris.

What Has Happened/What Can Happen

ConocoPhillips Glenpool Pipeline Terminal
Glenpool, Oklahoma
7 April 2003

An 80,000-barrel capacity storage tank exploded and burned at the ConocoPhillips Company's Glenpool South tank farm in Glenpool, Oklahoma, on 7 April 2003. This incident was investigated by the US National Transportation Safety Board (NTSB), which discovered that the floating roof tank exploded and burned while being filled with diesel at a high initial fill rate and with the roof still on the legs with a vapor space under the roof. The tank was also being switch-loaded, being filled with a combustible product, in this case, diesel, over a flammable heel, in this case, gasoline.

The resulting fire burned for about 21 hours and damaged two other storage tanks in the area at a cost of about $2.5M. There were no injuries or fatalities; however, nearby residents were evacuated, and nearby schools were closed for two days. See Figure 8.8.15 for photos of the resulting rank fire and Figure 8.8.16 for a photo of the destroyed storage tank.

Key Lessons Learned

This incident violated several key precautions discussed in Chapter 8.22, "Controlling Static Electricity." Certainly, filling at high fill rates or high velocities during the initial fill generates a lot of turbulence, leading to a lot of static electricity being generated. Also, filling with high fill rates before the floating roof is floating makes available space under the roof for vapors to accumulate rapidly. Switch loading produces the flammable atmosphere needed to have an internal explosion inside the tank and under the roof. All these practices were not followed in this case and contributed to the destroyed tank and cargo and some community disruption.

These issues should be developed into written procedures, and those procedures must be rigorously followed for each tank-filling operation each time a tank is being filled.

Figure 8.8.15 Photos of the ConocoPhillips Glenpool Storage Terminal tank during the firefighting effort. Source: Photo courtesy of the US National Transportation Safety Board (NTSB).

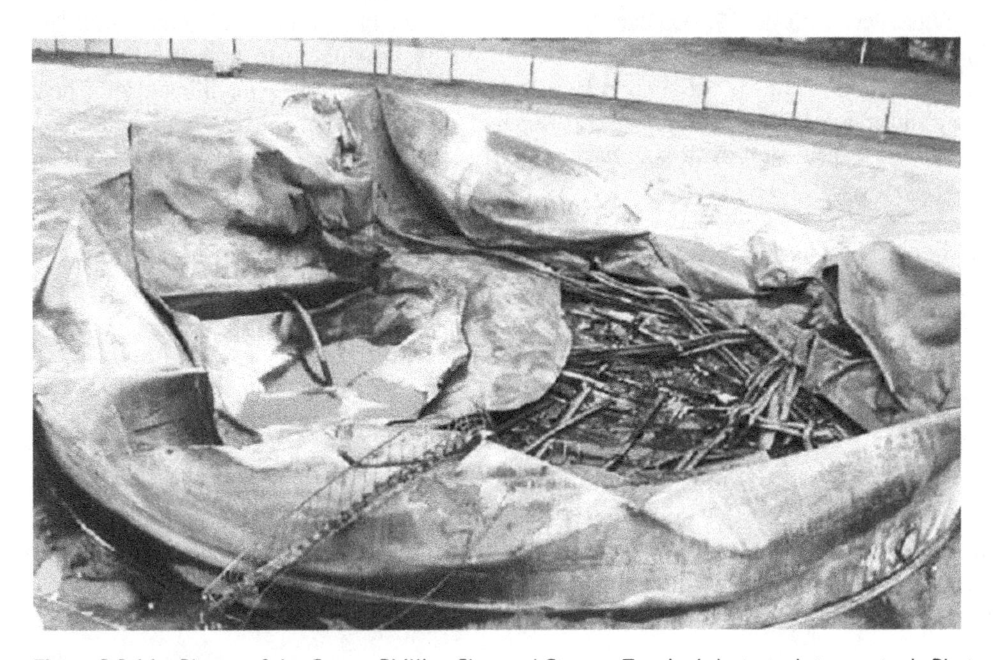

Figure 8.8.16 Photos of the ConocoPhillips Glenpool Storage Terminal destroyed storage tank. Photo courtesy of the US National Transportation Safety Board (NTSB).

What Has Happened/What Can Happen

REPSOL Refinery, Puertollano, Spain
Large Vapor Cloud Explosion (nine Fatalities)
14 August 2003

We briefly discussed this in Chapter 8.7, the "Maintaining Safe Product Rundown Control" chapter. We will add more on this incident in this chapter since the rundown significantly affected the tank farm operations. This very large refinery explosion and fire involved eight storage tanks and several process units, essentially destroying the crude unit. The incident occurred during an FCC Unit startup when the operator's attention was focused on the maintenance of the catalyst circulation. While focused on the catalyst circulation, the Operators missed a critical alarm on the Debutanizer bottom stream, indicating low Debutanizer tower temperature.

The low Debutanizer tower temperature allowed the light ends from the overhead system to drop into the bottoms and go out with the bottom's product to a floating roof storage tank. The unstabilized naphtha was released as a vapor cloud from the seal of the floating roof tank and was ignited by a passing contractor's vehicle, resulting in a devasting explosion. A "wave" of burning naphtha from one of the off-site storage tanks flowed into the atmospheric crude unit, resulting in catastrophic damage to the crude unit. The explosion also resulted in seven additional tank fires (eight total). It required two days to extinguish the fires.

Nine people were killed in this devastating incident, which paralyzed the refinery production.

"The alarm system, gauges, and other unit-panel warnings were not attended...," Repsol said in a statement that concluded weeks of investigation into the accident, which killed nine people and paralyzed refining and chemicals production at the sprawling site, located 250 kilometers (about 156 miles) south of Madrid.

What Has Happened/What Can Happen

TEPPCO Gasoline Terminal
Garner, Arkansas (three Fatalities)
12 May 2009

We discussed the TEPPCO terminal incident earlier; as a reminder, at this site, contractors were preparing to install a gauging pole containing a gauge float device inside a 57,000 Barrel, 100-foot diameter, 49-foot-tall floating roof tank. The tank design included an internal floating roof and a geodesic dome cover. The tank was out of service, with all products removed, and the internal floating roof was resting on the support legs. The three contractors were planning to cut a hole using an acetylene torch in the top and bottom plates of the internal floating roof to accommodate the new gauge pole. Refer to Figure 8.8.13 for an image of this tragic incident provided courtesy of the US Chemical Safety and Hazard Investigation Board.

A confined space entry and hot work permit were issued, with a valid gas test, at the beginning of the shift at 7:00 a.m. However, there was no indication that a second gas test was completed or that the contractors had struck an arc before the explosion at 2:30 p.m. When they started the hot work, the tank exploded, hurling the internal floating roof and the geodesic dome cover completely off the tank. All three of the contractors were killed instantly.

The following was extracted from the OSHA report on the incident:

> "C & C Welding was contracted by TEPPCO, INC to install a remote fuel level gauging rod in tank 1303, a 57,000-barrel capacity gasoline storage tank, located at the McRae, Arkansas terminal. The tank was equipped with two roofing systems; an interior, hollow "floating" roof was approximately 32 inches thick, 100 feet in diameter, and was resting upon 6-foot-tall legs at the bottom of the 49-foot-high tank. The exterior roof, a geodesic dome, consisted of an aluminum skeleton paneled with Plexiglas polygons. The task was to cut an 18-inch-diameter hole through the floating roof's top and bottom surfaces. Shortly after the workers entered the tank to begin this process, the tank exploded, ejecting the top half of the roof and the three workers up and through the geodesic dome to the exterior grounds surrounding the tank".

One of the key lessons learned in this incident is that work atmospheres can change rapidly; gas monitoring needs to be conducted immediately prior to and during hot work activities to ensure that workers are constantly aware of the potential development of an explosive atmosphere. This incident also highlights the importance of continuous gas monitoring. The atmosphere inside a storage tank can change dramatically during the heat of the day as the temperature change can drive hydrocarbon vapors out of residual sludge or scale, creating a flammable atmosphere within the tank.

The US Chemical Safety and Hazard Investigation Board safety spotlight, highlighting hazards of conducting hot work in and near storage tanks, reported, "The torch-cutting activity most likely ignited flammable vapor within the tank." This incident highlights the value of continuous gas monitoring when conducting hot work inside storage tanks and other similar types of hot work. Continuous gas monitoring or ensuring a gas test is carried out immediately before conducting the hot work may have prevented this incident.

What Has Happened/What Can Happen

Thai Oil Refinery
Sriracha, Thailand (8 Fatalities, 13 Injuries)
2 December 1999

A 115 feet (35 meter) diameter floating roof gasoline tank was overfilled, and the resulting spill was ignited by the off-site operator pick-up truck that was responding to the overfill. The Operator reportedly drove into the vapor cloud and was the ignition source. The remains of the truck were located very near the overfilled tank. The spreading fire quickly involved an additional four tanks.

The fire destroyed the nearby fire truck shed, three fire trucks, and foam storage. There were eight fatalities (two off-site operators, two Firemen, three Security Guards, and one other person) and fourteen injured. The fire burned for thirty-five hours, and damage was estimated to be about $22.3M. However, as we say so often, the real tragedy is that eight lives were lost, and many people were affected forever by this incident.

The fire occurred when a storage tank was overfilled, destroying five gasoline storage tanks and 250,000 barrels of gasoline. Even though high-level alarms sounded, the tank continued to be filled due to the opening of the wrong valve. Firefighting foam was flown in from Singapore after local stocks were exhausted. The personnel toll would have been much higher had the explosion not occurred when it did, at 11:30 p.m.

Thai Oil Company was blending product onsite when an operator manually opened a valve to fill a tank already filled with product. The tank quickly began to overfill, resulting in the loss of containment. The rising liquid level set off two safety alarms at an off-site control room, but the control room operator did not hear the alarms. See Figure 8.8.17 for a photo of the Thai Oil Company Off-site Fire and Figure 8.8.18 for an image truck used by the Operator who was responding to the overfilling tank. The Operator was one of the fatalities associated with this incident.

This incident highlights the importance of Operator training. Of course, we never want to have a storage tank overfilled. However, should this happen, the Operator should *NEVER* drive a vehicle (an ignition source) toward an overfilling storage tank. Operators should be trained to stay clear of the tank, stop traffic, and isolate the tank remotely by stopping product flow to the tank (block the valves remotely or stop the pumps or product transfer remotely).

Key Lessons Learned

Tank work involving cleaning, gas freeing, internal and external hot work, dismantling, and rebuilding is all high-hazard work. Work must conform to tank cleaning and maintenance procedures and practices. The proximity to other hot work or ignition sources must be tightly controlled, especially when significant vapors may be present, for example, during the

Figure 8.8.17 Photo of Thai Oil Company offsite fire (8 fatalities, 13 injured). Source: Image with the kind courtesy of the American Institute of Chemical Engineers (AICHE), The Center for Chemical Process Safety (CCPS), and the Beacon. http://www.aiche.org/CCPS/Publications/Beacon/index.aspx.

Figure 8.8.18 Thai Oil operator's pickup truck (following the explosion and fire). Source: Image with the kind courtesy of the American Institute of Chemical Engineers (AICHE), The Center for Chemical Process Safety (CCPS), and the Beacon. http://www.aiche.org/CCPS/Publications/Beacon/index.aspx.

tank cleaning operation. Static control procedures must also be in place and rigorously followed. Ensure that continuous gas monitoring is included in the work plan and that the equipment is in place and used during tank maintenance. Remember, continuous gas monitoring saves lives.

Ensure that there are layers of protection in place to help prevent the overfill of a storage tank. Tanks in flammable or toxic service should normally be equipped with a standard remote reading level gauge backed by a completely independent high-level alarm. Technology is evolving to include a third shutdown or diversion device in the event of a high level in the tank. The independent high-level alarm and the shutdown device (when installed) should both be categorized as SHE critical devices and subjected to periodic testing to ensure functionality. Ensure tank-level instrumentation systems and alarms are properly functioning (especially those classified as SHE critical). The failure of a tank-level instrument must be determined promptly, and actions must be taken to properly return the device to full online service.

Deviations from rundown limits should be immediately reported to the process area receiving the stream and to the Shift Superintendent.

Unit rundown streams must be in continuous compliance with the operating limits established for the streams, including the type of tank the stream is routed to. This includes limits for high temperature, high vapor pressure, water, or other contaminants. For example, sending wet streams into hot >212 °F (100 °C) tanks, hot material into cold tanks containing wet products, and light material to atmospheric tanks. Other examples include sending hot oil into a cold tank that has a water bottom that can result in overheating the water, resulting in a boilover, or sending light products to a floating roof tank not designed to contain products over 13 psia TVP. Either can result in loss of containment to the atmosphere, a vapor release, and a vapor cloud explosion.

Ensure that cone roof storage tanks are continuously protected against overpressure or vacuum and that the pressure vacuum vent is not fouled with sludge, wax, or insects. The P/V vent to the atmosphere should normally not be disabled for any reason other than periodic maintenance, but if it is, ensure that alternative protection is in place to prevent tank overpressure or vacuum.

In wet climates, where rain is frequent, periodically verify that the center roof drains are free and clear and can drain rainwater from the roof. In more arid climates, where the roof drains are maintained closed, be sure to open the drains following a rain event to drain the rainwater from the roof. Otherwise, water can enter the pontoons, resulting in a sunken floating roof.

Train Operators that in the event of a tank overfill, they should never enter the vapor cloud and never drive a vehicle toward an overfilling storage tank. Evacuate the area and isolate the flow remotely by cutting power or stopping the transfer.

Ensure that ALL work around or near storage tanks is considered high-hazard work, that plans are developed in advance, and that potential hazards are identified, addressed, or mitigated before work begins. This includes all hot work; ensure that hot work such as welding or cutting with a torch near storage tanks begins immediately following a valid gas test or confirm that continuous gas monitoring is in place before initiating hot work inside storage tanks or any confined space. This is also true for using any motorized equipment near storage tanks. All motorized equipment should be permitted by a valid work permit and only after all sewers are properly sealed, and a valid gas test has been completed, generally within two hours of work beginning or with continuous gas monitoring.

What Has Happened/What Can Happen

The National Fire Protection Association report indicates that fires at outside storage tanks have decreased markedly over the past three decades. In 2011, an estimated 275 reported fires involving storage tanks, a 76% decrease from 1,142 estimated fires in 1980. However, they still occur with some regularity in the industry, indicating that we must continue to emphasize reducing or eliminating these potential hazards.

The following is a Listing of Recent Incidents Involving Storage Tanks (examples only; not intended to be a complete list):

Date	Location	Company	Details or notes
10/15/2019	Rodeo, California	NuStar Energy Co.	Two Ethanol tanks were destroyed. The cause is still undetermined; there was an earthquake in the area 15 hours earlier.
3/18/2019	Deer Park, Texas	Intercontinental Terminals Co.	Eleven storage tanks burned; the preliminary report indicates release from an open valve started the fire.
1/11/2019	Aden, Yemen	Aden Refinery	Civil defense forces in Yemen's port city of Aden fought to extinguish a fire at an oil refinery on Friday, sparked hours earlier by an explosion, the cause of which was still unknown, the refinery company said. The news article details rebel action in the surrounding area at the time of the incident.
7/5/2018	Terengganu, Malaysia	Kemaman Bitumen Company	Crude oil tank farm fire affecting three storage tanks. One tank was destroyed (one injury).
4/26/2018	Superior, Wisconsin	Husky Energy Co.	Shrapnel from the FCC explosion struck an asphalt tank, resulting in a major refinery fire. The cause of the initial explosion was an unintended process flow on FCC, which occurred during the planned shutdown (26 injuries and large-scale evacuation of the adjacent community).
3/20/2018	Singapore	Tankstore Pulau Busing terminal	Fuel oil storage tank fire.
11/3/2017	Selangor, Malaysia	Kuala Garing Factory	A diesel storage tank exploded (three fatalities).
10/17/2017	Glenville, New York	Mohawk Asphalt Emulsions	Tank fire containing kerosene cleaning solution (two fatalities).
10/6/2017	Butcher Island, India	Port of Mumbai	Diesel tank fire.
6/14/2017	Oaxaca, Mexico	Salina Cruz Refinery	Crude oil tank fire (one fatality, nine injuries).
12/25/2016	Haifa, Israel	Bazan Oil Refinery	A benzene storage tank burned.
8/18/2016	Puerto Sandino, Leon, Nicaragua,	Puma Energy Plant	A fire broke out in a fuel-storage tank following an explosion. A second tank ignited the next day. Puma Energy and Williams Fire and Hazard Control let the fire burn itself out.

Date	Location	Company	Details or notes
4/22/2016	Jingjiang, Jiangsu Province, China	Chemical storage warehouse and storage tanks	A chemical warehouse and several storage tanks exploded and burned (one fatality).
4/20/2016	Jurong Island, Singapore	Jurong Aromatics Corp.	131-foot (40 meter) diameter storage tank fire.
4/4/2016	Apia, Samoa	Matautu Wharf	A fuel-storage tank exploded during maintenance; thousands were evacuated from the surrounding area (one fatality, one injury).
2/11/2016	Karachi, Pakistan	Keamari Oil Terminal	Chemical tank explosion and fire (two fatalities).
7/14/2015	Berre-l'Etang, France	LyondellBasell Petrochemical Plant	One gasoline tank and one naphtha tank burned along with a portion of the petrochemical plant, reported being the act of a saboteur.
6/27/2015	Aden, Yemen	Aden Refinery	Two fuel storage burned following an attack by Houthis rebels, an act of war (one fatality).
4/2/2015	Port of Santos (Brazil)	Ultracargo Fuel Storage Facility	The fire spread to four storage tanks, and the surrounding area was evacuated.
12/26/2014	Ras Lanuf, Libya	Port of Es Sider	A pro-Islamist coalition fired rockets at Libya's largest oil port, igniting an oil storage tank which spread to seven tank fires. All seven tanks were destroyed in a fire that lasted more than a week (an act of war; 19 fatalities).
11/19/2014	La Crosse, Wisconsin	Midwest Fuel Plant	A malfunctioning gauge led to a storage tank containing an asphalt-diesel fuel mixture.
9/27/2014	Sicily, Italy	Milazzo Oil Refinery	A fire broke out during nighttime maintenance work. The fire burned for 14 hours until the product had been consumed.
5/9/2014	Mohammedia, Morocco	SAMIR Refinery	During maintenance operations, a gas release occurred, resulting in two storage tank fires (one fatality and two injuries).
3/22/2014	Mendoza, Argentina	YPF Cerro Divisadero Plant	Large tank farm fire involving six storage tanks. Operations at this site were halted as a result of the fire.
9/12/2013	Lamezia, Italy	Storage facility owned by fuel producer Ilsap Biopro	Explosion at oil storage facility during maintenance (three fatalities).
8/11/2013	Anzoátegui, Venezuela	Puerto La Cruz Refinery	Lightning hit a storage tank, igniting a fire that forced authorities to evacuate the surrounding area.
1/5/2013	Ichapur, India	IOC Hazira Terminal	A fire broke out in a storage tank at this Indian Oil Corporation terminal and quickly spread to a second petroleum tank (three fatalities).
8/25/2012	Punto Fijo, Venezuela	Amuay Refinery	Due to a leak on a propane pump, a ferocious fire ravaged Venezuela's biggest refinery. Eleven storage tanks were destroyed or damaged along with much of the refinery infrastructure (48 fatalities, at least 151 injuries, and substantial damage in the adjacent community)
7/26/2012	South China Sea island of Labuan	Petronas terminal	Explosion on Malaysia oil tanker (five fatalities).
5/5/2012	Map Ta Phut, Thailand	BST Elastomers Factory	An explosion occurred while cleaning a storage tank with Toluene. (12 fatalities, 129 injured, and thousands evacuated from the surrounding communities.)
4/14/2012	Suez, Egypt	Nasr Oil Company	An explosion and fire raged through one of Egypt's biggest refineries, and two storage tanks were destroyed (two fatalities).
9/28/2011	Singapore	Shell Refinery	Gasoline blending manifold fire during preparation for maintenance due to static electricity.

(Continued)

Date	Location	Company	Details or notes
6/2/2011	Pembroke, Wales	Chevron Pembroke Refinery	Storage tank fire during a maintenance operation. Two other tanks were damaged during the fire (four fatalities).
5/31/2011	Gibraltar	Algeciras Bay Fuel Depot	Welders performing maintenance work in an empty diesel storage tank ignited fuel residues. The resulting fire spread to a second diesel tank destroying both tanks. Fuel also spilled into Algeciras Bay, contaminating several beaches (15 injured, 2 critically).
3/11/2011	Sendai, Japan	JX Nippon Refinery	An earthquake and tsunami devastated this facility resulting in a fire that destroyed a gasoline tank, asphalt tanks, molten-sulfur tanks, and the oil-shipping facility.
3/11/2011	Chiba, Japan	Cosmo Refinery	Multiple spheres BLEVE'd due to loss of containment caused by an earthquake
9/8/2010	Bonaire, Dutch Antilles	BOPEC Terminal	Two tanks caught fire during an electrical storm. Firefighters battled the blaze for three days but could not extinguish it until the tanks burned out. Dutch investigators later reported that the plant's firefighting system was not working due to improper maintenance.
6/13/2010	Greensboro, North Carolina	Colonial Pipelines	A gasoline tank was struck by lightning and burned for six hours in spite of efforts by firefighters. The fire eventually burned itself out.
10/29/2009	Mohanpura, India	IOC Terminal Jaipur	Storage tank overfilled while transferring gasoline from the depot to a pipeline; 12 Tanks were reported destroyed (12 fatalities, at least 300 injured).
10/22/2009	Catano, Puerto Rico	Caribbean Refining Co.	Storage tank overfilled while receiving gasoline from a ship; 20 Tanks were reported destroyed, and another 12 tanks were damaged. The fire raged for 66 hours until no more material was left to burn.
5/12/2009	Garner, Arkansas	TEPPCO Gasoline Terminal	Gasoline tank explosion during maintenance (three fatalities).
8/19/2008	Ras Lanuf, Libya	Ras Lanuf Oil Complex	Hot work ignited a crude oil storage tank during routine maintenance operations. About 1,000 firefighters managed to contain the blaze to a single tank but could not extinguish it for more than nine days.
8/17/2008	Johor, Malaysia	Tanjung Langsat Port	A fire started in a gasoline storage tank and spread to an adjacent tank containing naphthalene the next day. Both tanks were destroyed.
2/18/2008	Big Spring, Texas	Alon USA Refinery	Propylene vapors were released from a storage tank and ignited, resulting in an explosion. Firefighters controlled the blaze about an hour after it started, but the blast severely damaged the facility, homes, and businesses in the surrounding area (five injuries).
12/6/2007	Durban, South Africa	Engen Refinery	Lighting hit a tank containing refined petrol (gasoline), causing the floating roof and seals to collapse into the tank. The tank was heavily damaged, and 60% of the product was lost.
4/27/2007	Wynnewood, Oklahoma	Gary-Williams Refinery	Lightning struck a naphtha storage tank resulting in a fire that spread to a diesel tank later that evening. Both tanks were destroyed.
12/11/2005	Buncefield, England	Buncefield Terminal	Gasoline tank overfill; the resulting fire spread to 20 storage tanks. One hundred eighty firefighters fought the fire for four days (244 people required medical treatment). Damages were estimated at GBP 1 billion (USD 1.5 billion). The terminal has not been rebuilt.
1/28/2004	New South Wales, Australia	Port Kembla Industrial Site	Hot work during maintenance resulted in an explosion in a tank containing ethanol. The blast blew the top off the tank, and it took firefighters 20 hours to extinguish the resulting blaze.

Date	Location	Company	Details or notes
1/20/2004	Gresik, Indonesia	Petrowidada Chemical Plant	An overheated machine touched off a massive explosion and a fire that engulfed at least two chemical tanks. Authorities evacuated some 250 people from the surrounding area (2 fatalities, 68 injuries).
5/3/2003	Gdansk, Poland	Gdansk Oil Refinery	An explosion triggered this full surface fire in a gasoline storage tank. The blast blew the cone lid off the tank and caused the internal roof to sink (three fatalities).
3/8/2003	Guwahati, India	Digboi Refinery	A terrorist fired a mortar into a petroleum storage tank, causing a large tank fire. Firefighters prevented the fire from spreading to other adjacent storage tanks.
12/8/2002	Hagatna, Guam	Mobil Oil Terminal Cabras Island	A buildup of static electricity sparked this fire as Super Typhoon Pongsona pounded the island. The flames persisted for six days; three storage tanks were destroyed, prompting officials to ban gasoline sales to the public.
11/25/2002	Mohammedia, Morocco	SAMIR Refinery	Heavy rains triggered a flood that engulfed the Mohammedia refinery resulting in waste oil floating on the water's surface and igniting after contacting the hot refinery equipment. This resulted in a massive blaze (two fatalities, three more reported as missing).
5/5/2002	Malopolska, Poland	Trzebinia Refinery	A lightning bolt triggered this full surface fire at a 99-foot (30 meter) diameter tank with an internal floating roof. A semi-fixed extinguishment system began applying the foam, but the process had to be terminated immediately due to system damage. It took 362 firefighters five hours to put out the flames by blanketing them with 109.5 tons of foam, using 13 foam monitors supplied by 35 vehicles.
7/10/2001	Granite City, Illinois	Petroleum Fuel and Terminal Oil Co.	Disaster struck this facility twice during the summer of 2001. An unexplained fire burned 400,000 gallons of asphalt fuel in an asphalt storage tank. A month later, a second fire started as workers tried to warm up the remaining asphalt to transfer it into a new tank. This second fire consumed 378,000 gallons of asphalt.
6/7/2001	Norco, Louisiana	Orion Refinery	Lightning ignited this blaze, which holds the Guinness World Record for extinguishing the largest tank fire. After 12 hours of preparation, firefighters defeated the flames in 65 minutes by continuously spraying foam while pumping residual gasoline out of the 270-foot (82 meter) diameter tank.

Additional References

American Petroleum Institute (API)

- RP 2003 "Protection Against Ignitions Arising out of Static, Lightning, and Stray Currents."
- API Std 653 "Tank Inspection, Repair, Alteration, and Reconstruction."

National Fire Protection Association, NFPA 30 "Flammable and Combustible Liquids Code", (Basic Requirements for Storage Tanks).

US Chemical Safety and Hazard Investigation Board, Safety Bulletin, "Seven Key Lessons to Prevent Worker Deaths During Hot Work In and Around Tanks", February 2010. https://www.csb.gov/seven-key-lessons-to-prevent-worker-deaths-during-hot-work-in-and-around-tanks/

US Chemical Safety and Hazard Investigation Board, Final Investigation Report Caribbean Petroleum Tank Terminal Explosion and Multiple Tank Fires, Report No. 2010.02.I.PR. https://www.csb.gov/caribbean-petroleum-refining-tank-explosion-and-fire/

UK Health and Safety Executive, "Explosion and Fire: Chevron Pembroke Refinery, 2 June 2011". https://www.hse.gov.uk/comah/chevron-refinery.htm

UK Health and Safety Executive, "Buncefield response - Reports and recommendations arising from the Competent Authority's response to the Buncefield incident". https://www.hse.gov.uk/comah/buncefield/index.htm

The 100 Largest Losses 1972–2001. Marsh and McLennan, Large Property Damage Losses in the Hydrocarbon-Chemical Industries. (See page 4 – Report on the Thai Oil Explosion, December 2, 1999, Sriracha, Thailand).

Incidents that define process safety. (See page 111 – Report on the Repsol Incident, 14 August 2003, Puertollano, Spain), John Atherton, Frederic Gil, Wiley Inter-Science.

Press Release by the Industry Council of Castilla La Mancha "Puertollano accident investigation blames Repsol" (Report on the Repsol Incident, August 14, 2003, Puertollano, Spain). https://www.thinkspain.com/news-articles/puertollano#p:/news-spain/1719/puertollano-accident-investigation-blames-repsol

Investigation Committee's Report on the Puertollano Refinery Accident (Issued by the Industry Council of Castilla La Mancha) (Report on the Repsol Incident, August 14, 2003, Puertollano, Spain). https://www.sec.gov/Archives/edgar/data/847838/000112528203005356/b327193_6k.htm

Pressurized Instant Foam website (incident listing). https://pifoam.ch/#incidents

American Petroleum Institute (API), RP 2003 "Protection Against Ignitions Arising out of Static, Lightning, and Stray Currents".

National Transportation Safety Board (NTSB), Investigation Report: "Storage Tank Explosion and Fire in Glenpool, Oklahoma 7 April 2003".

AP Archive "Thailand; Explosion at Oil Refinery". http://www.aparchive.com/metadata/youtube/f03b9fa65209871ff2f2cd629f5db23d

Compilation of Pressure-Related Incident Summaries. US Department of Energy, Argonne National Laboratory. https://indico.fnal.gov/event/13755/sessions/1912/attachments/14044/17881/AccidentsCompilation_June16.pdf

Tank Fires, Review of fire incidents 1951–2003, BRANDFORSK Project 513-021. http://www.diva-portal.org/smash/get/diva2:962266/FULLTEXT01.pdf

Chapter 8.8. Storage Tank Safety

End of Chapter Quiz

1. Name the different types of storage tanks discussed in this chapter.
2. Do the safe operations guidelines for storage tank safety apply only to tank field personnel?
3. What type of products can be safely stored in external floating roof tanks?
4. What is the primary reason that a cone roof tank would be converted to an internal floating roof?
5. What services are cone roof storage tanks more typically used for?
6. What are some of the limitations or restrictions associated with cone roof tanks?
7. Can a cone roof tank be used in hot services such as asphalt or hot residuum?
8. What are spheres or horizontal storage vessels, sometimes referred to as bullets, used for?
9. What is the primary hazard when working on or around the spheres or bullets?

10 What are some of the safety features associated with spheres or pressurized horizontal storage vessels? Name all the safety features that you can.

11 What are some of the hazards associated with storage tanks? Name as many as you can.

12 Can you give at least one example of a site that appears to have lost its "license to operate" due to a significant tank overfill and the resulting explosion and fire?

13 To prevent overfill, what kind of device are tanks in low flash service, such as naphtha and gasoline, equipped with?

14 What are two potential causes of a storage tank boilover and devastating fire?

15 What are the immediate response steps an Operator should take in the event of a tank overfill?

16 What are the potential consequences if the tower internal temperature on a Debutanizer is allowed to go low due to mechanical failure or Operator error?

17 For storage tanks in flammable service and during normal operations, tank procedure generally avoids pumping the level down to the point where the roof is "landed" on the roof support legs. What is the safety concern with landing on the roof?

18 Why do we always say that ALL work around or near storage tanks should be considered high-hazard work?

8.9

Injection and Mixing Point Failures

In petroleum refining and petrochemicals processes, internal corrosion and unexpected rupture of piping circuits downstream of injection points or mixing points have been an issue that has not been all that uncommon. Some of these incidents have been catastrophic as they occur unexpectedly, and the resulting rupture results in a large loss of containment of the process, sometimes releasing volatile or toxic materials into the atmosphere, causing a large explosion.

Several failures have occurred at the first piping ell downstream of the injection or mix point or at other similar locations where the flow suddenly changes direction or turbulence exists. Other piping failures have occurred due to the chemical continuing to flow, even after the main process flow was either stopped or flowing at reduced rates, resulting in increased chemical concentration.

Injection Point Failures

Injection point failures are particularly elusive due to internal corrosion or erosion. In most piping circuits with a constant flow, iron sulfide deposits can form over time on the pipe internals, a passivation layer protecting the pipe internals from further corrosion or erosion. However, the resulting turbulence caused by the water or chemical injection can remove the passivation layer, causing continued corrosion or erosion and leading to the pipe rupture. The turbulence can be very severe in small, isolated areas; therefore, the corrosion is localized and difficult to detect unless the Inspector is looking in the right place.

This chapter provides information on these types of failures and examples of incidents where these occurred. We will also examine precautions and mitigation to help prevent this from happening in your unit or area.

Crude oils typically contain salt compounds and other impurities, which can lead to internal corrosion and plugging of process equipment. These compounds are usually removed by injecting water into the crude streams and subsequently coalescing and removing the impurities in the crude unit desalting equipment. However, small amounts of salts (primarily sodium chlorides) continuously escape the desalter and flow into the process equipment in other downstream units, where they may form salts or combine with other impurities such as ammonia to form salts (sodium or ammonium chlorides), resulting in fouling or plugging of exchangers or other processing equipment. Salts typically go overhead in process equipment and can lead to fouling or plugging of overhead condensers. This fouling or plugging is typically identified by a high-pressure differential across the exchanger or process equipment and almost always results in reduced throughput rates.

An effective way to remove the salting and restore the pressure drop to normal is to inject a small amount of water into the process stream, thereby dissolving the salts. The salts may then flow to a downstream condensate drum, where they may be safely removed from the process. While this effectively dissolves the salt fouling, it can result in significant internal corrosion in piping immediately at or just downstream of the water injection point. These locations have had several catastrophic failures due to internal erosion/corrosion. Generally, these failures have occurred at pipe bends (elbows) or other piping geometry changes, causing the salt-laden stream to change flow direction and impinge on the pipe wall.

Similar failures have also occurred at mixing point failures, including temperature mix points, different phases (vapor and liquid), acid or caustic mix points, and even mixing of different process streams. For example, failures have occurred at temperature mixing points where the temperature may vary, subjecting the adjacent piping or fittings to temperature-induced stresses. These fatigue failures have also resulted in catastrophic containment loss and fires or explosions.

For example, on Hydrocrackers, Reformers, and other similar units where Hydrogen is injected into process feed circuits, it is not uncommon to have a significant temperature differential between the injected gas stream and the process. In this

Process Operations Safety: The What, Why, and How Behind Safe Petrochemical Plant Operations, First Edition. M. Darryl Yoes.

case, relatively small changes in the flow rates of the gas streams can result in significant temperature deviations in the piping at the injection point. These temperature variances can result in temperature-induced thermal stresses that have led to piping fatigue and failure. When these failures occur, they are typically catastrophic, resulting in a full-bore type of failure and occurring totally without warning.

Mixing Point Lessons Learned the Hard Way, Including Design Considerations

Designers attempt to limit the differential temperature between the two streams to a maximum of about 100 °F (38 °C) or design the mix points with an internal sleeve to mitigate the anticipated thermal stresses. For some other mixing points, such as acid mix points, the corrosion or erosion may be mitigated using special alloy piping components. Figure 8.9.1 shows an example of a temperature mix point, and in this case, it is a mixture of hot Hydrogen at about 900 °F (482 °C) mixed with another Hydrogen stream at 100 °C (38 °C). These two streams with a large temperature differential created significant thermal stresses in the mixing tee. Even though the mixing tee was designed with an internal thermal sleeve intended to mitigate the temperature stresses, the tee eventually failed catastrophically, resulting in a large process unit fire and extended downtime. Figure 8.9.2 illustrates how the piping tee failed due to the piping stresses.

When designing a piping mixing or injection point, the designer may specify an inlet quill to be installed in the piping and pointed downstream, which requires a straight run of piping immediately downstream of an injection point or mixing point. For example, pipe designers may require 10 pipe diameters downstream of a mixing point before a change in the flow direction to stabilize the flow and become more laminar before the next piping ell or bend. The quill may require internal bracing to prevent fatigue failure, which could result in flow impingement on the piping internals. These requirements help minimize the amount of internal turbulence directly downstream of the mixing point that can contribute to internal erosion or corrosion.

It is critical that inspection identifies all mixing and injection points, including water injection and chemical injection, such as acids or caustic, temperature, or phase mixing points in the site inspection strategy. These injections and mixing points should be inspected more frequently than standard piping, especially in the downstream areas where the flow changes direction. Inspection should also consider the potential cracking in temperature mix points in their inspection strategy.

During the material selection process for injection points or mixing points, considerations must include local erosion, corrosion/erosion, and corrosion due to high turbulence or rapid changes in the process flow or temperature. The specific chemical or temperature may require piping with an internal liner to be designed to resist the corrosion/erosion of the chemical and operating temperature. Be aware that some internally lined piping systems require a venting system to release accumulated pressure between the liner and the piping system.

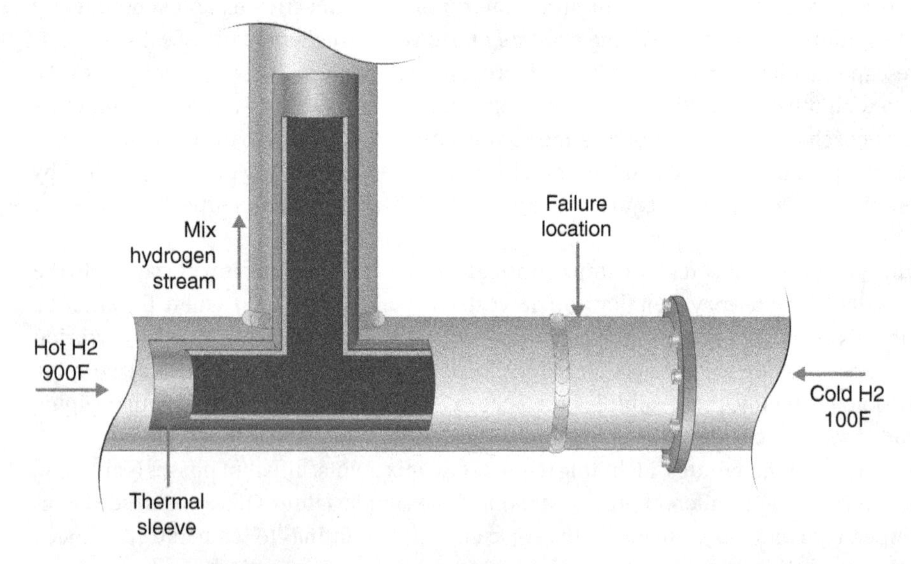

Figure 8.9.1 Hydrogen temperature mix point failure. These images were created to illustrate the hazards of mixing points. Example of failure of a temperature mixing point where 900 °F hydrogen was mixing with 100 °F hydrogen. This resulted in piping stress and the failure of the pipe segment. Source: Image courtesy of Ecoscience Resource Group Visual Artists.

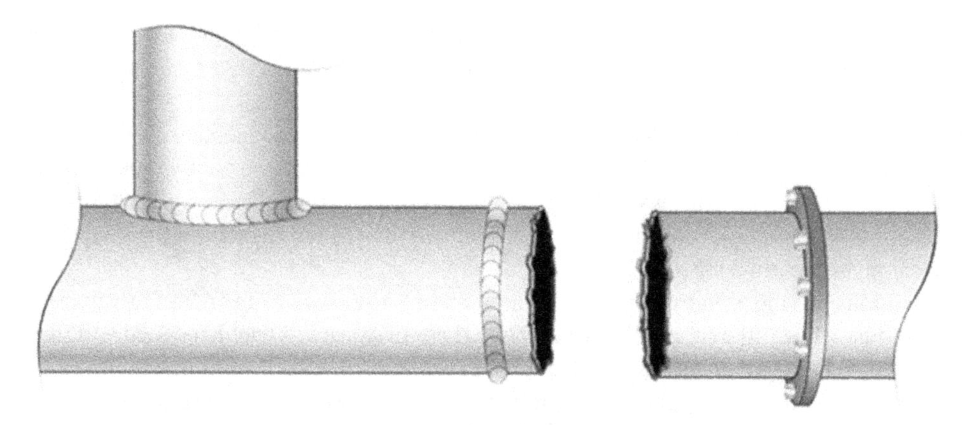

Figure 8.9.2 Hydrogen mix point failure. The fatigue failure occurred at the temperature mix point in the hydrogen line near the heat-affected zone of a weld. The temperature mix point failure resulted in a total loss of containment of hot hydrogen. This pipe segment operates at nearly 1,800 PSI, resulting in a large refinery fire when it failed. The failure occurred even though the mixing point was designed with a thermal sleeve on the piping interior. Source: Image courtesy of Ecoscience Resource Group Visual Artists.

Inspection programs for mixing pints should include water and chemical injection points such as amines, inhibitors, acids, caustics, etc. They should also include process gas mixing points such as H_2 or N_2 injection points and mixing different process streams with significantly different temperatures.

Other areas of concern include locations where chemicals or process additives are introduced into a process stream. Other examples are corrosion inhibitors, desalter additives such as demulsifiers, neutralizers, process antifoulants, oxygen, hydrogen scavengers, caustic, and water washes, and all types require specific design considerations and aggressive inspection programs.

Process Operations Guidance

Facilities managers should ensure that plant personnel are aware of the concerns with mixing points and injection point systems and the associated hazards. For operations with the noncontinuous flow of the injected chemical, the process flow must be started first and allowed to stabilize before starting the injected chemical. When stopping the injected chemical, always ensure to stop the injected chemical before shutting down the main process flow. In both cases, following this procedure prevents a concentrated chemical solution from contacting the host piping, thereby preventing increased corrosion. A checklist-type procedure should be required to help ensure procedure compliance.

Management of Change Requirements

When the installation or a change in an injection point or mixing point is being considered, it is important to implement the change using the Management of Change process. This will help ensure the change is engineered and any potential hazards are identified and mitigated as an integral part of the design and change.

What Has Happened/What Can Happen

Motiva Refinery
Port Arthur, Texas
2 June 2012

Severe pitting and cracking of new stainless-steel piping in the new crude unit.

Fire and unplanned outage to repair damaged Crude Distillation Unit.

A celebration was held by CEOs representing the Royal Dutch Shell and Saudi Aramco companies in Port Arthur, Texas, on 31 May 2012, noting the startup of the major expansion just completed at the Port Arthur Motiva refinery. The celebration was to mark the completion of a long and challenging five-year project that doubled the refinery's daily capacity to more than 600K B/D capacity. However, two days later, it was apparent to refinery Operators that something was wrong. After two fires and a failure in one of the new unit's atmospheric furnaces, it was decided to bring the unit to a safe position to investigate the leaks and failures.

Fresh crude oil contains chlorides which undergo a reaction at high temperatures in the preheat exchangers and furnaces, resulting in corrosive and destructive Hydrochloric Acid (HCL), which can lead to severe corrosion to the crude column overhead piping and equipment. One method to control this is to inject a dilute caustic into the circulating crude oil to react with the hydrochloric acid to form a more stable sodium chloride. The chlorides can then be removed from the bottom of the crude distillation columns, lessening the potential for severe overhead corrosion.

On 22 June 2012, Motiva reportedly confirmed the source of internal contamination to the press: "The preliminary inspection indicates that parts of the new unit have been contaminated with elevated levels of caustic." What happened was that caustic typically used to inject into the circulating crude oil only while the unit was on stream, had continued to flow into the unit while the unit was on slow hot circulation during the outages to repair the leaks. This also increased the caustic concentration, resulting in stress corrosion cracking (SCC) of the unit's miles of brand-new stainless-steel piping.

Caustic concentration and temperature play a huge role in stress cracking. The higher caustic concentration (higher pH) and an increase in temperature play a significant role in the susceptibility of stress cracking. The caustic continued to flow into the crude unit during a low flow period, increasing its concentration. The resulting stress cracking occurred when the unit was restarted, and the temperature returned to the typical very high temperature.

This incident resulted in significant unplanned unit outages and prevented the Motiva refinery from processing crude. It also resulted in expensive repairs to the unit.

Key Lessons Learned

Personnel should be trained on the hazards of chemical injection points and the critical nature of chemical injection. Concentrations of chemicals should be considered, and procedures should be in place to ensure the chemical injection rates are clearly understood and reinforced.

Procedures for chemical injection should also be developed and rigorously enforced. For example, procedures for chemical injection points should be included in all unit startup and shutdown procedures. They should be in a checklist format with signatures indicating when a chemical injection is started and stopped.

Generally, to ensure that concentrated chemicals are not left in the piping and other equipment, the procedures should specify that the process flow should be started first and that the chemical should be started after the process flow has stabilized. When stopping the chemical injection, it should be stopped before the process flow is stopped. The process flow should continue until after the chemical has been displaced from the system. This procedure ensures that the chemical concentration is not above critical limits.

For a critical application such as the case discussed here, a flow instrument with an alarm for the console operator may be appropriate to ensure that the operator is aware that the chemical is flowing, even though the unit is not at normal rates.

If a process piping system may be exposed to a higher concentration of caustic or chlorides, stainless steel, which is more resistant to SCC, may be considered. The austenitic stainless steels are the most susceptible due to their relatively low nickel content, with stainless steels 304/304L and 316/316L being the most susceptible. Austenitic grades with higher nickel and molybdenum, such as alloy 20, 904L, and the 6% molybdenum super austenitic grades, have much-improved chloride SCC qualities. The ferritic grades of stainless steel, such as type 430 and 444, are very resistant to chloride SCC.

What Has Happened/What Can Happen

ConocoPhillips Refinery
Humber, UK
16 April 2001

Injection point failure, loss of containment, and resulting explosion/fire.

A catastrophic loss of containment and resulting explosion occurred on 16 April 2001, at the ConocoPhillips Humber (UK) Refinery. No one was severely injured, but the refinery experienced severe damage to process equipment due to the explosion and resulting fire. Numerous buildings were damaged on-site, and damage extended to the surrounding communities of Immingham and South Killingholme. The incident occurred on holiday (Easter Monday) and at shift change time; therefore, the number of people who would have otherwise been exposed was far less. The Competent Authority investigated this incident in the United Kingdom, and a public report summarizing the investigation results was also prepared and released by the Health and Safety Executive. This summary is based on this report.

The pipe failure occurred downstream of a water injection point on the deethanizer tower overhead system, which is part of the saturated gas plant. The injection point was used to control salting in the Deethanizer overhead condensers and had been formed by connecting the water to an existing bleeder. No injection quill or other dispersion device was used to control or direct the water flow. The failure occurred in a 6″ diameter pipe ell located just 670 millimeter (26 inches) downstream of the water injection point in the outboard sweep of the ell. The six-inch diameter overhead pipe contained flammable gas at high pressure (primarily ethane, propane, and butane). The Health and Safety Executive reported that the piping in the ell had thinned from 7–8 millimeter (0.3 inch) to a minimum of 0.3 millimeter (0.01 inch) in the area of the failure. The piping thinned to the point where it could no longer hold the pressure, and when it ruptured, it resulted in a "full bore" release of the flammable gas into the atmosphere.

Additional piping systems failed following the initial explosion and resulting fire due to the impinging fire, resulting in secondary loss of containment and escalation of the event. The metallurgical examination of the piping following this incident revealed the following:

- The severe corrosion/erosion area was directly associated with the water injection.
- Uncorroded pipe sections were internally coated with black iron sulfide, which provided a "passivation" layer protecting the steel from further corrosion.
- This protective layer had been washed away by the water injection in the area of severe corrosion/erosion.
- This area was left open to attack by corrosive agents in the gas stream.
- The result was the severe thinning of the pipe wall and its eventual failure.

Figures 8.9.3–8.9.4 illustrate the significant damage that occurred due to this explosion. Figure 8.9.5 shows how the water injection was tied into an existing ¾-inch bleeder valve; Figure 8.9.6 illustrates the failed 90 piping ell.

Note the bleeder at the top of the pipe just upstream of the failure where the water injection was connected. Also, notice the complete rupture of the ell just downstream of the water injection.

Figure 8.9.3 ConocoPhillips saturate gas plant following April 2001 explosion. Contains public sector information published by the health and safety executive and licensed under the https://www.nationalarchives.gov.uk/doc/open-government-licence/version/3/. Source: Reproduced by the kind permission of HSE. HSE would like to make it clear that it has not reviewed this product and does not endorse the business activity of Safety Consulting International, LLC.

(a)

(b)

Figure 8.9.4 Figure (a) illustrates the damage to an adjacent unit and Figure (b) shows a close-up of SCP damage. Contains public sector information published by the health and safety executive and licensed under the https://www.nationalarchives.gov.uk/doc/open-government-licence/version/3/. Source: Reproduced by the kind permission of HSE. HSE would like to make it clear that it has not reviewed this product and does not endorse the business activity of Safety Consulting International, LLC.

Figure 8.9.5 Location of water injection point vs. 90 ° pipe ell. Contains public sector information published by the health and safety executive and licensed under the https://www.nationalarchives.gov.uk/doc/open-government-licence/version/3/. Source: Reproduced by the kind permission of HSE. HSE would like to make it clear that it has not reviewed this product and does not endorse the business activity of Safety Consulting International, LLC.

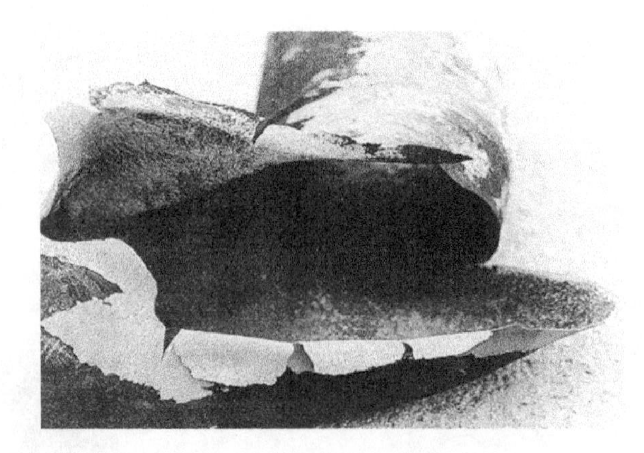

Figure 8.9.6 Rupture of 90 ° piping ell. Contains public sector information published by the health and safety executive and licensed under the https://www.nationalarchives.gov.uk/doc/open-government-licence/version/3/. Source: Reproduced by the kind permission of HSE. HSE would like to make it clear that it has not reviewed this product and does not endorse the business activity of Safety Consulting International, LLC.

What Has Happened/What Can Happen

Shell Refinery
Norco, Louisiana (7 Fatalities, 42 Injuries, and Destruction of an FCC Unit)
5 May 1988

Injection point failure, loss of containment, and resulting explosion/fire.

On 5 May 1988, a similar failure occurred at the Shell Refinery located at Norco, LA., when a piping section immediately downstream of a water injection mixing point failed on an FCCU gas plant. The loss of containment resulted in a major flammable vapor release, leading to a catastrophic fire and explosion, resulting in 7 fatalities and 42 injuries in the blast. Windows were blown out up to 30 miles from the site, and widespread damage was on both sides of the mile-wide Mississippi River. The FCCU main fractionating column was blown completely over onto its side. The FCCU (a primary gasoline-producing unit) was out of service for an extended period due to the blast.

Information is limited in this case, but this was reported to be a water injection point in the overhead system of a depropanizer tower, resulting in the catastrophic failure of the 8″ overhead vapor line. Otherwise, the case appears very similar to the ConocoPhillips incident at Humber (UK). The following paragraphs are from the OHSA report on this incident:

Accident Investigation Summary

At 3:37 a.m. on 5 May 1988, an explosion occurred in an oil and gas refinery's catalytic cracking unit (CCU). The explosion apparently resulted from the corrosion of an 8-inch vapor line. Under 270 psi pressure, this vapor line ran from a 10-inch header that originated as the main overhead vapor line from the depropanizer column. The apparent instantaneous line failure released approximately 17,000 pounds of hydrocarbon vapor for about 30 seconds. A possible ignition source could have been the unit's superheater furnace. The damage pattern indicated that the explosion was probably an aerial explosion with an epicenter between the depropanizer and the CCU control room. Eighteen people were killed in this event.

Employees #1, #2, #4, #5, and #7 were fatally injured inside the CCU control room as a direct result of the blast. Employee #3 was fatally injured approximately 30 feet outside the west side of the CCU control room as a direct result of the explosion. Employee #6 was fatally injured while exiting the GO-1 South control room. The negative pressure wave of the blast apparently created a vacuum that caused the east brick wall of GO-1 South to be "sucked" toward the CCU and then to fall on Employee #6. Employee #8 was on the northwest side of the reactor-regenerator vessels in the CCU and was critically injured as a result of the blast. The rest of the employees, #9 through #26, were in different units in the refinery at the time of the explosion. All received varying injuries to various degrees due to the explosion.

Key Lessons Learned

Mixing points require special attention to identify if any rapid and localized internal corrosion is occurring. The inspection program should ensure that mixing points are identified and that a systematic and comprehensive inspection protocol is in place. Inspections should include piping geometry changes, such as piping elbows downstream of mixing points. Inspection should include the outer sweep of bends or elbows downstream of injection or mixing points.

Inspection programs for mixing pints should include water and chemical injection points such as amines, inhibitors, acids, and caustics. They should also include process gas mixing points such as H_2 or N_2 injection points and mixing different process streams with significantly different temperatures.

Process designers should specify straight piping run between the injection point and the pipe bend or ell downstream of the injection point, typically, about 10 pipe diameters. This reduces the turbulence and allows time for the flow to change from turbulent to laminar before the flow reaches the change in direction. Likewise, designers should typically require injection points to have an injection quill to inject the water or chemical into the center of the process stream and direct the flow downstream. This also helps minimize turbulence, helps avoid pipe wall impingement, and promotes mixing before the stream reaches the downstream piping ell or other similar equipment where turbulence can occur and could otherwise result in internal corrosion or erosion. Quill design should also include flow calculations to help identify potential fatigue due to vibration or other frequency issues. Quills with beveled ends may help improve chemical dispersion and prevent corrosion or fouling.

This clearly illustrates the hazard of simply tying into an existing bleeder as a water or chemical injection point. This can result in simply dumping the injected material into the flow without regard for pipe wall impingement or internal turbulence, which can result in internal corrosion or erosion.

When considering an +, ensure that all the Safety Management Systems are fully used to identify if hazards are being introduced and properly mitigated. For example, Management of Change always drives a Process Hazard Analysis to identify hazards and address those hazards. The Pre-startup Review (PSSR) ensures that hazards identified are mitigated before the revised system is placed in service. See Chapter 6 for more details on Safety Management Systems.

Additional References

American Petroleum Institute (API)

- **API 570.** Piping Inspection Code: In-service Inspection, Rating, Repair, and Alteration of Piping Systems.
- **API RP 571.** Damage Mechanisms Affecting Fixed Equipment in the Refining Industry.

National Association of Corrosion Engineers, NACE 34101: Refinery Injection and Process Mixing Points.

NACE SP0114-2014: Refinery Injection and Process Mix Points.

ASTM F1545 Standard Specification for Plastic-Lined Ferrous Metal Pipe, Fittings, and Flanges.

What is Piping? A Blog for Piping Engineers, Dipak Samanta, Injection Points and Mixing Points for Process Streams.

Occupational Safety and Health Administration (OSHA), Inspection Detail; Report ID: 0625700 Shell May 5, 1988 Explosion. https://www.osha.gov/pls/imis/establishment.inspection_detail?id=100478866

Occupational Safety and Health Administration (OSHA), Inspection: 100478866 – Shell Oil Company, (OSHA report of an explosion at Shell Oil Refinery, which occurred on 5 May 1988, near New Orleans, LA). (8 Fatalities/18 Injured). https://www.osha.gov/pls/imis/establishment.inspection_detail?id=100478866

UK Health and Safety Executive Public Report, Public Report of: The Fire and Explosion at, The ConocoPhillips, Humber Refinery, 16 April 2001. https://www.hse.gov.uk/comah/conocophillips.pdf

Tests for Stress Corrosion, ASM International Documents, Gerhardus H. Koch, CC Technologies Inc., Dublin, Ohio. https://www.asminternational.org/documents/10192/1755223/amp15908p036.pdf/026e7c61-4606-424e-9ade-3455865aba71

Chapter 8.9. Injection and Mixing Point Failures

End of Chapter Quiz

1. From an Inspection perspective, what is the challenge with Injection Points or Mixing points?

2. What are some examples of types of failures that involved or could involve mixing points or injection points? Name all you can.

3. What additional steps may pipe designers consider when they realize that the concentration of caustic or chlorides may be higher than expected?

4. Which piping areas are most vulnerable to severe internal corrosion as a result of mixing points or injection points?

5. How can procedures in a checklist format requiring Operator signoff on each critical step in the procedure help ensure against operator error and/or equipment failure?

6 When it is necessary to start a new water wash to remove salting from an exchanger or condenser or start a new chemical injection to a unit, is it OK to tie the new connection directly into an existing bleeder? Please explain your answer and the considerations related to this question.

7 Where can we find additional information on Mixing Point inspection procedures and failures involving Mixing Points or Injection Points?

8 Temperature mixing points are less susceptible to piping failures than water or chemical injection points.
- True
- False

9 What management process is critical when installing a new chemical injection point on a processing unit?

10 What are just a few issues with connecting to an existing bleeder to start a new chemical injection?

8.10

Hazards of Piping "Dead Legs"

A dead leg is "a section of pipe where there is normally no flow, and which is filled with the stagnant process fluid, and pressurized by the system pressure." Dead Legs may also be referred to as "stagnant zones." Four types of failures have occurred in piping systems containing a dead leg, leading to loss of containment and resulting in catastrophic fires and vapor cloud explosions (VCEs). Dead leg piping systems are subject to (1) severe internal and uniform corrosion and pipe rupture; (2) freezing of free water that has accumulated in the dead leg resulting in pipe rupture; (3) in polyethylene chemical plants, dead legs may rupture from an uncontrolled Butadiene chemical reaction resulting in piping failure due to the growth of "popcorn" polymer causing extreme internal pressure, and (4) severe external corrosion due to corrosion under insulation (CUI) on relative long runs of cold dead leg piping.

Failures in piping dead legs have resulted in sudden, unexpected, and catastrophic releases of flammable vapor and resulted in fires and explosions. A piping dead leg can occur in piping systems due to piping modifications or "jump overs," which leave a section of process piping with no flow or little flow, but still connected to and pressurized by the process. Some piping dead legs result from piping geometry or may be created by the control valve, exchanger, or relief valve bypasses or by a dead-ended extension of a pipe section used for piping support. Some unit startup and shutdown lines may also meet the dead-leg criteria as they are connected to the piping system and are only infrequently used. However, most dead legs in piping are created by an uncontrolled change in process piping where sections are abandoned in place and where temporary piping is installed for a short-term application and not removed after the temporary operation is complete. Generally, a piping dead leg is defined as a section of piping without flow but connected to the "live" piping circuit. A rule of thumb to use in characterizing a dead leg is if the section of pipe without flow is longer than about 1.5 times the piping diameter.

Severe internal corrosion can occur in dead-leg piping because of water precipitation and sediments such as iron sulfide scale when operating at a temperature and pressure below the dew point of water when water can condense. This results in severe under-deposit internal corrosion in the piping, normally at the bottom of the pipe where the water has accumulated. Dead leg internal corrosion typically results in uniform pipe thinning due to internal corrosion and can result in the catastrophic release of hydrocarbons. Piping dead legs are particularly susceptible to this internal corrosion in sour service, that is, services containing hydrogen sulfide or other sulfur compounds in a stream with small amounts of water. In sour service, the H_2S combines with the water and iron scale, creating corrosive sulfide deposits on the pipe's interior. Services such as Light Ends fractionation column overheads, Crude and Coker unit overhead streams, sour water streams, and other similar sour streams are particularly susceptible to internal corrosion in piping dead legs.

Piping dead legs are also susceptible to freezing in cold climates where the ambient temperatures may be below freezing. In these cases, the free water will precipitate into low points in the dead leg of the piping and freeze, causing the piping to rupture. As the ambient temperature rises and the system thaws, the ice melts, releasing hydrocarbon or toxic vapor from the ruptured piping system. This has led to major fires and explosions in process units immediately following a hard freeze.

Also, as mentioned above, polyethylene chemical plants or other facilities handling Butadiene (BD), Isoprene, and similar reactive chemicals have a unique failure mode concerning dead legs. Butadiene is very chemically reactive, and in stagnant piping, zones have been known to react, resulting in the formation of "popcorn polymer," which has caused the piping failure and loss of containment. Popcorn polymer is more common in BD concentrations above 90% but has been seen in concentrations down to 50%. Operating procedures to minimize popcorn polymer should be developed and followed to prevent ingress of O_2, and the use of antioxidants; (e.g., tertiary butyl catechol) (TBC) should be considered if popcorn polymer has occurred in the past. Injection of methyl/ethyl mercaptan has also demonstrated effectiveness in mitigating the formation of popcorn polymer.

Process Operations Safety: The What, Why, and How Behind Safe Petrochemical Plant Operations, First Edition. M. Darryl Yoes.

What Has Happened/What Can Happen

Dead Leg Piping Failure at a Middle Eastern Oil and Gas Production Facility

Presented at the National Association of Corrosion Engineers (NACE) International 2007 Conference and Expo

At this production facility, gas is injected into the production well to increase well pressure to aid in producing hydrocarbon liquids. The multi-train gas injection system utilized gas compressors to inject the gas. A glycol mixture was also injected to prevent hydrates from forming in the injection system. The system also typically included some H_2S and CO_2 in the injection stream, and therefore, the equipment included multiple corrosion inhibitor injection points to control internal corrosion.

The compressor inlet drum and compressor were protected from overpressure by a pressure relief valve connected by an 8″ diameter pipe section. This pipe was installed in an orientation that was not self-draining and allowed condensed water (sour water due to H_2S) to accumulate in the line to the relief valve. Also, due to the piping orientation and location, the piping design did not allow the corrosion injection system to protect the safety valve piping.

After about a five-year run, the 8′ safety valve line failed catastrophically due to severe internal corrosion along the bottom section of the piping.

A similar incident occurred on a natural gas pipeline operated by the Tennessee Pipeline Company on December 8, 2010. In this incident, a dead-ended section of the pipeline failed catastrophically due to internal corrosion. The failure occurred in a section of 24″ diameter piping that was originally a .500″ wall thickness pipe in a lateral connected to the station discharge header piping near East Bernard, Texas. The piping was pressurized during normal operation, although no gas was flowing through this section of piping. The piping was designed for a maximum operating pressure of 750 psig and was operating normally at 720 psig when the failure occurred.

When the piping failed, a 12-foot section of the ruptured 24″ pipe was ejected 295' from the origin of the incident, resulting in a catastrophic loss of containment of high-pressure natural gas into the surrounding area. Fortunately, no explosion or fire was associated with this incident, and the pipeline company isolated the system without incident. Due to the pressure and the size of the system, it took about six hours to depressure the system. Several homes were evacuated, and nearby roads were closed until the system was depressurized. An estimated 70mm cubic feet of natural gas was released due to the rupture.

The DOT investigation noted that this section of the pipeline was identified on the station facility plan, and annual monitoring for dead legs was part of the Tennessee Pipeline Company's overall integrity plan.

This incident was investigated by the US Department of Transportation, and a detailed report was issued. Figures 8.10.1a and 8.10.1b are from this report and illustrate the hazards associated with dead-leg piping.

(a)

(b)

Figure 8.10.1 (a) Ruptured section of the Tennessee Pipeline Company 12″ diameter pipeline. This section was identified as a dead-leg section of the pressurized pipe, which failed while in service at 720 psig. A failed 12-foot section of the pipeline was ejected 295 feet from the failure site. (b) Ruptured end section of the Tennessee Pipeline Company 12″ diameter pipeline. This section had been identified as a dead leg in the company integrity plan and was subject to annual inspection. Source: US Department of Transportation investigation/ Public Domain.

What Has Happened/What Can Happen

Valero Energy Corporation, McKee Refinery (Sunray, Texas)
16 February 2007

The US Chemical Safety and Hazard Investigation Board (CSB) investigated this incident, and a report was issued with recommendations to help prevent a reoccurrence (CSB Report No. 2007-05-I-TX July 2008). The CSB also produced an excellent training video on the key lessons learned from this incident.

The Valero Refinery at Sunray, Texas, experienced a significant fire on 16 February 2007, when a piping elbow in a dead leg pipe in liquid propane service froze, and the piping fractured due to internal icing. A piping system control valve station on the propane deasphalting unit had been isolated for about 15 years before the incident, creating the dead leg and allowing this section of the line to accumulate free water in the dead leg portion of the pipe circuit. The area experienced freezing weather for several days before the incident, with a low temperature of 6 °F (−14 °C) reported nearby the day before. The ambient temperatures rose above the freezing point on the morning of the incident, allowing the ice to thaw. As the ice thawed, the propane was released from the ruptured pipe and was quickly ignited by nearby process equipment. The resulting propane jet fire impinged on other process equipment, causing secondary losses of containment and rapidly escalating into a major refinery fire.

Three workers sustained severe burn injuries due to this incident, and the entire refinery was evacuated due to concerns of further fire or explosion. This incident represents a characteristic dead leg failure due to the freezing of precipitated free water.

The CSB has completed a full Investigation Report, which includes much greater detail of the incident and a summary of other key lessons learned. A well-documented video of the incident is also available from the CSB and is a valuable training tool. The report and video are available for download from the CSB website (csb.gov).

Thanks to the CSB for making this information available to the industry. See Figures 8.10.2–8.10.4 for photos of the incident and the fractured pipe (courtesy of the CSB final report on this incident).

Figure 8.10.2 Valero McKee Refinery C3 dead leg piping failure. Photos of the incident and the fractured pipe. Source: Courtesy of US Chemical Safety and Hazard Investigation Board.

Figure 8.10.3 Piping dead leg failure (freezing of trapped water). Source: Courtesy of US Chemical Safety and Hazard Investigation Board.

Figure 8.10.4 Valero McKee Refinery resulting Fire Damage. Photos of the incident and the fractured pipe. Source: Courtesy of US Chemical Safety and Hazard Investigation Board.

What Has Happened/What Can Happen

Total Refinery, La Mede, France
9 November 1992
VCE (6 Fatalities [Operators], 38 Injured)
The French Ministry of Environment issued a report on this incident.
"Gas explosion in the cat cracking and gas plant units of a refinery 9 November 1992."

On 9 November 1992, the Total refinery at La Mede, France, experienced a major VCE and resulting fire at the FCC when an 8″ exchanger bypass line for the absorber Deethanizer intercooler operating at approximately 150 psig failed due to severe internal corrosion. The VCE destroyed the control room and significantly damaged the FCC and surrounding equipment. Six Operators were killed in the incident, including three who were in the control room at the time of the incident. The incident report indicated that the piping was exposed to sour water, ammonia, H_2S, and cyanides and that it was not being routinely inspected because it "is only a bypass."

The French Ministry of Environment investigated this incident, and a detailed report with photos of the devastating fire was released. Unfortunately, approval to use the photos in this book was not obtained; however, they can be viewed in the incident report at the following web address:

https://www.aria.developpement-durable.gouv.fr/wp-content/files_mf/FD_3969_La_Mede_1992_ang.pdf

The resulting failure resulted in an instantaneous pipe failure roughly equivalent to a 4-square-inch diameter hole. The gas plant, FCC unit, and associated control building were essentially destroyed. Two new process units under construction, scheduled to come into operation in 1993, were also seriously damaged. Total losses were about $1 billion in today's dollars.

A metallurgical analysis was performed on the failed absorber deethanizer bypass piping and revealed that the severe internal corrosion resulted from a dead leg where water could accumulate in an environment containing H_2S and CO_2. The analysis also indicated that corrosion monitoring was not ongoing at the locations of the highest potential corrosion; therefore, there was no advance indication of the corrosion. Finally, the analysis indicated that the key factor leading to the internal corrosion was the accumulation of water in this dead leg, highlighting that attention should be paid to piping design to ensure dead-ended piping is avoided or designed to be self-draining to prevent the accumulation of water that can lead to internal corrosion.

Following this incident, the Court found the President of Total (in position from 1988 to 1993) guilty of involuntary manslaughter and sentenced him to 18 months in prison. Two inspection managers were sentenced to suspended prison terms of 18 months, and two plant inspectors were given four-month suspended prison sentences.

What Has Happened/What Can Happen

Syncrude Canada Ltd.
14 March 2017
The Alberta Energy Regulator investigated this incident, and a report was published.
Investigation Summary Report
2017-012: Syncrude Canada Ltd.
Authorization No. 8573N

At about 1:36 p.m. on 14 March 2017, Syncrude identified a treated naphtha leak, resulting in a fire and subsequent explosion in one of their hydrotreater units (Unit 13-1). The failed pipeline was the idled 6″ (150 millimeter) naphtha recycle line, which ran from unit 13-1 stripper tower back to the unit feed surge drum. The Syncrude investigation determined that water had precipitated into this section of piping over the 2016–2017 winter and that the electric heat tracing had become inoperable on this section of pipe. The water froze in the piping, and the expanding ice ruptured the pipe, releasing naphtha. Due to the large size of the split piping, a significant amount of naphtha was released, which overwhelmed the sewer drain system, resulting in an accumulation of naphtha and vapors in the area. The vapor cloud expanded and reached an ignition source in the facility's operating units, resulting in the flashback (explosion) and subsequent pool fire.

One Operator was severely burned in this incident and was transferred to the Northern Lights Hospital and subsequently to the University of Alberta Hospital in Edmonton for further treatment. A large portion of the Syncrude operation was shut down, and about 1200 Syncrude workers were evacuated during the fire. The Naphtha Hydrotreater sustained significant damage.

During the investigation, it was determined that it was common for water to accumulate in the naphtha piping system. The water is normally maintained hot by an electric heat tracing system; however, the heat tracing on this section of piping was inoperable, allowing the water to freeze and resulting in the piping rupture. The rupture went undetected until the ambient temperature rose above freezing, which melted the ice, releasing the naphtha into the atmosphere.

Unfortunately, the photos of this incident were not obtained for copy in this textbook. However, they are available at the Alberta Energy Regulator web address below, including an image of the failed piping component due to the freezing and the expansion of trapped water. It is almost unbelievable how much pressure can be exerted by freezing water.

2017-012_ISR_Syncrude_20181116.pdf (aer.ca)

TPC Group Chemical Plant Butadiene Unit
Explosion and Fire caused by Popcorn Polymer
Occurred on 27 November 2019 in Port Neches, Texas

This incident occurred on the night shift at 12:54 a.m. when the piping from the bottom of the Butadiene fractionator column suddenly and unexpectedly ruptured. The pipe rupture occurred in a section of 16″ pipe about 35 feet in length between the bottom of the column and the isolation valve at the column bottom's pump. This rupture led to the immediate release of the contents of the column, butadiene, with a purity of about 98%. The hot butadiene immediately vaporized, forming a large vapor cloud and the initial explosion. At about 1:48 in the afternoon, a second, even more massive explosion occurred, propelling one of the columns into the air and landing into the facility, causing even more damage. During the event, four columns toppled, and the butadiene unit was destroyed. The fires burned for over a month before they were declared extinguished. This incident resulted in $450 million in on-site damages and another $153 million in off-site damage. This event had three minor injuries; all were treated and released at the local hospital. The TPC Group filed for bankruptcy on 1 June 2022.

There was also a significant impact on the community. Officials in Jefferson County, where the incident occurred, declared the county a state of disaster and issued a 4-mile radius evacuation area, affecting the communities of Port Neches, Groves, Nederland, and a portion of Port Arthur. The evacuation orders also affected local schools, which were damaged and impacted by debris from the incident. Additionally, restrictions were placed on the usage of the Sabine-Neches Waterway, a major US marine waterway, affecting the local communities and the transportation of goods and services to other parts of the country.

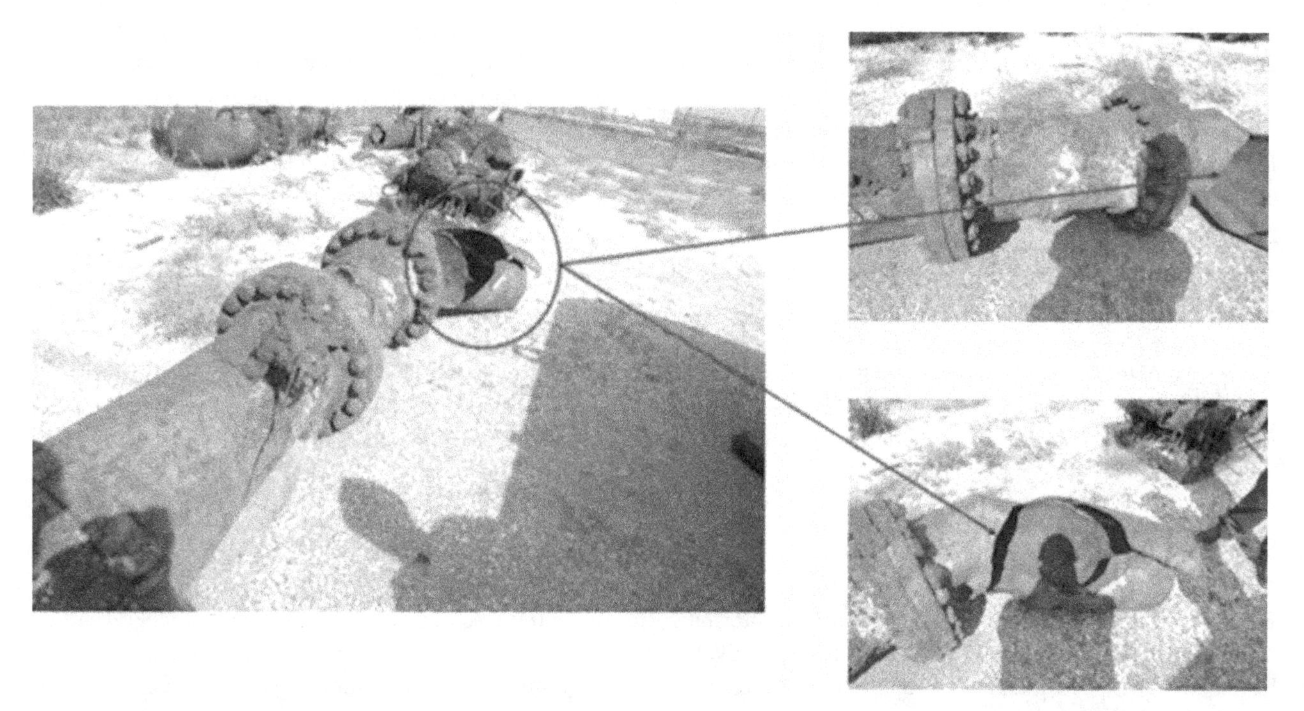

Figure 8.10.5 Photograph of the Failed Section of 16″ Piping. Source: Courtesy of US Chemical Safety and Hazard Investigation Board.

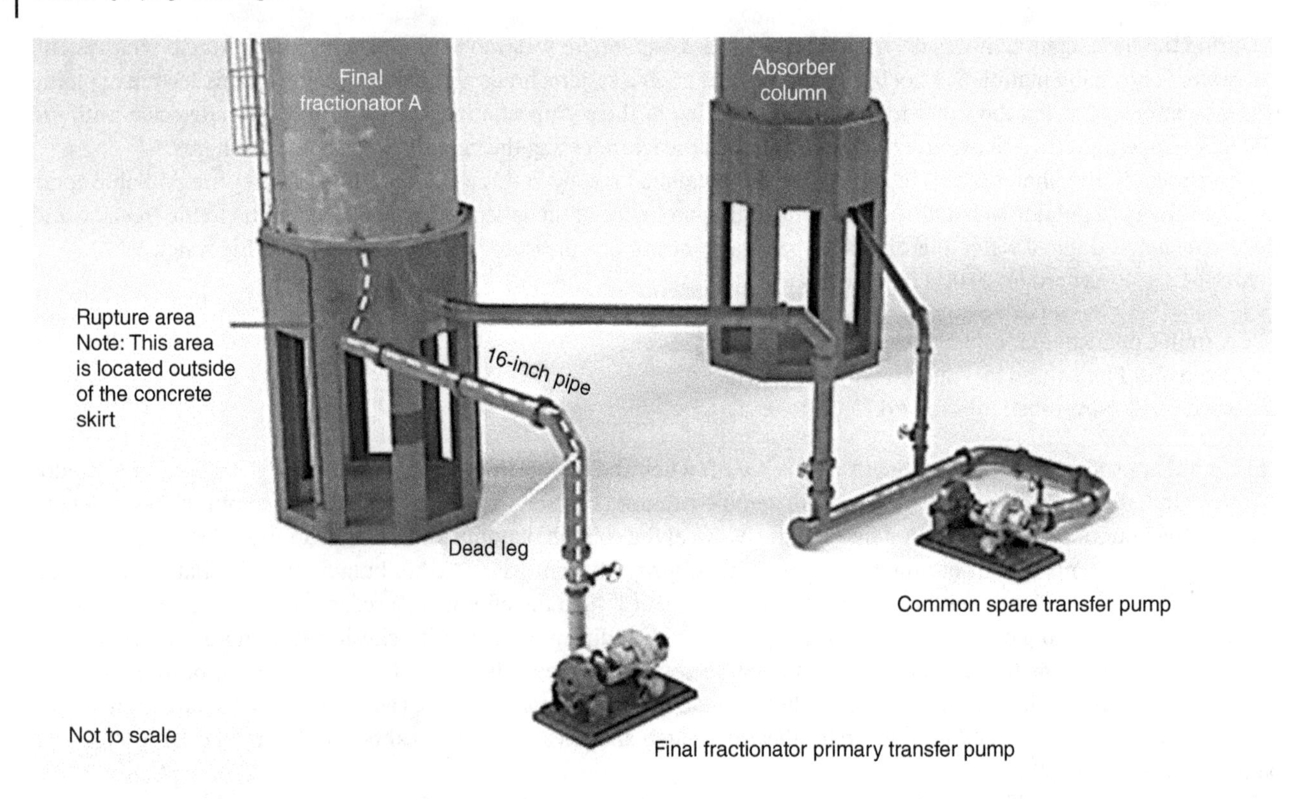

Figure 8.10.6 Image Indicating the Location of the Failed Section of 16″ Piping. Source: Courtesy of US Chemical Safety and Hazard Investigation Board.

This incident was investigated by the US Chemical Safety and Hazard Investigation Board (CSB), and an in-depth report was completed and posted on the CSB website. The CSB concluded that the tower bottoms pump had been isolated for maintenance for 114 days before the failure. Butadiene, especially in higher concentrations, can polymerize and form "popcorn polymer", especially in dead-ended piping and other stagnant areas. This was discussed earlier in this section, and previous incidents involving butadiene polymerization have occurred. In the CSB investigation, the CSB found that although procedures existed for routine flushing of the pumps to prevent the formation of the polymers, they were unable to confirm that anyone at the TPC site was familiar with the hazard associated with the isolation of a pump creating this hazard.

See Figure 8.10.5 for a photograph of the section of piping that failed due to the formation of popcorn polymer. Figure 8.10.6 illustrates the location of the failed section of piping. The closed isolation valve on the pump suction created a dead-leg section of pipe between the fractionator and the pump.

Exxon Refinery, Baton Rouge, Louisiana
24 December 1989
Massive explosion and Fire due to thermal expansion in an LPG piping system.

See the separate report and photographs in Chapter 8.14 of a massive explosion at the Exxon Baton Rouge Refinery on Christmas Eve of 1989. This incident occurred due to thermal expansion in an isolated LPG piping system. There were two fatalities in this event.

Key Lessons Learned

Piping dead leg failures can occur due to severe internal corrosion or freezing and rupture of the pipe in cold weather. Common corrosion protection programs are ineffective in preventing internal corrosion in dead legs, and a failure can be sudden, unexpected, and catastrophic. This can result in a large flammable or toxic vapor release and an explosion or fire. Piping dead legs should be identified through a rigorous piping inspection program and systematically removed if possible. Dead legs created by equipment bypasses or by a unit startup or shutdown line should be included in the piping inspection process. These bypass and startup lines should be monitored at regular intervals to guard against the potential for unexpected failure.

The first objective for managing dead legs is removal. Identify the dead legs, get them on your turnaround list, and see that they are removed and replaced with a self-draining pipe circuit so that sediments and moisture cannot settle into the pipe. Until the pipe is removed, we must ensure a rigorous inspection program for all dead legs, with the resulting inspection data published for the equipment owners. Verify that the Inspectors are looking in the right place, at the bottom of the pipe where water can accumulate. This data must be tracked until the removal of the dead legs.

The piping design for new installations and revamps should eliminate dead legs to the full extent possible. Piping design should preclude using dead-ended extensions for pipe supports and other similar piping modifications that would create a piping dead leg. For the remaining few that cannot be eliminated, the piping design should ensure that dead legs are self-draining to prevent the possibility of water accumulating in the piping, which can lead to internal corrosion or freezing in cold climates. For example, control valve bypass and exchanger bypass piping should extend up and over to ensure that water cannot accumulate in the low points in the piping. Evaluation for potential dead-leg piping should be a routine part of the site's process for managing change.

Also, it is essential that the existing dead legs are clearly on the winterization program and that they are protected against freezing during cold weather. Trapped water will freeze, expand, and easily rupture the piping or valves if water is not removed and insulation and heat tracing are not applied. Piping dead legs susceptible to freezing weather should be identified and protected by steam or electrical tracing and insulation and, if feasible, by frequent draining of any accumulated water before or during freeze events. An alternative means of protecting piping during winter is maintaining a small flow of warm fluids through the pipe, especially when the ambient temperature is below freezing.

Likewise, developing and implementing a process task is essential to ensure the periodic flushing of the control valve and exchanger bypasses. Flushing removes free water and sediments that may have accumulated in the bypass piping and can result in a catastrophic failure of the piping. This is a simple process of the Operator cracking open the bypass valve on a scheduled task (monthly or quarterly) to flush the accumulated water or sediments from the piping system. This has shown effectiveness in mitigating bypass hazards. Some plants have implemented dead leg flushing procedures as a process task item for equipment bypasses, such as exchanger or control valve bypasses, to help ensure they are not accumulating stagnant water or sediments.

When evaluating piping dead legs, remember that the potential for severe corrosion typically occurs on the piping interior and is almost always at the bottom quadrant of the piping, where the precipitated water and sediments accumulate. The piping exterior may appear completely normal with no indications of a problem right up to the point of catastrophic failure. Also, since the corrosion is uniform, it occurs over a large area. When failure occurs, it usually results in a large area of piping failing instantaneously. This is generally referred to as a "blowout," an unexpected catastrophic failure.

There may also be cases where a section of a dead-leg pipe may become cold since it is isolated from the process flow. In this case, if the dead leg is insulated, it may also experience severe external corrosion due to moisture accumulating under the insulated portion of the pipe (Corrosion Under Insulation or "CUI"). In this case, the corrosion will be on the piping exterior under the insulation and possibly around the full circumference of the pipe. CUI can also lead to a catastrophic loss of containment, resulting in fire or explosion.

For chemical plant units that have experienced issues with Butadiene popcorn polymer, consider using a chemical inhibitor to help guard against the polymerization of the Butadiene. Tertiary butyl catechol (TBC) and methyl/ethyl mercaptan have demonstrated effectiveness in mitigating the formation of popcorn polymer at some units.

Additional References

US Chemical Safety and Hazard Investigation Board, Investigation Report – LPG Fire at Valero – McKee Refinery, 4 Injured, Total Refinery Evacuation, and Extended Shutdown, Report No. 2007-05-I-TX July 2008. https://www.csb.gov/valero-refinery-propane-fire/

Marsh Risk Consulting; "The Top 100 Losses 1972–2001", 20th Edition, February 2003, 09 November 1992 Explosion – La Mède, France.

French Ministry of Environment – DPPR/SEI/BARPI, Gas explosion in the cat cracking and gas plant units of a refinery 9 November 1992, La Mède [Bouches du Rhône] France. https://www.aria.developpement-durable.gouv.fr/wp-content/files_mf/FD_3969_La_Mede_1992_ang.pdf

Paper No. 07667 "Corrosion of a Gas Injection Station – Case History", NACE 2007 Corrosion Conference and Expo., Daryoush Masouri, Mahmoud Zafari.

"Incidents That Define Process Safety", Pages 112–121; FCCU Explosion at Total La La Mède France, November 9, 1992, John Atherton, Frederic Gil.

Alberta Energy Regulator Report, Investigation Summary Report, 2017-012: Syncrude Canada Ltd., Authorization No. 8573N. http://www1.aer.ca/compliancedashboard/investigations/2017-012_ISR_Syncrude_20181116.pdf

US Chemical Safety and Hazard Investigation Board, Investigation Report – Popcorn Polymer Accumulation Pipe Rupture, Explosions, and Fires at TPC Group Chemical Plant Butadiene Unit, 3 Injured, Butadiene Unit was Destroyed, and the Company's Bankruptcy, Report No. 2020-02-1.

Chapter 8.10. Hazards of Piping Dead Legs

End of Chapter Quiz

1. How do we define the term "Dead Legs"?

2. Severe internal corrosion can occur in dead leg piping as the result of the precipitation of ___________ and ___________such as iron sulfide scale when operating at a temperature and pressure below the dew point of water when water can condense.

3. Name the most typical failure modes for piping with dead legs. Please name all you can.

4. Failures in piping dead legs have resulted in _______, _________, and __________ releases of flammable vapor and resulted in fires and explosions.

5. How can dead legs be avoided or mitigated?

6. See how many examples of piping design or configurations you can name that may meet the definition of a dead leg.

7. Piping dead legs are particularly susceptible to this internal corrosion in _______ service, that is, services containing hydrogen sulfide or other sulfur compounds in a stream with small amounts of water.

8. When is a major release likely to occur when a process pipe fails due to freezing water trapped in a dead-end piping section?

9. What is a major unit hazard in units processing high concentrations of Butadiene and Isoprene?

10. What is an effective practice for handling control valve or exchanger bypasses to prevent free water accumulation in the dead leg piping?

8.11

Brittle Fracture

Brittle Fracture is a type of metallurgical failure that occurs to process vessels (drums, tanks, fractionation columns, and other similar vessels and can also occur in heavy wall piping and fittings) and piping and occurs instantly and without warning. It can occur when the equipment is in service, out of service for maintenance, or on standby. The failure mode can be a large crack in the steel, but it is often a total pipe or vessel failure. When this occurs with the vessel in service, the resulting loss of containment results in the release of the contents, often leading to an instantaneous and catastrophic explosion and fire.

The most susceptible vessel materials of construction are carbon and low alloy steels, but they can also include 400 series stainless steel, like the 12 Cr varieties. Generally, carbon steel is ductile at a temperature above its Minimum Design Metal Temperature (MDMT). In this circumstance, and when stress is applied, the metal will bend, bulge, or flex until it eventually fails (a ductile failure). A ductile failure generally provides an advance warning by showing signs of bending, bulging, or other forms of deformation before the failure.

This is a desirable characteristic as we can see the failure occurring, and we have the opportunity for intervention. However, the characteristics of steel change as the vessel temperature is reduced to below the vessel MDMT. The steel changes from ductile to brittle. When the vessel is brittle (the vessel temperature is below the MDMT) and when a form of stress is applied, the vessel can fail suddenly and without warning due to brittle fracture. Stress in process equipment is often internal pressure but can be other forms of stress, such as pipe strain or impact.

Therefore, a brittle fracture is defined as a sudden and often catastrophic fracture of a material with little or no plastic deformation (no bending or distortion before the failure). It is generally characterized as a failure with minimal energy absorbed in the material before fracture and rapid crack propagation through the material. In a brittle fracture failure, the crack propagates near the speed of sound and can result in the steel material, such as a pressure vessel, breaking into many shards similar to that of broken glass.

The temperature where the metal properties change from ductile to brittle is characterized as the MDMT (also sometimes referred to as the "transition temperature," "ductile-to-brittle-transition temperature" [DBTT], the "nil-ductility transition temperature," or the "15 ft lb. transition temperature").

For our discussion, we will use the MDMT since we are dealing primarily with process piping and vessels, and the MDMT is the term most often used on vessel data plates. It is important to note that this temperature may be above ambient for carbon steels, some low alloy steels, and 400 series stainless steel. This is especially true for vessels with thick shells (i.e., 1/2″ or greater). When the equipment is in service (or anytime the equipment is pressurized), the vessel temperature must be maintained above the MDMT to avoid the potential for brittle fracture. See Figure 8.11.1 for an example of the ductile-to-brittle-transition temperature.

The potential for Brittle Fracture should be considered during equipment design for scenarios where the vessel may experience cold temperatures due to climate (ambient temperature) or the flash vaporization of LPG such as Ethane or Propane. In this case, a suitable material, such as 300 series stainless steel, should be considered for vessel fabrication. Stainless steel has very low MDMT and is an inherently safer material for cold services. Another method is to modify the carbon steel microstructure by post-weld heat treatment. Post-weld heat Treating (PWHT) (sometimes called stress relieving) lowers the transition temperature and can also help guard against the potential for brittle fracture. However, PWHT is not a guarantee against brittle fracture.

Process Operations Safety: The What, Why, and How Behind Safe Petrochemical Plant Operations, First Edition. M. Darryl Yoes.

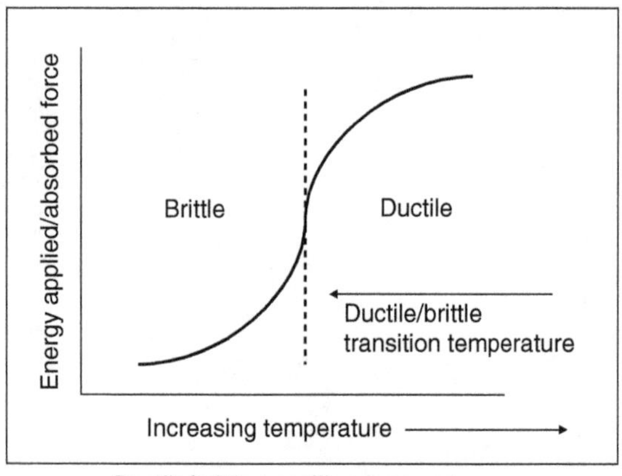

Figure 8.11.1 Ductile-to-brittle-transition temperature.

Key Brittle Fracture Terminology

Some other key Brittle Fracture terminologies that we will discuss here, and almost any time brittle fracture is being discussed, include the following:

- **Critical Exposure Temperature (CET):**
  - The CET is not a design parameter. It is the lowest temperature the piping or vessel is expected to see during its lifetime. This can be either the ambient temperature or the temperature of the process, whichever is the lowest. The process temperature may be flash vaporization or auto-refrigeration of the process fluids.
  - The CET is derived from normal and abnormal operating conditions and the lowest ambient temperature.
  - The CET is used to develop the process vessel's design criteria before vessel fabrication.
- **Minimum Design Metal Temperature (MDMT):**
  - The MDMT is a metallurgically determining minimum safe working temperature at the vessel's maximum design pressure. It is based on the type of steel used in the vessel, the wall thickness, and knowledge of the vessel's manufacture, including the type of welding, heat treatment, and material tests undertaken by the manufacturer, and further defined in API-579 "Fitness for Service."
  - The MDMT is the lowest temperature at which the vessel is designed to see if the vessel pressure is at the Maximum Allowable Working Pressure (MAWP), determined by the material(s) of construction and equipment design.
  - The MDMT is found on the process vessel's code stamp or data plate after fabrication.
  - Industry experience has shown that pressurization of the vessel at more than 25% of the MAWP when the temperature is below the MDMT may result in a catastrophic brittle fracture.
- **Minimum Safe Operating Temperature (MSOT):**
  - The MSOT is the Minimum Safe Operating Temperature at a corresponding pressure (typically a temperature and pressure curve indicating safe operating conditions).
  - Used most often only for a small population of process vessels, those that are very thick wall carbon steel construction. The MSOT curve is often used for equipment where the temperature and pressure must be slowly increased or decreased simultaneously, for example, during startup or shutdown.
  - Deviating from the MSOT curve, that is, increasing pressure ahead of increasing temperature, can result in catastrophic brittle fracture.
- **Maximum Allowable Working Pressure (MAWP):**
  - The maximum pressure the vessel may experience to keep within Code-Allowable Stress.
  - The MAWP is set by vessel design.
  - The MAWP is typically found on the vessel code stamp or data plate.

    Note: The MAWP can be affected by service life or the number of cycles and may need to be reanalyzed. The Mechanical Engineer or Fixed Equipment Engineer should be consulted if there are concerns about service life.
- **Post-weld Heat Treatment (PWHT):**
  - Post-welding thermal heat treatment of the piping or vessel to refine the welding metallurgy and reduce residual stresses.

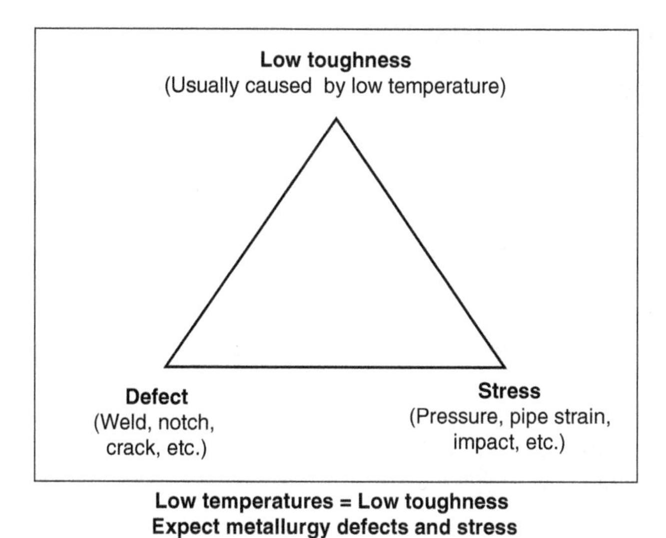

Figure 8.11.2 The brittle fracture triangle (low toughness [cold temperature] + Defect + Stress).

Before a brittle fracture can occur, the following three elements must be present (See Figure 8.11.2) (Brittle Fracture Triangle):

1. Low toughness. Generally, this occurs in carbon steel at a temperature below the MDMT. Toughness can be tested in the laboratory using the Charpy[1] impact test.
2. A flaw or defect in the steel or a weld. For example, the size, shape, and stress concentration from a flaw such as weld defects or a machined notch. Internal lamination-type defects in the steel can also introduce a flaw.
3. Sufficient residual or applied stress (above some minimum). This is often internal pressure in process vessels, but it could be piping stress induced by thermal shock, misaligned pipe flanges, stress at welded joints, an external mechanical impact, etc.

1. Low Toughness (typically CS >1/2″ Thick, exposed to cold – temperatures below the MDMT).
2. **Defect:** Likely already present, for example at a weld, machine notch, steel flaw, or lamination (PWHT can help).
3. Stress (internal pressure, thermal shock, pipe stress, mechanical impact).

When these three come together, the result is frequently a catastrophic brittle fracture failure and a total and instantaneous loss of containment.

A new vessel should have a code stamp, data plate, and a technical design datasheet specifying the design parameters (including MDMT and MAWP). New designs should ensure that the CET is above the MDMT. Operating procedures should be in place for older vessels to ensure the vessel temperature is at or above the MDMT before the vessel is pressurized.

A rule of thumb is to maintain operating pressure below 25% of MAWP until the vessel temperature is at or above the MDMT. A brittle fracture assessment may need to be performed for existing equipment, especially that manufactured before 1987.

Carbon steel piping generally has an MDMT of −20 °F (−29 °C) and is subject to brittle fracture failure at or below −20 °F (when stress is present). Certain alloys, such as stainless steel, will have an MDMT well below that of carbon steel.

When hydrotesting equipment, it is important to consider the temperature of the water or other testing medium used to ensure that it is at a temperature above the MDMT of the equipment being tested. Many brittle fractures occurred during hydrostatic testing when the water temperature was below the MDMT of the tested vessel. In some cases, it may be necessary to preheat the water with temporary facilities BEFORE increasing the pressure of the vessel being tested.

Also, flash vaporization and auto-refrigeration of some light hydrocarbons such as Propane or Butane can result in temperatures as low as −40 °F (−40 °C) and, therefore, certainly have the potential to result in brittle fracture of vessels or piping. See Appendix C for an excellent reference article on the effects of auto-refrigeration ("Auto-Refrigeration: When Bad Things Happen to Good Pressure Vessels" written by Francis Brown P.E. (National Board of Boiler and Pressure Vessel Inspectors) (NBBI). This article was reprinted with permission of the NBBI.

1 The Charpy impact test can help define the toughness of steel to be used in a new vessel. The results of the Charpy tests are valuable indications of how the material might behave in service. It is a destructive test and is not a simulation of material in service.

Other potential causes of brittle fracture failures in process piping and vessels include:

- Startup and Shutdown of equipment where the temperature drops while pressure is maintained (best if temp and pressure drop together).
- Emergency depressurization of LPG causing auto-refrigeration.
- Safe parking or squatting a unit at pressure for a longer period, particularly over a cold night where the temperature can drop below the MDMT.
- Hydrotesting with cold water (water below the MDMT for the vessel) or when the vessel shell is at a cold ambient temperature.
- Failure to remove nitrogen before introducing liquefied light ends results in the hydrocarbons flashing into the nitrogen. This results in auto-refrigeration of the LPG and produces temperatures much colder than when nitrogen is not present.
- Anytime equipment ices up, and it does not usually ice up.
- Using liquid nitrogen to speed up the cooling of the catalyst (for example, when vaporizing liquid nitrogen from a truck).
- The loss of pressure in compressor suction drums and flash vaporization of LPG during startup.
- Pressurizing compressor discharge drums or process vessels more than 25% of the MAWP when the vessel temperature is below its MDMT.
- Loss of heat input into a pressurized pressure vessel.
- Overflow or transport of cold liquids to downstream equipment.
- Nitrogen filling vessels containing liquid LPG (due to the significant difference in partial pressures, the LPG will rapidly vaporize in the nitrogen environment, resulting in auto-refrigeration).

Another excellent example of brittle fracture occurred with some regularity during World War II when US merchant ships were ferrying soldiers, munitions, and other war materials across the North Atlantic to support Europe during the war. Many of these ships broke apart in the cold waters of the North Atlantic and sunk. It was some time before we truly understood the nature of these failures. Most were initially thought to have been attacked by German U-boats; however, many were due to the poor quality of the steel, and they failed at sea due to brittle fractures in the cold waters of the North Atlantic.

Figure 8.11.3 from the Naval Materials Science and Engineering Course Notes, Chapter 11, Fracture of Materials, US Naval Academy, are images of liberty ship failures at sea.

The British Health and Safety Executive has a good discussion on avoiding the risk of brittle fracture in the initial design and the operation of process equipment. See the Reference section at the end of this chapter for the link to the article.

A startling example of a brittle fracture is shown in Figure 8.11.4. This was reportedly a new process vessel that was being hydrotested in preparation for being placed in service. Unfortunately, the test was being done with water below the minimum design metal temperature (MDMT for the process vessel). The vessel failed catastrophically due to the stresses imposed during the test, while the vessel metal temperature was below the ductile–brittle transition temperature.

In the photograph, there is no bending or tearing of the steel, and it failed as if it were broken glass. When this happens, there will be shards of broken metal with no bending or stretching of the steel. A brittle fracture is easily identified as the cause of the failure by the "chevron" marks in the broken edges of the steel.

(a)

(b)

Figure 8.11.3 US Naval Academy Science and Engineering, fracture of materials. Images of liberty ship failures at sea due to brittle fracture. Source: The image is courtesy of the US Naval Academy.

Figure 8.11.4 A process vessel brittle fracture that occurred while being hydrotested with water at a temperature below the vessel MDMT. Source: US Department of Energy, Argonne National Laboratory, Compilation of Pressure-Related Incident Summaries. (https://indico.fnal.gov/event/13755/sessions/1912/attachments/14044/17881/AccidentsCompilation_June16.pdf).

What Has Happened/What Can Happen

Brittle Fracture Failure
Alberta, Canada production well site
November 2006

Information in this chapter was gleaned and interpreted from a report (Marsden 2007) presented at the API Inspector's Summit held in February 2007 at Galveston, Texas. The report details the investigation of a brittle fracture failure that occurred at a sweet-gas well site located approximately 37 miles (60 kilometers) east of Calgary, Alberta, Canada, near the town of Strathmore.

Although no one was injured during this incident, the potential for severe injury or even the loss of life was real. Information from the detailed report is shared here to help prevent similar incidents from occurring.

Ambient winter temperatures in this part of Alberta can occasionally drop to −40 °F (−40 °C) and last for days or weeks. These periods are often interrupted by warm, dry winds known locally as "chinooks." Chinooks can cause ambient temperatures to rise from these low temperatures to above-freezing temperatures in less than 24 hours! This incident occurred during an extended period when ambient temperatures hovered around −31 °F (−35 °C). However, a chinook was in the weather forecast within a few days.

The well site was part of a larger natural gas project consisting of approximately 200 wells. The entire project was staffed using only three operators. Figure 8.11.5 provides a simplified schematic of the equipment associated with each well site. The key elements and the function of each are listed in Figure 8.11.6.

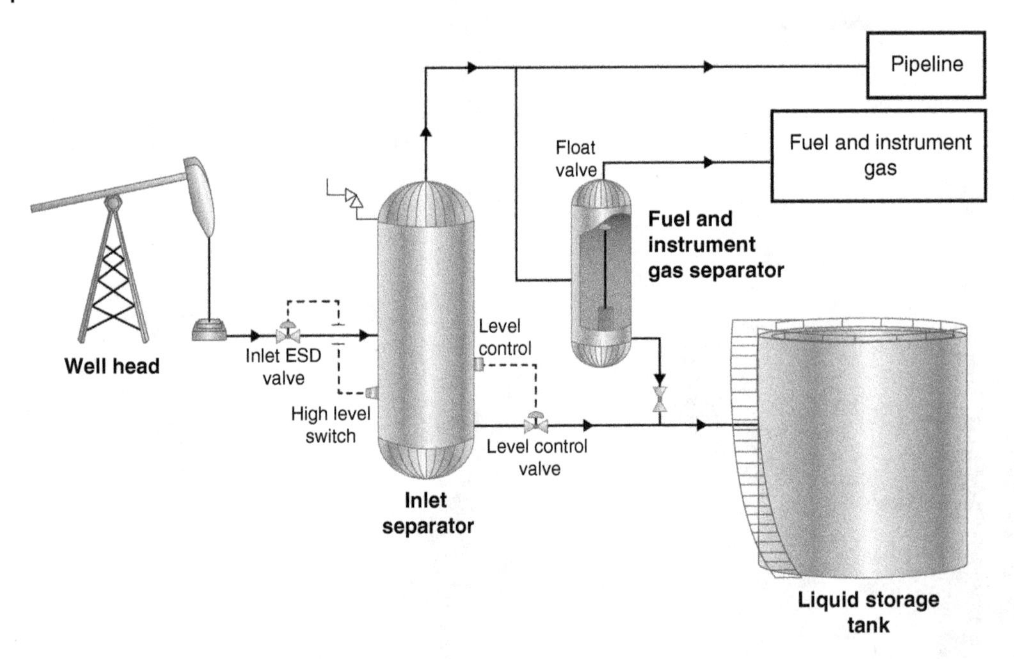

Figure 8.11.5 Alberta, Canada, gas well, liquid/gas separation unit. Source: Simplified Site Schematic (Image courtesy Ecoscience Resource Group Visual Artists, with edits courtesy of Marsden, R.).

Schematic key elements	Function/Comments
Wellhead	Liquids from the well (primarily water) were pumped to the inlet separator using a pumpjack (not shown) powered by an internal combustion engine. Produced fluids (e.g., natural gas and reservoir liquids including water) co-mingled in a 2″ flowline to the separator package.
Inlet ESDV (Emergency shutdown valve)	The inlet ESDV was configured to fail in the closed position upon the loss of instrument gas pressure. As designed, this would cause the ESDV to close automatically.
Inlet separator vessel (*the failed vessel*)	The inlet separator was 58″ tall × 16″ diameter (1.5 m × 0.41 m diameter) and was rated for 1440 psig. The wall of the SA-106 Grade B shell material was 1″ (25.4 mm) thick. The vessel separated the two-phase flow into individual gas and liquid streams. Liquids were automatically dumped into the storage tank using a level controller (LC) and a level control valve (LCV). Gas was sent to a pipeline gathering system except for a small amount used for instrumentation and fuel gas.
High-level switch	The operational purpose of the high-level switch was to initiate the closure of the inlet ESDV in the event of a high liquid level in the inlet separator. The switch controlled the instrument gas pressure going to the ESDV actuator. As designed, a high liquid level in the separator would lift the switch's float, causing the instrument gas pressure to be vented from the ESDV actuator.
Liquid level control valve (LCV) and level controller (LC)	The LCV and LC automatically controlled the separator's liquid level by dumping liquids into a storage tank.
Gas pipeline (to offsite)	The underground gas pipeline network transported gas to a local compressor station.
Fuel gas and instrument gas (FG/IG) separator vessel	The FG/IG separator removed any remaining liquids from the gas used for instrumentation controls, fuel gas for building heaters, or as motive gas for the methanol injection pumps. The vessel was rated for 150 psig and had a shell fabricated from SA-106B pipe.
FG/IG high liquid level float valve	A float-actuated shutoff valve was incorporated into the outlet of the FG/IG separator to stop the flow of gas if excessive liquids accumulated in the vessel. As designed, this would cause the ESDV to close. Liquids were manually drained to the storage tank using a manual valve (not shown).
Liquid storage tank	The tank provided on-site storage for separated liquids.
Site building (not shown on schematic)	A small building was used to house the inlet separator, FG/IG separator, and associated instrumentation. The building was heated using two catalytic heaters.

Figure 8.11.6 Schematic key elements and function.

The Sequence of Events Leading up to the Incident

Understanding the sequence of events is important to understand the causal factors of the incident.

Three days prior to the incident, an operator visited the site to investigate indications of erratic flow rates. The operator determined that the pumpjack had automatically shutdown due to a *low level* of engine coolant. Bubbling sounds were heard from inside the inlet separator, indicating the liquid level was above the inlet connection for the separator and, therefore, also above the high-level switch. This meant that the high-level switch had failed to cause the ESDV to close as designed. The operator determined that the 1″ dump line was frozen despite the two building heaters operating at maximum output. Methanol was, therefore, manually injected into the dump line to remove the ice and reestablish level control. The pump jack was restarted after coolant was added to the engine.

Two days prior to the incident, an operator visited the site again to investigate indications of erratic flow rates. The operator determined that the pumpjack had automatically shutdown again, but this time it was due to *high pressure* in the flowline. It was later concluded that ice had formed inside the 2″ flowline, which increased the resistance to flow and, therefore, the line pressure. The temperature inside and outside the separator building was −17 °F (−27 °C). The bubbling heard the previous day was no longer apparent, and the two catalytic heaters were not functioning. It was later concluded that the high-level switch failed to function properly, which resulted in liquids continuing to fill the inlet separator and, eventually, being carried over into the FG/IG separator. The float-actuated valve in the FG/IG separator also failed to function properly, allowing liquids to enter the fuel gas system and extinguishing the building heaters. It was later determined that the float-actuated valve assembly had been assembled incorrectly. The operator was unsuccessful in restoring the system to normal operations and left with the intention to return in a few days when the forecasted chinook would arrive.

There were no visits made to the site one day prior to the discovery of the incident. The ambient temperature continued to drop, reaching −22 °F (−30 °C) during this period.

Two instrumentation technicians who had arrived to perform scheduled preventative-maintenance tasks at the site discovered the incident. Upon arrival, they discovered the inlet separator failure and the door of the building open and damaged. Despite the gaping hole in the separator, there were no indications of fire nor a significant loss of fluids since everything was frozen solid. However, there was evidence of a tremendous amount of energy being released. A failure of the flowline was also found. Once the required notifications to management and regulatory authorities were made, a root cause failure analysis began. Figure 8.11.7a,b are photographs taken during the investigation.

Initial observations identified the failure mode as a brittle fracture. The vessel fragments had distinctive tear-drop shapes and had no indications of ductility at the fracture face. Most significantly, the fracture faces had the tell-tale brittle fracture "chevron" features, which pointed to the fracture initiation location. See Figure 8.11.8, which clearly shows the characteristic chevron features.

Tremendous pressures are possible when ice forms in a closed or nearly closed system (Rezvani Rad et al. 2018). As ice formed in the inlet separator, gas continued to flow for a period, resulting in gas bubbles being trapped in the ice. As the ice continued to form, the pressure continued to increase and compressed the gas bubbles. Eventually, the pressure

(a)

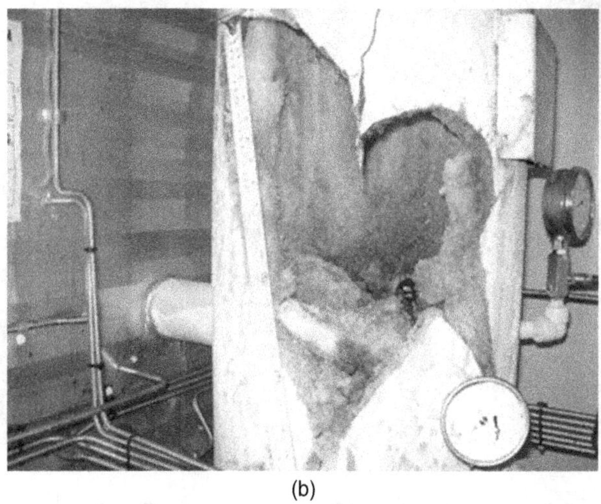

(b)

Figure 8.11.7 Failed inlet separator vessel with large segments of vessel shell missing. Source: Photos are courtesy; Marsden, R.

Figure 8.11.8 Cross section of the failed inlet separator vessel interior, showing "chevron marks," a clear indication of brittle fracture failure. Source: Photo is courtesy; Marsden, R.

was sufficiently high to instantaneously collapse the stainless-steel float for the high-level switch. Tests performed on an identical float determined the collapse pressure of the float was 3,500 psig (24,131 kPa). The energy associated with the instantaneous collapse of the float exceeded the fracture toughness of the vessel's shell and initiated the brittle fracture (See Figure 8.11.10 for an image of the collapsed float). The release of the energy stored in the compressed gas bubbles caused the three shrapnel pieces to be propelled violently from the vessel (See Figure 8.11.9a). The largest piece was found penetrating the building wall (see Figure 8.11.9b). The Separator vessel collapsed float ball is shown in Figure 8.11.10.

A detailed investigation was completed and included witness statements, a systematic approach to gathering operations and weather data, photographs of the scene, determining the collapse pressure of the float, computer simulation of energy to collapse float, and metallurgical analysis of damaged equipment. As previously noted, the investigation was comprehensive, with detailed documentation of lessons learned and the completion of a Root Cause Failure Analysis (RCFA). The RCFA focused primarily on physical root causes but acknowledged human and organizational (latent) root causes that might have contributed to the incident. This incident resulted in a research collaboration with the University of Alberta spanning several years (McDonald et al. 2014).

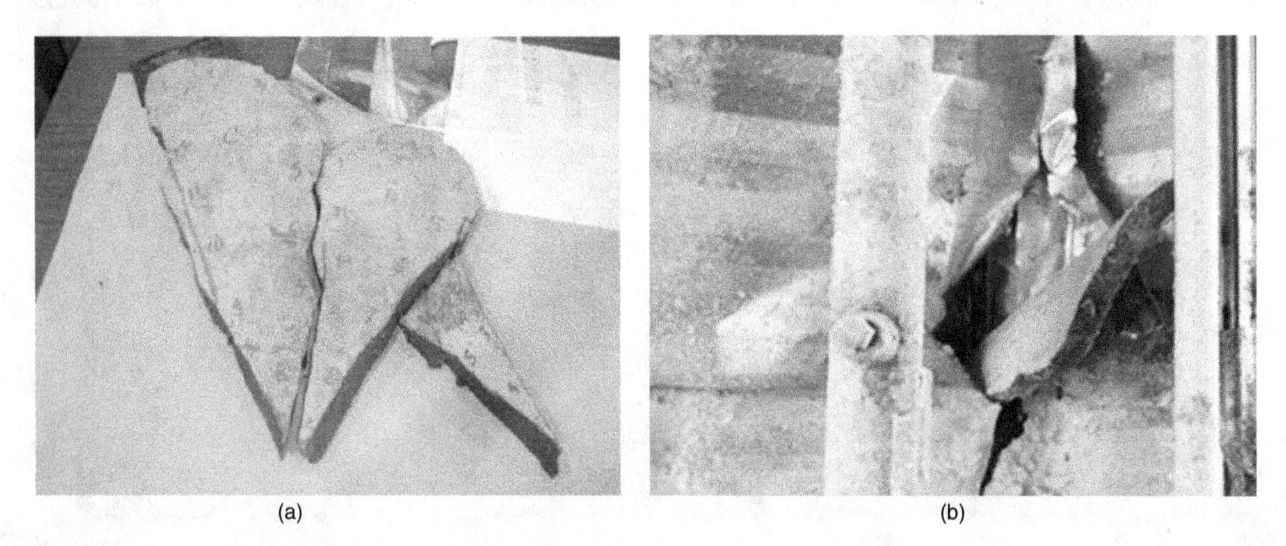

Figure 8.11.9 (a) Three metal fragments from the failed inlet separator vessel. (b) One projectile that penetrated the building wall. Source: Photos courtesy; Marsden, R.

Figure 8.11.10 Photograph of the collapsed float ball inside the inlet separator vessel. Source: Photo is courtesy; Marsden, R.

Below are several significant "findings" or lessons learned to help prevent a recurrence. Causal factors are also identified where applicable. However, this is not intended to be a complete listing of those in the report and is not in any special order. Additional recommendations to help prevent a reoccurrence are included in the summary of this chapter on brittle fracture.

- The inlet separator and associated piping were frozen, with a significant amount of ice still attached to the interior of the separator vessel.
- The flowline separated at a threaded connection due to internal freezing of the line and inadequate thread engagement. Notably, there was no significant fluid release, indicating that the line was frozen at the time of the pipe failure.
- The two separator vessels were fabricated from a carbon steel material not rated for low temperatures (Physical Causal Factor).
- The thermostat control for the building heaters was located too far above the floor level (Contributing Cause).
- Piping supports for the dump line were thermodynamically connected directly to the cold metal floor, creating a large heat sink to draw heat away from the dump line (Contributing Cause).
- Metallurgical analysis of the fragments from the inlet separator vessel confirmed that the inlet separator failure was consistent with an instantaneous brittle fracture.
- Analysis of the high-level switch determined that an internal "O" ring was missing resulting in the failure of the switch to operate as designed and vent the ESDV actuator (Physical Root Cause). This allowed gas and liquids to continue to flow into the separator, resulting in liquids carrying over into the FG/IG separator vessel.
- The float-actuated valve on the FG/IG separator was assembled incorrectly (Physical Root Cause).
- The building heaters were reliant on produced gas. Any interruption to the flow of gas or liquids in the fuel gas would extinguish the heaters (Contributing Cause).
- Material selection during the design phase is very important and should consider the operating environment, including ambient temperature and process temperature during normal and abnormal operations. In this case, a more suitable material choice would be SA-333 Grade 6, which has proven toughness properties at the ambient temperatures of the wellsite.

Let us look at how this incident fits into the brittle fracture triangle as discussed in the introduction to this chapter (the three requirements needed for a brittle fracture to occur):

- **Low toughness (susceptibility to brittle fracture):** Low toughness resulted from the carbon steel vessel being pressurized while at very low temperatures, temperatures well below the MDMT for the steel pipe used during the fabrication of the vessel. Note the ice formed in the images above.
- **Defect (or initiation point for the brittle fracture):** Determined to be a poor-quality weld and weld splatter near the welded coupling used for the level float.
- **Stress (or initiating event):** Determined to be the shock created when the level float ball collapses due to excess internal pressure.

What Has Happened/What Can Happen

Brittle Fracture Failure
Esso Longford Gas Plant; Longford, Australia
25 September 1998
2 Fatalities/8 Injuries

This facility has been described as "one of the most important industrial facilities in Australia." At the time of the incident in 1998, it was the major supplier of natural gas to this area of Australia. The fire resulted in two deaths and eight severe injuries, and the facility shutdown for 20 days due to the resulting equipment damage. The loss of natural gas significantly impacted the region, including some industries shutting down during this period. Also, natural gas was unavailable for home heating, cooking, and other residential and industrial purposes.

Following the fire at Longford, an Australian Royal Commission was appointed to investigate the incident, resulting in a detailed report being issued in June 1999. This comprehensive report provides significant insight into the complexities of the operation, including the heat and product integration between the three gas plants, the Gas Plant-1 (GP-1), the Crude Oil Stabilization Plant, and the Long Island LPG plant. Much of the information in this chapter is extracted from this report and condensed, and key lessons learned are shared to help ensure an event like this one is not repeated.

The Longford Gas Plant consists of three separate gas separation plants to process the incoming natural gas from the platforms in the Bass Strait. Gas Plant-1 (GP-1), the first plant built at Longford, uses lean oil (similar to kerosene) in counterflow in a liquid extraction tower to absorb the heavier hydrocarbons (Ethane, Propane, and Butane) from the incoming natural gas stream. The resulting "rich" oil is then fractionated in a Deethanizer tower and a lean oil fractionator to recover the heavier components (the LPG) from the natural gas. The lean oil is then cycled back to the extraction towers to support the continuing process. The natural gas and LPG are then sent to consumers via off-site pipelines. This process is standard in the industry and is used in many oil refineries and petrochemical plants to remove heavier ends from gas streams.

The other two plants (GP-2 and GP-3) were built later and used cryogenic technology to separate the incoming light materials from the natural gas. The initial fire occurred at Gas Plant 1 (GP-1), although the remaining two plants (GP-2 and GP-3) were also affected and shutdown due to the fire. The site also contained a Crude Oil Stabilization Plant to process oil from the Bass Strait (less involved in this incident). Additionally, the three gas plants (GP-1, 2, and 3) are integrated with a separate facility known as Long Island Point, which receives the raw LPG (ethane, propane, butane) and processes the raw LPG into saleable products.

On Friday, 25 September 1998, at about 12:26 p.m., a major fire occurred at the Esso Longford Gas Plant in Victoria, Australia. The fire resulted in the deaths of two Esso employees and injuries to eight others. Supplies of natural gas were disrupted throughout the state of Victoria, including to most industrial facilities, for several weeks. The incident occurred when the Deethanizer bottoms reboiler heat exchanger failed due to a brittle fracture, causing the release of large amounts of hydrocarbon vapor and the subsequent series of explosions and a major fire. Secondary fires resulted when other interconnecting piping failed during the fire. The remaining operating crew had difficulty isolating the leaking piping due to the many interconnecting lines between the three gas plants and other interconnected facilities.

The brittle fracture occurred when the Operators responded to a plant upset due to the carrying over of LPG (liquid ethane and propane) into the low-pressure section of the unit. As the liquid LPG vaporized, it auto-refrigerated, resulting in extreme cold conditions in equipment not designed for these conditions and the loss of the lean oil pumps due to vaporized ethane and propane. During the upset, one of the exchangers started leaking hydrocarbons at both ends of the exchanger.

During the unit upset, the Operators called for mechanical support to tighten the bolts on the leaking exchanger heads. They realized the exchangers were very cold. The temperatures in the lower portion of the Deethanizer reached about $-54\,^{\circ}F$ ($-48\,^{\circ}C$), contributing to the leak of the exchanger flange. The Operators felt restarting lean oil circulation would bring temperatures closer to normal and stop the exchanger leaks. When the lean oil pumps were restarted and circulation

Figure 8.11.11 Gas plant deethanizer reboiler exchanger. Photo from the Australian Royal Commission report on the Esso Longford gas plant incident. The Royal Commission determined the origin of the crack was at the 8:00 o'clock position in the tube sheet to channel weld (see photograph below). Source: Photo is with kind courtesy of Det Norske Veritas (annotation by investigation team).

reestablished, the warm lean oil entering the extremely cold equipment resulted in thermal shock and rapidly increasing internal pressures. This caused GP-905 (the Deethanizer reboiler exchanger) to fail catastrophically due to a brittle fracture. See Figure 8.11.11, photo of failed GP-905 exchanger.

The massive vapor release from the ruptured exchanger was ignited within a few minutes from a fired reboiler nearby. This resulted in a large jet fire impinging directly onto an adjacent pipe rack, causing secondary failures of in-service piping and rapid escalation of the incident.

The detailed metallurgical analysis supports the conclusion of a cold, brittle fracture failure. A quote from the Royal Commission Report states, "On the balance of probabilities, the additional stress required to cause the failure arose from the temperature differences between the channel and shell. The higher temperature in the shell was due to the introduction of hot lean oil resulting from the restart attempts of the GP1201 pumps."

Findings by the Commission included the following *(condensed)*:

- The facility's safety culture was more oriented toward personnel safety.
- A HAZOP had not been conducted on the heat exchange system.
- Previous excursion incidents were not adequately reported to management.
- Isolation of equipment was difficult due to facility design.
- Personnel training in normal operating procedures was less than adequate.
- Excessive alarms caused operator distractions from the primary event.
- Plant engineers had previously been relocated to Melbourne and were not on-site to support the Operators.
- Less than adequate communication between shifts.

An independent review noted that the Longford plant was a well-run facility with a world-class safety record. However, the operators were not sufficiently trained to recognize the potential for brittle fracture or the proper response to bring the unit back to safe conditions following the extreme cold excursion.

Let us look at how the Longford event fits into the brittle fracture triangle as discussed in the introduction for this chapter (the three requirements needed for a brittle fracture to occur):

- **Low toughness (susceptibility to brittle fracture):** Low toughness resulted due to the carbon steel exchanger shell being at very low temperatures due to internal auto-refrigeration of light hydrocarbons (around −54 °F) (−48 °C). The temperatures were well below the MDMT for the steel pipe used during the fabrication of the vessel.
- **Defect (initiation point for the brittle fracture):** Analysis determined the origin of the failure was in a weld between the channel and tube sheet on the Deethanizer reboiler (GP905) – some embedded slag was found in the weld.
- **Stress (or initiating event):** Determined to be a thermal shock that occurred when the lean oil pumps were restarted, sending warm lean oil into the very cold exchanger.

What Has Happened/What Can Happen

Dutch State Mines (DSM)
Naphtha Steam Cracker Brittle Fracture Failure; Explosion and Fire
Beek, Limburg (Holland)
7 November 1975 (14 Fatalities / 107 Injuries)

This tragic explosion and fire resulted in the deaths of 14 people and injuries to 107 others, some very seriously. In this devastating fire, the naphtha steam cracker and much of the infrastructure were essentially destroyed (See Figures 8.11.12 and 8.11.13).

The Dutch State Mines plant was an Olefins steam cracker feeding naphtha and producing ethylene for the plastics and chemical industries. Many records were lost in the explosion and fire, but investigators believe the unit was in startup mode following a maintenance downtime. During the startup, the propylene compressor tripped offline. The incident occurred while the unit was returned to service following the compressor problem. Compressed gas was sent to the low-temperature system, including the Deethanizer tower, at about 6:00 a.m. during the restart. This resulted in an upset in the light ends systems, cold temperatures, and high pressures in process vessels not designed for these conditions. Just before 10:00 a.m., witnesses reported seeing a large vapor cloud in the area of the Depropanizer feed drum, followed within a few minutes by the catastrophic explosion and fire. As indicated above, 14 lives were lost, and 107 people were injured. A significant portion of the unit and infrastructure was destroyed. Three of those injured were outside the facility. The resulting fires, including several large storage tanks, burned for several days.

Figure 8.11.12 Photos of Dutch state mines naphtha cracker after the November 1975 explosion and fire contains public sector information published by the Health and Safety Executive and licensed under the https://www.nationalarchives.gov.uk/doc/open-government-licence/version/3/. Source: Reproduced by the kind permission of HSE. HSE would like to make it clear that it has not reviewed this product and does not endorse the business activity of Safety Consulting International, LLC.

Figure 8.11.13 Photos of Dutch state mines naphtha cracker after the November 1975 explosion and fire contains public sector information published by the Health and Safety Executive and licensed under the https://www.nationalarchives.gov.uk/doc/open-government-licence/version/3/. Source: Reproduced by the kind permission of HSE. HSE would like to make it clear that it has not reviewed this product and does not endorse the business activity of Safety Consulting International, LLC.

An investigation was conducted to identify the cause. However, the investigation appears complicated due to the loss of records and the significant damage. It is believed that the upset in the light ends section, primarily the Deethanizer and Depropanizer towers, and the resulting cold temperatures led to a cold, brittle fracture. The depropanizer feed drum normally operates at 149 °F (65 °C). The process upset in the Deethanizer tower resulted in the Depropanizer feed stream being very cold, about 32 °F (0 °C), and the feed stream contained an unusually high percentage of ethane. This resulted in LPG flash vaporization and auto-refrigeration in the depropanizer feed drum, with resulting drum temperatures as low as 14 °F (−10 °C).

The failure occurred at the 1.5 inch (40 mm) carbon steel diameter pipe to the safety valve on the Depropanizer feed drum and was consistent with a cold brittle fracture. Investigators reported that the depropanizer feed drum metallurgy should have been suitable for temperatures as low as −4 (−20 °C). However, the investigation team also believed the failure occurred on this piping riser at a weld. The report also indicated that this weld could fail at temperatures as high as 32 °F (0 °C) due to its age and length of time in service. Investigators concluded that the feed drum relief valve opened, causing vaporization of the light material, auto-refrigeration, and possibly vibration, all contributing to the brittle fracture.

Let us look at how the Dutch State Mines event fits into the brittle fracture triangle as discussed in the introduction for this chapter (the three requirements needed for a brittle fracture to occur):

- **Low toughness (susceptibility to brittle fracture):** Low toughness resulted due to the carbon steel safety valve riser pipe being at very low temperatures due to internal auto-refrigeration of light hydrocarbons.
- **Defect (initiation point for the brittle fracture):** Analysis determined the origin of the failure was in a weld in the piping.
- **Stress (or initiating event):** Thought to be associated with LPG flash vaporization, auto-refrigeration, and vibration of the safety valve.

What Has Happened/What Can Happen

Vessel Hydrotesting Failures (Testing the Vessel Integrity with Pressurized Water)

Many equipment brittle fracture failures occurred during pressure testing with water to ensure vessel integrity before bringing the equipment back into service (hydrostatic testing). Hydrotesting at ambient temperature results in high stresses in the

Figure 8.11.14 Vessel failures due to hydrostatic testing with water at a temperature below the vessel's Minimum Design Metal Temperature (MDMT). Source: Photo kind courtesy of TWI Global, Ltd.

piping and vessel (due to internal pressure), especially when the vessel steel has low toughness (due to low temperatures). This results in high susceptibility to brittle fracture. In most failures, the water temperature or vessel MDMT was unknown and not verified before the pressure test. Most of these failure reports state that the vessel was being tested with "cold water," which likely means the water was below the MDMT.

The water and vessel temperature should always be verified during the planning stage for the vessel hydrostatic test to ensure the temperatures are above the verified MDMT for the vessel being tested (verified MDMT data from the vessel data plate). This ensures that the vessel and water temperatures are well above the vessel MDMT to avoid the potential for brittle fracture during the pressure test. Special procedures may be required when a hydrostatic test is being performed during cold weather periods or when the ambient temperature is below the vessel MDMT. This may include building temporary shelters with portable heaters or other means of protecting the vessel against the potential of brittle fracture due to cold conditions.

Let us look at how a hydrostatic test failure fits into the brittle fracture triangle as discussed in the introduction of this chapter (the three requirements needed for a brittle fracture to occur):

- **Low toughness (susceptibility to brittle fracture):** Low toughness results when the vessel (generally a carbon steel vessel) is at a temperature less than the vessel MDMD during the hydro test.
- **Defect (initiation point for the brittle fracture):** This is often identified as a weld or machined notch or possibly a lamination in the steel of a similar flaw. We should anticipate that an initiation point for the brittle fracture will be present.
- **Stress (or initiating event):** In most cases, stress is the internal pressure associated with the hydrostatic test.

See Figure 8.11.14 for an image of an exchanger that experienced a catastrophic failure while being hydrostatically test using water at a temperature below that of the exchanger (MDMT).

What About Pneumatic Pressure Testing (Testing the Vessel with Pressurized Nitrogen, Air, or Another Pressurized Gas)

The hazards of pressure testing process vessels and piping using a pneumatic such as Nitrogen or air is covered in more detail in Chapter 8.15. However, it is briefly covered here due to the potentially catastrophic consequence of a brittle fracture failure during a pneumatic pressure test. The discussion here involves a pneumatic test to verify the integrity of the equipment (when done instead of the typical hydrostatic test).

Pneumatic testing is particularly hazardous, regardless of the failure mechanism. In a hydrostatic test (pressure testing with water), the energy is released instantaneously when a leak or failure occurs. Water is noncompressible, and therefore the initial leak instantly releases the pressure. However, a pneumatic is compressible, and energy must be applied to compress the gas to reach the designated test pressure. In the event of a failure, this "stored energy" is instantly released in the form of an "explosion." Not an explosion that we usually think about with a fireball but an explosive release of compressed gas. This sudden release of stored energy can propel shrapnel and other materials long distances.

In a brittle fracture, the consequences can be catastrophic due to the vessel failing in "shards" or large fragments of broken steel. The stored energy can propel these fragments with great force for hundreds of feet. The shrapnel can strike personnel, resulting in the potential for severe injury or causing secondary damage to process equipment and potentially the loss of containment of flammable or toxic materials.

A pneumatic pressure test should typically be the method of last resort for pressure testing process vessels. Anytime a pneumatic pressure test is proposed or considered, a detailed and thorough hazard analysis should be completed by personnel with a good knowledge of the brittle fracture potential. The review team should always ensure that the hazard review considers the brittle fracture potential and that safeguards are enforced during the test. The assessment should be reviewed and approved or endorsed at the senior management levels at the site before allowing a pneumatic test of process vessels. This is particularly important on vessels constructed with heavy wall designs due to the brittle nature of these materials.

Key Lessons Learned

The good news is that brittle fractures do not frequently happen due to the factors that must all be present simultaneously for the incident to happen. However, as indicated by these three case studies, when a brittle fracture does happen, it often occurs without warning. It can be catastrophic regarding the impact on people, the environment, and the facilities. Therefore, the potential for brittle fracture should always be considered a higher risk during equipment design, commissioning, operations (normal and abnormal), risk assessment (HAZOP), and equipment inspection.

Due to the nature of the brittle fracture, Operators and others need to understand when we are susceptible and what actions to take to minimize the risk and return the unit or equipment to a safe operating condition. While not intended to be a complete list, the following are examples of situations or operations when we may risk a catastrophic brittle fracture failure.

- Transient operations such as unit upset, startup, and shutdown of equipment where the temperature is reduced while pressure is maintained (best if temperature and pressure are reduced together).
- For thick wall carbon steel vessels, it may be necessary to closely control the vessel heat-up and cooldown rates and, in some cases, the "soak temperatures" to ensure the vessel is evenly warmed without temperature gradients in the vessel. Thick metals may have temperature gradients; measured at the surface, they will appear fine, but without adequate time or soak, they may still be below the transition temperature (MDMT) inside or within joints.
- Any emergency depressurization of LPG resulting in auto-refrigeration.
- Safe parking/squatting a unit while it is pressurized for a longer period, particularly over a cold night.
- Hydrotesting with cold water (water at a temperature below the MDMT for the vessel being tested) or when a vessel is at a cold ambient temperature.
- An isolated pressure vessel at a temperature below the MDMT and in a full water condition (no vapor space) is susceptible. A very small increase in temperature will result in a large pressure increase and potential brittle fracture in this state.
- The failure to remove nitrogen from a vessel before introducing liquefied light ends (LPG). The hydrocarbons flash into the nitrogen, producing temperatures much colder than when nitrogen is not present.
- Anytime equipment ices up, especially when it does not normally ice up.
- The use of liquid nitrogen to speed up the cooling of catalyst or process equipment. Nitrogen auto-refrigerates when it is flashed from a liquid to vapor, and the cold Nitrogen can reduce the temperature of the equipment to below the MDMT.
- Leaking LPG equipment or exchanger internals (such as ethane or propane) results in cold spots due to auto-refrigeration
- The stopping or diverting of the warm stream in an exchanger.
- The fractionator reboilers stop functioning and allow cold liquid to dump into the associated fractionator.
- The failure to follow approved operating procedures without a thorough review (applies to normal and emergency procedures).

- The bypassing of low-temperature trips or controls.
- Introducing abnormal materials or substituting one for another without formal review and approval.
- The relief valve fails open, resulting in a rapid pressure drop, followed by placing the spare relief valve in service and the system rapidly repressured (e.g., from an upstream higher-pressure system).
- Depressurizing liquids through a system designed for vapor (e.g., on vessel overfill).
- Flashing liquid LPG across a relief valve results in auto-refrigeration and upstream or downstream carbon steel piping failure.
- Shock chilling – introduces large additional stresses.

Sites or process units that process LPG or where they operate in climates where equipment can be exposed to cold conditions should have procedures in place to help Operators identify the brittle fracture hazard and, in the event of an excursion, the necessary steps to protect people and safely address the risk (more detail in Chapter 8), ("Importance of Operating Procedures"). It is equally important that Operators are familiar with the required procedures, including periodic refresher training on the procedures and participate in routine tabletop drills and exercises involving brittle fracture scenarios.

Due to the variety of operations, these procedures must be developed specifically for each process or facility. The following are generic response guidelines for recognizing the brittle fracture potential. These guidelines are intended only as examples for training and discussion purposes and should not be used for an actual response for an excursion resulting in temperatures outside the safe operating conditions (i.e., temperature deviation below the MDMT).

Generic Guidelines to Respond to a Potential Brittle Fracture Scenario

- Recognize that a brittle fracture with catastrophic consequences is possible.
- Take appropriate steps to minimize personnel exposure (evacuate personnel from the area around the equipment at risk).
- Stabilize the unit operations – do not make sudden changes in operations variables (temperature, pressure, flow, or level).
- If possible, slowly drain or remove the remaining LPG liquid from the vessel remotely (do not send people to the vessel to do this).
- Very slowly reduce the vessel pressure while closely monitoring the vessel temperature:
  - The objective is to reduce the pressure to 25% or less of the maximum allowable working pressure (MAWP).
  - Recognize that this may take many hours or even days, especially in LPG vessels or vessels containing LPG (due to the continued auto-refrigeration of the LPG).
- During this operation, the vessel pressure should not be increased. Increasing the pressure while the vessel temperature is below the MDMT may result in a brittle fracture.
- Also, do not add a "dry" gas (e.g., Nitrogen) during this period. Adding nitrogen or any dry gas to a vessel containing liquid LPG can result in flash vaporization of the LPG into the dry gas, leading to auto-refrigeration and driving the temperatures even lower, which may result in brittle fracture.
- After the vessel pressure is below 25% of the vessel MAWP, a thorough examination of the vessel for cracks, especially on welds on nozzles, bolts/studs, flanges, and fittings, should be completed by the Materials Engineer or Metallurgist before the vessel is returned to service.

Guidelines to Prevent a Potential Brittle Fracture Scenario

- Actively participate in emergency response drills and exercises. Practice equipment isolation from remote isolation valves. What can we do to help prevent a catastrophic brittle fracture?
- We have all probably heard the motto, "design it right, build it right, operate it right, maintain it, and inspect it." Nothing could be more accurate when it comes to brittle fractures. This includes consideration for the proper metallurgy and PWHT to ensure the equipment is suitable for the minimum temperatures to which it may be exposed (temperatures including the lowest ambient weather conditions, normal process temperatures, and low temperatures that can result from abnormal operations or upsets). Guidance is readily available in API RP 920, API RP 579-1, and ASME FFS-1.
- Ensure that operating limits are established to prevent brittle fractures and that these are rigorously enforced. Deviations should be reported and analyzed, and corrective actions should be taken to help eliminate reoccurrences. Operate within pressure/temperature limits.

- Unit procedures should also include appropriate cautions and warnings regarding deviations to critical operations parameters or operating limits, especially for transient operations such as startups and shutdowns.
- Ensure operators (field and panel operators) are trained on brittle fractures, including recognizing the potential and steps to take to return to safe operations.
- A periodic inspection ensures the equipment is in good condition and free from defects that may become initiation points for brittle fractures.
- Any changes to vessels should always include a detailed MOC with reviews by the materials engineer or metallurgist.
- The potential for brittle fracture should always be considered during unit Process Hazards Analysis (PHA/HAZOPS) and mitigations identified and implemented.
- If the vessel temperatures are decreasing and are approaching the MDMT, the pressure should be reduced to 25% or less of the MAWP (See prior paragraph for more detail). Verify that critical brittle fracture data is included in operating procedures and readily available to console and field Operators (e.g., MDMT/MSOT).
- Ensure that brittle fracture potential is considered during the following operations:
  - Hydrotesting pressure vessels.
  - Transient operations such as startup or shutdown.
  - Cold temperature excursions such as auto-refrigeration of LPG.

What About After the Cold Exposure; Is it Safe to Return the Equipment to Service if the Vessel Has No Apparent Damage?

Following a low-temperature excursion resulting in the vessel temperature experiencing temperatures below the MDMT, the vessel should be inspected to ensure fitness for return to service. Most importantly, the equipment specialists should be involved in this analysis. They should be involved in the decision to return to service before attempting to increase the temperature or pressure on the vessel. In some cases, the vessel may require extensive inspection before returning to pressurized service.

It is also very important that immediately following a low-temperature excursion, operations should be aware and prevent any effort to rapidly heat the vessel or allow a "hot" stream into the cold vessel. Experience has shown that this adds significant thermal stress and can result in an instantaneous brittle fracture (Brittle Fracture Case Study 2 in this chapter).

Recommendations from The National Board of Boiler and Pressure Vessel Inspectors (NBBI) http://www.nationalboard.org) include the vessel being slowly warmed in a non-pressurized condition and the vessel being thoroughly inspected for cracks before returning to service. The recommendation provides good guidance on the types of inspection that may be required. More detail is available on the board's website (the board's white paper is also included as a reference in Appendix C ("Auto-Refrigeration: When Bad Things Happen to Good Process Vessels"). Reprinted with permission from the NBBI).

Additional References

Marsden, R. (2007). Brittle fracture case study and root cause failure analysis. *Proceedings of API Inspector's Summit*, (6–10 February 2007). Galveston, Texas.

McDonald, A., Bschaden, B., Sullivan, E., and Marsden, R. (2014). Mathematical simulation of the freezing time of water in small diameter pipes. *Applied Thermal Engineering* **73**: 140–151.

Rezvani Rad, M., McDonald, A., and Marsden, R. (2018). Testing and analysis of freezing phenomenon in conventional carbon steel pipes. *Proceedings of The Canadian Society for Mechanical Engineering International Congress 2018*. http://dx.doi.org/10.25071/10315/35356

British Health and Safety Executive, Report on Dutch State Mines Steam Cracker Explosion; November 7, 1975, (Brittle Fracture). https://www.hse.gov.uk/comah/sragtech/casebeek75.htm

"The Esso Longford Gas Plant Accident Report of the Longford Royal Commission", The Honorable Sir Daryl Michael Dawson, AC KBE CB-Chairman, Mr. Brian John Brooks, BE FIEAust FAIP FAIE FIE-Commissioner, June 1999, By Authority. Government Printer for the State of Victoria, No. 61- Session 1998-9. https://www.parliament.vic.gov.au/papers/govpub/VPARL1998-99No61.pdf

"Lessons from Esso's Gas Plant Explosion at Longford", Andrew Hopkins PhD, Australian National University. http://www.futuremedia.com.au/docs/Lessons%20from%20Longford%20by%20Hopkins.PDF

The National Board of Boiler and Pressure Vessel Inspectors, Note: Includes guidance for vessel inspection following a cold temperature excursion. http://www.nationalboard.org

Article: "Brittle Fracture": A quick primer on brittle fracture. https://www.myodesie.com/wiki/index/returnEntry/id/3060

Lessons from the Longford Gas Explosion and Fire (White Paper), R. B. Hutchison (A), D. M. Boult (B), R. M. Pitblado (C), G. D. Kenny (D).

A: Safety and Risk, Det Norske Veritas, Level 19, 100 Miller St, North Sydney, 2060, Australia

B: Process Marketing and Sales, Det Norske Veritas, 16340 Park Ten Place, Suite 1000, Houston, TX 77084, U.S.A.

C: District Manager, Det Norske Veritas, Palace House, 3 Cathedral St, London, SE1 9DE, U.K.

D: Vice President, Division Americas, Det Norske Veritas, 16340 Park Ten Place, Suite 1000, Houston, TX 77084, U.S.A

http://bowtieconsulting.com.au/pdfs/Lessons%20from%20the%20Longford%20Gas%20Explosion%20and%20Fire.pdf

When Bad Things Happen to Good Process Vessels, Francis Brown, P.E., National Board of Boiler and Pressure Vessel Inspectors. https://www.nationalboard.org/PrintPage.aspx?pageID=164&ID=249

UK Health and Safety Executive, "The assessment of pressure vessels operating at low temperature" (HS(G)93, 2nd edition). https://www.hse.gov.uk/pubns/priced/hsg93.pdf

Compilation of Pressure-Related Incident Summaries, US Department of Energy, Argonne National Laboratory. https://indico.fnal.gov/event/13755/sessions/1912/attachments/14044/17881/AccidentsCompilation_June16.pdf

Chapter 8.11. Brittle Fracture Failures

End of Chapter Quiz

1 Brittle Fracture is a catastrophic type of failure that has resulted in devastating incidents. What types of vessels or equipment can be subjected to this kind of devastating failure?

2 What is the main cause of brittle fracture?

3 How is brittle fracture defined?

4 The potential for Brittle Fracture should be considered during equipment ________ for scenarios where the vessel may experience cold temperatures due to climate (ambient temperature) or the flash vaporization of LPG such as Ethane or Propane.

5 What are the three things that must be present for a vessel to pipe to experience a brittle fracture?

6 How is the Maximum Allowable Working Pressure defined?

7 How is Critical Exposure Temperature defined?

8 A new vessel should have a code stamp and data plate, plus a technical design datasheet that specifies the design parameters (including ________ and ________). New designs should ensure that the ________ is above the MDMT.

9 How is brittle fracture managed for older vessels?

10 Briefly describe post-weld heat treatment or sometimes referred to as "stress relieving."

11 Name three operations tasks where the potential for brittle fracture should always be considered.

12 Ensure that __________ _________ are established to prevent brittle fracture and that these are rigorously enforced. Deviations should be reported and analyzed, and corrective actions taken to help eliminate reoccurrences. Operate within pressure/temperature limits.

13 What about after the vessel exposure to cold temperatures; is it safe to return the equipment to service if the vessel has no apparent damage?

14 What are the key steps to take if a vessel is discovered to be operating at pressure and a temperature below the MDMT?

15 What recommendations are from the National Board of Boiler and Pressure Vessel Inspectors following a pressure vessel exposure to temperature and pressure below the Minimum Design Temperature, a near miss involving a potential brittle fracture failure?

16 Why must a Management of Change (MOC) review be completed after weld repairs on a certified API 510 pressure vessel?

17 What is the hazard associated with conducting a pneumatic pressure test on a process vessel or any large vessel?

18 Flash vaporization and auto-refrigeration of some light hydrocarbons such as Propane or Butane can result in temperatures as low as –___ °F (–___ °C) and, therefore, certainly have the potential to result in brittle fracture of vessels or piping.

8.12

Hazards of Furnaces and Heaters

Due to the critical nature of furnace operations and the high incident rate involving furnaces, we will spend a little time discussing furnace operations and the types of incidents that have occurred. In furnace operations, we purposely bring fuel, air, and an ignition source together inside a confined space (the furnace box); we do that with oil or gas flowing through the furnace tubes. It is easy to see how this can get away from us, resulting in an explosion or significant loss of containment. Unfortunately, there have been many furnace explosions and fires in our industry, and they continue to occur. Many people have been seriously injured, and lives have been lost in these incidents. For this reason, I consider this topic to be one of the most important in this book.

A few of the main topics we will cover in this chapter include:

- Furnace procedures, including startup, shutdown, and routine procedures.
- Failure consequences (furnace tube failure involving loss of containment).
- Emergency shutdown system, especially operating with the system bypassed or inoperable.
- Failure to follow prescribed procedures during startup or shutdown.
- Loss of flow in a single pass in a multiple-pass furnace.
- Operating a furnace in a fuel-rich (flooded) state (a significant hazard).
- Coking inside the furnace tubes.
- Afterburning.
- Operating the furnace with high tube metal temperatures.
- Flame impingement on the furnace tubes.

Furnace Procedures are Extremely Important and Must Be Followed in Detail

Procedures, including startup, shutdown, and normal operating procedures, should be in checklist format with signed and dated logs for each critical step. This helps ensure that steps are not omitted or executed in the wrong sequence. And most important, the procedures must be followed in the field. If a procedure is incorrect or outdated, the procedure should be updated by redlining the procedure to insert the corrections. The redlined procedure should be approved by plant supervision and the appropriate management of change documents. Then, the revised procedure should be followed by signing each step as it is executed.

Refer to Figure 8.12.1 for an example of a furnace explosion that occurred while the burners were being adjusted.

Furnace Safety Hazards

Furnace/Heater Failure Consequences – Furnace Tube Failure

In the event of a furnace tube failure, there will be a loss of containment of the process, resulting in fire, environmental impact, equipment damage, and downtime for repairs. This can also result in an explosion with potential personnel injury and more extensive equipment damage. The most common cause of tube failure is loss of flow in a single pass or the complete furnace. The process flow through the tubes provides the cooling to keep the furnace tube at a safe operating temperature. Loss of flow results in the tube metal overheating to the point where the tube fails, resulting in the loss of containment. Any indication of loss of flow through a single pass or the complete furnace is a very serious issue and

Process Operations Safety: The What, Why, and How Behind Safe Petrochemical Plant Operations, First Edition. M. Darryl Yoes.

Figure 8.12.1 Furnace explosion; furnace destroyed and fractionator column damaged. Source: Image with the kind courtesy of the American Institute of Chemical Engineers (AICHE), The Center for Chemical Process Safety (CCPS), The Beacon, and NOVA Chemicals Corporation. http://www.aiche.org/CCPS/Publications/Beacon/index.aspx.

demands immediate attention from the operators. Indications may be low pass flows, high tube metal temperatures, or even visual indications of hot spots or bulges on furnace tubes. These require immediate action to reduce the firing rate or shut the furnace down before failure occurs.

Loss of furnace flow should be protected by the furnace Emergency Shutdown Device, which should be programmed to trip the furnace in the event of loss of pass flows. These devices must be in service anytime the furnace is in service. If they are defeated for any reason, for example, for preventative maintenance or repairs, control of defeat and alternative mitigation should be in place while they are disabled (Refer to Chapter 8.4 for information on the Control of Defeat work process).

Operating with Emergency Shutdown Systems Bypassed

The Emergency Shutdown Systems protect the furnace from low process flow, high bridge wall pressures, low pilot or fuel gas pressures, loss of induced draft or forced draft fans, and high temperatures to the induced draft fan, indicating a potential fire in the induced draft fan. If these critical devices are disabled during furnace operation, the furnace is not protected in the event of an abnormal event.

The Emergency Shutdown System should be classified as safety-critical and receive periodic function checks to ensure its availability for service. The trips should remain in service except for preventative maintenance or repairs and then be defeated with alternate protection provided by the Control of Defeat process. An alternate protection plan must be in place that replaces the device's functionality while it is defeated.

Failure to Follow Prescribed Procedures During Startup/Shutdown

Always ensure that a check of the critical instruments and the furnace emergency shutdown system is completed and that all instrumentation is fully functional before returning the furnace to service.

Before the startup of a furnace, the firebox must be thoroughly purged to remove any residual fuel before attempting to light pilots or burners. Attempting to light pilots or burners without a thorough purge may result in an explosion, personnel injury, or loss of life. A good practice is to purge the furnace and then perform a flammable gas test at the peep doors and near the burners to ensure no fuel is present before attempting to light the burners.

If the box is equipped with forced air blowers, the purge can be accomplished quickly by running the induced draft and forced draft fans. Otherwise, if the furnace is equipped with steam injection, this is also an effective way to purge the furnace. An alternative is an air educator installed on the furnace stack. Otherwise, you can simply open the damper and the air registers on the burners and allow the furnace to self-purge using a natural furnace draft. In this case, the box should be purged for a minimum of 30 minutes before gas testing for flammable vapor before lighting the pilots or a burner. If the pilots and burners are extinguished during the furnace startup, you must thoroughly purge the firebox before resuming the relight.

Ensure the feed is circulating through the tubes before attempting to light the burners. The tubes can be severely damaged if the burners are placed in service before a flow is established in the tubes. If the box is a multi-pass design, ensure feed is flowing through all the furnace passes.

Never light a burner from another burner or a hot wall. There is no assurance that the burner will light, which may release unburned fuel into the firebox and result in an uncontrolled explosion.

Operating Furnace Above Maximum Allowable Tube Metal Temperatures (TMT)

Furnaces should never be operated with the tube metal temperatures in alarm status. The TMTs are, at best, a spot check on the tubes. TMTs should never be ignored and must be supplemented by periodic Operator observations looking for hot spots on the tubes. Any indication of a hot spot is a serious condition and deserves immediate attention, as this can lead to tube failure.

Loss of Flow in a Single Pass of a Multiple-pass Furnace

Loss of flow in a single pass of a multiple-pass furnace is a serious concern and deserves immediate attention to prevent tube rupture and loss of containment. If the flow cannot be returned in minutes, the furnace must be shut down to prevent tube failure.

Operating a Furnace in a Flooded (Fuel-rich) State

The symptoms of a fuel-rich state, called furnace flooding or bogging, are:

- Low O_2 (<0.5%) and/or high combustibles (>1000 ppm) are usually the first indications of a potential firebox flooding situation.
- Pulsating flames.
- Local draft gauge swinging widely.
- Hazy appearance in the firebox or smoke from the stack (severe bogging).
- Smoking stack.

Furnace flooding is a very serious situation that must be promptly addressed. When the firebox is fuel-rich, we no longer have the correct ratio of fuel-to-air needed for good combustion (the stoichiometric mixture), and the coil outlet temperature starts going below the setpoint. When the coil outlet temperature decreases, the furnace control will attempt to bring the coil outlet temperature back to its temperature setpoint by opening the fuel gas valve. Adding more fuel gas to an already flooded furnace makes the flooding worse. The furnace cannot correct itself without intervention, and severe flooding can result in a total flameout and an explosion in the firebox. When the burners flame out, the furnace goes to natural draft, pulling in ambient air. The fresh air mixes with the hot fuel in the presence of embers on the tube surface, and the furnace can explode.

The only way to address a flooding furnace is to; (1) Get people away from the furnace. Do not send anyone to the furnace; instead, get people away from the furnace. (2) Place the damper in manual control to prevent the controller from operating the damper. (3) Break the cascade on the fuel gas control and manually reduce fuel to the firebox by 25%. (4) Be patient and wait for the firebox to start making heat again. If the coil outlet temperature does not increase in 8–10 minutes, reduce the fuel again. Continue doing this until the furnace starts making heat again. When the firebox temperature starts recovering and generating heat, place the controls back into automatic, returning the firebox to normal operation.

In a flooding situation, do not trip the furnace. Tripping the firebox puts the firebox in a natural draft, pulling additional air into the fuel-rich firebox. Due to the hot firebox and a fuel/air mixture, the box may explode.

Do not do anything that may increase air to the firebox in a flooding situation. Do not adjust the burners, do not open the peep doors, do not adjust the damper, or anything similar that could bring in additional air into the firebox. This can also result in an explosion in the firebox.

See John Zink Hamworthy's presentation on furnace flooding at the HSE Conference in Austin, Texas, in February 2017. The web address is included as a reference in this chapter.

Coking Inside the Furnace Tubes

Coking is where a layer of coke builds up on the inside of a furnace tube. Coking can occur due to hot spots on the tubes, low furnace flow rates, or a dirty burner with flame impingement on the tubes. The coke acts as an insulator between the radiant heat inside the firebox and the fluid or vapor flowing through the tubes, cooling the metal tubes. Furnace tube coking can be very localized. As a result, the steel gets hotter and can cause the tube to bulge and fail. Internal coke buildup results in rapid tube overheating and accelerated tube failure.

Coking can be prevented by monitoring the tube metal thermocouples and, most importantly, by periodic observation of the tube condition by the operators. Operator observation can also identify dirty burners or burners that are out of adjustment with flames impinging on the tubes. These need prompt attention to prevent coke buildup inside the tubes. See Figures 8.12.2 and 8.12.3 for examples of coke build up inside a furnace tube.

Afterburning

Afterburning is when unburned fuel (fuel that has not been burned at the burner) comes into contact with air in the convection section and burns in the convection section. Radiant tubes are designed to withstand the intense heat from a nearby flame. Convection tubes are not! Convection tubes subjected to a flame due to afterburning will soon become soft and fail. Typical causes of afterburning are unbalanced burners (some rich, others lean), unbalanced boxes (for multi-box heaters), holes in the furnace casing (lower O_2 at burner vs. analyzer), or inadequate or poorly positioned oxygen analyzers. As a result, the fuel combusts after the burner (in the convection section or furnace stack). As a result, the tube design temperatures are exceeded, leading to a failure (tube, tube hanger, or air preheat equipment).

Figure 8.12.2 Example of coke build up inside a furnace tube.

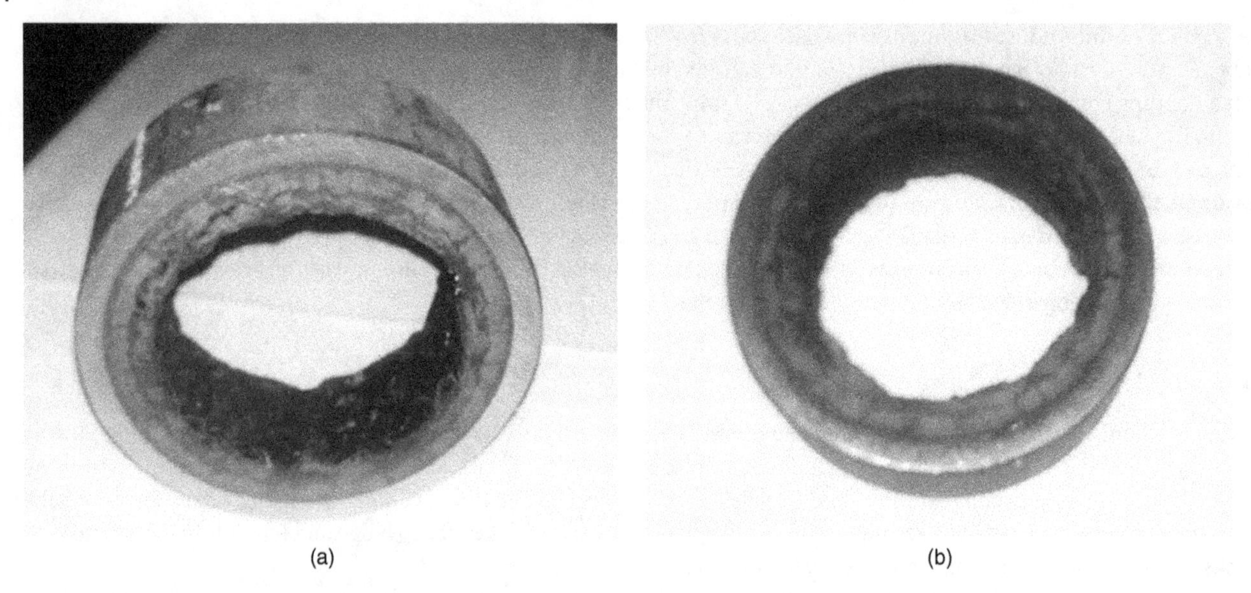

(a) (b)

Figure 8.12.3 Examples of coke buildup inside furnace tubes. Both images highlight the thickness and uniform nature of the coking in this example.

Dirty burners or burners out of adjustment can also cause afterburning. The unburned fuel flows up into the convention section, where it finds tramp air leaking into the firebox from casing leaks in the firebox. The gas is very hot and auto-ignites in the convection section, where the flame impinges on the convention tubes and the tube supports. As a result, the tube hangers can fail to allow the tubes to drop into the convention section. See Figure 8.12.4 for an illustration of damage to convection tubes due to afterburning.

Operating with High Tube Metal Thermocouples (TMTs)

The maximum TMT limits are based on achieving the desired tube life considering tube corrosion and creep rupture. Designers typically design furnace tubes for 100,000 hours; however, if the tubes are operated within the original design temperatures, the tube life can be extended well beyond 100,000 hours. The TMT data should be reviewed daily and constantly monitored by DCS alarms. TMTs should also be supplemented by observation of combustion performance at least twice daily, including a visual inspection for localized hot spots. Periodically, infrared scans of the furnace tubes should validate the TMT data. Remember, the TMTs are, at best, a spot check of the tube metal temperatures. However, they should never be ignored. See Figure 8.12.5a and b for examples of a furnace tube infrared scan.

Flame Impingement onto the Furnace Tubes

When a burner is dirty or out of adjustment, the flames may contact and touch the heater's furnace tubes or other internal components. This can result in furnace tube failure due to hot spots, accelerated corrosion and coke formation at the furnace tubes, and failure of furnace tube supports or refractory.

This is an example of what the Operator should be looking for during their twice-daily observation inside the firebox. Any indication of flame impingement should be promptly addressed, as this can lead to coking inside the tube and an unscheduled shutdown. Dirty burners should be taken out of service until cleaning can be scheduled. Occasionally, a piece of furnace refractory or other material may be lying on the burner. This, too, will require the burner to be removed from service until cleaning or adjustment can be arranged.

A useful technique to determine burner flame pattern and indicate flame impingement is using Sodium Bicarbonate. Sodium Bicarbonate injected into the furnace burners (air inlet) will help make flame patterns visible. This may help identify flame impingement, burner patterns, afterburning, and other issues with burner performance. Ensure combustion engineers are involved when injecting Sodium Bicarbonate due to potential impacts on environmental controls and other furnace equipment. See Figure 8.12.6 for examples of dirty burners or burners out of adjustment and impinging on the furnace tubes.

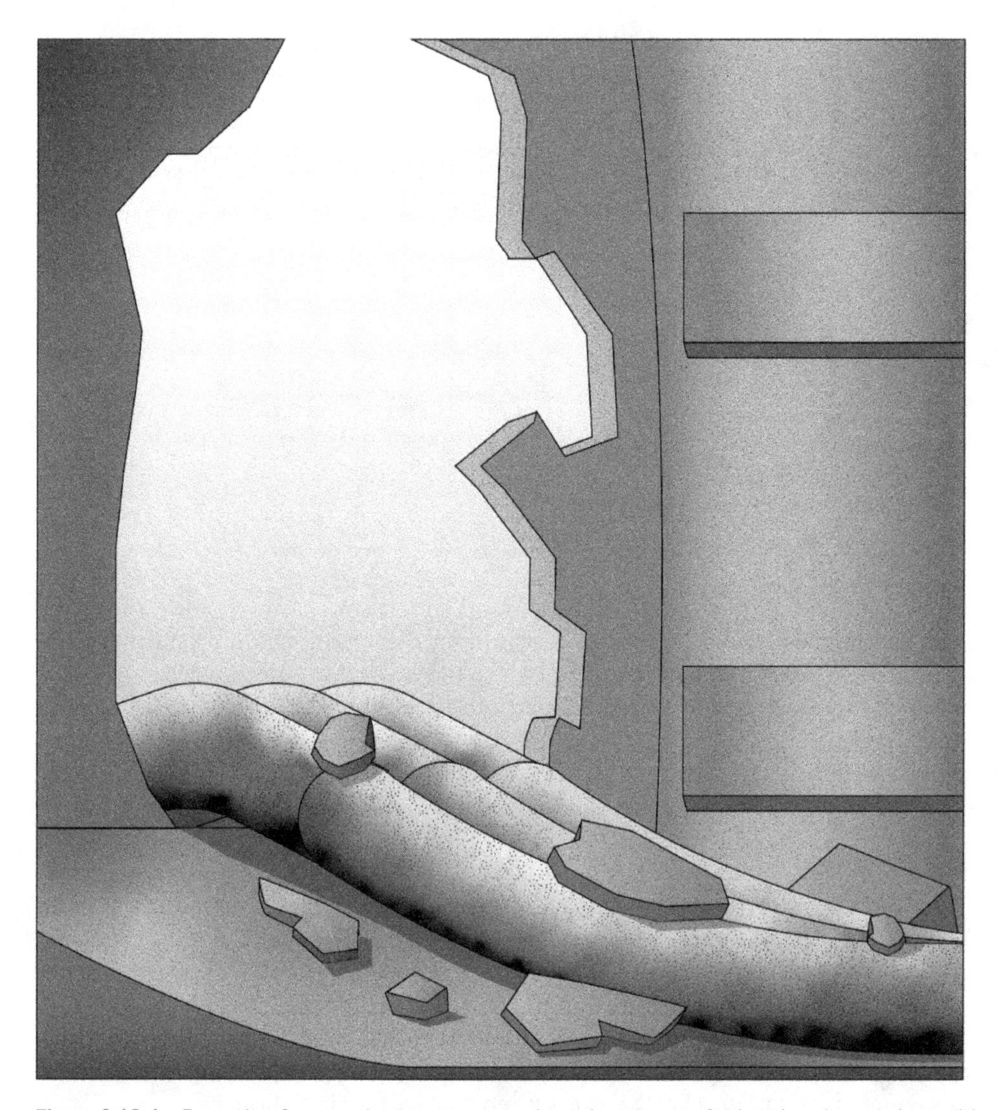

Figure 8.12.4 Example of severe damage to convection tubes due to afterburning. An artist's rendition of furnace convection tubes that were severely overheated due to afterburning. Notice that the tube hangers and supports are completely burned away, and the tubes have sagged due to overheating. This damage results in a complete tube replacement. Source: Image is courtesy of Ecoscience Resource Group Visual Artists.

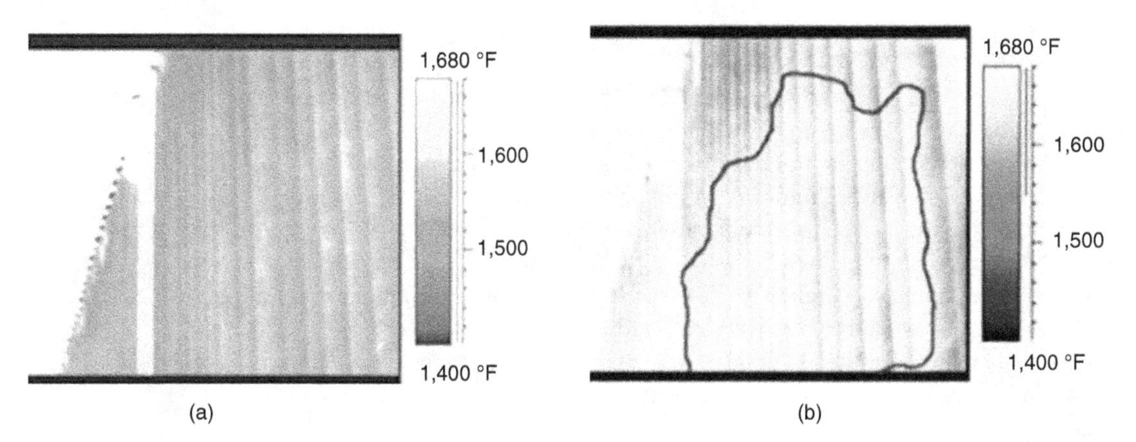

Figure 8.12.5 Examples of a reformer furnace tube infrared scan. The left thermal scan indicates good burner patterns without tube flame impingement. The right scan clearly shows burner flame impingement on the adjacent furnace tubes. Note that the tube metal temperatures are near the top of the temperature scale in this area (white). Source: Photo courtesy Sonny James, Thermal Diagnostics Limited. (http://www.tdlir.com/).

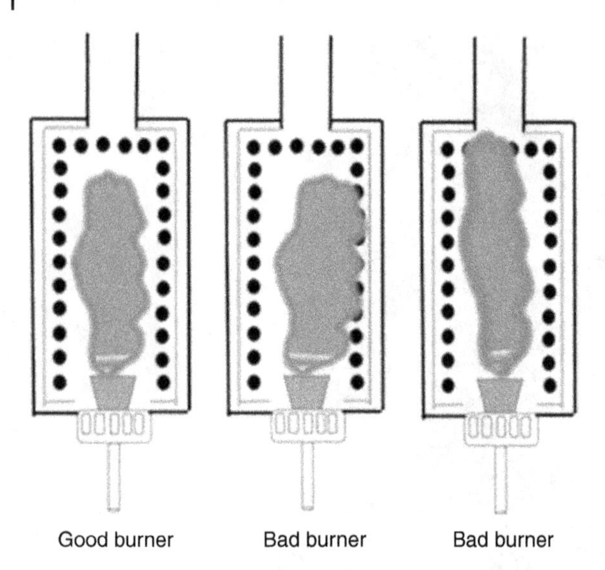

Figure 8.12.6 Examples of dirty burners or out of adjustment resulting in the flame pattern impinging on the furnace tubes. The left illustration is a good burner with no impingement on the tubes. However, the flame patterns illustrated on the two remaining burners are pinging on the tubes. They are either dirty or in need of adjustment. This is a serious condition as it can lead to the coking of the tubes.

Failure to be Aware of Other Furnace Safety Concerns

Be aware of operating at low excess O_2 targets without minimum prescribed instrumentation and controls. Refineries prefer to operate furnaces at very low excess O_2 targets for energy conservation to avoid heating the excess air. However, it is essential to have proper instrumentation and analyzers (O_2 and combustibles) to help identify potential flooding situations when operating at very low excess O_2 targets.

Entry into furnaces (during downtimes) must be properly controlled. Entry into furnaces requires a confined space entry permit and all the other confined space entry requirements (gas testing, hole watch, tracking of people in and out of the confined space, rescue plan).

Preparing vertical tube furnaces for hot work also represents a significant hazard. It is very difficult to remove all the hydrocarbons from the bottom of tall vertical furnace tubes, and there is the potential for "trapped" hydrocarbons in the bottom section of tall vertical furnace tubes. Ensure a method for verifying the tube preparation before issuing a hot work permit for hot work on the bottom section of the tubes. One method is drilling into the bottom of the tube to verify the tube is hydrocarbon free.

Be constantly aware of the potential for fires in air preheaters, especially the induced draft preheater. Unburned fuel from the combustion box can be carried over into the air preheater, where it finds air and can self-ignite in the air preheater. An indication of high preheater temperatures may indicate a fire that can severely damage the air preheater.

Access onto furnace structures (walking off the hard walkways or ladders) should not be done unless the Inspectors have reviewed the walking path and approved the integrity of the furnace roof. There is the potential for failed refractory and damage to the furnace roof, which may not be visible.

Furnace Key Concepts

- ESD system should not be bypassed except for testing and maintenance.
- Closely monitor all critical flows/temperatures during startup/shutdown situations.
- Critical instrument check of Emergency Shutdown System before startup.
- Never attempt to light pilots or burners without first purging the firebox.
- Never light a burner from another burner or a hot wall.
- If all burners and pilots are extinguished, re-purge the firebox before relighting any pilots or burners.
- Install pilot and fuel gas blinds promptly after a furnace shutdown.
- Operating above the maximum allowable TMTs will significantly reduce tube life.
- Timely problem-solving is needed when a sudden change in TMT occurs.
- Never ignore alarms indicating loss of pass flow.
- A prompt and significant reduction in firing should accompany the loss of pass flow.
- The same rules govern furnace entry as with other confined spaces.
- Vertical tube furnaces require special testing near tubes needing hot work.
- Watch out for fires in air preheaters.
- Special precautions are required for nonroutine jobs on furnace structures.

And Finally – A Personal Safety Reminder

Always Wear the Proper Personal Protective Equipment When Near Fired Heaters:

- Hard Hat.
- Fire Retardant Clothing (FRC).
- Eye, Hand, and Foot protection.
- Face shield when under the furnace.
- Hearing Protection.

What Has Happened/What Can Happen

Chevron Refinery
Pascagoula, Mississippi
15 November 2013 (1 Fatality)

An Operator was killed attempting to restart a process unit furnace after the unit tripped due to a power interruption. During the relight process, the furnace exploded, killing the process operator. See Figure 8.12.7 for a photo of the Chevron explosion.

Following this incident, OSHA cited Chevron for failing to provide operators with clear instructions for proper furnace startup after an emergency shutdown. The OSHA report indicated that proper furnace purging procedures were not performed during the attempted restart.

Key Lessons Learned

The furnace relight procedures should be detailed, and specific instructions for a relight vs. a cold startup should be provided. During a relight, it is essential that if the furnace trip is complete, the pilots and burners have tripped, and the firebox must be thoroughly purged to ensure all residual fuel has been eliminated. Attempting a relight without completely re-purging the firebox can ignite any remaining residual fuel, causing an explosion in the firebox.

Figure 8.12.7 Photo of the Chevron furnace explosion. Source: Photo is with the kind courtesy of Matt Defelice.

This should be clear in the procedures, and the procedures should be in a checklist format. This should also be included in the Operator training to help ensure Operator awareness.

What Has Happened/What Can Happen

NOVA Chemical Plant
Bayport, Texas
11 June 2003

An explosion occurred in a Furnace serving as a reboiler for an ethylbenzene fractionator at the Nova Chemical Plant in Bayport, Texas, on 11 June 2003. Fortunately, there were no injuries during this incident; however, an operator had been at the furnace a few minutes before the incident, adjusting the airflow to the burners and adjusting the excess oxygen in the flue gas by adjusting the furnace damper. This furnace had recently been retrofitted with low-NOx burners, and the Operator was attempting to respond to burner instability by adjusting the airflow to the burners and adjusting the excess O_2. See Figure 8.12.1 for a photo of the damage to the furnace and distillation column from this incident.

Shortly after these adjustments, a loud puff was heard in the furnace, quickly followed by the furnace explosion. Fire immediately came from the radiant section of the furnace and the inlet header and continued being fed from the ethylbenzene release from the return piping, which was connected to the ethylbenzene fractionator. The ethylbenzene continued to fuel the fire for about 3½ hours, significantly damaging the furnace and the ethylbenzene fractionator. The reboiler furnace was destroyed, and the fractionator was heavily damaged during this incident.

In a report issued by NOVA Chemicals, it was reported that the cause was linked to plugging in the burner nozzles by iron oxide (rust) from corrosion of the carbon steel burner nozzles, coke deposits from the decomposition of the natural gas due to the high temperatures that were created due to the flow restrictions, and construction debris that had not been removed prior to commissioning the new burners. An inadequate fuel-to-air ratio resulted in a flameout in the furnace and an explosion.

Similar incidents have occurred across the industry with different causal factors. When a furnace goes fuel-rich, when the percentage of fuel exceeds the ratio of fuel-to-air required for efficient combustion, the furnace can become very unstable. If the Operator fails to recognize the cause of the burner instability as furnace flooding (fuel-rich) and attempts to correct the instability by increasing air to the burners, the furnace can explode.

Taking any steps to increase air into a flooded furnace (fuel-rich) can result in a devastating explosion. For example, attempting to increase air to the burners, opening the damper to increase the draft, opening the furnace peep doors, or any similar steps that result in increased airflow to the furnace can result in an explosion in a flooded furnace.

Key Lessons Learned

At this site, NOVA reported several significant changes to address the burner plugging that resulted in this incident, including changes to their project management guidelines, specifically to these low-NOx burners and other new burners to include the following:

- Thorough hazard reviews must be conducted to identify hazards associated with new technology.
- Strictly follow Management of Change and include the identification of new technology.
- Consider using stainless-steel headers and burner nozzles instead of carbon steel for new low-NOx burners.
- Consider a natural gas filtration system upstream of the burners to prevent debris, coke, and iron oxide scale from entering the small-diameter burner nozzles.

Since this incident is closely related to incidents involving furnace flooding, the following should also be considered at other industry sites:

- Ensure Operators are trained to recognize the indications of a furnace that is fuel-rich (flooded). These include low coil outlet temperature, low bridge wall temperature, low excess O_2, pulsating draft (draft swinging between positive and negative), furnace panting, and fuel gas rate increasing but coil outlet temperature dropping.
- Operators should be trained to take the proper steps to a furnace with symptoms of flooding, including the following:
  - Get people away from the furnace.

- Reduce the fuel rate to the furnace break the cascade to the firing control, place the fuel gas on manual control, and manually reduce the firing rate.
- Break the cascade on the furnace damper and place the damper in manual control (to prevent the damper from opening, which could increase the draft and increase the airflow to the furnace).
- Do not do anything to increase the air to the furnace (do not adjust the burners, do not adjust the damper, do not open the peep doors, do not attempt to adjust the burners).

What Has Happened/What Can Happen

Taiwan Petrochemical Company
Taoyuan City, Taiwan
29 January 2018

A devastating explosion occurred at the Taiwan Petrochemical Company at 6:42 a.m. local time. The explosion occurred in a furnace in startup mode, and the resulting fire engulfed the desulfurization unit. The fire bureau dispatched 93 personnel, 37 fire appliances, and 3 ambulances to the fire scene. No reported casualties were associated with this event, although damage appears significant. According to preliminary reports, the explosion and fire occurred when a pipe ruptured, and a boiler exploded as workers attempted to restart a furnace at the second hydrodesulfurization plant. The incident triggered protests among residents, including the mayor, who demanded the relocation of the refinery. The blast resulted in a spectacular refinery fire. A video of this incident is available on the RT News website.

Key Lessons Learned

The cause of the Taiwan incident is unknown at the time of this writing. However, the startup of a refining or petrochemical furnace should always be considered a high-hazard operation. The point is that there have been many furnace incidents across the industry. However, the causes and learnings from these incidents are rarely shared outside the company involved. In this regard, I believe we (the industry) "have a learning disability," to quote Dr. Andrew Hopkins, Emeritus Professor of Sociology at Australian National University. I also believe that improved sharing of lessons learned across the industry can lead to fewer incidents and, therefore, fewer lives lost and fewer workers being treated in hospital burn units.

Having up-to-date procedures is critical, and startup procedures must be followed. These procedures should be considered "SSHE Critical" and be developed in a checklist format requiring the Operator's signature for each step in the procedure as that step is executed. This ensures that a critical step in a procedure is not missed or omitted and that the steps are completed correctly.

When starting a furnace or relighting pilots or burners, following a furnace trip, it is critical that *the furnace firebox be thoroughly purged* to ensure no residual fuel exists in the firebox before attempting to light any pilots or burners. Likewise, it is critical to establish a flow through all the furnace tubes before the furnace is started. These critical steps should be included in Operator training and the furnace startup procedures.

As already stated, the operators should also be trained to recognize a flooded furnace (fuel-rich) quickly, react to protect people near the furnace, and take the correct steps to return the furnace to safe operation.

The key learnings from incidents involving furnaces or heaters should be shared broadly to prevent a reoccurrence of the event.

Additional References

American Petroleum Institute, API 560; Fired Heater for General Refinery Service, (A standard developed by the American Petroleum Institute specifies requirements and gives recommendations for the design, materials, fabrication, inspection, testing, preparation for shipment, and erection of fired heaters and air preheaters (APHs), fans, and burners for general refinery service. The most recent edition (5th Ed.) was released in February 2016.)

American Petroleum Institute, API RP 573; Inspection of Fired Boilers and Heaters, (A standard covers the inspection practices for fired boilers and process heaters (furnaces) used in petroleum refineries and petrochemical plants).

Fired Equipment Safety in the Oil and Gas Industry. A review of changes in practices over the last 50 years. Jacques Dugué, Published by Elsevier Ltd. https://www.sciencedirect.com/science/article/pii/S1876610217327170

Hydrocarbon Processing, September 2018, Special Focus: Refining Technology – A Game-Changing Approach to Furnace Safeguarding. https://www.hydrocarbonprocessing.com/magazine/2018/september-2018/special-focus-refining-technology/a-game-changing-approach-to-furnace-safeguarding

The Chemical Engineer. https://www.thechemicalengineer.com/news/fire-engulfs-taiwan-refinery/

NOVA Chemicals, Key Findings Incident at Bayport, Texas, site – 11 June 2003. http://www.georgemihalovich.info/DOCS/Bayport.pdf

CCPS Process Safety Beacon, The article "Avoid Improper Fuel-to-Air Mixtures" (Article on the Bayport, Texas furnace incident). https://www.aiche.org/sites/default/files/2004-01-Beacon-English_0.pdf

Training presentation on Furnace Flooding by John Zink Hamworthy Combustion Presented at HSE Conference, Austin, Texas – February 2017. http://content.4cmarketplace.com/presentations/FiredHeaterFlooding.pdf

Chapter 8.12. Hazards of Furnaces or Heater Operations

End of Chapter Quiz

1. Why are furnace procedures deemed critical procedures and in checklist format for the Operator to sign off each step as they are completed?

2. What are the correct actions to take if a procedure is outdated or incorrect?

3. What is one of the most important steps in lighting furnace pilots or burners from a cold startup or immediately following a furnace trip?

4. Please explain what is meant by furnace flooding or bogging.

5. When a furnace is operating in a flooded state, without Operator intervention, the fuel gas control valve will continue to _______, increasing _______ to the flooded furnace.

6. The process flow through the tubes provides the _________ to keep the furnace tube at a safe operating ___________. Loss of flow results in the tube metal overheating to the point where the tube fails, resulting in the loss of containment.

7. What are the indications or symptoms of a flooded or bogging furnace?

8. What steps can we take to correct a furnace flooding or bogging situation?

9. The safe response to a furnace operating in a flooded state is to quickly trip the furnace offline.
   - True
   - False

10. Why do vertical furnace tubes represent special hazards when doing hot work inside the furnace, for example, during turnarounds?

11. What can happen if the flow is lost in a single pass of a multi-pass furnace?

12. Which of the following is typically protected by the furnace emergency shutdown system? What is the effect of disabling this critical system?

13 The furnace emergency shutdown system should not be _________ except for testing and maintenance.

14 Ensure the feed is circulating through the ______ before attempting to light the burners. The tubes can be severely damaged if the burners are placed in service before a _______ is established in the tubes. If the box is a multi-pass design, ensure feed is flowing through all the furnace _________.

15 What can cause flame impingement from the furnace burners onto the adjacent tubes?

8.13

Corrosion Under Insulation and Under Fireproofing

Corrosion under insulation (CUI) and corrosion under fireproofing (CUF) can result in sudden and unexpected loss of containment due to severe external corrosion in piping and other process equipment. This type of corrosion is hidden from view due to the insulation and can cause failures in areas that are not normally of primary concern to an inspection program. CUI can result in sudden, hazardous leaks with potentially catastrophic consequences, especially when toxic materials, flammable fluids, LPG, very high temperatures, or pressure vessels are involved. There have also been severe corrosion cases under fireproofing that have resulted in the catastrophic failure of structural members, such as vessel support beams, causing structural failure and collapse of the equipment.

The failures are often the result of uniform corrosion in carbon steel, resulting in large failures. However, CUI is not limited to carbon steel, and it is possible to have pitting-type corrosion or stress corrosion cracking, especially in 300 series stainless steel. CUI/CUF can also result in an extended downtime or outage of the process equipment.

CUI is common in carbon steel and 300 series stainless steel, operating in temperature ranges of about 25 to about 350°F (−4 to 175 °C). CUI is most severe in carbon steels operating at about 200°F (93 °C). CUI in stainless steel generally results in pitting or stress cracking but is nonetheless a loss of containment. CUI is exacerbated by low-PH water and is very severe, with a PH of about four or less. Water can be introduced by breaks in the insulation barrier, leaking steam tracing, internal system leaks, ineffective waterproofing, or improper maintenance of the insulation systems. The water can be from rainwater, cooling tower drift, fire protection systems such as a deluge system, or sweating from temperature cycling or low-temperature operation from liquified petroleum gas (LPG) or refrigeration units. Susceptibility to CUI can be significantly increased by cyclic temperatures or with the presence of dead legs (piping with little or no flow).

CUI can be exacerbated by the chemical content of water, such as by chlorides or acids that may be present in the insulation or from minor process leaks. In cases where moisture becomes trapped on the metal surface by insulation, these corrosive constituents, such as chlorides and sulfuric acid, can concentrate and accelerate the corrosion rates, such as chlorides from cooling tower drift. The corrosion rate under wet insulation can be up to 20 times greater than corrosion in the ambient atmosphere. This high rate of corrosion, coupled with the fact that the metal surfaces are "hidden" from view by the insulation, makes this one of the more significant problems in insulated systems.

CUI and CUF are difficult to find because they are normally covered by insulation or fireproofing, masking the problem until a leak or pipe rupture occurs and a significant loss of containment occurs. Inspection can be difficult and expensive due to the need to remove the insulation or fireproofing for inspection, especially if asbestos insulation is involved (due to the remediation costs of the asbestos).

Risk-based Inspection

Currently, the only way to combat the challenges with CUI and CUF is with a rigorous inspection program, an effective pipe coating regime, and an aggressive program to eliminate or minimize water intrusion into the pipe insulation. Several technologies are evolving to inspect CUI, including profile radiography and ultrasonic spot readings. Eddyfi Technologies, in the article *"Corrosion Under Insulation: The 7 Inspection Methods You Must Know About,"* discusses seven methods that may be used to inspect for CUI and CUF (see References below for information on how to access this article). However, for now, the complete removal of the insulation or fireproofing is the only surefire inspection method, although it is the most expensive.

Process Operations Safety: The What, Why, and How Behind Safe Petrochemical Plant Operations, First Edition. M. Darryl Yoes.

The API-570 Piping Inspection Code: *"In-service Inspection, Rating, Repair, and Alteration of Piping Systems"* provides inspection guidance on the areas that may be susceptible to CUI. This information should be invaluable to piping inspectors. The following areas are specified in the code:

- Areas exposed to mist overspray from cooling water towers.
- Areas exposed to steam vents.
- Areas exposed to deluge systems.
- Areas subject to process spills, ingress of moisture, or acid vapors.
- Carbon steel piping systems, including those insulated for personnel protection, operate between 25 and 350°F (−4 and 175 °C). CUI is particularly aggressive where operating temperatures cause frequent condensation and re-evaporation of atmospheric moisture.
- Carbon steel piping systems that normally operate in service above 350°F (175 °C) but are in intermittent service.
- Dead legs and attachments that protrude from insulated piping and operate at a temperature different than the active line.
- Austenitic stainless-steel piping systems operating between 150 and 400°F (60 and 204 °C). These systems are susceptible to chloride stress corrosion cracking.
- Vibrating piping systems that tend to inflict damage to insulation jacketing and provide a path for water ingress.
- Steam-traced piping systems may experience tracing leaks, especially at the tubing fittings beneath the insulation, resulting in additional moisture beneath the insulation.
- Piping systems with deteriorated coatings and wrappings.
- Locations where insulation plugs have been removed to permit thickness measurements on insulated piping should receive special attention to ensure the plug has been appropriately resealed.

For the reasons stated above, each petroleum refinery and petrochemical plant should have a rigorous risk-based inspection program to address the CUI/CUF potential. CUI is a challenging adversary and one of the most difficult corrosion processes to find and prevent. It seems that regardless of the steps taken to prevent CUI, water will still find its way into the insulation in areas where it is almost impossible to find until a leak or catastrophic release occurs. Inspection data tells us that the most common cause of leaks and releases from process piping is CUI, which is the highest at sites in coastal areas or other areas with high humidity. Undoubtedly, billions of dollars are spent on mitigation, repairs, downtime, and lost production due to CUI.

Each site should evaluate the piping systems using the criteria in API 570 as a guide. Where these issues exist at the facility, the site should ensure inspection of those piping systems for the potential of CUI or CUF and ensure remediation based on the inspection's findings.

Piping Inspectors should be trained in the procedures and techniques to inspect for the potential for corrosion under insulation.

API RP 583, "Corrosion Under Insulation and Fireproofing," is also available and covers design, maintenance, inspection, and mitigation practices to address corrosion under insulation on pressure vessels, piping, and all types of storage tanks. The RP examines the various issues related to the CUI/CUF and presents guidelines and practices for preventing external corrosion or stress cracking under insulation and fireproofing. It also addresses the maintenance practices to avoid additional damage and spells out the inspection practices to detect and assess CUI damage. It also provides guidance for conducting risk assessments on equipment or structural steel that CUI may impact.

What Has happened/What Can Happen

The sphere in Figure 8.13.1 was being filled with water during a routine hydrostatic testing procedure when the support legs collapsed, resulting in fatal injuries to one person involved in the test.

Filling the sphere with water will expose the sphere and the support legs to about twice the weight as if the sphere were filled with LPG. Water has a specific gravity of 1, whereas Propane has a specific gravity of 0.495 (at 77°F or 25 °C). Water is about twice the weight of liquid propane.

An investigation found that the support legs were severely corroded due to external corrosion under the fireproofing material. Apparent water intrusion into the fireproofing contributed to the CUF and the resulting failure. Figures 8.13.2 and 8.13.3 are examples of typical CUI-type failures in refinery and petrochemical plant piping systems.

Figure 8.13.1 Support legs collapsed during hydrotest due to corrosion under fireproofing. Source: Russeltech.com (Russell NDE Systems, Inc. – from public images).

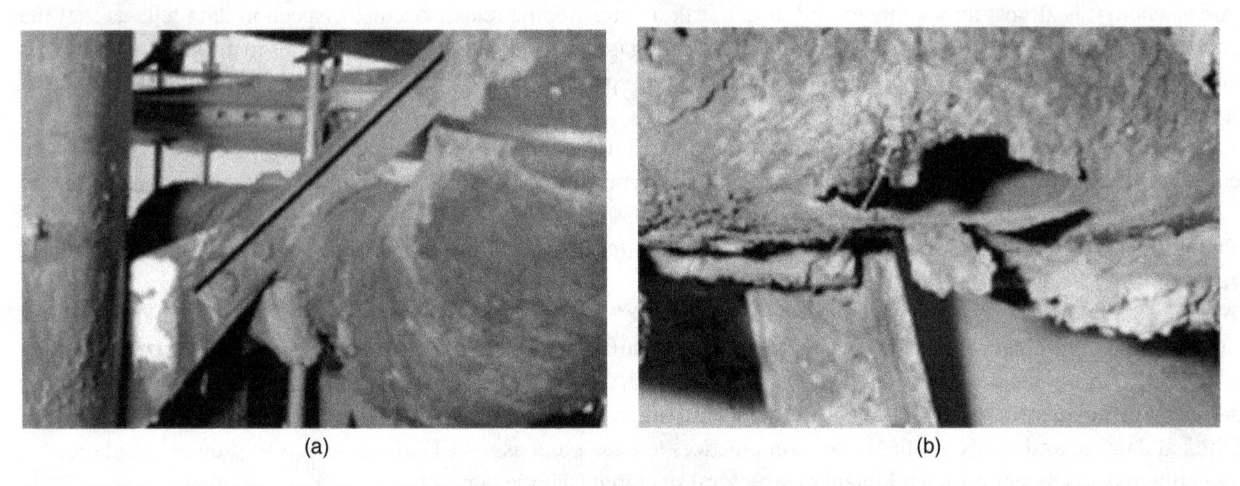

(a) (b)

Figure 8.13.2 (a) PetroPlus This figure illustrates the Coryton refinery Dehexanizer tower walkway support brace located under the piping insulation at the point of piping failure. This is where the water was trapped adjacent to the Dehexanizer feed line. Figure (b) shows the catastrophic failure which occurred due to the CUI. Contains public sector information published by the health and safety executive and licensed under the https://www.nationalarchives.gov.uk/doc/open-government-licence/version/3/. Source: Reproduced by the kind permission of HSE.HSE would like to make it clear that it has not reviewed this product and does not endorse the business activity of Safety Consulting International, LLC.

Shell LPG Distribution/Bottling Plant at La Goulette, Tunisia. On 12 May 2000, the LPG sphere collapsed while being filled with water during a hydrostatic test. The support legs under fireproofing were found to be severely corroded. One person was under the sphere at the time of collapse and was killed. Despite other spheres being pressurized, there was no LPG release as the interconnecting blinded lines remained intact. The sphere reportedly collapsed while being filled with water for a planned hydrostatic test. The support legs failed due to external corrosion (Corrosion Under Fireproofing).

(a)

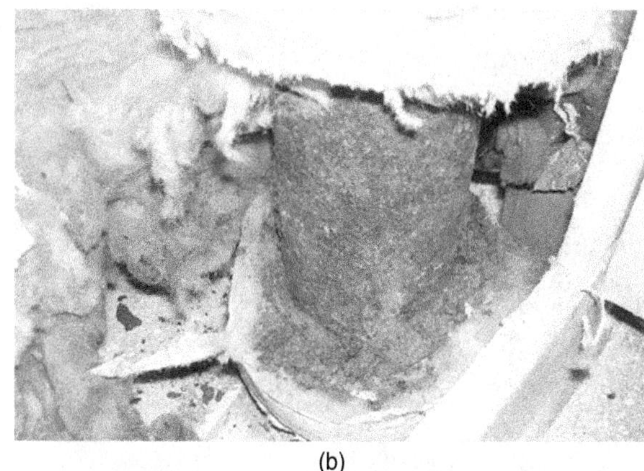

(b)

Figure 8.13.3 Figures (a) and (b) are both example s of corrosion under insulation, sometimes created due to leaking steam tracing under the insulation or moisture leaking into insulation due to broken insulation covering. Source: Russeltech.com (Russell NDE Systems, Inc. – from public images).

What Has Happened/What Can Happen

PetroPlus Coryton Refinery Fire (United Kingdom)

On 31 October 2007, a major fire occurred at the PetroPlus Refinery near Coryton, a hamlet and civil parish in the West Devon district of Devon, England. The incident occurred on the C5/C6 Isomerization Unit from an insulted pipe feeding the Dehexanizer column, approximately 82 feet (25 meters) above ground level. The leaking naphtha quickly found an ignition source, resulting in a very large fire.

At the time of the fire, the unit was temporarily offline, and the feed was bypassed to allow repairs to the control system. While the work on the control system was underway, the programmable logic controllers (PLCs) failed twice. Between the two failures, the flow control valve between the failed pipe and the Dehexanizer column was manually reset to 15% open, which was the typical operating setting while the unit was in service.

The investigation team reported that this resulted in a pressure surge in the piping during startup. Although the pressure surge was within the pipe design, this resulted in a pipe failure very close to a brace adjacent to the pipe, which was hidden under the insulation. The pipe ruptured catastrophically in this location due to corrosion under the insulation.

The investigation team determined that the brace interfered with the piping insulation and resulted in the entrapment of water under the insulation and against the pipe. This resulted in corrosion under the insulation, resulting in the piping failure. The brace was placed in that location to support an adjacent personnel walkway. Unfortunately, the corrosion progressed over time and went unrecognized by the refinery's ongoing integrity management systems and inspection routines. CUI had previously been identified as endemic throughout the company's operations and a significant threat. A program was in place; however, this location had not yet been identified. Following this incident, the company formalized an inspection and management program to address CUI across the refinery.

Gulf Coast Refinery Close Call (Corrosion Under Insulation)

To further illustrate the serious nature of CUI and the associated risk, a CUI pipe failure at a Gulf Coast refinery several years ago could have easily escalated to a catastrophic explosion with the potential for fatalities and major damage.

This incident occurred when an eight-inch diameter Ethylene line completely separated while in service due to CUI. This failure resulted in a very large Ethylene vapor release directly into process operating areas where numerous ignition sources could have been present. As expected with this kind of failure, the large vapor cloud quickly developed; however, no ignition occurred.

The failure occurred in a vertical section of the Ethylene line where water had penetrated the insulation, resulting in external corrosion to the eight-inch piping. This refinery had an ongoing CUI inspection and prevention program, and it was apparent that this section of piping was missed.

This incident is why it is so important to have a systematic CUI Inspection and prevention program to ensure that all critical piping systems are evaluated for the potential of corrosion under insulation and fireproofing. All areas that are identified as potentially exposed to CUI should be prioritized and risk assessed with pending actions tracked until they are completed.

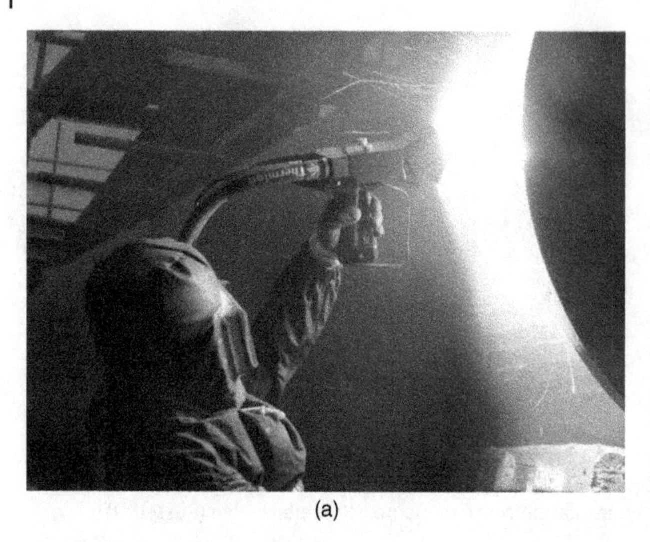

(a) (b)

Figure 8.13.4 Figures (a) and (b) illustrate workers applying thermal spray aluminum (TSA) metalized coating to address corrosion under insulation (CUI). Source: Photo courtesy Thermion Inc. (https://www.thermioninc.com/).

Key Lessons Learned

A risk-based inspection program should be established to help identify and evaluate piping and vessels operating in the temperature range for potential CUI/CUF, with a focus on equipment operating in light hydrocarbon, toxic, high-temperature, or high-pressure service. It is essential to ensure the integrity of the insulation systems to prevent water intrusion into circuits operating in the CUI temperature range. Also, applying insulation materials to avoid high levels of corrosive impurities such as chlorides is critical to reducing corrosion under insulation. Insulation with wicking properties, such as mineral wool, should also be avoided as they will absorb moisture and hold the moisture against the pipe or vessel, resulting in aggressive corrosion.

Applying paint or coating on new piping and equipment before the insulation or fireproofing is installed can help protect from CUI/CUF. Consider installing external insulation in system loops to prevent water vapor migration through the system. Also, installing water drains at low points can help drain any accumulated water from the insulation system. To prevent damage to the insulation water barrier, avoid mistreatment of insulated systems (such as walking on the insulated piping). Damaged insulation should be promptly addressed to prevent water intrusion into the insulation. This includes prompt repair of insulation damage when performing UT inspection for piping thickness.

Current technology to address CUI includes applying Thermal Spray Aluminum coating (TSA), an aluminum coating bonded onto the clean (sand and grit blasted) surface at elevated temperatures using an electric arc or flame spray. These thermal spray coatings protect against corrosion by excluding moisture and acting as a barrier coating (like paints, polymers, and epoxies). But unlike a typical barrier coating, they also provide sacrificial anodic protection. The coating acts as anodic protection, whereas the aluminum corrodes preferentially protect the underlying carbon steel or 300 series pipe or vessel. Images of workers applying the Thermal Spray Aluminum coating are available in Figure 8.13.4a,b.

Caution should be used before attempting any on-stream leak repair involving CUI. Due to the potential for uniform corrosion, what initially appears to be a minor leak of a flammable, toxic, or hot hydrocarbon may result in unexpected and catastrophic loss of containment with the potential for explosion or fire. In some cases, an unexpected loss of containment has occurred while attempting to remove the external insulation.

Additional References

American Petroleum Institute:

- API-570: In-service Inspection, Rating, Repair, and Alteration of Piping Systems
- API RP-583: Corrosion Under Insulation and Fireproofing

Article: "Corrosion under the insulation of plant and pipework v3", British Health and Safety Executive. https://www.hse.gov.uk/foi/internalops/hid_circs/technical_general/spc_tech_gen_18.htm

Article: "Preventing Corrosion Under Insulation", The National Board of Boiler and Pressure Vessel Inspectors, V. Mitchell Liss, Engineering consultant, La Grange, TX. https://www.nationalboard.org/Index.aspx?pageID=184

Eddyfi Technologies, Corrosion Under Insulation: The 7 Inspection Methods You Must Know About. https://eddyfi.com/en/blog/corrosion-under-insulation-the-7-inspection-methods-you-must-know-about

European Commission Report – PetroPlus Coryton Fire (Corrosion Under Insulation). https://emars.jrc.ec.europa.eu/en/emars/accident/view/20454cd4-25ec-804c-53b5-27596696deb3

Blog of several significant refinery fire reports, including some with CUI as the cause. https://www.rbiclue.com/blog/

Russell NDE Systems, Inc., Mitigating The Risks Of Corrosion Under Insulation With Russell NDE Systems. https://www.russelltech.com/News/ArtMID/719/ArticleID/255/Mitigating-The-Risks-Of-Corrosion-Under-Insulation-With-Russell-NDE-Systems

Chapter 8.13. Corrosion Under Insulation and/or Under Fireproofing

End of Chapter Quiz

1. What is the primary concern with corrosion under insulation or fireproofing?

2. At what temperature range is corrosion under insulation an issue at our refining and chemical plant facilities?

3. The most typical source of corrosion is due to water penetrating the insulation or fireproofing barrier. How does the water get into the insulation, and what are some potential sources of this water? Name as many as you can.

4. All areas that are identified as potentially exposed to CUI should be __________ and risk ________ with pending actions tracked until they are completed.

5. Insulation with ________ properties, such as mineral wool, should also be avoided as they will absorb moisture and hold the moisture against the pipe or vessel, resulting in aggressive corrosion.

6. Why is particularly hazardous to attempt a repair of what appears to be a minor leak involving CUI?

7. What are some additional factors that can greatly increase the corrosion rate of CUI?

8. Can you explain the current technology used to address the CUI issue?

9. Currently, the only way to combat the challenges with CUI and CUF is with a rigorous _________ program, an effective pipe ________ regime, and an aggressive program to eliminate or minimize __________ intrusion into the pipe insulation.

10. Why are special precautions required when responding to what appears to be a minor leak on piping with external insulation or fireproofing?

8.14

Hazards of Thermal Expansion

Most of us consider thermal expansion a de minimis issue, not a significant process safety hazard. For example, when I started working in the refinery, I remember opening large block valves in gasoline service during the summer against built-up pressure in the system. The valves were very hard to "break off the seat," when they first opened, I could hear the squeal of liquid flowing under the valve seat. We called this "sun pressure." The built-up pressure in the system resulted from thermal expansion and liquid heating in the long gasoline transfer lines. We may dismiss thermal expansion as an insignificant process hazard during the process hazard analysis (PHA); however, thermal expansion is a serious process safety hazard and must be recognized.

Several significant process safety incidents have occurred because of thermal expansion, which directly resulted from the thermal expansion of trapped liquids exposed to very modest temperature increases. The facts are that with liquid-filled process piping or vessels with no vapor space, extreme pressures can be generated within those systems with exposure to very modest temperature increases. For example, gasoline in a long loading pipeline subjected to only a 1 °F temperature rise can result in a pressure increase of 98 psi.

Using the gasoline line as an example, if the pipe is liquid-filled and blocked overnight during the coolest part of the day, the temperature can rise by more than 20–30 °F by midafternoon. This can increase pressure in this trapped system by nearly 2,000–3,000 psi. These 150# pipe class systems are typically hydrostatically tested at 450 psi (per the ANSI B16.5 pipe standard). Think about this: a hydrostatically tested pipeline at 450 psi can be pressurized to nearly 3,000 psi from daytime heating!

So extreme pressures can be generated in trapped piping or vessels that are liquid-filled and blocked in with no overpressure protection. Some piping or vessels are protected with small thermal relief valves. These are very effective and prevent inadvertent overpressure when blocked in. A controlled release of the overpressure takes only a very small flow to prevent pressure build-up in the system. However, when these devices are removed for maintenance, we must ensure that overpressure protection is provided while the device is removed or otherwise disabled.

The consequences of unprotected thermal expansion can be severe. Most often, gasket failures are typical since they are the weakest link. However, thermal expansion can also result in flange or piping ruptures with catastrophic consequences, particularly with light hydrocarbon lines, due to the potential for vapor cloud formation.

The risk is overpressure if the equipment is blocked in with no relief path provided. Examples of equipment that can be subjected to overpressure include the following:

- Piping in intermittent service blocked in at both ends.
- Vessels are blocked when removed from service (or returning).
- Heat exchangers blocked while preparing for cleaning.
- Heat-traced piping or vessels (if the equipment is liquid-filled).
- Thermal relief valves were removed for servicing.
- Heat exchangers with the hot side flowing while the cold side is blocked.

Protecting From Thermal Expansion Piping or Vessel Failures

The most effective way to protect against thermal expansion is to install a thermal relief valve on lines required to be blocked in. Otherwise, we must develop and follow procedures to ensure that pipelines or equipment are left with an open valve and floating at one end to provide a relief path. Another alternative is to employ procedures or facilities to ensure that

Process Operations Safety: The What, Why, and How Behind Safe Petrochemical Plant Operations, First Edition. M. Darryl Yoes.

blocked-in equipment does not remain liquid-filled; for example, a drilled check valve prevents inadvertent isolation of the system by the check valve.

Thermal Expansion Hazards Associated With Heat Exchangers

Heat exchangers are notorious for creating thermal expansion. They have a hot side and a cold side, and the purpose of the exchanger is to exchange heat between the tube side and the shell side (between the hot side and the cold side). If the cold side is blocked in and the flow is started on the hot side, or flow continues on the hot side, the heat is transferred to the cold site, resulting in extreme pressures from thermal expansion on the cold side (the side that is blocked in).

This has resulted in ruptures on the cold side and loss of product containment, leading to explosions and fire. Most sites require either a safety relief valve on the cold side to protect it from inadvertent overpressure or at least a warning sign to remind the operators to never block in the cold side with the hot side flowing, coupled with the use of procedures for commissioning an exchanger or for safely removing the exchanger from service.

Thermal Expansion Hazards Associated With Double-seated Block Valves

One hazard that is not well known is the thermal expansion hazard of double-seated block valves. This hazard may be overlooked by safety professionals and Operators alike. Some double-seated block valves may have thermal expansion hazards associated with the internal cavities above and below the gate. Liquid can be trapped in these cavities with the valve either in the fully open or closed position making these valves significant process safety hazards when the valve is in liquid service. Especially if a hot liquid flows through the valve or we attempt to steam through the valve. The hot process flow can heat the trapped liquid in these cavities resulting in extreme internal pressures and causing the valve to fail in service. These failures have resulted in the loss of containment of the process fluids. Unfortunately, Operators who were in the vicinity of these failures lost their lives when this occurred.

Certain models of double-seated block valves have been reported with this unique hazard. Please verify this with your valve manufacturer, the manufacturer's catalog, or the Mechanical Engineer to determine if this hazard exists at your site.

The valve manufacturers recognized this hazard and fitted these double-seated valves with either a small thermal expansion valve or bleed valves to depressurize the cavities, thereby relieving the pressure. However, these devices must be maintained and have procedures for the Operators to follow in the field. Otherwise, failure of these valves can result in loss of containment and a resulting explosion and fire. My experience is that the thermal relief valves are frequently seen as a nuisance and are removed and plugged over time. This becomes the perfect condition for a tragic process safety incident. See Figure 8.14.1 for an image of a double-seated block valve with this unique thermal expansion hazard.

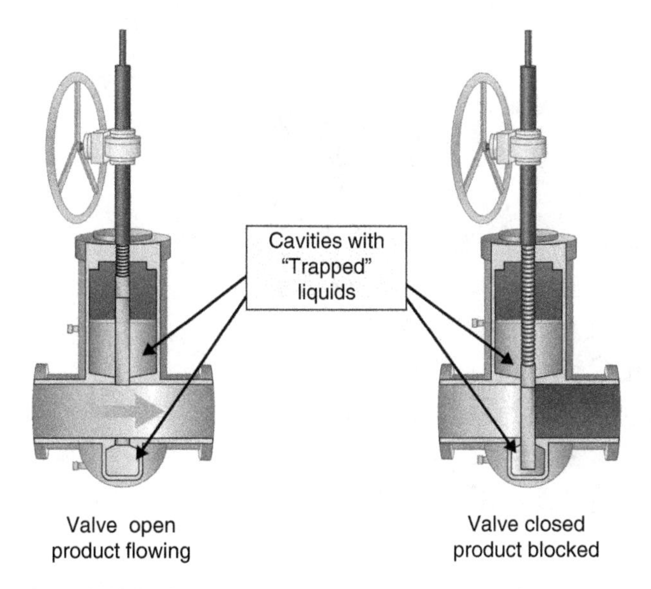

Figure 8.14.1 Double-seated block valve – image of a generic double-seated block valve with cavities above and below the seat. Source: The image was created courtesy of Ecoscience Resource Group Visual Artists.

These images were created to illustrate the hazard associated with trapped liquids in double-seated block valves. These images do not represent any certain valve manufacturer and only illustrate the hazard associated with double-seated block valves.

These valves present a unique hazard of trapped liquid in cavities above and below the valve gate. This can result in excess pressure when exposed to a hot stream or even during daytime heating. For example, when a hot product is flowing through the valve or when steaming through the valve.

- Potential overpressure of valve body cavity due to Internal Thermal Expansion of trapped liquids. These valves have ruptured due to overpressure from thermal expansion in the valve cavities.
- Excess pressure can occur in the valve bodies in the fully open or fully closed position or between travel.
- A vent or PRV is typically used to protect the valve body, preventing valve damage.

A quick review of the valve manufacturer's catalogs will illustrate the valve's internal cavities and how the different valve manufacturers mitigate these hazards (vents or external relief devices designed to relieve the internal pressure).

Signs to Warn Operators of Thermal Expansion Hazards

It is good to provide and post warning signs on exchangers and double-seated block valves to remind the Operators of the severity of the thermal expansion hazards associated with this equipment, especially if the exchangers and valves are not equipped with relief devices. The signs should also be reminders of the proper sequence of valve operation when isolating exchangers from service and when returning exchangers to service, for example, after maintenance.

See Figure 8.14.2 for an example of warning signs that should be placed on each exchanger to help warn the operators of the hazards of thermal expansion. Similar signs should be placed on double-seated block valves. These are examples of signs to warn personnel to provide a vent or drain before isolating the cold side of exchangers to prevent thermal expansion of trapped liquids. The signs can also remind the operators of the correct sequence to operate isolation valves when removing the exchanger from service or returning it to service. These are actual employee warning signs posted on units to ensure the operators know the hazards of trapped liquids in either the shell or tube sides of exchangers. While the two signs provide different ways to avoid trapped liquids, they serve as reminders to operators to avoid the hazards of thermal expansion in exchangers. Figure 8.14.2a provides a procedure; 8.14.2b reminds operators to never block in an exchanger without providing a vent.

What Has Happened/What Can Happen

Williams Olefins Plant
Geismar, Louisiana
13 June 2013

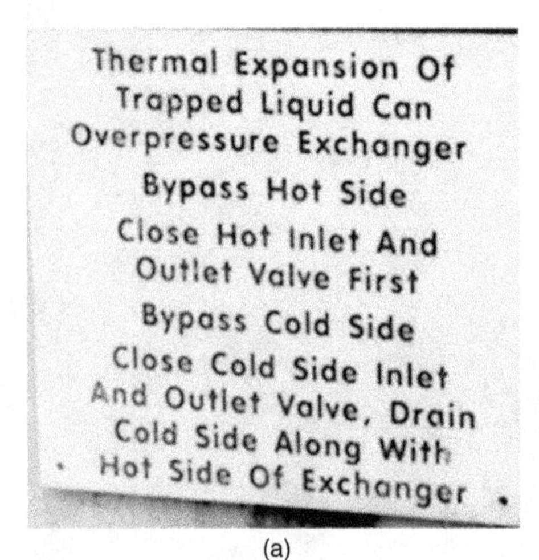

(a)

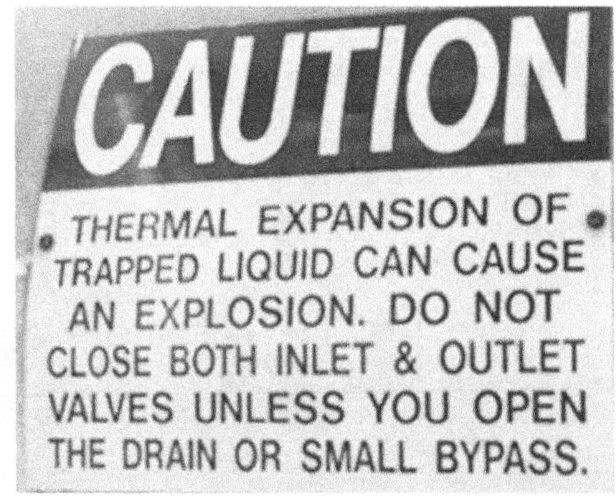

(b)

Figure 8.14.2 Examples of warning signs that should be placed on each exchanger to help warn the operators of the hazards of thermal expansion.

This incident was investigated by the U.S. Occupational Safety and Health Administration (OSHA) and the U.S. Chemical Safety and Hazard Investigation Board (CSB). The CSB completed a detailed investigation report and an excellent process safety training video indicating the key lessons learned from the incident.

The Williams Olefins Plant in Geismar, Louisiana, experienced a catastrophic explosion on 13 June 2013, as a result of process flow started on the tube side of a Propylene fractionator reboiler exchanger with the shell side blocked in. Two Operators were killed, and 167 people reported injuries due to this incident.

This incident was a direct result of the thermal expansion of the shell side of the exchanger due to the hot process flow on the tube side (hot side) of the exchanger. On the day of the incident, the supervisor was troubleshooting the exchanger; he suspected fouling on the tube side of the exchanger that was in service and attempted to start a small flow of hot water through the tube side of the standby exchanger. However, he failed to realize that the exchanger shell side was full of liquid (primarily propylene), and the shell side of the standby exchanger was isolated from the relief valves. Shortly after starting the flow of hot water, the exchanger failed catastrophically due to the rapid overpressure of the shell side from the thermal expansion of the trapped liquid. The supervisor was killed along with an Operator who was working nearby.

The CSB report indicated several failings of Williams's process safety management work processes, including the Process Hazard Analysis, the Pre-startup Safety Review (PSSR), and the Management of Change process. See the photos in Figure 8.14.3 for an image of the fire and damage photo in Figure 8.14.4.

Key Lessons Learned

Operators should be trained, and procedures should be readily available to return exchangers to service following an outage. Flow should never be started on the hot side of an exchanger unless the cold side is open to the process and is flowing. Otherwise, the temperature from the hot side will transfer to the cold side resulting in extreme pressure increases on the cold side from thermal expansion.

Procedures for this type of operation should normally be in checklist format with signed and dated logs for specific tasks. This ensures procedures are correctly followed without omitting a critical step or executing steps out of sequence.

Figure 8.14.3 Photo of explosion and fire at Williams Olefins Plant (13 June 2013). Source: Photo courtesy of the U.S. Chemical Safety and Hazard Investigation Board.

Figure 8.14.4 Photo of damage at Williams Olefins Plant (13 June 2013). Note that the exchanger shell is completely laid open due to the extreme overpressure on the shell side of the exchanger. Source: Photo courtesy of the U.S. Chemical Safety and Hazard Investigation Board.

Block valves associated with pressure relief valves should be car-sealed when the valves are in the open position. Car-sealed valves should always be respected by the Operators and others. They should never be closed unless there is an approved control of defeat or similar authorization for impairment with proper documentation in place. See Chapter 8.34 (Criticality of Car Seal Valve Management Plan) for additional details on car seal procedures and Chapter 8.4 (Ensuring the Availability of SSHE Critical Devices) for additional details on the control of defeat process.

The Car Seal Valve Management plan also requires a "car seal valve line up procedure" and car seal valve audit process to ensure the safety, security, health, and environmental (SSHE) valves are in the correct position (fully opened or fully closed) and car sealed. Additional details can be found in Chapter 8.34 (Criticality of Car Seal Valve Management Plan).

Operators should be trained to "walk down" the systems on exchangers (and other similar equipment) before returning them to service to ensure that proper overpressure protection is fully functional and that all valves are in the appropriate position. This expectation should be documented in procedures and enforced by site management.

What Has Happened/What Can Happen

Exxon Refinery
Baton Rouge, Louisiana
24 December 1989

This incident was investigated by the U.S. Occupational Safety and Health Administration (OSHA), and a report is available from the agency.

The Exxon Refinery, located in Baton Rouge, Louisiana, experienced a catastrophic vapor cloud explosion on Christmas Eve, 24 December 1989, due to extreme overpressure of an 8-inch Ethane/Propane pipeline over 20 miles long. This catastrophic incident resulted in a total power failure to the site (for a short duration), the total loss of steam pressure, and fire water pressure. It also resulted in sixteen tank fires, including two large heating oil/diesel tanks. Two people were killed due to the explosion, and the 500K B/D refinery and the adjacent chemical plant were essentially out of service for several months due to the significant damage from the explosion. Please refer to Chapter 5 (Vapor Cloud Explosions), Figures 5.5 and 5.6, for amazing images of this incident.

During a severe freeze with the refinery in the third consecutive 24-hour period, the incident occurred at temperatures well below freezing. The area reached a low temperature of 8 °F (13 °C) during the early morning hours preceding the incident. The following afternoon, the ambient temperature reached about 30 °F (−1 °C) at 1:20 p.m. (the time of the incident).

Investigation revealed that the 8-inch pipeline had been blocked overnight during the freezing weather and remained blocked on both ends until it failed at about 1:20 p.m. the following afternoon. This line was subjected to extreme overpressures because of daytime heating resulting in failure. There was no overpressure protection on this pipeline. Gasoline and naphtha lines have the most severe coefficient of thermal expansion; however, LPG is a close second, and no doubt this line saw extreme internal pressures due to the rapid change in ambient temperatures.

Freeport LNG Liquefaction Plant
Quintana Island, TX.
8 June 2022

A similar incident to the 1989 Exxon, Baton Rouge case occurred at the Freeport LNG Liquefaction facility on 8 June 2022. The Freeport LNG plant, located on Quintana Island, just South of Galveston, TX., experienced a loss of containment of LNG and a resulting explosion and fire on 8 June 2022. No one was injured in this incident; however, debris and shrapnel were spread over the facility. The company contracted with a firm to investigate and understand why and how this happened. Freeport LNG and the contact firm have been very open with the incident findings and shared the learnings with the local press and community.

At Freeport, they found that a pressure relief valve on one of the LNG transfer lines had been removed for maintenance sometime before the current incident occurred. Several days before the incident, an Operator manipulated valves and inadvertently isolated a section of the LNG piping, leaving the line blocked. Over these several days, the line was exposed to daytime heating, and the LNG expanded to the point where the piping failed due to overpressure.

As a result of this incident, the liquefaction plant was out of service for several months during a time when LNG was in high demand worldwide. This incident clearly illustrates the importance of ensuring that relief valves are maintained and, when defeated, must have mitigation in place for the entire duration of the defeat until the device is returned to full online service. This is discussed in more detail in Chapter 8.4, "Ensuring the Availability of SSHE Critical Devices." Additional discussion is included in Chapter 8.34, "Car Seal Management Plan" (of SSHE Critical Devices).

Key Lessons Learned

Changes in ambient temperatures can result in extreme pressure increases in petroleum pipelines or vessels if the equipment is liquid-filled and blocked in with no opportunity for the release of thermal expansion. The daily temperature change from nighttime lows to daytime highs can easily exceed 30 °F (or about 17 °C), which can easily result in increases of between 2,000 and 3,000 psig (138–207 bar) in equipment that is blocked in. Very few systems are designed to see these kinds of pressures.

Our role is to ensure that long pipelines are equipped with a thermal relief valve to prevent overpressure when the line is blocked. An alternative is to ensure that procedures are developed and rigorously followed to establish a flow path on one end of the long pipelines to relieve pressure and prevent overpressure of the piping or equipment.

What Has Happened/What Can Happen

Goodyear Chemical Plant
Houston, Texas
11 June 2008 (One fatality, six injuries)

An incident almost identical to the one at Williams Olefins occurred at Goodyear Chemical Plant in Houston on 11 June 2008. In that incident, one person was killed, and six received injuries.

Figure 8.14.5 Photo of exchange failure due to overpressure at Goodyear Chemical Plant, Houston, TX. Source: Photo courtesy of the U.S. Chemical Safety and Hazard Investigation Board.

The exchanger at Goodyear was protected by a ruptured disk and a relief valve. The rupture disk was installed below and in line with the relief valve. However, both were blocked the day before the incident to allow maintenance personnel to replace the rupture disk. Maintenance personnel completed the replacement of the rupture disk; however, the isolation valve below the rupture disk remained closed.

The following day, a Goodyear employee started the process of returning the exchanger to service, connected a steam hose, and started steaming the hot side of the exchanger (with the cold side blocked in and the relief valve and rupture disk still isolated). The pressure built up rapidly, resulting in the failure of the shell side (the cold side), releasing ammonia and shrapnel into the area.

Five employees were exposed to ammonia, and one was injured while escaping the area. Later, another employee's body was discovered under the debris, having been killed in the incident.

The US Chemical Safety Board investigated this incident and prepared a detailed report. See Figure 8.14.5 for a photo of the Goodyear Chemical Plant exchanger.

Key Lessons Learned

Ensure that detailed startup procedures are developed with critical steps in a checklist format that requires signed and dated logs for each task. Operators should be trained to "walk down" the system to ensure proper valve alignment and verify that overpressure protection is functional before beginning the return to service.

Operators should be trained in the hazards of thermal expansion and the risks associated with returning equipment to service. It is a good practice to have warning signs placed on exchangers and double-seated block valves to remind the Operators of the proper procedures to prevent isolation of the cold side with the hot side flowing.

Isolation valves under protective devices, such as pressure relief valves and rupture disks, should be designated as "car seal" valves. Car seal valves should be painted a unique color, and no other valves should be painted the same distinctive color. They should be full port valves and subjected to periodic audits to ensure they are fully open with car seals in place.

It is essential that these are not just viewed as recommendations but that they become part of the organizational culture and are ingrained in every Operator at the site. They should be viewed as how things are done for every task at the site.

Additional References

U.S. Chemical Safety and Hazard Investigation Board, Investigation Report and training video "Blocked In": Williams Olefins Explosion. https://www.csb.gov/williams-olefins-plant-explosion-and-fire-/

U.S. OSHA report and citations: Williams Olefins Explosion. https://www.osha.gov/news/newsreleases/region6/12112013

U.S. Chemical Safety and Hazard Investigation Board report of the Houston Goodyear Chemical Plant. https://www.csb.gov/goodyear-heat-exchanger-rupture/

ASME Section VIII (The Boiler and Pressure Vessel Code) and ANSI Codes (B31.3) recognize the danger of overpressure by thermal expansion and provide guidance for the prevention of overpressure by thermal expansion.

API RP-520 "Sizing, Selection, and Installation of Pressure-Relieving Devices in Refineries," and API RP-521 "Guide for Pressure-Relieving and Depressuring Systems" also provide guidance on preventing overpressure by thermal expansion.

Occupational Safety and Health Administration (OSHA), Report of Exxon Baton Rouge Refinery 1989 Offsite Explosion. https://www.osha.gov/pls/imis/establishment.inspection_detail?id=101477560

Baton Rouge Channel 9 News video article: A LOOK BACK: 30 years since Exxon explosion in Baton Rouge. https://www.wafb.com/2019/12/24/look-back-years-since-exxon-explosion-baton-rouge/

Chapter 8.14. Hazards of Thermal Expansion in Closed or "Trapped" Systems

End of Chapter Quiz

1. The facts are that with liquid-filled process piping or vessels with no ______ space, extreme ________ can be generated within those systems with exposure to very modest temperature increases.

2. When returning equipment to service after an outage for repair or replacement, the Operator is expected to _______ the system down to ensure all __________are complete and the _____________is ready to return to service.

3. List some examples of equipment that can be subjected to overpressure if a relief path is not made available.

4. If a process system is blocked in and a relief device is not available, what must be available and followed to prevent piping or vessel overpressure?

5. Anytime the thermal expansion relief valves are removed for inspection or maintenance, we must ensure that overpressure __________ is provided while the device is removed or otherwise disabled.

6. Another serious thermal expansion hazard exists with double-seated block valves. These valves may be equipped with external drains or thermal expansion relief valves. What is the hazard associated with these valves, and how is the hazard controlled?

7. Extreme pressures can be generated in trapped piping or vessels, pipe or vessels that are liquid filled and blocked in with ___ ______________ protection.?

8. Procedures used to return equipment such as exchangers to service should be in __________ format, and each relevant step filled out __________ as the procedure is being followed.

9. Isolation valves under protective devices, such as pressure relief valves and rupture disks, should be designated as _____ _____ valves and should be painted a unique _______, and no other valves should be painted the same distinctive __________. They should be full port valves and subjected to periodic __________ to ensure the valves are fully open with car seals in place.

10. What is the unique hazard associated with heat exchangers?

11. What can be done to remind the Operators of the extreme safety hazard associated with thermal expansion in exchangers and double-seated block valves?

8.15

Hydrostatic Testing vs. Pneumatic Testing

This chapter refers to the strength testing required for vessels and piping for new construction or after modifications such as welding on existing piping or vessels. Strength testing is usually performed using water or other liquid to ensure the integrity of the equipment, typically completed at 130% of the design pressure or the vessel's maximum allowable working pressure (MAWP). This is a hydrostatic test and is the preferred testing method. However, occasionally testing with water or other liquid is not possible due to the weight of the liquid, the presence of a catalyst, or other reasons. In this case, the testing can be done with a pneumatic, such as air, nitrogen, or other gas. When using a pneumatic, pressure is normally limited to 110% of the MAWP, and other significant precautions are required. This chapter is intended to emphasize the safety difference between a hydrostatic test and a pneumatic test. While either process can be used successfully to conduct the required test, there is a significant safety difference between the two, and pneumatic testing normally requires additional procedures, safety evaluation, and a higher level of approval.

In conducting a hydrostatic test, the water or other fluid is noncompressible; therefore, the pressurization energy is directly applied to the equipment being tested. In the pneumatic test, the gas must first be compressed to apply internal pressure to the equipment. Therefore, in the pneumatic test, a very large amount of energy is stored in the compressed gas. Since the liquid is not compressible, the pressure is immediately released should an equipment failure occur while conducting a hydrostatic test. The opposite is true for a pneumatic test. A failure of equipment containing compressed gas, for example, during a pneumatic test, will result in a large release of stored energy due to the compression of the gas. This sudden and instantaneous release of stored energy can cause severe damage to surrounding equipment and injury or death to personnel. A failure during a pneumatic test also can propel projectiles long distances causing secondary damage and secondary loss of containment.

Testing with water or other hydrostatic testing has additional benefits as well. It is much easier to locate small leaks with a hydrostatic test as the pressure loss will immediately be evident on the pressure gauge. It may also be much easier to repair small leaks when found with the hydrostatic test, whereas, with a pneumatic test, the small leak is more likely to result in significant equipment damage before the pressure can be released. A common failure mode with a pneumatic test failure is a complete rupture and rapid pressure release with the appearance of an explosion.

For these reasons, strength testing using compressed gas (air, nitrogen, or other compressible gas) should be avoided if possible; however, it does have a place and occasionally will be required. For example, the testing of piping, such as refinery or chemical plant overhead flare piping, where the piping is very large, and the pipe supports are not designed to support the weight of the piping when filled with a liquid. Other examples where pneumatic tests may be required include testing vessels where the water or other liquid would result in product or catalyst contamination or may be reactive with the process chemicals.

If the pneumatic testing process is used, the process should receive higher-level safety and management review and approvals, including a risk assessment to ensure all hazards are identified and mitigated before being carried out. The process should have a well-thought-out safety plan which includes barricading the area and excluding all nonessential personnel before applying pressure to the equipment. Controls should be in place to guard against the equipment's potential overpressure, such as a relief valve and appropriate pressure gauges and alarms. Air should only be used for pneumatic testing when the equipment is known to be completely hydrocarbon-free and free of iron sulfide scale or other deposits or materials that may be pyrophoric.

Consideration should also be given to the potential for brittle fracture before filling the process equipment with fluid or gas for testing (See Chapter 8.11 for additional discussion of brittle fracture). Catastrophic brittle facture failures have been known to occur with water and nitrogen during pressure testing operations when the test medium temperature was below the equipment's minimum design metal temperature (MDMT). Remember, the MDMT for heavy wall carbon steel vessels

Process Operations Safety: The What, Why, and How Behind Safe Petrochemical Plant Operations, First Edition. M. Darryl Yoes.

Comparing a hydrostatic test to a pneumatic test

- A hydrostatic test using water or other incompressible liquids is carried out at 30% above the design pressure as per ASME Sec VIII, Div. 1, UG 99. A pneumatic test using air or other compressible gas is done at 10% above the design pressure as per ASME Sec. VIII, Div. 1, UG 100.
- The hydrostatic test is to verify the equipment integrity (strength) whereas the pneumatic test is typically used to ensure tightness after the equipment integrity has been verified.
- Pneumatic testing also generally requires dedicated safety procedures, special safety precautions and close monitoring by professionals. Whereas hydrostatic testing can generally be performed by trained craftsmen and does not require specialized support.
- A pneumatic test requires a relief valve or other method to ensure against inadvertent overpressure. Overpressure protection is recommended for a hydrostatic test.
- Due to the compressibility of the pneumatic, an equipment failure can have catastrophic consequences with extensive damage. An equipment failure during a hydrostatic test can also have damage but is generally not catastrophic.
- Hydrostatic testing generally does not require specialized personnel to conduct the testing. Pneumatic testing generally requires engineering and safety support with approval from senior management.
- Barricading of a large area around the test site to ensure unauthorized personnel are excluded is required for a pneumatic test. A limited area is required to be barricaded for a hydrostatic test.
- A hydrostatic test using liquid will require a thorough cleaning to remove the moisture. This is especially true in cryogenic service or where catalyst or other similar material is used. A pneumatic test does not introduce moisture so generally no cleaning is required.
- Hydrostatic testing can test larger piping and equipment segments in one test. Pneumatic testing generally is limited to smaller segments of these systems to minimize the potential for equipment failure during the test.

Figure 8.15.1 A table comparing hydrostatic testing and pneumatic testing.

may be significantly above normal ambient temperature. In these cases, the test medium may require heating to carry out the pressure test.

Procedures are readily available in ASME code PCC-2: "Repair of Pressure Equipment and Piping." Section 6.2 covers Pneumatic Testing of Pressure Vessels and Piping, including detailed considerations and precautions and a formula for calculating the stored energy potential during the test.

Figure 8.15.1 provides a table that compares the differences between a hydrostatic Test and a pneumatic test.

What Has Happened/What Can Happen

- Midcontinent Express Pipeline Company
- Smith County, Mississippi
- 14 July 2009

Frequently in these cases and since the hazards of pneumatic testing are not intuitive, the owner or contractor will make the decision to conduct a pneumatic pressure test without consulting with safety or other professionals before conducting the test. This happens more frequently than we may be aware of, and occasionally things can go terribly wrong, and people get caught up in the aftermath of a devasting explosion. Not an explosion involving hydrocarbons and generally there is no fire involved. The explosion is the sudden and unexpected pressure release and the release of the "stored energy" associated with a pneumatic pressure test. A couple of brief examples follow which both involved the loss of life due to the unexpected release of the stored energy. This explains why we prefer to test with water, called a hydrostatic test. Water is noncompressible, therefore, there is minimum "stored energy" with a hydrostatic test.

The first incident occurred in rural Smith County, Mississippi, and was investigated by the US Occupational Safety and Health Administration (OSHA), and a report is available on the OSHA website. The Administrator for the Mississippi Public Service Commission told me that the incident did occur and involved Kinder Morgan Pipeline Company. He explained that the incident was under the federal jurisdiction and was investigated by the US Department of Transportation, Pipeline Safety and Hazardous Material Administration (PHSMA). There is no report of the incident on the PHSMA website, and after contacting the agency multiple times, I received no response about this incident.

The reports also indicate that the US National Transportation Safety Board was notified; however, there is no report of this incident in the NTSB database. However, the images of the accident site are posted on a blogger's website ("Chemical and Process Technology") and they clearly show the carnage associated with a failure during pneumatic testing. The images include large-diameter piping bent and twisted like a pretzel. Unfortunately, I do not have permission to use the images here, but they can be easily found online.

According to the OSHA report, the incident occurred during a pressure test being conducted with nitrogen at a new meter station at the Midcontinent Express Pipeline station. On 14 July 2009, according to the OSHA investigation, a Midcontinent Express contractor and subcontractor were conducting a nitrogen pressure test on metering facilities at a Midcontinent Express delivery meter station that was under construction. The pressure test involved piping and two pressure vessels being tested at the same time. During the test, the equipment being tested failed, and an unexpected pressure release occurred resulting in a sudden and massive pressure release of the pressurized nitrogen. This caused failure of the piping and vessels and resulted in flying missiles and debris across the work site. One contract employee was killed, and four other employees of the contractor or subcontractor were injured. Kinder Morgan, Inc. and Midcontinent Express Pipeline were later named in a legal suit arising from the incident. OSHA cited three of the subcontractors for "failing to protect their workers." The citations included three "willful" violations, the most serious OSHA violation.

OSHA reported that when the equipment reached the required test pressure of 2,225 psig, one of the employees observed that the pressure had dropped to 2,205. As the employee walked away, the door on a PECO separator (a pressure vessel) blew off, releasing pressurized nitrogen gas that sent projectiles flying. The Smith County Sheriff said that the workers were "literally right on top" of the explosion and that the injuries were caused by pressure, not by fire or heat. The report also indicated that one of the workers was injured when a section of the pipe fell on him. The explosion caused a pipeline that was connected to a massive separator unit to be bent and hurled several yards.

This incident clearly illustrates the awesome force that can occur when the stored energy is suddenly released. We will talk more in the Key Lessons Learned section below about what precautions should be in place before attempting to perform a pneumatic pressure test on process equipment.

What Has Happened/What Can Happen

Brazil
1 January 2006

A freak accident occurred in Brazil (another report indicated that this happened in China) while the maintenance crew was conducting pneumatic testing of new piping associated with two new atmospheric storage tanks on a process unit. The piping was not fully isolated from one of the tanks, although there was a valve in the pipeline; however, this valve was not completely closed or blinded. This allowed air used to conduct the pneumatic test to flow into the storage tank, over-pressuring the tank. Cone roof storage tanks are designed to fail at the top roof seam when over-pressured; however, they do not always fail there; this tank failed at the bottom seam, and the tank rocketed off the foundation and landed on top of the process unit. See photos in Figure 8.15.2a,b of the tank in its new location.

While freak in nature, this incident clearly shows the tremendous force associated with a pneumatic test when things go wrong. Fortunately, no one was injured, but anyone can see the potential in an incident like this.

Kawasaki Japan
1 June 2008

Workers at this chemical plant site were pressure testing a reactor piping system following mechanical repairs using high-pressure nitrogen as the test medium. The pipe was pressurized to 2,900 psig (200 BARG) as planned, the pressure test was completed, and the workers opened valves to vent the pressure from the piping. They then removed a blank flange, and when the flange suddenly failed, the piping moved violently in the pipe rack. The sudden movement caused the pipe to strike two workers, resulting in the death of one worker and injury to the second worker.

The workers attempted to depressure the reactor piping system but failed to recognize that a check valve was installed in the small piping being used to vent the pipe. This check valve prevented the depressurization, and therefore, the nitrogen pressure was not vented from the piping system. Consequently, the workers started removing the blank flange while the piping was still fully pressurized. The stored energy in the pressurized piping system was suddenly and unexpectedly released. The sudden pressure release was sufficient to cause the horizontally oriented pipe to rotate nearly 90° up in the pipe rack into an almost straight-up vertical position. The piping then rotated back down almost 180° to a nearly straight-down vertical position, striking the two workers.

This incident illustrates the considerable effect of "stored energy" in pneumatic testing and the sudden and catastrophic release of this energy should a failure occur. The energy potential should be evaluated when workers are involved in pressure testing equipment, especially when a pneumatic test is considered. In this example, one worker lost his life, and another was seriously injured.

(a)

(b)

Figure 8.15.2 Photos of the failed storage tank following the pneumatic pressure test. Figure (a) illustrates how the tank was propelled to the top of the process unit by the sudden release of the pneumatic pressure. Figure (b) shows the remaining tank foundation with the broken studs, illustrating the amount of force was generated by the pneumatic test. Source: Courtesy of the American Institute of Chemical Engineers (AICHE), The Center for Chemical Process Safety (CCPS), and the Beacon.

What Has Happened/What Can Happen

Shanghai LNG Terminal
February 2009 (1 Fatality, 15 Injured)

Details on this incident are sketchy; however, government reports from Shanghai, China, indicated that workers at the new Shanghai Liquefied Natural Gas Terminal were completing construction of the facility and testing the equipment. The report stated that they were pressurizing one of the gasifiers with air for the pressure test when the piping system suddenly failed catastrophically.

Figure 8.15.3 Shanghai LNG Terminal; February 2009, Piping Failure during Pneumatic Pressure Test. Source: Courtesy of the American Institute of Chemical Engineers (AICHE), The Center for Chemical Process Safety (CCPS), and the Beacon.

The resulting release of "stored energy" severely damaged the piping systems and caused cement support beams to buckle and fail. Major process vessels were dislodged from their foundation. One worker was killed when he was struck by a piece of shrapnel while working in a warehouse some distance from the LNG terminal. Another 15 workers were injured, and the new LNG plant was completely destroyed.

This incident also illustrates the impact of an equipment system failure during a pneumatic pressure test and the resulting energy release. See Figure 8.15.3 for a photo of the Shanghai LNG Terminal piping system failure, which occurred during a pneumatic pressure test. There are many additional images of this incident available at the following link: https://www.wermac.org/misc/pressuretestingfailure1.html.

What Has Happened/What Can Happen

PetroChina West/East Natural Gas Pipeline
Failure During a Pneumatic Pressure Test
8 May 2011

"The world's longest natural gas pipeline at 2,500 miles (4,000 kilometer)" failed at a flange weld. The failure occurred while the pipe was being pressure tested using nitrogen to conduct the test (a pneumatic test).

Portions of the pipe were propelled through the end of the building into the surrounding area. See Figure 8.15.4a,b for photos of the pipeline failure during a pneumatic pressure test.

(a)

(b)

Figure 8.15.4 Photos of a natural gas pipeline failure during a pneumatic pressure test. Note how a section of the piping was blown completely through the building into an adjacent area. Figure (a) illustrates how a large section of piping failed during the pneumatic test when the flange failed. Figure (b) shows how a large section of pipe was blown completely through the wall of a building. After an exhaustive search, the original source for these images could not be found.

Key Lessons Learned

- Use the hydrostatic testing process when possible; avoid pneumatic testing if possible.
- If a pneumatic "strength" test is planned, involve the site management and safety personnel and ensure they understand the hazards.
- Ensure that a detailed safety plan has been developed and that hazards have been identified and mitigated.
- Ensure the equipment is protected against potential overpressure while performing the test.
- Be sure the potential for brittle fracture has been considered, especially if using vaporized nitrogen to perform the pneumatic test. Remember, it may be necessary to heat the vessel to a temperature at least equal to the MDMT before conducting the test.
- Barricade the area and keep nonessential personnel away while the pneumatic test is performed. Only personnel necessary to perform the test should be present, and they should minimize time inside the barricaded areas to only that necessary to conduct and monitor the test.

Strength testing with a pneumatic such as air or nitrogen can be done safely. However, due to the risks and consequences of failure, we should always consider pneumatic testing a method of last resort and only conduct a pneumatic test after a thorough analysis and mitigation of the hazards. Ensure that only those directly involved in conducting the test are in the vicinity of the test while the test is underway.

Additional References

ASME PCC-2: "Repair of Pressure Equipment and Piping": ASME PCC-2 provides methods for repairing equipment, piping, pipelines, and associated ancillary equipment within the scope of ASME Pressure Technology Codes and Standards after being placed in service. These repair methods include relevant design, fabrication, examination, and testing practices and may be temporary or permanent, depending on the circumstances. The methods provided in this Standard address the repair of components when a repair is deemed necessary based on appropriate inspection and flaw assessment.

Included in ASME PCC-2 is a description of pneumatic testing of process equipment, including considerations and precautions for pneumatic testing and the method of calculating potential stored energy in a pneumatic test (Article 5.1, Appendix II).

"Hazards and Safety Concerns During Pneumatic Testing of Pressure Plants", Naveenraj, T. R., Mechanical Engineer, Chempro Inspections Pvt. Ltd. as Senior Manager (QA/QC). www.chemproindia.com/file.php/pdf/HAZARDS-Pneumatic Test.pdf

Safety requirements for pressure testing Guidance Note GS4 (Fourth edition). https://www.hse.gov.uk/pubns/gs4.pdf

US Department of Transportation, Pipeline and Hazardous Materials Safety Administration, 49 CFR 195: Current regulation with changes from the previous regulation noted. https://www.phmsa.dot.gov/sites/phmsa.dot.gov/files/docs/195_Master_195_100_Highlighted.pdf (See Subpart E).

Chapter 8.15. Hazards of Pneumatic Testing (For Testing the Integrity of Process Equipment) (Hydrostatic Testing vs. Pneumatic Testing)

End of Chapter Quiz

1 What is the most significant difference between a hydrostatic pressure test using water as the test medium and a pneumatic pressure test using air or nitrogen as the test medium?

2 What are some situations where a pneumatic test would be necessary to conduct the pressure test?

3 If the pneumatic testing process is used, the process should receive higher level ________ and __________ review and approvals, including a ________ assessment to ensure all hazards are identified and mitigated before being carried out.

4 Where are procedures readily available that cover Pneumatic Testing of Pressure Vessels and Piping, including detailed considerations and precautions for conducting the test?

5 What are the key lessons learned from this module?

8.16

Cold-wall Vessel and Piping Failures

Process vessels such as reactors and piping on some units operate at temperatures exceeding the vessel's design temperature or piping metallurgy, and the pipe or vessel would quickly fail if not otherwise protected. In most cases, these vessels and piping systems are internally lined to protect the metal from excessive temperatures and the resulting degradation and failure. This is generally referred to as a "cold-wall design," meaning that the materials of construction (the vessel or piping) are protected by internal insulation (usually refractory). Therefore, the steel operates at a much lower temperature than the temperature of the process. In a cold-wall design, the steel of the vessel or piping is protected from excess temperature by the internal insulation, thus "cold-wall design." With a hot-wall design, the internal temperature of the process is in direct contact with the steel.

The internal refractory or lining may be gunite, a concrete, ceramic, or composite type. Some systems, such as furnace stacks, may be lined with bricks to protect the pipe walls. Cold wall-designed piping and vessels are lined with refractory inside to protect against very high process temperatures and have no external insulation. The designers prefer that the steel used for these designs radiate the heat away as quickly as possible.

Using cold-wall design in refineries and chemical plants is a beneficial technology. It allows the use of carbon steel for the pipe wall or vessel wall in high-temperature services that otherwise would require very expensive metallurgy and still require replacement at relatively frequent intervals. Carbon steel in these high-temperature services, with the application of internal refractory, can achieve a service life of 30–50 years without replacement. However, the refractory must be maintained in good condition, requiring periodic inspection and verification of its condition. Any indication of a suspected failure must receive prompt attention to mitigate the hot spot.

Refractory, when exposed to extreme temperature changes and other stresses imposed by the process conditions, can fail in service, and refractory failure due to cracks or spalling will subject the pipe and vessel wall to extreme process temperatures. If this internal lining fails, the vessel and piping are subjected to process temperatures exceeding metallurgy limits. This can lead to an unexpected and catastrophic failure of the vessel or piping system, resulting in loss of containment, explosion, and fire.

It is extremely important to closely follow the refractory's heating and cooling rate during unit startup and shutdowns. This ensures that the refractory is allowed time to expand and contract slowly, avoiding damage to the refractory. It is equally important to allow time for any trapped water to evaporate without explosive expansion, which can destroy the refractory. These precautions should be built into the unit operating procedures in a checklist format and rigorously followed. Likewise, be aware that unplanned emergency shutdowns should be avoided to the extent possible due to the severe temperature changes that can occur inside the refractory-lined equipment.

Monitoring Internal Refractory

An infrared camera and a well-trained Unit Inspector can detect and monitor potential refractory failure. Inspections should be done during and after unit startup, and a baseline should be established for ongoing inspections of critical areas. Inspections should then be conducted at an established interval during the unit run to determine the condition of the internal refractory lining. Indications of a "hot spot" should be managed promptly as this may lead to piping or vessel failure. See Figure 8.16.1a,b for photos of an infrared scan to help identify internal refractory failure and the resulting hot spot. Figure 8.16.1a is what we may see when observing a section of cold wall designed pipe. Note there is no indication of refractory failure and no "hot spots". However, Figure 8.16.1b is the same section of pipe using an infrared camera. The hot spots are clearly visible using the infrared.

Process Operations Safety: The What, Why, and How Behind Safe Petrochemical Plant Operations, First Edition. M. Darryl Yoes.

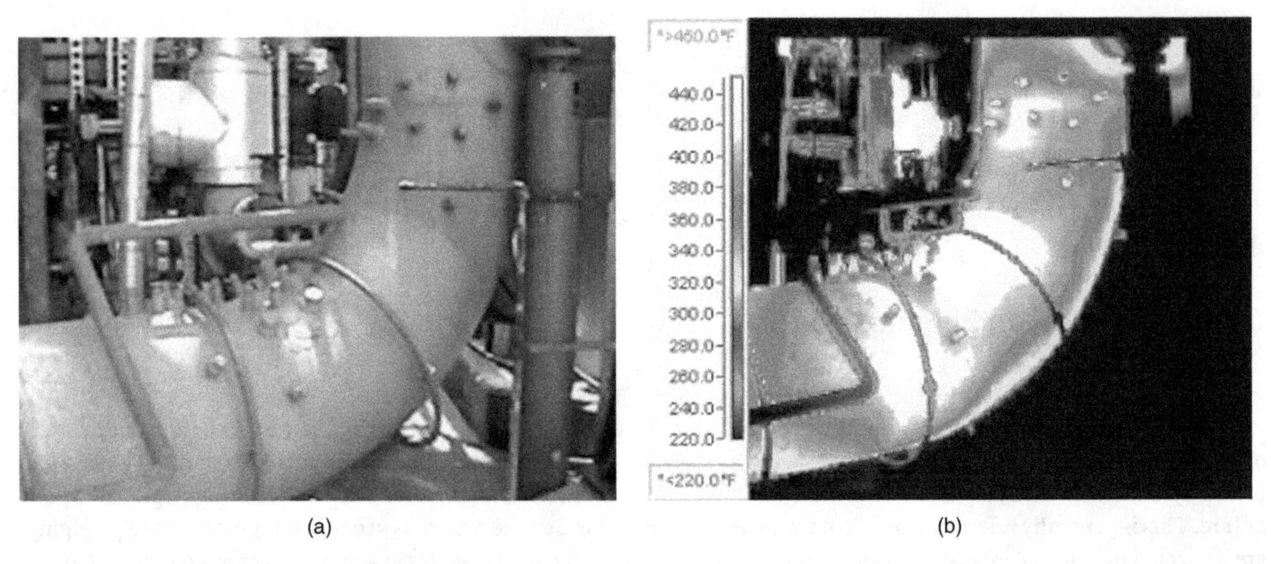

Figure 8.16.1 Photograph of typical refractory-lined transfer pipe. The photo on the left is "as-viewed," while the photo on the right is an infrared image. The hot spots cannot be detected visually but can be seen with the infrared image. This is an indication of a refractory failure or possibly the result of "hot gas bypass," the internal channeling of the hot process gasses behind the refractory. Source: Photos courtesy of Sonny James, Thermal Diagnostics Limited (http://www.tdlir.com/).

Temperature-indicating paints also offer another attractive method for identifying potential refractory damage and hot spots. Thermal Paint is a temperature-sensitive paint-based product that creates an irreversible visual picture of the temperature contour patterns. It is an effective tool to help monitor internally lined equipment. Paint suppliers advertise that Temperature Indicating Paint will:

- Provide an early warning indicator of process vessel overheating due to gas bypassing or refractory failure.
- Provide an early warning indicator of temperature conditions conducive to hydrogen attack of low carbon allow steels in high-pressure/high-temperature refinery and petrochemical processes utilizing hydrogen-rich atmospheres.
- The paint offers accurate results between the temperatures of 118–2,282 °F (48–1,250 °C).

Hot spots indicating potential refractory damage should be immediately reported to the Equipment Owner and the Mechanical Engineer for follow-up.

Temporary Mitigation of "Hot Spots"

We can take action on a temporary, short-term basis to control and mitigate metal "hot spots" due to potential internal refractory damage. Depending upon the severity of the refractory damage, some hot spots can be controlled by applying steam, air lances, or spargers. In this application, we are positioning a steam lance or sparger to blow live steam or refinery air directly onto the hot spot. The steam is hot but much cooler than the process temperature inside the pipe or vessel. Therefore, it cools the steel and keeps it at a safe temperature. This is illustrated in Figure 8.16.2a,b. The infrared thermal images clearly indicate how the temperature rises when the steam lances are turned off, and how the temperature returns to a safe zone when the steam is restored.

Once installed, these lances or spargers should be considered a critical application to prevent pipe or vessel wall failure. These devices should be considered safety-critical and subject to controls such as Management of Change. To help ensure the refractory failures are not propagating, piping and vessels where the temporary steam application is controlling hot spots should be periodically monitored by the equipment inspector using thermography or other similar inspection techniques.

The Hazard of Installing Temporary Insulation on Cold-wall Piping or Vessels

Most operators do not understand the hazard of installing temporary insulation on the exterior of piping or vessels with internal refractory (cold-wall design). It does happen periodically, with the very best of intentions. However, installing

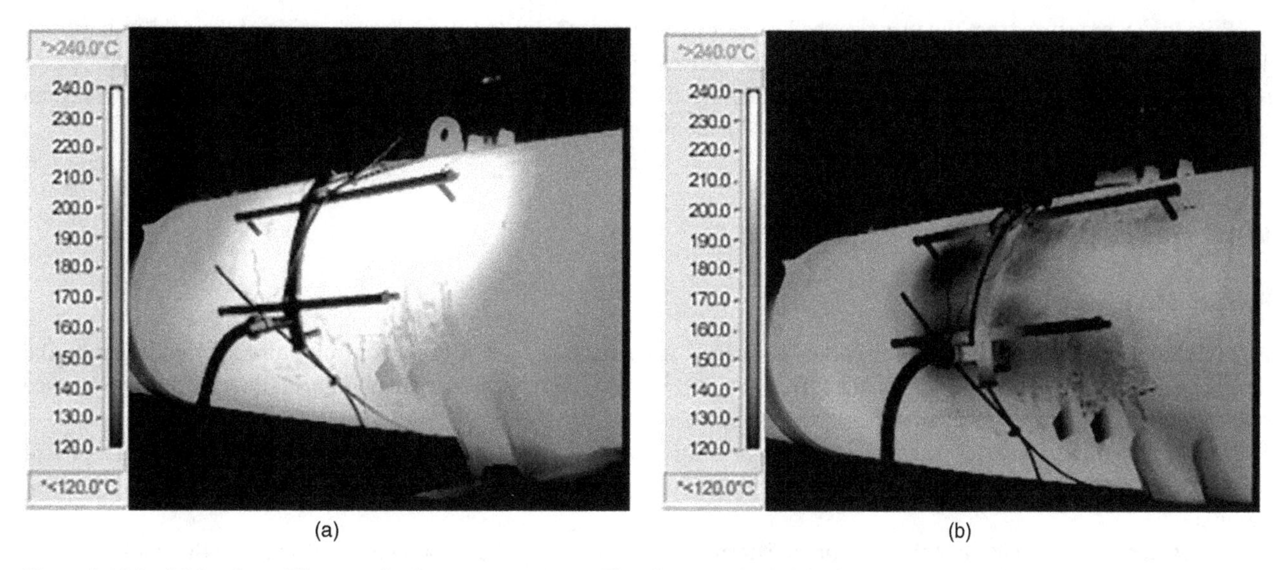

Figure 8.16.2 Mitigation of "hot spot" using steam spargers. The photo on the left is the hot spot with the steam cooling turned off, while the right image was taken with the steam sparger on. Note the significant difference in temperature. Source: Photos courtesy of Sonny James, Thermal Diagnostics Limited (http://www.tdlir.com/).

external insulation over equipment with internal refractory can result in catastrophic failure of the cold-wall equipment. The external insulation can result in severe damage to the internal refractory and rapid overheating of the pipe or vessel steel, leading to a catastrophic failure or rupture.

Caution

Never place external insulation, such as an insulating blanket, on a pipe or vessel with cold-wall-designed equipment (with internal refractory). This can lead to sudden and unexpected failure of the piping or vessel. The external insulation can trap the heat, which will rapidly overheat the metal wall of the pipe or vessel, leading to a catastrophic failure and loss of containment of the process, leading to a fire or explosion.

External insulation should not be placed on the exterior of cold-wall-designed equipment (piping or vessels) unless a detailed analysis by Mechanical Engineers has been completed and it has been determined it is safe to do so and authorized by a documented Management of Change.

A section of piping or vessel that does not usually glow and is glowing is potentially a very serious situation and should be addressed immediately by removing heat, pressure, or both.

What Has Happened/What Can Happen

Methane Reformer Unit Fire

A major unit fire occurred on a Methane Gas Reformer Hydrogen Unit at a West Coast refinery due to a failure of the furnace outlet line. The product in the furnace outlet line normally operates from approximately 1,450—1,480 °F and 400 psig; however, with the internal refractory, the outlet line's normal skin temperature ranges from 150 to 250 °F (65–121 °C). However, the line had experienced minor internal refractory damage, and the resulting pipe wall hot spots were managed with steam lances.

Meanwhile, some mechanical work was in progress in the area under the furnace adjacent to the furnace outlet pipe. It was extremely hot under the furnace, and the team was working close to the furnace outlet piping. Also, their glasses were fogging due to the steam from the steam lances. To improve this, they redirected the steam lances away from where they were working and from the outlet line; however, it was still very hot adjacent to the outlet line, so they wrapped a section of the line near their work area with an insulating blanket.

While they were away from the furnace, the line suddenly ruptured due to internal overheating of the cold-wall piping. This resulted in rapid depressurization of the furnace tubes and a major fire under the furnace. The depressurization also severely damaged the internal refractory and resulted in a discharge of the catalyst from the furnace tubes over a wide area in the refinery. While there were no injuries in this incident, the potential for a catastrophic incident existed.

What Has Happened/What Can Happen

Furnace Stack Near Failure

At a US Gulf Coast refinery, mechanics worked to repair a stuck furnace stack damper control. Thinking the issue with the damper was due to the excess heat from the furnace stack, they installed a temporary insulating blanket around the furnace stack to shield the damper mechanism from the excess heat.

Unknown to them, the stack internal refractory had failed, resulting in the excess heat near the damper, which caused overheating of the furnace stack metallurgy and most likely resulted in the binding of the damper shaft and the stuck damper.

As a result of the application of the temporary external insulation, excess heat was generated, causing the stack piping to fail. When the external insulation was removed, the furnace stack in the damper area was completely melted, and one side of the stack had a large diameter hole in the area where the blanket was installed. Fortunately, the stack did not collapse. Although no one was injured and damage was minimal, this could have resulted in the stack toppling from the furnace and impacting other operating equipment in proximity.

What Has Happened/What Can Happen

FCCU Reactor Riser Pipe

At another Gulf Coast refinery, mechanics were working in the structure of an FCC unit adjacent to a 68″ (1.7 M) FCC catalyst riser pipe equipped with a cold-wall design and operating at around 1,000 °F (537 °C). Attempting to mitigate the excessive temperatures in the area near where they were working, they called for insulators to install temporary insulation on the huge catalyst riser pipe to help shield them from the high temperature.

This insulation was left on the pipe for several days until an Operator making a normal process round observed the temporary insulation crumbling, with some of the insulation falling onto the adjacent platforms. The Operator quickly called for his supervisor to come up and help him understand what was happening. Upon arrival, the supervisor immediately determined that the pipe was a cold-wall design and should not have any installed external insulation. They had an insulation crew remove the temporary insulation immediately.

Upon insulation removal, they discovered that the 68″ (1.7 meter) diameter reactor pipe was found to be deformed (bulged) by 14 inches (35 centimeter) in circumference over a 4-foot (1.2 meter) long section. The unit was immediately

Figure 8.16.3 Artist rendition of failed cold-wall pipe when covered by insulation. This image is courtesy of Ecoscience Resource Group Visual Artists. It was created to illustrate the hazards of coldwall-design equipment when applying external insulation. The hazard is that the metal quickly overheats and can fail catastrophically and without warning.

shut down for repairs to the pipe and the internal refractory. This close call had the potential for a major loss of containment had the catalyst riser failed, leading to a potential explosion and major unit fire.

Consequences of Placing External Insulation on Cold-wall Equipment

Figure 8.16.3 illustrates what can happen if a vessel or piping system designed with cold-wall design, with refractory on the interior and no insulation on the exterior, is subjected to the addition of external insulation. This can quickly lead to rapid overheating of the steel and an unexpected vessel or pipe rupture. Of course, this can result in rapid depressurization of the contents and a large explosion or fire.

Key Lessons Learned

- The consequences of applying exterior insulation to cold-wall carbon steel equipment may result in piping and vessel failure and loss of containment. Even "small" deviations in the installation of insulation can result in huge impacts, for example overlapping external insulation over the cold wall to hot-wall transition.
- Pay close attention to hot/cold transition areas. When conducting potential problem analysis of unusual observations, consider the possibility of accelerated metal degradation (overheating of carbon steel piping or vessels).
- Installation of temporary steam or air lances to help control metal temperatures should be considered safety-critical and adequately managed through a Management of Change (MOC). The same process should also be used for the removal of temporary cooling after it is installed.
- Be alert for indications of hot spots on cold-wall equipment and immediately ensure the involvement of the equipment owner, mechanical engineers, and equipment inspector if any are discovered.
- Ensure piping and equipment designed with internal refractory ("cold-wall design") are identified in the field and on the unit drawings. This is to help prevent the accidental application of external insulation, which could lead to piping or vessel failure.
- Ensure Operators are aware of the hazards of piping or vessels in cold-wall service (if not properly managed).
- Ensure periodic temperature surveys of cold-wall equipment to identify potential hot spots and the installation of any external insulation.

Additional References

"The Fundamentals of Refractory Inspection with Infrared Thermography", Sonny James, Level I, II, and III Infrared Thermographer, Managing Director Thermal Diagnostics, LTD. http://tdlir.homestead.com/papers/Refractory.pdf

Fluid Catalytic Cracking (FCC) Transfer Line Flexibility – Analysis and Design Considerations, Chuck Becht. https://becht.com/becht-blog/entry/considerations-in-fcc-transfer-line-flexibility-analysis-and-design?webapp=1

Protecting Refractory Linings in FCC Units. https://catcracking.com/protecting-refractory-linings-in-fcc-units/

Chapter 8.16. Hazards of Cold Wall-design Piping and Vessels (Refractory-lined Equipment)

End of Chapter Quiz

1 What is meant by the term "cold-wall design"?

2 Cold wall-designed piping and vessels are lined with ___________ inside to protect against very high process temperatures and have no external insulation.

3 What happens if the refractory starts to wear or fail?

4 How can hot spots or small refractory failures be identified?

5 Depending upon the severity of the refractory damage, some hot spots can be controlled by applying _________, air lances, or spargers.

6 After the steam lances are installed on a hot spot, they should be regarded as _______-_______ and subject to controls such as Management of Change.

7 What is one of the primary concerns with cold wall-design equipment?

8 We understand that external insulation should not be placed on cold wall-designed equipment, and when done, it is typically without an understanding of the hazards. What are a couple of reasons why this may be done?

9 External insulation placed on cold wall-designed piping or equipment can result in severe damage to the internal refractory and rapid ____________ of the pipe or vessel steel, leading to a catastrophic failure or rupture.

10 To help ensure the refractory failures are not propagating, piping and vessels where the temporary steam application is controlling hot spots should be periodically _________ by the equipment ___________ using thermography or other similar inspection techniques.

8.17

Metallurgy Matters: Positive Materials Identification (PMI)

Many significant process safety incidents have started with metallurgy, specifically the wrong metallurgy for the service or the corrosion mechanism. Several of these have had severe consequences, including loss of life. For example, in October of 2006, at the Mažeikiu Nafta refinery located in Lithuania, it was reported that piping near the bottom of a vacuum tower failed catastrophically, resulting in a major fire. During the extensive fire, the vacuum tower collapsed nearly on top of the heat exchanger train, destroying it and its associated equipment. The cause of the pipe failure was reported to be inadequate positive materials identification or PMI. The failed piping was reported to be 5% chrome; however, the circuit contained a section of carbon steel pipe welded into the chrome piping circuit, which failed catastrophically in service. If this is what happened, it is most certainly a high-temperature sulfidation corrosion failure. These are uniform corrosion, and failure is almost always a rupture. The lack of fireproofing reportedly contributed to the collapse of the vacuum tower. See Figure 8.17.1 for a photo of the collapsed vacuum tower.

Unfortunately, this incident is not all that uncommon. There have been numerous PMI incidents in our industry similar to the one that occurred in Lithuania. When leading the Safe Operations Training, I always mention that carbon steel in chrome piping circuits is our "Achilles' heel" regarding PMI. The issue is that it is impossible to distinguish the two metals apart using only our senses. They have the same appearance and will both corrode or rust when left in the elements. The welder cannot tell one from the other as they both have the same characteristics during the welding.

When we modify a chrome piping circuit or install new chrome or other alloy piping, the replacement pipe, or the new pipe, is laid out by craftsmen and then prefabricated in the shop. After fabrication, the new piping is delivered to the field for installation. If the layout or fabrication is off by a few inches or even several feet, it has not been uncommon for the field craftsman to pick up a section of pipe and weld it into the circuit to make up the difference. The results can be catastrophic if that piece of pipe happens to be carbon steel and it gets welded into a chrome circuit.

Unfortunately, when this happens, the chances that our inspection process will discover the short section of carbon steel is nil. If we do not PMI the pipe as it is installed, we will not find it until it fails; and that will not be for several years. The on-stream thickness monitoring process for pipe thickness is a periodic random ultrasonic thickness (UT) measurement. The chance that this random placement of the UT probe would locate a small section of thinning carbon steel pipe in a long or complex chrome piping circuit is extremely low. The result is that the corrosion is much more aggressive in piping and vessels fabricated from carbon steel than in similar vessels fabricated from chrome.

Carbon steel has little resistance to many corrosion mechanisms and corrodes much faster. For example, we typically use chrome piping in circuits where the primary corrosion mechanism is due to high-temperature sulfidation operations above about 450 °F (232 °C) in the presence of sulfur. Under these conditions, carbon steel pipe will fail within a few years, whereas chrome piping will last much longer, generally several decades. Another issue with Carbon steel, when installed in an alloy circuit, is the corrosion mechanism and the type of corrosion. When exposed to high-temperature sulfidation, carbon steel pipe corrodes uniformly; every section of the carbon steel pipe that "sees" the process corrodes at the same rate, and when it fails, it is most often a complete rupture. This almost always results in an unexpected and total loss of containment and an explosion or major fire. See Figure 8.17.2 for an example of sulfidation corrosion in carbon steel vs. chrome, and pay attention to the uniform nature of the corrosion. The thickness of the carbon steel component, the center section in the photograph, was the same as the chrome when it was first installed.

Thus far, I have discussed our Achilles' heel: a carbon steel section of pipe installed in a chrome piping circuit. However, the PMI process should be applied to all alloy piping and other alloy equipment that can "see" the process pressure. For example, stainless steel is produced in many different grades, and the corrosion resistance and other properties vary widely by grade. It is impossible to distinguish among the various grades of stainless and many other metals without PMI.

Process Operations Safety: The What, Why, and How Behind Safe Petrochemical Plant Operations, First Edition. M. Darryl Yoes.

Figure 8.17.1 Major refinery fire and the collapse of the vacuum tower at the Mažeikiui Nafta refinery located in Lithuania. The crude unit vacuum column collapsed onto the heat exchanger train. The furnace transfer lines have separated from the column. Figure shows the bottom of the vacuum tower completely separated from the supporting foundation.

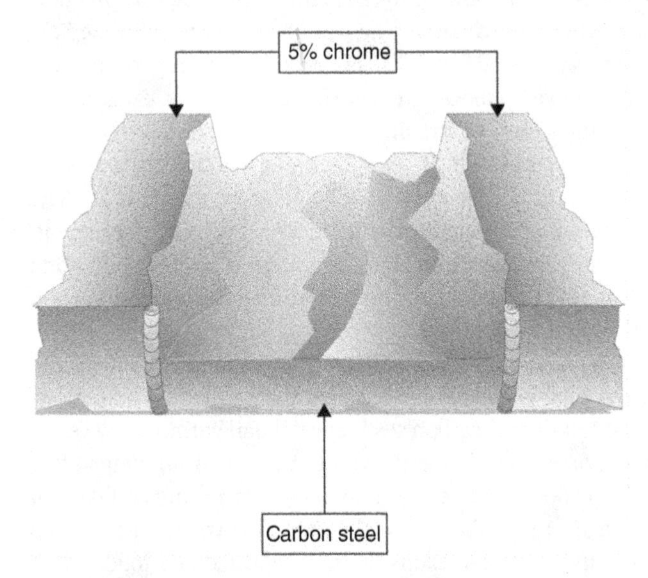

Figure 8.17.2 Illustrates two sections of 5% chrome pipe with a small section of carbon steel welded between the two. Note how high-temperature sulfidation corrosion affected the Carbon Steel and the uniform nature of the corrosion. The CS pipe was severely corroded, while the chrome was resistant to the sulfidation corrosion. This resulted in a significant fire on a crude unit due to a failure on the opposite side of the pipe. Source: This image is courtesy of Ecoscience Resource Group Visual Artists. It was created to illustrate the hazard of mixing metallurgies or using the incorrect metallurgy for pressure containment.

More recently, we have rediscovered that it is also essential to understand the type and composition of carbon steel piping used in hot service where sulfur is present. On 6 August 2012, the Chevron refinery in Richmond, California, experienced a loss of containment event and a major refinery fire. It was rediscovered in that incident that corrosion is much more aggressive in Carbon Steels with a low silicon content (less than 0.10 wt% silicon). This is certainly true when operating at high temperatures with sulfur compounds in the piping circuit. Due to this incident, many refineries are conducting PMI on carbon Steel piping circuits in hot service to verify the silicon content of the piping.

So, What Is PMI?

PMI is field verification to ensure that the correct metallurgy has been installed in alloy equipment. PMI is covered by API 578 ("Material Verification Program for New and Existing Alloy Piping Systems"). This recommended practice provides the guidelines for material control and verification programs on ferrous and nonferrous alloys during the construction, installation, maintenance, and inspection of new and existing process piping systems covered by the ASME B31.3 and API 570 piping codes. Carbon steel components specified in the new or existing piping systems are not specifically covered under the scope of the API Recommended Practice.

Why Do We Conduct PMI?

Unfortunately, mistakes happen, the material can be mislabeled, and, as mentioned above, some alloys look the same or very similar. Sometimes the mill certificates or shipping papers may be incorrect, or materials are mixed up in shipment or material storage bins. We know from experience that the wrong metallurgy materials installed in the field can lead to unexpected failures.

When Do We Conduct PMI?

PMI should always be conducted following the new installation of pressure-containing components in alloy circuits or the replacement of alloy components during maintenance. We also conduct PMI for quality control in the warehouse and when materials are delivered to the field, also for quality control.

Who Is Responsible for PMI?

Conducting PMI is a sitewide responsibility. While responsibilities should be well defined and understood, everyone should feel responsible for ensuring that PMI is done, especially the final PMI after installation or construction is complete. For example, Operators should feel empowered to ask to see the PMI documentation before placing an alloy piping circuit in service following maintenance or modifications.

How Is PMI Conducted?

PMI is conducted using completely portable X-ray Fluorescence Technology (XRF Spectrometry). In this process, elements of an alloy are excited by X-rays and emit a definitive wavelength, resulting in characteristic X-rays specific to that element's atomic structure. This phenomenon is known as X-ray fluorescence. The handheld analyzer definitively identifies the spectral signatures of each detectable element. Lighter elements, such as carbon, cannot be detected by XRF technology. However, most metals can be positively identified by the composition of primary alloying elements. During field analysis, most of the newer devices provide the technician with a direct readout of the metallurgy. These handheld devices can quickly distinguish between carbon steel and alloy piping materials and various other metallurgies without damaging the base metal. See Figure 8.17.3 for a photo of a Handheld X-ray Fluorescence PMI analyzer.

Conducting PMI is relatively easy. However, recording and tracking the data is much more difficult due to many individual data points. Each component must be analyzed to ensure it is the correct metallurgy. Components may include piping,

Figure 8.17.3 Conducting PMI with a handheld X-ray fluorescence PMI analyzer. Source: Photo kind courtesy of the Evident Scientific Corporation (EvidentScientific.com).

fittings, welds, instruments, or any other component exposed directly to the process fluid. This means that in a typical piping circuit, every section of pipe, flange, small pipe component or small connection, valve, vent or drain, and weld must be analyzed. Even with a relatively short piping run, this can quickly add up to hundreds of data points. See Figures 8.17.5 for example, pipe sections and the corresponding PMI data points that must be recorded and managed.

Equipment Degradation Mechanisms Related to PMI

The following equipment degradation mechanisms can be directly related to PMI. The use of inadequate or wrong metallurgy in circuits where these mechanisms exist can result in catastrophic piping or other equipment failures and lead to catastrophic loss of containment, explosion, and fire. Remember that dissimilar metallurgies are affected in different ways when subjected to different chemicals and temperatures. Dissimilar metals react differently in types of corrosion, corrosion rates, and failure mechanisms:

- High-Temperature Sulfidation.
- Naphthenic Acid Corrosion.
- High-Temperature Hydrogen Attack.
- Brittle Fracture.
- Creep.
- Ammonium Salt Corrosion and Under Deposit Corrosion.
- HCl, H_2SO_4, Organic Acid Corrosion.

The number of PMI locations and data points can quickly add up. Note the simplified piping layout in Figure 8.17.4. Where are the locations that will require PMI?

See the companion sketch in Figure 8.17.5 for the required PMI locations. Remember, it only takes one to be missed to result in an incident.

PMI is not easy but must be done on all alloy materials replacements or installations, during all new construction, and after modifications to circuits containing alloy piping or other alloy equipment. Each test is not complicated; however, all insulation must be removed.

As evidenced by the above sketches, data tracking is the most difficult. Incidents have occurred due to data being misplaced or incorrectly recorded.

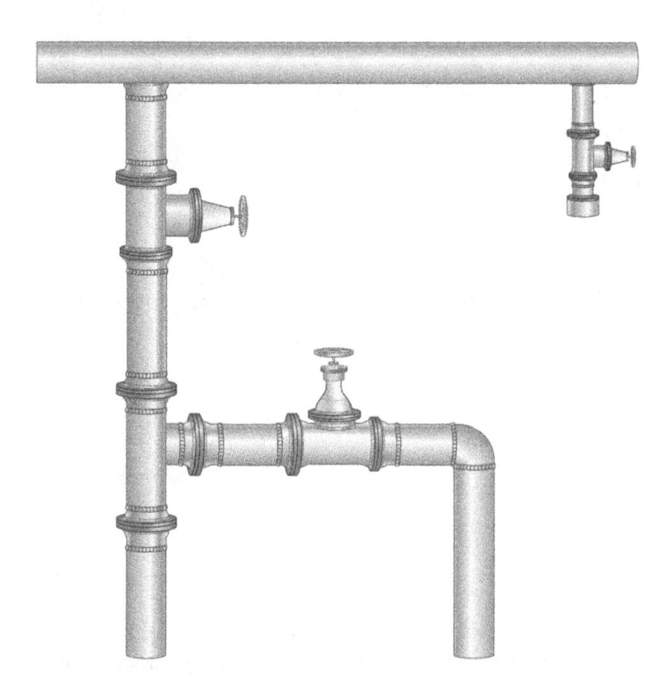

Figure 8.17.4 Simplified piping layout requiring 100% PMI. Source: Image courtesy of Ecoscience Resource Group Visual Artist. These images were created to illustrate the hazards associated with PMI and are not representative of any company.

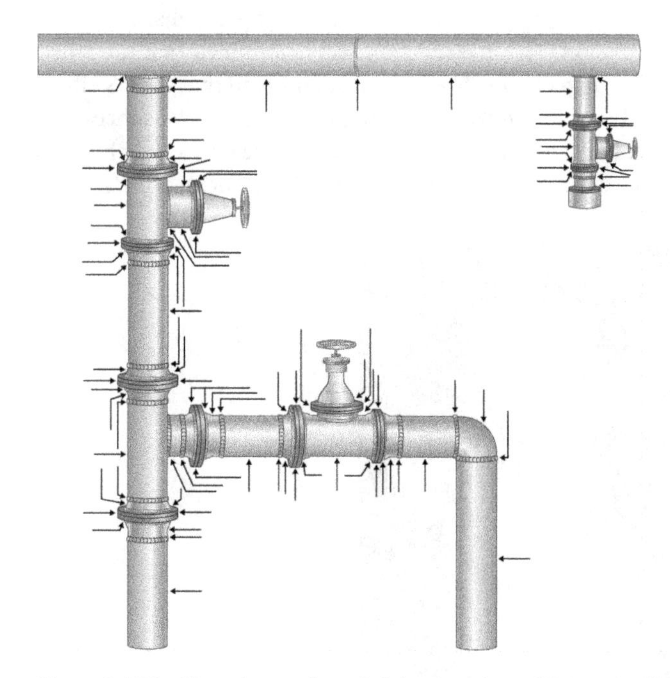

Figure 8.17.5 Note the number of piping locations that require PMI testing to ensure 100% PMI. Each arrow indicates a required PMI to ensure 100% PMI. Source: Image courtesy of Ecoscience Resource Group Visual Artist.

Typical metallurgies used in a petroleum refinery or petrochemical plant (not intended to be a complete list, and not each metallurgy mentioned here is used in every refinery or chemical plant):

- Carbon Steel.
- Carbon ½ Mo.
- 1¼ Cr-½ Mo (P11).
- 2¼ Cr-1 Mo(P22).
- 5 Cr-½ Mo (P5).
- 9 Cr-1 Mo (P9).

- 9 Cr-1 Mo – V (P91).
- Low Nickel steels: 3.5Ni, 5Ni, 9Ni steels.
- 13 Cr (410 SS, CA6NM).
- 300 Type SS: 304(L) SS, 309SS, 316(L) SS, 321SS, 347SS.
- Duplex Stainless Steel (2101, 2205, 2507, CD3MN, CD4MCuN).
- Nickel Alloys (Alloy 20, Monel 400, Incoloy 800, 825, Hastelloy).
- Titanium, Zirconium, Tantalum.

All these metallurgies have different characteristics and will corrode at different rates when exposed to the process or the elements. Many of these metallurgies are impossible to distinguish apart without conducting PMI.

What Has Happened/What Can Happen

Exxon Baton Rouge Refinery
Delayed Coker Fire (Three Fatalities)
2 August 1993

On 2 August 1993, a catastrophic loss of containment of hot residuum feed occurred on the Baton Rouge Refinery East Coker Unit, a four-drum delayed Coker unit, resulting in a major unit fire that partially destroyed the large process unit. Unfortunately, three people were killed in the incident due to the loss of containment and fire. Two Operators and a contract employee were lost in this incident.

The investigation determined that a 6″ diameter (15 centimeter), 45° piping elbow on the hot feed pump discharge piping circuit failed catastrophically due to internal sulfidation corrosion. The investigation also determined that the failed discharge ell was carbon steel and was installed in a 5% chrome piping circuit. The investigation found that the CS elbow was installed in 1963 during the unit's original construction and that Exxon was unaware that metallurgy was incorrect. The bottom line is that the fitting could have failed at any time, but tragically, it failed during a coke drum switch while people were working in the structure. See Figure 8.17.6 for a photo of the refinery fire and Figure 8.17.7 for an artist's image

Figure 8.17.6 Exxon Baton Rouge Refinery, 2 August 1993. Major fire in the Delayed Coker Unit. Source: Photo courtesy WAFB Channel 9 News.

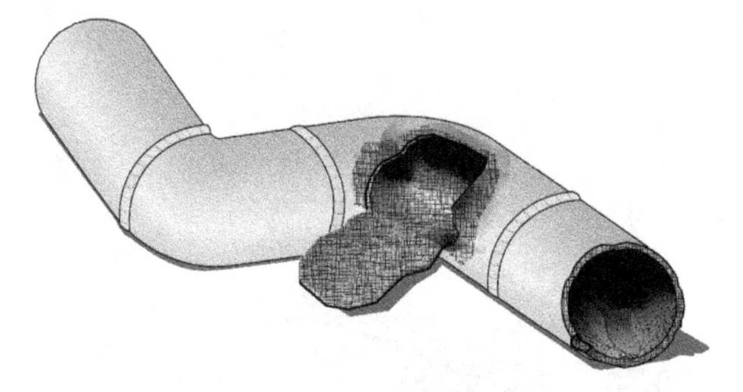

Figure 8.17.7 An artist's rendition of the failed 6-inch diameter carbon steel 45 ° piping elbow on the delayed Coker hot feed pump discharge. Source: This image courtesy of the Ecoscience Resource Group Visual Artists.

of the failed delayed Coker carbon steel piping elbow. The 45 ° elbow failed due to internal sulfidation corrosion, and the resulting fire caused considerable damage to the Coker structure.

The response to this incident was a significant investment in a renewed and greatly expanded PMI program. As an integral part of that effort, the company undertook a retroactive PMI process. They returned to all other operating units and conducted a comprehensive PMI examination for all alloy piping and alloy pipe components and fittings from scratch.

This depiction illustrates the type of failure that can occur when a single section of pipe is installed using incorrect metallurgy. In this example, with this one exception, the entire piping circuit was the correct 5% chrome. When the failure occurred, pipe thickness measurements confirmed that the remainder of the chrome piping was essentially the original thickness when installed about 30 years earlier. However, the failed section, a 45-degree elbow, was carbon steel and was affected by high-temperature sulfidation corrosion. When the pipe failed, it was a catastrophic loss of containment of residuum feed operating at about 700 °F (370 °C) and around 500 psig. A massive fire resulted, which destroyed the 4-drum delayed Coker structure. As stated earlier, three people lost their lives in this incident.

What Has Happened/What Can Happen

BP Texas City
Resid Hydrotreater Fire
28 July 2005

This incident occurred at the same site, where just four months earlier, a devastating explosion resulted in the loss of 15 and over 180 people injured. This incident was also investigated by the US Chemical Safety and Hazard Recognition Board, and a PMI safety bulletin was issued on the incident. On the BP Texas City Refinery Resid Hydrotreater, contractors were called to remove the reactor feed/effluent exchanger bundles for cleaning. Following cleaning, the exchanger tube bundles were reinstalled. Each exchanger had an 8″ (20 centimeter) diameter pipe elbow on the inlet and outlet of the exchanger. These were removed when the tube bundles were removed for cleaning and then reinstalled before startup.

After the exchangers were returned to service and following a run of only about three months, 1 of the 8″ outlet piping elbows failed and resulted in a significant loss of containment of hot recycle gas, which was primarily hydrogen under approximately 3,000 psi, and operating at a temperature above 500 °F (260 °C). The release quickly ignited, resulting in a large jet fire with an estimated path 75 feet westward from the flange and significant damage localized along the fire path. The fire and resulting damage resulted in an extended shutdown for repairs.

After investigation, it was revealed that contractors were unaware that the exchanger inlet and outlet piping elbows were different metallurgy. The elbows on both the inlet and outlet appeared to be the same. However, the inlet to the exchangers (the cold side) was carbon steel, while the outlet elbows (the hot side) were 1.25% chrome. These piping elbows were inadvertently switched during the prior routine maintenance job. The carbon steel elbows were placed on the outlet of the exchangers, where they were exposed to very high process temperatures causing a piping failure resulting in the large process unit fire. The carbon steel piping elbow failed due to a high-temperature hydrogen attack (HTHA). chrome is inherently more resistant to HTHA, whereas carbon steel is not. See Figure 8.17.8a,b for photos of the failed piping elbow.

In this incident, there was no PMI of the elbows or other materials, and the piping elbows were not marked or identified as to their location on the unit. This exemplifies that one cannot go by appearances; we must conduct PMI to determine the correct metallurgy.

(a) (b)

Figure 8.17.8 Photos of BP Texas City Resid Hydrotreater failed carbon steel piping elbow. Figure (a) illustrates how the carbon steel fitting failed due to HTHA and separated from the exchanger while in service. Figure (b) shows the effect of HTHA on the carbon steel ell. Source: US Chemical Safety and Hazard Recognition Board Safety Bulletin.

What Has Happened/What Can Happen

LyondellBasell Houston Refinery
Houston, Texas
8 April 2016

In their delayed Coker unit, a large fire occurred at the LyondellBasell refinery in Houston, Texas, on 8 April 2016. According to an investigation report, a 4″ heavy gas oil pipeline to the wash oil sprays in the main fractionator failed adjacent to the tower. The hot gas oil flashed a few seconds after the release, probably due to autoignition. Following the fire, the line was discovered to have a classic fish mouth failure due to obvious general thinning or uniform internal corrosion. The metallurgical analysis indicated that the pipe was the incorrect metallurgy, and the severe internal corrosion was consistent with high-temperature sulfidation. Although the resulting damage was mitigated by water deluge systems and prompt emergency response, this damaging fire resulted in an extended downtime of the Coker unit for repairs. Fortunately, there were no injuries reported as a result of this incident.

What Has Happened/What Can Happen

Philadelphia Energy Solutions Refinery
South Philadelphia, Pennsylvania
21 June 2019

A section of pipe failed in the Alkylation unit at 4:00 a.m. on Friday, 21 June 2019, resulting in a catastrophic vapor cloud explosion (VCE) and massive fire. During the event, approximately 3,200 pounds of highly toxic chemical hydrofluoric acid (HF) was released into the atmosphere, according to the US Chemical Safety and Hazard Investigation Board, which is currently investigating the incident. According to the Chemical Safety and Hazard Investigation Board (CSB) in a "Factual Update" report on the incident, the failure occurred due to the pipe metallurgy not meeting current industry standards. The CSB reported that the failed pipe section was not being inspected for corrosion and was found to be as thin as about ½ a credit card in sections. The investigation by the CSB is ongoing; however, the Factual Report contains a good overview of the incident and its cause.

The preliminary CSB report has found that the Alkylation unit pipes were installed in 1973. During this period, the industry standards were not as developed and did not specify how much nickel or copper the pipe should contain. However, these standards were updated in 1995 and specify that pipes in flammable hydrocarbon service should have a maximum of 0.4% of each nickel and copper. The ruptured pipe at PES contained 1.74% nickel and 0.84% copper, above the current pipe specifications, and may be vulnerable to internal corrosion at a higher rate than expected. In May of 2013, the American

Figure 8.17.9 Video still of the Philadelphia Energy Solutions refinery explosion 21 June 2019. Source: US Chemical Safety and Hazard Recognition Board and NBC-10 (Philadelphia). From the CSB Final Report "Fire and Explosions at Philadelphia Energy Solutions Refinery Hydrofluoric Acid Alkylation Unit." Includes annotations by the US Chemical Safety Board (annotations are by the CSB).

Petroleum Institute (API) issued a Recommended Practice (RP) covering "Safe Operation of Hydrofluoric Acid Alkylation Unit." This report indicated that "HF corrosion has been found to be strongly affected by steel composition, and localized corrosion rates can be subtly affected by local chemistry differences."

This catastrophic explosion hurled a process vessel and other refinery hardware for hundreds of feet. The Alkylation unit was reported to be destroyed in the explosion. Fortunately, there were no fatalities; however, there were five minor injuries, and the release of hydrofluoric acid is a serious concern. Thanks to a panel Operator's extremely quick actions, most of the HF was diverted, and a potential disaster was averted. Philadelphia Energy Solutions shut down the refinery and filed for bankruptcy after about a month. They have announced that the 335,000 B/D refinery will not be returned to service. Please see the CSB Factual Update and the final CSB report, when issued, for additional details.

Please refer to Figure 8.17.9 for a video still of the Philadelphia Energy Solutions refinery explosion, which occurred on 21 June 2019.

Update:

The Chemical Safety Board Final Report for the Philadelphia Energy massive explosion and fire was issued on 11 October 2022. I encourage all individuals involved with petroleum refining or petrochemical operations to review this report. It has some great photographs of the event and the aftereffects on the refinery and the community.

The report also contains some great recommendations that will significantly improve process safety if implemented in our industry.

Key Lessons Learned

High-temperature sulfidation is one of the most common PMI failures, and in the case of the LyondellBasell Delayed Coker fire, this is exactly the mechanism that caused the failure. This was also the cause of the three fatalities at the Exxon Baton Rouge delayed Coker. High-temperature sulfidation is a very aggressive corrosion mechanism, especially in carbon steel piping, when operating at high temperatures in the presence of sulfur. The only way the piping can be confirmed as the correct metallurgy and with the expected resistance to the corrosion mechanisms present in the system is through a rigorous PMI process. This means that every component that is installed in the alloy piping circuits must be confirmed with a rigorous PMI process. This means every pipe, flange, small connection, valve, and weld section must be confirmed as the correct metallurgy.

I prefer to do the PMI after the welding is complete because we know it only takes one wrong component to result in a catastrophic loss of containment. It is easy for a craftsman to insert a carbon steel pup piece or other carbon steel component into a chrome circuit with the best intentions, such as correcting a minor piping layout error or pipe fabrication error. This would be detected with a rigorous PMI that is done **after** the welding is complete.

The following should also be in place at each facility and rigorously enforced:

- Following maintenance or all new installations of piping or components in alloy systems, PMI must be conducted on all alloy components in their "installed location" in line with API-578 ("Material Verification Program for New and Existing Alloy Piping Systems"). A good practice is to conduct the final PMI after all welding is complete, and each weld should be verified as an integral part of the PMI process.
- PMI is normally carried out on alloy materials received in the warehouse as a quality control check.
- PMI is a sitewide responsibility, and all parties must clearly define and understand roles and responsibilities: Operations, Maintenance, Project, Inspection, or QC.
- Pressure-containing components shall be supplied with a traceable mill certificate in accordance with EN 10204 3.1.

Note: EN 10204 3.1 Certification. 3.1 is a certificate issued by the mill declaring that the plates comply with the specification and include the test results. Test results are validated by the mill's in-house test department, which has to be independent of the manufacturing department.

Additional References

American Petroleum Institute (API)

- API-578 "Material Verification Program for New and Existing Alloy Piping Systems."
- RP 941, "Steels in Hydrogen Service at Elevated Temperatures and Pressures in Petroleum Refineries and Petrochemical Plants."

US Occupational and Health Administration (OSHA), Incident Report on Exxon Baton Rouge East Coker Fire (2 August 1993). https://www.osha.gov/pls/imis/establishment.inspection_detail?id=107635211

US Chemical Safety and Hazard Recognition Board (CSB), Safety Bulletin No. 205-04-B, Issued October 12, 2006, "Positive Material Verification: Prevent Errors During Alloy Steel Systems Maintenance". https://www.csb.gov/userfiles/file/rhubulletin.pdf. https://www.csb.gov/assets/1/20/rhubulletin1.pdf?13897

US Chemical Safety and Hazard Recognition Board (CSB), Factual Update – Philadelphia Energy Solutions Investigation. https://www.csb.gov/assets/1/20/pes_factual_update_-_final.pdf?16512

Hydrocarbon Processing, Article: "Fire-Destroyed Refinery: What went wrong", 16 October 2019.

Article on Philadelphia Refining Refinery explosion and fire. https://www.hydrocarbonprocessing.com/news/2019/10/fire-destroyed-refinery-what-went-wrong

Chapter 8.17. Metallurgy Matters (The Importance of Positive Materials Identification – PMI)

End of Chapter Quiz

1. Many significant process safety __________ have started with metallurgy, specifically the wrong metallurgy for the service or the corrosion mechanism. Several of these have had severe consequences, including loss of life.

2. Using only your senses, is it possible to distinguish the difference between carbon steel pipe and a similar-sized section of chrome pipe?

3. What are the differences between carbon steel and chrome that make proper identification of the materials used important?

4. Why is chrome piping used in piping circuits where the which operate above about 450 °F (232 °C) in the presence of sulfur?

5 Carbon steel in chrome piping circuits is our "Achilles' heel" regarding PMI. The issue is that it is impossible to ___________ the two metals apart using only our senses.

6 Why do we Conduct PMI?

7 The PMI process should be applied to _____ alloy piping and other alloy equipment that can "see" the process pressure. For example, stainless steel is produced in many different grades, and the corrosion resistance and other properties vary widely by grade. It is impossible to distinguish between the various grades of stainless and many other metals without PMI.

8 When do we conduct PMI?

9 How is PMI Conducted?

10 More recently, we have rediscovered that it is also essential to understand the type and composition of carbon steels. On 6 August 2012, Chevron Richmond refinery loss of containment and fire, it was rediscovered that corrosion is much more aggressive in Carbon Steels with low _________ content.
Select the correct answer.
  a. Chrome
  b. Moly
  c. Silicon
  d. Carbon steel
  e. Copper

8.18

Hazards of Nitrogen and Inert Entry

Nitrogen (N_2) is used throughout the petrochemical industry for a variety of purposes, including the air freeing and tightness testing of process equipment during prestart-up before the introduction of hydrocarbons, pressurization of tank trucks and tank cars for offloading products, blanketing for storage tanks. Nitrogen is also used in other process vessels to prevent air from entering the tank vapor space and in the preparation of equipment for hot work. Mechanical work such as catalyst replacement is sometimes done using nitrogen to provide an inert atmosphere (this requires highly specialized equipment and personnel with extensive training and expertise).

There have been numerous nitrogen incidents in the industry, resulting in fatalities. According to a study by A Suruda J. Agnew at the Division of Occupational Medicine, Department of Environmental Health Sciences, Johns Hopkins University School of Hygiene and Public Health, Baltimore, Maryland, "a review of 4,756 deaths investigated by the Occupational Safety and Health Administration (OSHA) in 1984–1986 found 233 deaths from asphyxiation and poisoning, excluding asphyxiations from trench cave-ins. The highest rates were in the oil and gas industry and utilities."

The US Chemical Safety and Hazard Investigation Board also conducted a study of Nitrogen-related incidents in the United States. They identified 85 nitrogen asphyxiation incidents that occurred in the workplace between 1992 and 2002, in which 80 people were killed, and 50 were injured. These studies indicate that nitrogen is a serious hazard, especially in the oil and gas industries, and additional focus is needed to minimize the hazards of nitrogen.

A few recent examples of nitrogen asphyxiation incidents are highlighted below, and these are the ones we know about:

- **2001 Lavera Refinery, France**: An employee was found dead at the bottom of the hydrocracker reactor under a nitrogen blanket for catalyst unloading. The employee had been making a gas test at the top manhole of the reactor, was overcome, and fell inside.
- **2001 Texas City Refinery, US**: The contractor was adjusting the ladder from the top manhole of a reactor being purged with nitrogen. He was overcome and fell inside, and died of asphyxiation.
- **2002 Merak Polyethylene Plant, Indonesia**: The analyzer technician died inside the analyzer house after analyzer instrument air purges had been changed to nitrogen.
- **2003 Belgium**: fitter found dead inside decene railcar which had been purged with nitrogen in preparation for repair of a leak.
- **2005 Valero Delaware City**: Contract fatality when a worker entered or fell into a reactor under nitrogen purge. A second fatality occurred when another worker attempted rescue.
- **2011 Singapore**: Inert entry contractor fatality due to breathing air cylinder contaminated with Argon (One cylinder in a bank of sixteen was determined to be Argon).

Properties of Nitrogen

Nitrogen makes up about 78% of the air we breathe; therefore, there can be a misconception that it is benign. The CSB study prompted a Safety Bulletin on nitrogen Asphyxiation Hazards and a safety training presentation. Unfortunately, the high rate of nitrogen incidents and deaths is continuing indicating the need for enhanced worker training on the hazards of nitrogen and continued improvement in Nitrogen-related work practices.

The hazard with nitrogen is that when nitrogen is increased in a confined space, it automatically reduces the available Oxygen. This can result in Oxygen deprivation with a relatively small increase in nitrogen. We normally consider concentrations below 19.5% Oxygen unfit for human occupancy, even for short periods. The real hazard with nitrogen is that it

Process Operations Safety: The What, Why, and How Behind Safe Petrochemical Plant Operations, First Edition. M. Darryl Yoes.

only takes one or two breaths of concentrated nitrogen, and you collapse and pass out. There is no odor. The person simply goes to sleep and will die unless rescued quickly. There is absolutely no warning!

Nitrogen is a clear, odorless, and colorless gas at normal working temperatures. It is very effective at displacing oxygen in confined spaces and near openings or vents/drains on equipment under a nitrogen atmosphere. Nitrogen can only be detected by specialized gas detection equipment. Most fatalities involving nitrogen asphyxiation occur because the person(s) was not aware that they were entering an Oxygen-deficient atmosphere or were not aware that the atmosphere had changed and had become deficient in Oxygen. Another cause of nitrogen asphyxiation is when one worker attempts to rescue a coworker without adequate rescue training or rescue equipment such as breathing air apparatus. Unfortunately, these cases usually result in the rescuer's death and the person being rescued also being killed. Unofficial data indicates that 50% of nitrogen deaths are related to attempting rescue without protective equipment.

The concentration of Oxygen in normal atmospheric conditions (at sea level) is about 20.9% Oxygen (the remainder being nitrogen [78%] and trace amounts of other gases). OSHA requires employers to maintain workplace Oxygen levels between 19.5 and 23.5% (except when specialized breathing equipment is being used). Oxygen concentration below 19.5% can result in Oxygen deprivation; Oxygen concentration greater than 23.5% results in an extreme fire hazard.

Nitrogen Hazards

When exposed to nitrogen, the Oxygen in the brain is quickly displaced, causing unconsciousness and death. *It only takes one or two breaths of concentrated nitrogen, and there is no warning before being overcome!*

Effects of oxygen deficiency on the human body

Atmospheric oxygen concentration (%)	Possible result
20.9	Normal
19.0	Some unnoticeable adverse physiological effects
16.0	Increased pulse and breathing rate, impaired thinking and attention, reduced coordination
14.0	Abnormal fatigue upon exertion, emotional upset, faulty coordination, poor judgment
12.5	Very poor judgment and coordination, impaired respiration that may cause permanent heart damage, nausea, and vomiting
<10	Inability to move, loss of consciousness, convulsions, death

Source: Compressed Gas Association, 2001

In the petrochemical industry, most nitrogen-related incidents and fatalities occur in or near confined spaces such as reactors, drums, and towers. These vessels may be nitrogen blanketed for process reasons or associated mechanical work. The deaths occur when personnel enter the confined space without following proper work permit procedures or when the conditions in the confined space change. Deaths also occur when employees are near access openings or vents/drains of equipment under a nitrogen purge or a nitrogen atmosphere. In some cases, the person became unconscious and fell into the vessel, which had been inerted by nitrogen.

Vessels under nitrogen purge must be properly identified with signage and hard barricades to keep personnel out of the area. Personnel should be excluded from the work decks where nitrogen is being exhausted out of the vessel unless they are wearing supplied air-breathing equipment such as hose line supplied respirators or self-supplied breathing air packs (SCBA) and have a backup, also in breathing air and ready to respond in the event of an incident involving Oxygen deprivation.

It is recommended that cylinders of breathing quality air, industrial-grade air, and nitrogen be equipped with distinctly different and incompatible connections that cannot be cross-connected. All breathing air cylinders should be clearly labeled and color-coded as breathing air, and no other cylinders should have similar labeling. I am not a fan of "blended air" or "manufactured air," where nitrogen and Oxygen have been blended to make breathing air, as incidents have occurred with these cylinders. I prefer that only compressed air that has been verified as containing 21% Oxygen should be used as breathing air. Each breathing air cylinder should be tested to verify that the cylinder contains breathing air BEFORE it is used for breathing air.

A common practice in our past was to back up the site instrument air system with nitrogen. However, we know now that this is not a good idea as instrument air may be exhausting into control rooms, analyzer shelters, instrument rooms, etc. When using nitrogen as backup, the instruments may exhaust nitrogen into spaces normally occupied by personnel resulting in Oxygen deficiency in these areas. Fatalities have occurred from this practice. It may be a good idea to check your site to ensure nitrogen is not used as a backup for the instrument air supply.

Cryogenic Hazards of Nitrogen

Liquid nitrogen expands to 700 times the volume when it is vaporized. For this reason, we frequently transport nitrogen as a compressed liquid since we can efficiently transport very large quantities. However, it boils at −320°F (−196 °C); therefore, contact with liquid nitrogen can cause significant cold burns. Cold nitrogen is heavier than air, so it accumulates at ground level. When liquid N_2 is exposed to air, the cloudy vapor you see is only the condensed moisture (water vapor) from the air, not the N_2 gas. Remember, nitrogen gas is invisible, and this is the Danger! Remember, do not bend over into the gas cloud, as you may be overcome by Oxygen deficiency.

Vaporizing nitrogen from a tank truck into process equipment may also drive the temperature of the piping and vessels below the minimum design metal temperature and may result in a potential brittle fracture of the equipment. The process temperature must be very closely controlled to prevent this from happening.

Other gases have also been known to result in Oxygen deprivation and loss of human life. For example, incidents have occurred with Carbon Dioxide (CO_2) used in some fire suppression systems and Argon used in welding applications, Helium, Methane, or natural gas. Ensure you have a valid gas test before peering into a confined space, entering a confined space, or even working near an opening to a confined space.

Confined Space Entry and Inert Entry Hazards

OSHA 29 CFR 1910.146 Permit-Required Confined Spaces for the general industry requires employers to identify all confined spaces in their workplace and determine if any of those are permit-required confined spaces. A permit-required confined space contains a hazardous atmosphere, an engulfment hazard, an entrapment or asphyxiation hazard, or other serious safety and health hazards.

OSHA requires confined spaces with elevated nitrogen concentrations (i.e., reduced oxygen concentrations below safe levels) to qualify as permit-required confined spaces. The OSHA standard establishes the requirement that a confined space "containing an actual or potential atmospheric hazard qualifies as a permit-required confined space".

To warn workers of nitrogen-enriched atmospheres and other permit-required confined spaces, OSHA 29 CFR 1910.146(c)(2) requires the posting of a warning sign, e.g., "Danger: Permit-Required Confined Space, Do Not Enter" or "any other equally effective means." See Figure 8.18.1 for an example of an analyzer shelter with this sign attached to both entry doors.

The Chemical Safety and Hazard Investigation Board (CSB) also recommends installing flashing lights, audible alarms, and auto-locking entryways to prevent access. In addition, individual personal monitors can warn workers via an audible or vibration alarm of low oxygen concentrations. The CSB also emphasizes the application of continuous atmospheric monitoring of confined spaces when occupied to help detect an area unfit for breathing before entry, or it may change over time. The CSB says the atmosphere in the entire confined space should be tested and confirmed safe before workers enter the space and should be continuously monitored while workers are in the space.

OSHA 29 CFR 1910.146(d)(5)(ii) and (iii) explain: "Test or monitor the permit space as necessary to determine if acceptable entry conditions are being maintained during the course of operations; and when testing for atmospheric hazards, test first for oxygen, then for combustible gases and vapors and then for toxic gases and vapors."

OSHA 1910.146(d)(6) also requires employers to have an attendant outside the permit-required confined space at all times while a worker is inside. The attendant's job is to use instruments to monitor the conditions within the space, to remain in contact with the entrant in case of emergency (and to alert rescuers, if necessary, or perform a nonentry rescue), and to know the hazards of the space and the signs or symptoms of exposure to the space's hazards, among other duties. The attendant should never enter the confined space unless the attendant is part of a rescue team and has been relieved by another attendant.

Figure 8.18.1 Image of a typical petroleum refinery analyzer shelter. These generally have pressurized nitrogen piped to the instruments located in the shelter (a confined space). Always be aware of this and ensure the shelter is equipped with an oxygen analyzer reading in the safe zone before entry; otherwise, a gas test for O_2 should be done, or the entrant should wear a self-contained breathing apparatus before entry. Source: Photo courtesy of the Chemical Safety and Hazard Investigation Board. https://www.csb.gov/assets/1/17/swpicture_smaller.jpg?14290.

The CSB also emphasizes continuous ventilation with fresh air whenever workers enter a confined space or a small or enclosed area without wearing a supplied air-breathing apparatus. The CSB emphasizes that ventilation with fresh air is very important when recently purged with nitrogen, carbon dioxide, or some other gas, and the area has been brought to the minimum safe breathing level of 19 1/2% oxygen.

In contrast, fresh-air ventilation is not an option when workers perform an inert entry, such as entering a pure nitrogen environment, such as when workers are changing a catalyst in a reactor. In this case, nitrogen would likely protect the catalyst from pyrophoric ignition or from becoming contaminated by oxygen or moisture; however, workers in an inerted environment should have specialized breathing equipment such as redundant forced air respirators with nonremovable clamshell helmets and other specialized precautions (see more in this section on Inert Entry).

CSB, in its 2003 safety bulletin, also makes an important observation in that several of the nitrogen asphyxiation cases they have investigated involved people who were not working in the nitrogen-enriched space, room, or enclosure but were working near vessels that were open and being purged with nitrogen. This has occurred in the United States and other countries as well. For example, in 2001, two very similar incidents occurred in the industry. One occurred at the Lavera Refinery in France, where an employee was found dead at the bottom of the hydrocracker reactor, which was under a nitrogen blanket for catalyst unloading. The employee had been making a gas test at the top manway of the reactor. He was overcome and fell inside. Also, in 2001, at the BP Texas City Refinery, a contractor was adjusting the ladder at the top manhole of a reactor being purged with nitrogen. He was also overcome and fell inside and died of asphyxiation.

In the CSB investigation of the Valero incident (discussed in more detail later in this chapter), the first contract employee was likely overcome by nitrogen being vented from the open manway and fell inside the reactor. The nitrogen venting from the open manway created a hazardous environment outside the confined space.

The bottom line is that all confined spaces and the breathing zones around confined spaces should be treated as if they contain a deadly toxic atmosphere or an atmosphere that will not sustain life until they are proven otherwise by testing. The "Technical Advisory on Working Safely in Confined Spaces" is an excellent resource for additional information on confined space entry procedures and practices. See Additional References at the end of this chapter.

Inert Entry

Inert entry is a special type of confined space entry involving personnel entry into a confined space, typically a reactor, which contains an atmosphere that is immediately dangerous to life and health (IDLH). These tasks often involve the replacement of catalysts to help preserve the catalyst for regeneration and reuse. Most of the catalysts used in petroleum or petrochemical reactors are pyrophoric and, therefore, will self-ignite if exposed to air, even in small quantities. Also, the air in the atmosphere contains moisture, which will damage most catalysts. Thus, tasks like removing the catalyst or working with the catalyst, catalyst support systems, or any reactor internals must be done in an inert atmosphere, typically nitrogen, if the catalyst is going to be regenerated and reused.

The typical requirements for confined space work must be followed. The vessel must be isolated from sources of hazardous energy and materials. Electrical sources should be isolated and locked out to prevent inadvertent reenergizing. Entry by persons not involved in the inert entry process should be eliminated. Any access to the vessel access area should be highly restricted to only those personnel engaged in the inert entry task, and they should be in full PPE, including supplied breathing air. The following should be considered when preparing for inert entry:

- Ensure adequate warning signs or inert entry work to prevent access to the work area.
- Establish barricades and personnel exclusion zones (no access to areas near the manway other than catalyst technicians in full PPE).
- Ensure adequate intrinsically safe lighting in anticipation of night work.
- Ensure that all pneumatic tools are powered by nitrogen if the tools are to be used inside the inert atmosphere (pneumatic tools should not be powered by air – discharge of air into the inert space will enrich the oxygen concentration and may result in fire).
- Establish adequate firefighting equipment at strategic locations.
- Ensure electrical bonding and grounding for equipment near vessel openings or used inside vessels.
- Ensure access for rescue operations.

Working inside a process vessel under inert atmosphere conditions involves a specialized contractor company with highly trained catalyst support staff and specially designed equipment and personnel protective equipment (PPE) for their technicians. The atmosphere inside the vessel is typically about 99% nitrogen with an oxygen concentration of about 1%. (Typically, 4% maximum). Most companies require a special work permit that establishes the limits for Oxygen (%), flammability (LEL, %), Hydrogen Sulfide (H_2S, ppm), Carbon Monoxide (CO, ppm), total hydrocarbon (ppm), and Benzene (ppm). The temperature in the confined space is also monitored with a maximum allowable of typically 100°F (38 °C). Normally the testing for flammables (LEL) is conducted with the inert gas purge turned off, and testing for Oxygen is done with the inert gas flowing.

The contractors are required to have specialized PPE, including specially designed fully encapsulated helmets with a triple-redundant breathing supply. The helmets contain two separate cylinder-supplied hose lines delivering air supplies and an emergency escape breathing supply. The helmet is a type that is locked (or bolted) in place so the catalyst technician cannot remove the helmet while working inside the vessel. The catalyst technician inside the vessel is in radio communication with a "hole watch" attendant in the triple air-supplied breathing apparatus and stationed at the top manway. Breathing air cylinders are typically mounted in banks of cylinders connected and monitored by a third catalyst technician, also in radio communication with the technicians (inside the vessel and the hole watch). The technician working inside the vessel will also be fitted with a special rescue harness with an attached retrieval line to aid in rescue if necessary. A separate rescue plan will be developed and ready for implementation if this becomes necessary (more on the rescue plan below).

Specialized Inert Entry Procedures

Detailed and specialized procedures must be developed covering the work inside an IDLH atmosphere and provided to the catalyst technicians, and these procedures must be enforced during the field execution. The following are only examples of the types of specialized procedures that should be in place before and during inert entry execution:

- The procedures should ensure the catalyst bed is continuously monitored for potential temperature rise. A rise of 5°F (3 °C) in 15 minutes may indicate a potential issue with the catalyst bed or oxygen intrusion into the vessel.

- A significant amount of compressed breathing air is used during a typical inert entry procedure. One way this is done is to connect many smaller cylinders of compressed air in parallel and then use one cylinder at a time for breathing air supplied to the catalyst technicians. The procedure should specify and enforce requirements for each cylinder to ensure the cylinder is tested for breathing air quality before the work begins. In some past incidents, cylinders have been shipped, labeled, and painted, with papers certifying them as containing breathing air but containing another gas or had a high percentage of nitrogen. This has resulted in on-site accidents and the deaths of catalyst technicians.
- The procedure should recognize the potential of a crust or hardened layer forming on the catalyst, which can suddenly fail due to pressure below the catalyst from the inert gas purge. This requires special precautions such as the continuous monitoring of pressure below the catalyst bed and a temporary pressure relief valve on the inert gas line below the catalyst set at very low pressure or maintaining the vessel in an inert state by adding nitrogen into the vessel with a hose above the catalyst bed or similar precautions.
- Another potential issue for consideration is catalyst migration below the catalyst surface. Should the catalyst migrate, or if some catalyst has been dumped below the catalyst surface, a void can be created below the catalyst, creating a fall hazard for the catalyst technicians. The procedure should recognize these hazards and provide additional fall protection for the catalyst technicians. More on this is in the "What Has Happened" discussion below.

Confined Space and Inert Entry Rescue Plan

It is understandable why confined space incidents all too often involve multiple fatalities. If a worker sees their buddy lying still inside a vessel, it is natural to want to rescue their coworker. Their first thought likely is that their coworker has experienced a heart attack or has fallen, and they want to help. However, if a toxic gas has overcome the coworker, or if the atmosphere has changed due to nitrogen, they too can be overcome. As in the two cases mentioned above, the Valero Texas case and the Canadian refinery case, both co-workers died when attempting to rescue a coworker. In both cases, they were killed by Oxygen deprivation due to a nitrogen-enriched atmosphere. Nitrogen is truly a silent killer.

Workers MUST be trained; if they are faced with a coworker who is down, they must STOP and immediately call for help. They must know that they can be overcome, sometimes with only one breath, and they will die. This is training – workers must know what to do in the event someone is down (call for help) and what not to do (never enter a confined space to rescue unless they are trained as responders and they have the proper breathing apparatus, and then only with a backup person).

According to OSHA 29 CFR 1910.146, employers can provide rescue with in-house personnel or by calling outside emergency services, but either way, the responsibility is to ensure the person is rescued before any long-term harm comes to that person.

The CSB safety bulletin emphasizes that a rescue plan should be developed in advance and implemented during the confined space entry. The plan may include placing the workers in a body harness with a lifeline and wristlets or anklets for workers who are entering a confined space to accommodate rescue by the attendant in the event of an emergency. Another potential response plan may include having trained response personnel standing near the job site with rescue equipment such as a hoist tripod of similar equipment at the ready. The plan should always include an effective communication plan between personnel inside the space to the attendant and response personnel. Under no circumstances should anyone enter the confined space without proper PPE. This includes emergency response personnel.

What Has Happened/What Can Happen

Valero Refinery
Delaware City, DE.
5 November 2005 (2 Fatalities)

This incident was investigated by the US Chemical Safety and Hazard Investigation Board, and a detailed report and a great training video were released.

On 5 November 2005, two fatalities occurred at the Valero Refinery in Delaware City, Delaware. The contract crew was working on a Hydrocracker reactor under a nitrogen purge and was preparing to install the reactor inlet elbow on top of the reactor when one of the workers noticed a roll of duct tape lying inside the reactor. Since this would cause the reactor to fail the cleanliness inspection, the contract employee attempted to remove the duct tape. The contractor intentionally entered the reactor vessel or, while trying to retrieve the roll of tape, was overcome by the nitrogen and accidentally fell into the reactor. The second contract employee immediately went to his aid and entered the reactor to attempt a rescue, and he was also overcome. Both employees were recovered within 10 minutes but were pronounced deceased when they reached the hospital. Please see Figure 8.18.2 for images of the work area, including the location of the open nozzle at the top of the reactor. Figures 8.18.3 and 8.18.4 illustrate the roll of tape the employee was attempting to retrieve and the wire used for this purpose.

The work permit indicated "Nitrogen purge N/A" and "LOTO N/A." There was a sign to make the contractors aware of the confined space. However, no signs or other warnings indicated the use of nitrogen purge or the potential for an Oxygen-deficient atmosphere.

The US Chemical Safety and Hazard Recognition Board (CSB) investigated this accident and prepared a detailed report and an excellent training video on the hazards of nitrogen and the dangers of nitrogen asphyxiation (Oxygen deprivation). The CSB also issued an excellent nitrogen safety training presentation and a safety bulletin on these hazards. These documents are available in Appendix Ha and Hb.

Unfortunately, a similar incident occurred at a Canadian Refinery in January of 1981, resulting in two fatalities. In that incident, a craftsman dropped a clamp used in a stress-relieving job. The clamp went directly into the open-top manway on a guard bed reactor under nitrogen purge for catalyst replacement. The contract employee immediately jumped into the reactor to retrieve the clamp without discussing the incident with his coworker or anyone else. His coworker saw the contractor jump into the reactor and ran over to the vessel, looked in, and realized his partner had collapsed. He, too, jumped into the reactor to rescue his partner and quickly collapsed. Both contractors were rescued by specially trained personnel but were pronounced dead when they reached the hospital.

Figure 8.18.2 Image of the work area on top of the hydrocracker reactor. The manway surrounded by large studs was open and venting nitrogen from the reactor purge when the incident occurred. The sign reads "Danger, Confined Space." There was no warning about the nitrogen purge and associated hazards. The red barricade tape and plastic covering were added before the work began and removed by the workers to install the pipe elbow. Source: Photo courtesy of the US Chemical Safety and Hazard Investigation Board. This image is from the CSB Report "CASE STUDY, Confined Space Entry – Worker and Would-be Rescuer Asphyxiated." https://www.csb.gov/valero-delaware-city-refinery-asphyxiation-incident/.

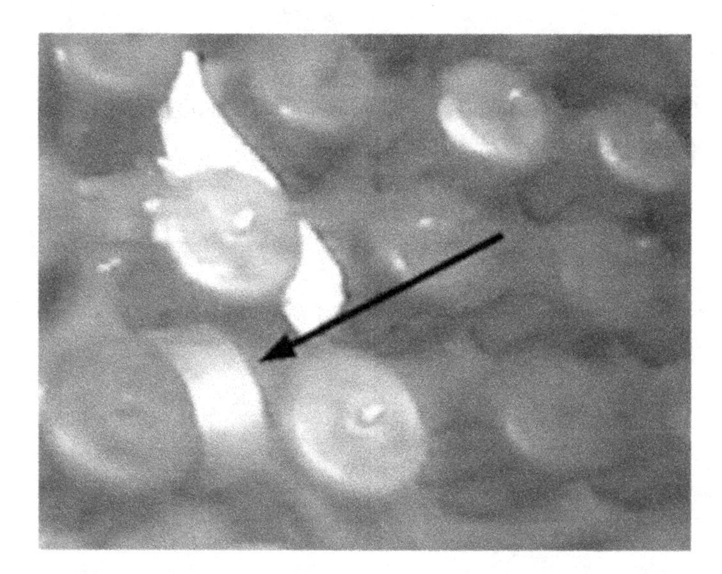

Figure 8.18.3 Image of the roll of duct tape lying on the reactor inlet distributor. Source: Photo courtesy of the US Chemical Safety and Hazard Investigation Board. This image is also from the CSB Report.

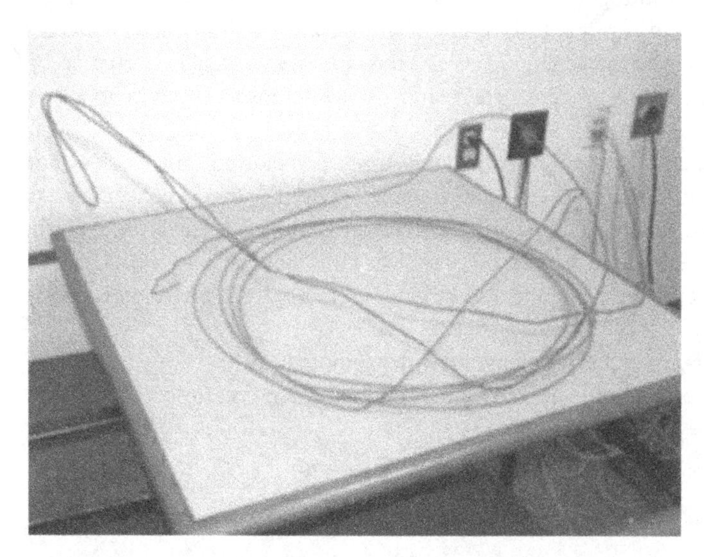

Figure 8.18.4 Image of roll of wire with a hook that was used by one of the workers to attempt to retrieve the tape. Source: Photo courtesy of the US Chemical Safety and Hazard Investigation Board. This image is also from the CSB Report.

Key Lessons Learned

- Anytime vessels or piping are being purged with nitrogen, ensure that proper barricades and warning signs are placed to warn workers of nitrogen asphyxiation and oxygen deprivation hazards. *A rule of thumb to follow is that no one should be allowed on a work deck where nitrogen is venting from a process vessel or a nitrogen purge is underway unless they are in breathing air and with a backup that is also in breathing air.*
- Ensure that all Confined Space Entry procedures are in place and rigorously followed anytime entering a process vessel or space. This includes the following:
  - Confined space training for all personnel who are entering the vessel.
  - Proper respiratory protection when nitrogen or any other asphyxiant is used (for example, SCBA or hose line supplied breathing air).
  - Full-time hole watch attendant to track personnel in and out of the vessel.
  - Continuous gas test monitoring of the confined space during entry (for O_2, LEL, H_2S, and other potential chemicals specific to the process).

- Clear communications between the hole watch attendant and the personnel inside the confined space.
- A predefined rescue plan to remove personnel from the confined space should there be an incident.

- Work permits and other similar documents (Job Hazard Analysis, Last Minute Risk Assessment, Job Site Planning Documents, etc.) should include reference to potential nitrogen exposure and Oxygen deficiency and provide steps to ensure protection to personnel.
- Ensure that personnel working near where a nitrogen purge or nitrogen blanket is underway or where nitrogen is exhausted, such as manway openings, open pipes, railcars, tank truck hatches, etc., are protected with a hose line or self-contained breathing apparatus (SCBA).
- *Never attempt to rescue a colleague unless you are well-trained in rescue techniques and you are protected with a self-contained breathing apparatus. Always call for help and wait for the professionals.*

What Has Happened/What Can Happen

Augusta, Georgia – 5 February 2017
Sperm Bank (1 Fatality, 4 injuries)
Reported in CHEST Annual Meeting (Toronto, Canada)

In this asphyxiation incident, an employee was treated for nitrogen cold burns and other related nitrogen asphyxiation injuries (Oxygen deprivation) sustained while attempting to repair a leak of liquid nitrogen at a sperm bank. Unfortunately, the report also makes it clear that a police officer who was trying to rescue the employee was also overcome by Oxygen deprivation and died in this incident. Three other police officers were treated for symptoms of exposure to a lack of Oxygen (O_2).

In the report, the authors clearly state that inert gases such as nitrogen, Helium, and Argon are colorless and odorless, but they can cause significant harm without warning. The report reinforces that a person may experience loss of consciousness and subsequent death without warning signs or symptoms. As reported in this incident review, liquid nitrogen is particularly hazardous since it rapidly expands at nearly 700 to 1 when released under atmospheric conditions at 68°F (20 °C). This incident report indicates that the employee and the emergency responders were exposed to an atmosphere with an O_2 content of about 0.2% (a lethal atmosphere).

In addition to the asphyxiation hazard, liquid nitrogen is an inert cryogenic fluid with a temperature of −320°F (−196 °C). Any contact with liquid nitrogen will result in extreme frostbite and cold burns to the skin and exposed flesh.

Key Lessons Learned

You do not need to enter a confined space to be overcome by nitrogen. Workers were overcome while working near the open manway where nitrogen was exhausted out of the vessel and then fell into the vessel and died. For example, in 2001, a worker at the Lavera Refinery in France was found dead at the bottom of the hydrocracker reactor under a nitrogen blanket for catalyst unloading. The employee had been making a gas test at the top manway of the reactor, was overcome, and fell inside. Also, in 2001 at the BP Texas City Refinery, a contractor was adjusting the ladder from the top manway of a reactor being purged with nitrogen. While doing so, he was overcome and fell inside, and died of asphyxiation. In these two examples, neither worker intended to enter the vessel. However, they were both overcome by the Oxygen-deficient atmosphere near the open manway, fell into the open vessel, and died.

Evaluate the use of nitrogen during the work planning process and ensure that procedures are in place and limits are established on where and how nitrogen is to be used. The work plans should establish barriers (hard barricades) to keep other personnel off work platforms where nitrogen will be venting into the atmosphere. Only personnel who are properly outfitted with hose lines or SCBA-supplied breathing air should be allowed on the platforms or near the vessel openings.

Verify that your site has a process for verification of breathing air quality. Each breathing air cylinder on the site should be individually tested to ensure breathing air quality. These tests should normally be done by someone other than the company that supplied the breathing air, and the test documents should be readily available for inspection. This is a very important part of the site safety and health program. We have experienced several examples where cylinders were labeled as breathing air but contained a different toxic gas. A breathing air certification process that includes testing each cylinder can ensure this will not happen at your site.

It has been estimated that about 50% of nitrogen-related fatalities are personnel attempting to rescue someone else. It is a human urge to try and rescue a team member or coworker who has been overcome. However, this is training; never attempt to rescue someone who is already down unless you are a qualified rescue team member AND have the proper PPE, including breathing air. Call for help and then direct the rescue or medical team to the site.

Do not use nitrogen as a backup to the site instrument air system. This can allow nitrogen to enter occupied areas such as control rooms located in remote areas, analyzer shelters, instrument enclosures, etc. It can impact personnel. This happened at a Polyethylene Plant in Indonesia in 2002. An analyzer technician died inside an analyzer shelter after the site instrument air system had been switched to nitrogen backup. The analyzer technician was called out after the radio call, notifying all personnel that the site was operating on nitrogen for instrument air backup. Unfortunately, he entered the analyzer shelter, a confined space that contained a nitrogen atmosphere from the instrument purges.

What Has Happened/What Can Happen

Waste Oil Refinery
Mandya, India
6 March 2014 (5 Fatalities)

I felt a need to share the following incident to show what can happen if procedures do not exist or are not followed. In this tragic incident, five workers were asphyxiated when they entered a waste oil refining unit rectifier at Tubinakere Industrial Area on the outskirts of Mandya (India). The police said the refinery did not have adequate facilities to combat emergency situations following this incident.

Local media reported that the factory supervisor had instructed the workers to clean the rectifier, and the workers entered the tank one by one and fell unconscious. When the response brigade arrived, they found the workers unconscious in the reactor. The brigade personnel used gas cutters to open the reactor and extricate the bodies.

Key Lessons Learned

This tragic incident, where five lives were lost, shows what can happen if procedures do not exist or if they are not followed. This clearly shows why we have detailed confined space entry procedures, which require a gas test for Oxygen and any other contaminant that may be present before anyone is allowed to enter the confined space. It shows why we have a dedicated "hole watch" person stationed at the entryway with no responsibility other than staying in touch with the workers inside the vessel. The hole watch must be trained on how to alert responders and call for help in an emergency.

This incident also shows why we have a detailed accounting system to track exactly who is in the confined space, including when they go in and when they leave the confined space. I believe this incident also shows why continuous monitoring, although not currently required by the OSHA standard, is so very important to continuously monitor the Oxygen in the confined space and the continuous monitoring for hydrocarbons (LEL or lower explosive limit) and any other gas or contaminant which may be present.

What Has Happened/What Can Happen

Petroleum Refinery
Sakai, Osaka, Japan
19 February 1991 (Fatality)

The following was extracted from a report issued by the Japanese Association for the Study of Failures (From the Failure Knowledge Database):

On 9 February 1991, a kerosene floating roof tank base plate inspection was scheduled for sandblasting work. At about 10 a.m., one worker was outside the tank to monitor the work in the tank, and four other workers were inside the tank to start the sandblasting work. Three of the workers donned their breathing hoods and almost immediately collapsed. They were rescued from the tank and immediately sent to an emergency hospital by an emergency response team. One worker was pronounced dead at the hospital, and the two others were hospitalized for additional treatment and observations.

Breathing air "protectors" or hoods are necessary for sandblasting work because of the large amount of minute dust particles present. The report stated: "Imperfect field validation and imperfect action indications could be indirect causes of this accident. The cause of this incident was that the breathing air hose for their breathing hoods was mistakenly connected to nitrogen gas piping. The piping arrangement was complicated, and nitrogen piping was close to oxygen piping. Imperfect field validation and imperfect action indications could be indirect causes of this accident". Countermeasures to prevent human error were not taken. It is apparent that there would be an accident if nitrogen were used instead of breathing air. Instrument safety measures such as fixing the color or shape of piping and nozzles for a specific gas and using individual couplings corresponding to each gas are a minimum requirement and a duty of management.

Key Lessons Learned

This incident happened long ago, and a lot has changed since then. I included this incident to share what has occurred in our industry in the past and highlight what has changed. Many of these incidents are the basis for some of the standards and requirements we have in place today.

It is evident that the workers routinely attached their breathing air hoses to pipeline-supplied utility air in this incident and did so with management and plant approval. It is also obvious that they were not using distinctively different hose quick connections for the separate utilities (air, steam, nitrogen, and water). They inadvertently connected their breathing air hoses directly to a nitrogen utility connection, resulting in one death and two injuries requiring hospitalization.

Today, our standards should never allow connections to pipeline-supplied utility air as breathing air. Breathing air hoses should be supplied either from a hose line cylinder, which is certified as breathing air or a self-contained breathing air (SCBA) apparatus should be used. All breathing air cylinders should be individually tested to confirm that the air meets breathing air standards of at least 19.5% O_2 and is void of other contaminants such as hydrocarbons and Carbon Monoxide (CO).

The report also mentions connections to oxygen piping. I believe this is an issue with the Japanese translation, and I think they were referring to the refinery utility air header. Regardless, oxygen should never be used in place of breathing air; even in 1991, this was not allowed and is certainly not permitted today. Oxygen under pressure when contacting hydrocarbons (oil and grease) can react violently, resulting in explosions, fire, injury to personnel, and property damage. Also, high oxygen concentrations can lead to irreversible injury and death.

What Has Happened/What Can Happen

Foundation Food Group
Gainesville, Georgia (28 January 2021)
6 Fatalities, eleven hospitalized (3 critically injured)

Although this incident occurred at a chicken processing plant, there are many lessons to be learned from it. The facility utilizes liquid nitrogen as a refrigerant in food processing operations. Liquid nitrogen is a significant safety hazard on two fronts. Of course, it is an asphyxiant that works by replacing Oxygen in the breathing space. Contact with liquid nitrogen can also result in severe burns to the skin and body. This incident was investigated by the US Chemical Safety and Hazard Investigation Board, and a detailed report has since been released.

The CSB report found that a single-level bubbler instrument in a liquid nitrogen freezer failed which allowed the liquid nitrogen to overflow from the freezer into an enclosed space where workers were present. Two workers were asphyxiated outright and were undetected for about 45 minutes or so. While attempting to locate the deceased employees, another worker found a dense white vapor cloud approximately four to five feet deep inside the enclosure where the freezer was located and reported this to management. Management immediately started an evacuation of the facility; however, as the workers and managers evacuated, some responded to the freezer room to aid the downed workers and respond to the release. Fourteen of these were exposed to nitrogen, resulting in four more deaths and additional injuries, including three seriously injured. A total of six workers and managers were killed, and eleven were hospitalized, three of whom had severe injuries.

The CSB reported several significant contributors to this tragic incident:

- The bubbler-level instrument in the freezer did not have any redundancy. Also, a clamp or bracket to support the instrument had been omitted during manufacturing, allowing it to be bent during maintenance.

- There were no environmental hazard detectors in the enclosed space where nitrogen was being used in the process. There were no alarms or trip devices to alert employees of the presence of nitrogen and to stop the flow of liquid nitrogen to the equipment.
- There was a severe lack of training in the hazards of nitrogen and response to a release or loss of nitrogen containment. Employees were not aware of the asphyxiation hazards of nitrogen.
- There was no PPE such as Self-Contained Breathing Air (SCBA) or hose line supplied breathing air for employees to use in the event of an emergency.
- There was no documented Safety Management Policy or Process Safety Management System in place. The position of safety manager had been vacant for a year prior to this accident.
- The supplier of the liquid nitrogen continued to provide the product even though they had been made aware of shortcomings regarding industry practices and procedures. The CSB said in their report that had the product been made unavailable, possibly this incident would not have occurred.

Key Lessons Learned

Again, it would be easy for us to say that this incident occurred in a food processing plant, so it does not apply to me. I believe that would be shortsighted. Anytime we are working with hazardous materials or equipment, we must be aware of the industry practices, guidelines, and regulations that are intended to help keep us safe and allow us all to go home at the end of the workday fully intact and uninjured.

Liquid nitrogen is certainly a hazard that requires our full attention to how it is handled and processed within our facility. A loss of containment of liquid nitrogen can be a major event like the one that happened in this incident. Six fatalities and eleven injuries due to exposure to liquid nitrogen or to its vapors certainly fit in the category of catastrophic.

Identification of single points of failure and the potential consequence of that failure is integral to the Process Hazard Analysis (PHA). In a PHA, the team goes through the process flow diagram point by point and asks the question of what could go wrong. Then, they ask what the consequence of the failure is, and at the end, they identify those failures that have high consequences and a plan for addressing the identified hazard.

This is only one small example of what should be in place for a detailed and proactive Process Safety Management System. I believe that had a Process Safety Management System been in place, the hazard would have been identified and corrected. Please refer to Chapter 6 (An Introduction to Process Safety Management Systems) for more information.

What Has Happened/What Can Happen

BP Lavera Refinery
Marseilles, France
Date of incident uncertain; Incident communicated on 22 January 2004 (1 Fatality)
BP reported this incident in an incident-sharing communication through the CCPA.

I ran across this report and felt it was certainly worth sharing here. The incident occurred at a refinery near Marseilles, France, and reinforces the point that an individual does not need to be inside a vessel to be overcome by nitrogen. The incident report said an employee was making a gas test at the top manway of a reactor under nitrogen purge for catalyst unloading. This was obviously an inert entry-type job in which the vessel is maintained with an inert atmosphere to protect the catalyst and prevent spontaneous ignition of the pyrophoric catalyst.

The incident report states that the employee was found dead at the bottom of the reactor. At the time the report was written, the cause of death had not been conclusively determined; however, the writer indicated that it was most likely due to Oxygen deficiency near the manway that resulted in the employee being overcome and falling into the reactor. The writer points out that a similar near-miss occurred at another refinery in July 1999 when a safety specialist accessed the top of a reactor to collect a gas sample during a permit update. In this near-miss, the oxygen sensor on his meter indicated an oxygen-deficient environment before he even got to the opening of the vessel.

In the safety sharing, the writer expresses, "While it is widely known that human exposure to excessive amounts of nitrogen inside purged equipment can result in swift death, people can be equally affected while standing near openings of nitrogen-purged equipment may be less understood. This situation must not be underestimated."

The safety sharing discusses preventative measures, which included posting danger signs at all access paths to vessel openings, controlled access, and an accountability system using ground personnel and sign-in/sign-out logs.

The safety sharing also mentions the implementation of a buddy system required to access the deck, a manway lockout device to prevent someone from falling in the reactor when it is unattended, and the designation of an employee to control nitrogen flow in the reactor. All of which are significant improvements.

Key Lessons Learned

The writer is correct, and it is not well understood that one can be overcome when standing or working near an open manway or any pipe or vessel being purged with nitrogen or other inert gas. There have been many examples of this happening in our industry. For instance, in the Valero reactor incident we discussed earlier, the CSB developed two scenarios of how the first employee entered the reactor. In the first scenario, he intentionally entered the reactor to remove the tape; in the second scenario, he was overcome and fell into the reactor while working directly over the open manway. We will never know for sure, but we know that there was a high concentration of nitrogen, and he was working right at the reactor opening. BP also reported another case in which an employee was adjusting the ladder at the top manhole of a reactor under nitrogen purge, was overcome, and fell inside the reactor.

The BP report made several very significant observations that are worth repeating here:

- If a person enters an atmosphere of nitrogen, they can lose consciousness without any warning symptoms in as little as 20 seconds; death can follow in less than 3–4 minutes.
- When overcome by nitrogen, the effect is the same as a person being struck by a blow on the head.
- All personnel must understand that one breath of 100% nitrogen will be fatal!
- Breathing is stimulated and controlled by the presence of carbon dioxide (CO_2) in the lungs.
- A high concentration of nitrogen will displace carbon dioxide completely. When the CO_2 is displaced by nitrogen, the absence of CO_2 sends a signal to the brain, and breathing stops.
- Personnel working in areas where the atmosphere may contain less than 19.5% Oxygen should wear a self-contained breathing apparatus or breathing air mask. This includes rescue personnel.

To add to what has already been said in this report, after reviewing so many incidents involving nitrogen asphyxiation (really Oxygen deprivation), I feel that warning signs and barriers should be placed on the ladders at the ground level to alert all personnel in the area that they may be exposed to Oxygen depravation in the work area in addition to those located at the entry point.

The warning signs should alert personnel that nitrogen is in use and entry without breathing air can lead to asphyxiation. No one should be allowed onto the top deck or where nitrogen is being vented unless they are in full breathing air protection.

What Has Happened/What Can Happen

Conoco Refinery
Billings Montana
6 October 1994 (1 Fatality)

In this incident, inert contractors trained in inert entry procedures were performing catalyst maintenance on a hydrotreater reactor that had experienced a high-pressure differential across the catalyst bed. This catalyst bed delta drop was because the catalyst had developed a crust or hard crusty layer on the top of the catalyst bed. The contractors performed an inert entry into the reactor and vacuumed this crust from the catalyst. In preparation for the inert entry, the refinery had established a flow of nitrogen through the reactor to inert the reactor. The nitrogen flowed into the bottom of the reactor through the catalyst bed and exited from the manway near the top of the reactor. As the contractors were in the reactor vacuuming the catalyst, the catalyst suddenly and unexpectedly erupted, propelling the contractor through the open manway, where he fell to his death. This tragic incident was a wake-up call to the industry that pressure could form below the catalyst during internal catalyst work and can result in fatal injuries to workers inside the reactor.

The following is directly from the OSHA report on this incident in 1994:

"Employee #1 was removing the catalyst from a reactor at the Conoco Petroleum Refinery. The reactor was over-pressurized with nitrogen, and when he broke through the crust layer at the top of the catalyst, the pressure release forced the employee out of the reactor. He suffered multiple trauma injuries and died."

Key Lessons Learned

This is the formula: pressure times area = force. The pressure applied under the catalyst bed was relatively small compared to the vessel-rated pressure. However, the surface area of the catalyst bed is very large, which resulted in significant force on the catalyst bed, and when the crust layer failed, the resulting force resulted in the contractor's unfortunate death. In cases where the catalyst below the crust has been dumped, the catalyst crust may also be strong enough to support the catalyst technician working inside the vessel. Then, suddenly and without warning, the crust failed, resulting in the catalyst technician falling inside the vessel or being propelled from the vessel.

During the catalyst inert entry work planning, the possibility of catalyst crust must be incorporated into the plan, and precautions must be included in the work plan. For example, continuous monitoring of pressure below the reactor bed and installing a pressure relief valve on the purge gas supply line below the catalyst bed. Also, if the catalyst has been removed below the crust, the plan should ensure the catalyst technician is protected from a fall by working either with a harness and support line or from a boatswain chair or equivalent.

Additional References

Occupational Safety and Health Administration, 29 CFR 1910.146; "Permit-Required Confined Spaces for General Industry."

U.S. Chemical Safety and Hazard Recognition Board, Nitrogen Case Study: "Valero Refinery Asphyxiation Incident". https://www.csb.gov/valero-refinery-asphyxiation-incident/

U.S. Chemical Safety and Hazard Recognition Board, Nitrogen Training Presentation: "The Hazards of Nitrogen Asphyxiation", Available as Appendix Ha or from the following web address. https://www.csb.gov/hazards-of-nitrogen-asphyxiation/

U.S. Chemical Safety and Hazard Recognition Board, Nitrogen Safety Bulletin: "Hazards of Nitrogen Asphyxiation", Also available as Appendix Hb or the following web address. https://www.csb.gov/hazards-of-nitrogen-asphyxiation/

An excellent article on Nitrogen safety: Published by EHS Today (ehstoday.com). It is available on their website below: https://www.ehstoday.com/safety/confined-spaces/ehs_imp_38471

"Five workers suffocate to death in Mandya refinery." News article – Bangalore, 6 March 2014, DHNS. https://www.deccanherald.com/content/390280/five-workers-suffocate-death-mandya.html

"Deaths from asphyxiation and poisoning at work in the United States 1984–1986", A Suruda, J Agnew, From the Division of Occupational Medicine, Department of Environmental Health Sciences, Johns Hopkins University School of Hygiene and Public Health, Baltimore, Maryland, USA. https://www.researchgate.net/publication/20379918_Deaths_from_asphyxiation_and_poisoning_at_work_in_the_United_States_1984-6

"Inert Gas Asphyxiation: A Liquid Nitrogen Accident", CHEST Journal; Official Publication of the American College of CHEST Physicians. https://journal.chestnet.org/article/S0012-3692(17)31922-0/fulltext

EHS Today, "Nitrogen: The Silent Killer". https://www.ehstoday.com/industrial-hygiene/article/21909148/nitrogen-the-silent-killer

Science Focus, Why does breathing pure oxygen kill you?. https://www.sciencefocus.com/the-human-body/why-does-breathing-pure-oxygen-kill-you/

Occupational Safety and Health Administration (OSHA), Inspection Detail, Inspection: 102800265 – Conoco, Inc. https://www.osha.gov/pls/imis/establishment.inspection_detail?id=102800265

Technical Advisory on Working Safely in Confined Spaces. https://www.wshc.sg/files/wshc/upload/cms/file/2014/cs2.pdf

BP Incident Sharing to CCPA, BP Fatality at Manhole of Nitrogen Purge Vessel, 22 January 22 2004.

U.S. Chemical Safety and Hazard Recognition Board, Fatal Liquid Nitrogen Release at Foundation Food Group – 28 January 2021, Gainesville, GA, Final Report of Nitrogen Asphyxiation incident|Incident. csb.gov/investigations

6 Killed In Liquid Nitrogen Leak At Georgia Poultry Plant, 29 January 2021 by Jaclyn Diaz. https://www.npr.org/2021/01/29/961923732/6-killed-after-liquid-nitrogen-leak-at-georgia-poultry-plant

Chapter 8.18. Hazardous of Nitrogen and Inert Entry into Nitrogen Filled Confined Spaces

End of Chapter Quiz

1. See how many different uses you can name for how Nitrogen is used in refining or petrochemical plant operations. Name all you can.
2. Describe the characteristics of Nitrogen. What are the hazards of a Nitrogen atmosphere?
3. Oxygen concentration in a normal atmosphere is 20.8%. We know that as we increase nitrogen concentration in a confined space, the Oxygen concentration available for breathing drops. Below about what percentage of Oxygen do we start feeling the negative effect of lack of oxygen?
4. Are there other gases, other than Nitrogen, that can also result in Oxygen deprivation incidents at the plant or in the mill?
5. Where are there good training resources for Nitrogen safety?
6. Do you have to be inside a process vessel to be overcome with Nitrogen?
7. What specialized response plan must be developed and available BEFORE an entry is approved for any space that meets the definition of a confined space?
8. Statistically, about half of those who die in Nitrogen-related incidents died while attempting to rescue someone else. How can this be prevented?
9. Who is allowed to carry out the duties associated with an inert entry job; for example, catalyst work inside a reactor prepared for inert entry?
10. When is the only time that Nitrogen should be used to power pneumatic tools? Why?

8.19

Hazards of Hydrogen Sulfide (H_2S)

Hydrogen sulfide is a dangerous and unique hazard because it is highly toxic and can kill you in seconds at high concentrations. It is particularly toxic when inhaled and affects the central nervous system, interrupting breathing. H_2S is Immediately Dangerous to Life or Health (IDLH) at concentrations of 100 ppm. This includes acute respiratory exposure that poses an immediate threat to life, immediate or delayed irreversible adverse health effects, or acute eye exposure that would prevent escape from a hazardous environment. Knockdown (rapid unconsciousness) often results in falls that can seriously injure or kill the worker. According to the US Bureau of Labor Statistics, H_2S is one of the leading causes of workplace gas inhalation deaths in the United States. They reported 60 fatalities in the US between 2001 and 2010 due to H_2S exposure! Figure 8.19.1 details the effects of hydrogen sulfide on the human body. An H_2S warning sign is illustrated in Figure 8.19.2.

Hydrogen sulfide (formula H_2S) generally comes into a refinery either directly in the raw materials (crude oils and natural gas) or is generated from the sulfur compounds in the crude oils. High sulfur crudes will typically have higher levels of H_2S in the vapor space and products produced from these crudes (unless treated). H_2S can dissolve in water, hydrocarbon, and liquid sulfur. When process streams containing H_2S are opened to the atmosphere, H_2S is released. H_2S can also be present in sewer gas, rotting vegetation, volcanos, animal manure, and other natural causes.

H_2S is flammable, slightly heavier than air, and can explode if exposed to an ignition source when in higher concentrations. It can be detectable by odor in extremely low concentrations (as low as 0.47 ppb) and has a distinctive rotten egg odor. The greatest fire and explosive potential are from releases where large concentrations are handled or processed. It has a wide flammable range (4% LEL–44% UEL) and an autoignition temperature of 500 °F (260 °C).

Since it is heavier than air, escaped H_2S gas may accumulate in low-lying areas, such as:

- Bottom of vessels
- Ditches or trenches
- Tank levees
- Dike walls
- Drains
- Sewers
- Containment areas

The OSHA 8-hour exposure limit is 10 ppm, and the short-term (15 minutes) exposure limit is 15 ppm. However, the American Conference of Governmental Industrial Hygienists (ACGIH®) recommended exposure limits for hydrogen sulfide for the 8-hour exposure limit is 1, and 5 ppm for the short-term exposure limit (15 minutes). As a result, several companies are reducing the 8-hour exposure limit to well below the OSHA limit of 10 ppm.

H_2S is highly corrosive to most metals. It reacts to form metal sulfides that may cause metal fatigue. Streams containing H_2S can leave a sulfur-containing residue on internal surfaces that can lead to the generation of a sulfur dioxide (SO_2) compound, and residue can form during welding on internal metal surfaces or during thermal cutting activities. When planning for acid cleaning, check for the possibility of iron sulfide deposits which may react with the acids and produce H_2S.

Process Operations Safety: The What, Why, and How Behind Safe Petrochemical Plant Operations, First Edition. M. Darryl Yoes.

Concentration (ppm)	Symptoms/Effects
0.00011–0.00033	Typical background concentrations
0.01–1.5	Odor threshold (when rotten egg smell is first noticeable to some). Odor becomes more offensive at 3–5 ppm. Above 30 ppm, the odor is described as sweet or sickeningly sweet.
2–5	Prolonged exposure may cause nausea, tearing of the eyes, headaches, loss of sleep, and airway problems (bronchial constriction) in some asthma patients.
20	Possible fatigue, loss of appetite, headache, irritability, poor memory, dizziness.
50–100	Slight conjunctivitis ("gas eye") and respiratory tract irritation after 1 hour. May cause digestive upset and loss of appetite.
100	Coughing, eye irritation, loss of smell after 2–15 minutes (olfactory fatigue). Altered breathing, drowsiness after 15–30 minutes. Throat irritation after 1 hour. Gradual increase in severity of symptoms over several hours. Death may occur after 48 hours.
100–150	Loss of smell (olfactory fatigue or paralysis).
200–300	Marked conjunctivitis and respiratory tract irritation after one hour. Pulmonary edema may occur from prolonged exposure.
500–700	Causes staggering, collapse in 5 minutes. Severe damage to the eyes in 30 minutes. Death after 30–60 minutes.
700-1000	Rapid unconsciousness, "knockdown" or immediate collapse within one to two breaths, breathing stops, death within minutes.
1,000–2,000	Nearly instant death.

Figure 8.19.1 Effects of hydrogen sulfide on the human body. Source: US Occupational Safety Health Administration (OSHA).

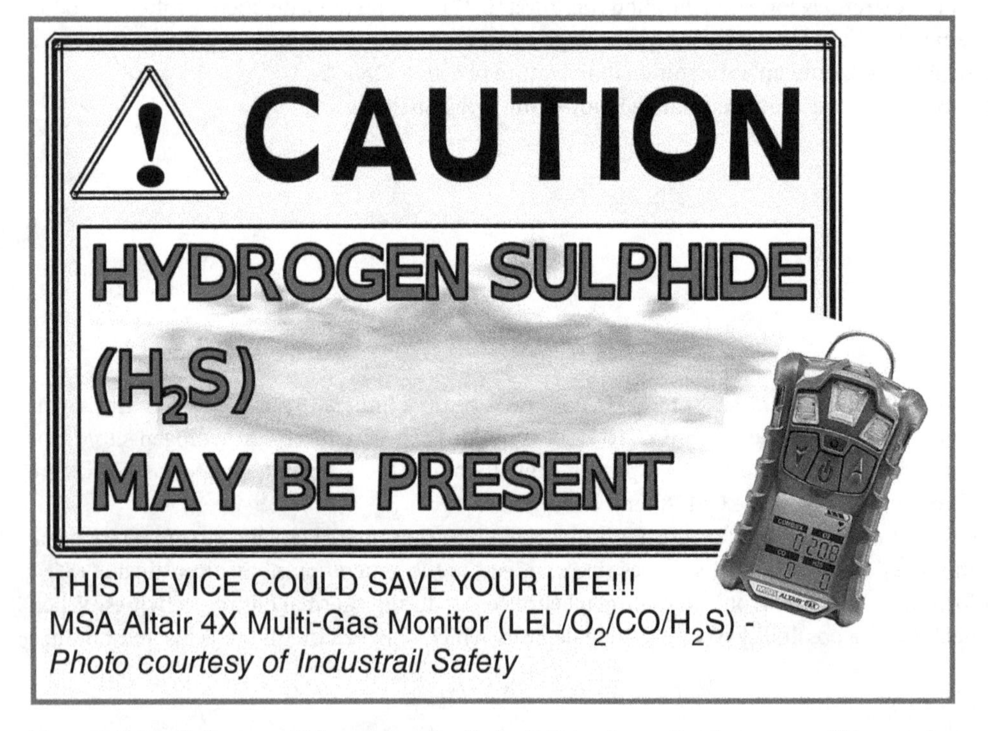

Figure 8.19.2 Hydrogen sulfide warning sign (Note: H_2S may be spelled hydrogen sulfide or hydrogen sulphide). Source: Photo kind courtesy of IndustrialSafety.com (industrialsafety.com). These are actual signs posted to provide Operators information about the potential hazards associated with Hydrogen Sulfide.

Typical Refinery Locations with Relatively High Concentrations of H_2S (>1.0%)

- Sulfur-Recovery Units
- Sour Water Stripping Units
- Hydrocracker (1st stage and H_2S Strippers)
- Reformers (Pretreater Reactor and Off Gas Absorbers)
- CHD or Hydrotreating units
- Unit H_2S Absorbers using MEA or DEA
- Coker/Coker Gas Plant
- FCC and FCC gas compressor trains
- Crude Units
- Crude tankage
- Sour Naphtha Tankage
- Foul MEA or DEA
- Flare System
- Crude Ships or Barges

Hydrogen Sulfide Generated from Chemical Reactions

H_2S can also be generated in significant quantities by inadvertently mixing acids with sulfur-containing materials like sour water. Sour water contains significant amounts of sulfur compounds, including H_2S. If sulfuric acid contacts sour water, the resulting chemical reaction can release significant quantities of H_2S and overcome unsuspecting personnel in the area.

Similar reactions should be anticipated between other chemicals containing sulfur compounds and acids. For example, various forms of hydrosulfide acids, when mixed with other acids, result in the rapid generation of significant quantities of sulfides, including potassium hydrosulfide, sodium hydrosulfide, and ammonium hydrosulfide. People working around or near these materials should be aware of these hazards. More information can be found on the MSDS for your site's specific chemical(s).

Actions to Reduce or Eliminate the Risks Associated with H_2S

- Provide initial and periodic refresher training for Operators and Craftsmen.
  Personnel must be trained on the hazards of H_2S and how to respond to H_2S releases/incidents, including initial detailed training and periodic (typically annual) refresher training. Training should emphasize the use of H_2S monitors, SCBA for short-term use, respect for barriers, and rescue techniques.
- Ensure adequate warning and demarcation controls (signs, barriers, and access controls):
  Provide signs or barriers (barricades, painted areas, etc.) to indicate operating areas designated as "H_2S Posted Areas". At a minimum, areas where process streams contain >2% H_2S or poorly ventilated areas containing sour streams, should be posted as H_2S areas and require an H_2S personal monitor or a second stand-by person. Additional required personal protective equipment may include escape respirators or hoods, two-way radio, etc.
  Examples:
  - Analyzer houses (due to analyzer purges, tubing leaks of Nitrogen or H_2S, etc.).
  - Walk-in acoustic enclosures (such as gas turbine enclosures).
  - Roof of floating roof tanks (especially if the roof is within four feet of the shell).
  - Poorly ventilated compressor houses etc. (also areas below grade or enclosed pipeways).
    Strict controls where H_2S may be present include red barricade tape with an appropriate flagging tag to indicate a hazardous area. Only employees participating in the specific task are permitted to enter with this tape and tag.
- Provide and enforce work procedures, PPE requirements, and respiratory protection:
  Site rules regarding proper work procedures (work permit, JLA/JSA, SOC, Standby, etc.) and PPE requirements (breathing apparatus, personal H_2S detectors, etc.), in particular when removing 50% of bolts on a flange in H_2S service before opening up, or opening flanges for blinding, de-blinding, removal of valves, etc. Working on safety valves and other

connections to flare/blowdown/sewer de-sludging work, chemical cleaning, etc. Procedures should require that a safety standby person be present when using breathing apparatus for mechanical work.
When the H_2S concentration is expected to exceed 5 ppm or is unknown, initial gas tests must be done by process personnel wearing a Self-Contained Breathing Apparatus (SCBA).
When opening process equipment, if there is >100 ppm H_2S directly behind any closed, unblinded isolation valve, the minimum requirement is to wear a pressurized breathing apparatus as isolation valves may leak, and the conditions behind the isolation valve can change. Work must be conducted with air-supplied breathing equipment if the potential exposure exceeds the 8-hour exposure limit. No entry should be allowed if the H_2S concentration exceeds 100 ppm.

- Ensure emergency response plan preparedness – conduct drills and exercises:
  To increase and maintain awareness of the H_2S hazards, employees should be familiar with respiratory protection and H_2S monitors. They should practice using the self-contained breathing apparatus (SCBA) and breathing air equipment and be knowledgeable with the fixed H_2S monitors. Employees should also be familiar with their personal H_2S monitors and ensure they are calibrated and bump tested at regular intervals. SCBA and 5-minute escape packs should be strategically located in the unit, including at elevated areas where H_2S streams are equal to or above 10 vol%, at frequently accessed work and operating platforms, and in congested areas that make emergency evacuation difficult. Entrants to the H_2S posted area should be familiarized with the locations of these escape sets. SCBA and five-minute breathing air sets must be serviced and checked regularly (at prescribed intervals and with a checklist of items to verify as functional).
  All employees should constantly take note of the wind direction and be prepared to evacuate crosswind in the unlikely event of a release using an escape set if needed. They should be aware of how to sound the alarm during a release.
  Do not attempt a rescue. If a colleague should go down, call for help immediately, evacuate the area, and be prepared to direct the responders toward the person. Fifty percent of fatalities in confined space incidents are people attempting to rescue others.
- Ensure a pre-job meeting with all workers to review the hazards and job plan before starting work:
  Team leaders will conduct a pre-job meeting with all workers working in a "hydrogen sulfide" area or on a job with the potential of H_2S exposure over 10 ppm. The following topics must be covered:
  - Review the appropriate MSDS.
  - Orientation of the job's specific rules and regulations.
  - H_2S hazards and where they can be found.
  - Safety requirements such as the following:
    - Flagging and warning signs.
    - Flagging tags.
    - Backup and standby personnel requirements.
    - Evacuation procedures.
    - Communication procedures.
    - Gas monitoring requirements.
    - Unit/area alarms.
    - Emergency Assembly Area and Emergency Meeting Points.
    - Emergency procedures in case of an H_2S leak.
    - Eyewash stations and emergency shower locations.

A summary of a quick review of the OSHA database for workplace fatalities caused by hydrogen sulfide exposure for the period 2019 through early 2023 is available at the end of this Chapter.

What Has Happened/What Can Happen

Georgia–Pacific Naheola Mill
Pennington, Alabama (two Fatalities, eight Injuries)
16 January 2002

Two contractors were killed at the Georgia–Pacific Naheola mill in Pennington, Alabama, and eight were injured due to a hydrogen sulfide poisoning incident on 16 January 2002. The US Chemical Safety and Hazard Investigation Board investigated this incident and issued a detailed report. This summary includes information available in the CSB report.

Burkes Construction employees were working on a construction project at the Naheola mill near the tank truck unloading station, where various chemicals could be unloaded. Sodium hydrosulfide (NaHS) was in the process of being unloaded on

15 and 16 January. Per the sodium hydrosulfide Safety Dats Sheet, NaHS solution contact with acids or acidic materials will cause highly toxic Hydrogen sulfide vapors to be released.

Unknown to the contract crew, previous modifications had been made to the sewer system, which tied the acid sewer to the sewer that serves the truck unloading station. On the day of the incident, sulfuric acid was added to the acid sewer to control pH downstream in the effluent area. NaHS from the oil pit and the collection drain drained into the sewer and reacted with the sulfuric acid to form H_2S. Within five minutes, an invisible cloud of H_2S gas leaked through a gap in the seal of a manway in the area near the Burkes Construction workers. Two contractors near the manway were killed by H_2S poisoning; seven other Burkes employees and one Davison Transport driver were injured due to H_2S exposure.

The CSB found that good engineering and process safety practices were not followed when connecting the drain from the truck unloading station and the oil pit to the acid sewer. No formal MOC analysis was conducted and there was no consideration for mixing the chemicals and/or the potential consequences. Mill personnel were not trained on the specific hazards of NaHS, such as handling spilled material or keeping it separate from acid. The fiberglass manway was not adequately designed or sealed to keep the sewer closed. The injured contractors did not have adequate training to understand the hazards of hydrogen sulfide (H_2S).

This is a summary; please see the complete CSB report for additional information on this tragic incident.

What Has Happened/What Can Happen

Release of toxic hydrogen sulfide into Poza Rica, Mexico
Oil Field Sulfur-Recovery Plant – 1950
22 Fatalities, 320 Hospitalized

A catastrophic exposure episode involving the release of large quantities of hydrogen sulfide into a small community was reported by McCabe & Clayton (1952). This occurred in 1950 in Poza Rica, Mexico, a city of 22,000 people located about 210 kilometers northeast of Mexico City. Poza Rica was then the center of Mexico's leading oil-producing district and the site of several oil field installations, including a sulfur-recovery plant. An early morning malfunction of the waste gas flare resulted in the release of large quantities of unburned hydrogen sulfide into the atmosphere.

The unburned gas, aided by a low-level temperature inversion and light early morning breezes, was carried to a residential area adjacent to the plant area. Residents of the area were overcome while attempting to leave the area and assisting stricken neighbors. Within 3 hours, 320 persons were hospitalized, and 22 died. The most frequent symptom was the loss of the sense of smell. More than half of the patients lost consciousness; many suffered signs and symptoms of the respiratory tract and eye irritation, and nine developed pulmonary edema.

Key Lessons Learned

The Poza Rica tragedy provides ample evidence that the accidental release of hydrogen sulfide into a community can be expected to cause severe impacts of varying severity, including multiple deaths.

A similar event involving a significant loss of high-concentration H_2S, for example, from an Amine Regenerator or Sour Water Stripper overhead system, could have similar consequences in or adjacent to a refinery. We must have a robust and practiced emergency response plan and a notification process to ensure adjacent areas are quickly evacuated or sheltered in place in the event of a site emergency involving the loss of containment of H_2S into downwind areas.

What Has Happened/What Can Happen

An Aghorn employee and his wife die after exposure to hydrogen sulfide
Ector County, Texas (two Fatalities)
26 October 2019

Amber Stegall of KCBD reported the following tragic story in Lubbock, Texas. The account was adapted to fit within this format. However, this incident clearly shows that hydrogen sulfide does not discriminate and is a fast-acting and highly toxic gas in high concentrations, and it will kill you.

Jacob and Natalee Dean died on 26 October 2019, after being poisoned by hydrogen sulfide gas. Jacob (44) and Natalee were husband and wife, and Jacob received a call from his employer, Aghorn Energy requesting him to check a pump house. After Jacob's wife, Natalee, did not hear from him, she went to the pump house to check on him. She took their two children with her and left them in the car when she went to check on Jacob. As she walked from the car to the pump house, she became overcome by H_2S.

An Aghorn supervisor later called law enforcement to check on Jacob since they had not heard from him. Deputies arrived within a few minutes and found that Jacob and Natalee had both died due to H_2S exposure. The children were removed from the car, decontaminated, and taken to the hospital for further treatment. The employee's H_2S detector was found in his pickup truck.

(a)

(b)

Figure 8.19.3 Small remote gas plant where this incident happened. Note the gathering tanks and the small pumphouse give the appearance of a small benign facility. Due to the high concentration of hydrogen sulfide, the lack of safety awareness, and complacency where H_2S was concerned, two people lost their lives here. Source: Photos is courtesy of the US Chemical Safety and Hazard Investigation Board. These images are from the CSB Report "Hydrogen Sulfide Release at Aghorn Operating Waterflood Station." Annotations by the CSB.

This incident was investigated by the US Chemical Safety and Hazard Investigation Board and a report was issued. This is a great report with some very significant lessons learned about the hazards of hydrogen sulfide and the precautions that should be taken to prevent another tragic incident like this from occurring. I strongly recommend that readers take the time and go to the CSB website and review this incident report.

The images below are taken directly from this report and show how at first glance the facility appears to be quiet, peaceful, and almost benign. Of course, due to the presence of highly toxic H_2S, appearances can often be deceiving. At high concentrations, only one breath can knock a person to the ground and breathing ceases after a brief few minutes (Figure 8.19.3a and b).

Key Lessons Learned

When working around hydrogen sulfide, we must all be aware of becoming complacent. It is easy to become familiar with smelling H_2S and accept the odor as a routine, almost expected, part of our job. At least to me, a lot is unknown about what happened to Jacob in this incident. Why was he not wearing his H_2S detector for this unplanned visit to the pump house? If he had it should have provided him with an alarm. Did he have access to respiratory protection, for example, a Scott Air-Pak, or equivalent? Why did he go to the pump house alone? Was this a practice? What training did Jacob have in H_2S awareness? I do not have any of these answers, but I am not sure I would be happy to hear them.

My point is that this tragic accident should not have happened. Had Jacob been adequately trained and had he used his H_2S detector, and had he had a partner with him, it is possible that it would not have happened. Be aware of complacency where H_2S is concerned. Ensure that even small leaks are promptly attended to. We should not be able to smell H_2S in our work areas.

ALWAYS, ALWAYS wear your H_2S detectors when you are out and about and in the process areas, even if you only expect to be there a couple of minutes. Ensure your H_2S detector is bump tested on schedule, and treat all H_2S alarms as real. Leave the area when the H_2S detector or alarms are alerting, and do not return until you have properly donned an air-pack and have a workmate to back you up who is also wearing an air-pack. Remember, you can die at H_2S concentrations of about 500–1,000 ppm. A typical foul Amine Regenerator overhead gas stream can easily be 98% H_2S, or 980,000 ppm! The OSHA 8-hour exposure limit is only 20 ppm (10 ppm for construction, and most sites set 10 ppm as their 8-hour exposure limit, some are even lower). Do not let this happen to you.

What Has Happened/What Can Happen

Saipan Sewer Lift Station
Hydrogen Sulfide exposures – three Fatalities
7 July 2017

The following is directly from the OSHA investigation report for this incident:

At 11:00 a.m. on 7 July 2017, Employee #1 attempted to dislodge a 24-inch rubber plug from a 2-foot diameter sewer pipe inside a 24-foot deep wet well. The workers were outside the well pulling on a 1/4-inch nylon rope attached to the 24-inch diameter plug. The plug was lodged inside a T-shaped PVC fitting from the force of the wastewater emptying into the well.

Without conducting atmospheric testing of the workspace, Employee #1 climbed down the ladder with a crowbar to dislodge the deflated 24-inch diameter rubber plug about 8 feet below the top of the well. He had difficulty releasing the plug with the crowbar and started to climb the ladder. He lost consciousness about 2 feet from the top of the well and fell into the 24-foot deep well. Employee #2 descended the ladder to provide emergency rescue but lost consciousness and went underwater. The wastewater level was about three feet deep at this point. Employee #3 climbed down the ladder to provide emergency rescue but also lost consciousness. All three workers were overcome and died due to hydrogen sulfide (H_2S) gas.

Key Lessons Learned

Ensure work plans recognize the potential for H_2S and that no one enters a confined space until the appropriate gas tests are completed, the entry permit has been issued, and only after a hole watch has been established. Each site should develop and rigorously enforce detailed confined space entry procedures, including verifying the gas test before entry.

In the event of an H_2S release and a person down, do not attempt a rescue. Call for help and immediately evacuate the area. Remain in the adjacent roadway to help direct the emergency responders to the person and ensure the responders are aware of the potential for H_2S in the area.

A Quick Review of the OSHA Website for Fatalities Due to Exposure to Hydrogen Sulfide (Only Years 2019 – early 2023)

2/24/23: Employee was potentially exposed to H_2S and was found dead inside a small pump shack known to have H_2S. https://www.osha.gov/ords/imis/accidentsearch.accident_detail?id=154242.015

7/5/23: Employee is overcome by H_2S and falls into sewer manhole and dies. https://www.osha.gov/ords/imis/accidentsearch.accident_detail?id=139143.015

8/22/22: Four employees operating motorized equipment at a waste treatment site were overcome by vapors suspected to be H_2S. One died, and three were treated at the hospital.

9/19/22: Employee was overcome by H_2S while pumping wastewater and died. https://www.osha.gov/ords/imis/accidentsearch.accident_detail?id=149915.015

5/14/22: Three employees were seriously burned when H_2S was ignited while working near a tank truck. One died, and two were hospitalized for burns. https://www.osha.gov/ords/imis/accidentsearch.accident_detail?id=146481.015

9/12/21: Two employees killed working in a pit (H_2S) https://www.osha.gov/ords/imis/accidentsearch.accident_detail?id=138824.015

8/29/21: Two employees killed while cleaning a tank truck (H_2S) https://www.osha.gov/ords/imis/accidentsearch.accident_detail?id=139073.015

8/30/21: An employee was killed cleaning a tank truck containing animal parts (H_2S) https://www.osha.gov/ords/imis/accidentsearch.accident_detail?id=138824.015

6/27/22: Two employees are working at the top of an open-top frac tank (H_2S). Both died. https://www.osha.gov/ords/imis/accidentsearch.accident_detail?id=147970.015

6/8/21: One employee killed, and one was hospitalized while cleaning tanker due to H_2S https://www.osha.gov/ords/imis/accidentsearch.accident_detail?id=136167.015

6/7/21: Two employees killed by H_2S https://www.osha.gov/ords/imis/accidentsearch.accident_detail?id=138824.015

6/8/21: Employee killed by H_2S while working in sewer manhole https://www.osha.gov/ords/imis/accidentsearch.accident_detail?id=136124.015

5/26/21: Two employees entered sewer vault to retrieve cell phone 6/8/21: and were both killed by H_2S https://www.osha.gov/ords/imis/accidentsearch.accident_detail?id=136332.015

7/24/20: Four employees were overcome by H_2S while working in a sewer – two died https://www.osha.gov/ords/imis/accidentsearch.accident_detail?id=128141.015

7/9/20: Two employees were exposed to H_2S while working in a sewer manhole – both died. https://www.osha.gov/ords/imis/accidentsearch.accident_detail?id=127680.015

5/29/20: One employee was killed and one injured by exposure to H_2S while working in a sewer manhole. https://www.osha.gov/ords/imis/accidentsearch.accident_detail?id=126541.015

10/26/19: Employee entered pumphouse due to alarm, was overcome by H_2S, and died. https://www.osha.gov/ords/imis/establishment.inspection_detail?id=1440890.015

5/24/19: A farmer cleaning a storage tank was overcome by H_2S and died. https://www.osha.gov/ords/imis/accidentsearch.accident_detail?id=116597.015

4/10/19: Employee entered a vault manhole, was overcome by H_2S, and died. https://www.osha.gov/ords/imis/accidentsearch.accident_detail?id=115356.015

Additional References

U.S. Chemical Safety and Hazard Investigation Board, Hydrogen Sulfide Poisoning, Georgia-Pacific Naheola Mill, Pennington, Alabama 16 January 2002, Report No. 2002-01-I-AL. https://www.csb.gov/georgia-pacific-corp-hydrogen-sulfide-poisoning/

U.S. Chemical Safety and Hazard Investigation Board, Safety Bulleting covering the hazards of Sodium Hydrosulfide in the workplace, CSB Releases Safety Bulletin Warning of Dangers of Sodium Hydrosulfide (NaHS) in the Workplace; Outlines Safe Practices to Prevent Harm – Investigations – News | CSB.

International Program on Chemical Safety (IPCS), Report on Hydrogen Sulfide: http://www.inchem.org/documents/ehc/ehc/ehc019.htm

OSHA Investigation Report: Saipan Sewer Incident (three Fatalities): https://www.osha.gov/pls/imis/establishment.inspection_detail?id=1246546.015

Excellent article on H_2S Safety: "Co-worker fatalities from hydrogen sulfide". https://www.researchgate.net/publication/8670870_Co-worker_fatalities_from_hydrogen_sulfide

News Article by Amber Stegall of KCBD News: "Gas company employee, wife die after Hydrogen Sulfide gas poisoning." https://www.kcbd.com/2019/10/28/gas-company-employee-wife-die-after-hydrogen-sulfide-gas-poisoning/

U.S. Chemical Safety and Hazard Investigation Board, Two H_2S fatalities at the Aghorn Operating Water Flood Station on October 26, 2019 (a worker and his wife were killed by Hydrogen Sulfide poisoning). https://www.csb.gov/aghorn-operating-waterflood-station-hydrogen-sulfide-release-/

CSB Aghorn H_2S Safety Video: https://www.youtube.com/watch?v=jh2HWT8gPeY

Chapter 8.19. Hazards of Hydrogen Sulfide (H_2S)

End of Chapter Quiz

1 Where does the hydrogen sulfide come from?

2 Hydrogen sulfide is a dangerous and unique hazard because it is highly toxic and can _____ you in seconds at high concentrations. It is particularly toxic when inhaled and affects the central nervous system, interrupting breathing. H_2S is Immediately Dangerous to Life or Health (IDLH) at concentrations of ____ ppm.

3 H_2S is__________, slightly heavier than air, and can explode if exposed to an ignition source when in higher concentrations.

4 It can be detectable by odor in extremely low concentrations (as low as 0.47 ppb) and has a distinctive ______ _______ odor.

5 What is the US OSHA eight-hour working limit for H_2S in ppm?

6 Hydrogen sulfide is a dangerous and unique hazard because it is highly toxic and can kill you in seconds at high concentrations. It is particularly toxic when inhaled and affects the central nervous system interrupting the ability to breathe. At what concentration is H_2S deadly?

7 Since H_2S is heavier than air, it will most likely accumulate and concentrate in low-lying areas. What are some examples where higher concentrations of H_2S may be found?

8 You are preparing the agenda for an upcoming job site safety briefing that will be done in a work area that has H_2S potential. Please list as many of the topics that should be covered as you can.

9 Employees should practice the use of the self-contained __________ __________ and be knowledgeable with the fixed __________ monitors.

10 Employees should also be familiar with their personal _____ __________and ensure they are calibrated, and bump tested at regular intervals.

8.20

Hazards of Product Sampling and Drawing Water from Process Equipment

Unfortunately, many people in the petroleum and petrochemical industries have been seriously injured or even lost their lives while collecting hydrocarbon or chemical samples for laboratory analysis. These incidents have occurred while sampling large petroleum storage tanks or collecting rundown samples on process units; they both have their hazards which will be covered in this chapter.

When collecting samples from process vessels, including storage tanks, ships, barges, tank cars or tank trucks, and similar conveyances, we have the "relaxation period." The relaxation period allows static electricity to dissipate before sampling or gauging. No sampling or gauging should occur during this relaxation period, and the tank or vessel should not be moving; that is, should not have production into or out of the vessel. For small volumes, less than 10,000 gallons (38 Cu Meters), this relaxation period should be 5 minutes; for volumes larger than 10,000 gallons, which most of our samples will involve, the relaxation period is 30 minutes.

During this relaxation period, the vessel should be completely still with no movement in the tank. The mixer should be turned off if the vessel is equipped with a mixer. The relaxation period allows the static electricity that has accumulated in the vessel or the product to dissipate to the shell of the tank and from there to the earth. More detail is available in Chapter 8.22, "Controlling Static Electricity," and Chapter 8.8, "Storage Tank Safety."

All overhead sampling of storage tanks (sampling through the vapor space) should be done using the "stilling well" or "sounding tube." This gauging pipe helps insulate the sampler from any static electricity accumulated in the rest of the tank. Also, only approved sampling and gauging equipment should be used for this purpose. The approved sampling equipment will be conductive material and an all-bonded design. It will have a bonding cable that should be attached to the tank or conveyance *before* anything goes into the tank vapor space. This is important as it will quickly dissipate any static electricity built up in the gauging equipment or static differential between the equipment and the tank. If, for some reason, it is required to use a nonconductive sample or gauging device, the only acceptable nonconductive rope is a dry cotton rope. Never use nylon, polypropylene rope, or synthetic rope, as they generate static electricity.

Sampling with metal cans or smaller containers requires adherence to these same rules. Always ensure the bonding and grounding of the smaller containers; the bonding and grounding cables should be the first thing that goes on and the last thing that comes off and should remain attached throughout the operation. For small containers, the relaxation period is only five minutes, but it is important to allow the static discharge to relax.

Collecting samples on operating units also has hazards and can result in fires and explosions. Some sample streams can ignite because of autoignition (ignites upon immediate contact with air), light hydrocarbons or hot vapors, or uncontrolled liquid releases can find a nearby ignition source resulting in an explosion or fire. And, of course, unit samples are also very susceptible to ignition from static electricity.

Operators have also been seriously burned while taking samples when coming into contact with hot process streams or hot condensate or cold burns from exposure to cryogenics like LPG. Chemical burns have also been sustained from acids, or caustic and toxic exposures can result from exposure to H_2S, chlorine, nitrogen, etc.

A good rule of thumb when collecting samples is to know the hazards before attempting to catch the sample:

- Know the pressure and temperature of the sample.
- Know the physical and chemical properties.
- Catch the minimum volume for testing.
- Choose the proper container (depending on the product and temperature).
- Leave a vapor space for the expansion of liquids.

Process Operations Safety: The What, Why, and How Behind Safe Petrochemical Plant Operations, First Edition. M. Darryl Yoes.

Sample points should be installed in accessible, unobstructed areas to facilitate safe access and egress. Most sample points should be "closed-loop" designed to help ensure a representative sample. Sample points handling high melting point products such as wax and asphalt should be heat traced and insulated, and sample points for hot streams should be equipped with sample coolers to cool the stream before sampling. Occasionally, a sample such as a bleeder will be needed from an "in situ" sample point. In this event, a special procedure should be developed to detail the safety precautions required to collect the sample.

At first glance, product sampling might appear to be a basic, low-risk, routine activity, but some basic rules must be followed to prevent incidents. We should always ensure the correct PPE is worn by Operators and others responsible for collecting samples of potentially hazardous materials. For example, we should always wear the proper PPE for each sampling task, such as chemical goggles AND a face shield and long chemically resistant gloves for acid or caustic; a face shield and long protective gloves for asphalt and bitumen; and breathing apparatus when H_2S exposure can exceed ten ppm.

Ensuring the correct container for the specific product being sampled is always necessary. As a rule of thumb, we use sample cylinders for LPG and unstabilized streams (vapor pressure >13 PSIA) (90 kPa)[1]; we do not use plastic containers for process streams >100 °F (40 °C)[1]; and cans rather than glass bottles for hot oil (>150 °F) (65 °C)[1].

It is important to only use the designated sample locations for each process stream sampled. These are typically quick-connected cylinder connections for LPG and unstabilized streams with vents and purges to safe locations, closed-loop connections for most hydrocarbon samples, and water coolers for hot oil streams. Do not use homemade or make-shift sample points, which are unsuitable for taking samples. These include high-temperature locations without coolers or locations close to ignition sources like furnaces or hot piping. Never attempt to collect a sample from a ¼ turn valve or plug valve not meant for throttling. We should preferably use needle or globe valves that provide much better control.

Some streams may contain Benzene. For those streams, a special Benzene label is required for all containers and equipment with 0.1% or greater by volume of benzene (not currently required for piping). Sampling Benzene also requires specific respiratory protection (currently half face with organic vapor cartridge for concentrations 0.5–5 ppm, full face with organic vapor cartridge for concentrations 5–25 ppm, and supplied-air respirator for concentrations greater than 25 ppm).

Figure 8.20.1 Photo of piston-type sample valve (Used to collect samples of residual type materials such as asphalt, where a representative sample may otherwise be difficult).

1 These are only typical examples - actual site rules for PPE and sample containers required should be followed.

Never attempt to collect samples of low flash or higher volatility products like gasoline or naphtha if welding or other hot work is in progress nearby. Never attempt to sample a stream operating above its autoignition temperature unless a cooler is provided.

Asphalt or bitumen samples are best taken using a piston-type positive displacement sample valve whereby the sample is displaced from the sample valve by an internal shaft. This prevents the asphalt from solidifying inside the valve and plugging it, preventing it from being used for the following sample. Also, waste is minimized with this type of valve. See Figure 8.20.1 for a photo of a piston-type positive displacement sample.

Remember, samples should be transported in open pickup truck beds, never in the cab or inside a vehicle, and the transportation of samples in personal vehicles is strictly prohibited.

Collecting Samples from LPG Spheres or Horizontal Storage Vessels (Bullets)

When collecting samples from LPG storage vessels (spheres, bullets, tank cars, etc.), the procedure is almost the same as drawing water from LPG storage vessels. The sample location on the sphere or bullet will have two valves close to the vessel, an upstream and a downstream valve. After connecting the cylinder sample container using the quick connect, open the cylinder valve first. Then fully open the upstream valve (or valve nearest the vessel), and then crack the downstream valve (valve away from the vessel) to a throttle position to collect the sample into the sample cylinder. After the cylinder has reached the fill mark, close the downstream valve and then close the upstream valve, close the cylinder valves, and disconnect the cylinder from the quick connect.

The rationale behind this procedure is to prevent the auto-refrigeration and potential freezing of the sample valves in the open position. If the upstream valve is accidentally throttled, a pressure differential across the upstream valve will result in a pressure differential. LPG subjected to a pressure drop will auto-refrigerate, freezing the upstream valve AND the downstream valves in the open position. Should this happen, the Operator may not be able to close the valves, which could lead to a loss of containment incident. If there are questions about this procedure, please see the section below covering water drawing on LPG vessels for additional discussion and sketches illustrating the procedure.

Sampling Sour Water/Rich DEA

Sampling sour water or foul Amines (MEA OR DEA) requires special precautions due to the relatively high concentration of hydrogen sulfide (H_2S). Operators have been overcome by the H_2S when precautions are not taken or sampling procedures are not followed.

For routine sour water or foul amine samples, the sample point should be enclosed in a special vapor-tight box that is vented overhead. The sample valve should be located approximately five feet from the box and equipped with a spring-loaded close valve. This protects the Operator and prevents potential exposure to the high concentration of H_2S while collecting the sample.

A nonroutine sampling of these sour streams (sour water or foul amines) requires special procedures since the closed box and the spring-loaded valve will not be available. In these cases, at a minimum, the Operator should be wearing a personal H_2S monitor set to alarm at 10 ppm, with radio communication while operating the sample valve and observed by a second person with fresh air readily available. If these requirements cannot be met, the Operator must wear a supplied air respirator to catch the sample.

Precautions for Static Electricity while Sampling

Static electricity is covered in more detail in Chapter 8.22, "Controlling Static Electricity," however some basic rules to control static electricity follow:

- Never sample a barge, ship tank, or cone roof tank through the roof while it is being filled or within 30 minutes of completing the filling (unless a stilling well is provided).
- Always observe the relaxing period before sampling (30 minutes for large volumes, volumes greater than 10,000 gallons (38 Cu meters), or 5 minutes after filling before sampling a tank truck or railcar (volumes less than 10,000 gallons) (38 Cu meters).

- Use all conductive sampling equipment bonded to the vessel shell or piping or,
- If you must use a nonconductive gauging bob or sample apparatus, use a dry cotton rope as the nonconductive rope.
- Never mix a conductor and a nonconductor when sampling or gauging hydrocarbons.

Always Adopt a Defensive Behavior when Taking Samples

- Stand upwind, away from the vapors.
- Avoid inhaling vapors.
- Anticipate possible splashing of liquid.
- Anticipate a potential increase in flow as viscosity lowers or any blockages clear.
- Be aware of dropping greasy, filled glass bottles.
- Use only special containers to carry multiple samples (sample carriers designed for this purpose).
- Insert plugs in bottles and other containers on a sturdy, level surface.
- Be aware of the locations of safety showers and eyewash fountains.
- Be aware of hot streams (asphalt, Coker feed, bitumen, and others).
- Never transport samples inside vehicles or allow anyone else to do so.

Hazards of Drawing (Draining) Free Water from Process Tanks and Vessels

This chapter will discuss the safety implications of draining or drawing water from process vessels in preparation for shipment to a customer or feeding to a process unit. We draw water from a wide variety of products from crude oil, gasoline, naphtha, and liquefied petroleum gas (LPG) like Propane. Many water draw systems for atmospheric tankage (cone roof and floating roof tanks) use a closed system to prevent contaminants from being released onto the soil or the wastewater treatment facility. Occasionally, we use tank trucks or vacuum trucks to collect the water bottoms and truck the bottom's sediment and water for treatment.

Static electricity is a continuing concern anytime a product is being moved through hoses into trucks or other conveyances for transport. This is covered in more detail in Chapter 8.22 "Controlling Static Electricity," but the most important consideration is the electrical bonding and grounding to prevent the ignition from static electricity. The other common denominator when drawing water is the constant attention required during the entire period the valves are open. It requires diligence and close observation to prevent draining hydrocarbons to the sewer or into a tank truck. An Operator must be present the entire time the water draw valves are open and ready to close the valve the instant the stream changes from water to product.

Precautions of Hydrogen Sulfide (H_2S)

The hazards of H_2S are covered in more detail in Chapter 8.19 (Hazards of Hydrogen Sulfide); however, you should always be aware that many of the products contained in storage tanks have a significant amount of hydrogen sulfide (H_2S). Always check the material safety data sheets (MSDS) to verify the H_2S concentration. The Operator should wear a personal H_2S monitor set to alarm at 10 ppm and always stand upwind of the water draw. In higher concentrations of H_2S, the Operator will be required to wear a positive air respirator and have a "buddy" standing by in a respirator if the vapors exceed ten ppm.

Drawing Water from Liquefied Petroleum Gas (LPG)

Drawing water from or sampling LPG such as propane, butane, pentane, and the whole family of LPGs represents a very special hazard. When LPG flashes from higher pressure to lower pressure, it goes through a change in state from a liquid to a vapor. The other change that occurs is the temperature of the product. When LPG flashes from a liquid to a vapor, it auto-refrigerates and gets very cold. For example, flashing propane can result in −40 °F (−40 °C) and below temperatures. This auto-refrigeration property of LPG has resulted in devastating consequences during water draw or sampling operations.

As a result of this challenge, when drawing water from an LPG sphere (e.g., Propane or Butane), we always use two valves and a special procedure for the water draw operation. The water draw line comes directly from the bottom of the sphere, and there are two valves in the piping from the sphere to the sewer or the closed system where we are drawing the water. The first valve from the sphere, the upstream valve, is typically a quarter-turn valve, and the second, the downstream, generally is either a gate or globe valve. We open the first valve from the sphere (the upstream quarter-turn valve) wide open, and then we start the flow by slowly throttling the second valve (the downstream gate or globe valve).

This procedure avoids accidentally creating an auto-refrigeration that may freeze both valves. If both valves become frozen, we would be unable to stop the propane flow, which could result in a devastating event. By fully opening the first valve, there is no pressure differential across this valve and, therefore, no auto-refrigeration. The pressure differential is across the downstream valve. The upstream valve can still be closed to isolate the system if the downstream valve is frozen. Should this procedure be reversed, and the flow was controlled by throttling the upstream valve, both valves could become

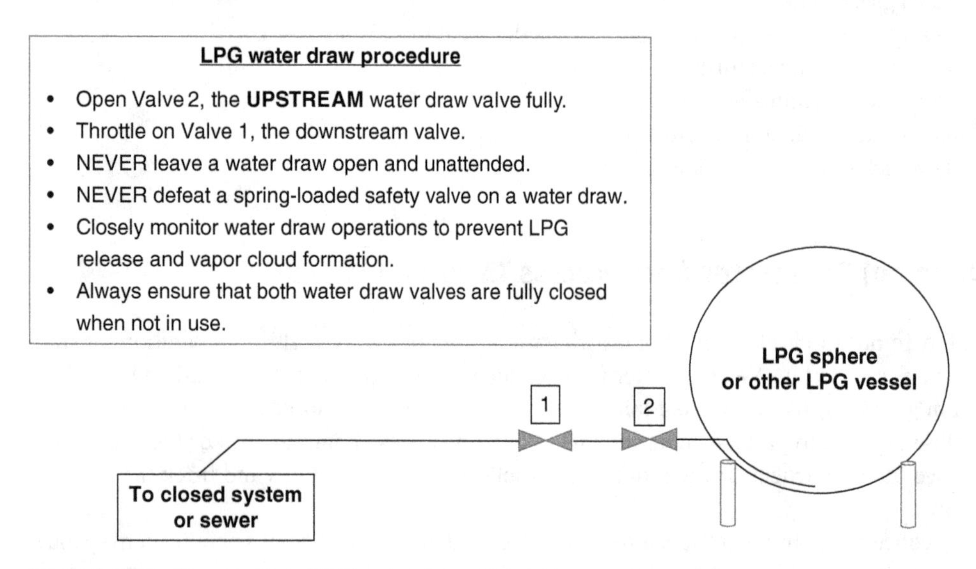

Figure 8.20.2 Recommended LPG water draw procedure.

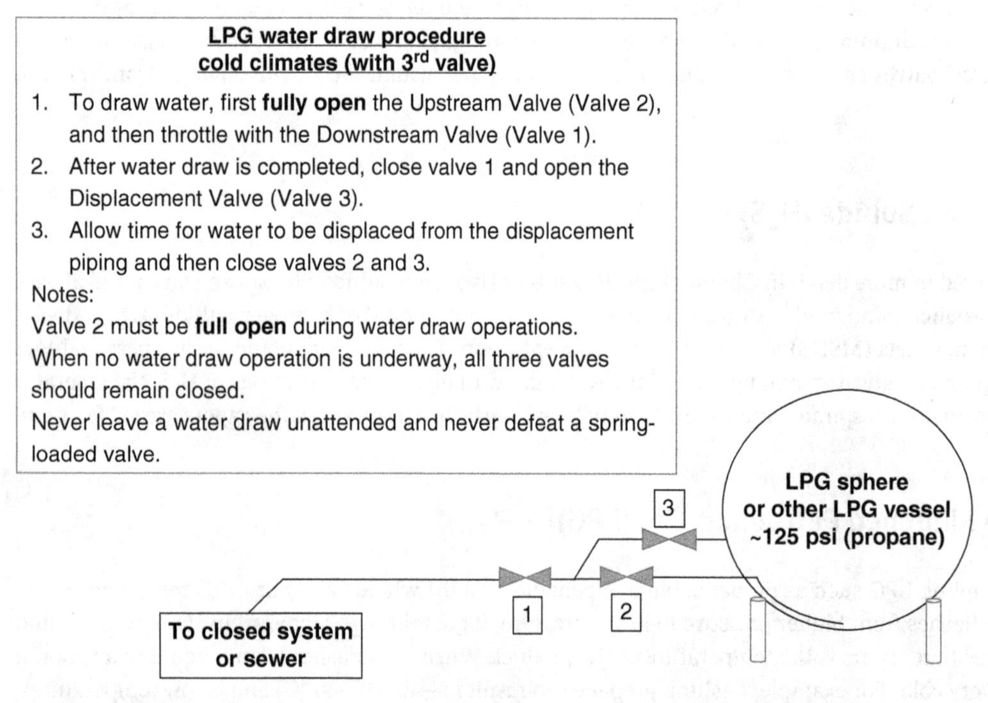

Figure 8.20.3 Recommended LPG water draw procedure when configured for cold climates.

frozen and inoperable. This is one of the most straightforward procedures to prevent a significant incident, but it is also one of the most important. This is one that we simply cannot afford to get wrong.

Always open the upstream valve fully and throttle on the downstream valve. When the water draw operation is complete, close both valves and leave both valves fully closed. See Figure 8.20.2 for a recommended LPG water draw procedure.

When a refinery or chemical plant is configured for cold climates, the water draw system may include a displacement valve used at the end of the water draw operations to displace the water from the piping near the sphere or other light ends vessel. Please refer to Figure 8.20.3 for the procedure to follow when using this displacement piping.

What Has Happened/What Can Happen

The Feyzin Refinery, Feyzin, France
A spectacular series of BLEVE explosions occurred on 4 January 1966
18 Fatalities, 81 Injured, and extensive damage
This tragic incident occurred when a Shift Supervisor, a Plant Operator, and a laboratory technician were draining water from a propane sphere to take a sample of the propane from the water draw. At Feyzin, this was a regular occurrence, with water samples drawn about every three to five days due to the formation of sodium hydroxide solution (caustic soda) separating from the LPG during storage. In this case, the Propane was also high in C_2 content, requiring extra samples and laboratory analysis.

The sphere water draw valves consisted of two 2″ (50 mm) plug valves in series from the sphere bottom separated by a short spool. The short spool in the center between the two 2″ valves had a 3/4″ (20 mm) drain valve leading to a sample point where they intended to collect the propane sample.

The refinery staff had prepared a procedure for the Operators to follow, which included the following:

1. Place a valve spanner wrench on either of both valves.
2. Fully open the upper valve (upstream valve nearest to the sphere).
3. Adjust the draw rate as required by operating the lower downstream valve or the ¾″ sample valve.

This procedure was designed to allow the Operators to control the cooling at the downstream valve and prevent any freezing of the upper valve. The valve handles were not permanently attached to the valves.

The Operator's only had a single valve spanner (operating lever), and the Operator placed it on the lower or downstream valve and proceeded to open it fully in the reverse sequence from the procedure prepared in advance. He then slightly opened the upper or upstream valve, and a small amount of caustic soda solution was released, followed by a bit of gas. He then closed the valve and reopened it, and suddenly a very powerful jet release of Propane gushed out.

The Operator received cold burns from the Propane and fell backward away from the valve; however, he pulled the valve spanner from the valve when falling. The Operator and a fireman in the area tried desperately to replace the valve wrench but were unable to get it back on the valve. Ice quickly formed on the valve, making this impossible.

The men turned the water sprays on the sphere and then set out on foot to report the release and seek help, afraid to start their vehicle. Shortly after 7:00 a.m., the fire truck arrived, and the fireman also attempted to close the valves but to no avail. A large Propane vapor cloud quickly formed and was ignited by a vehicle at about 7:15 a.m.

At about 8:40 a.m., sphere 443 exploded, a Boiling Liquid Expanding Vapor Explosion or BLEVE. The sphere ruptured into five large fragments creating a massive fireball resulting in eighteen fatalities, and many people were seriously injured. One of the fragments knocked the supports from under sphere 442, tipping it over, and another fragment tipped over another sphere releasing butane into the fire area. Another large fragment damaged the piping that connected the refinery with the storage area, starting several large fires. One other tank fragment damaged piping near four floating roof tanks and started even more extensive fires.

At about 8:55 a.m., the firefighting efforts were abandoned, and then at about 9:30 a.m., sphere 442 also exploded (BLEVEed), and sphere 441 dumped its contents, adding to the intensity of the fire. Three other butane spheres ruptured, adding to the fire's intensity.

The investigation concluded that the cause of this incident was the Operator opening the lower or downstream valve fully and operating the upper or upstream valve by throttling, directly opposite the written procedure. This allowed a differential pressure across the upper valve, creating flashing and auto-refrigeration of Propane resulting in either an ice plug or propane hydrate in the upper valve. As the Operator attempted to manipulate the valve, this ice plug blew out, resulting in the large and unexpected release of propane directly into the atmosphere.

This was compounded by only having a single valve spanner wrench. When the release occurred, the Operator jumped back and slipped and fell. When the Operator fell, he pulled the valve spanner wrench off the valve. Although the Operators tried desperately to replace it, they could not, most likely due to ice buildup on the valve body. Plug valves can induce high-pressure drop when flowing and are not well suited for throttling service and should normally not be used for this service.

When I review this incident, I cannot help but wonder why they were attempting to collect samples from the water draw. It is possible that the sphere was not equipped with a dedicated sample location. Collecting samples from the water draw introduces multiple hazards, including the potential freezing of the water draw valves.

The result was five spheres BLEVEed in addition to multiple tank and pipe rack fires. Damage to the refinery and the surrounding community was extensive and significant product was lost in the ensuing fires. However, the real cost in human terms was the loss of 18 lives and 81 people injured. Many lessons were learned from the Feyzin incident that carries over to today's procedures to help prevent a reoccurrence of a similar event, particularly the procedure for drawing water and collecting samples from LPG vessels.

Please see Figures 8.20.4 and 8.20.5 for images of the Feyzin BLEVE incident (Photos are with kind courtesy of DRIRE 69 (rights are reserved/copyrights DRIRE 69) (©DRIRE 69) (France). A couple of additional images of this event are available in Chapter VI (Boiling Liquid Expanding Vapor Explosion (BLEVE).

Key Lessons Learned

Sampling Procedures

- Always know the hazards of the product before attempting to catch a sample.
- Personnel responsible for product quality samples should be trained in sample techniques and procedures.
- Ensure static electricity precautions are integral to the sampling procedure (relaxation period and bonding and grounding).
- Always be alert for H_2S and other toxic/corrosive materials during sampling and water drawing, and always wear the proper PPE for the hazard when sampling.
- Use the proper sample container and the sample location and ensure the sample containers are correctly labeled.
- Always leave the sample point locations double-blocked when you are finished.
- If special sample points are unavailable, the special sampling procedure must be used; if those requirements cannot be met, then FRESH AIR MUST BE WORN TO TAKE THE SAMPLE!

Water Drawing Operations

- Personnel should be trained to draw water from LPG-containing vessels and other products that contain toxins like Benzene and Hydrogen Sulfide.
- LPG-containing equipment should be equipped with two water draw valves in series, with the upstream valve normally spring-loaded to close the quarter-turn valve. This spring-loaded valve should never be defeated for any reason. The second or downstream valve should be either a gate or globe valve designed for throttling.
- Always leave the water draw lines double-blocked when you have completed drawing water.

Additional References

Institution of Chemical Engineers (IChemE), Loss Prevention Bulletin 077 October 1987, The Feyzin Disaster. https://www.icheme.org/media/1278/lpb251_digimag.pdf

PetroWiki, "Fluid Sampling Safety Hazards". https://petrowiki.org/Fluid_sampling_safety_hazards

Occupational Safety and Health Administration (OSHA/NIOSH) Safety Bulletin, "Health and Safety Risks for Workers Involved in Manual Tank Gauging and Sampling at Oil and Gas Extraction Sites". https://www.osha.gov/Publications/OSHA3843.pdf

(a)

(b)

Figure 8.20.4 Images from the Feyzin Refinery BLEVEs. The fire resulted in 18 deaths and destroyed or severely damaged the LPG storage area and an adjacent liquid hydrocarbon storage facility, including 11 tanks (5 spheres, 2 horizontal tanks and 4 floating roof tanks). Source: Photo used with kind courtesy of DRIRE 69 (rights are reserved/copyrights DRIRE 69) (©DRIRE 69) (France).

(a)

(b)

Figure 8.20.5 Image of one of the LPG spheres that ruptured and BLEVEed at the Feyzin Refinery. Source: Photo used with kind courtesy of DRIRE 69 (rights are reserved/copyrights DRIRE 69) (©DRIRE 69) (France).

ASTM D4057 – 12, "Standard Practice for Manual Sampling of Petroleum and Petroleum Products". https://www.astm.org/DATABASE.CART/HISTORICAL/D4057-12.htm

Strahman Valves, Article and introduction to sampling valves. https://www.strahmanvalves.com/products-en/process-valves/sampling-valves/

Chapter 8.20. Sampling and Drawing (Draining) Water from Process Vessels

End of Chapter Quiz

1 When collecting samples from process vessels, including storage tanks, ships, barges, tank cars or tank trucks, and similar conveyances, we have the "relaxation period." The relaxation period allows static electricity to dissipate before sampling or gauging. No sampling or gauging should occur during this relaxation period, and the tank or vessel should not be moving; that is, should not have production into or out of the vessel. How long is this relaxation period for small volumes and large volumes?

2 During this relaxation period, the vessel should be completely ______ with no movement in the tank. The mixer should be turned ______ if the vessel is equipped with a mixer.

3 What is the purpose of this relaxation period?

4 The approved sampling equipment for sampling ships, barges, and storage tanks will be ___________ material and an all-bonded design. It will have a bonding cable that should be _________ to the tank or conveyance before anything goes into the tank vapor space.

5 To prevent a static discharge, what type of rope should never be used to catch a petroleum sample?

6 What product characteristics and other information about the sample should you be aware of before attempting to collect a sample of the product? Answer all that you are aware of.

7 Describe the type of facilities required to collect sour water or foul amine samples.

8 In addition to the hydrocarbon vapor and potential H_2S hazards, what other significant process hazard must be managed when collecting samples?

9 When collecting a sample of LPG like Butane or Propane, a special _________ is required to prevent the possibility of auto-refrigeration.

10 What is the main hazard with collecting samples of sour water or foul MEA/DEA?

11 How many key points can you remember about taking a defensive posture while collecting product samples? List as many as you can.

12 After a water draw operation is completed, we always leave the water lines in what condition before walking away?

13 When draining water on an LPG sphere, why is it important to open the upstream water draw valve fully and control the operation by throttling the downstream water draw valve?

14 What was the primary cause of the series of catastrophic BLEVEs at the Feyzin Refinery in France that resulted in 18 fatalities, 81 injuries, and catastrophic damage to the facility?

8.21

Pyrophoric Ignition Hazards in Refining and Chemical Plants

The word pyrophoric is derived from the Greek for "fire-bearing."

This chapter will discuss the hazards of pyrophoric materials in refining and chemical plant operations and review examples of where pyrophoric materials may be found in our plants. We will also review incidents involving ignition from pyrophoric materials and lessons learned and review methods/practices to minimize the potential for ignition of pyrophoric materials.

A pyrophoric material is liquid or solid that, even in small quantities and without an external ignition source, can ignite after air contact. This is not the real hazard; the hazard is that a pyrophoric can ignite other flammable or combustible materials that may be nearby or in the area.

Pyrophoric ignition is a concern in many petroleum refining or petrochemical processes, especially where sulfur is present in low Oxygen environments. They are typically found in process equipment containing Hydrogen Sulfide in concentrations >1% (but can form in lower concentrations). Corrosion products such as iron scale or rust combine with the sulfur compounds, creating iron sulfide scale (FeS), which is pyrophoric. This pyrophoric iron sulfide lies dormant in the equipment for years until the equipment is shut down and opened for maintenance.

When opening the equipment for maintenance, the pyrophoric material in the presence of air creates heat and may ignite other flammable or combustible materials such as:

- Residual hydrocarbons, coke, or sludge that may remain in the equipment.
- Internal demister screen or internal tower packing materials.
- Scaffold boards, rags, or other combustible materials.

A pyrophoric ignition in a confined space may also reduce the Oxygen concentration below safe limits.

Some process chemicals and catalysts may also be pyrophoric and must be controlled. This includes alkali metals such as triethylaluminium or TEAL, which are used in most plastics manufacturing plants as co-catalysts in the production of polyethylene and polypropylene. These are normally contained in an inert atmosphere to prevent ignition. Other highly pyrophoric materials include:

- Nickel carbonyl in some catalysts.
- Phosphorus.
- Other metal alkyls, such as alkyl magnesium, etc.

Process Operations Safety: The What, Why, and How Behind Safe Petrochemical Plant Operations, First Edition. M. Darryl Yoes.

The Chemistry of the Pyrophoric Reaction

The pyrophoric compound (typically iron sulfide scale) + O_2 (typically air) results in an Oxide of the compound + heat and Sulfur Dioxide (SO_2).

This can be several intermediate reaction steps. The reaction can be very reactive or very slow to react, and the reaction can vary with conditions such as humidity, temperature, particle size, degree of disbursement in air, etc. The resulting heat can ignite other materials (flammable or combustibles). White smoke (SO_2) can be mistaken for steam.

The bottom line is that pyrophoric materials can be elusive and require our attention to prevent uncontrolled spontaneous combustion inside or near process equipment.

More on the chemistry:

Pyrophoric compound (typically iron sulfide scale) + O_2 (typically air) = Oxide of the compound + heat + SO_2.

Typical chemical reactions: the result is heat and Sulfur Dioxide gas:

- $Fe_2O_3 + 3H_2S = 2FeS + 3H_2O + S$
- $4FeS + 3O_2 = 2Fe_2O_3 + 4S + \text{heat}$
- $4FeS + 7O_2 = 2Fe_2O_3 + 4SO_2 + \text{heat}$

We should expect to find pyrophoric materials in process equipment and areas where sulfur or sulfur compounds like Hydrogen Sulfide (H_2S) will be present. For example, H_2S will go overhead, so it is not a surprise that pyrophoric materials will be found in overhead systems like flare systems. However, there are many other areas where we should expect to find significant amounts of pyrophoric materials. Other examples are listed below (not a complete list):

- Flare and blowdown system piping, drums, and associated equipment.
- Process vessels in sour service
- Fractionation columns (trays, demister screens, internal packing)
- Reactors (scale traps are a common occurrence)
- Coke drums
- LP and HP separator drums
- Gas compressors and compressor KO drums.
- Storage tanks in sour service
- Crude oils
- Asphalt
- Sour Water
- Sour Naphtha
- Heat exchangers in sour service
- Crude, light naphtha, etc.
- Process sewer systems
- API Separators
- Portable tanks and totes, when used in sour service
- Marine tankers and barges (crude, sour naphtha, and other sour services)

Modern Fractionation

A modern approach to fractionation includes using demister screens to help remove the liquid droplets from rising hot vapors or random or structured packing in fractionation columns to improve the fractionation and the distillation end points. The internal packing dramatically improves the contact between the rising hot vapors and the falling liquids and improves fractionation. The packing materials are extremely dense and are designed with various metallurgies suitable for the operating temperature inside the columns. They must also be designed for the corrosivity of the process fluids, operating pressures, etc. In some cases, the packing may be made of plastic material. While demister screens and internal packing will significantly improve fractionation, the packing is very dense and requires special procedures when opening process equipment for maintenance.

The dense demister and packing can attract sulfur compounds and coke materials, which can adhere to the packing and be pyrophoric. For example, pyrophoric materials such as iron sulfide scale or coke particulates with sulfur compounds

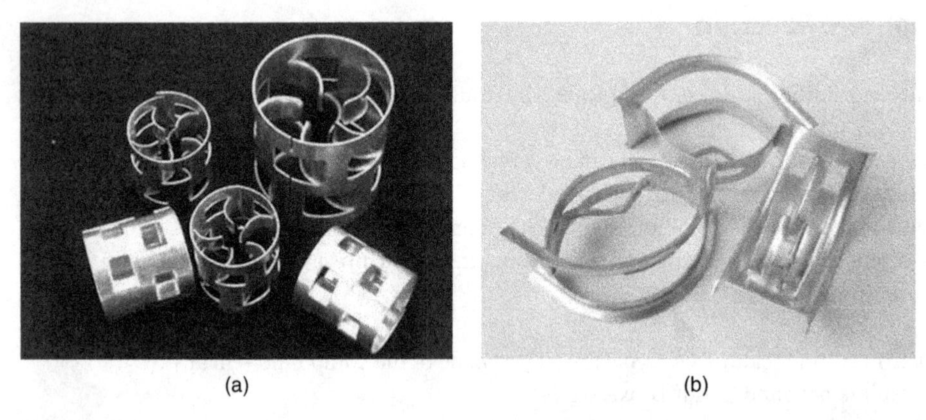

(a) (b)

Figure 8.21.1 Images of *random packing* materials that may be installed in sections of modern fractionation columns to help improve fractionation. Ransom packing is available in many designs and configurations. Two designs of random packing are shown in the figure. Remember, the issue is not with the internal packing, but with how we prepare the equipment before and during the opening of the vessel. Source: Images are courtesy of demisterpads.com (https://www.demisterpads.com/demister-pad/random-packing.html).

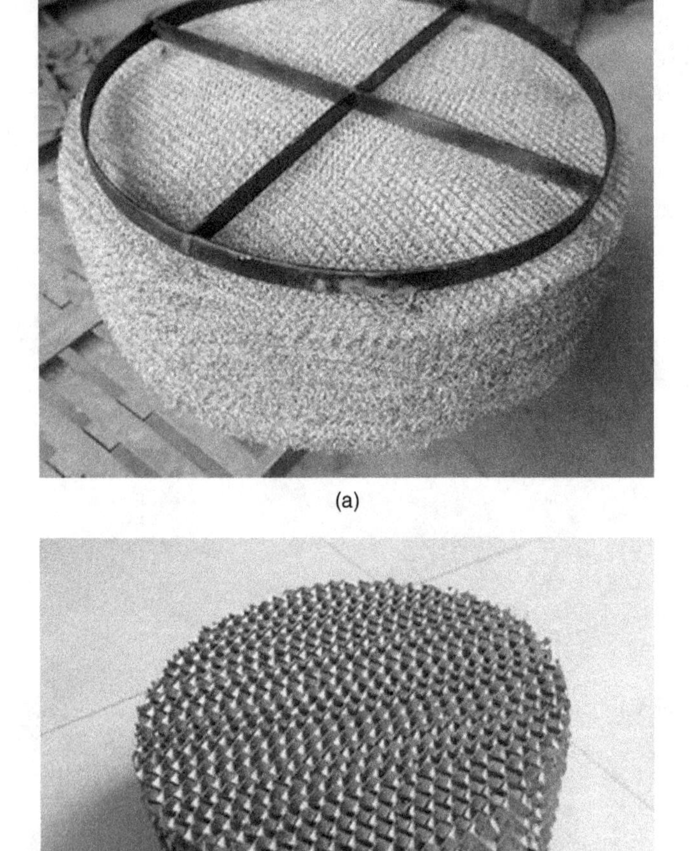

(a)

(b)

Figure 8.21.2 Images of *demister screen and structured packing* that may be installed in sections of modern fractionation columns to help improve fractionation. Due to the potential for pyrophoric ignition in the packing, special procedures are required before and during vessel opening. Source: Images are courtesy of demisterpads.com (https://www.demisterpads.com/demister-pad/stainless-steel-demister-pad.html) – (https://www.demisterpads.com/demister-pad/structured-packing.html).

may be embedded in demister screens and random or structured packing in fractionator columns. It lies dormant there for years while the tower operates and does not cause any issues. The packing materials are very dense, and it is difficult to remove the pyrophoric materials during normal preparation for opening. Steaming, water washing, and even chemical cleaning are not very effective in removing these materials.

When the fractionation tower is removed from service and opened to the atmosphere, these pyrophoric materials can ignite when coming in contact with air and ignite the metal packing or the metal demister screen. Metal fires burn at extremely high temperatures, and the total surface area of a bed with structured packing can easily exceed a million square feet. It may be virtually impossible to extinguish if an internal fire is not detected early and extinguished in the first few minutes. As a result, large fractionator columns have completely toppled over due to the internal fire burning through the shell or weakening the shell.

Images of the internal demister screen and packing materials are available in Figures 8.21.1 and 8.21.2.

Mitigating Pyrophoric Hazards?

A practical approach is ensuring the equipment is water wet when open to the atmosphere. Water flooding is generally not required; however, periodic water spraying should be considered to ensure the pyrophoric materials do not dry.

If work is planned near the demisters or internal packing, the preference is to remove these materials from the fractionator or ensure periodic water spray so that it does not dry. Another good practice is to position a charged fire hose at each manway where there is a possibility of ignition by pyrophoric materials and have someone with the responsibility to periodically walk up and down the tower spraying water into each manway. The key to preventing ignition is to keep the humidity high inside the vessel with the periodic water spray. Early fire detection can be done by continuously monitoring the temperature and carbon monoxide levels in the column while it is open to the atmosphere.

If hot work is to be done inside the tower and near the packed bed, it may be prudent to include the removal of the packing in the work plan. Otherwise, plan to build internal barriers to protect the packing against accidental ignition by the hot work. Metal fires have occurred due to spontaneous combustion from increasing the temperature inside the tower; therefore, if the packing is not removed, the periodic wetting must be continued throughout the maintenance operation.

Chemical Neutralization

Another effective approach is chemical neutralization before opening the equipment. During the final water wash step, this is typically done with a 1% potassium permanganate solution. Potassium Permanganate is reactive; ensure you are working with a reputable chemical vendor and experienced chemical engineer and that hazards have been thoroughly reviewed before use. Potassium permanganate should not be used in acid service or equipment containing low pH materials (refer to MSDS and chemical vendor).

The chemical vendors will circulate and monitor the color of the Potassium Permanganate solution (purple). Neutralization may also be done following chemical cleaning/de-oiling (e.g., Zyme-Flow). Chemical cleaning, or de-oiling alone, has proven ineffective as a neutralization agent for pyrophoric compounds. This is especially true for the pyrophoric materials deeply embedded in the demister screens or towers with internal packing. If neutralization is not used, periodic water spray will be required to prevent the potential for pyrophoric ignition (do not allow the pyrophoric materials to dry in the presence of air).

Reactors and Catalysts

For reactors and catalysts where the circulating chemical solution or water wetting is not possible, use nitrogen or other inert gases to keep oxygen out. Nitrogen hazards (potential for low oxygen environments) must be recognized and managed. Tanks, reactors, columns, and exchangers in high-sulfur service must be adequately blanketed with N_2 during idle periods.

The drums should also be inerted to displace air when loading spent catalyst into drums. A common practice is to N_2 purge the drum and then insert a piece of dry ice into the drum before it is closed to prevent oxidation of the pyrophoric material.

Sludge and Scale

Enforce the requirement that all deposits (scale/sludge, etc.) be removed from the vessel immediately after opening. Keep all materials wet until they are removed from the vessel and placed in a safe place. Remove sludge, debris, old demister screens, and packing materials to a safe place away from the unit (in the event combustion does occur when they dry). Ensure people handling material removal know that the material may be pyrophoric and understand handling procedures (keep it wet!).

Our History

At one time or another, most petrochemical sites have experienced the spontaneous ignition of iron sulfide on the ground or inside equipment, some with catastrophic consequences. Ensure mitigation procedures are included in the planning process and rigorously followed during equipment opening. Ensure that deposits that may be pyrophoric are kept wet until removed from the vessel and the unit.

What Has Happened/What Can Happen

Citgo Refinery
Lemont, Illinois
17 August 2001

The Citgo refinery, located in Lemont, Illinois, experienced a fire on the bottoms circuit of the crude distillation unit vacuum tower. The crude unit was shut down to assess and repair the damages, and the atmospheric and vacuum towers were opened to the atmosphere. Three days later, a fire started inside the bottom section of the atmospheric column, obviously due to ignition by pyrophoric material.

Coke and residual oil burned internally from the draft of air through breached piping, which ignited the internal structured packing material. The internal metal fire burned very hot, weakening the vessel shell and resulting in the total collapse of the 183-foot-tall (56-meter) atmospheric fractionation tower. No one was injured in this spectacular incident, but the unit sustained heavy damage.

Key Lessons Learned

This incident reinforces the need to remove the packing from the tower, neutralize the pyrophoric material with Potassium Permanganate, or maintain periodic water wetting of tower internals to prevent ignition by pyrophoric materials (do not allow the packing material to dry when in the presence of air). To reiterate, pressurized water hoses should be available for wetting tower internals as soon as manways are opened. Ignitions have occurred within minutes of opening the vessel, or sometimes not for hours or even days later.

What Has Happened/What Can Happen

Singapore Refinery
Internal fire during sour naphtha tank cleaning operation
21 August 2003

At this Asian refinery, a fire occurred inside a sour naphtha tank while being cleaned and prepared for inspection and maintenance. In preparation for tank cleaning and repairs, the sour naphtha was pumped down to a very low level, the tank blinds were set, and the manways were removed. Tank cleaning personnel were removing the remaining small amount of sour naphtha from the tank using vacuum trucks.

The fire started when pyrophoric material that was embedded in the tank's internal walls, and under the floating roof started smoldering, burning the paint from the tank exterior. The embers started dropping into the naphtha, resulting in several low-level explosions inside the tank and fires at the open manways. Fortunately, there were no injuries and no serious damage to the tank.

Key Lessons Learned

The refinery team had identified pyrophoric as a potential hazard during the development of the Job Safety Analysis (JSA); however, they did not go to the next step and discuss how the hazards would be mitigated. Mitigating identified hazards should be an integral part of the JSA process; keeping the humidity in the tank with water spray may have prevented the pyrophoric material from igniting.

What Has Happened/What Can Happen

Imperial Oil Refinery
Sarnia, Canada
2 April 2019

A 150-foot-tall (46 meters) fractionation column at the Imperial Oil Sarnia, Ontario refinery toppled over on 2 April 2019. The cause was determined to be an ignition in the column's bottom section due to pyrophoric materials.

The Sarnia Refinery Manager was quoted in the local media as saying the cause of the vessel collapse has been attributed to pyrophoric materials that can ignite when exposed to air. He said a design change "unknowingly increased the risk of pyrophoric," and their investigative team "was able to verify with absolute certainty that the cause was not related to the age, inspection, or maintenance of the tower." A photo of this failed column is available in Figure 8.21.3.

Key Lessons Learned

The packing materials have certainly helped with improved fractionation. However, this incident reminds us that these events can still happen today and in our facilities. Good practices such as neutralizing the pyrophoric, ensuring the equipment is maintained wet, or removing the internal packing to prevent a subsequent ignition inside the equipment must be understood and rigorously enforced. The key message is to maintain a water spray at each open manway until one of the above has been completed.

Figure 8.21.3 Photo of 150-foot-tall (46 M) fractionation column at The Imperial Oil Refinery, Sarnia (2 April 2019). Source: Photo courtesy Blackburn Media Sarnia photo by Dave Dentinger.

Figure 8.21.4 ExxonMobil Baton Rouge refinery offsite piping fire, 11 February 2020. Source: Photo courtesy of WBRZ.com.

What Has Happened/What Can Happen

ExxonMobil Refinery
Baton Rouge, Louisiana
11 February 2020

More recently, on 11 February 2020, a very large fire erupted at the ExxonMobil Refinery in Baton Rouge, Louisiana, resulting in significant damage and the resulting shutdown of several of the process units. This event happened at about 11:30 p.m. and was so large that it turned the night sky a brilliant orange that could be seen from miles away. There was also a large plume of smoke released from the fire. No injuries were reported from this incident. An image of this incident is available in Figure 8.21.4.

The release and resulting fire occurred adjacent to and between some of the process units, although unit damage was minimal. Most of the damage was reported in the massive pipe rack and associated refinery infrastructure. The refinery reported that the cause was determined to be an air purge into one of the unit's hydrocarbon rundown lines where the air mixed with hydrocarbons, resulting in ignition inside the piping. More than likely, the piping contained iron sulfide scale, which is pyrophoric, completing the fire triangle and leading to the massive fire.

On the crude units, the refinery had a practice of clearing the lube oil product lines with air between products to prevent contamination caused by mixing the various grades of lube oil. In this case, there may have been lighter material in the piping, along with iron sulfide scale, and the hydrocarbons ignited when air was injected into the pipe. The line quickly failed, damaging additional surrounding piping in the pipe rack, resulting in this very large fire.

The local news reported that ExxonMobil has agreed to ensure that the valves associated with air ingress are locked in a closed position to prevent recurrence.

A Short List of Other Known Incidents Involving Ignition by Pyrophoric Material

- BASF Total Petrochemicals, Port Arthur, TX. September 2023
  A fractionator column toppled on the pyrolysis gasoline unit at the refinery on 5 September 2023, around 10:45 a.m. The refinery reported that the investigation is ongoing; however, in all probability, this was the result of a pyrophoric fire inside the column which led to the weakening of the column and its collapse. The entire refinery was shutdown due to this incident.
- Valero Refinery; Wilmington, CA. January 2004

FCC Fractionator experienced an internal packing fire 14 hours after manways were opened. The fire distorted the tower shell and damaged the packing in two beds.

- ExxonMobil Chemical; Baton Rouge, LA. January 2001
  Titanium packing ignited by a pyrophoric catalyst destroyed a unit feed stripper tower.
- Chevron Phillips Plant; St. James, LA. February 2001
  250 Ft. column experienced internal fire resulting in the total collapse of the column. The fire was in internal packing and was reported to be started by hot metal slag. Pyrophoric material may have also been a contributor.
- Citgo Refinery, Lemont, Il. August 2001
  Fire on the Vacuum tower bottoms piping resulted in the crude unit shutdown. Three days after opening manways on the Vacuum and Atmospheric towers, a fire occurred in the Atmospheric column (in the metal packing). The 183 Ft tall fractionation column collapsed.
- Exxon Refinery; Fawley, England 1989
  Several hours after manways were opened, a fire started inside an Amine absorber. Iron sulfide on pall ring packing ignited, causing internal tower damage.
- Exxon Refinery; Baton Rouge, LA. 1974
  Fire started inside Vacuum Crude tower while personnel was working inside the vessel. All personnel escaped, no injuries, and tower damage was limited.

Additional References

Mary Kay O'Connor Process Safety Center; 2003, "Best Practices in Prevention and Suppression of Metal Packing Fires". http://psc.tamu.edu/files/library/center-publications/white-papers-and-position-statements/metalfires.pdf

Koch-Glitsch; 2001, "Minimize the Risk of Fire During Distillation Column Maintenance". http://folk.ntnu.no/skoge/prost/proceedings/distillation10/DA2010%20Sponsor%20Information/Koch%20Glitsch/Technical_Articles/Packing/Minimize_Risk_of_Fire.pdf

Work Safe BC – Work Safe Bulletin. http://www.ecolog.com/daily_images/1003890388-1003892086.pdf

"CHEMICAL CLEANING AND DECONTAMINATION OF REFINERY AND PETROCHEMICAL DISTILLATION EQUIPMENT", B. Otzisk1 and Dr. M. Urschey2, 1 Kurita Europe, Industriering 43, 41751 Viersen, Germany (otzisk@kurita.de) 2 Kurita Europe, Niederheider Strasse 22 (Y20), 40589 Duesseldorf, Germany. http://docshare01.docshare.tips/files/27264/272648301.pdf

Causes and Prevention of Packing Fires (White Paper), Chemical Engineering, By Design Practices Committee Fractionation Research Inc. | 1 July 2007, Causes and Prevention of Packing Fires – Chemical Engineering | Page 1 (chemengonline.com).

WBRZ News Report "ExxonMobil reveals cause of February blaze at Baton Rouge refinery.", 13 April 2020, 2:29 p.m. in News. https://www.wbrz.com/news/exxonmobil-reveals-cause-of-february-blaze-at-baton-rouge-refinery/

Mach Engineering Structured Packing Article, "Selecting a Suitable Structured Packing" and "Tower Internals". https://www.machengineering.com/selecting-a-suitable-structured-packing/

Demister pad for removing liquid droplets from gas or vapor streams, Wire Mesh Demister Pads for Liquid and Gas Separating Problems.

Chapter 8.21. Pyrophoric Ignition Hazards, Including Fractionation Columns with Internal Packing

End of Chapter Quiz

1 What is the definition of a pyrophoric material?

2 Is this the real hazard of a pyrophoric, or is it something even more significant?

3 Pyrophoric ignition is a concern in many petroleum refining or petrochemical processes, especially where _________ is present in low Oxygen environments.

4 Pyrophorics are typically found in process equipment containing Hydrogen Sulfide in concentrations >1% (but can form in lower concentrations). Corrosion products such as iron scale or rust combine with the sulfur compounds creating _____ _______ _______ (FeS), which is pyrophoric.

5 What is another less-known hazard involving a pyrophoric fire in a confined space?

6 We should expect to find pyrophoric materials in process equipment and areas where sulfur or sulfur compounds like Hydrogen Sulfide (H_2S) will be present. For example, H_2S will go overhead, so not a surprise that pyrophoric materials will be found in overhead systems like flare systems.

7 Where can pyrophoric materials typically be found? Name all you can.

8 How is the simplest and most effective way to control pyrophorics?

9 What about fractionation column internal packing or demister screens?

10 The bottom line is that pyrophoric materials can be __________ and require our attention to prevent uncontrolled spontaneous combustion inside or near process equipment.

8.22

Hazards of Static Electricity

Static Electricity has resulted in or contributed to many significant process safety incidents in petroleum refineries and petrochemical plants, and static has been identified as the ignition source in many fires and explosions. In many cases, static electricity was not the primary cause of the event but contributed as the ignition source to ignite the resulting vapor cloud following an accidental loss of containment. For example, static was identified as the primary cause of an 80,000-barrel floating roof tank fire at the ConocoPhillips Glenpool, Oklahoma pipeline terminal near Glenpool, Oklahoma, on 7 April 2003. The Glenpool Fire destroyed the tank, damaged two other nearby tanks, and resulted in the evacuation of the surrounding community. Schools were closed for two days because of the incident.

Many hydrocarbons and other products will generate static electricity. Some hydrocarbon products will accumulate static electricity in the product. For those products, a spark can occur when an object, such as a gauging tape or sample device at a different electrical potential, is lowered into the product. When this happens, and when the vapor space is in the flammable range, the ideal ratio of fuel and air can result in a catastrophic explosion and a significant fire.

The following two sentences help frame the hazards associated with static electricity:

- Static electricity is generated when liquids come in contact with other materials. It is a common occurrence when liquids are moved through nozzles, pipes, mixed, poured, pumped, filtered, or otherwise agitated.
- Other causes include the settling of solids or an immiscible liquid through a liquid, the ejection of particles or droplets through a nozzle, and the splash or spray of a liquid against a solid surface, for example, in a storage tank or railcar.

This is what we do in a petroleum refinery or a petrochemical plant. We transfer products, blend products, load products into barges, ships, railcars, and other conveyances, and many other operations where controlling static electricity is extremely important.

This module will cover static electricity hazards, discuss several additional significant incidents involving static electricity, and discuss how static electricity can be effectively managed in oil refineries and chemical plants.

The good news is that we have two reference documents that help us understand the static electricity hazards and provide the procedures and tools to help us manage these hazards.

The National Fire Protection Association code 77 (NFPA-77) "Recommended Practice on Static Electricity" states that "under certain conditions, particularly with liquid hydrocarbons, static may accumulate in the liquid," with the danger of subsequent sparking in a flammable vapor-air mixture. This code goes on to provide the practices to help manage static electricity.

The American Petroleum Institute Recommended Practice 2003 (API 2003) "Protection Against Ignitions Arising out of Static, Lightning, and Stray Currents" presents the technology and know-how used to prevent hydrocarbon ignition by static electricity, lightning, and stray currents.

In the United States, API RP-2003 has been adopted by reference to the US Code of Federal Regulations (49CFR 195.405(a) or the transportation of hazardous materials by pipeline) and by the USCG 33CFR 127.1101(h) or piping systems transporting flammable liquids or vapors. Regarding these facilities in the United States, API RP-2003 is the law. Similar regulations are in place in many countries around the world. Note: Many of the requirements of API-2003 are being replaced by API-545 Recommended Practice of Lightening Protection of Above Ground Storage Tanks for Flammable or Combustible Liquids. Please refer to the new standard for these requirements.

Process Operations Safety: The What, Why, and How Behind Safe Petrochemical Plant Operations, First Edition. M. Darryl Yoes.

Static accumulators	Non accumulators
Non-conductive liquids	*Conductive liquids*
• Jet fuel • Solvents • Diesel (especially ULSD) • Naphtha • Gasoline • Hexane • Benzene • Toluene • Xylenes • Other refined products such as lube base stocks	• Black oils • Crude oils • Residual fuel oil • Bitumen / asphalt • Alcohols, ketones • Water

Figure 8.22.1 Conductivity of typical petroleum products.

Electrical Conductivity Adds to Static Accumulation in the Liquid

The amount of static accumulation in liquids is directly related to the electrical conductivity of the liquid. Electrical conductivity measures how well a product accommodates the transport of an electric charge, which is the movement of charged particles through a product in response to an electric field. Conductive liquids typically do not represent significant static hazards as they can rapidly dissipate a static charge through the liquid to the tank's steel or conveyance and the earth. However, nonconductive liquids will generate static and accumulate a significant static charge in the liquid. Nonconductive liquids cannot quickly discharge the static charge; therefore, the static charge will accumulate in the liquid. Static build-up in the liquid can result in a static discharge to the tank's steel or another object and ignite a flammable atmosphere.

Electrical conductivity is measured in Pico siemens/meter (pS/m), typically indicated on the product Material Safety Data Sheet (MSDS). The prefix "p" represents pico and equals one trillionth or 10^{-12}. For example, the conductivity of typical diesel fuel is about 0.5–50 pS/m. A liquid is typically considered nonconductive if its conductivity is below 100 pS/m and semi-conductive if its conductivity is below 10,000 pS/m. Whether a liquid is nonconductive or semi-conductive, the precautions are the same. Several factors, for example, liquid temperature, the presence of contaminants, and the use of anti-static additives, can significantly influence the conductivity of a liquid. Figure 8.22.1 indicates whether typical petroleum products are conductive or nonconductive based on current MSDS data.

Static in Mists, Aerosols, and Dust Clouds

Mists, aerosols, and dust clouds are also static accumulators, and static can also play a significant role in explosions involving accumulations or clouds of combustible dusts. Static electricity accumulated in a dust cloud can result in an arc that can ignite the dust cloud. Any combustible dust can explode when dispersed into the air in fine particulates. Dusts that you may think could not possibly explode can be extremely explosive under the right conditions. Of course, any other arcing or sparking device or process can ignite a cloud of combustible dust; however, the cause is frequently suspected of static electricity.

Examples of types of dust that are combustible and may represent significant static and explosion hazards include:

- Petroleum coke dust or particulates.
- Plastic dust or particulates.
- Coal dust or particulates.
- Metal dust.
- Grain dust (sugar, corn, rice, wheat, etc.).
- Wood dust.

The molecules are separated by air or vapor at the molecular level in a dust cloud, making a combustible dust cloud particularly hazardous. The cloud is conductivity-independent, and the method of charge decay is by gravity settling or coalescing. The static charge cannot readily dissipate, and a dust cloud can retain a static charge for as long as five hours. While the charge is present, there is a real possibility of a dust explosion.

In the petroleum and petrochemical industry, coke dust and plastic dusts are special hazards, and many incidents have occurred, especially with plastic dust. Explosions and fires have occurred due to plastic dust build-up, with the cause frequently being static electricity due to a breakdown in electrical bonding and grounding.

High Flow Rates (Velocity) and Micron Filters = Higher Static Charge

Higher flow rates or velocities can result in a higher static charge build-up in a flowing hydrocarbon system. For example, a micron-size filter placed in a flowing pipeline system can significantly increase the filter's velocity and result in about 500 times the static build-up in the system. We frequently use micron filters in flowing hydrocarbon systems, such as jet fuel delivery systems.

Generally, three classifications of filters may be used in flowing hydrocarbon systems as follows (Source: International Safety Guide for Inland Navigation Tank-barges and Terminals; Chapter 3 "Static Electricity"):

- Course (greater than or equal to 150 microns):
  These do not typically generate a significant charge and require no additional precautions if kept clean.
- Fine (less than 150 microns, greater than 30 microns):
  These can generate a significant charge and typically require a minimum of 30 seconds of residence in the piping downstream of the filter. Flow velocity must be controlled to ensure this residence time.
- Microfine (less than or equal to 30 microns):
  These filters are a significant static hazard. The residence time to allow enough time for the charge to relax downstream of microfine filters is typically a minimum of 100 seconds before the liquid enters the tank or conveyance. Again, flow velocity must be controlled to ensure adequate residence time.

A Low-level Charge Can Ignite Flammable Hydrocarbon and Air Mixtures

An extremely small electrical charge or spark can ignite a flammable hydrocarbon and air mixture. The electrical energy is measured in Joules, or Millijoules, which is one-thousandth (10^{-3}) of a Joule (symbol: mJ). One Joule is equivalent to the energy dissipated as heat when an electric current of one ampere passes through a resistance of one ohm for one second. A Millijoule is equal to about one-thousandth of this energy, or an extremely small amount of electrical energy.

The Minimum Ignition Energy (MIE) of most flammable gases such as gasoline or petrol, methane, ethane, propane, butane, and benzene is usually about 0.1 mJ (according to Lewis and von Elbe; authors of Combustion, Flames, and Explosions of Gases, third edition) (see references below). A mixture of Hydrogen and air requires an even smaller electric charge or MIE, requiring only 0.019 mJ or only about 1/10 of other flammable gases. Materials with a MIE less than 30 mJ are susceptible to static discharge, which in turn causes flash fires and explosions.

The bottom line is that an extremely small amount of electrical energy is required to ignite a flammable mixture, less than what you feel when you have an electrical spark from your fingertips when you walk across the carpet and touch a doorknob. The MIE required to ignite a flammable hydrocarbon and air mixture is available in most MSD sheets; however, the electrostatic discharge energy required for ignition varies significantly over the flammable range (LEL to UEL) and is at its lowest at the midpoint of the flammable range. See Figure 4.10, an image that illustrates how little static charge is required to ignite a mixture of hydrocarbons when they are in the flammable range.

Electrical Bonding and Grounding

Electrical bonding and grounding are relatively simple practices to eliminate the charge between objects and avoid arcing and sparking. Bonding is the cable placement between two objects to ensure no electrical potential. An example of bonding is the cable placement between a railcar loading line and the railcar, eliminating the opportunity for an arc between the two. Grounding, sometimes referred to as Earthing, connects a conductive cable between an object, such as a railcar, and the earth. Grounding is typical to a rod driven into the ground. Grounding effectively dissipates the electrical charge into the earth. I always prefer to do both bonding and grounding to ensure the charge is eliminated, and there is no electrical potential between the two objects that can result in an arc or spark.

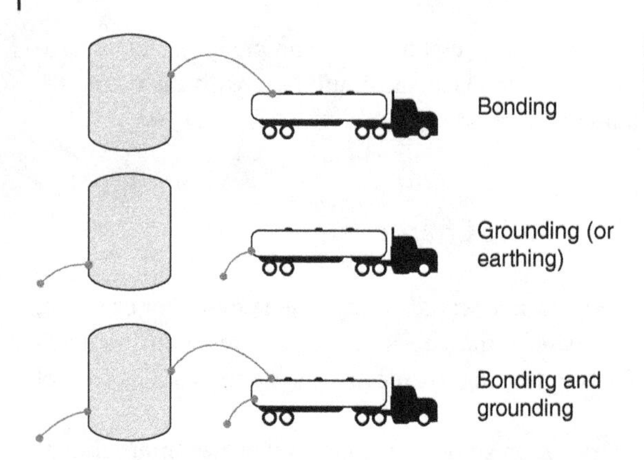

Figure 8.22.2 Illustration of proper electrical bonding and grounding. Source: Illustration courtesy of U.S. Chemical Safety and Hazard Investigation Board (annotations are by the CSB).

For example, when a tank truck or railcar arrives at the loading rack to be loaded or unloaded with a petroleum product, the bonding and grounding cables should be the first things that go on and the last things to be removed. These cables should always go on before any tanks or compartments are opened, and they should remain in place and undisturbed for the loading or unloading until the tanks are securely closed. This same concept is true for other operations involving transfer, loading, filling, or other movements involving hydrocarbon products. Figure 8.22.2 from the US Chemical Safety and Hazard Investigation Board illustrates proper bonding and grounding.

Are we perfect in this area? A review of 310 accidents by the Japanese chemical industry found that improper grounding caused 70% of all static electricity accidents. These incidents involved the petroleum industry, specifically hydrogen, natural gas, and liquefied petroleum gas. We may find room for improvement in bonding and grounding worldwide if the data were available.

Hazards of "Switch Loading" (Loading a Combustible Product onto a Flammable Heel)

Switch loading is defined as loading a high-flash product such as kerosene, jet fuel, or diesel after loading a low-flash product such as gasoline or naphtha or loading onto a low-flash heel. Switch loading has been defined as one of the primary causes of storage tank or vessel fires due to static charge build-up in the liquid and the flammable atmosphere inside the vessel. For example, switch loading along with splash filling has been estimated by ASTM (ASTM International, formerly known as the American Society for Testing and Materials) to be the fundamental cause of approximately 80% of loading rack fires. The problem created by switch loading is that a relatively safe fuel, like diesel fuel, kerosene, or heating oil, with relatively high-flash points can become contaminated with a low-flash product like naphtha or gasoline. The accumulation of static electricity is also accelerated in a switch-loading operation. The hazard with switch loading is that the atmosphere inside the vessel passes through the flammable range very slowly, and should an arc occur from static electricity, the flammable atmosphere can explode, or a fire can result.

So how does switch loading result in ignition inside a vapor space? Most petroleum cargo vessels are divided into separate compartments or tanks, with some compartments handling flammable products like gasoline and others handling combustible products like diesel fuel. Due to vapor pressure and the low-flash point, gasoline vaporizes at ambient temperatures. Therefore, the atmosphere inside the gasoline tanks is "too rich" to ignite, and although arcing from static electricity can occur, the product cannot ignite. On the other hand, diesel fuel has a much lower vapor pressure and a much higher flash point and is not vaporizing at ambient temperatures; therefore, the atmosphere inside the diesel tanks is "too lean" to ignite. Again, in this case, ignition cannot occur even in the event of arcing or sparking due to static discharge; ignition cannot occur. However, when switch loading, a flammable product is in the tank; therefore, the atmosphere contains flammables.

During switch loading, as the load begins, the air is brought in with the higher flash product, and some of the low-flash vapors are displaced from the compartment. This results in the tank atmosphere in the vapor space quickly going into the flammable range and an explosive environment in the vapor space where the fuel-to-air concentration is between the Lower Explosive Limit (LEL) and the Upper Explosive Limit (UEL). The issue with switch loading is that the vapor space stays in the flammable range throughout the loading operation. Any arcing or sparking due to static discharge can result in an explosion or fire.

Switch loading should normally be avoided and is restricted by law in some jurisdictions. This may require the tank truck or barge to be cleaned before loading. If switch loading is considered without tank cleaning, some additional considerations should be in place before loading, for example, reduced loading rates to minimize the splash and spray, which can result in static charge build-up, purging the vessel or tank atmosphere with an inert gas such as Nitrogen, or use of an anti-static additive. However, even with these additional precautions, the risk exists when switching loading. Refer to ASTM Standard D-4865 ("Standard Guide for Generation and Dissipation of Static Electricity in Petroleum Fuel Systems") for additional information on switch loading operations.

Insulated Cargo Transfer Piping or Hoses

Generally, the cargo transfer piping from the onshore terminal should be electrically insulated before the piping crosses onto the offshore loading facility, such as a marine terminal. This is to help ensure that any electrical potential between the onshore and offshore facilities is eliminated and current that could ignite flammable vapor is not allowed to pass between the two.

The primary reason for the insulated flanges is that many cargo vessels like ships and barges are equipped with onboard electrical power generators, and in the event of a short circuit, the electric energy could be passed to onshore facilities. Another reason is that most offshore facilities, like pipelines, are equipped with electrical cathodic protection, which could also transmit an electrical charge to the onshore facility. The insulated flanges also help prevent ignition from static electricity.

Insulating flanges, joints, or sleeves are required by US Coast Guard Regulation 35.35-4 for flammable liquid or vapor cargoes. These divide the cargo piping or hoses into electrically isolated halves (onboard and shoreside), allowing the vessel and shoreside facilities to be at different electrical potentials. The regulation states "The insulating flange must be inserted at the jetty end and take all reasonable measures to ensure the connection will not be disturbed. The hose must be suspended to ensure the hose to-hose connection flanges do not rest on the jetty deck or other structure that may render the insulating flange ineffective or short-circuited by contact with external metal or through the hose handling equipment. The insulating flange must be inspected and tested at least annually, or more frequently if necessary due to deterioration caused by environmental exposure, usage, and damage from handling." The testing requirements are provided in more detail in the regulation.

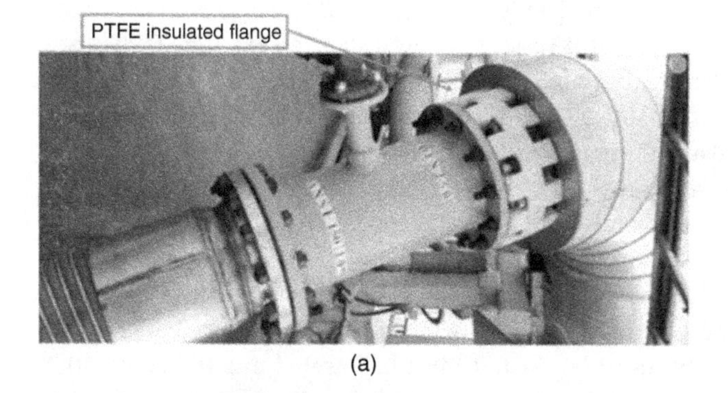

(a)

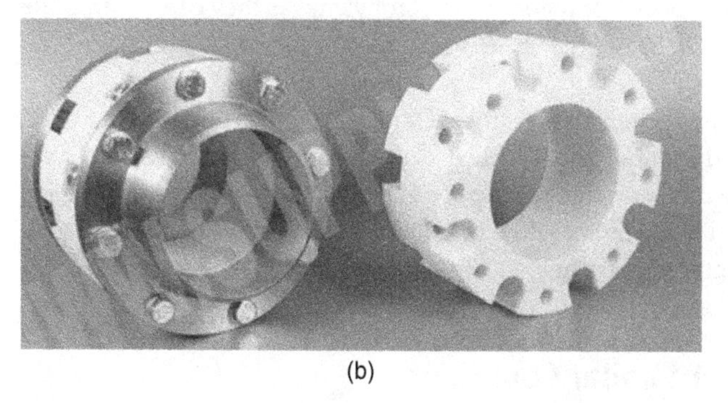

(b)

Figure 8.22.3 Examples of insulated flanges for metal cargo hoses and other forms of petroleum or petrochemical product transfers. Note that the bolts do not penetrate the nonconductive material, therefore: the cargo hose is completely insulated from the vessel being loaded or offloaded. Source: Images courtesy of TMS Sia. (www.supplier.lv) and Ftoroplasts.lv.

The insulated flange should be readily identifiable with signage or tags, and the insulated flanges' integrity should not be compromised during maintenance. See Figure 8.22.3a and b for photos of insulated flanges designed for marine terminals handling hydrocarbon cargos.

Precautions for Loading or Product Transfer (to Minimize Static Buildup or Static Discharge)

Bond and ground the loading line or arm to the conveyance and from the conveyance to the ground. The bonding and grounding cables should be the first thing that goes on before any cargo tanks are opened and the last thing to come off after the cargo tanks are closed.

Avoid switch loading, loading a combustible liquid like diesel onto a flammable heel like gasoline.

When loading into an empty barge, tank car, tank truck, or another tank, minimize the potential for splash and spray and the resulting static charge by limiting the initial loading rate until the fill line is well covered. The loading or transfer procedure should ensure the fill line is covered by at least 2 feet (61 centimeters). This is usually accomplished by loading by gravity flow until the fill line is well covered by about 2 feet of liquid. For floating roof tanks, the fill rate should be maintained at a low velocity until the roof is floating. Again, gravity flow normally accomplishes this until the roof is floating.

The product flow should be limited to a maximum of 3 feet per second (1 meter/second) at the tank inlet for the whole operation unless the product is "clean." In this context, a "clean" product is defined as one that contains less than 0.5% by volume of free water or other immiscible liquid and less than 10 mg/liter of suspended solids.

Avoid splash filling by using bottom connections when loading railcars and tank trucks. If bottom connections are unavailable, avoid splash filling by using a long fill tube extending to the tank bottom with a fill diffuser nozzle in contact with the tank bottom.

Precautions for Sampling and Gauging Storage Tanks and Conveyances (Ships, Barges, Tank Cars, etc.)

Procedures should not allow sampling or gauging during loading or immediately after loading, as this may produce a static spark. The procedures should specify a "relaxation" period for the tank or vessel to allow a static charge time to dissipate and prevent an arc or spark during the sampling or gauging. The relaxation period should be 30 Minutes for tanks, marine vessels, and tank cars/trucks with volumes >10,000 gallons (38 Cu Meters) or 5 minutes for smaller volumes. Tank mixers, if equipped, should also be idled during the relaxation period and during sampling or gauging.

The procedures should ensure that gauging is conducted within the approved tank "stilling well" or "sounding tube." This pipe is designed to shield any potential static build-up in the tank from the sampling and gauging equipment. The sampling and gauging equipment should be all conductive and bonded to the tank before it is placed in the vapor space. It is best always to use only commercially available sampling and gauging equipment that is all conductive and has a bonding cable attached. The bonding cable should be connected to the tank or conveyance before anything is allowed to enter the vapor space. If a nonconductive sample and gauging equipment must be used, it must be lowered into the tank with a nonconductive rope (e.g., dry cotton rope). Never use synthetic ropes like nylon or polypropylene, as they can accumulate a static charge and can result in arcing or sparking during use. See Figure 8.22.4 for an example of an improper tank gauging tape.

Ensure there are no loose objects in the tank, and do not drop things in the tank. If you drop something into the tank, make sure it is reported. Tanks containing loose objects should normally not be filled as the object can accumulate a static charge and arc or spark during filling operations (e.g., sample cans or other sample containers, float balls, etc.). The object will be charged from the fluid, and there is the potential for a static spark when the object de-energizes against the tank wall or other tank internals. Metal objects will often give a high-energy spark.

Sampling or Draining Using Small Pails or Other Similar Containers

Most of the above guidelines apply to sampling or draining hydrocarbons into small containers. Metal pails or containers should be grounded and bonded before starting the flow of hydrocarbons. Do not use a metal can or bucket with a plastic or

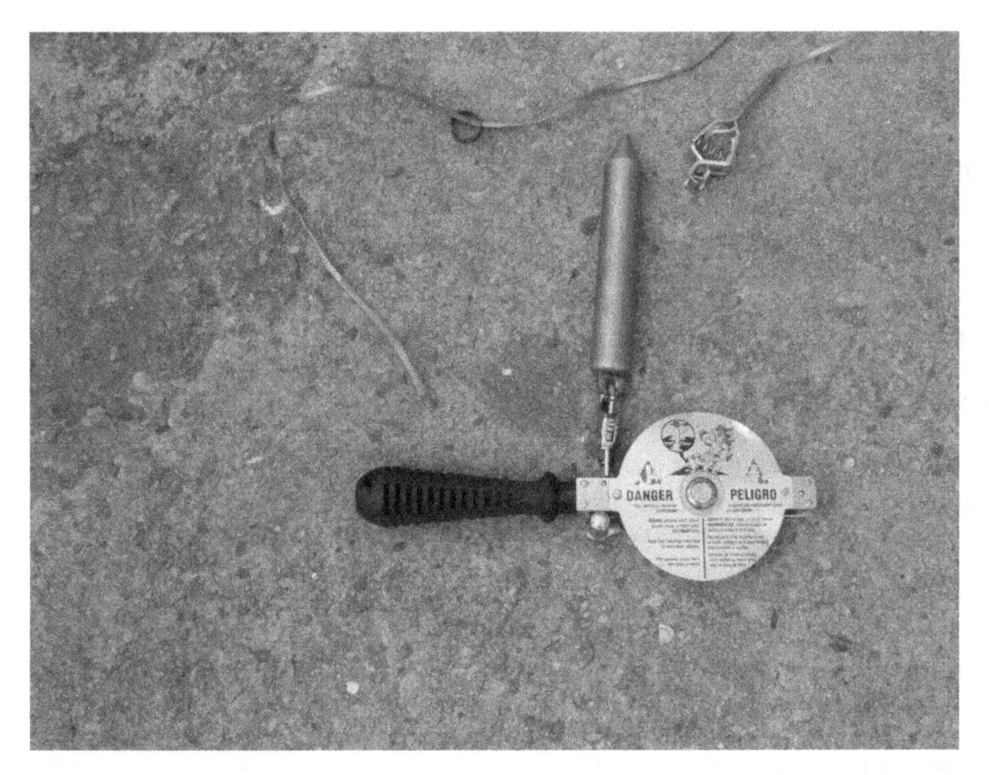

Figure 8.22.4 An improper tank gauging tape. Note that the bonding cable has become detached from the gauging tape. Source: Use of this device without a bonding cable can be deadly!

other nonconductive handle and ensure the relaxation time is followed before sampling or gauging (5 minutes for volumes less than 10,000 gallons (38 Cu Meters) and 30 minutes for larger volumes).

To minimize the splash and spray that can generate static electricity, the National Fire Protection Association Recommended Practice (NFPA 77) recommends that small tanks such as totes be filled with a bottom fill connection. If a bottom fill connection is unavailable, filling with a dip pipe extending to the container's bottom is recommended. The container should be filled at a velocity not to exceed 3.3 feet per second (1 meter per second) or less until the end of the dip pipe is covered by 6 inches (150 millimeters).

Plastic or synthetic containers like polypropylene should never be used as drain pans or for other containment of flammable hydrocarbons. Some sites have developed a practice of using plastic child's swimming pools as drain pans. These types of plastic containers can generate static electricity and can serve as the ignition source for flammable vapors that may be present in process areas.

Medium Volatility Products; Another Static Hazard

Low volatility products such as diesel or kerosene do not evolve enough vapors to create an explosive fuel/air flammable mixture at normal tank temperatures, so the vapor space, when at ambient temperature, is not at risk of explosion or fire. Gasoline and other high-volatility products rapidly create a vapor-rich environment, even at ambient temperature, and therefore are too rich to ignite, so the vapor space is not at risk of an explosion or fire. However, medium volatility products may create the perfect atmosphere with the correct ratio of fuel to air for an explosion or fire to occur, even at normal tank temperatures. Therefore, the vapor space above a medium volatility product may be just the right atmosphere for an explosion to occur when a small arc or spark occurs from static electricity. This atmosphere can occur from things not in our control, such as the ambient temperature, atmospheric pressure, temperature of the product, etc. Examples of medium volatility products include medium boiling range naphtha's like paint and varnish naphtha's, jet A, and xylene.

Medium volatility products, especially those with low conductivity, require special safety precautions, such as storage in a floating roof tank to eliminate the vapor space. Additional mitigation should be considered if a medium volatility product is stored in a cone roof tank or any vessel with a vapor space. This includes tank blanketing with nitrogen or natural gas to eliminate the potential for a flammable atmosphere, anti-static additives or conductivity improvers to minimize the potential for static discharge, limiting pumping or transfer rates, and ensuring all tank internals are properly bonded. We can safely handle medium volatility products with the proper precautions and mitigation.

Lightning, Another Significant Static Hazard

Lightning is a naturally occurring phenomenon that can result in a devastating event under the right circumstances. Lightning is static electricity on steroids! It is estimated that 95% of floating roof storage tank rim-seal fires result from lightning strikes. Data suggests that 0.16% of all storage tanks with rim seals will experience a rim-seal fire in any given year (Source Breitweiser, C., "AST Lightning Protection – API 545 Update," American Petroleum Institute Tank Conference Proceedings, 2008). 0.16% is not a very large number, but think about how many floating roof tanks there are worldwide, which becomes a very large number. See Figure 8.22.5 for a photo of a floating roof storage tank fully involved in a fire following a lightning strike.

Storage tanks must have supplemental grounding if they are not "direct grounded" (API 2003). This is a critical recommendation since many storage tanks are resting on a concrete or asphalt pad and are not directly grounded. The storage tank grounding and bonding cables should be inspected periodically, including conductivity and resistance testing. Ground cables found defective, damaged, or loose must be repaired or replaced.

Floating roof tanks must have roof shunts to electrically bond the floating roof to the tank shell (NFPA-780). Each steel shoe section shall be electrically bonded to the roof with stainless steel shunts spaced at periodic intervals. The shunts should be inspected periodically to ensure they are not corroded or damaged and are in contact with the tank shell. The floating roof seal must be tight-fitting against the shell. Note: Many of the requirements of API-2003 are being replaced by API-545 Recommended Practice of Lightening Protection of Above Ground Storage Tanks for Flammable or Combustible Liquids. Please refer to the new standard for these requirements.

Officials reported a lightning strike ignited a storage tank fire at the Puerto La Cruz, Venezuela, refinery. The massive tank fire resulted in a significant release of black smoke and an evacuation of the surrounding community.

At the Orion Norco, Louisiana Refinery on 7 June 2001, lightning struck this massive 270-foot diameter (82 Meter) floating roof gasoline storage tank resulting in a full-surface tank fire with large plumes of black smoke. The incident occurred during Hurricane Allison, which also struck the refinery with heavy winds and deluge rain amounts.

Figure 8.22.5 Floating roof storage tank fully engulfed in a fire following a lightning strike. Source: Photo courtesy of Williams Fire and Hazard Control.

This still stands as the largest full-surface tank fire to be extinguished, which was done by the Williams Fire and Hazard Control company.

What Has Happened/What Can Happen

Shell Singapore Refinery Fire
Pulau Bukom Manufacturing Site
28 September 2011

On 28 September 2011, a devastating fire occurred in the Shell Singapore Pulau Bukom Refinery. According to local media reports, the fire occurred in the "pumphouse," or gasoline blending area. Apparently, Shell employees and contractors were preparing for work to be performed on the gasoline blending equipment. The spectacular fire was visible for miles around the site, forcing the large refinery to shut down operations. Shell declared force majeure because the company could not provide products to their customers.

The Singapore Ministry of Manpower's investigation identified lapses in workplace safety in three areas: flammable naphtha vapor had accumulated because of the naphtha drainage process, the use of plastic trays to collect naphtha resulted in static charges that could have sparked the naphtha vapor, and the failure to provide gas monitors to detect dangerous levels of naphtha vapor.

Shell prepared a safety bulletin highlighting what led to the incident, the valuable lessons learned, and images of the aftermath of the fire, including the damaged gasoline blending area. I felt this was a fantastic way to communicate the key learnings from this incident to the rest of the industry. Unfortunately, I could not obtain permission from Shell to use the information or images in this book. I was told that the information was for "internal use only." After several attempts, I was forced to rely on the Ministry of Manpower, local media, and my experience to describe what could go wrong during an operation of this type.

At most refineries, the gasoline blend area is comprised of multiple large-diameter pipes containing highly flammable naphtha, which come together to form the gasoline blend headers. This means that when preparing for maintenance, these large pipes must be drained and prepared for opening. This can be done in several ways, including draining into drain pans and collecting the material with portable pumps to closed systems or by using vacuum trucks to collect the product and properly dispose of it in tanks or other closed systems, or by a combination of these.

This can introduce multiple hazards into the worksite, mainly the potential accumulation of flammable vapors, and most importantly static electricity from flowing liquids. When flammable vapors are allowed to accumulate, it only takes a small spark to result in a devasting fire. A spark can come from an operating engine, such as a vacuum truck or pump motor, or a static discharge caused by the flowing liquids.

When carrying out operations of this type, it is important to continuously monitor for the accumulation of flammable vapors by conducting continuous gas monitoring using instruments equipped with both audible and visual indicators, typically horns and lights, providing a warning to the workers when flammable vapors are present. Also, it is essential that all hoses used for product transfer are equipped with internal bonding cables to protect against static sparks. All equipment, including transfer hoses and the vacuum truck if one is used, should be grounded to help carry away any accumulated static electricity.

Plastic trays or plastic containers of any type should not be used to drain flammable products. Plastic, being a nonconductor, will accumulate an electrical charge and can result in arcing or sparking, which is a potential source of ignition. An industry best practice is to only use metal containers for the accumulation of flammable products and to ensure that each container is properly electrically bonded and grounded to prevent static accumulation and to drain away any static that does accumulate.

Key Lessons Learned

This incident provides a good reminder about using plastic or any synthetic material to drain hydrocarbons or to accumulate or store hydrocarbons. Plastic materials cannot be bonded or grounded since they are nonconductive, and they will generate and accumulate static electricity and serve as an unwanted insulator in the system. In my experience, I have been made aware of Operators using plastic "kiddie" pools (plastic pools designed as swimming pools for children) as drain pans for hydrocarbons. This should not be allowed or tolerated due to the significant safety issues due to static electricity. Draining flammable product into polyethylene drums has the same type of hazards and should not be used for this purpose.

Open drain systems where flammable vapors are released into the atmosphere should be avoided to the extent possible. Draining light hydrocarbons into closed systems can eliminate any accumulation of vapors in the area and significantly reduce the possibility of ignition or fire.

Ensuring periodic inspection and maintenance of emergency equipment is essential to ensure equipment availability when needed. Fire monitors, turrets, water deluge systems, and fixed foam systems must function properly when called upon.

Ensure flammable hydrocarbon detectors are in place at key locations during this type of operation. Flammable hydrocarbons are heavier than air and will tend to accumulate in low areas. Ensure flammable detectors are in service near the vacuum trucks and especially in the lower elevations where the hydrocarbons would tend to accumulate. Hydrocarbon detectors should be equipped with both audible and visual indications of flammable vapors being present.

What Has Happened/What Can Happen

ConocoPhillips Pipeline Terminal
Glenpool, Oklahoma
Storage Tank Explosion and Fire
7 April 2003
Investigated by the U.S. National Transportation and Safety Board

An 80,000-barrel storage tank at ConocoPhillips Company's Glenpool South tank farm in Glenpool, Oklahoma, exploded and burned as it was filled with diesel. The tank was being "switch loaded" in that the prior cargo was gasoline which had been removed from the tank earlier in the day. The resulting fire burned for about 21 hours, damaging two other nearby storage tanks and resulting in the evacuation of residents in the surrounding area and closing of the local schools for 2 days. There were no injuries from this incident. The National Transportation Safety Board investigated this incident (NTSB), and the report "Storage Tank Explosion and Fire in Glenpool, Oklahoma, 7 April 2003" (Pipeline Accident Report NTSB/PAR-04/02) was issued with recommendations to help prevent a recurrence of the incident.

At 8:33 p.m., Explorer Pipeline Company started a batch of 24,500 barrels of diesel to tank 11 at an initial delivery rate of 24,000–27,500 barrels per hour. Tank 11 exploded at 8:55 p.m. and erupted into a major fire. At the time of the explosion, the tank contained around 55 barrels of gasoline in the tank sump and another 7,500 barrels of diesel, which was filled on top of the gasoline. The tank was being filled at a high velocity, with the floating roof on the legs and a vapor space under the roof. The tank was being switch loaded, loading diesel after a cargo of gasoline.

Filling at high rates with the roof on the legs results in splash and spray under the roof and can generate a significant amount of static electricity. When this occurs during switch loading, the vapor space under the roof is in the flammable range, the ideal ratio of fuel to air that can result in an explosion or fire. The NTSB determined that the probable cause of the explosion and fire was the ignition of a flammable fuel–air mixture within the tank by static electricity.

Please see Chapter 8.8, "Storage Tank Safety," for images of the ConocoPhillips Pipeline Terminal storage tank fire.

Key Lessons Learned

Switch Loading

When a tank is filled with a volatile high vapor pressure product like gasoline at ambient temperatures, the product rapidly vaporizes and results in a vapor space that is too rich to ignite. Even though static electricity may be present, ignition cannot occur. When a tank is filled with low vapor pressure and higher flash product such as diesel, the diesel does not vaporize at ambient temperatures due to the higher flash point; therefore, the vapor space is too lean to ignite. In either case, when filling the tank with gasoline or diesel, ignition cannot occur. However, filling with a high-flash combustible product like diesel over a low-flash flammable product like gasoline results in a vapor space that goes into the flammable range in the first few minutes of the loading and remains in the flammable range throughout the filling operation. Should static occur during switch loading, an explosion or fire can result.

Procedures should be in place and rigorously enforced to prevent switch loading, where possible, and outline procedures to follow in the event switch loading does occur, including limiting filling rates, the use of a static inhibitor, or inerting the tank vapor space, especially when the floating roof is on the legs with a vapor space under the roof.

Limiting Fill Velocity During the Initial Fill

To guard against such explosions, the American Petroleum Institute Recommended Practice (API RP-2003) recommends limiting the fill line and discharge velocity of the incoming liquid stream to 3 feet per second (1 meter/second) until the fill pipe is submerged by either two pipe diameters or 2 feet (61 centimeters), whichever is less. In the case of a floating roof tank (internal or open top), observe the 3 feet/second (1 meter/second) velocity limitation until the roof becomes buoyant (until the roof is floating).

Landing a Floating Roof on the Legs

In a refinery or terminal such as this one, avoiding "landing" a floating roof on the roof legs and creating a vapor space under the roof is good. When landing on a roof, the roof vent opens and allows air into the vapor space to break the vacuum and prevent damaging the roof. This can result in a fuel and air mixture within the flammable range in the vapor space under the roof. Opening the roof vent also results in venting hydrocarbon into the atmosphere and may violate the environmental permit for the site. The combination of flammable range and static electricity is a recipe for disaster.

What Has Happened/What Can Happen

Fire and Explosion of Highway Transport Trucks
Stock Island; Key West, Florida
29 June 1998
The incident was investigated by the National Transportation Board (NTSB)
Hazardous Materials Accident Report NTSB/HZM-99/01

On 29 June 1998, at about 5:14 a.m., a Dion Oil Company transport driver was on the top of a fuel transport tank truck, checking the contents of the truck's cargo compartments in preparation for a cargo transfer when an explosion occurred in the tank. The explosion threw the driver from the top of the truck to the ground, resulting in injuries. The subsequent fire and several more explosions destroyed the truck and a tractor, the front of a semitrailer, and a second truck cargo tank, resulting in additional burn injuries to the driver. The driver sustained a broken left knee and second and third-degree burns to his hands, arms, and legs. Damage was estimated at more than $185,000. This was a serious incident, but the consequences could have been much more severe.

The NTSB's investigation focused on the adequacy of the transport company's product transfer procedures and employee training and the oversight of the Federal Highway Administration and the State of Florida on training and fire safety for loading and unloading hazardous materials. This chapter will only focus on the company's training and procedures.

On the morning of the accident, the transport driver arrived at the marina just before 5 a.m. and was preparing to transfer a load of diesel fuel from a temporary storage tank to one of the cargo tanks on his transport truck. He remembered collecting what he believed to be a mixture of gasoline and diesel spilled from previous loads from below the temporary storage tank into a plastic pail and taking it to the top of his transport truck. He ascended to the top of the cargo tanks and opened the three cargo tank hatches. He believed the two rear compartments contained diesel and the front tank had gasoline based on finding no pressure on the two rear tanks and pressure on the hatch on the front tank when they were opened. He stated that he might have been pouring the mix of gasoline and diesel from the plastic container into the front compartment when the explosion occurred.

NTSB Findings – Procedures

The NTSB determined that the Dion Oil Company had no written cargo transfer procedures before this accident, nor did they have procedures for handling the containers of mixed fuels. Also, procedures that required the drivers to ground and bond the cargo tanks during fuel transfer operations did not exist, and no grounding rods or other means of grounding the temporary storage tank were in place before this accident. Additionally, the transport truck operated by the driver and the temporary storage tank were not equipped with bonding straps or cables to facilitate bonding or grounding. The driver stated that he did not bond his vehicle while transferring fuel. The transfer hose that the driver used was equipped with an internal or embedded conductive wire; however, the 1996 Hose Handbook, published by the Rubber Manufacturers Association, does not recommend using embedded wires for bonding, as they may break, and this may not be detected visually.

NTSB Findings – Training

The NTSB found that Dion Oil Company trained its drivers with an American Trucking Association training video and required them to work with experienced Operators as an integral part of their training. There was no specific training on loading or unloading cargo tanks. During the driver interviews, it was also apparent that Dion's drivers routinely "switched loaded" cargo in the compartments of their trucks as needed to make deliveries. As specified by the NFPA (NFPA-30) and API (API-2003) standards, switch loading is extremely hazardous as a flammable range can easily be created in the vapor space of the compartment and can easily be ignited by static electricity. Switch loading is defined as loading diesel fuel or any other combustible after loading gasoline or other flammable material.

NTSB Findings

The NTSB found that the probable cause of this accident was the transport company's lack of procedures and driver training. The agency found in their investigation that the incident occurred when the driver poured a mixture of gasoline and diesel fuel from a plastic bucket into one of the truck's cargo tanks. The cargo tank was determined to have contained a mixture of explosive vapors (gasoline and air in the flammable range).

Key Lessons Learned

Pouring a nonconductive product from a plastic pail or any other plastic container into a storage tank or compartment containing a flammable mixture with a vapor space in the flammable range is extremely hazardous. Plastic is well known for generating a static charge and can readily ignite a flammable atmosphere and result in an explosion. Plastic containers are not conductive and cannot be bonded and grounded to prevent static sparks.

Ensure that all transfers of nonconductive petroleum fuels are properly bonded and grounded to prevent the arcing and sparks that can occur from static electricity. This includes tank trucks, railcars, and smaller containers like drums and pails. All conveyances should be conductive, and they should be bonded and grounded before any hatches or compartments are opened and remain connected and grounded until the hatches or compartments are fully closed.

Switch loading of petroleum products should be avoided. If a conveyance such as a tank truck compartment is scheduled to load a combustible product such as diesel, kerosene, jet fuel, or any other combustible product with a flash point above 100 °F (or 38 °C), and the previous load was gasoline or light naphtha, or any other flammable with a flash point below 100 °F (or 38 °C). In that case, the product's compartment should be stripped entirely and cleaned before loading begins.

Ensure all employees involved in handling, loading, blending, or transporting petroleum products are trained in the hazards of static electricity and procedures are in place for employees to follow for the transfer of all petroleum products. The procedures should specify the grounding and bonding of all cargo tanks and loading lines before starting the fuel transfer. The procedures should also be reviewed periodically to ensure they are accurate and current.

Although not a contributor in this incident, the initial loading or transfer of a nonconductive petroleum product should be at low rates (low velocities) until the fill line is fully covered (by 2 feet or 61 centimeters); this is normally accomplished by gravity flowing; no pumps should be started until the fill line is well covered. If the transfer is to a floating roof tank, the transfer rate should be maintained at a low rate (by gravity flow) until the roof is floating. This reduces the splash and spray, resulting in static charge build-up and ignition by static discharge.

Additional References

National Transportation Safety Board (NTSB), Storage Tank Explosion and Fire in Glenpool, Oklahoma, 7 April 2003, Pipeline Accident Report NTSB/PAR-04/02; http://ntsb.gov/investigations/AccidentReports/Reports/PAR0402.pdfPB2004-916502 Notation 7666

National Transportation Safety Board (NTSB), Fire and Explosion of Highway Transport Trucks, Stock Island; Key West, Florida, 29 June 1998, Hazardous Materials Accident Report NTSB/HZM-99/01. https://www.ntsb.gov/news/events/Pages/Dion_Oil_Company_cargo_tank_explosion_and_fire_Key_West_Florida_June_29_1998.aspx

Shell "Learning from Incidents; Awareness Alert", DSM-AW-201201, Downstream Manufacturing, January 2012, Pump House Fire – September 28, 2011.

National Fire Protection Association code 77, NFPA-77 "Recommended Practice on Static Electricity".

American Petroleum Institute, API Recommended Practice 2003 (API 2003), "Protection Against Ignitions Arising out of Static, Lightning, and Stray Currents", Note: many of the requirements of API-2003 are being replaced by API-545 Recommended Practice of Lightening Protection of Above Ground Storage Tanks for Flammable or Combustible Liquids. Please refer to the new standard for these requirements.

International Safety Guide for Inland Navigation Tank-barges and Terminals, Chapter 3 "Static Electricity".

Combustion, Flames, and Explosions of Gases, 3rd edition, Lewis and Von Elbe, Academic Press, Published Date: 4 June 1987.

ASTM International; International Standards Organization, ASTM: D4865 "Standard Guide for Generation and Dissipation of Static Electricity in Petroleum Fuel Systems."

American Petroleum Institute, API Tank Conference Proceedings, 2008. "AST Lightning Protection – API 545 Update," Breitweiser, C.

National Fire Protection Association, NFPA-780 "Standard for the Installation of Lightning Protection Systems".

American Petroleum Institute, API Recommended Practice 2219 "Safe Operation of Vacuum Trucks Handling Flammable and Combustible Liquids in Petroleum Service".

American Petroleum Institute, API Publication 1003 "Precautions Against Electrostatic Ignition During Loading of Tank Motor Vehicles".

ANSI/API Standard 2015 "Requirements for Safe Entry and Cleaning of Petroleum Storage Tanks".

"1996 Hose Handbook", Published by the Rubber Manufacturers Association.

National Fire Protection Association, NFPA 30 "Flammable and Combustible Liquids Code".

Horizon PSI Newsletter January 2016, "Static Electricity Can Lead to Plant Explosions.", Static Electricity Can Lead to Plant Explosions | HorizonPSI.

US Coast Guard Regulation 35.35-4. "Insulating flange joint or nonconductive hose—TB/ALL".

Chapter 8.22. Controlling Static Electricity

End of Chapter Quiz

1 Static electricity is frequently not the primary cause of an event but contributes as the ________ _________ ignition source to ignite the resulting vapor cloud following an accidental loss of containment.

2 Static electricity is generated when ___________ move in contact with other materials and is a common occurrence when _________ are being moved through nozzles, pipes, mixed, poured, pumped, filtered, or otherwise agitated.

3 Other causes include the ________ of solids or an immiscible liquid through a liquid, the ejection of particles or droplets through a nozzle, and the _______ or spray of a liquid against a solid surface, for example, in a storage tank or railcar.

4 Many hydrocarbons and other products will generate static electricity. Some hydrocarbon products will _________ the static electricity in the product. For those products, a spark can occur when an object, such as a gauging tape or sample device at a different electrical potential, is lowered into the product.

5 When a ________ charge occurs in the vapor space of a storage tank or a barge, and when the vapor space is in the __________ range, the ideal ratio of fuel and air, it can result in a catastrophic explosion and a significant fire.

6 What three reference documents help us understand and manage static electricity?

7 The amount of static accumulation in liquids is directly related to what characteristic of the liquid.?

8 Name several products that are low conductivity and therefore suspectable to static hazards.

9 What are the two most effective ways to control static electricity?

10 Describe "switch loading" as it relates to static electricity.

11 How long is the relaxation period, when the tank or vessel must be still, before sampling or gauging a tank or vessel involving a large quantity (greater than 10,000 gallons)?

12 The sampling and gauging equipment should be all conductive and should be ___________to the tank before it is placed in the vapor space.

13 When gauging storage tanks, to shield any potential static build-up that may be in the tank, the procedure should ensure that gauging is conducted within the approved tank ______ _________ or ______ _______.

The sampling and gauging equipment should be all conductive and should be ___________to the tank before it is placed in the vapor space.

14 When initiating a load or transfer of a nonconductive petroleum product, the initial transfer should be at _____ _____ or______ _______until the fill line is fully covered (by 2 feet or 61 centimeters). This is normally accomplished by gravity flow with no pumps started until the fill line is well covered.

If the transfer is to a _______ ________ tank, the transfer rate should be maintained at a low rate (by gravitation) until the roof is floating.

15 Mists, aerosols, and dust clouds are also ________ accumulators, and static can also play a significant role in explosions involving accumulations or clouds of combustible dusts.

8.23

Hazards of Hydrofluoric Acid

We start this discussion with a question: What is Hydrofluoric Acid, known as HF, in the industry? HF is a strong and highly corrosive acid with a strong, irritating odor, and it is highly toxic, even at very low concentrations. HF is a contact poison with the potential for skin burns and irritation that is painless initially but can result in excruciating burns and significant damage to the underlying tissues and bones. In low concentrations, HF can result in systemic toxicity and eventual fatality by interfering with the body's ability to metabolize calcium.

HF is also highly toxic to the respiratory system, and breathing HF can damage lung tissue and cause lung swelling and fluid accumulation (pulmonary edema). Small splashes of high-concentration hydrogen fluoride products on the skin can be fatal. Skin contact with HF may not cause immediate pain or visible skin damage (signs of exposure).

HF is highly volatile and vaporizes at room temperature, creating a large vapor cloud that can travel long distances. Therefore, HF must be handled with extreme care, and procedures must be in place and executed correctly in the event of loss of containment. This discussion will attempt to cover all aspects of handling HF, including the emergency response procedures.

Hydrofluoric Acid (HF) is typically used in some petroleum refineries as a catalyst in the Alkylation process to produce a high-octane gasoline component known as Alkylate. Of the 148 petroleum refineries in the United States, approximately 50 are known to use HF as a catalyst in their Alkylation units. HF is also used in Europe and other parts of the world. There are three types of HF available in the industry: anhydrous HF (at higher concentrations, anhydrous HF is a colorless gas or a fuming liquid), modified HF (modified HF contains an additive to reduce the volatility and minimize vaporization in the event of a release to the atmosphere), and aqueous HF (HF in solution with water). Those refineries that use HF as a catalyst typically use anhydrous HF as the catalyst in the alkylation process. Four refineries in the US use Modified HF in their alkylation units (including two in California).

For emergency responders, HF may be known as Hydrogen fluoride (UN 1052) or Hydrofluoric acid (UN 1790). Identification numbers are CAS number 7664-39-3, UN: 1052, or RTECS: MW7875000. The primary HF manufacturers and HF importers are DuPont (US), Allied (US), and Honeywell (US).

The only other acceptable catalyst currently in Alkylation is sulfuric acid, which is used in many of the remaining refineries in the US and worldwide (those refineries not using HF). The significant difference between HF and Sulfuric Alkylation Units is that those units using sulfuric acid require many times the amount of acid. Additionally, sulfuric acid requires acid regeneration, generally offsite, which means a large amount of acid is transported to/from regeneration by pipeline or truck. Only strong acids can catalyze the alkylation; therefore, the acid strength for both sulfuric and HF acids must be maintained above 88%; otherwise, the olefins may polymerize. This means that acid requires almost continuous regeneration for the sulfuric process to maintain acid strength.

Is there a new, safe Alkylation catalyst on the horizon? Until now, concentrated sulfuric and hydrofluoric acids were the only commercial catalysts to produce high-octane gasoline. However, new technology is forthcoming as Chevron and UOP have recently licensed a new ionic liquid alkylation catalyst. The desired alkylation reactions are forming C8 carbonium ions and the subsequent formation of Alkylate for high-octane gasoline.

A new ISOALKY catalyst has been retrofitted into the Chevron Salt Lake City refinery for demonstration purposes. However, the new technology catalyst does require unit modifications to convert the process to accommodate the new catalyst, and the full scope of these modifications remains undefined. This new technology catalyst, which can be regenerated on-site, uses a nonaqueous liquid salt, or ionic liquid, at temperatures below 212 °F (100 °C) to convert typical FCCU olefins into a high-octane gasoline blending component.

Alkylation Units have two main and inherent hazards: large volumes of highly flammable light hydrocarbons that are potentially explosive if released into the atmosphere and a highly corrosive catalyst (sulfuric acid or hydrofluoric acid) (HF).

Process Operations Safety: The What, Why, and How Behind Safe Petrochemical Plant Operations, First Edition. M. Darryl Yoes.

Both are highly corrosive and toxic. However, HF represents a much more significant risk due to the acid's volatility and toxicity. HF has a boiling point of 67 °F (19 °C) and will vaporize rapidly at room temperature if released into the atmosphere.

HF is also very toxic to humans and has an IDLH of only 30 ppm. A toxic cloud of HF can travel a long distance (miles). Due to the hazards of HF, the American Petroleum Institute (API) has issued a recommended practice specific to HF units, which recommends restricted access to HF units. API Recommended Practice 751, "Safe Operation of Hydrofluoric Acid Alkylation Units," also recommends that refineries audit the safety of HF alkylation operations every three years and provide the detailed elements to be included as part of a comprehensive audit plan. A copy of the API 751 checklist is available for purchase from the American Petroleum Institute.

Scientific Testing of an Unmitigated Release of HF in the Nevada Desert in 1986

The US Department of Energy, sponsored by the AMOCO Corporation, performed an unmitigated release demonstration in the Nevada desert in 1986. Department spokesperson Jim Boyer was quoted in the press that the 7-minutes controlled test released HF 10–15 feet high and that the cloud was visible as far as 4,000 feet from the spill area.

Boyer said the test was designed to help gather data for predicting the movement of HF vapors in the atmosphere in case of an accidental leak. While conducting these scientific tests using HF, the research team was reportedly surprised when 100% of the released HF was fully vaporized. The vaporized HF created a large, rolling, dense white cloud of extremely toxic gas that traveled a very long distance. It was reported that HF in toxic concentrations was measured at 3–6 miles (about 5–10 kilometer) from the release point.

The Laurence Livermore Laboratories performed the test at Frenchman's Flat, NV, in 1986. The release involved a 1,000-gallon release of HF from a Tank Truck through a 4″ line in about 2 minutes. The release occurred at 110 psi and 104 °F operating conditions, at conditions similar to an HF alkylation unit's operating temperature and pressure. The initial cloud size was ¼ mile in length, and the HF cloud tended to hug the ground all along the long distance of the test. The test reportedly used 3,000 sensors, including two different types of sensors located at various distances and elevations.

HF Toxicity (for reference): OSHA IDLH = 30 ppm (Immediately Dangerous to Life and Health), OSHA TWA = 3 ppm (8 hours) (Time Weighted Average), OSHA STEL = 6 ppm (15 minutes) (Short Term Exposure Limit).

During the test, specific readings of 2,900 ppm HF occurred 1/4 mile away within 30 seconds. Other toxicity readings during the test included 362 ppm at 1.86 miles from the release point, and lethal concentrations were detected about five miles from the release point (downwind). Overall test readings were reported to be from 0.3 to 42,000 ppm (IDLH of 30 ppm) during the test (depending on location and elevation).

This test confirmed that a similar *unmitigated release*, if it occurred inside a refinery, could place a large number of workers and members of the community in harm's way. Similar testing has been performed using sulfuric acid, and no vaporization has occurred in those tests. This reinforces the importance of precautions and strict adherence to operating and emergency response procedures, especially in release prevention and emergency response, and communication downwind during an accidental release.

Please refer to the Final Report prepared by the US Chemical Safety and Hazard Investigation Board of the 21 June 2019 Philadelphia Solutions Refinery explosion for additional detail and the amazing images of the HF release demonstrations test that was done in the Nevada desert by the US Department of Energy during August 1986.

There have been several recent releases and near releases of HF over the past several years. For example, in March of 2009, a release of HF occurred at the Philadelphia Sunoco Refinery, resulting in 13 workers receiving "precautionary" medical treatment for potential exposure to HF. In this incident, 22 lbs. of hydrogen fluoride HF (the gas from HF acid) was released. During this same period, the US CSB announced it was investigating two other releases of HF, one at the ExxonMobil Joliet Refinery in Illinois and the other at the Citgo Refinery in Corpus Christi, Texas. In the Citgo incident, the company reported that 43,000 lbs. of HF were released during an Alkylation unit fire that occurred on 9 December 2009; however, they reported that all but 30 lbs. were captured by the water mitigation system. Again, on 5 March 2012, Citgo reported an HF release due to leakage at pipe flanges on the alkylation unit. See Figures 8.23.1 and 8.23.2 for photos of a flange with evidence of HF leakage.

Additionally, a significant release of HF occurred in an explosion and fire at the Philadelphia Energy Solutions refinery in South Philadelphia on 2 June 2019, and another near loss occurred at the ExxonMobil Torrance Refinery when an explosion occurred there on 18 February 2015. Shrapnel from the Torrance explosion reportedly narrowly missed striking HF-containing equipment on the HF Alkylation unit. The US CSB reported this near loss of HF, which could have potentially impacted the local community.

(a)

(b)

Figure 8.23.1 (a and b) Photos of 2012 Citgo Alkylation Unit flange showing evidence of HF leakage (paint that changes color in the presence of HF). Figure (b) is a closeup image illustrating the distinctive color paint which is designed to change color in event of an HF acid leak. Source: Photos courtesy of the US Chemical Safety and Hazard Investigation Board.

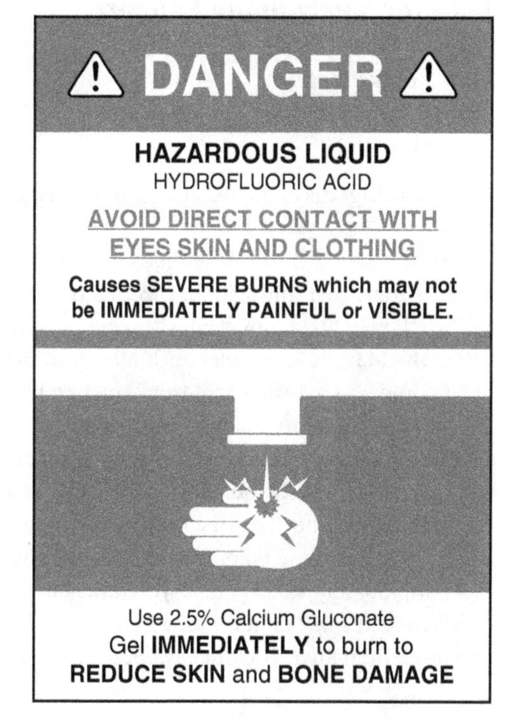

Figure 8.23.2 Example of an employee warning poster in the event of direct contact with hydrofluoric acid (HF). Source: This image is courtesy of the Ecoscience Resource Group Visual Artists. It was created to illustrate the hazards of hydrofluoric acid and is not representative of any company. Similar signs should normally be posted in all locations where Hydrofluoric Acid is present or in use.

The National Response Center (NRC) recorded approximately 400 incidents at industrial plants (including refineries) involving HF from 1990 to 2005. In April 2019, The CSB issued a letter to the EPA requesting the EPA to review its existing HF study to determine the effectiveness of existing regulations and the viability of utilizing inherently safer alkylation technologies in petroleum refineries. Also, The US PIRG Education Fund (US Public Interest Research Group) issued a study on HF in 2005 titled "Needless Risk," and the United Steelworkers Union issued a similar study in April 2013 titled "A Risk Too Great," also challenging the petroleum companies to develop a safer alternative to HF for gasoline production.

What Has Happened/What Can Happen

Hube Global Company
Gumi National Industrial Complex

South Korea
27 September 2012

On 27 September 2012, at the Hube Global Company at the Gumi National Industrial Complex in Southern Korea, five workers were killed by Hydrofluoric Acid (HF) Exposure. This incident occurred while workers attempted to unload HF from a truck when the hose or hose coupling failed. The workers were right in the path of the HF when the release occurred and were not wearing any type of protective equipment or respiratory protection.

Many people were exposed to HF during this incident, including residents, local workers, news journalists, and emergency responders, many without protective equipment. Police believe it might have happened while acid was being transferred to a storage tank. The HF quickly traveled to the surrounding area where more than 3,000 nearby residents had to be treated for respiratory and skin problems.

The release impacted areas as far as 4 miles (6.5 kilometer) from the leak site, including people, animals, and vegetation. Questions remain about the long-term health effects of the exposure. A video of this incident, which was taken as it was happening, is posted on YouTube. If you watch the video, you will see the work crew completely engulfed by the HF vapor cloud as they connected the tank truck. Five of these guys died from the exposure to HF. There are also photographs of the large white vapor cloud as it enveloped the tank truck and the area within the facility and beyond into the local community. Some were in the local news; however, I could not secure those for this book.

Marathon Petroleum Refinery

Marathon Refinery
Texas City, Texas
30 October 1987.

The following was extracted and summarized from an article published by the British Health and Safety Executive on their website:

Probably one of the most serious releases of HF occurred on 3 October 1987 at the Marathon Refinery in Texas City, Texas, when a crane carrying a 50-foot section of a convection heater dropped its load onto an anhydrous hydrogen fluoride tank within the HF alkylation unit, shearing two lines leading to the top of the tank. This resulted in an atmospheric release of HF at the Marathon Petroleum Company refinery in Texas City.

One line was a 4-inch acid truck loading line, and the other was a 2-inch tank pressure relief line. The tank was at the normal operating pressure of approximately 125 psi. The tank originally contained 35,700 gallons of AHF, of which about 6,548 gallons were released over a 44-hour period. Most of the release occurred during the first two hours as the tank depressurized. The cloud of HF-containing some light hydrocarbons (primarily isobutane) plus water vapor moved with the prevailing wind. The first mitigation action was to place stationary fire monitor nozzles and erect a water spray curtain about 10 feet downwind of the release to control the HF acid vapor plume.

Approximately 4,000 people were evacuated from the residential areas threatened by the plume, and the three area hospitals treated 1,037 patients, of which nearly 100 were hospitalized. There was extensive damage to trees and vegetation in the residential area. Fortunately, there were no fatalities.

Key Lessons Learned

HF Release Prevention and Mitigation

We have talked primarily about the downside of HF in our facilities. Now, let us talk about what we do to ensure that HF is handled safely to protect our people, community, environment, and reputation. As mentioned earlier, it is very important to understand what is at stake and ensure that the procedures and practices are followed each time we enter the unit. This is especially true for emergency response procedures. If you work on an Alkylation unit, regardless of whether the unit uses HF or sulfuric acid, it is very important that you are well-trained and know what to do in the event of a loss of containment of a flammable or acid. We will generically talk through those procedures as it would be impossible to discuss your specific procedures in this book. However, you should periodically review your specific procedures and follow the review by going to the field and locating the valves you would operate if the event were happening. Refer to Chapter 9, "Importance of Operations Emergency Response Drills and Exercises," for more details.

Rapid De-inventory System to Quickly Remove the HF to Safe Storage

In the unlikely event of a significant release of HF into the atmosphere, one of the first steps is to quickly remove the HF inventory from the unit. Limiting the inventory available effectively limits the duration and, therefore, the impact of the release. The HF Alkylation unit's largest HF inventory is held in the HF Settler. Many Alkylation units around the world include a means of rapidly dumping the HF acid inventory from the settler vessel to another vessel, typically an HF acid storage tank, in the event of a release to minimize the material that can be released. This system is commonly referred to as a Rapid Acid De-Inventory System (RADS) or Rapid Acid Transfer System (RATS). RADS is designed to quickly remove the HF inventory to a safe place, typically into an HF storage tank. There may be either a transfer pump or the system may be designed to use unit pressure (a pressure-based system), whereby simply opening a valve (or a series of valves) empties the entire inventory of HF to a separate and safe storage vessel. The RADS can either be manually or detector activated, and the total transfer time will depend upon the unit size and capacity, the amount of HF in inventory, and the transfer mechanism. However, it is designed to eliminate the hazard quickly.

In the event of an apparent large release of HF, you should never be intimidated about using the RADS system to quickly remove the HF inventory from the unit. Some automated systems may incorporate a time delay to allow operations personnel to visually identify the release location and allow aborting activation. The operator might abort activation if the detection alarm is considered spurious or the operations personnel realize that the release is coming from another part of the unit unaffected by activating the acid dump.

Multiple HF Gas Detectors are Placed Around the Unit

In the unlikely event of a release, these sensors would detect the release and sound alarms in the control room and the unit. To ensure reliability, there are typically two or more different gas sensors designed to simultaneously detect the loss of containment. Sensors are classified as safety-critical equipment and are periodically tested to ensure they remain fully functional. Any indication of a release by one of the sensors should be considered an actual release until independently verified otherwise.

Water Mitigation Systems

Please refer to Appendix E, a great brochure from BakerRisk on HF Water Mitigation for HF releases. This brochure titled HF Water Mitigation Techniques and Design Considerations contains great details and images of how water can help mitigate a potential HF release. Included are photos of water spray curtains or water walls, water cannons, and additional physical barriers to prevent a release from spreading. Additionally, this brochure provides details on equipment layout and design and the strategies and tactics that can be used in the event of a release. We will elaborate on the variety of tactics that can be used here.

Firewater Wall Mitigation, Water Curtain System Mitigation, and Aim-and-Shoot Mitigation

There are multiple benefits from a very strong water deluge or firewater curtain-type system. First and foremost, the firewater can readily absorb HF, effectively removing the HF from the toxic cloud. This provides significant protection for anyone located downwind from the release. The second benefit is the turbulence created by the water curtain, which results in more substantial mixing and the addition of additional air to the release cloud. The air reduces the concentration of available HF.

Three types of engineered water sprays can be used: Water Wall mitigation, Water Curtain mitigation, and Aim-and-Shoot mitigation (using fixed and portable water monitors and water cannons). It is not uncommon that all three may be employed in a significant release.

Water Wall Mitigation

The water wall system is a dispersed and water contact-type system and involves an arrangement of firewater monitors to encapsulate the HF release, regardless of where the release emanates from on the unit. This system produces an extensive water fog, which creates a wall of water. The water wall system is typically located along the perimeter of the HF Alkylation unit to achieve the spacing required for effective absorption and maximize the scenarios it can effectively mitigate. Activation of the water wall may be automated and based on a voting logic involving multiple detectors or manually initiated once a release has been confirmed.

The Water Curtain System

The water curtain system is also a disperse and contact approach and is typically applied to high-risk parts of the unit, such as high inventory vessels or high-risk HF pumps, rather than the unit as a whole. Using strategically located monitors and rows of fog nozzles, the area is encircled with a high-flow, small drop-size wall of water. A higher flow rate is required than the water wall approach as an HF release will have limited opportunity to disperse and thus be at a higher concentration and momentum before contacting the water mitigation. The water curtain systems can also be activated automatically based on the detector(s) or manually, like the water wall system.

For example, a site can employ multiple fog nozzles surrounding a high inventory HF Settler, forming a water wall to quickly absorb and dilute the HF. Another tactic is the aim-and-shoot mitigation, where the name says it all. This mitigation involves several firewater monitors with a cone-shaped pattern aimed directly at the point of HF release. Although this approach is effective at knocking down the release, it requires a degree of accuracy. The monitors can be fixed manually or remotely operated. Fixed monitors are usually pre-aimed at likely sources of release (pump seals, etc.), remotely operated monitors often with a "home" position based on a predetermined likely release point, and manually operated monitors are usually located on the edge of the unit.

A water deluge is a localized and concentrated deluge of water located strategically through risk assessment or aimed at multiple vital areas. Water deluge involves multiple strategically placed water deluge monitors that can be fixed or remotely operated, typically preset, and directed at higher-risk equipment.

Closed-circuit TV Cameras on all Unit Critical Areas

Almost all HF Alkylation units have multiple closed-circuit recording cameras to monitor higher-risk areas and operations on a 24/7 basis. They generally monitor higher-risk equipment such as pump seals, truck operations, large-volume equipment, etc. These cameras actively display recording activities on the unit, including monitoring for any hydrocarbon vapors or HF release. This continuous monitoring supplements routine rounds by Operators and helps enable rapid response to any situation.

Reducing the Vapor Pressure of the Released HF

Several refineries use a particular type of HF called "Modified HF." Modified HF is blended with a proprietary chemical designed to reduce the vapor pressure of the HF should it be released into the atmosphere. The special inhibitor is designed to reduce the amount of airborne HF, thereby reducing the risk of downwind contact and vapor inhalation. The result is that the released HF is less volatile and, therefore, less likely to create a large vapor cloud when released.

Availability of Safety Equipment on Alkylation Units

All alkylation units should have ample availability of functioning safety showers and eyewash fountains, and any safety shower or eyewash fountain found not to be working should be repaired on a priority basis. Charged water hoses should also be readily available in the unlikely event it is required to wash someone down following an acid release. In HF alkylation

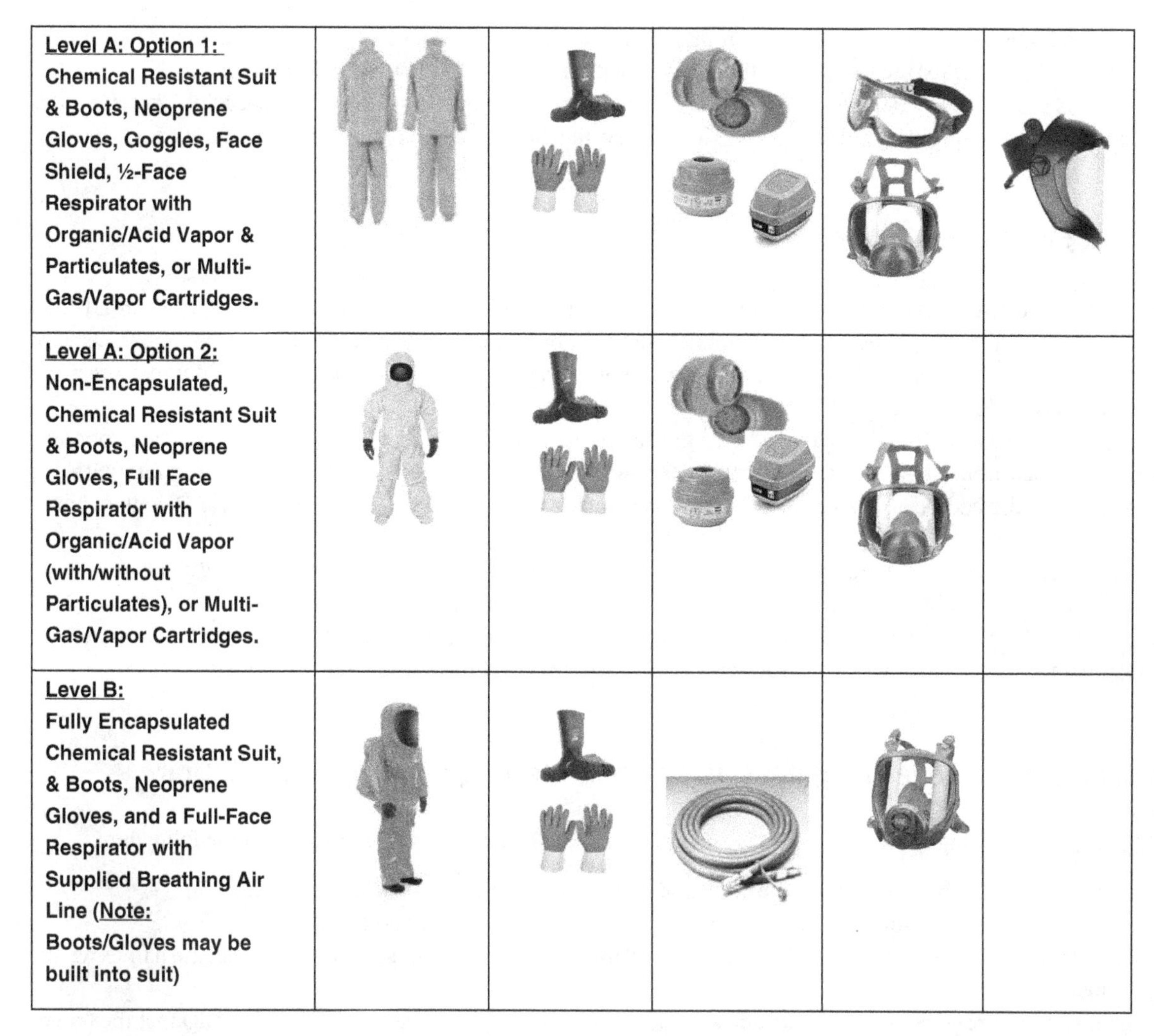

Level A: Option 1: Chemical Resistant Suit & Boots, Neoprene Gloves, Goggles, Face Shield, ½-Face Respirator with Organic/Acid Vapor & Particulates, or Multi-Gas/Vapor Cartridges.					
Level A: Option 2: Non-Encapsulated, Chemical Resistant Suit & Boots, Neoprene Gloves, Full Face Respirator with Organic/Acid Vapor (with/without Particulates), or Multi-Gas/Vapor Cartridges.					
Level B: Fully Encapsulated Chemical Resistant Suit, & Boots, Neoprene Gloves, and a Full-Face Respirator with Supplied Breathing Air Line (Note: Boots/Gloves may be built into suit)					

Figure 8.23.3 Examples of the types of protective PPE that should be worn while working or visiting an HF Alkylation unit. These are examples only; the site Industrial Hygienist and/or Safety Manager should approve PPE for use in hazardous areas. Source: Images and text are from the Industrial Safety website and are provided courtesy of IndustrialSafety.com (industrialsafety.com).

units, it is also essential to have a ready supply of 2.5% Calcium Gluconate Gel for immediate treatment of personnel in the event of contact with HF. Please see Figure 8.23.3 for an example of an employee warning poster in the event of direct contact with HF.

Use of the "Shelter-In-Place" Procedure in the Event of an Accidental Release

Most refineries have implemented the "Shelter-In-Place" procedure during a vapor or toxic gas release. This is due to the number of personnel affected by a vapor release and the ease of quickly protecting many people during an emergency. The Shelter-In-Place procedure is a documented and practiced process wherein, in the event of a toxic gas release, personnel who downwind enter buildings designed as "safe havens" close the doors and windows, and turn off the air conditioning or air handling equipment. They may also place tape around the doors and windows to help ensure an effective seal against the vapors leaking into the building. They then simply wait until the leak or release is secured or until the emergency response team notifies them that it is safe to exit the building.

While this may sound like an archaic process, it has proven to be quite effective in real life and has been widely recognized as a legitimate response strategy by many safety professionals and response groups. Even in the event of HF releases, the industry has experienced the effective use of the Shelter-in-Place process. During the Marathon, Texas City HF Acid release, no one who elected to shelter-in-place required any treatment following the release. Also, following a tank car rupture of

Oleum in Richmond, CA, employees of the Chevron facility, located about 3,000 feet away and in the direct path of the vapor cloud, sheltered in place for the duration of the incident with no required follow-up or treatment.

Shelter-in-place is a recognized response strategy by OSHA and The Department of Homeland Security. Therefore, it is only appropriate to keep this effective tool in our tool kit in the event of a significant loss of containment of a toxic like HF.

Characteristics of Hydrofluoric Acid

HF is a very small molecule similar in size to helium, has a low boiling point (67 °F), and freezes at −11 °F. An HF acid leak can produce a cloud that resembles a steam cloud. It will result in a dense white cloud of HF vapors that reacts with moisture in the air. HF is very soluble in water and in most hydrocarbons. HF atomizes when released under pressure, creating tiny droplets of HF.

HF is a highly corrosive acid; it is the only acid that is used to etch glass. HF has a PH of 0 and can attack concrete, rubber, leather, and cast iron. The Fluoride ions can penetrate skin and flesh, de-calcify bones, and interfere with the calcium metabolism in the body, causing significant harm. People have died from very small UNTREATED contact burns with HF.

Generally, with a small HF contact burn, there may be no pain for the first 24–48 hours. This can lead to what initially appears to be a minor contact burn, which later becomes very serious if there is no treatment with the antidote (2.5% Calcium Gluconate). I have seen one report of a fatality of a person who had a quarter-size burn that went untreated. This small burn resulted in the fatality. Any burn suspected to be caused by HF requires immediate treatment.

Special Personal Protective Equipment (PPE) when Working on or Near an HF Unit

Standard PPE, anytime working on or simply visiting an HF acid unit, should include a minimum of the following:

- Acid-resistant coveralls and boots to protect the body and feet from the potential of acid exposure.
- Face shield PLUS acid-resistant goggles to ensure absolute protection for the face and eyes.
- Acid-resistant gloves or gauntlets (Neoprene or Nitrile (22 mil) or other HF-resistant gloves) – to protect the hands, wrists, and lower arms.
- Air-supplied respirator or self-contained breathing apparatus with supplied compressed air to protect against the potential of breathing HF vapors or fumes.

Other personnel protective equipment is illustrated in Figure 8.23.4.

Human Safety and Health Hazards of HF

HF is extremely toxic to humans through inhalation of vapors, skin absorption, ingestion, and eyes.

The following first aid advice is extracted from an HF Material Safety Data Sheet:

For Any Route of Contact

Detailed First Aid procedures should be planned before beginning work with HF. In all cases of contact, regardless of the degree of contact, immediately call a POISON CENTER or doctor/physician.

The effect of HF, that is, the onset of pain, particularly in dilute solutions, may not be felt for up to 24 hours. Persons working with or near HF must have immediate access to an effective antidote even when they are away from their workplace so that first aid treatment can be commenced immediately (Note: Antidote is 2.5% Calcium Gluconate).

In the Event of Inhalation

Get medical help immediately. If the patient is unconscious, give artificial respiration or use an inhalator. Keep the patient warm and resting, and send them to the hospital after first aid is complete.

In the Event of Skin Contact

1. Remove the victim from the contaminated area and immediately place him or her under a safety shower or wash with a water hose, whichever is available.
2. Remove all contaminated clothing. Handle all HF-contaminated material with gloves made of appropriate material, such as PVC or neoprene.
3. Keep washing with large amounts of water for a minimum of 15 minutes.
4. Have someone arrange medical attention while you continue flushing the affected areas with water.
5. If the following materials are available, limit the washing to five minutes and immerse the burned area in a solution of 0.2% iced aqueous *Hyamine 1622 (Tetracaine Benzethonium Chloride) or 0.13% iced aqueous **Zephiran Chloride (Benzalkonium Chloride).
   If immersion is not practical, towels should be soaked with one of the above solutions and used as compresses for the burn area. Ideally, compresses should be changed every 2 minutes.
   Alternately, 2.5% calcium gluconate gel should be massaged into the affected area.
6. Seek medical attention as soon as possible for all burns, regardless of how minor they may appear initially.

Calcium gluconate gel is a topical antidote for HF skin exposure. Calcium gluconate works by combining with HF to form insoluble calcium fluoride, thus preventing the extraction of calcium from tissues and bones. Keep calcium gluconate gel nearby whenever you are working with HF. Calcium gluconate can be ordered through scientific supply companies. Calcium gluconate has a limited shelf life and should be stored in a refrigerator if possible and replaced with a fresh supply after its expiration date. Use disposable exam gloves to apply calcium gluconate gel. Even after applying calcium gluconate, a medical evaluation must be made.

Hazards of Hydrofluoric Acid – Our Role and Areas of Emphasis

- Rigorously follow the PPE requirements for working on the HF Alkylation unit – every time you enter the unit.
- Help enforce these requirements for others you work with on your unit, especially visitors who may not be familiar with HF work requirements.
- Know the emergency procedures for loss of containment of HF and periodically participate in drills and exercises.
- For example, know how to activate the HF Rapid De-inventory and water wall systems.
- Always know where your Calcium Gluconate is and how to use it if needed.
- Should you feel any skin irritation or discomfort after leaving work, immediately seek medical attention and report this to your management immediately.
- Know your safety showers and eyewash stations and ensure they are always fully operational.

More Recent HF Recommendations by the US Chemical Safety and Hazard Investigation Board (CSB)

As a result of the Philadelphia refinery explosion and the resulting HF release, the CSB, in 2019, recommended that the API update its standards on the safe operation of HF alkylation units. Specifically, the CSB recommended that API require that critical safeguards and associated control system components be protected from fire and explosion hazards, including radiant heat and flying projectiles.

They also have consistently recommended that API look for inherently safer designs. Recognizing that of the 155 US petroleum refineries currently in operation in the United States, 46 operate Alkylation units using HF catalyst, and the remainder operate units with Sulfuric Acid as the catalyst.

HF is highly toxic and is one of the eight most hazardous chemicals regulated by EPA's Risk Management Program (RMP). New alternative alkylation technologies have been developed in recent years, such as a solid acid catalyst and new ionic liquid acid catalyst alkylation technology. Replacing highly toxic chemicals such as HF with less hazardous chemicals meets the definition of "inherently safe design." Some refinery alkylation units use sulfuric acid as a catalyst instead of HF. Although sulfuric acid is highly corrosive and will result in skin burns when it contacts the skin, sulfuric acid does not vaporize when released and, therefore, does not potentially affect the downwind areas.

Conversely, when HF is released, as demonstrated by the Nevada dispersion test, it results in concentrations in air for miles from the release point in excess of 30 ppm, which is Immediately Dangerous to Life and Health (IDLH). Sulfuric acid remains a liquid upon release and does not present the same risk to surrounding communities as HF, which vaporizes upon release and has the potential to travel offsite.

The CSB also made a recommendation to Citgo Corpus Christi, TX, and Lemont, IL. Refineries after a release of HF occurred at the Corpus Christi refinery on 19 July 2009. That recommendation was related to improvements to Citgo's water mitigation systems for the HF alkylation units. During the July 2009 incident, CITGO nearly exhausted its stored water supply for fire suppression and HF mitigation on day one of the multi-day incident response.

Additional References

Press Release: "Honeywell UOP Introduces Ionic Liquids Alkylation Technology". https://www.uop.com/?press_release=honeywell-uop-introduces-ionic-liquids

U.S. Chemical Safety and Hazard Investigation Board, Philadelphia Energy Solutions (PES) Refinery Fire and Explosions, Final Report issued October 12, 2022, The incident occurred on 21 June 2019.

U.S. Chemical Safety and Hazard Investigation Board, Urgent Recommendation: Citgo Refinery, Corpus Christi, Texas, July 19, 2009 Alkylation unit hydrocarbon gas release, fire, and hydrogen fluoride (HF) release. https://www.csb.gov/assets/recommendation/urgent_recommendations_to_citgo_-_board_vote_copy.pdf

U.S. Chemical Safety and Hazard Investigation Board, Media Report; Update by CSB Investigation Team Leader, Citgo Refinery 5 March 2012 Alkylation unit hydrofluoric acid gas leak. https://www.csb.gov/assets/1/17/statement_-_final_to_post_3_16_2012_-_2.pdf?14356

U.S. Chemical Safety and Hazard Investigation Board, Letter to U.S. EPA Administrator: Re: Encouraging the EPA to initiate a review and update of its 1993 HF study to determine if the US petroleum refineries risk management plans for HF to determine whether these refineries' existing risk management plans are sufficient to prevent catastrophic releases; and to determine whether there are commercially viable, inherently safer alkylation technologies for use in petroleum refineries. https://www.csb.gov/assets/1/6/letter_to_epa_4.23.2019.pdf

Marathon Refinery HF Release, Texas City, Texas (October 30, 1987), British Health and Safety Executive. https://www.hse.gov.uk/comah/sragtech/casemarathon87.htm

"Needless Risk – Oil Refineries and Hazard Reduction", U.S. PIRG Education Fund, August 2005. https://uspirg.org/sites/pirg/files/reports/Needless_Risk_USPIRG.pdf

"A Risk Too Great – Hydrofluoric Acid in US Refineries", United Steel Workers Union, April 2013. http://assets.usw.org/resources/hse/pdf/A-Risk-Too-Great.pdf

Video of Nevada Hydrofluoric Acid dispersion test. https://www.facebook.com/BanMHF/videos/1986-hydrofluoric-acid-release-test-goldfish/290182224646882/

"Questions remain after huge hydrofluoric acid leak.", HF Release; Hube Global chemical plant in Gumi, South Korea, September 27, 2012, Chemistry World. https://www.chemistryworld.com/news/questions-remain-after-huge-hydrofluoric-acid-leak/5611.article

CITGO Refinery Hydrofluoric Acid Release and Fire, Final Report. https://www.csb.gov/citgo-refinery-hydrofluoric-acid-release-and-fire/

CSB Issues Urgent Recommendations to CITGO; Finds Inadequate Hydrogen Fluoride Water Mitigation System during Corpus Christi Refinery Fire Last July. https://www.csb.gov/csb-issues-urgent-recommendations-to-citgo-finds-inadequate-hydrogen-fluoride-water-mitigation-system-during-corpus-christi-refinery-fire-last-july/

"CSB Will focus on Hydrofluoric Acid in refining, Kulinowski says". https://fluoridealert.org/news/csb-will-focus-on-hydrofluoric-acid-in-refining-kulinowski-says/

Baker Risk brochure: "HF Water Mitigation Techniques and Design Considerations." (See Appendix E). https://www.bakerrisk.com/wp-content/uploads/BakerRisk_Best_Practices-HF_Water_Mitigation-FNLv3-WebRes_Spreads-1.pdf

Chapter 8.23. Hydrofluoric Acid Hazards (And How We Manage HF Safely in Our Facilities)

End of Chapter Quiz

1 HF is a strong and highly _________acid with a strong, irritating odor, and it is highly ________, even at _______ _____ concentrations. HF is also a contact poison with the potential for deep and initially painless burns and can cause significant damage to the underlying tissues and bones.

2 HF is also highly toxic to the respiratory system, and breathing HF can damage lung tissue and cause lung swelling and fluid accumulation (pulmonary edema).
- True
- False

3 How is a small contact burn with HF treated?

4 Hydrofluoric Acid (HF) is typically used in some petroleum refineries as a __________ in the Alkylation process to produce a high-octane gasoline component known as Alkylate.

5 What are the two types of acid used as catalysts for alkylation units?

6 HF is highly _______ and _________ at room temperatures creating a large vapor cloud that can travel long distances. Therefore, HF must be handled with extreme care, and procedures must be in place and executed correctly in the event of loss of _____________.

7 What are the three types of HF currently used in industry?

8 HF is very _______ to humans and has an IDLH of only 30 ppm. A ________ cloud of HF can travel a long distance (miles). Due to the hazards of HF, the American Petroleum Institute (API) has issued a recommended practice specific to HF units which recommends _________ access to HF units.

9 What are the most important emergency response equipment associated with an HF Alkylation unit?

10 Why is it so important to immediately flood a release of HF with deluge amounts of water?

11 What are the OSHA exposure limits for working around HF vapors?

12 Many refineries have implemented the ________ ___ ________ procedure which is documented and practiced at regular intervals. This procedure is designed to protect many people in the event of a loss of containment of HF.

13 In addition to the de-inventory and water wall water deluge systems, what other HF response equipment is readily available on most HF Alkylation units?

8.24

Safe Use of Utilities; Connecting Utilities to Process Equipment

Utilities (air, water, steam, nitrogen) are essential in unit operations. We use utilities for many purposes, including preparing equipment for mechanical work, preparing equipment for return to hydrocarbon service, cleaning or purging process equipment, powering pneumatic-powered tools, and many other applications. However, many incidents and even fatalities have occurred by not following basic safety precautions when connecting utilities to process equipment.

Regarding the use of plant utilities, there are several "Cardinal Safety Rules" as follows:

1. Never use plant air as breathing air.
2. Never connect potable water (drinking water) directly to process equipment.
3. Ensure the hose being used is rated for the pressure and temperature and compatible with the material before it is connected to the equipment.
4. Ensure that check valves are installed at the connection to the process and the utility to prevent backflow into the hose or the utility system.
5. Ensure the utility stations are equipped with unique quick disconnects or fittings for each utility service (especially for plant air, nitrogen, service water, and steam).
6. Instrument air should not be used in place of plant air.
7. Do not use Nitrogen as a backup for Instrument air (due to the asphyxiation hazards of Nitrogen).

Potable Water

Occasionally, potable water will be required to fill process equipment or used in process equipment. For example, potable water may be used for the hydrostatic testing of stainless-steel process equipment where clean water with very low chloride content is required. If potable water use is required, a "break tank" should be used to allow the water to fall freely from the utility system into the water supply tank. The water is then pumped from the break tank into the process. The physical "break" ensures that any backflow from the process cannot enter the potable water system, causing contamination of the drinking water system.

Water Filling of Process Piping or Equipment

Water is relatively heavy, weighing about 8.5 lbs. (3.85 kg) per gallon or 62.5 pounds (28.35 kg) per cubic foot (0.028 Cu m). Some process equipment is not designed to support water weight and can be severely damaged or destroyed during water-fill operations. For example, flare headers and flare piping are typically large-diameter piping and are not normally designed to support the weight of water. Some fractionating columns or overhead exchangers are also not designed to support the weight if water-filled. Failure of the piping or structural collapse can result if these systems are inadvertently filled with water and may result in a major fire or safety incident by breaking hydrocarbon containment or causing secondary equipment failures. When filling process equipment with water for testing or maintenance work, always verify with the Mechanical Engineer to ensure that the foundation and structure are designed to support the weight of the water.

Another significant concern is the potential for brittle fracture during water-fill operations. Procedures should include precautions and considerations for the water temperature relative to the piping or vessel's minimum design metal temperature (MDMT) (see Chapter 8.11 on Brittle Fracture for additional detail and information).

Process Operations Safety: The What, Why, and How Behind Safe Petrochemical Plant Operations, First Edition. M. Darryl Yoes.

Connecting Water to Hot Process Equipment

Remember that water expands about 1700 times its volume when it flashes to steam. A small amount of water introduced into HOT process equipment, such as a fractionating column, can severely damage the internals of the tower. The rapid overpressure may result in loss of containment by opening relief valves to the atmosphere. Also, the overpressure may result in equipment leaks and vessel rupture in extreme cases (see Chapter 8.1 on "Returning Equipment to Service – Safe Startups and Other Transient Operations").

Water accidentally introduced into a hot vacuum tower will flash at a lower temperature because of the vacuum, and the expansion rate is much higher, as high as 4,800 times at full vacuum. We must be careful not to allow water to enter any hot equipment, especially those operating under vacuum conditions.

Utility Hose Rating

The hose rating should be provided by the hose vendor and should be clearly stamped or embossed on the hose. The utility hose must be rated at a temperature and pressure higher than the utility being used AND rated higher than the temperature and pressure of the process line or vessel being connected (in the event of unexpected backflow). Of course, the hose must also be designed and rated for the type of service. For example, never use a hose designed for Nitrogen in steam service, etc.

Use of Check Valves

Check valves at the process connection and the utility to prevent a backflow of the process fluid into the utility hose or system. A reverse flow of light hydrocarbon into the refinery nitrogen or air systems or hydrocarbons to the firewater system can result in serious safety incidents. Should the hose fail, a significant loss of containment of the process can occur if the Operators have not installed a check valve in the proper location. Many incidents of this type have already occurred.

It is also desirable to have bleeder valves installed at the hose connections to help depressurize the hose after use and to check for the possibility of reverse flow. An example of this utility connection is provided in Figure 8.24.1.

Unique Utility Hose Fittings

Utility stations should be standardized for the site so that all units comply and use the same unique coupling for each utility. A unique type of coupling should be required for each utility, steam, plant air, nitrogen, and utility water. This is to help

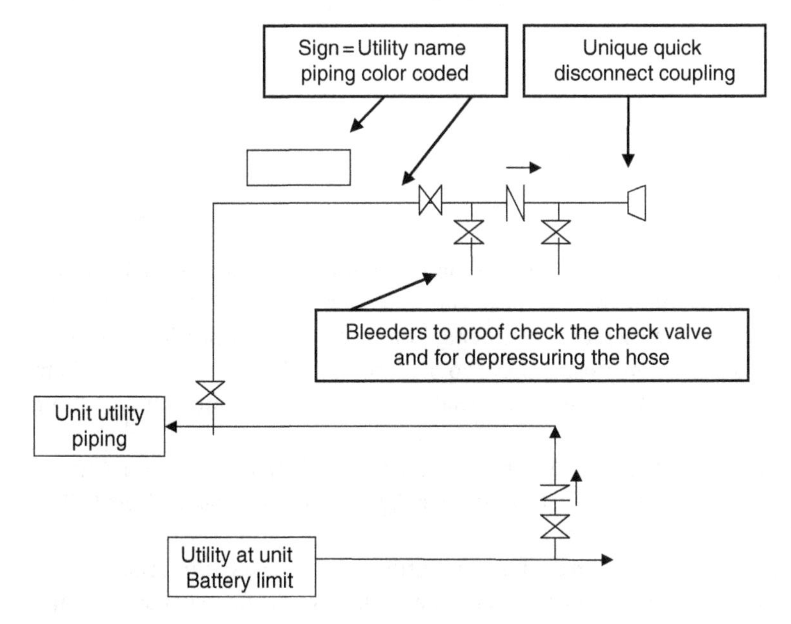

Figure 8.24.1 Example of a safe utility connection for temporary connections to process equipment (helps prevent process backflow into the utility system).

prevent the inadvertent connection of a nitrogen hose to utility air or vice versa. Also, to help avoid connection to the wrong utility, all utility headers or connections should be labeled and color-coded using a scheme that is also consistent across the site.

Plant Air

Plant air connections should be limited to utility use only and marked with signs/warnings that plant air shall not be used as breathing air. Fatalities have resulted from contract employees connecting breathing air hoses and sandblast hoods to what they thought were plant air but were nitrogen systems. Plant air is not adequately prepared or tested as breathing air and should never be used for that purpose. Breathing air should *only* be provided by compressed air cylinders certified as breathing air or air compressors designed to provide breathing air (oil-free compressors certified for breathing air use).

Plant air should not be connected directly to process equipment unless the equipment has been assured to be free of hydrocarbons (by venting and steaming, water filling, or by Nitrogen purge and subsequently gas tested to ensure they are hydrocarbon free).

The OSHA Safety Standard Regulation 29CFR CHXVII Paragraph 1926.302(b)(7) requires that all compressed air hoses exceeding 1/2-inch inside diameter shall have a safety device at the source of supply or branch line to reduce pressure in the event of a hose failure. This is to help prevent dangerous air hose whips and accidents. These compressed air safety shut-off valves are designed to provide simple but efficient protection to pneumatic systems in the event of a broken compressed air hose or pipe.

Nitrogen

Nitrogen is an asphyxiant and should only be used with good air circulation. Extreme care should be used when purging or inerting process equipment with Nitrogen. Vents, drains, and access ways should be barricaded and marked to keep personnel away during all purging or venting operations when using Nitrogen. See Chapter 8.18, which covers the "Hazards of Nitrogen and Inert Entry."

Steam

Many good process Operators have felt the pain of steam or hot condensate burns. To prevent blow-out of the steam hose, ensure that the correct hose and fittings are used for steam service and be aware of the potential for hot condensate being trapped in the steam connections or the hose. A good practice is to equip each steam hose coupling with a "whip check device." This device will not prevent a steam release should the coupling fail, but it will help prevent the hose from flailing, thereby reducing the potential for personnel to be sprayed by the hot steam or condensate or struct by the flailing hose.

What Has Happened/What Can Happen

Tosco Avon Refinery
Martinez, CA.
22 March 2000

A fire occurred at the alkylation unit. A contractor was welding in the alkylation unit with a fire watch, spraying down any errant sparks when the firewater caught on fire when flammable hydrocarbons backed into the firewater system. This incident reportedly occurred due to a separate connection between the firewater system and a hydrocarbon line without a check valve installed. This allowed hydrocarbons to back into the firewater system and exit directly below the welder from the fire watch. The hydrocarbons were quickly ignited from the welding overhead, resulting in two contractors receiving burn injuries.

This incident was highlighted in the OSHA database for industrial injuries and the listing of "Major Accidents at Chemical/Refinery Plants in Contra Costa County," published by Contra Costa County. (See Figure 8.24.2) (Extracted from OSHA injury database).

This incident highlights the importance of using a check valve when connecting a utility to a process system. Had a check valve been installed in this system, it would have prevented the hydrocarbons from backing into the firewater system, and this incident and the resulting burn injuries could have been prevented.

Tosco (now Tesoro) Avon, CA March 22, 2000	A fire occurred at the alkylation unit. A contractor was welding in the alkylation unit with a fire watch spraying down any errant sparks. The firewater caught on fire because flammable hydrocarbons backed into the fire waterline.	Two contractors were injured.

Figure 8.24.2 Excerpt from the OSHA incident report (Tosco Refinery; Martinez, CA., March 2000). The following text was copied directly from the OSHA incident report from the OSHA website.

Importance of Field Observation and Hazard Recognition

The following are only examples of the types of job site hazards that may be in your work areas. These are intended to represent observations of issues you should be alert for and observations you can identify during routine process rounds in the field. These are actual observations in refineries and chemical plants involving utility connections. The following does not represent any company or refinery and were collected over many years of personal observations at many refining and petrochemical locations worldwide.

Always intervene should you observe these types of situations on your equipment. Please refer to the following images:

Figure 8.24.3 – An unauthorized adapter used to operate pneumatic tools on Nitrogen.

Figure 8.24.4 – Using Nitrogen to purge hydrocarbons from a crude unit charge pump without a check valve to prevent reverse flow.

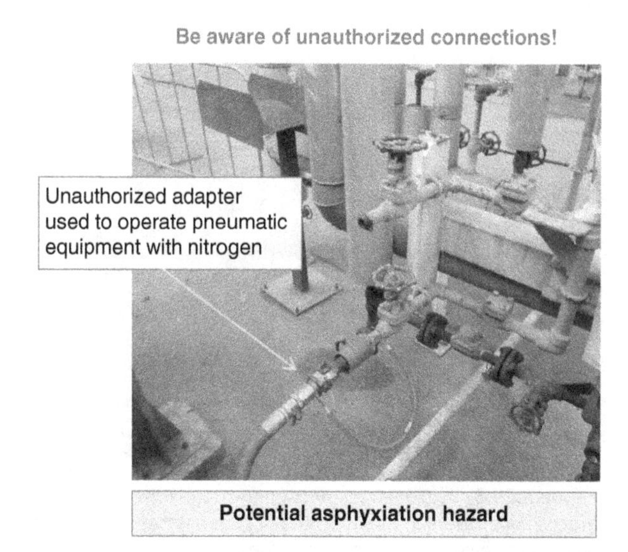

Figure 8.24.3 An unauthorized adapter used to operate pneumatic impact guns on nitrogen. A potential asphyxiation hazard!.

Figure 8.24.4 Nitrogen to purge hydrocarbons from a crude unit charge pump – no check valve to prevent backflow of hydrocarbons into the nitrogen.

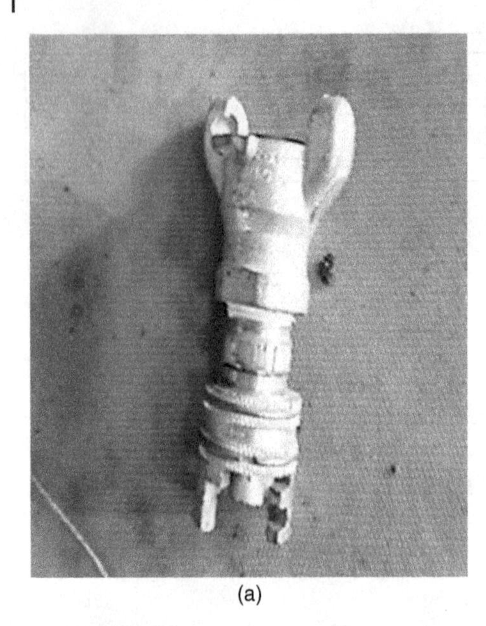

(a)

(b)

Figure 8.24.5 (a and b) Photographs of an unauthorized adapter were discovered during the unit turnaround. This is designed to cross-connect between two different utility connections. Please note that Figure (b) is connected to a hose and cross connects to a completely different utility connection. This should not be allowed.

Figure 8.24.6 Nitrogen connection to hot tap machine with no check valve to prevent backflow. Higher pressure in the hot tap process line reversed flow into the nitrogen hose, causing the hose to fail. Hydrocarbons were released from a failed nitrogen hose, spraying two contract employees with hot oil. The end of the nitrogen hose is visible in the photo (no check valve).

Figure 8.24.5 – Two unauthorized adapters used cross-connect between two different utility connections.
Figure 8.24.6 – A nitrogen connection to a hot tap machine without a check valve to prevent backflow.
Figure 8.24.7 – A utility connection with two different quick-connect fittings (utility air and Nitrogen, no labeling, and no check valve).

Key Lessons Learned

The following should be considered "Cardinal Rules" for connecting utilities to process equipment:

- Ensure that check valves are used when connecting any utility to process equipment. The check valve should be installed where the utility hose is connected to the process piping or vessel.
- Check the utility hose pressure and temperature rating and ensure the hose is designed for the service *before* connecting to process equipment.
- Never use a hose designed for one service for a different service.

Figure 8.24.7 Another example of a utility connection with two different quick-connect fittings (utility air and nitrogen, no labeling, and no check valve). Which utility is this (no label or color code)?

- Never connect potable water directly to process equipment unless a "break" tank is used to prevent the backflow of the process fluids into the potable water system.
- Never use plant air as breathing air.
- Always consider guidelines to prevent a brittle fracture before filling process equipment with water. Remember: Some equipment's MDMT is considerably higher than ambient temperature and may fail catastrophically if water at a temperature below the MDMT is used.
- Ensure process equipment is designed to support the weight of water before water-filling equipment.

Additional References

Chemical Processing, "Maintain Inert Gas Purity | Chemical Processing." https://www.chemicalprocessing.com/articles/2003/108/

Contra Costa Health Services, Contra Costa County, California, USA, "Major Accidents at Chemical/Refinery Plants in Contra Costa County", (Includes the Tosco Refinery incident mentioned in this chapter). https://cchealth.org/hazmat/accident-history.php

Piping Engineering, "Utility Stations Piping Layout". https://www.pipingengineer.org/utility-stations/

Chapter 8.24. Safe Use of Utilities and Connecting Utilities to Process Equipment

End of Chapter Quiz

1 Many incidents and even fatalities have occurred by not following basic safety precautions when connecting utilities to process equipment. What are examples of how utilities are used in refining and petrochemical operations? Name all you can.

2 What are the "cardinal rules" of utilities and connecting utilities to process equipment?

3 How much does water weigh? What is the concern about the weight of water, especially when water-filling process equipment prior to startup or for hydrostatic testing?

4 If potable water use is required, a "_____ _______" should be used to allow a free fall of the water from the utility system into the water supply tank.

The physical "break" ensures that any __________ from the process cannot enter the potable water system causing contamination of the drinking water system.

5 How many times will a water stream expand when introduced into a hot fractionation column operating above the boiling point of water?

6 How do we control what type of utility connections are allowed?

7 Another significant concern is the potential for _______ ________ during water-fill operations. The water temperature should be above the minimum design metal temperature (MDMT) for the vessel before filling the vessel. In some cases, this may require ________ the water.

8 What special precautions should be in place for purging equipment with Nitrogen when preparing equipment for mechanical work?

9 The utility hose must be rated at a temperature and pressure ________ than the utility being used AND rated _________ than the temperature and pressure of the process line or vessel being connected (in the event of unexpected backflow).

10 The hose rating should be provided by the hose ________ and should be clearly _________ or embossed on the hose.

11 Utility connects should be standardized with a _______ quick connection and utility station and hose ________ for each of the main utilities (_______, ________, ________, and _________). All site utility connections should meet this standard.

12 Utility air should not be used for which of the following:

a. Reformer regen air.
b. Sulfur unit combuster air.
c. Breathing air.
d. For powering maintenance impact guns.

13 Why is extreme care required when working with or purging equipment with Nitrogen?

8.25

The Hazards of Steam and Hot Condensate

My experience in the numerous process roles over my career tells me that most people consider steam and hot condensate a process utility, not a hazardous commodity. Many people within our industry have been seriously injured or killed due to steam or hot condensate exposure. In their regular jobs or routine tasks, some were not even aware that they were working near high-pressure steam or hot condensate systems.

A quick review of the OSHA database for a 10-year period, considering only US data and only the refining industry, indicates 26 hospitalizations and four fatalities (during the 10 years). Burn injuries are excruciating, with long treatment periods before the employee can return to regular work duties if they can ever return. Burn injuries also have the potential for disablement and disfigurement as well. See Appendix G for a review of the OSHA injury database for a summary of burns involving steam and hot condensate in the refining industry in the United States.

This chapter will help readers understand the potential exposures to burn injury and the severe consequences of a steam or condensate (or hot oil) burn.

Steam is invaluable to process operations, and we use steam in many ways. Steam is excellent for heat transfer, and we use it to apply heat to the process with steam coils in tanks or other process vessels, steam heat tracing, steam reboilers, etc. We use steam for purging equipment before shutting down or startup, for example, purging hydrocarbons during shutdown and air-freeing equipment before startup. We use steam to operate process equipment, steam turbines to drive pumps and compressors, steam stripping in fractionators for product quality control (typically superheated steam), and some firefighting operations. Steam and Steam Condensate can potentially impact Personnel Safety and Process Safety, and both will be discussed in this chapter.

Personnel contact with Steam or Hot Condensate has resulted in severe burn injuries, permanent disablement, disfigurement, and death. How Does Contact Typically Occur?

The most common injuries occur from one of the following:

- Contact burns from steam tracing (a very painful burn).
- Leaking steam tracing or faulty steam traps may result in pools of scalding water ("hot holes").
- Open-ended vertical pipes connected to steam systems may result in scalding water and the "geyser effect."
- The potential release of scalding water when breaking containment of process equipment due to inadequate isolation and trapped steam or hot water ("stored energy") and,
- The release of scalding water from a steam turbine exhaust, a safety valve outlet, or other similar steam vents.

We will discuss each of these in more detail in this chapter.

The steam or hot condensate temperature is directly related to the system pressure. The higher the system pressure, the hotter the steam or condensate. See Figure 8.25.1 for a chart indicating this pressure/temperature relationship for saturated steam. Steam removed from the presence of water can be "superheated" to much higher temperatures than shown in this chart.

With a steam or condensate burn, the severity of the burn is directly related to the temperature of the steam or hot condensate and the contact time. Even low-pressure steam and standing hot water can result in catastrophic burn injuries. Also, the Nomex or fire-retardant clothing does not protect against a steam or condensate burn. It makes the burn more severe because the fire-retardant clothing absorbs and holds the heat against our bodies for a longer time. This temperature/contact time relationship is illustrated in Figures 8.25.2 and 8.25.3.

Exposure to hot condensate produces more severe burns than steam because of its higher heat content in the liquid. Prevention against contact by positive equipment isolation must always be the primary objective, and where isolation is not possible, more robust personnel protection must be considered.

Process Operations Safety: The What, Why, and How Behind Safe Petrochemical Plant Operations, First Edition. M. Darryl Yoes.

System pressure	Temperature	
	(°F)	(°C)
20 Inch Hg (Vac)	161	72
5 Inch Hg (Vac)	203	95
0 psig	212	100
25 psig (1.7 Bar)	267	131
100 psig (6.9 Bar)	338	170
150 psig (10.3 Bar)	366	186
400 psig (27.6 Bar)	448	231
1500 psig (105.5 Bar)	597	314

Figure 8.25.1 Pressure/Temperature Relationship for Saturated Steam.

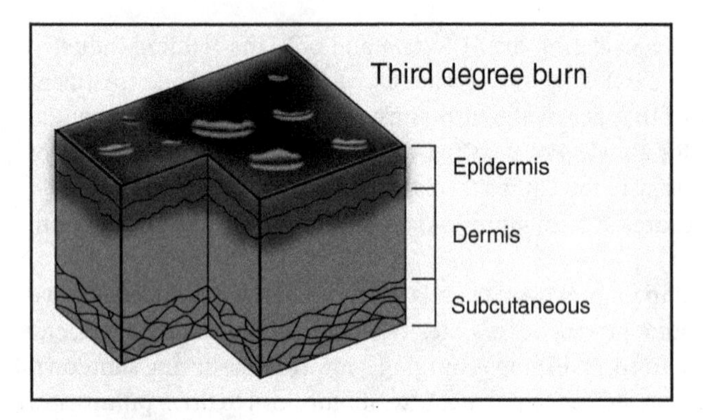

Figure 8.25.2 Third-burn. Source: Image credit Wikimedia Commons (File: Burn Degree Diagram. Svg). Wikimedia image includes Wikimedia annotations.

Time and temperature relationship to severe burns (Third degree burns)

Water temperature		*Time for a third degree burn to occur*
155° F	(68° C)	1 Second
148° F	(64° C)	2 Seconds
140° F	(60° C)	5 Seconds
133° F	(56° C)	15 Seconds
127° F	(52° C)	1 Minute
124° F	(51° C)	3 Minutes
120° F	(48° C)	5 Minutes
100° F	(37° C)	safe temperature for bathing

Remember:

- ***150 psi steam or condensate = 366°F/186°C***
- ***A "Hot Hole" with live steam = 212°F/100°C***

Remember:

- **Steam and hot condensate in our worksite has temperatures of 300 to 600F!**
- **Even low-pressure steam and standing hot water in "hot holes" or steam from steam trap discharge can result in catastrophic burns to personnel**

Chart source: American Burn Association "Scald Injury Prevention Educators Guide"

Figure 8.25.3 Chart indicating the time and temperature relationship to severe burns. Source: American Burn Association - SCALD INJURY PREVENTION Educator's Guide. A Community Fire and Burn Prevention Program Supported by the United States Fire Administration Federal Emergency Management Agency.

Types of Burns

In addition to burns from steam and hot water (thermal burns), there are at least three other types of burns that can affect our bodies. They are mentioned here for completeness:

- A thermal burn is due to exposure or contact with external heat or extreme cold, which can damage the skin and the deeper underlying tissues. A thermal burn can cause tissue damage, charring, or even death under extreme circumstances. Examples are contact with hot water, steam, hot oils, or hot metals. Thermal burns can also occur when working with hot materials, such as burning and welding.
  Inhalation of hot/cold gases or vapors can result in severe internal organ injuries and potential loss of life. The hot gases cause burns to the airway passages and can cause swelling that blocks airflow to the lungs.
- A chemical burn can result from contact with the skin or eyes with strong acids, alkalis, detergents, or some solvents. For example, this category includes contact with sulfuric acid, caustic such as sodium hypochlorite, or most hydrocarbons.
- An electrical current can severely damage the skin and underlying layers, depending on the voltage and current (amps) of the contact. This can happen with either alternating current or direct current (AC or DC current).
- The fourth category is burn due to radiation and is less of a concern in refining and petrochemical operations, but it is not zero. Radiation burns can occur from prolonged exposure to sources of radiation such as X-rays. This explains why we limit personnel from areas where inspectors are X-raying piping or vessels.

The Makeup of the Skin

The skin has three layers: the epidermis, the dermis, and the subcutaneous or fat layer. The epidermis is the outer layer or first layer of skin. The dermis is the thick layer of living tissue below the epidermis, which forms the true skin. It contains the blood capillaries, nerve endings, sweat glands, hair follicles, and other structures. The deepest layer of skin is the subcutaneous or fat layer. This layer serves many functions, including cushioning to protect our internal organs.

Burn Classification

Burns are classified as first, second, or third-degree, depending on the severity of the injury. A first-degree burn is very painful and affects only the outer exposed layer of the skin (known as the epidermis). See Figures 8.25.4–8.25.6 for illustrations of burn injuries for first, second, and third-degree burns.

Fahrenheit	Celsius	1st degree burns	2nd degree burns reversible	2nd degree burns irreversible	3rd degree burns
124°	51°	≤ 1 Minute	2 Minutes	3–4 Minutes	≤ 5 Minutes
131°	55°	5 Seconds	17 Seconds	25 Seconds	30 Seconds
140°	60°	2 Seconds	3 Seconds	3–5 Seconds	5 Seconds
148°	64°	≤ 1 Second	≤ 1 Second	≤ 2 Seconds	2 Seconds
155°	68°	≤ 1 Second	≤ 1 Second	≤ 1 Second	1 Second

Figure 8.25.4 Estimated Contact Time for 1st, $2^{nd,}$ and 3rd degree burns to occur to exposed skin. The chart also indicates contact time for Reversible vs. Irreversible Damage for 2nd-degree burn injuries. Source: American Burn Association - SCALD INJURY PREVENTION Educator's Guide. A Community Fire and Burn Prevention Program Supported by the United States Fire Administration Federal Emergency Management Agency.

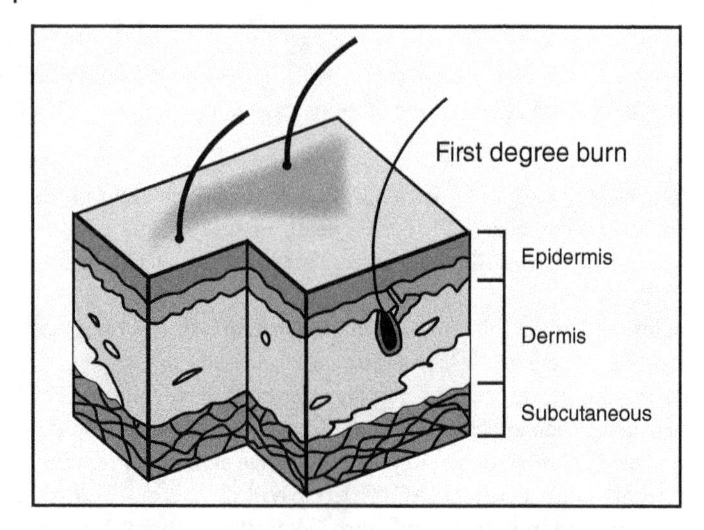

Figure 8.25.5 First-Degree Burn. Source: Image credit Wikimedia Commons (File: Burn Degree Diagram. Svg). Wikimedia image includes Wikimedia annotations.

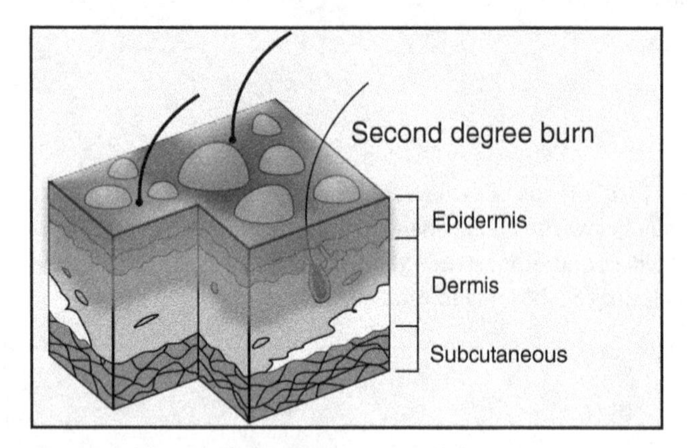

Figure 8.25.6 Second-Degree Burn. Source: Image credit Wikimedia Commons (File: Burn Degree Diagram. Svg). Wikimedia image includes Wikimedia annotations.

First-degree Burn

A first-degree burn will be red in color with a dry feel and no blistering. Most of us have experienced a mild sunburn or a mild burn from hot water without blistering. This is characteristic of a first-degree burn. There is generally no lasting damage to the skin. A first-degree burn will generally heal in three to six days with no lasting damage to the skin. Note the reddened area on the top layer of skin in the artist's sketch below.

Second-degree Burn

The burn has impacted the outer layer and underlying skin (the dermis). A second-degree burn is very painful and generally requires skin grafts and a prolonged recovery period.

A second-degree burn is also very painful. These are also sometimes referred to as a partial thickness burn and not only affect the epidermis, but a second-degree burn will also impact the underlying dermis. The burn area will be red and is almost always blistered. It may also be swollen and very tender to the touch. A second-degree burn may have some minor scarring after healing.

Third-degree Burn

The burn has severely impacted the underlying skin (the dermis) and the fat layer. Generally, there is little pain due to damage to the nerves. Consequences can be death if a large area is involved. If survived, grafting and prolonged recovery is required. Heavy scaring can also occur.

A third-degree burn is generally not very painful; this is because the nerves are severely damaged by the burn. A third-degree burn will affect the fat layer that lies beneath the dermis. The skin may be stiff with a white or tan leathery-type appearance. Most third-degree burns require skin grafts, and most do result in some scarring. A third-degree burn requires an extended time in the burn unit for treatment and can be life-threatening, depending on the percent of body exposure.

There are also fourth-degree, fifth-degree, and sixth-degree burn classifications as well. These are catastrophic burn injuries that generally result in amputations or death. We will not cover these burn injuries here.

Inhalation of hot gases such as high-temperature steam can result in severe internal organ injuries and potential loss of life. The hot gases cause burns to the airway passages and can cause swelling that blocks airflow to the lungs. These burns often occur in poorly ventilated spaces such as boiler rooms, pipeways, or steam turbine enclosures.

It is noteworthy that a person can sustain a third-degree burn with exposures less than 155 °F (68 °C) in less than one second. Figures 8.25.4–8.25.6 show that regular 150 psi (10.3 bar) utility steam and condensate have a temperature of 366 °F (185.5 °C). Exposure can result in catastrophic burn injuries.

Stepping into a hot hole at the boiling point of water (212 °F or 100 °C) can also result in catastrophic third-degree burns in less than a second. I believe this explains why steam and hot condensate burns are so severe.

Personnel Hazards When Breaking Containment

Opening process equipment in steam or hot condensate service has unique hazards that must be managed. Steam and Condensate have severe scalding potential and must always be considered hazardous materials. Complete energy isolation must comply with the site's energy isolation practices for breaking containment. Be aware of trapped hot water, for example, above a check valve. Failure to achieve energy isolation has resulted in the loss of containment and severe injury to personnel from scalding water.

Caution

Failure to achieve total isolation can be deceiving. The initial leakage/release may appear to be only cold water. However, as soon as the water is displaced, it will be followed by steam and hot condensate (scalding water).

Hazards of "Hot Holes"

"Hot Holes" are created in unpaved areas when steam traps discharge into the soil. Over a prolonged period, the steaming results in a pool of scalding water, creating a serious hazard. Personnel stepping into "hot holes" have sustained disabling burn injuries to their feet and ankles.

Steam trap discharge should be routed to avoid creating the potential for "hot holes." The release from steam traps should be routed to a closed system or secured into a sewer catch basin. Steam traps should be directed away from walkways and other areas where the discharge of scalding water can spray personnel. Any identified steam trap concerns should be promptly barricaded and reported or corrected.

The Steam "Geyser Effect"

The "Geyser Effect" results from an accumulation of hot condensate in an open-ended vertical pipe (open at the top) caused by steam either released or leaking into the pipe over a period. Over a prolonged period, the steam heats the liquid causing it to flash from water to steam resulting in a sudden and unexpected discharge or "geyser" of scalding water and steam from the open-ended pipe. The geyser may catch unsuspecting workers off guard, spraying them with scalding water. Steam turbine atmospheric exhaust vents, steam safety valve outlets, and any other type of atmospheric steam vent have the potential to create the geyser effect.

Warning signs should be placed near atmospheric vents to alert workers of the hazard. Any of the following is the minimum acceptable work practice for LOTO jobs on steam/condensate systems, especially to prevent or mitigate the geyser effect:

1. Approved and Certified single block with open low-point drain.

2. Double block with a bleeder valve in the open position between the blocks.
3. Blinds to isolate the system(s) being worked on (the preferred approach).

Additional PPE (i.e., rain suits, face shields, rubber gloves/boots) should be a standard requirement for any work on steam or condensate systems that cannot be depressurized in a controlled fashion.

Be Aware of Atmospheric Steam Vents

Hot condensate released from atmospheric steam vents has severely burned many people. Atmospheric vents should have bottom drains to prevent condensate accumulation, which can result in the geyser effect. Ensure people are cleared from the immediate area when vents are being commissioned. For work near atmospheric vents, ensure one or more of the following:

1. Follow OPE and ensure implementation and control of Hazardous Energy Practices.
2. Ensure there is an open low-point drain.
3. Use enhanced PPE such as heavy rain suits, face shields, etc.

An image of a typical steam vent is illustrated in Figure 8.25.7.

Precautions for Sampling Hot Condensate

Sampling hot condensate can potentially spray the person collecting the sample with hot condensate, resulting in serious burn injuries. Before sampling, the individual performing the task should become familiar with the hazards associated with

Figure 8.25.7 Photo of an atmospheric vent (potential for geyser effect or hot condensate release).

hot condensate/steam. The required PPE should always include a face shield and protective gloves. The condensate stream should be adequately condensed and cooled before attempting to collect the sample. A temporary cooler should be used if a permanent cooler is unavailable.

Emergency Treatment for Steam or Hot Water Burns

Call for professional medical help immediately. Burn injuries can be deceiving; the injury may not appear serious initially; therefore, unless you are a trained medical professional, call for a medical response as soon as possible.

Depending on the severity of the burn and the body parts affected, burn injuries can range from minor first-aid treatment to life-threatening. Call for emergency medical help and emergency transport (do not transport unless advised by medical professionals).

Cool the affected areas with cool running water for scalding injuries until medical help arrives (cold water or ice is not recommended). Remove tight clothing and jewelry as soon as possible, as burns will swell quickly. Once you have cooled the burn, cover it with a clean, dry dressing or a clean damp cloth.

All injuries must be reported immediately to the Team Leader and the plant medical staff.

Process Hazards Associated with Steam and Condensate

Thermal Expansion Damage Caused by Flashing Water to Steam

When water flashes to steam, the volume expands by 1,600 times at atmospheric pressure, almost 5,000 times in a vacuum, such as in a vacuum tower. Water entering hot fractionator towers has the potential for severe internal damage to the trays, resulting in the loss of containment. Often, this water is the result of condensing steam inside the vessel. Water or steam condensate entering hot process equipment during steaming operations may result in severe equipment damage and the potential for loss of containment.

Storage Tank Boilovers

Most petroleum storage tanks often have water in the bottom of the tank. During a tank fire, it is possible to overheat the water, flashing it to steam and resulting in a pressure wave inside the tank, causing the product to be expelled over the top of the tank. Usually, when this happens, the tank roof is blown clear, and the propelled product results in an explosion and fire. This is known as a tank boilover. Boilovers have devasting results, and many firefighters and others have been killed by the rapidly expanding product and the resulting fire.

Tank overheating can also result in the flashing of water bottoms, resulting in a tank boil over or froth-over. Tank contents should not exceed 200 °F (93 °C) unless designated as a "hot tank." Hot tanks must be water-free and operate above 265 °F (130 °C) to ensure trace amounts of moisture do not accumulate in the tank that could later flash, causing a boilover. See Figure 8.25.8 for an example of a storage tank boilover. Refer to Chapter 8.8 "Storage Tank Safety," for more details on how a storage tank boil over can occur and the effects of a boilover.

Condensing Steam can Create a Damaging Internal Vacuum

Condensing steam in an isolated or blocked-in vessel can generate a vacuum, resulting in equipment damage and loss of containment. For example, blocking equipment following steaming without providing a vent may result in vessel collapse or severe equipment damage. See Figure 8.25.9 for an example of a tank car damaged by a vacuum from condensing steam.

Inadvertent Heating Resulting in Thermal Expansion and Overpressure of Equipment

Another hazard of steam and hot condensate is the inadvertent heating and thermal expansion of liquids in blocked systems. Steam heating of blocked-in equipment or steam tracing of a section of pipe full of liquid can generate extreme pressures if venting or overpressure protection is not provided. This overpressure may result in piping or vessel failure and loss of containment, leading to explosion and fire. This is also covered in more detail in Chapter 8.14, "Thermal Expansion of Liquids in Closed Systems."

Figure 8.25.8 A storage tank boilover due to heating the water bottoms during the tank fire. Image of boil over of Milford Haven (UK) crude oil storage tank T011. Additional information on this incident is available in Chapter 8.8. Source: Image courtesy of Mid and West Wales Fire and Rescue Service.

Figure 8.25.9 Damage to a tank car caused by internal vacuum because of condensing steam. Source: Image with the kind courtesy of the American Institute of Chemical Engineers (AICHE), The Center for Chemical Process Safety (CCPS), and the Beacon. http://www.aiche.org/CCPS/Publications/Beacon/index.aspx.

Returning Steam Lines to Service

Unique hazards exist when returning steam lines to service. Flashing of free water in the pipe or equipment can result in steam and water hammering, causing severe equipment damage, loss of containment, and personal injury. Another more destructive process is called condensate-induced water hammer and occurs when a steam bubble in the pipe is enveloped in cooler water. The almost instant collapse of the steam creates a vacuum, causing the remaining water in the system to rush into the vacuum, resulting in the water hammer. Therefore, steam lines must be free of water before steam is introduced. All low-point drains should be cracked open to drain water before introducing steam to the equipment.

Condensate-induced Water Hammer

When water flashes to steam, it expands by 1,600 times when heated to the boiling point (1,600 times when at atmospheric pressure, less when pressurized, and more under a vacuum). We have had many steam explosions due to this basic fact.

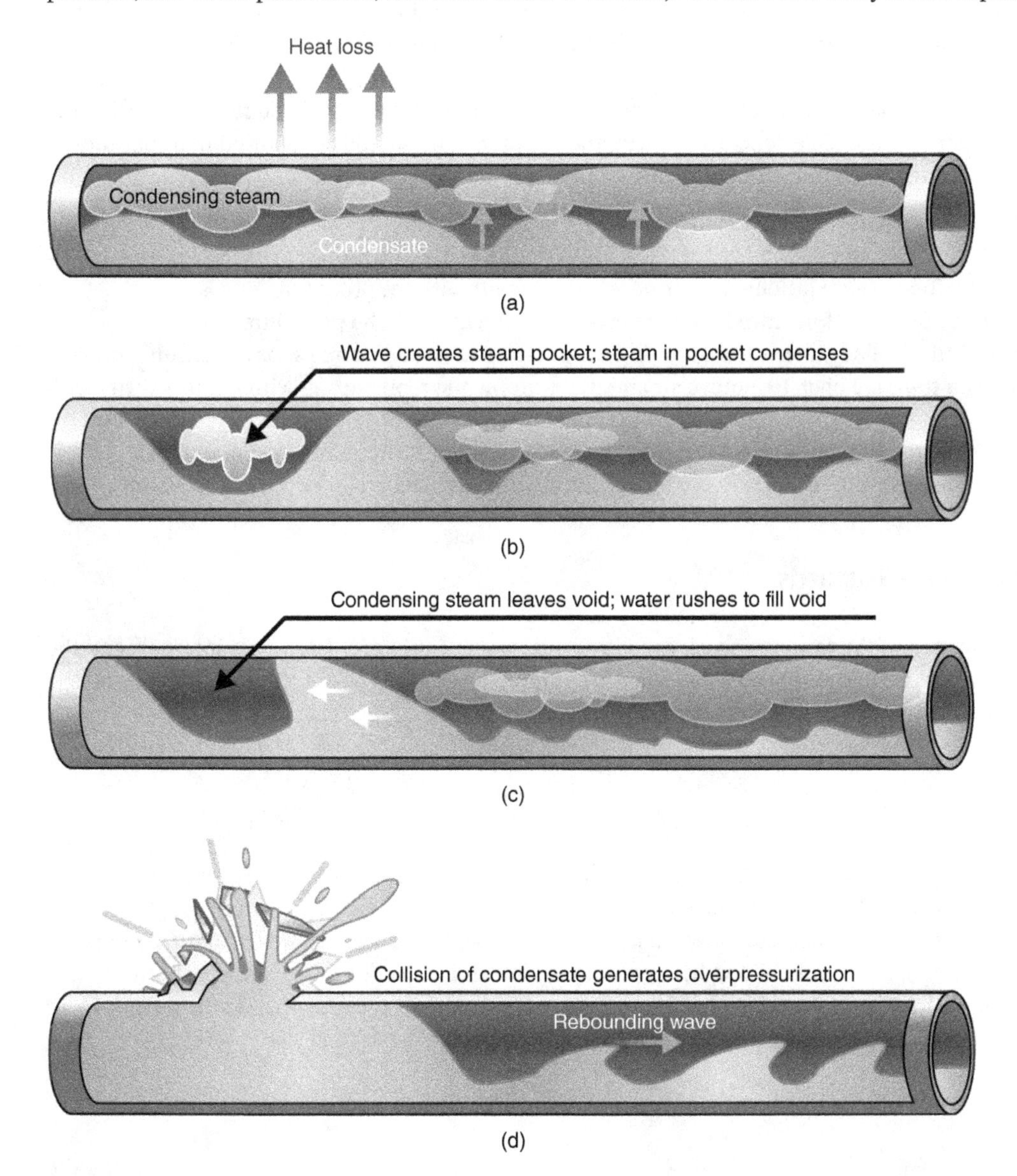

Figure 8.25.10 The Mechanism for Condensate-Induced Water Hammer. These four images illustrate the sequence of events that can lead to condensate induced water hammer. From steam flowing over condensate (water) in the first illustration, to a large wave of condensate in the fourth image resulting in pipe or vessel rupture and loss of containment. Source: Images courtesy of Ecoscience Resource Group Visual Artists. They were created to illustrate the hazards associated with condensate-induced water hammer and are not representative of any company or specific incident.

For example, when a storage tank is burning, and if it contains a water bottom, the water can be heated above its boiling point, resulting in a tank boilover, a devastating event.

When steam condenses, it can result in a powerful vacuum. We have had this occur when steaming an atmospheric tank or railcar, turning off the steam without providing a vent, resulting in tank collapse due to the vacuum from the condensing steam.

Something very similar can happen when commissioning a steam pipeline. If all of the condensate (water) is not drained or removed before placing a steam pipeline into service, the hot steam rapidly cools as it flows over the colder liquid. When this happens, the steam condenses, creating a vacuum above the condensate, which pulls in more steam, which also condenses, creating even more vacuum.

The vacuum creates a wave of water inside the piping, which can seal off a section of the pipe. When this happens, the wave is even greater, and the rushing water (condensate) creates a strong pressure wave slamming into the piping and fittings, resulting in a very loud water hammer. Severe damage can result from condensate-induced water hammer.

Condensate Induced Water Hammer is illustrated in more detail in Figure 8.25.10.

The British Health and Safety Executive has a good discussion on Condensate Induced Water Hammer on their website located at the following web address: https://www.hse.gov.uk/safetybulletins/phenomenon-of-condensate-induced-water-hammer.htm.

Ensure that a procedure specific to the equipment being placed in service has been developed and followed during the recommissioning. The commissioning should be done very slowly, and the procedure should be followed in its entirety.

The critical elements of a procedure for commissioning a steam line:

- Ensure a detailed procedure has been developed for commissioning the steam line.
  - The procedure should list all the steam traps and water drains and be followed as the procedure is implemented.
- Ensure all steam traps are functional before introducing steam – verify the list with the procedure.
- Ensure steam lines are drained of water before introducing steam – verify the list with the procedure.
- Introduce steam very slowly until the line reaches the operating temperature to avoid thermal shock or steam hammer.
- Verify that all drains and steam traps are open to remove air and free water while repressuring with steam – verify with the procedure.
- Slowly repressure until the temperature is consistent with the saturated steam table (pressure vs. temperature).
- Close the drains after the line temperature is consistent with the saturated steam table (temperature vs. pressure).

Other Steam and Condensate Hazards

Steam can generate a static charge, and the resulting spark can ignite a flammable mixture. The vessel internals and the steam nozzle should be bonded and grounded when using steam for purging or cleaning vessels. The steam hose nozzle should be grounded by a separate grounding wire or a metal braided steam hose.

High-pressure Steam

Steam leaking from a high-pressure system cannot be seen but can cut through solid objects and result in severe injury. Use all your senses when a leak is suspected. Operators have been observed using a broom to locate a leak from a high-pressure steam system. The Meet-In-The-Field should include a physical walk of the line and isolation with an Operator. Be satisfied that the equipment has been depressurized and drained before the work permit is issued. Ensure that a last-minute risk assessment (LPSA/FLRA) is completed before working on the equipment, and ensure a risk assessment using a JHA or JLA is completed for higher hazard tasks.

Whip Checks

A Whip Check is a safety device that, when properly installed, secures the steam hose and steam manifold or between two steam hoses or between the manifold and hose. The purpose of the Whip Check is to prevent personnel injury from a flailing hose in the event the connection fails. The Whip Check does not prevent loss of containment of steam but may help direct the release away from personnel. See Figure 8.25.11 for a properly installed whip check.

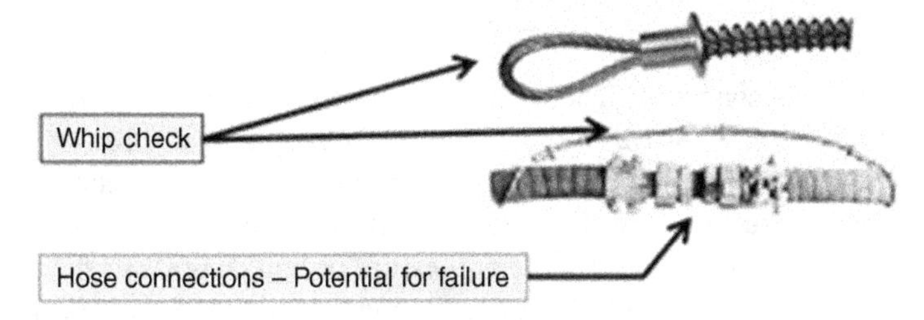

Figure 8.25.11 Properly installed whip check to prevent a flailing hose in the event of hose failure.

What Has Happened/What Can Happen

BP Texas City Refinery
Texas City, Texas
2 September 2004

The information for this incident was derived from a BP HSSE Communication, a Preliminary Incident Investigation Update. Thanks to BP for sharing the key learnings from this incident.

A hot condensate loss of containment occurred at the BP Texas City Refinery Ultraformer #3 unit on 2 September 2004. The incident happened during a routine maintenance task, replacing the check valve on the discharge side of a condensate (hot water) pump. This incident resulted in the deaths of two mechanics, and a third received severe burn injuries when drenched with scalding water.

In this incident, the Scalding water was released from the 10″ (25 cc) 300# check valve while opening the flange between the check valve and the discharge valve. The 12″ discharge valve was partially open due to an internal obstruction. The check valve had inadvertently become the isolation and was holding pressure between the check and discharge valves. Energy isolation was not confirmed between the check valve and the discharge block valve ("trapped energy"). Figure 8.25.12 shows the process configuration of the pump and its associated valves and the check valve to be replaced.

A review of the OSHA database discovered another incident at the same site on 26 April 1989, when Amoco owned the site. A Pipefitter was drenched with hot water when a pipeline was opened to replace a valve. The Pipefitter sustained thermal burns over a large portion of his body and died from his injuries. The lack of positive energy isolation also appears to have been the root cause of that incident.

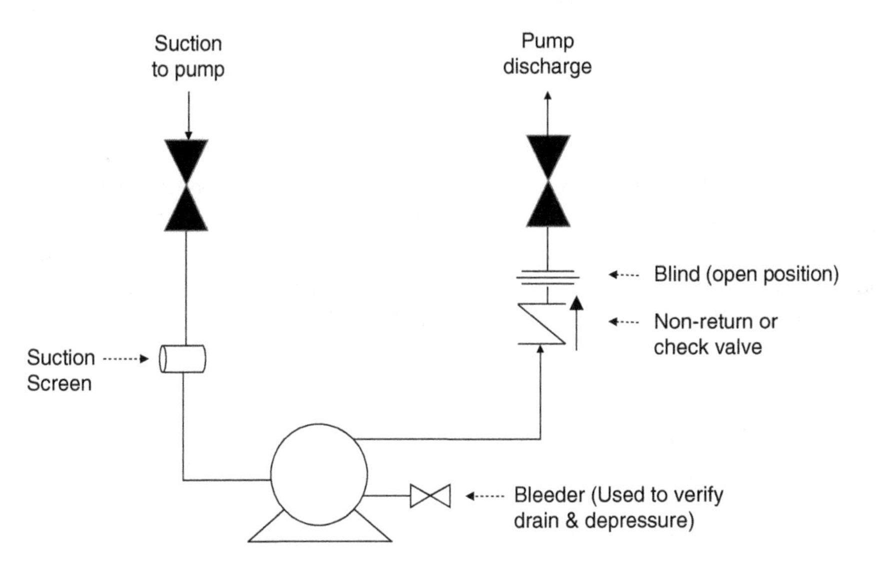

Figure 8.25.12 This sketch was recreated to illustrate how this incident occurred. Source: Image courtesy of Ecoscience Resource Group Visual Artists. The following is a representation of a process hot water pump being prepared for maintenance.

Key Lessons Learned

In the BP incident, it appears that positive energy isolation was not confirmed. There appears to have been no consideration of a leaking discharge valve that could hold pressure between the check valve and the pump discharge piping circuit.

Other factors to consider for a task such as this one include the following. Consider the position of the workers relative to the flange they are opening. Ensure they are not below the flange where they can encounter residual process fluids or pressure while opening the flanges. Always open the flanges slowly, and if there is pressure or liquids present, then immediately reclose the flanges and call the process personnel back to the job site. Consider the type of PPE the workers are wearing. For a task such as this, they should wear heavy-duty rain gear with face shields and heavy gloves. The rain gear will shed the hot water away from their clothing, helping to minimize the burns. Otherwise, the hot water will soak into their clothing, holding the heat and resulting in more severe burns.

What Has Happened/What Can Happen

Goodyear Tire and Rubber Company
Houston, Texas
11 June 2008

This incident occurred at the Goodyear Tire and Rubber Company in Houston, TX, on 11 June 2008. The root cause was thermal expansion caused by heating liquids in closed systems. See Chapter 8.14 for more discussion on the "Hazards of Thermal Expansion in Closed Systems."

Employees closed valves to isolate a rupture disk protecting the shell side of an Ammonia exchanger to replace the failed disk. This effectively isolated the shell side of the exchanger from its overpressure protection. The shell side was filled with pressurized liquid anhydrous ammonia. Later, steam heating was applied to the tube side of the exchanger, which heated the ammonia on the shell side, resulting in the violent rupture of the shell and release of the contents. One Operator was passing through the area and was killed in this incident. The shell side isolation valves and relief systems were found closed. This incident was investigated by the US Chemical Safety and Hazard Investigation Board. This incident was also discussed in Chapter 8.14 (Hazards of Thermal Expansion). Please refer to Figure 8.14.5 for a photo of the ruptured Ammonia heat exchanger at Goodyear Tire and Rubber Company on 11 June 2008.

A very similar incident occurred at the Williams Olefins Company facility located in Geismar, LA, on 13 June 2013. Two employees were killed in that incident, and 167 employees and contract personnel reported injuries. The CSB also investigated the Williams Olefins incident and published a report and video on the incident. The Williams Olefins incident is also discussed in Chapter 8.14, "Hazards of Thermal Expansion in Closed Systems."

Key Lessons Learned

Be aware of isolation or disablement of relief valves, even for a few minutes. Process temperatures and pressures can change dramatically in a very short period, especially if the overpressure protection is isolated. Ensure that a process is in place to car seal isolation valves in the correct position and that management approval is required for any defeat of overpressure protection. Also, an alternate relief path or similar protection must be in place before defeating or closing the overpressure protection. The alternate protection should replace the functionality of the defeated device. See Chapter 8.14 for more discussion on disablement or defeating safety-critical devices ("Ensuring the Availability of SSHE Critical Devices").

What Has Happened/What Can Happen

MEG Energy Corp. Steam Pipeline Failure and Hydrocarbon Release
Christmas Lake Regional Project located about 93 miles (150 kilometers) south of Fort McMurray, Canada, 5 May 2007

At 5:35 a.m. on Tuesday, 5 May 2007, a new 24″ diameter high-pressure steam pipeline failed catastrophically due to a condensation-induced steam/water hammer. The pipeline was fabricated from schedule 160 pipe, with 2.34″ wall thickness.

When the steam pipeline failed, sections of the piping were propelled into a broad surrounding area; one section 100 feet long and weighing over 50,000 lbs was propelled about 2,600 feet (800 meters). Some of those sections impacted adjacent hydrocarbon piping, releasing hydrocarbon to the surrounding environment; others took down adjacent electrical power lines, resulting in power outages.

The Energy Resources Conservation Board (ECRB) investigated this incident, and a detailed report was written to document the investigation. A summary of the results of the ERCB investigation is included here. The report found the failure was due to mechanical overload (due to the water hammer) and not from weld or material defects. The report indicated that the procedures were written at a high level without sufficient detail, especially related to heating, monitoring, and pressuring the steam header. The procedure left the detailed steps to the discretion of the Operators. Also, modifications had been made to the piping system during construction, eliminating some drains and drain tanks. The pipeline was placed into service without adequate controls and instrumentation to eliminate potential hazards.

A condensation-induced steam/water hammer is sometimes also referred to as steam bubble collapse because that is what it is. This is a violent action caused by trapped liquid, usually steam condensate, in a steam line. When this liquid accumulates in a steam line, it causes some steam to condense, creating more liquid and causing the steam to give up heat. As the steam bubble collapses into liquid, it leaves a void, resulting in steam rushing in to fill the void. The steam flowing over the liquid creates a "wave" of liquid inside the pipe, referred to as "the Bernoulli effect." As this progresses, it worsens as the liquid fills the pipe on one end, effectively sealing off the pipe internally. As the steam bubbles collapse and the steam rushes in to fill the void, it results in a violent hammering and can destroy the pipe, as in the Christmas Lake incident.

I have seen photographs of this incident but was unable to obtain permission to use them in this book. The scene looked much like the children's game of "pick up sticks," except the sticks were 50,000-pound (22.7 metric tons) (mt) sections of 24″ (~61 centimeter) diameter heavy wall piping. It was an unbelievable scene. It could have been a very bad day if this happened on a regular process unit with people around.

Key Lessons Learned

The investigation into this incident determined that some design modifications had been made during construction, and the facilities to drain condensate were not present in the constructed piping. Therefore, there was no practical way to drain the liquid from the new steam piping during startup and keep the liquid drained while heating the piping. Also, the procedures were written at a high level and were not detailed to ensure the piping was continuously drained as the piping was being heated to place it in service.

The investigation confirmed that the piping integrity was intact and did not play a role in the failure. This was a straightforward lack of drains to remove water and a lack of detailed procedures for Operators to follow while commissioning the new steam piping.

Remember, as indicated in this chapter, it takes incredible patience to successfully return a steam line to service. To reiterate some key steps in this process, always ensure that the steam line is completely drained of free water and that all steam traps are functional before introducing steam. Introduce the steam very slowly and continuously drain water from the drains as the line heats up. All drains should remain open as the line is heating and allow condensing water to drain. Slowly repressure the steam line until the line temperature is consistent with the temperature on the saturated steam table (pressure vs. temperature). The drains can be closed at this point, and the line is considered in service.

What Has Happened/What Can Happen

Sonatrach Skikda LNG Liquefaction plant
Skikda, Algiers, 20 January 2004
Plant Explosion with 27 Fatalities

The Sonatrach Skikda LNG plant and refinery (known locally as the GL-1K complex) began production at a site about 310 miles (500 kilometer) east of Algiers (in the country of Algeria) in the early 1970s. The state-owned oil and gas company's Sonatrach complex had grown to six trains, with the last train commissioned in 1981. All six trains had been upgraded to current plant specifications and had an LNG capacity of about 7.7 million tonnes per year.

On 20 January 2004, a fired boiler exploded, resulting in the deaths of 27 workers; 72 others were injured, and seven were reported to be missing. The explosion resulted in extensive damage to the facility, with three of the six LNG trains destroyed.

The explosion and fire also damaged an electrical power plant and other adjacent industrial facilities. A marine terminal designed for the loading of LNG tankers was also heavily damaged in the event.

Immediately before the massive explosion, unusual noises and high vibrations were heard from the steam boiler, leading to speculation that the boiler had experienced slug or two-phase flow. It is known that slug flow can result in extreme vibration and has resulted in piping and boiler failure at other sites. Other reports attributed the explosion to an inadequate purge of the firebox. However, after a more extensive study, the cause was determined to be the failure of a cold box (heat exchanger) designed to exchange heat between the refrigerant (typically LPG) and the LNG. In either case, the 1,500 PSI boiler exploded with extreme force, causing further loss of containment of adjacent LNG-containing equipment and a

Figure 8.25.13 Image of the Sonatrach Skikda LNG Liquefaction plant after the January 2004 explosion and fire. Source: Photo courtesy of Richard A. Hawrelak (Presented to CBE 497, CBE 317 at UWO in 1999).

Figure 8.25.14 Image of the Sonatrach Skikda LNG Liquefaction plant after the January 2004 explosion and fire. Source: Photo courtesy of Richard A. Hawrelak (Presented to CBE 497, CBE 317 at UWO in 1999).

disastrous explosion. The 335K B/D refinery and the LNG plant were shut down due to the damage from the explosion and to facilitate the investigation.

LNG Liquefaction train six restarted in May 2002, followed by trains five and ten in September. The remaining trains (trains 20, 30, and 40) were destroyed. Property damage was estimated to be nearly 1 billion in current dollars. The original fired steam boilers have been replaced with safer and more efficient gas-fired turbines and compressors. The Skikda LNG explosion had numerous impacts on the design, construction, and operation of LNG projects and facilities the world over.

Chapter 8.15, "Hazards of Pneumatic Testing," discusses the extreme hazards of pneumatic testing equipment using air, nitrogen, or other pneumatic as the medium. The Skikda explosion of the 1,500 steam boiler illustrates the powerful force of pressurized equipment when a pneumatic is involved as the medium. In the Skikda incident, the investigation concluded the cause of the initial explosion was due to the failure of the cold box exchanger and the resulting loss of containment of LPG and/or LNG, causing the initial explosion. However, it is clear that the force of the explosion was made much worse by the failure of the highly pressurized steam boiler. High-pressure (1,500 psig) steam boilers have extreme potential or "stored" energy, and if suddenly released into the environment, they can easily be equivalent to many pounds of TNT. This type of damage is clearly illustrated in the images in Figures 8.25.13 and 8.25.14. This is the very definition of a BLEVE (a Boiling Liquid Expanding Vapor Explosion).

Key Lessons Learned

Pressure vessels operating at high pressures, especially when pressurized by a pneumatic, have extremely high energy potential. The formula is pressure times area = force. A pneumatic such as air, nitrogen, or any other gas, including steam, has a large amount of "stored" energy. If this energy is suddenly released, whatever the cause, it can result in devastating consequences. However, a vessel pressurized by a noncompressible liquid will quickly depressurize as soon as a leak occurs. In the Skikda incident, the vessel was pressurized by steam, which is a compressible vapor. When the explosion occurred, the steam drum ruptured violently, releasing a large amount of stored energy.

What Has Happened/What Can Happen

The above incidents are certainly tragic and involve loss of life. However, there have been many other incidents related to steam or water hammer. A great website for additional incident descriptions and details is located at Kirsner Consulting Engineering (https://www.kirsner.org/pages/forensicResAlt.html). A couple of additional incidents from this report are described below.

Incident 1: The New York State Public Service Commission investigated and reported on one such incident, which occurred on a steam incident that occurred in New York State. The report indicated that the failure was likely caused by a condensate-induced steam event that ruptured the high-pressure steam piping. This incident was reported by Kirsner Consulting Engineers and is posted on their website. Please refer to Figure 8.25.15 for an image of the damage associated with this incident.

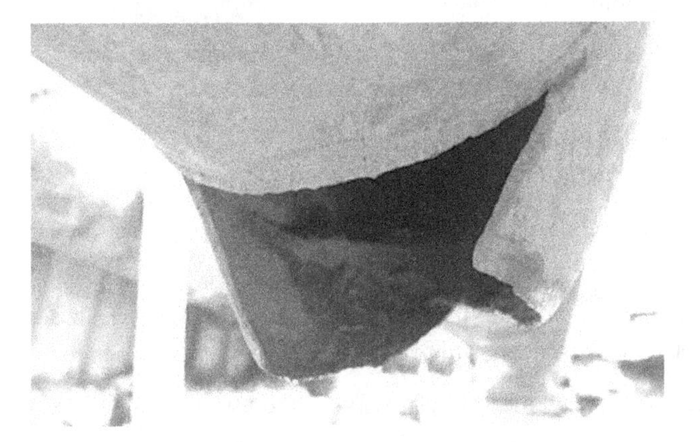

Figure 8.25.15 Rupture High-Pressure Steam Piping reported by New York State Public Service Commission. Source: Image courtesy of Kirsner Consulting Engineering.

Figure 8.25.16 Failed High-Pressure Steam Block Valve due to failure to continuously drain condensate from the piping during the commissioning process. Source: Image courtesy of Kirsner Consulting Engineering.

Incident 2: A steam worker was killed while commissioning a branch steam line at the Industrial Campus. OSHA issued a citation and fine, contending that the OSHA lockout and tagout standard had not been followed. The actual cause was determined by Kirsner Engineering to be related to the commissioning procedures lacking a requirement for the opening of a drain valve and maintaining continuous draining of the steam piping during the heating process. See Figure 8.25.16 for a photo of the failed valve due to a condensate-induced steam hammer.

Additional References

The Engineering Toolbox, "Properties of Saturated Steam – Imperial Units". https://www.engineeringtoolbox.com/saturated-steam-properties-d_273.html

Wikipedia, Burn Severity Chart Signs and Symptoms: https://en.wikipedia.org/wiki/Burn

American Burn Association, "Scald Injury Prevention", Educator's Guide, (Includes chart "Time and Temperature Relationship to Severe Burns."). http://ameriburn.org/wp-content/uploads/2017/04/scaldinjuryeducatorsguide.pdf

Energy Resources Conservation Board (ERCB), ERCB Investigation Report, "MEG Energy Corp. Steam Pipeline Failure", September 2, 2008. https://www.aer.ca/documents/reports/IR_20080902_MEG_Energy.pdf

Mayo Clinic, "Burns" (Symptoms and Treatment). https://www.mayoclinic.org/diseases-conditions/burns/symptoms-causes/syc-20370539

ABSA: The Pressure Equipment Safety Authority (Alberta, Canada), "Condensate-Induced Water Hammer Safety Principle & Recommendations". https://www.absa.ca/media/1490/ab-503_waterhammer_card.pdf

The Process Piping (A good article on Condensate Induced Water Hammer). https://www.theprocesspiping.com/water-hammer/

Steam Accidents and Forensic Investigations. https://www.kirsner.org/pages/forensicResAlt.html

Chapter 8.25. Hazards of Steam and Hot Condensate

End of Chapter Quiz

1 Steam and Hot Condensate are utilities, but they are also very _______ _________.

2 How many ways that steam is used in a refinery or petrochemical plant can you list? List as many as you can.

3 Personnel contact with Steam or Hot Condensate has resulted in serious burn injuries and even death. How Does Contact Typically Occur? Name the most common types of burn injuries.

4 Steam leaking from a high-pressure system, operating above about 400 psi, can cut through solid objects and can result in severe injury.

Can steam leaking from a high-pressure system like this be seen?

5 What is a Whip Check, and why are they commonly used on steam hoses?

6 Following steam out of a process vessel, rail car, or storage tank, what additional step is essential before allowing the vessel to cool?

7 What is one of the most predominant issues with returning high-pressure steam lines to service?

8 The steam or hot condensate temperature is directly related to the system pressure. As the system pressure is increased, does the temperature rise or fall?

9 Explain the most common way a "hot hole" is created.

10 Explain the key elements of the procedure for commissioning a high-pressure steam line that has been out of service.

11 How long of contact time is required for catastrophic burn injuries if a person comes into direct contact with live steam or hot condensate?

12 How can these painful injuries and suffering be prevented?

13 What causes steam hammer in steam piping systems?

8.26

Electrical Hazardous Area Classification System

Using household electrical appliances or even household electrical extension cords in an industrial setting where flammable hydrocarbons may exist can readily ignite most flammable hydrocarbons and lead to a disaster. Standard household appliances create electrical sparks and have more than enough energy to ignite a flammable atmosphere. In industry, this is addressed by the Electrical Hazardous Area Classification System, a requirement of the National Electrical Code, or NFPA 70. The Electrical Hazardous Area Classification System ensures that equipment suitable for the surrounding atmosphere is used and that arcing and sparking electrical equipment is only used in areas where no flammable atmosphere exists or is used only in enclosed protective equipment designed to contain an explosion inside the equipment.

The Occupational Safety and Health Administration (OSHA) requires the electrical area classification system, and compliance is enforced. The definitions for electrical hazardous area classification can be found in OSHA 29 CFR 1910.307 and 1910.399. There are two systems used to classify hazardous areas: The Class/Division system and the Zone system. The Class/Division system is used predominately in the refining industry in the United States, whereas most other countries typically use the Zone System. For example, in Canada, similar requirements are in place; the Canadian regulation is referred to as the Canadian Electrical Code, which uses the Class/Zone system.

The National Electric Code (NEC), NFPA 70, is published by the National Fire Protection Association and covers the details and requirements of area classification. Article 500 describes the NEC division classification system; articles 505 and 506 define the NEC zone classification system. The NEC zone classification system was created to provide companies involved in international operations with a system that could correspond with the NEC classification system used in the United States.

The US Electrical Hazardous Area Classification System classifies hazardous areas by Class, Division, Group, or Class and Zone. Using this system helps identify areas that will or may have flammable gases, combustible dusts, and ignitable fibers present and which may be ignited by arcing and sparking electrical equipment. The OSHA regulations also define the classified areas and describe the various protective systems allowed in each classified area.

The following is a general description of the NEC Electrical Hazardous Area Classification Class/Division System:

Class

The Class defines the general nature (or properties) of the hazardous material in the surrounding atmosphere, which may or may not be in significant concentrations.

a. **Class I:** Locations in which flammable gases or vapors may or may not be in concentrations to produce explosive or ignitable mixtures.
b. **Class II:** Locations in which combustible dust (either in suspension, intermittently, or periodically) may or may not be in enough quantities to produce explosive or ignitable mixtures.
c. **Class III:** Locations in which ignitable fibers may or may not be in enough quantities to produce explosive or ignitable mixtures.

☆ Note: This chapter is provided as an educational aid (like the rest of this book) and should *not be used* in actual field applications. Please refer to the appropriate regulations and to the relevant engineering guides for field use.

Process Operations Safety: The What, Why, and How Behind Safe Petrochemical Plant Operations, First Edition. M. Darryl Yoes.

Division

The Division defines the probability of the hazardous material being able to produce an explosive or ignitable mixture based on its presence in a mixture with air.

a. **Division 1:** Indicates that the hazardous material is highly likely to produce an explosive or ignitable mixture due to it being present continuously, intermittently, or periodically or from the equipment itself under normal operating conditions.
b. **Division 2:** Indicates that the hazardous material has a low probability of producing an explosive or ignitable mixture and is present only during abnormal conditions for a short period.

Class/Division

A Class I, Division 1 location is one:

- In which ignitable concentrations of flammable gases or vapors may exist under normal operating conditions; or,
- Where ignitable concentrations of such gases or vapors may frequently exist because of repair or maintenance operations or because of leakage; or,
- In which breakdown or faulty operation of equipment or processes might release ignitable concentrations of flammable gases or vapors and cause simultaneous failure of electric equipment.

A Class I, Division 2 location is a location:

- In which volatile flammable liquids or flammable gases are handled, processed, or used, but in which the hazardous liquids, vapors, or gases will normally be confined within closed containers or closed systems from which they can escape only in the event of accidental rupture or breakdown of such containers or systems, or as a result of abnormal operation of equipment; or,
- In which ignitable concentrations of gases or vapors are normally prevented by positive mechanical ventilation and which might become hazardous through failure or abnormal operations of the ventilating equipment; or,
- That is adjacent to a Class I, Division 1 location and to which ignitable concentrations of gases or vapors might occasionally be communicated unless such communication is prevented by adequate positive-pressure ventilation from a source of clean air and effective safeguards against ventilation failure are provided.

A Class II, Division 1 location:

- This classification may include areas of grain handling and processing plants, starch plants, sugar-pulverizing plants, malting plants, hay-grinding plants, coal pulverizing plants, areas where metal dusts and powders are produced or processed, and other similar locations that contain dust-producing machinery and equipment (except where the equipment is dust-tight or vented to the outside). Under normal operating conditions, these areas would have combustible dust in the air in quantities sufficient to produce explosive or ignitable mixtures.
- Combustible dusts that are electrically nonconductive include dusts produced in the handling and processing of grain and grain products, pulverized sugar and cocoa, dried egg and milk powders, pulverized spices, starch and pastes, potato and wood flour, oil meal from beans and seed, dried hay, and other organic materials which may produce combustible dusts when processed or handled.
- Dusts containing magnesium or aluminum are particularly hazardous, and extreme caution is necessary to avoid ignition and explosion.

A Class II, Division 2 location is a location where:

- Combustible dust will not normally be in suspension in the air in quantities sufficient to produce explosive or ignitable mixtures, and dust accumulations will normally be insufficient to interfere with the normal operation of electric equipment or other apparatus, but combustible dust may be in suspension in the air as a result of infrequent malfunctioning of handling or processing equipment; and
- Resulting combustible dust accumulations on, in, or in the vicinity of the electric equipment may be sufficient to interfere with the safe dissipation of heat from electric equipment or may be ignitable by abnormal operation or failure of electric equipment.

A Class III, Division 1 location is a location in which easily ignitable fibers or materials producing combustible flying's are handled, manufactured, or used.
A Class III, Division 2 location is a location in which easily ignitable fibers are stored or handled other than in the process of manufacture.

Division vs. Zone

Some sites use Zone instead of Division, which is allowed by US OSHA and API RP-505. Employers may use the zone classification system as an alternative to the division classification system for electric and electronic equipment and wiring for all voltage in Class I, Zone 0, Zone 1, and Zone 2 hazardous (classified) locations where fire or explosion hazards may exist due to flammable gases, vapors, or liquids (29 CFR 1910.307) (g) (1). The classifications are similar; however, the process areas are classified as Zone 0, Zone 1, and Zone 2 as follows:

a. **A Class I, Zone 0 location is a location in which one of the following conditions exists:**
   - Ignitable concentrations of flammable gases or vapors are present continuously, or,
   - Ignitable concentrations of flammable gases or vapors are present for long periods of time.
b. **A Class I, Zone 1 location is a location in which one of the following conditions exists:**
   - Ignitable concentrations of flammable gases or vapors are likely to exist under normal operating conditions; or,
   - Ignitable concentrations of flammable gases or vapors may frequently exist because of repair or maintenance operations or because of leakage; or,
   - Equipment is operated, or processes are carried on of such a nature that equipment breakdown or faulty operations could result in the release of ignitable concentrations of flammable gases or vapors and cause simultaneous failure of electric equipment in a manner that would cause the electric equipment to become a source of ignition; or,
   - A location adjacent to a Class I, Zone 0 location from which ignitable concentrations of vapors could be communicated unless communication is prevented by adequate positive-pressure ventilation from a source of clean air and effective safeguards against ventilation failure are provided.
c. **A Class I, Zone 2 location is a location in which one of the following conditions exists:**
   - Ignitable concentrations of flammable gases or vapors are not likely to occur in normal operation and, if they do occur, will exist only for a short period; or,
   - Volatile flammable liquids, flammable gases, or flammable vapors are handled, processed, or used, but the liquids, gases, or vapors are normally confined within closed containers or closed systems from which they can escape only as a result of accidental rupture or breakdown of the containers or system or as the result of the abnormal operation of the equipment with which the liquids or gases are handled, processed, or used; or,
   - Ignitable concentrations of flammable gases or vapors normally are prevented by positive mechanical ventilation but may become hazardous as the result of failure or abnormal operation of the ventilation equipment; or,
   - A location adjacent to a Class I, Zone 1 location, from which ignitable concentrations of flammable gases or vapors could be communicated unless such communication is prevented by adequate positive-pressure ventilation from a source of clean air and effective safeguards against ventilation failure are provided.

Division and Zone Classification

- In Class I locations, an installation must be classified as using the division or zone classification system.
- In Class II and Class III locations, an installation must be classified using the division classification system (1910.307) (a)(4).

Group

The Group defines the type of hazardous material in the surrounding atmosphere. Groups A, B, C, and D are for gases (Class I only), while groups E, F, and G are for dusts (Class II) and ignitable fibers (Class III).

a. **Group A:** Atmospheres containing acetylene.

b. **Group B:** Typical gases include hydrogen, butadiene, ethylene oxide, propylene oxide, and acrolein. Flammable gases in this category may create atmospheres containing a flammable gas mixture, flammable liquid-produced vapor, or combustible liquid-produced vapor whose maximum experimental safe gap (MESG) is less than 0.45 millimeter or minimum ignition current (MIC) ratio is less than 0.40.
c. **Group C:** Typical gases include ethyl ether, ethylene, acetaldehyde, and cyclopropane. Flammable gases in this category may create atmospheres containing a flammable gas mixture, flammable liquid-produced vapor, or combustible liquid-produced vapor whose MESG is greater than 0.45 millimeter but less than or equal to 0.75 millimeter or MIC ratio is greater than 0.40 but less than or equal to 0.80.
d. **Group D:** This is the most common Hazardous Area Classification in petroleum refineries. Typical gases include acetone, ammonia, benzene, butane, ethanol, gasoline, methane, natural gas, naphtha, and propane. Flammable gases in this category may create atmospheres containing a flammable gas mixture, flammable liquid-produced vapor, or combustible liquid-produced vapor whose MESG is greater than 0.75 millimeter or MIC ratio is greater than 0.80.
e. **Group E:** Atmospheres containing combustible metal dusts such as aluminum, magnesium, and their commercial alloys.
f. **Group F:** Atmospheres containing combustible carbonaceous dust with 8% or more trapped volatiles such as carbon black, coal, or coke dust.
g. **Group G:** Atmospheres containing combustible dusts not included in Group E or Group F. Typical dusts include flour, starch, grain, wood, plastic, and other chemicals.

Electrical Hazardous Area Drawings and Classification of Electrical Equipment

Guidelines for developing the Electrical Hazardous Area Classification drawings are presented in API RP 505. Electrical classification drawings shall be produced as part of any new design, plant modification, or maintenance that may change or impact the hazardous area classification. The electrical classification drawings shall indicate all electrically classified areas both vertically, including above and below grade, and horizontally via plan, elevation, and section views.

The electrical classification drawing shall be referenced during all plant or area modifications and maintenance activities to ensure that the work does not change the electrical classifications. Also, all equipment placed in the classified areas must meet the electrical classification indicated on the drawings. Refer to API Recommended Practice 500 and 505 for additional details and other information that should be included on the electrical classification drawings.

Classification of areas and selection of equipment and wiring methods shall be under the supervision of a qualified registered professional engineer (OSHA 29CFR 1910.307) (g)(4)(i).

Equipment shall be approved not only for the class of location but also for the ignitable or combustible properties of the specific gas, vapor, dust, or fiber that will be present (OSHA 29CFR 1910.307) (c)(2)(i).

Each room, section, or area shall be considered individually in determining its classification (1910.307) (a)(1).

A Class I, Division 1, or Division 2 location may be reclassified as a Class I, Zone 0, Zone 1, or Zone 2 location only if all of the space that is classified because of a single flammable gas or vapor source is reclassified (1910.307) (g)(4)(iii).

Equipment shall be marked to show the class, group, and operating temperature or temperature range, based on operation in a 40 °C ambient, for which it is approved. The temperature marking may not exceed the ignition temperature of the specific gas or vapor to be encountered. However, the following provisions modify this marking requirement for specific equipment (OSHA 29 CFR 1910.307) (c)(2)(ii).

- Equipment of the nonheat-producing type, such as junction boxes, conduit, and fittings, and equipment of the heat-producing type having a maximum temperature not more than 100 °C (212 °F) need not have a marked operating temperature or temperature range.
- Fixed lighting fixtures marked for use in Class I, Division 2, or Class II, Division 2 locations only need not be marked to indicate the group.

Electrical Hazardous Area Drawings and Classification of Electrical Equipment (Continued):

- Fixed general-purpose equipment in Class I locations, other than lighting fixtures that are acceptable for use in Class I, Division 2 locations, need not be marked with the class, group, division, or operating temperature.

- Fixed dust-tight equipment, other than lighting fixtures that are acceptable for use in Class II, Division 2, and Class III locations, need not be marked with the class, group, division, or operating temperature, and
- Electric equipment suitable for ambient temperatures exceeding 40 °C (104°F) shall be marked with both the maximum ambient temperature and the operating temperature or temperature range at that ambient temperature and,
- Safe for the hazardous (classified) location. Equipment that is safe for the location shall be of a type and design that the employer demonstrates will provide protection from the hazards arising from the combustibility and flammability of vapors, liquids, gases, dusts, or fibers involved.

Please refer to API RP-500 and RP-505 for the recommended Class I, Division 2 recommended distances (Note: API RP 500 covers Division I and II areas; API RP-505 covers Zones 0, 1, and 2) (not shown here). In both documents, the authors emphasize that the documents are to be used as a guide and that additional engineering and judgment should be used, depending on the circumstances involved.

Numerous other examples and recommendations for safe distances for hydrocarbon sources, including other drawings available in API RP-500 and API RP-505. Please review and use these important and readily available resources when evaluating Electrical Hazardous Classifications in process areas.

Electrical Hazardous Area Safeguards

Hazard Elimination

A variety of methods or technologies are used to provide protection and help minimize the risks of ignition of a flammable vapor by arcing and sparking electrical devices. The most obvious of these is the elimination of arcing and sparking devices from areas considered to be hazardous, physically isolating the hazard, or placing or relocating the normal arcing and sparking equipment to a nonhazardous area.

Elimination of arcing devices should always be considered when evaluating hazardous areas. However, elimination is not always possible; therefore, let us look at other considerations to minimize the threat of ignition of flammable vapors by electrical equipment. Each of the devices mentioned here should be certified and have a code stamp indicating that the equipment is compatible with the electrical area classification where it is to be used.

Nonincendive Equipment

Some electrical devices are considered "nonincendive" or incapable of producing a spark with enough energy to ignite a flammable atmosphere under normal conditions. Generally, hearing aids or ordinary battery-operated wristwatches fall into this category. They are generally considered nonincendive devices incapable of igniting a flammable atmosphere under normal conditions. A code stamp is not required.

Explosion-proof Equipment

An explosion-proof device generally contains the arcing electrical equipment, such as relays or solenoids, inside an enclosure that can withstand an internal explosion of flammable vapors, thereby preventing the ignition of explosive gas or vapor that may be in the surrounding area. The enclosure contains the explosion and cools the resulting hot vapors to a temperature below the ignition temperature of vapors in the surrounding area before releasing those vapors. All bolts must be in place and tight to contain the internal explosion that may occur should gas or vapor reach the internals. Refer to Figures 8.26.1 and 8.26.2 images of explosion-proof equipment.

Intrinsically Safe Equipment

An intrinsically safe device under normal or abnormal conditions is incapable of releasing enough electrical or thermal energy to ignite vapors in the surrounding atmosphere, even in its most easily ignitable concentration. Devices classified intrinsically safe will have a stamp or seal of approval by one of the certification agencies (FM – Factory Mutual Research Corporation, UL – Underwriters Laboratories, or in Canada by the Canadian Standards Association, and by the CENELEC in the European Union). Process radios and flashlights are generally designed and certified to be intrinsically safe devices. Refer to Figure 8.26.3 for an image of intrinsically safe equipment.

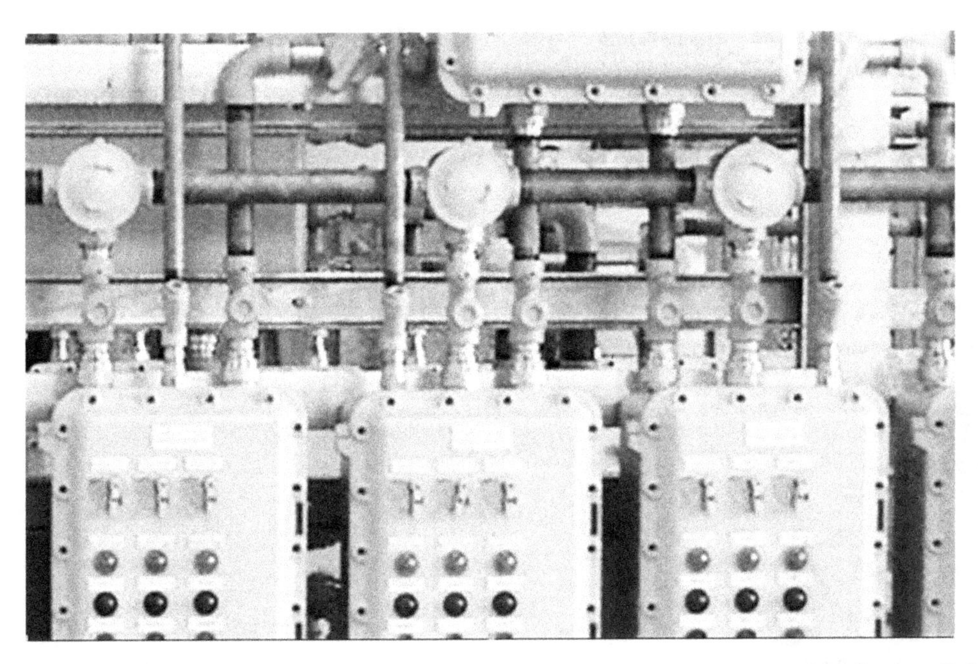

Figure 8.26.1 Example of explosion-proof electrical enclosures. Note: For these to be effective, all the bolts, plugs, and covers or caps must be installed and tight.

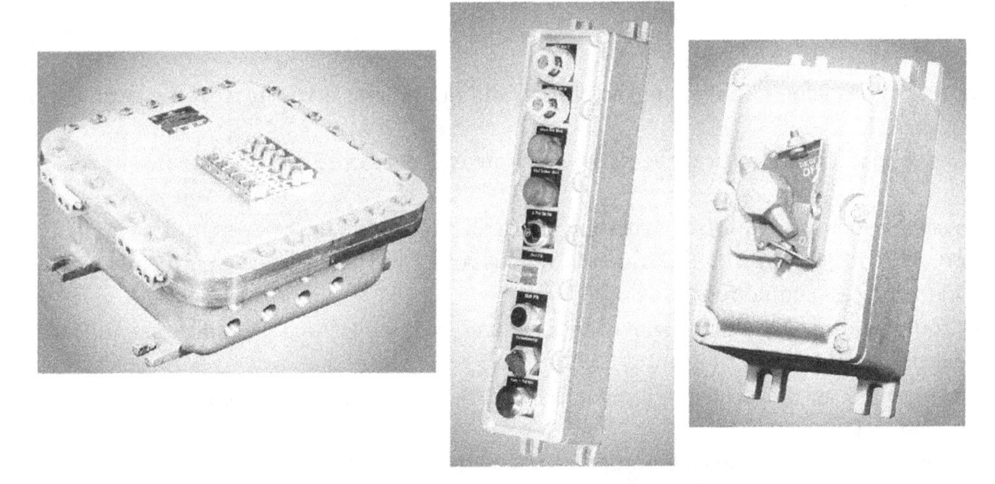

Figure 8.26.2 Random examples of other types of explosion-proof (Division II) electrical equipment.

However, "simple devices" such as thermocouples, resistive sensors, LEDs, and switches may be employed in a hazardous area without certification, provided the device does not generate or store more than 1.2 V, 0.1 A, 20 μJ, and 25 mW. This IEC (International Electrotechnical Commission) definition is now used in the USA and Canada.

Dust Ignition-proof

The dust ignition-proof device is a type of protection that excludes ignitable amounts of dust or amounts that might affect performance or rating when the device is installed and used per the original design for the equipment. A device approved for use in areas of ignitable dust such as Class III, Division 1 area, will not ignite dust in the atmosphere or accumulate dusts. The device controls or prevents the ignition from arcs, sparks, or heat otherwise generated inside the enclosure to cause ignition.

Figure 8.26.3 Flashlight with UL code stamp for use in hazardous areas. Resent to Akron Electric with ongoing correspondence – we should receive this one. Suggest we obtain Approval for this image or use a stock image of a Division II Electrical Equipment.

Key Lessons Learned

- Resources and references are readily available to help with the classification of process areas to meet the regulatory requirements.
- Ensure that all process areas are reviewed, and electrical hazardous areas are well developed and documented by a Professional Engineer.
- This documentation shall be available to those authorized to design, install, inspect, maintain, or operate electric equipment at the location (OSHA 29CFR 1910.307) (b). The Electrical Hazardous Area Drawings are posted in clearly visible process work areas (typically in the process control room).
- Ensure all maintenance and equipment modifications follow the Electrical Hazardous Area Classification (EHSC), including all newly installed electrical equipment.
- When using electrical equipment in process areas, ensure it complies with the EHAC. Otherwise, ensure a gas test is completed and a work permit is issued for its use.
- Ensure all bolts and plugs are tightly installed in explosion-proof enclosures in hazardous areas. Bolts or plugs should not be missing or loose.

Additional References

OSHA 29 CFR 1910.307 "Hazardous (classified) locations."

OSHA 29 CFR 1910.399 "Definitions applicable to this subpart."

NFPA-70 "The National Electrical Code"

NFPA-497 "Classification of Flammable Liquids, Gases, or Vapors and of Hazardous (Classified) Locations for Electrical Installations in Chemical Process Areas"

NFPA-499 "Classification of Combustible Dusts and of Hazardous Locations for Electrical Installations in Chemical Process Areas"

API RP-500 "Recommended Practice for Classification of Locations for Electrical Installations at Petroleum Facilities"

API RP-505 "Recommended Practice for Classification of Locations for Electrical Installations at Petroleum Facilities Classified as Class I, Zone 0, Zone 1, and Zone 2"

Chapter 8.26. Electrical Hazardous Area Classification System

End of Chapter Quiz

1. Where do the requirements for electrical area classification system in the United States come from?
2. Can you explain an Electrical Hazardous Area Classification of Class 1, Division 2, Group "D"?
3. What is meant by intrinsically safe?
4. How do you know if your device is intrinsically safe?
5. What is meant by explosion poof?
6. How many of the enclosure bolts can be left loose or missing, and the device still protects against an explosion outside the enclosure?
7. What is meant by a Class I, Division 1 classified area?
8. Devices classified ____________ _______ will have a stamp or seal of approval by one of the certification agencies (FM – Factory Mutual Research Corporation, UL – Underwriters Laboratories, or in Canada by the Canadian Standards Association, and by the CENELEC in the European Union).
9. What types of common process equipment and devices are intrinsically safe?
10. What are two additional good reference standards covering classification of electrical installations at petroleum refineries?

8.27

Hazards of Combustible Dust

In a detailed report on combustible dust hazards, the US Chemical Safety and Hazard Investigation Board identified 281 combustible dust incidents between 1980 and 2005, resulting in 119 fatalities and 718 injuries. The CSB report identified an average of 10 dust explosion incidents per year and indicated that this may be low due to the potential for underreporting in the earlier years. The CSB and OSHA have issued detailed incident reports on the hazards of combustible dust explosions. Yet they continue to occur with alarming frequency and result in many deaths, injuries, and catastrophic property damage. Most of these incidents are catastrophic, with fatalities or serious injuries occurring in 71% of the events.

Although there has been a significant amount of publicity, combustible dust hazards are still not recognized in most industries; several combustible dust explosions and fires have occurred at sites where the potential for a dust explosion was not recognized by those sites prior to the incident. At some sites, inspections had been completed prior to the incidents by safety and insurance professionals who also had not recognized the explosion potential of combustible dust. Some Material Safety Data Sheets (MSDS) do not identify or recognize the potential for explosions from combustible dusts.

Essentially, any material that will burn in the air as a solid form can become a combustible dust when reduced to a finely divided form such as granules or powders. Generally, combustible dust with a particle size of less than 420 microns (about that of fine sand) is explosive unless testing demonstrates otherwise. Relatively small amounts of combustible dust have resulted in catastrophic damage. The National Fire Protection Association (NFPA) has warned that amounts as small as 1/32nd of an inch in thickness over only about 5% of a room's surface can be a combustible dust hazard. This is roughly the thickness of a dime.

Any combustible dust that will burn can be explosive when dispersed into the air in fine particulates. Some examples of materials that can result in combustible dust include:

- Coal
- Petroleum coke
- Grains (rice, corn, wheat, sugar, etc.)
- Metals (e.g., aluminum, magnesium, iron, etc.)
- Plastics
- Resins
- Wood

The Hazards of Combustible Dusts

The traditional fire triangle requires the presence of Fuel, Oxygen, and an Ignition source for a fire or explosion to occur. This also holds true for a fire involving combustible dust. However, a dust explosion requires the simultaneous presence of these three elements and two others: dust dispersion and confinement. If either the dispersion or confinement is removed, a fire may still occur, but an explosion may be prevented. The elements needed for a dust explosion were illustrated in the CSB report on Combustible Dust (shown in Figure 8.27.1). The aeration of accumulated dust can cause dispersion through routine process activities or a sudden and unexpected "shock" wave. Combustible dust explosions have frequently occurred as a secondary explosion following another unrelated but adjacent smaller explosion.

The ignition source in a dust cloud explosion has been frequently identified as static electricity. All ignition sources must be controlled when dusts are present, but static electricity is a significant hazard since a dust cloud can accumulate a static charge and hold that charge for as long as five hours. See Chapter 8.22 for additional information on "Controlling Static Electricity."

Process Operations Safety: The What, Why, and How Behind Safe Petrochemical Plant Operations, First Edition. M. Darryl Yoes.

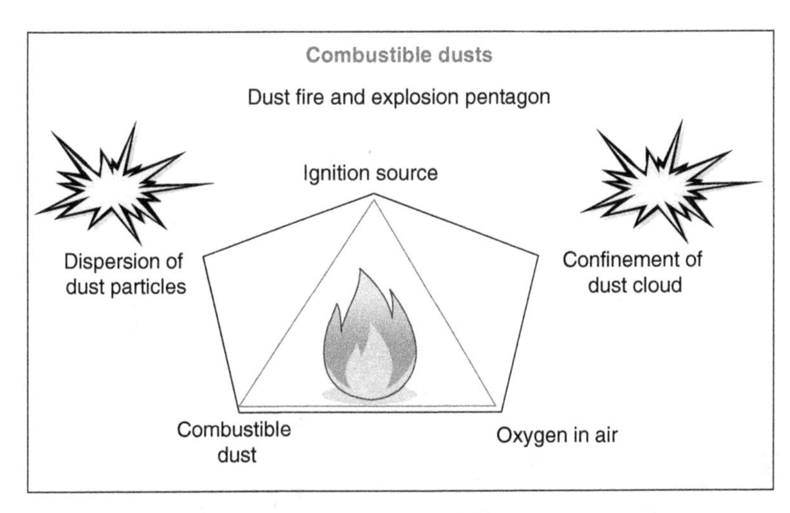

Figure 8.27.1 Elements required to support a combustible dust explosion. Source: Image courtesy US Chemical Safety and Hazard Investigation Board (annotations by CSB).

In direct response to a large number of incidents as outlined at the beginning of this chapter, and due to the consequence of these events, OSHA has enacted a Combustible Dust National Emphasis Program (NEP) where OSHA will be conducting audits of facilities for compliance with the appropriate standards (29 CFR 1910.22) (housekeeping) or, where appropriate, 29 CFR 1910.176) (c) (housekeeping in storage areas), among other OSHA standards. Information on this NEP is available on the OSHA website (refer to directive number CPL 03-00-008).

Additionally, the National Fire Protection Association (NFPA) has issued a standard covering the "Fundamentals of Combustible Dusts" (NFPA 652). The requirements of NFPA-652 are applied retroactively, and compliance is required by the regulation regardless of how long the facility has been operating.

NFPA-652 is a comprehensive standard and requires each site that generates combustible dust or may accumulate combustible dust to do the following:

- Determine the combustibility and explosibility hazards of materials they handle.
- Identify and assess any fire, flash fire, and explosion hazards or Dust Hazard Analysis (DHA).
- Manage the identified fire, flash fire, and explosion hazards.
- Communicate the hazards to Operators and other affected personnel
- Operating procedures must be documented, and practices implemented to prevent or mitigate dust explosions and fires.
- Training and hazard awareness is a significant part of NFPA-652 compliance; however, the type and depth of training expected by all personnel can vary depending on their responsibilities.
- Housekeeping becomes an even more important part of compliance with NFPA-652 due to the dynamics of a dust explosion. Layers of very fine dust can result in significant fires but, even more significantly, can result in catastrophic explosions. Housekeeping must be audited periodically to ensure it is effective.
- Hot Work requires well-thought-out procedures, including consideration for the potential ignition of accumulated dust.
- Emergency response procedures and training should consider the appropriate firefighting response and equipment to prevent the spreading of a dust cloud. Also, consider the type of dust to ensure the appropriate extinguishing medium is used.
- The incident investigation procedure should be documented to ensure qualified and experienced people are involved and that recommendations are developed to prevent a reoccurrence of the event.

What Has Happened/What Can Happen

West Pharmaceutical Services, Inc.
Kinston, North Carolina
29 January 2003 (6 Fatalities, 38 injured)

(a) (b)

Figure 8.27.2 West Pharmaceutical Services explosion and fire damage occurred on 29 January 2003 (6 fatalities and 38 injured).You can see from (a) that this was a catastrophic event and that many lives were changed forever. The damage is even more evident in (b). Source: US Chemical Safety and Hazard Investigation Board.

An explosion of fine plastic powder used in the manufacture of polyethylene products occurred at the West Pharmaceutical Services in Kinston, North Carolina, on 29 January 2003, resulting in the deaths of 6 employees and an additional 38 people injured. The pharmaceutical services facility was destroyed.

The US Chemical Safety and Hazard Investigation Board investigated this incident and concluded that it occurred "after polyethylene powder accumulated on surfaces above a suspended ceiling, providing fuel for a devastating secondary explosion." They noted that the production areas were extremely clean; however, few employees knew the dust had accumulated above the suspended ceiling. Additionally, the electrical equipment in the ceiling area was not suitable for an area with combustible dust (the National Electrical Code requires equipment rated for Class II where combustible dust is present).

Additional details on this incident can be found in the CSB "Combustible Dust Hazard Study" and a safety bulletin prepared by The North Carolina Department of Labor (Division of Occupational Safety and Health). See Figure 8.27.2a,b for photos of the West Pharmaceutical Services explosion and fire damage.

What Has Happened/What Can Happen

Imperial Sugar Company
Port Wentworth, Georgia
7 February 2008 (14 Fatalities, 38 injured)

This incident is very similar to the West Pharmaceutical Services event. This explosion was also investigated by the US Chemical Safety and Hazard Investigation Board, and they reported that this was the deadliest combustible dust explosion in the United States in decades. The CSB found similarities between the accumulation of combustible dust in the West explosion and the accumulation of sugar dust in sugar processing and packaging areas at Imperial Sugar Company. Sugar is a grain and combustible; therefore, sugar accumulation in powder form quickly becomes a significant combustible dust hazard.

This incident started with a small explosion at 7:15 p.m. as the initiating event, which released additional dust and rapidly propagated the incident to other areas of the facility. The explosions caused catastrophic damage to the facility, resulting in electrical power loss, loss of lighting, and damage to stairwells and other escape routes. Many of the escape routes were

Figure 8.27.3 The three photographs in Figure (a, b, and c) also clearly show the catastrophic damage which occurred at the Imperial Sugar Company explosion and fire. Fourteen lives were lost and thirty-eight were injured, some seriously in this tragic event. I realize that it is difficult to comprehend that dust can cause this much havoc, damage, and death. Source: US Chemical and Hazard Investigation Safety Board.

also reportedly blocked by the damage. The explosions killed eight people, and six more died in hospitals. The explosions and resulting fires injured 38. The facility was essentially destroyed.

The accumulation of combustible dust at Imperial appears to have several causes. The facility was equipped with dust evacuation systems; however, they were inoperable at the time of the incident, and operations practices appear to have accepted the spilling of sugar as a normal occurrence. The bottom line is that significant accumulations of sugar were present in several locations throughout the facility.

This incident highlights how attention to detail and prevention of dust accumulation can go a long way toward the prevention of a catastrophic dust cloud explosion. See Figure 8.27.3a–c for photos of the Imperial Sugar Company explosion and fire.

Key Lessons Learned

The potential for a combustible dust explosion is not well recognized or known, even by some safety and insurance professionals, and reliance on the product's MSDS for combustible dust explosion potential is uncertain. Continued awareness and training on the combustible dust hazard are needed.

A combustible dust explosion can occur when dust accumulates in confined areas and where a source of ignition is present, such as electrical devices or a static charge. A relatively small amount of dust is required as fuel for a combustible dust explosion; therefore, good housekeeping is essential to prevent dust build-up, including in areas that are usually not visible to workers, such as in suspended ceilings, air or product ductwork, shelves, or similar places where dust can accumulate.

Companies must develop and implement comprehensive dust control, housekeeping, and training programs. These programs should be exposed to periodic audits and reviews to ensure their effectiveness in preventing dust accumulation that can lead to catastrophic explosions. Dust accumulation should never be taken for granted; be aware of the hazard of complacency regarding dust accumulation. Dust control goes far beyond routine housekeeping!

Ignition sources should be controlled when dust is present in manufacturing operations, including properly rated electrical equipment and other arcing/sparking equipment. Static electricity may also be a factor when dust is present, and therefore, all equipment should be electrically bonded and grounded to help prevent a stray static charge.

Companies with the potential for generating or accumulating combustible dust should comply with the NFPA Combustible Dust Standard (NFPA-652) requirements to help ensure against a devasting explosion.

Some manufacturing where dust is usually present in closed processing equipment may require inert blanketing using Nitrogen or other inert gas to eliminate the presence of Oxygen. Of course, Nitrogen brings about many additional safety concerns that must be managed. See Chapter 8.18 for further discussion about the "Hazards of Nitrogen."

Ensuring effective written and practiced emergency and evacuation procedures is also essential. Drills and exercises should be conducted periodically to ensure workers are aware of how to escape in an event like those described in this chapter. See Chapter 9 for additional information on the importance of "Operations Emergency Response Drills and Exercises."

Please refer to the "Additional References" section below for additional resources on Combustible Dusts. Agencies such as the Chemical Safety Board (CSB), The National Fire Protection Association (NFPA), The North Carolina Department of Labor, and The Occupational Safety and Health Administration (OSHA) have all developed a wide range of Combustible Dust training and awareness bulletins, and readily available videos. The New York Times has even written an Op-Ed article on the Dangers of Combustible Dust. I am sure that there are many others also available on the web. Many of these are ready for in-house training and as a resource for developing procedures for protecting a production site against combustible dust hazards. I encourage those of you who may have this hazard at your workplace to take full advantage of these materials and put them to use.

Additional References

U.S. Chemical and Hazard Investigation Board, West Pharmaceutical Services Dust Explosion and Fire, Report No. 2003-07-I-NC September 2004. https://www.csb.gov/assets/1/20/csb_westreport.pdf?13815

U.S. Chemical and Hazard Investigation Board, Sugar Dust Explosion and Fire, Imperial Sugar Company, Wentworth, GA., Report No. 2008-05-I-GA September 2009.

National Fire Protection Association (NFPA), NFPA 652 "Standard on the Hazards of Combustible Dusts".

U.S. Chemical and Hazard Investigation Board, "Combustible Dust Hazard Study", Report No. 2006-H-1, November 2006. file:///C:/Users/Owner/AppData/Local/Packages/Microsoft.MicrosoftEdge_8wekyb3d8bbwe/TempState/Downloads/Dust_Final_Report_Website_11-17-06%20(1).pdf

U.S. Chemical and Hazard Investigation Board, "Combustible Dust Fact Sheet". https://www.csb.gov/assets/1/6/csb_2018_factsheet_combustibledust_05.pdf?16219

U.S. Chemical and Hazard Investigation Board (Videos), "Combustible Dust an Insidious Hazard", "Combustible Dust Solutions Delayed", "Inferno: Dust Explosion at Imperial Sugar".

North Carolina Department of Labor, Industry Alert on Combustible Dust. https://www.labor.nc.gov/safety-and-health/occupational-safety-and-health/occupational-safety-and-health-topic-pages/combustible-dust#hazard-overview

Occupational Safety and Health Administration (OSHA), Safety and Health Information Bulletin, "Combustible Dust in Industry: Preventing and Mitigating the Effects of Fire and Explosions". https://www.osha.gov/dts/shib/shib073105.html

Occupational Safety and Health Administration (OSHA), Safety and Health Information Bulletin, Combustible Dust: An Explosion Hazard. A list of OSHA bulletins and articles related to combustible dust hazards. https://www.osha.gov/dsg/combustibledust/guidance.html

Occupational Safety and Health Administration (OSHA), Combustible Dust National Emphasis Program (Reissued), Directive Number CPL 03-00-008. https://www.osha.gov/enforcement/directives/cpl-03-00-006

Occupational Safety and Health Administration (OSHA), Combustible Dust Awareness Training (PPT). https://view.officeapps.live.com/op/view.aspx?src=https%3A%2F%2Fwww.osha.gov%2Fsites%2Fdefault%2Ffiles%2F2018-12%2Ffy08_sh-17797-08_combustible_dust.ppt

Occupational Safety and Health Administration (OSHA), Fact Sheet "Combustible Dust Explosions". https://www.osha.gov/OshDoc/data_General_Facts/OSHAcombustibledust.html

Occupational Safety and Health Administration (OSHA), Pamphlet prepared by the Texas Engineering Extension Service (TEEX), "Combustible Dust Explosion Hazard Awareness". https://www.osha.gov/sites/default/files/2018-12/fy08_sh-17798-08_participant_manual.pdf

North Carolina Department of Labor, "A Guide to Combustible Dusts", PDF Document. https://files.nc.gov/ncdol/osh/publications/ig43.pdf

Plastic News article on NFPA 652, "NFPA 652: Beyond the Dust Hazard Analysis". https://www.plasticsnews.com/perspective/nfpa-652-beyond-dust-hazard-analysis

The New York Times (Op-Ed article), The Danger of Combustible Dust, By Rafael Moure-Eraso. https://www.nytimes.com/2014/08/23/opinion/the-danger-of-combustible-dust.html

Chapter 8.27. Hazards of Combustible Dust

End of Chapter Quiz

1. How would you describe a combustible dust?

2. The typical fire triangle has three sides: Fuel, Air or Oxygen, and Heat or an ignition source. What are the additional elements needed to form a combustible dust explosion?

3. Companies must develop and implement comprehensive ________ ________, ______________, and training programs.

4. What resource is readily available to provide guidance and support to industry to help minimize the opportunity for dust explosions?

5 Any combustible dust that will burn can be explosive when dispersed into the air in fine particulates. Please list as many materials that can result in combustible dust explosions as possible.

6 Most of these incidents are ___________, with fatalities or serious injuries occurring in 71% of the events.

7 Although there has been a significant amount of publicity, combustible dust hazards are still not ___________ in most industries; several combustible dust explosions and fires have occurred at sites where the potential for a dust explosion was not recognized by those sites prior to the incident.

8 The National Fire Protection Association (NFPA) has warned that amounts as small as 1/32nd of an inch in thickness over only about ___% of a room's surface can be a combustible dust hazard.

9 Where dust clouds are concerned, _________ electricity is a significant hazard since a dust cloud can accumulate a static charge and hold that _________ for as long as five hours.
Companies with the potential for generating or accumulating combustible dust should comply with the NFPA Combustible Dust Standard (NFPA-652) requirements to help ensure against a devasting explosion.

8.28

Hazards of Undetected or Uncontrolled Exothermic Reactions

An exothermic reaction pertains to a chemical change accompanied by a liberation of heat and is a term used to describe a reaction or process that releases energy in the form of heat. In the petroleum and petrochemical industries, exothermic reactions occur when hydrocarbons are exposed to heat, pressure, and catalysts in the presence of Hydrogen. For example, reactions such as hydrocracking and olefin saturation are exothermic reactions. However, uncontrolled exothermic reactions can occur in these and in other processes if reaction temperatures are not closely controlled.

Exothermic reactions can also occur by mixing two or more incompatible chemicals. For example, mixing acid and water will result in a violent exothermic reaction. Mixing incompatible chemicals is covered in Chapter 8.31, "Hazards of Chemical Incompatibility or Reactivity." This chapter focuses more on uncontrolled exothermic reactions associated with hydro-processing equipment, equipment where hydrocarbons are processed at high temperatures in the presence of catalyst and Hydrogen.

A simple example of an exothermic reaction is burning Methane in the air. This reaction is exothermic as it generates heat and produces CO_2 as a byproduct of the reaction. This energy change is illustrated in the following equation and generates about 215,000 Calories of heat:

$$CH_4 + 2O_2 \rightarrow CO_2 + 2H_2O + 213 \text{ K cal}$$

Of course, this is a very beneficial reaction, and we use this process to heat our homes, cook our food, and for many other good and valuable purposes. This module will talk about what can happen when the reaction occurs uncontrolled.

There are many examples of refining and petrochemical processes that are, by their nature, exothermic reactions. These processes involve hydrocarbons in the presence of heat, catalyst, and hydrogen. The following are a few of the processes that are exothermic (not intended to be a complete list):

- Hydrocracking
- Hydrotreating
- Sulfiding of Hydrotreating and Hydrocracking Catalyst
- Alkylation
- Isomerization
- Olefins Polymerization
- Claus Sulfur Plants

Undetected or uncontrolled exothermic reactions, which can lead to temperature excursions, otherwise known as a "thermal runaway," have occurred in hydrocracking and other processes, especially during startup and other transient operations. The hydrocracking reaction (and other similar reactions) generates heat and causes the reaction to accelerate. These reactions must be recognized and controlled to ensure temperatures remain in a safe range, especially during transient operations like startup and unit upsets. Uncontrolled reactions can lead to temperatures exceeding the design limits for equipment, resulting in equipment failures, loss of containment, fires/explosions, and unit or catalyst damage. An undetected or uncontrolled thermal runaway can create a catastrophic incident and loss of lives.

The British Health and Safety Executive (HSE) defines a thermal runaway as follows: "An exothermic reaction can lead to thermal runaway, which begins when the heat produced by the reaction exceeds the heat removed. The surplus heat raises the temperature of the reaction mass, which causes the rate of reaction to increase. This, in turn, accelerates the rate of heat production. An approximate rule of thumb suggests that reaction rate – and hence the heat generation rate – doubles with every 10 °C rise." Note: 10 °C is equal to about 18 °F; therefore, the rate of heat generation doubles with about every 18 °F increase in temperature.

Process Operations Safety: The What, Why, and How Behind Safe Petrochemical Plant Operations, First Edition. M. Darryl Yoes.

The following include some of the more common causes of exothermic reactions or thermal runaways and the resulting temperature excursions (examples):

- Uneven flow and heat distribution in the catalyst bed cause internal (catalyst) or external (vessel shell) hot spots. These hot spots can trigger additional and self-propagating exothermic reactions.
- Internal reactor failures lead to catalyst migration and dead or "stagnant" zones. This can interrupt reactor quench flow distribution, leading to hot spots and exothermic reactions. For example, the ineffective sealing of the catalyst support bed to the reactor shell allows the catalyst to migrate from one bed onto the distribution grid for the bed below. A change in reactor bed differential pressure (DP) may indicate catalyst migration, with the catalyst migrating through catalyst seals from an upper catalyst bed into a lower bed. Catalyst migration has been known to result in internal hot spots in the catalyst bed, damaging the reactor and resulting in loss of containment.
- Incomplete sulfiding of catalyst. Additional sulfurization of the catalyst is exothermic and can lead to uncontrolled exothermic reactions in the catalyst.
- Raising reactor temperatures too quickly when using a fresh, highly reactive catalyst. The higher temperatures and very active catalysts can rapidly trigger uncontrolled exothermic reactions.
- Uncontrolled reaction temperatures or failure to respond quickly to a rapid increase in reaction temperatures. Failure to respond to a rapid rise in reactor temperatures by failure to recognize the temperature rise or failure to increase quench rates can lead to uncontrolled exothermic reactions.
- Feed temperature too high can increase reactor temperatures leading to uncontrolled exothermic reactions.
- Loss of recycle gas. The recycle gas serves as a quench to the catalyst. The loss of compressors or failure of the control system can lead to a loss of quench, a rapid rise in reaction temperatures, and an uncontrolled exothermic reaction.
- Low recycle gas or oil flow rate. Again, this results in inadequate quench to the reactor catalyst and a rapid rise in reactor temperatures.
- Inadequate reserve quench gas capacity or operating with the quench gas control valve wide open. This can lead to a rapid rise in catalyst temperatures without the ability to quickly control the reaction.
- Improper control, Operator inattention, or overreaction to some process change. An increase in a highly reactive feed to the unit, for example, a Hydrocracker running a mix of virgin gas oil and FCC cycle oil that, experiences a sudden increase of cycle oil (reactive feed) without adjusting temperatures.
- Carry over the catalyst from the reactor to downstream vessels where H_2 may be present at high temperatures. This can inadvertently create another reactor with unreacted feed (hydrocarbons), high temperatures (heat), pressure, and Hydrogen, but without adequate temperature monitoring or controls. An exothermic reaction without temperature monitoring or cooling capability can result in extreme temperature excursions in the downstream vessel due to uncontrolled exothermic reactions.

Operator Training for Controlling an Out-of-Control Exothermic Reaction

Operators, especially panel operators, should be trained to recognize a potential exothermic reaction and the appropriate response actions. Operator training should emphasize the importance of closely tracking reactor catalyst bed temperature profiles and reactor skin temperatures. Training should emphasize awareness of one or more reactor temperatures in a common bed elevation deviating from the trend (+ or −) and the rate of change vs. the standard deviation for temperature indications in a common bed elevation. This includes emphasis on close monitoring of distributed control system (DCS) and temperature data for indications of temperature rise or unexplained change in catalyst bed DP. Changes in DP may occur over a short period or over a longer period; monitoring should be for both the short term and the long term. It is essential to emphasize the importance of a rapid response, including the Operator actions required to stop the reaction in the event of a potential exothermic reaction.

Inherently Safe Design

Another way to prevent an exothermic reaction is by inherently safe design. The concept of Inherently Safe Design is discussed in more detail in Chapter 16, but simply stated, inherently safe design is a method that avoids hazardous chemicals rather than controlling them. An example of an inherently safe design is selecting products that are not susceptible to runaway reactions or substituting a hazardous chemical with one that is less susceptible to temperature excursions. Another method of inherently safe design would include ensuring operating conditions and equipment that maintain temperatures well below the point where a temperature excursion would be possible.

Operating Limits Must be Established and Followed

Operating limits (OL), sometimes referred to as the operating envelope (OE), should be established below the minimum temperature or pressure where an exothermic reaction can occur. Once established, these limits must be rigorously enforced, with reporting required for all excursions above the established OL. "OL^s" or "OE^s" should include reactor catalyst bed maximum temperatures, temperature rate of change, and the differential pressure (DP) and DP rate of change. The minimum rate of a reactor quench or cooling medium may also be established as an OL to ensure adequate cooling for the catalyst bed.

Management must enforce strict adherence to Operating Envelopes in reactor operations (e.g., catalyst bed temperatures, quench rates, and differential pressures). Operators should ensure the OLs are followed and deviations or excursions are promptly addressed and reported. In the event of rapid temperature rise situations, Operators should be familiar with reactor depressurization procedures or other means to stop the reaction excursion, and they must ensure these procedures are followed.

An unexplained change in temperatures or DP may indicate an uncontrolled exothermic reaction. Operators must be trained not to second guess the instrumentation in the event of a rapid rise in reactor catalyst temperatures. Delays while troubleshooting unexplained reactor temperature or DP have resulted in the loss of containment events, including vessel or piping ruptures. Examples are Tosco Avon Refinery (Martinez, CA.) and Shell Chemical Company (Moerdijk, Netherlands), discussed below. Rapid reactor depressurization is required to stop the reaction.

Verify that procedures are in place for prompt response to indications of uncontrolled exothermic reactions, including unit evacuation and emergency depressurization in the event of a rapid reactor temperature increase.

Ensure Enhanced Process Control

Enhanced process control ensures that key process variables are continuously monitored for indications of an increase in reactor temperatures or other early signs of a potential exothermic reaction without Operator involvement or intervention. This can be advanced software on the computer DCS system tied to priority alarms and unit trips to alert the Operator that a potential temperature excursion is occurring and automatically trip the unit if an exothermic reaction is detected. In critical applications, this is more likely to be a triple modular redundant (TMR) computer that can immediately take control of the process and trip the reactor or return the process to a safe state, completely independent from the DCS and without Operator involvement.

An example is the Hydrocracker reactor temperature monitoring and shutdown controls. These systems typically consist of multiple temperature thermocouples in the catalyst bed and additional thermocouples on the reactor shell and outlet piping. These thermocouples feed the data to a TMR logic solver with voting logic independent of the DCS. The TMR computer is also connected to the reactor shutdown and depressurization controls. It can detect the temperature rise in the reactor catalyst bed, determine if this is a false signal or an actual event (via voting logic), and shut down and depressure the reactor, stopping the reaction without the involvement of the Operator.

Periodic Inspections

Ensure that process equipment on hydro-processing units subjected to exothermic reactions is inspected periodically (visual) for undetected damage, especially in areas that are not visible during routine Operator rounds, for example, under the reactor and high-pressure separator vessel skirts, process piping downstream of reactors, etc.

Design Changes

All design changes must follow the detailed Management of Change (MOC) and Process Hazardous Analysis (PHA) processes and include reviews by experienced process engineers and Materials engineers. Reviews should include the potential for the unanticipated contact of unreacted hydrocarbon and catalyst in the presence of hydrogen, potential dead zones, and potential catalyst migration. Experienced engineers qualified in hydro-processing equipment should also be involved in all design changes or modifications to hydro-processing equipment, especially the reactor catalyst systems.

After the construction is complete, the Pre-startup Safety Review (PSSR) should be done to ensure no safety issues have been introduced into the process and that all priority action items discovered during the project and pre-startup reviews have been addressed.

Reactor Inspection Following Maintenance or Before Catalyst Loading

Reactors require a detailed internal visual inspection before catalyst loading to ensure no gaps or voids exist that may lead to catalyst migration during operations. This includes the catalyst support screens and inert ball type, size, and placement. For reactors with interbed quench, a detailed inspection is required to ensure the interbed assemblies are tight and there can be no leakage around the interbed panels. Also, these interbed panels should be verified as clean (no fouling or other debris).

Ensure that personnel experienced in the functional inspection of reactor internals are involved in these inspections. Technical troubleshooting on critical processes should involve an in-depth risk analysis of the existing operations. Design changes should be proposed only after fully understanding existing problems and symptoms.

What Will I See as a Console or Field Operator?

Any indication of a hot spot on a reactor or piping associated with a reactor in hydro-processing equipment may indicate an uncontrolled exothermic reaction. As discussed earlier, reactor catalyst bed or vessel skin temperatures should be monitored closely for indications of a hot spot. Anomalies in reactor catalyst bed temperatures or outlet temperatures should be considered potentially serious situations. A bulging reactor or downstream vessel or outlet pipe is an extreme case and requires immediate evacuation of personnel from the unit and rapid reactor depressurization and removal of heat input.

Likewise, a change in catalyst bed differential pressure (DP) may indicate an uncontrolled exothermic reaction. This change can be rapid in DP over a short time, or it may be a very slow change in DP that occurs over a much longer period, which could be weeks or months and may indicate catalyst migration inside the reactor. Therefore, the continuous monitoring of catalyst bed DP must be based on short-term and long-term trends.

Even with the best design, the best facilities, and the best instrumentation, controls, and alarms, a thermal runaway is still possible, and our operators are the best line of defense. This means well-trained Operators who know the indications of a thermal runaway and continuously monitor for the signs. Operators must know the symptoms and how to quickly diagnose and determine if a thermal runaway is occurring and be able to take the actions to return the unit to a safe state.

What Has Happened/What Can Happen

Shell Chemical Plant
Belpre, Ohio
27 May 1994 (Three Fatalities)

At 6:25 a.m. on Friday, 27 May 1994, a temperature excursion and runaway occurred in a polymer reactor vessel at a Shell Chemical Plastics Plant near Belpre, Ohio. The temperature excursion resulted in a devastating explosion and secondary fires that destroyed the polymer unit. Shrapnel from the initial explosion punctured a styrene storage tank approximately 600 feet (183 meters) away, resulting in a fire in the tank. About 5 minutes later, another explosion occurred in or near the burning styrene storage tank, spreading the fire to five additional storage tanks. Three to five million gallons (11–18 million liters) of flammable products were burned in the fires. Other process units in the facility were also damaged by shrapnel from the explosions. Three workers were killed during the explosions. See Figures 8.28.1 and 8.28.2 for photos of the resulting fire and damage from this significant exothermic reaction explosion.

The Occupational Safety and Health Administration (OSHA) reported: "A catastrophic failure of a 15,000-gallon polymer reactor vessel was initiated by a runaway chemical reaction" in the company's Kraton-D polymer unit. "The reactor failure and resulting fire," explained OSHA, "caused the complete destruction" of that unit.

Figure 8.28.1 Photo of explosion and fire at Shell Belpre, Ohio polymer plant (three fatalities). Source: Photos are courtesy of Mackey's Antique Clock Repair. Parkersburg, WV.

Figure 8.28.2 Photo of damage as the result of explosion and fire at Shell chemical. Source: Photos are courtesy of Mackey's Antique Clock Repair. Parkersburg, WV.

The following is a direct quote from the OSHA report on this incident:

"On 27 May 1994, Employees #1 through #3 were working at the Shell–Belpre Chemical Plant in Belpre, OH. A catastrophic failure of a 15,000-gallon polymer reactor vessel was initiated by a runaway chemical reaction involving an abnormally high amount of 1,3-butadiene during the production of Kraton-D, the Shell trademark for polymer. The reactor failure and ensuing fire resulted in the complete destruction of the polymerization unit. Missile fragments from the failed reactor vessel damaged adjacent units in the plant. One fragment punctured a styrene storage tank approximately 600 feet away. This subsequently resulted in the burning of five styrene storage tanks containing approximately 3.5 million gallons of flammable products. Employees #1 through #3 were killed in the explosion."

Local newspapers reported that 1,700 residents were evacuated from the area due to the explosions and fired and contaminated the adjacent Ohio River with toxic chemicals for 20 miles downstream. Another newspaper reported, "Explosion, Fire Force Evacuations; Governor Declares State of Emergency in Belpre." The fire raged for nine hours before being brought under control, as dozens of fire companies and 150 firefighters from surrounding communities responded. At times, flames shot 300–600 feet into the air, and firefighters had to back away from their battle at various points due to the intense heat.

Key Lessons Learned

According to Independent Commodity Intelligence Services (ICIS), Shell reported that although the exact cause of this incident was not confirmed, they believe it was caused by one of the employees killed in this incident intentionally ignoring alarms and overriding the safety system. They also reported several instances of high pressures in the same reactor on the same day as the explosion, resulting in the automatic shutdown of the butadiene shutoff valve. They reported that they believed the employee's action was based on his belief that the reactor contained isobutane and not reactive butadiene. Shell has since implemented an improved system for automatically monitoring and controlling the ratio of isobutane to butadiene. The system also prevents a single employee from defeating or overriding these controls.

Shell's report on the incident confirmed that the explosion occurred in the polymer reactor due to excessive reactor pressure caused by a buildup of butadiene. The failed vessel, known as a STEP II reactor, is the second of two reactors in the polymerization section of the unit. The STEP I reactor prepares materials for the STEP II reactor, where butadiene and styrene are mixed with a catalyst to form an elastomer. Butadiene is a very reactive hydrocarbon, and reactions can result in very high temperatures and pressures.

This incident emphasizes the importance of Operator training and awareness of causal factors of uncontrolled exothermic reactions. Real-time monitoring of reactor temperatures and differential pressures is critical, and rapid response by Operators in response to increases in reactor temperatures is critical to preventing a repeat of this incident. For example, Operators should never bypass or override alarms or controls without a thorough analysis, a detailed review and approval by senior management, and a completed and approved MOC.

What Has Happened/What Can Happen

Tosco Avon Refinery – Martinez, CA
Hydrocracker Explosion and Fire – 21 January 1997
1 Fatality/46 Injuries
The US EPA investigated the incident (information provided here was extracted from the EPA's final report).

When this incident occurred, the Operators were faced with anomalies in the reactor temperature profile. The reactor thermocouples were erratic, with some readings at the bottom of the temperature scale and others indicating very high temperatures. It was evident to the investigators that the operators were questioning the data from the temperature indicators and troubleshooting the process when the incident happened. They had some indications of an exothermic reaction, but other indicators led them to believe that the temperature indicators were failing. The erratic reactor temperatures resulted in the Operators questioning the data and their loss of confidence in the reactor instrumentation. Therefore, they were not sure that an exothermic reaction was occurring. Later analysis found that the TIs were designed to return to the bottom of the scale if they reached an over-range condition, and this is what was happening. The reactor thermocouples went into an over-range condition and then quickly dropped to the bottom of the scale. This was compounded because most temperature readings were only available at a field panel near the reactor's base. See Figures 8.28.3 and 8.28.4, reactor temperature charts indicating a rapid increase in temperatures in catalyst beds and then a drop to zero.

At the time of the failure and the resulting explosion and fire, the Operators were attempting to troubleshoot the cause of the reactor temperature increase reflected on their instruments. One Operator was at the reactor base trying to recover the temperature data from a local temperature stamping instrument. At this time, the extremely high reactor temperature resulted in the reactor outlet piping failing catastrophically. This resulted in a major loss of containment, a rapid depressurization of the entire unit, and a massive explosion and fire.

The incident was caused by an undetected internal reactor exothermic excursion and catalyst bed temperatures exceeding the design limit of the reactor outlet piping. The Operator, who was at the local instrument near the base of the reactor,

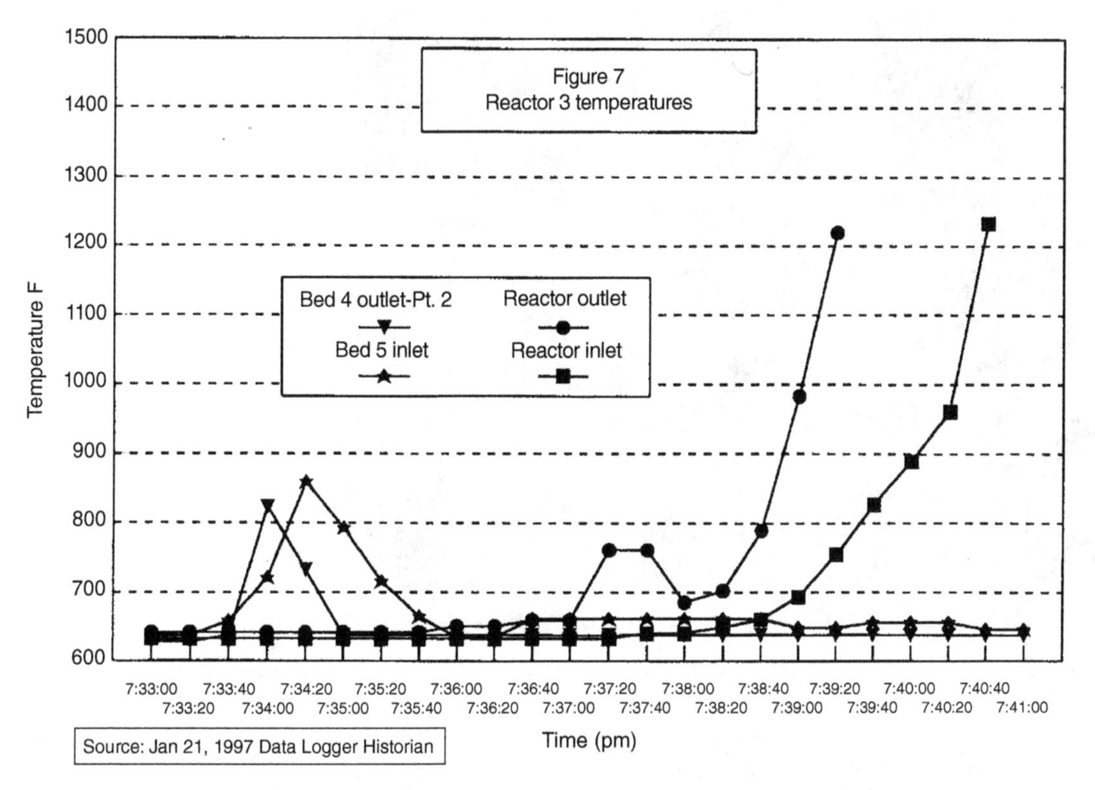

Figure 8.28.3 Reactor temperature chart showing rapid temperature rise in reactor beds. Source: The US Environmental Protection Agency (EPA annotations by the US EPA).

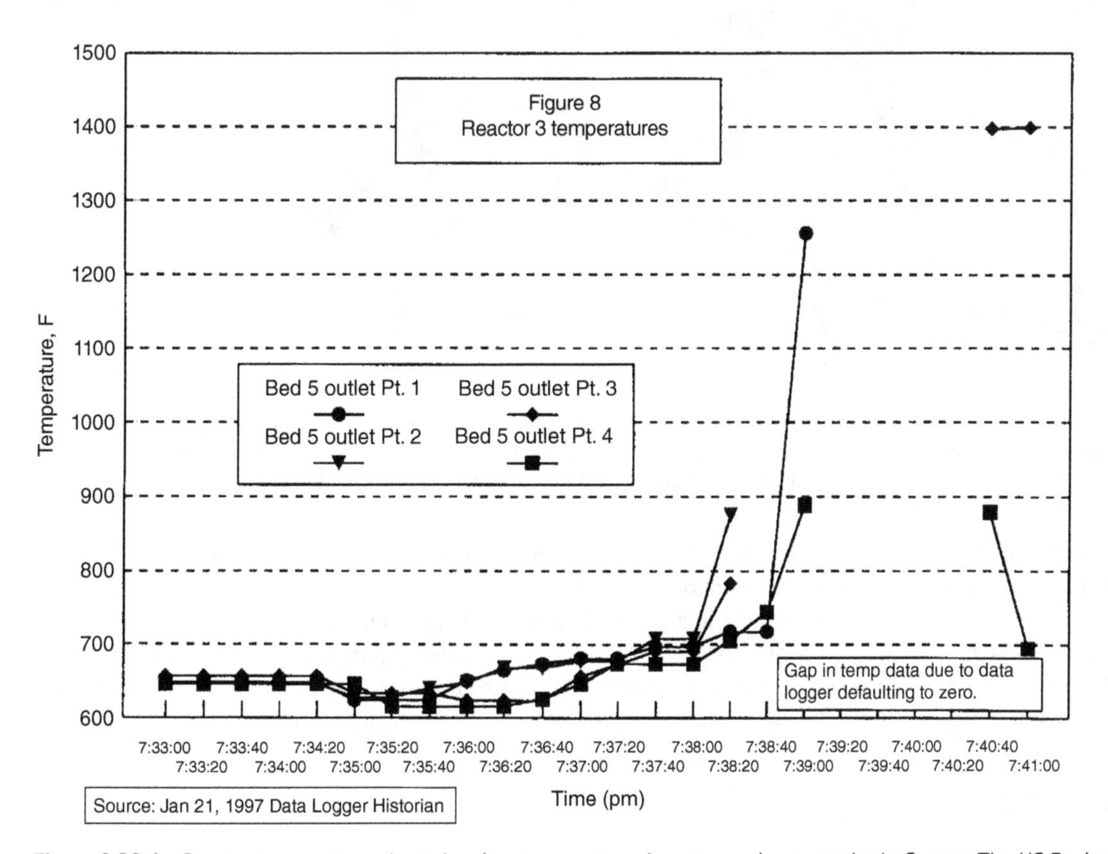

Figure 8.28.4 Reactor temperature chart showing temperature drop to zero in reactor beds. Source: The US Environmental Protection Agency (EPA) (annotations by the US EPA).

Figure 8.28.5 Location of external reactor temperature recording panel. Source: The US Environmental Protection Agency (EPA) (annotations by the US EPA).

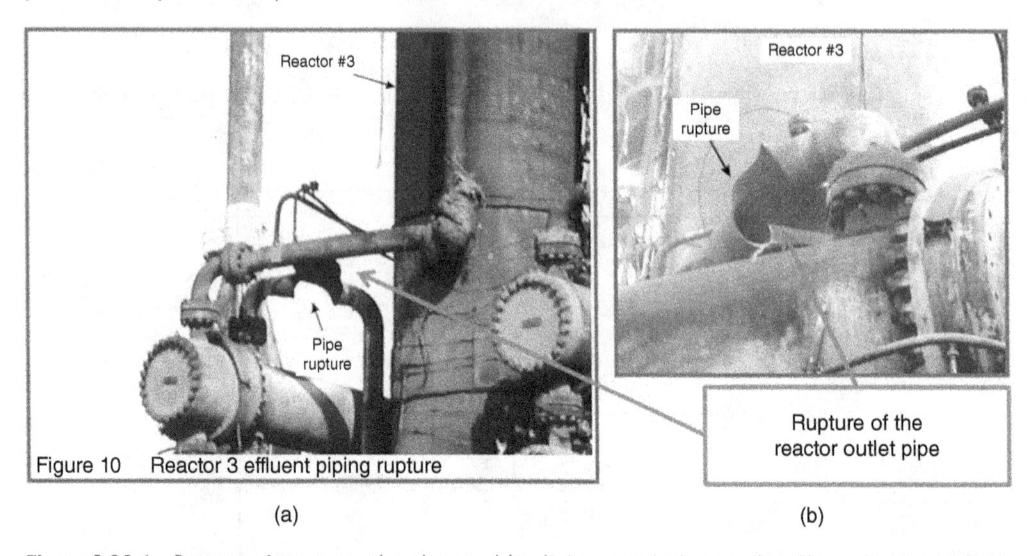

Figure 8.28.6 Rupture of reactor outlet pipe resulting in large explosion and fire. Figure (a) clearly illustrates the rupture in the section of pipe just above the exchanger. This released the hydrogen and hydrocarbons from the pressurized reactor and unit to the atmosphere. Figure (b) is a close up of this rupture. Source: The US Environmental Protection Agency (EPA) (image annotations by the US EPA, additional annotations by author).

was killed, and 46 others were injured during the explosion and fire. See Figure 8.28.5 for a photo of the location of the local temperature recording instrument where the Operator was attempting to recover the reactor temperature trend data.

The investigation team reported that the reactor was in an uncontrolled exothermic reaction. It was estimated that the reactor and outlet piping reached a temperature of 1,400 °F (760 °C) before the outlet piping failed catastrophically. When the reactor outlet piping failed, the entire unit rapidly depressurized from the normal pressure of 1,560 psi, resulting in a massive fire near the base of the reactor. See Figure 8.28.6a,b for photos of the failed reactor outlet piping.

Key Lessons Learned

When the exothermic reaction occurred, the reactor thermocouples initially indicated the high temperatures and would quickly drop to the bottom of the temperature scale and then back to normal. This was confusing to the Operators, and they were not sure if an exothermic reaction was occurring or not and were trying to diagnose the problem when the incident occurred. Instrumentation should normally be designed to fail in the "fail-safe" position and, in this case, should have failed at the high end of the scale. In this position, the Operators would have had a direct indication of a temperature excursion in the reactor and could have taken the correct response to stop the reaction. The Operator who was killed in this incident was standing near the local temperature instrument panel near the base of the reactor, attempting to collect the reactor temperature data for analysis. See Figure 8.28.3, Reactor temperature profile. Note how the catalyst bed temperatures went above 800 °F (426 °C) and then dropped to the bottom of the scale.

Operators should be trained to continuously monitor the reactor temperature and pressure profiles for any indication of an uncontrolled exothermic reaction and quickly act if one is detected. Trends should be continuously run for reactor catalyst bed temperatures and bed delta pressures, both short-term trends and longer-term trends. Procedures should be in place and followed in responding to reactor temperature or pressure anomalies. In the event of a rapid rise in temperatures, the Operator should be trained to quickly dump the reactor pressure, stopping the reaction and returning the reactor to a safe state. Most of the newer hydrocrackers and some hydrotreaters are controlled with TMR computer systems designed to monitor and automatically respond to an abnormally high exothermic reaction.

Operators should not be troubleshooting a reactor, indicating a potential exothermic reaction. Any indication of temperature excursion or high pressure in a reactor should signify a potential temperature runaway, and actions should be taken immediately to stop the reaction. There is always time to troubleshoot after the unit is returned to a safe state. Again, this should be entrenched in their training and procedures.

What Has Happened/What Can Happen

Shell Chemical Company – Moerdijk (Netherlands)
3 June 2014
Undetected exothermic reaction during startup resulted in explosion and fire and two injuries.
The Dutch Safety Board investigated this incident and made recommendations to help prevent a recurrence.

The Shell chemical plant located at Moerdijk in the Netherlands just out of Rotterdam experienced an explosion during the startup of the Propylene Oxide/Styrene Monomer reactor. After the turnaround, the two reactors had just experienced catalyst replacement and were undergoing unit startup. The resulting two explosions at 10:48 p.m. resulted in the catastrophic failure of one of the reactors and a separator vessel and two injuries. Reactor debris was spread as far as 0.5 miles (800 meters) from the unit, and sections of the reactor vessel were located over 800 feet (250 meters) from the unit. The resulting damage also resulted in a long-term outage of the unit for investigation and repairs.

The unit was in the start-up mode following a catalyst replacement in both reactors, and the Operators were circulating ethyl benzene following routine start-up procedures. As the unit was circulating, the Operators believed the operation was taking longer than they expected; therefore, they increased the reactor temperatures. An internal exothermic reaction occurred shortly afterward, resulting in a very rapid temperature and pressure increase in both reactors.

The Dutch Safety Board reported anomalies in the reactor temperatures and flow to the second reactor during startup. However, these were considered routine by the Operators, and they were ignored since they had seen similar excursions in prior startups. Operators proceeded with the startup even after critical alarms and automatic protection systems had been activated, including the automatic closure of the vent valve to the flare system. When the vent valve closed, the unit pressure increased rapidly to more than 2,600 psi (180 bar), more than 25 times the normal operating pressure. Within a few minutes, the second reactor ruptured, followed by the rupture of the first separator vessel within about 20 seconds. The Operators never realized that they should have aborted the startup. The explosion resulted in a very large fire and the release of hot catalyst pellets and other debris from the reactors. See Figures 8.28.7 and 8.28.8 for a photo of the fire and the resulting damage to the process unit.

The Dutch Safety Board found that ethylbenzene and catalyst hazards had been studied extensively in 1977. However, this study was not updated following modifications to both the unit and the catalyst system being used in 2014. There had not been a study on the new catalyst being used to determine its activity in the presence of ethylbenzene, temperature, and pressure. The catalyst reacted, resulting in a temperature excursion in the reactors and a temperature runaway.

Figure 8.28.7 Shell Moerdijk chemical plant explosion and fire. Source: Photo courtesy Marcel Otterspeer photography (Included in the Dutch Safety Board Investigation Report, which is open to the public). (https://www.otterspeer.com/beeldbank/136419-brand-shell-editorial-journalistiek).

Figure 8.28.8 Damage to Shell Chemical Company (Moerdijk, Netherlands – 3 June 2014). Source: The Netherlands Police LTFO (Included in the Dutch Safety Board Investigation Report, which is open to the public).

Key Lessons Learned

This incident highlights the importance of strict adherence to the MOC and ensuring that the new catalyst is thoroughly evaluated before starting a reactor with a new catalyst.

Highly active catalysts in the presence of reactive hydrocarbons can result in temperature excursions and temperature runaways. Start-up procedures may need modifications to consider a different, less active hydrocarbon for sulfiding or startup. Also, the heating-up period may require a slower heat-up to help prevent a temperature excursion.

Operators should be well-trained on how to monitor the unit safely during startup and when and how to abort quickly in the event of anomalies in reactor temperature and pressure profiles. In the event of a temperature excursion or other indications of an overactive thermal reaction, the Operators should be trained to react quickly and bring the unit back to a safe condition. This includes a proper response to unit alarms or trips.

Additional References

EPA Chemical Accident Investigation Report, EPA 550-R-98-009, November 1998, Tosco Avon Refinery, Martinez, California.

Dutch Safety Board Incident Report, "Explosions MSPO2 Shell Moerdijk". https://www.onderzoeksraad.nl/en/page/3438/explosions-mspo2-shell-moerdijk

Dutch Safety Board Video, Shell Moerdijk MSPO2 Explosion. https://www.youtube.com/watch?v=JcAFkMX8XJM

Independent Commodity Intelligence Services (ICIS), "Belpre blast alarm ignored, says Shell." 14 November 1994. https://www.icis.com/explore/resources/news/1994/11/14/39086/belpre-blast-alarm-ignored-says-shell/

The British Health and Safety Executive, "Chemical reaction hazards and the risk of thermal runaway". https://www.hse.gov.uk/pubns/indg254.pdf

Chapter 8.28. Hazards of Undetected or Uncontrolled Exothermic Reactions (Thermal Runaways)

End of Chapter Quiz

1. In the petroleum and petrochemical industries, exothermic reactions occur when hydrocarbons are exposed to ________, _________, and _______ catalyst in the presence of __________.
2. However, _____________ exothermic reactions can occur if reaction temperatures are not closely controlled.
3. Please list the exothermic refining processes. Please list all you can.
4. Exothermic reactions can also occur by the mixing of two or more ______________chemicals; for example, mixing acid and water will result in a violent exothermic reaction.
5. Undetected or uncontrolled exothermic reactions, which can lead to temperature excursions, otherwise known as a "__________ __________," have occurred in hydrocracking and other processes, especially during startup and other transient operations.
6. Uncontrolled reactions can lead to temperatures exceeding the ________ limits for equipment, resulting in equipment failures, loss of containment, fires/explosions, and unit or catalyst damage.
7. An undetected or uncontrolled thermal runaway can create a catastrophic incident and the potential for the loss of ________.
8. The console operator operating the Hydrocracker sees a steady rise in reactor temperatures and catalyst bed differential pressures. Should he troubleshoot to understand if there is a problem with the reactor thermocouples, or what other action should he take?
9. How many of the potential causes for an exothermic reaction can you list? List all you can.
10. What are the critical factors needed to help guard against an uncontrolled exothermic reaction leading to an explosion or loss of containment event?

    What are the indications of an incipient exothermic reaction and a temperature runaway in a reactor catalyst bed?

8.29

Responding to Minor Leaks and Releases

On 6 August 2012, the Chevron Refinery near Richmond, CA, experienced a loss of containment and a major fire that significantly impacted the surrounding community. This incident occurred while the refinery was attempting to strip insulation from the Atmospheric 4th sidestream draw to identify the source of a "dripping" leak to install a clamp to stop the leak, allowing the unit to continue operations. As a result of this loss of containment, 18 people were "enveloped" in a vapor cloud and escaped through the vapor. Fortunately, the eighteen workers escaped before the vapor cloud flashed, and one fireman who was trapped in a fire engine escaped in bunker gear through the fire without injury. The US Chemical Safety and Hazard Investigation Board investigated this incident and released a detailed report and an excellent training video.

This incident highlights the importance of improved risk assessment when responding to even small leaks. Unfortunately, the industry experienced similar events before the Chevron incident ("Organizational Memory"). This chapter presents the hazards of responding to even small leaks and provides strategies for assessing/managing the hazards.

The Chevron Incident

This was a small dripping leak from the 8-inch insulated Atmospheric #4 sidestream light gas oil draw piping circuit operating at 640 °F (338 °C), well above the autoignition temperature of about 410 °F (210 °C). There were no isolation valves on the #4 Sidestream draw. The team attempted to remove the insulation from the pipe and planned to install a temporary fitting known as a clamp to contain the leak when the line suddenly ruptured, resulting in a large release of hot gas oil, forming a major vapor cloud. All 18 people were caught in the vapor cloud and had to escape through it, some on their hands and knees. One fireman was trapped in a fire engine and escaped through the flames wearing bunker gear. Fortunately, no one on the unit was injured due to the vapor release or the fire; however, a dense plume of black smoke drifted into the adjacent communities from the facility.

Over the next several days, 15,000 people in the adjacent community reported to the hospital, most with breathing difficulty. The corrosion mechanism was found to be high-temperature sulfidation, a uniform corrosion mechanism, and the piping circuit was found to be carbon steel with very little silicon. This type of metallurgy with very little silicone is susceptible to this type of corrosion.

The US CSB investigated this incident and released a detailed investigation report and a very good training video (animation) of the incident and the causal factors that led to it. Please see Figures 8.29.1 and 8.29.2 for photos of the fire damage and Figure 8.29.3 for the impact on the community.

Key Lessons Learned

This incident highlights that a change in responding to even minor leaks is required. When a small leak is discovered, our instinct is to quickly do what is necessary to repair the leak and stop the loss of containment, for example, by attempting on-stream temporary repairs such as a clamp or leak box. The Chevron incident (among others) tells us we must have the "right" people involved and ensure careful analysis before attempting this temporary repair. We should ask what information is needed before we begin the installation of a clamp, leak box, or another type of on-stream repair and who has the information or data.

Process Operations Safety: The What, Why, and How Behind Safe Petrochemical Plant Operations, First Edition. M. Darryl Yoes.

Figure 8.29.1 Chevron Richmond Refinery; dmage sustained from the 6 August 2012, Fire. Source: Photo courtesy US Chemical Safety and Hazard Investigation Board and EHS Today.

Figure 8.29.2 Chevron Richmond Refinery; cooling tower damage sustained from the 6 August 2012, Fire. Source: Photo courtesy US Chemical Safety and Hazard Investigation Board and EHS Today.

Figure 8.29.3 Chevron Richmond Refinery fire (6 August 2012). Source: Photo courtesy of the US Chemical Safety and Hazard Investigation Board.

What information is needed before we begin the installation of a clamp, leak box, or other types of on-stream repair? The following are only examples of questions we should ask before attempting an on-stream temporary repair (such as installing a clamp).

- Are the "right" people involved in this decision?
  - Process management/supervision, Shift Superintendent.
  - Fixed Equipment Engineer or Materials Engineer, Unit Inspector.
  - Safety Department representative.
- Is the corrosion mechanism known?
  - The fixed equipment engineer or Inspector will have this information.
- Is the corrosion Localized or Uniform?
  - Uniform corrosion has the potential for rupture mode of failure.
- What is the nature of the material involved?
  - Hot (above autoignition temperature).
  - Light (LPG, Gasoline, or light naphtha, etc.).
  - Toxic/Corrosives (H_2S, Ammonia, Chlorine, Acids/Caustic, etc.).
- What are the operating temperatures and pressures?
- Can this work be done while minimizing personnel exposure in the event the equipment fails during the repair?
- Does the job plan detail how exposures will be minimized?
- Are there isolation valves to shut off the source quickly in the event the pipe or equipment suddenly fails?
- Is the potential release elevated?
- Do we have an isolation plan in case something goes wrong?
- Are there ignition sources in the area?
- Will a sudden release impact the community?
- Is the emergency response team on standby?

- Ready to respond should something go wrong?
- Are they stationed out of harm's way/out of the line-of-fire?

A detailed job plan should be developed and carefully followed to control the work and ensure hazards are being addressed during the repair, including the following:

- Identification of corrosion mechanism(s) and applicable repair strategies (including a go/no-go decision for repair strategy).
- Equipment isolation/shutdown procedure in the event of an unanticipated release.
- Monitoring of hazardous vapors (LEL, Toxics) during the repair.
- Ensure proper PPE for those performing the work.
- Minimizing all other personnel exposures (including clearing of nonessential personnel).
- Proper notifications before work begins (includes Site Shift Supt).
- Special emphasis for LPG/Toxics/Materials above Autoignition.

The CSB Final Report on the Chevron Refinery Fire contains a new "Leak Response Protocol" developed by Chevron, incorporating much of what is covered in this chapter.

See the "Reference" section of this chapter for the web address for the CSB and EHS Today reports.

What Has Happened/What Can Happen

Shell Oil Products
Martinez Refinery
Martinez, CA
8 November 2005

At 8:24 p.m. on 8 November 2005, a small leak was noticed at an isolation valve on an approximately 12-inch diameter filter on the FCC Unit after taking the filter out of service for cleaning. This occurred while the maintenance workers were placing the clean filter back into the filter housing. The leak quickly escalated into a large release of 700 °F (370 °C) slurry bottoms product mixed with lighter flushing oil. The stream of hot oil reached about 200 feet into the air above the filter housing. Fortunately, there was no fire; however, one contract employee received 1st, 2nd, and 3rd-degree burns and was treated at a local medical center. Due to the nature of the hot material and the difficulty isolating the system, the all-clear was not called until 2:54 a.m. the following morning. About 150 barrels of hot products were released during this period. The heavy residue of slurry oil and gas oil coated the surrounding equipment and some vehicles in the area, requiring a cleanup.

Key Lessons Learned

This incident is another example of how what first appears to be a small leak or release can quickly escalate into a large-scale event. In hindsight, this team was very fortunate that this release did not autoignite, resulting in a large fire potentially engulfing the entire team.

This incident also emphasizes that we should never take anything for granted when working with a small release. Before opening the equipment, our procedures should require a careful analysis of the material or product we are working with, including the product's temperature, volatility, and how the product line or equipment is isolated. Current procedures should require that equipment is equipped with double isolation valves with a vent in between to allow for double block and bleed to ensure complete isolation. If these are unavailable, procedures should require a risk assessment to ensure additional precautions and mitigation are identified and in place before work execution. Further details on opening process equipment are covered in Chapter 8.2 ("Safe Permit-to-Work – Breaking Containment and Controlling Work").

Remember

A release of materials that readily vaporize or are toxic can create large vapor clouds and have the potential for high-consequence events, including impact on the community.

The following require special emphasis due to these higher hazards:

- Products that will rapidly vaporize (LPG and process streams operating at temperatures above autoignition).
- Toxics such as H_2S (especially streams >2% H_2S, Anhydrous Ammonia, Chlorine, and HF Acid).
- Crude Oil and Gasoline or Light Naphtha Storage Tanks.

Don't let this happen at your site!

Additional References

US Chemical Safety and Accident Investigation Board, Final Investigation Report – Chevron Richmond Refinery Pipe Rupture and Fire, (The Final Report includes Chevron's "New Leak Response Protocol." See Section 5.3.4 in the CSB Final Report). Report No. 2012-03-I-CA January 2015. https://www.csb.gov/chevron-refinery-fire/

EHS Today News Article, "CSB: Pipe Corrosion Caused Chevron Refinery Fire" (25 February 2013). https://www.ehstoday.com/safety/article/21915578/csb-pipe-corrosion-caused-chevron-refinery-fire

News Article – SFGate, "Small refinery leak leads to big disaster." https://www.sfgate.com/bayarea/article/Small-refinery-leak-leads-to-big-disaster-3770451.php

Contra Costa Health Services, "Major Accidents at Chemical/Refinery Plants in Contra Costa County". https://cchealth.org/hazmat/accident-history.php

Chapter 8.29. Response to "Minor" Leaks and "Temporary" Repairs

End of Chapter Quiz

1. This lesson highlights that sometimes "we do not know what we do not know" about a situation or event. In event of a small dripping leak for one of our process units, who should be in attendance and consulted before we touch the pipe or equipment?
2. What information do we need to know about the situation described above?
3. Can you explain the difference between localized corrosion and uniform corrosion?
4. Is high-temperature sulfidation a localized or uniform corrosion mechanism?
5. In terms of these two corrosion mechanisms, which is subject to a catastrophic loss of containment?
6. When responding to what appears to be a minor leak, possibly only dripping, what could go wrong if we proceed before getting the right people involved and the details mentioned above?
7. Do we need to consciously make a go/no-go decision after carefully considering all the facts?
8. The US Chemical Safety and Hazard Identification Board (CSB) Final Report on the Chevron Refinery Fire contains a new "______ __________ __________" developed by Chevron, incorporating much of what is covered in this chapter.

8.30

Electrical Safety Hazards for Operations Personnel

In this chapter, we will discuss Electrical Hazards for Operations Personnel. Most operators and others are involved with electrically powered equipment as a routine part of our roles. We start and stop electrically powered equipment, and we are called upon to prepare this equipment for maintenance, including ensuring that the equipment is isolated correctly, including lock-out/tag-out. We occasionally operate electrical breakers and other similar tasks. We also perform similar tasks around the home involving electrical energy sources.

Many people do not understand the devastating effect of electrical exposure, even at relatively low voltages. This chapter will discuss these hazards, some incidents related to electrically powered equipment, and the critical lessons learned to help guard against injury to personnel. We will discuss three types of electrical hazards: (1) shock or electrocution by direct contact with a live circuit, (2) Arc flash/Arc blast, and (3) Ignition of flammable or other process materials by electrical equipment or circuits.

Electrical Shock

The first and most common hazard is an electrical shock when we come into contact with a live electrical conductor, such as a bare wire or other conductor types. These can be painful and can result in severe injury and the tragic loss of life, especially when the current level is sufficiently high and other conditions are present, such as an electrical current path through the body and to the ground. Many people have been killed while working with standard household voltage (110—120 volts) in the United States or 220–240 volts internationally. A quick review of the OSHA database indicates that the most common cause of death of trained electricians in the United States occurs while replacing the ballast on fluorescent light fixtures (household current).

Figure 8.30.1 illustrates the effect on the body with a corresponding energy level (current). Note that this chart is in milliamperes (mA) and that one mA equals 1/1,000th of an amp. Using this chart as a guide, it only takes 100 milliamperes or 0.1 ampere to stop the heart and take your life. Most of us have reset a 15-ampere breaker at home without much thought. When you recognize the current levels in a petroleum refinery or petrochemical plant, it quickly becomes clear why this chapter is included in this book.

A current as low as 100 milliamperes (or 0.1 ampere) can stop the heart, resulting in death 100 milliamperes (0.1 ampere) is the current required to light a typical bathroom or hallway night light.

A . Ground Fault Circuit Interrupter (GFCI), when properly used, will stop the current flow at about five mA before the current can have devastating effects on the body. GFCIs can save lives and will be discussed later in this chapter.

Protection Against Electrical Shock Hazards

Lock-out Tag-out

The first order of protection is to ensure that work on electrical systems is only done when the system is fully isolated and deenergized. This is done by electrical lock-out tag-out (LOTO). LOTO is done by operations personnel by opening the electrical breaker on the equipment scheduled for maintenance or other similar work. They also place a lock and a safety tag on the breaker handle indicating LOTO. The safety tag should include, as a minimum, supporting information, identification of the breaker, why the breaker is open, the date and time the breaker was opened, the name and signature of

Process Operations Safety: The What, Why, and How Behind Safe Petrochemical Plant Operations, First Edition. M. Darryl Yoes.

Effects of electric current on human body

Source: (NIOSH) Publication Number 2009–113

Current level (in milliamperes)	Probable effect on human body
1 mA	Perception level. Slight tingling sensation. Still dangerous under certain conditions.
5 mA	Slight shock felt - not painful but disturbing. Average individual can let go. However, strong involuntary reactions to shocks in this range may lead to injuries.
6–30 mA	Painful shock, muscular control is lost. This is called the freezing current or "let-go" range.
50–150 mA	Extreme pain, respiratory arrest, severe muscular contractions. Individual cannot let go. Death is possible
1000–4,300 mA	Ventricular fibrillation (the rhythmic pumping action of the heart ceases). Muscular contraction and nerve damage occur. Death is most likely.
10,000 mA	Cardiac arrest, severe burns and probable death

Note: One (1) ampere equals 1,000 milliamperes.

Figure 8.30.1 Effect of electrical shock on the human body. Source: OSHA Publication "Electrical Hazards in Construction book." Text and annotation by NIOSH and OSHA.

Figure 8.30.2 Electrical breaker with electrical breaker opened and locks and tag in place (LOTO).

the person who opened the breaker, and how they can be contacted. The lock should also be labeled or identifiable to the person who placed it on the breaker. Figure 8.30.2 shows an example of a proper electrical LOTO (lock and tag).

Once the breaker is open with the appropriate lock and tag, the operator should attempt to start or operate the protected equipment from each location or start/stop switch. This is called TryOut and is equally important for a couple of reasons: it is possible that the wrong breaker was opened (this has happened), and therefore the equipment is still energized. It is also possible that the internal mechanism in the circuit breaker is broken or failed, and the breaker is still energized, even though the breaker handle is in the open position (this has also happened). See Figure 8.30.3 for an alarming example of a breaker with a broken internal mechanism.

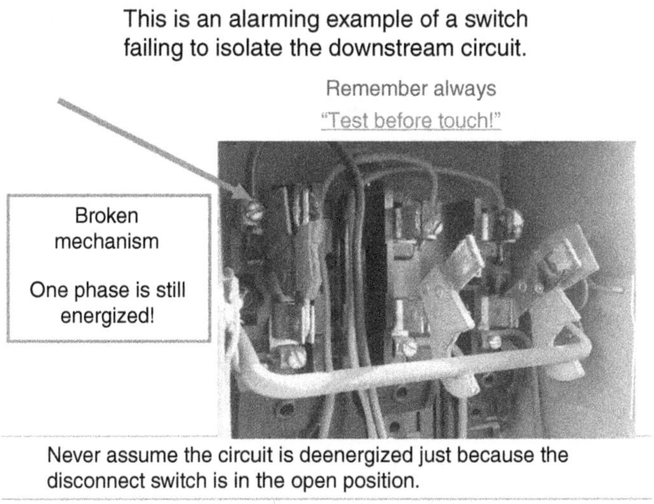

Figure 8.30.3 Electrical breaker with a broken internal mechanism. Note: The breaker is in the "open" position; however, one phase is still energized due to the broken internal mechanism. Be aware that this mechanism is located inside the breaker; therefore, you will not be able to see this when isolating the breaker. This is exactly why the craftsmen must verify energy isolation before beginning work.

Once the equipment is turned over to the craftsmen before maintenance begins, the mechanics should test for current in the isolated system and install their personal locks on the breaker. These locks remain in place during all periods their work is active. For more details, refer to OSHA 29CFR 1910.147 "The Control of Hazardous Energy (Lock-out/Tag-out)."

LOTO, in the context of this Chapter, refers to the isolation of electrical energy prior to work on circuits or other electrical equipment. LOTO is equally important for the isolation of other energy hazards, such when opening process equipment or where there is the potential for moving parts or other gravity, spring-loaded, thermal, or chemical hazards present.

In these cases, LOTO and TryOut should be done each time to ensure the potential hazard has been isolated.

Use of Ground Fault Circuit Interrupters (GFCI)

Ground Fault Circuit Interrupters (GFCI) have significantly reduced the number of electrocutions and serious electrical burn injuries in the workplace and at home. A GFCI is designed to disconnect the power source when it detects any unintentional contact between a protected energized conductor and a fault in ground. These can happen with simple unintended contact with an energized conductor or a damaged or defective appliance or tool. For example, an electrically operated device with internal damage may result in a live circuit within the device contacting the metal frame or case. In this example, you could be electrocuted if your body contacts the energized appliance frame and the ground. A GFCI, in this example, would immediately recognize the fault and disconnect the power, protecting you from potential electrocution. The National Electrical Code (NEC) specifies GFCI use for new installations or major upgrades or renovations.

The GFCI device works by monitoring the current on the incoming hot power lead flowing through the device to the powered equipment and comparing the return current on the neutral lead. When the GFCI detects a loss of about five MAs, it immediately opens the circuit. GFCIs are designed to operate before the electricity can affect your heartbeat. The GFCIs are equipped with a test button and should be periodically tested to ensure they function properly. Simply press the button, and the device should immediately trip. Once tested, simply press the reset button to restore functionality.

GFCIs can be installed in household-type electrical receptacles, whereas the device(s) plugged into the receptacle are protected by the GFCI. Another approach is to have GFCI devices installed in conjunction with circuit breaker panels, where the panel then serves to protect in the event of an electrical fault or electrical overload on any of the devices associated with that circuit. GFCI devices are also available as portable devices plugged directly into a power supply. GFCIs should always be attached directly to a power source and not to the end of an extension cord, as this would leave the extension cord unprotected. Figures 8.30.4 and 8.30.5 show examples of GFCIs.

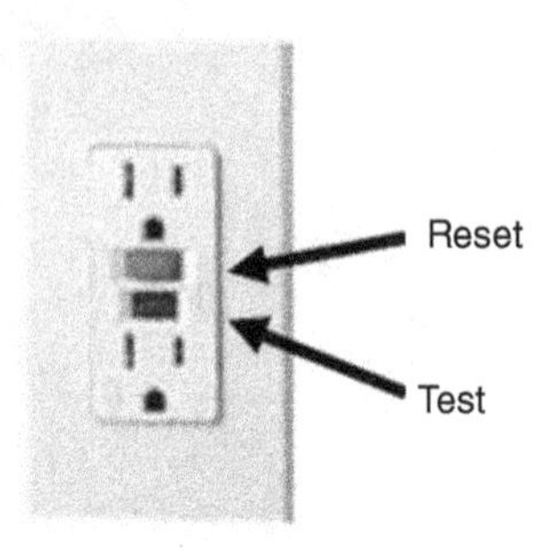

Figure 8.30.4 Household-type GFCI indicating test and reset buttons. *This is not for use in Class I, Division II areas.*

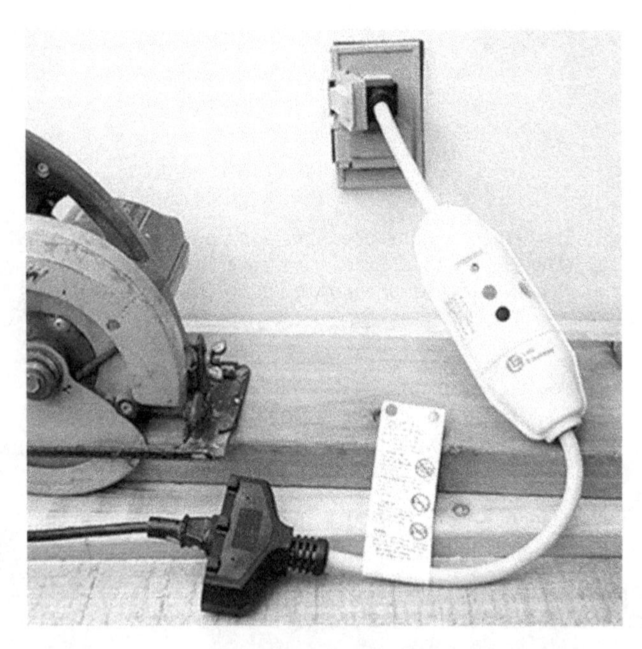

Figure 8.30.5 Example of portable type GFCI. Note: this type of GFCI is not for use in Class I Division II areas. *They must accompany a completed gas test and an approved hot work permit if used in Division II areas.*

Use of Electrical Equipment Inside Tanks or Other Process Vessels

Due to new technology and the equipment available to help expedite work, we see the more frequent use of electrically powered tools and equipment inside process vessels. This presents a unique safety hazard because the vessel is a direct ground, and any incidental contact between the power leads or other energized conductors and the vessel with people present can Corrresult in the potential for electrocution. Site guidelines or procedures should be available to help address this hazard. This chapter helps provide some insight into some of these guidelines.

The first approach is to avoid using electrically powered equipment in the first place, as this virtually eliminates the electrical safety hazard. Other alternatives are the use of pneumatically powered tools or low-voltage (less than 50 volts) equipment, for example, low-voltage lighting @50 volts maximum. OSHA regulations specify a limit of 12 volts maximum for wet locations.

If these alternatives are not feasible, another approach is to use 110 volts tools or equipment and ensure that each device is protected by a ground fault circuit interrupter (GFCI). The GFCI must be physically located outside the vessel, and the power cables must be protected from damage, especially where they pass through the vessel manway(s) or other similar pinch points. Using 110 volts-powered equipment inside process vessels should be considered an exception to routine work practices. The requirements for each use should include a hazard assessment, including a review by electrical specialists and approval by the appropriate site management representative(s).

Additional information and requirements can be found on the OSHA website, including the following regulations:

- OSHA 29 CFR 1910.304 (f)(5)(v)(C)(5, 7)
- OSHA 29 CFR 1910.334 (a)(4), 1926.404 (b)(1)(ii)

- o OSHA 29 CFR 1926.404 (f)(3), (f)(7)(IV)(C)(4, 6)
- o OSHA 29 CFR 1926.405 (a)(2)(ii), (G), (I) and (J)
- o OSHA 29 CFR 1926.404 (f) (3)

The National Electrical Code (NEC 250.45(d)) also includes relevant information on electrical work inside process vessels.

Motor Terminal Junction Boxes

Most industrial electric motors have at least two electrical circuits. The starting circuit is typically operated by a 110 volts circuit involving the motor start button near the motor and an electrical magnetic relay in the electrical substation. The second circuit is a high-voltage circuit from the breaker through the motor stator/rotor, which drives the motor. The high-voltage circuit can be very high, depending on the motor size and horsepower (240–13,800 volts or higher). When the operator presses the start button, the low-voltage circuit closes the magnetic relay, which activates current through the high-voltage motor circuit to start the motor.

The Motor Terminal Junction Box (MTJB) is mounted directly to the motor (generally on the side of the motor) and is the junction point for the three high-voltage motor leads (in a three-phase motor) coming in from the breaker and where they mate with the corresponding three leads going to the motor windings. These connections may be a terminal strip where each pair of leads is bolted together, or each lead may be bolted directly to the corresponding motor lead and then taped, or they may simply be wound together and taped. This depends upon the manufacturer and sometimes a decision by an individual craftsman. The three high-voltage phases are connected inside the explosion-proof box, which normally has a bolted or threaded cover.

During a motor start, these high-voltage connections are exposed to a current "in-rush" or "surge" of about six to seven times the typical motor running amps. This current in-rush is due to the energy required to get the motor from 0 rpm to 1,800 or 3,600 rpm in just a few seconds, depending on the motor design. This current in-rush results in a high load on these motor lead junctions during a motor start. If there is moisture inside the junction box or the electrical leads are not tight, there can be an explosion inside the box when the Operator presses the motor start push button. There have been cases where the junction box cover has been blown off, striking the Operator and resulting in serious injury.

Hazards of the Motor Terminal Junction Box (MTJB):

In cases where the motor start button is located directly in front of the MTJB, consideration should be given to having it relocated to an inherently safer location. As indicated in Figure 8.30.6, operators should also be trained on the safe body position out of the direct line-of-fire until these start buttons can be relocated. Figure 8.30.6 indicate body positioning from line-of-fire to one where the operator is in a safe position, out of the direct line-of-fire of the MTJB.

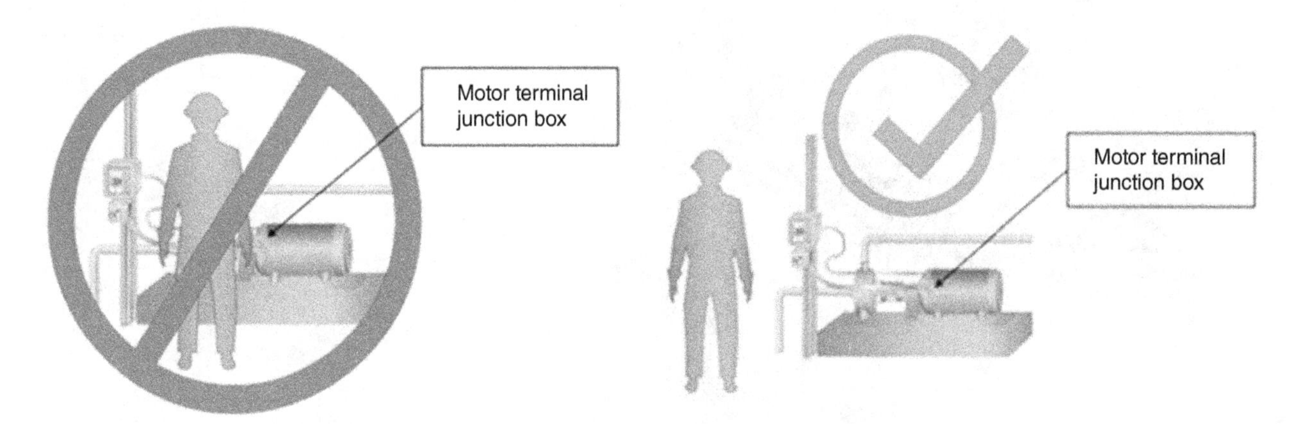

Figure 8.30.6 Shows the operator is in an unsafe position in the direct line-of-fire of the motor terminal junction box. Source: Ecoscience Resource Group Visual Artists. These images were created to illustrate the hazards of body placement in the line-of-fire of motor terminal junction boxes when starting electrically powered equipment.

Work Around or Near Overhead Power Lines

Other significant hazards at many of our sites are the overhead power lines. At most sites, these are the high-voltage incoming power ranging from 68 to 230 kilovolt and higher (1 kilovolt = 1,000 volt). Some sites also have intra-substation lines running overhead at voltages of 4.1–13.8 kilovolt or higher, most of which are in off-site areas and inside electrical substations. *Many of these overhead cables are uninsulated* and, therefore, represent serious hazards for overhead equipment, including cranes and other lifting equipment such as tipping trucks and during the transport of large equipment and overhead scaffolding. Refer to Figures 8.30.7 and 8.30.8 for images of high-power lines and associated warning signs.

These overhead lines must be clearly identified, especially when work near the energized cables is planned. In the US, OSHA regulates work around or near these overhead power lines (29CFR 1910.333(c)(3). The OSHA Regulations require maintaining clearances of 10′ up to 50 kilovolt, and an additional 4″ (10 centimeter) for each 10 over 50 kilovolt).

The preferred approach for work in proximity is to have the lines deenergized before the work begins and for the duration of the work. Any work near these overhead lines should include a hazard review involving the site electrical specialists and management approvals before the work begins.

The UK Health and Safety Executive has developed great training materials for working near overhead power lines. For example, see the material in the Reference section at the end of this chapter.

Figure 8.30.7 Clearly identified overhead power lines.

Figure 8.30.8 High-powered overhead power lines.

Electrical Arcs and Sparks as Sources of Ignition of Flammable or Combustible Vapors

This chapter will discuss how electricity can be a source of ignition when flammable vapors are present and how these hazards are managed. Equipment such as switches, circuit breakers, motor starters, pushbutton stations, or plugs and receptacles can produce arcs or sparks in normal operation when contacts are opened and closed. Other electrical equipment, such as motors or lighting systems, can produce temperatures above the ignition temperature of most hydrocarbons. Another potential ignition source is electrical insulation failure, especially wiring with splices, as these can break down or fail, resulting in arcing or sparking. A devastating explosion or fire may result if any of these occur in the presence of flammable vapors or combustible hydrocarbons. The NEC developed the Electrical Hazardous Area Classification (EHAC) to help identify hazardous work areas, and equipment placed in these areas is designed and controlled.

Electrical Hazardous Area Classification (EHAC) (Note: More Detail on the EHAC Is Included in Chapter 8.26)

In the United States, the National Fire Protection Association (NFPA) developed the Electrical Hazardous Area Classification as NFPA-70, also known as the National Electrical Code (NEC). The NEC is also approved as an American national standard by the American National Standards Institute (ANSI) and is identified as ANSI/NFPA 70.

These EHAC requirements are also detailed in the Code of Federal Regulations as OSHA CFR 29.1910.307. The following paragraph is directly from the OSHA regulation and helps define the requirements and scope of the NEC.

> "The National Electrical Code, NFPA 70, contains guidelines for determining the type and design of equipment and installations that meet this requirement. Those guidelines address electric wiring, equipment, and systems installed in hazardous (classified) locations and contain specific provisions for the following: wiring methods, wiring connections; conductor insulation, flexible cords, sealing and drainage, transformers, capacitors, switches, circuit breakers, fuses, motor controllers, receptacles, attachment plugs, meters, relays, instruments, resistors, generators, motors, lighting fixtures, storage battery charging equipment, electric cranes, electric hoists and similar equipment, utilization equipment, signaling systems, alarm systems, remote control systems, local loudspeaker and communication systems, ventilation piping, live parts, lightning surge protection, and grounding."
>
> These standards outline the Electrical Hazardous Area Classification system used in the United States (and adopted worldwide by most other countries, although the nomenclature may be different in other countries). The standards include the methodology for determining the types and severity of the hazards that may be present, including consideration of the materials' characteristics.

The NEC (NFPA 70E) provides a table for determining the EHAC by Class, Division, and Group as follows (also available in OSHA 29CFR 1910.399) (See Figure 8.30.9):

- Class defines the type of hazard that exists or may exist:
  - Class I locations are where flammable gases or vapors may be present in the air in quantities sufficient to produce explosive or ignitable mixtures.
  - Class II locations are those that are hazardous because of the presence of combustible dust.
  - Class III locations are those that are hazardous because of the presence of easily ignitable fibers or "flying."
- Division defines the frequency (or availability) of the hazard:
  - Division 1 indicates that the hazard is likely present during normal operations.
  - Division 2 indicates the hazard may be present during an accident or abnormal operations.
- The group determines the degree of the hazard or the explosiveness of the materials involved. Groups applicable to Class I, Division 1 and 2 are:
  - A. Acetylene.
  - B. Hydrogen.
  - C. Ethyl ether
  - D. Gasoline, Hexane, Natural Gas (Methane)

N.E.C/C.E.C	National Electric Code (NEC) Classification Chart (NEC 500, 501, 502, 503)	
CLASS	**DIVISION**	**GROUP**
I: GAS	**1. HAZARD EXISTS** An area where GASES or VAPORS are normally present. **2. POTENTIAL HAZARD** An area where GASES or VAPORS are handled or stored but are not normally confined or in closed container systems.	A. Acetylene. B. Hydrogen. C. Ethyl, Etc, Ether D. Gasoline, Hexane, Natural Gas Methane
II: DUST	**1. HAZARD EXISTS** An area where combustible DUST is always present. **2. POTENTIAL HAZARD** An area where combustible DUST is present in atmosphere.	E. Metal Dust, Aluminum, Magnesium, etc. F. Carbon Black, Coal Dust, Coke Dust. G. Flour, Grain.
III: FIBERS	**1. HAZARD EXISTS** An area where fibers are handled, manufactured, or used. **2. POTENTIAL HAZARD** An area where fibers are stored or handled, other than in the process of manufacture.	None

Figure 8.30.9 NEC electrical hazardous area classification.

- Groups applicable to Class II, Division 1 and 2 are:
  - E. Metal Dust, Aluminum, Magnesium, etc.
  - F. Carbon Black, Coal Dust, Petroleum Coke Dust.
  - G. Flour, Grain.

The most common Electrical Hazardous Area Classification encountered in a petroleum refinery or chemical plant is Class I (flammable vapor), Division 2 (flammable vapor is not present during normal operations), and Group D (materials like gasoline, hexane, or natural gas) (Methane) are processed. The chart from the NEC indicating Electrical Hazardous Area Classification areas is available in Figure 8.30.9.

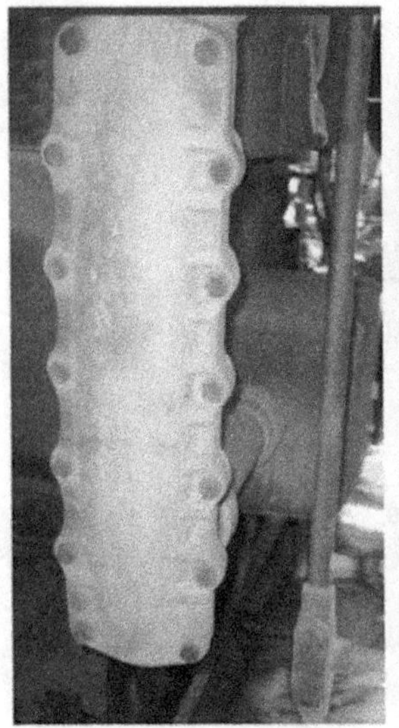

Figure 8.30.10 Example "Explosion-Proof" bolted enclosure electrical equipment.

The NEC provides guidelines for the design and type of electrical equipment suitable for the classified areas. For example, electrical equipment installed or operating in Class I, Division 1, or 2 areas must be "explosion-proof." Examples of "explosion-proof" electrical equipment are available in Figure 8.30.10.

As a reminder, all of the bolts must be installed and tight; otherwise, the integrity of the enclosure is jeopardized.

Fixed Electrical Equipment

Fixed electrical equipment must meet the EHAC classification requirements, including requirements for most electrical equipment, such as terminal boxes, light fixtures, motor starter boxes, etc. This means that the equipment must either be nonarcing or placed in process areas that are "unclassified," or the equipment must meet the "explosion-proof" requirements for classified locations. For example, in Class I, Division 1 and 2 locations, arcing devices such as relays and switches must be enclosed in an explosion-proof housing. The words "explosion-proof" are somewhat of a misnomer. The device is not explosion-proof but is designed to contain an explosion inside the device to the point that it will not ignite a flammable vapor in the adjacent atmosphere.

These devices are generally enclosed in bolted or threaded enclosures, as shown in Figure 8.30.10. As noted above, these devices are designed to contain an explosion inside the device, preventing ignition of a flammable atmosphere outside the enclosure, and operate at such an external temperature that the device or the enclosure will not ignite a surrounding

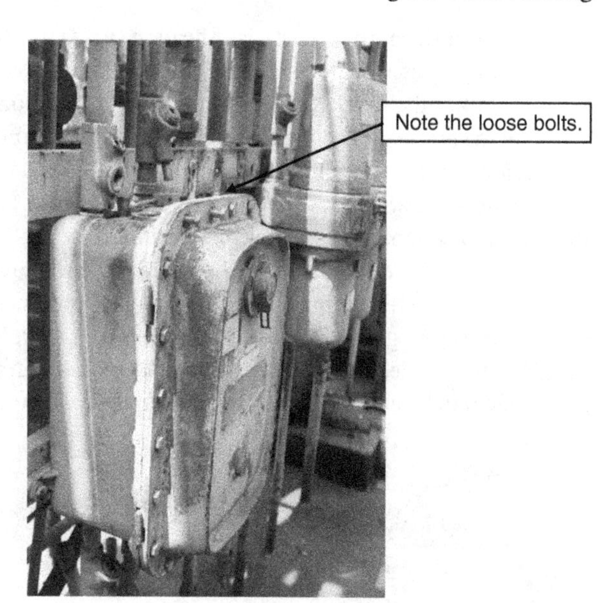

Figure 8.30.11 Example of bolted electrical enclosure with loose bolts.

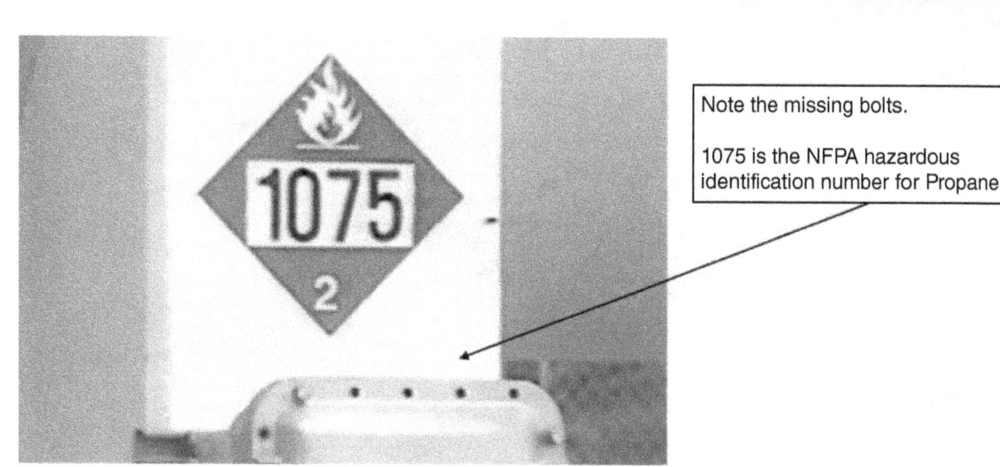

Figure 8.30.12 Example of bolted electrical enclosure with missing bolts.

flammable atmosphere. However, all the bolts must be in place and tight for these devices to function correctly. Maintenance must be notified if bolts are missing or the bolts are loose. Examples of explosion-proof electrical enclosures with loose and missing bolts are shown in Figures 8.30.11 and 8.30.12.

Portable Electrical Equipment

Portable electrical equipment must be approved as "intrinsically safe" or fully compliant with the electrical hazardous area classification before use in the field. Devices approved for use in hazardous locations will have a certification tag attached to them indicating the certification body or agency (UL, FM, ANSI, CSA, etc.). For example, flashlights or unit radios authorized in hazardous locations will each have a tag indicating certification. Otherwise, a permit-to-work should be authorized, including a gas test for flammable vapor, before the device is authorized for use. Examples of devices requiring permit-to-work in EHAC classified areas include cameras, analytical devices such as ultrasonic detection devices (for pipe or vessel inspection), handheld computers, etc. Cellular telephones are generally not permitted in hazardous classified areas at most refining and petrochemical sites.

Use of Electrical "Cheater Cords"

The Electrical Hazardous Area Classification (EHAC) required by the NEC establishes standards for the regulation of arcing/sparking devices that are allowed on or near process areas. Most process areas are classified as "Class I, Division II," which signifies the processing of flammable hydrocarbons or other materials and that these materials are normally contained in the processing equipment. The NEC requires arcing/sparking devices to be designed to contain an explosion inside the equipment. For example, an electrical outlet in a Class I, Division II area must contain any electrical spark inside the equipment. An example of these receptacles is shown in Figure 8.30.13. When the Operator inserts the male end of the cord into the connection, there is no electrical connection until the Operator rotates the male end to the right, closing the internal circuit. This energizes the attached cord and associated equipment. This design ensures no arcing or sparking in the process areas where flammable materials may be present.

In some cases, generally, as a convenience, the explosion-proof end of the cord is removed and replaced with a household-type electrical connection. This makes connecting standard (nonexplosion-proof) equipment to the unit receptacle much easier. Of course, this connection is now an arcing/sparking device and may become an ignition source

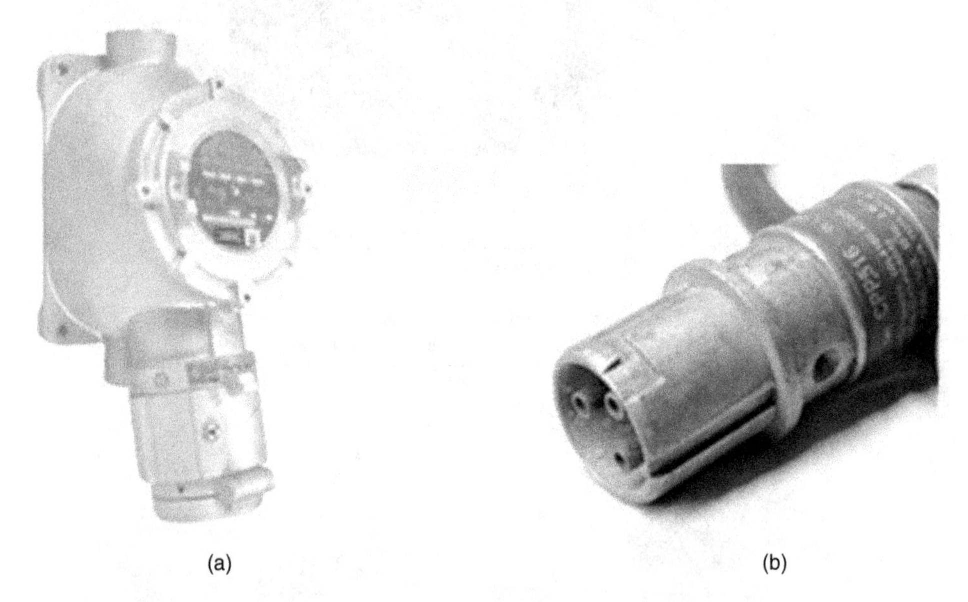

(a) (b)

Figure 8.30.13 Explosion-Proof receptacle (male and female ends. Designed to 'insert and turn to energize'). Please note that the female is the left image and the male connection is the right image. These connections are explosion proof and will not energize until the connection is completed in the enclosed housing and the connetion is rotated to energize.

Figure 8.30.14 Electrical "cheater cord" (Note that the explosion-proof female end has been removed and replaced with a regular household electrical connection, defeating the explosion-proof properties).

for flammable vapors if they are present when the connection is connected or disconnected. Figure 8.30.14 shows a cheater cord – note that the explosion-proof end has been removed and replaced with a standard household receptacle.

Cheater cords should not be permitted in process areas except for special circumstances, only after a hot work permit has been issued to cover the arcing/sparking that will occur during its use.

Arc-Flash/Arc-Blast Hazards

The other significant electrical hazards are arc flash and arc blast. Arc flash and arc blast (AF/AB) occur when the voltage level is sufficiently high, generally above 120 volts, and the contact is made between a charged conductor and ground or between two or more energized phases. In industry, many of these events have occurred when the voltage level is above 480 volts. While these happen almost instantaneously, they are two separate events and will be discussed as two events here.

An arc flash is a resultant fireball when a fault occurs in energized conductors. This is a bright electrical flash during the rapid release of energy due to an arcing fault between a phase bus bar and another phase bus bar, neutral or ground. In this event, the ionized air becomes the conductor. Arc-flash temperatures can reach 35,000°F (20,000 °C), resulting in critical burns to personnel and major equipment damage. The ultraviolet light generated by the arc causes burns to the skin and damages the eye's retina.

The arc blast is the resultant pressure wave and projectile emanation that occurs when a fault occurs between the energized conductors. Generally, AF/AB occurs at voltage levels above 110 volts and is most common in equipment at voltages over 480 volts. The tremendous temperatures of the AF/AB cause the expansion of the surrounding air and the metal in the arc path, resulting in sudden and catastrophic damage or rupture of the equipment. For example, copper conductors expand

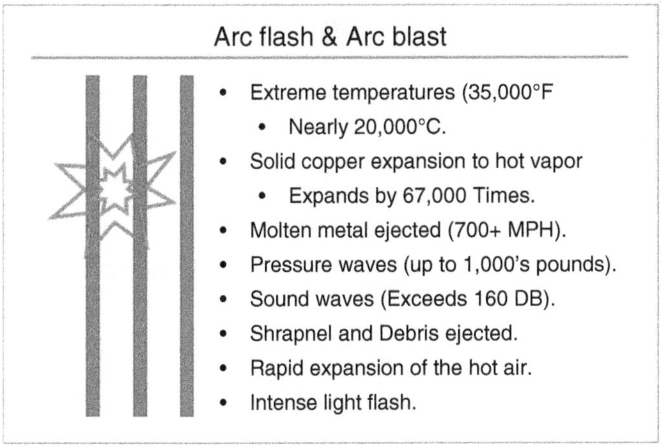

Arc flash & Arc blast

- Extreme temperatures (35,000°F
  - Nearly 20,000°C.
- Solid copper expansion to hot vapor
  - Expands by 67,000 Times.
- Molten metal ejected (700+ MPH).
- Pressure waves (up to 1,000's pounds).
- Sound waves (Exceeds 160 DB).
- Shrapnel and Debris ejected.
- Rapid expansion of the hot air.
- Intense light flash.

Figure 8.30.15 The effects of an Arc Flash and Arc Blast. This example illustrates arcing between two of the three high-voltage conductors.

approximately 67,000 times as they change from a solid to a gas. Such an explosion and the expulsion of hot gases can cause severe injuries, including burns, ruptured eardrums, collapsed lungs, injuries from flying debris, and other injuries caused by the pressure wave's sudden impact on the body. Figure 8.30.15 illustrates the effects of an arc flash/arc blast.

Requirements of NFPA 70E (Selection of Electrical PPE and Warning Workers of Hazards)

NFPA 70E provides the requirements for employee selection of electrical PPE and warning workers of potential electrical hazards. Employers are expected to adopt and follow NFPA 70E to comply with OSHA's general duty clause. Article 90 of NFPA 70E provides a good description of the requirements of NFPA 70E.

More specifically, Article 130.5 of NFPA 70E sets the expectation that employers will perform an arc flash risk assessment for electrical equipment, especially electrical breakers and other electrical switchgear. This is to ensure that all potential electrical hazards that may be present have been identified and the equipment labeled as a warning to workers. Article 130.5 requires specific arc flash labels to include system voltage, arc flash boundaries, and other information to determine the minimum arc rating for the required PPE. This helps ensure that the workers have available and use the appropriate PPE for arc flash and arc blast protection. Many sites have adopted the practice of posting the arc flash boundaries on the equipment labels and painting the warning boundaries on the pavement in front of the electrical equipment.

Of course, a better practice is to ensure that the equipment is completely deenergized anytime the breaker or enclosure doors are open.

Many people have been critically injured, and lives have been lost due to being near an electrical arc flash or arc blast. Industry standards and US regulations define specific distances required for personnel working near live circuits and detail requirements for providing safe distances (boundaries) for unprotected personnel. The standards require each site to define these "flash boundaries" and post them on each piece of electrical equipment susceptible to AF/AB.

Figure 8.30.16 Failed high-voltage electrical switchgear. Source: Photo courtesy Montgomery County (MD). Fire and Rescue Service.

Figure 8.30.16 is an example of 4160-volt electrical switchgear that has experienced an internal fault and arc-flash/arc-blast event. In this example, the arc blast blows the protective doors clear, exposing other equipment and personnel to the resulting force from the explosion. For this reason, most sites require that operations personnel remain clear of this type of equipment during electrical switching operations and that only qualified electricians are allowed in the area.

Sites can also provide additional protection for electricians by providing arc-flash protective gear (arc-flash suits, gloves, boots, etc.). Also, some sites use robotic or remote-switching equipment, which electricians can use to execute electrical switching of high-energy circuits from a safe distance.

High-Voltage PPE (Arc-Flash/Arc-Blast Protection)

The electricians are also typically required to wear specially designed arc-flash/arc-blast protective PPE. See Figure 8.30.17 for an example of a high-voltage arc flash and arc blast. Figures 8.30.18 and 8.30.19 illustrate examples of the specialized arc-flash/arc-blast PPE for this type of work. Industry and regulatory standards also require each site with arc-flash/arc-blast hazards to calculate potential energy levels and post approach boundaries on or near the equipment at risk.

The main message in this section for Operations personnel is to recognize the significant hazards present when the electricians are working with the high-voltage equipment. All other personnel, including operations personnel, should be clear of the area when the electricians conduct electrical switching, opening or closing high-voltage breakers, and other work of this type. The electricians should be well trained, including in the appropriate PPE for protection against hazards such as arc flash and arc blast; however, operations personnel can help, for example, when issuing work permits, to ensure the appropriate PPE is used.

Remote Closing Devices (Robotic Closers)

Newer technology includes remote opening/closing devices to operate electrical circuit breakers in high-voltage service. These breakers are usually operated (opened or closed) only when the equipment being fed is down, with no load on the breaker and no load expected when the breaker is closed. Opening or closing the breaker with load can result in an arc flash/arc blast and is therefore avoided. However, as a precaution, electricians are protected with specialized PPE should the event occur. To further protect the Electricians, remote-switching devices allow electricians to be entirely out of harm's way should an unexpected flashover occur. Examples of remote-switching devices are shown in Figures 8.30.20 and 8.30.21.

Figure 8.30.17 Example of arc flash/arc blast due to a high-voltage electrical fault. Source: Photo courtesy of Westex®: A Millken Brand, Spartanburg, S.C. (annotation by Westex®) (Leader in FR/AR Protective Fabric|Westex®: A Milliken Brand).

Figure 8.30.18 Example of arc-flash/arc-blast protective PPE used by electricians servicing high-voltage equipment. Source: Photo courtesy of Lee Marchessault at Workplace Safety Solutions, Inc. (Workplace Safety Solutions Experts In Electrical Safety Training and Arc Flash Analysis).

Figure 8.30.19 Other common protective equipment designed for use by electricians when working near high-voltage equipment. Source: Photo courtesy IndustrialSafety.com (Safety Supplies, Industrial Supplies, PPE, and MHE – IndustrialSafety.com).

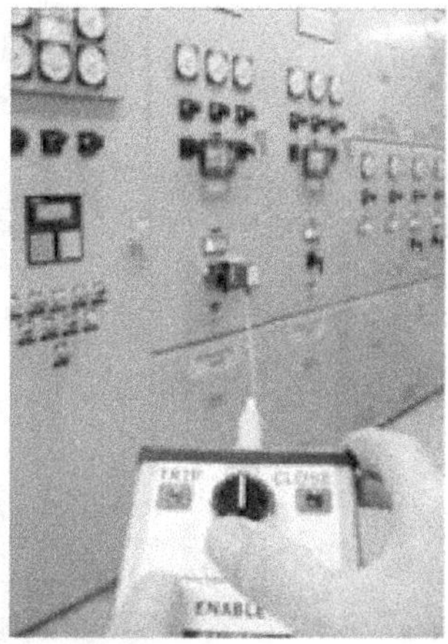

Figure 8.30.20 Remote breaker closing device sometimes called a "Chicken Switch." Source: Electrical remote-switching images are provided courtesy of CBS ARCSAFE (CBS ArcSafe Remote Circuit Breaker Racking) (wpengine.com).

Hazard Recognition

The following Figures are examples of some electrical safety issues you can help identify and correct in the field. These all have the potential for injury to personnel and may ignite flammable vapors or combustibles, leading to fire or explosion. If you see any of these, please be sure to have these addressed promptly or notify your supervisor. How many can you spot in your work areas?

Please refer to the following figures:

Figure 8.30.22 – Damaged electrical cords.
Figure 8.30.23 – Damaged electrical conduit and exposed wiring.
Figure 8.30.24 – Missing electrical seals.
Figure 8.30.25 – Missing conduit cover with exposed wiring.
Figure 8.30.26 – Household-type extension cord and electrical "cheater cord." Please note that these are not explosion proof and will arc and spark. They should not be used in hazardous locations where flammable materials may be present.

What Has Happened/What Can Happen

Work Near Overhead Power Lines
Jackson, TN (Near the airport)
23 July 2008

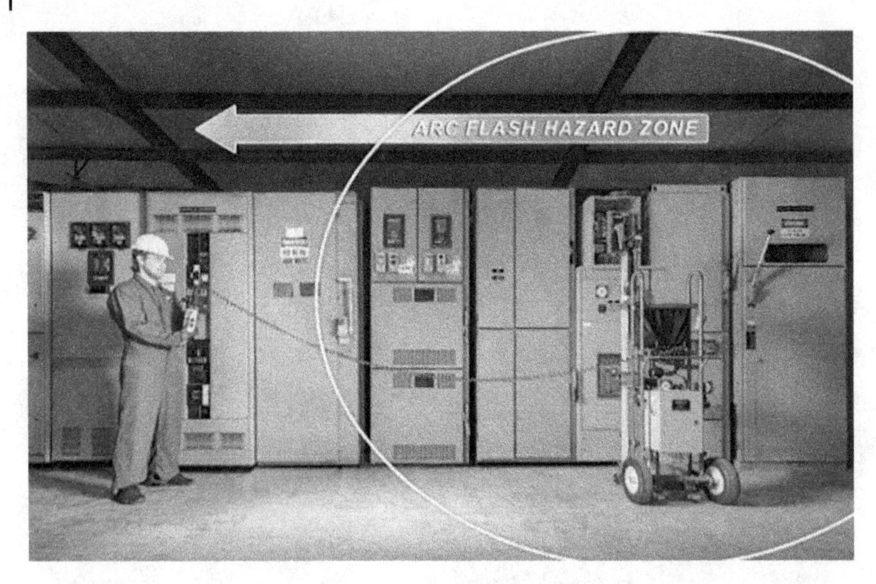

Figure 8.30.21 Remote breaker racking devices. Source: Electrical remote-switching images are provided courtesy of CBS ARCSAFE (annotation by CBS ARCSAFE). (CBS ArcSafe Remote Circuit Breaker Racking) (wpengine.com).

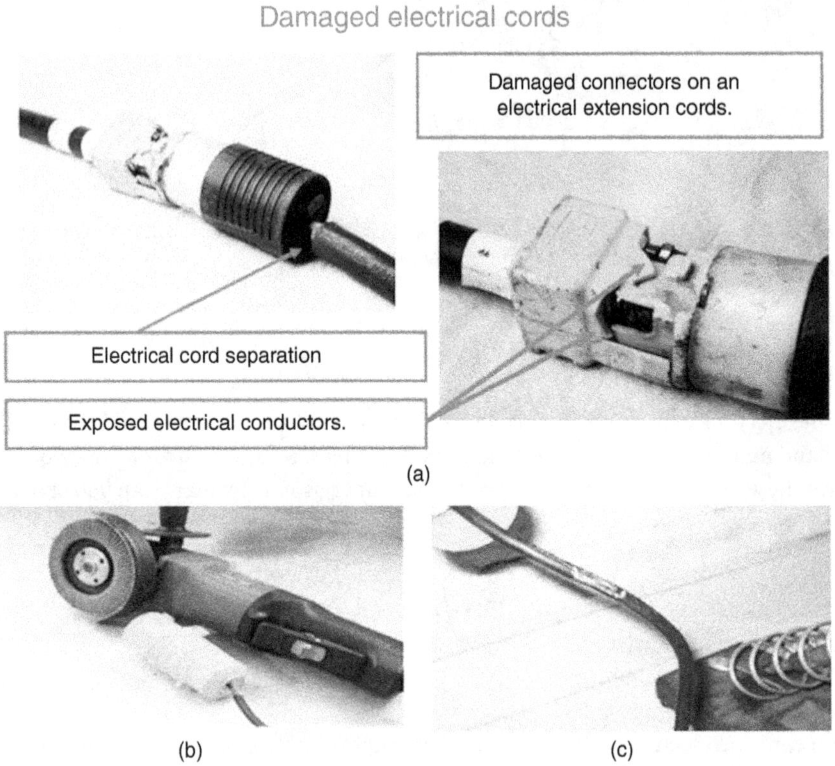

Figure 8.30.22 Damaged electrical cords (two shown here). Note: The first electrical extension cord illustrated below was located under several inches of snow. Note the exposed conductors.

This incident shows how hazardous it is to work directly below or near energized overhead electrical power lines. These lines are typically very high-voltage and contact can mean certain death and destruction of equipment and cause power loss to a refinery or chemical plant. This example also shows how easy it is to come into direct contact with these overhead power lines.

Figure 8.30.23 Damaged electrical conduit and exposed wiring.

Figure 8.30.24 Missing electrical seals.

Figure 8.30.25 Missing conduit cover with exposed wiring.

Figure 8.30.26 Household-type extension cord and electrical "cheater cord" in Class I, Division II Classified Area.

Figure 8.30.27 Photo of log truck in direct contact with an energized overhead electrical power line. This image was used with the permission of the company involved.

In this example, a log truck parked under an energized overhead power line, and the driver may not have realized the lines were there. As the driver attempted to secure the logs on the trailer, he threw a cable over the logs, intending to tie the cable over the logs and to the trailer on both sides. Unfortunately, the cable landed directly on the overhead power line, resulting in an electrical fault in the trailer and truck. Fortunately, the driver was reported to be uninjured; however, the potential certainly existed for severe injury or even loss of life. The truck, trailer, and a load of logs were all destroyed in this incident.

See Figures 8.30.27–8.30.29 for photos of the truck, trailer, and the energized power line.

Figure 8.30.28 Photo of log truck rigging cable across an energized electrical power line. This is the cable the driver was attempting to throw acoss the load to secure the load with. This driver was very fortunate to have survived. This image was used with the permission of the company involved.

Figure 8.30.29 Photo of destroyed tractor and log trailer due to contact with an energized electrical power line. This image was used with the permission of the company involved.

Key Lessons Learned

When working near overhead power lines, the safe practice is contacting the energy supplier and deenergizing the power lines. However, if work is planned near energized overhead power lines, the maximum allowable proximity to the overhead electrical power lines is covered in OSHA 29CFR 1910.333 "Selection and use of work practices." This OSHA standard details the safe distances that should be maintained from energized overhead power lines as follows, depending on the voltage of the power line. An excerpt from the regulation is included below:

> When an unqualified person is working in an elevated position near overhead lines, the location shall be such that the person and the most extended conductive object they may contact cannot come closer to any unguarded, energized overhead line than the following distances:
> "1910.333(c)(3)(i)(A)(1)
> For voltages to ground 50 kV or below: 10 feet (305 cm);
> 1910.333(c)(3)(i)(A)(2)
> For voltages to ground over 50 kV: 10 feet (305 cm), plus 4 inches (10 cm) for every 10 kV over 50 kV.
> 1910.333(c)(3)(i)(B)
> When an unqualified person is working on the ground in the vicinity of overhead lines, the person may not bring any conductive object closer to unguarded, energized overhead lines than the distances given in paragraph (c)(3)(i)(A) of this section."

In the case of the log truck driver, we do not know the voltage of the power line that he came into contact with. However, per the regulation cited above, a clearance of 10 feet must be maintained up to a voltage of 50,000 volts. An additional distance of 4 inches should be added for each additional 10,000 volts over 50,000 volts. For example, for a power line operating at 138,000 volts, the distance required from the longest available extension is 13 feet.

In high-voltage power lines, one does not need to touch the power line to be electrocuted; a person can be electrocuted simply by approaching the energized power line. The distances cited above are designed to help keep personnel safe while working near energized power lines.

What Has Happened/What Can Happen

Northwestern US Refinery
Electrical Enclosure Exploded

The operator was Struck in the chest by the electrical cover plate (severe injury).

In this example, the Operator was in the process of returning a centrifugal compressor to service. After lining up the compressor and completing all the pre-startup checks, he pressed the start button to start the motor driving the compressor. As soon as the motor energized, the large electrical cover plate blew from the MTJB, striking the concrete pavement and the Operator in the mid-chest area. I spoke with the Operator sometime after this incident, and he told me he spent nine days in intensive care due to the injuries received.

Key Lessons Learned

We talked earlier about the operator's body position relative to the line-of-fire from the MTJB. For an operator, this is an essential learning in this chapter. Operators should never stand directly in the line-of-fire of the MTJB. I am personally aware of several of these things exploding and projecting shrapnel. In this case, the operator was severely injured when the large metal electrical box cover plate blew off, ricocheted off the pavement, and struck him directly in the sternum. The operator survived, but this was a life-threatening injury.

This injury is precisely what the precautions in this chapter are intended to help prevent. Operators should be trained to avoid the line-of-fire of MTJB when starting electrical motors or similar equipment. The electrical terminal junction box is where the three-phase wiring between the electrical breaker and the motor is connected (the three wires are connected inside the MTJB). Sometimes, these wires are connected by bolting the leads from the breaker to the leads going to the motor and then taping the three connections. The electricians may twist the leads together for smaller motors and then tape the leads.

So, what is the hazard with these connections? When the motor is started, the initial current through these leads is about eight or nine times greater than when the motor is running. This electrical power surge is required to get the motor from 0 to 1,800 rpm or 3,000 rpm, depending on the motor design. This power surge places a huge electrical load on these connections. Should there be moisture in the terminal box, or if the electrical leads are loose, there can be an explosion, and the cover plate can be blown free. This is what happened to this Operator.

Prevention includes the following: (1) ensuring that all the bolts in the cover plates are correctly installed and that all bolts are tight, (2) training the Operators to stand out of the line-of-fire of the MTJB cover plates, and (3) relocating the start/stop switches if they are in direct line-of-fire.

What Has Happened/What Can Happen

Mine Safety and Health Administration
US Department of Labor
Incident Sharing – Fatality Alert
7 August 2019

A 42-year-old plant electrician with 15 years of mining experience was electrocuted on 7 August 2019 when he contacted an energized connection of a 4,160 VAC electrical circuit. This incident is reported here due to the significant lessons learned and shared in the fatality alert safety sharing.

The victim was in the plant's Motor Control Center (MCC), adjusting the linkage between the disconnect lever and the internal components of the 4,160 VAC panel supplying power to the plant feed belt motors. See Figure 8.30.30 for the location of this incident. The MCC was not electrically isolated, and the worker came into contact with the energized internal components (4,160 volts).

Key Lessons Learned

The following are considered electrical best practices and are shared to help ensure that an incident such as this one will not be repeated.

- Lock-out and Tag-out the electrical circuit yourself, and NEVER rely on others to do this for you.

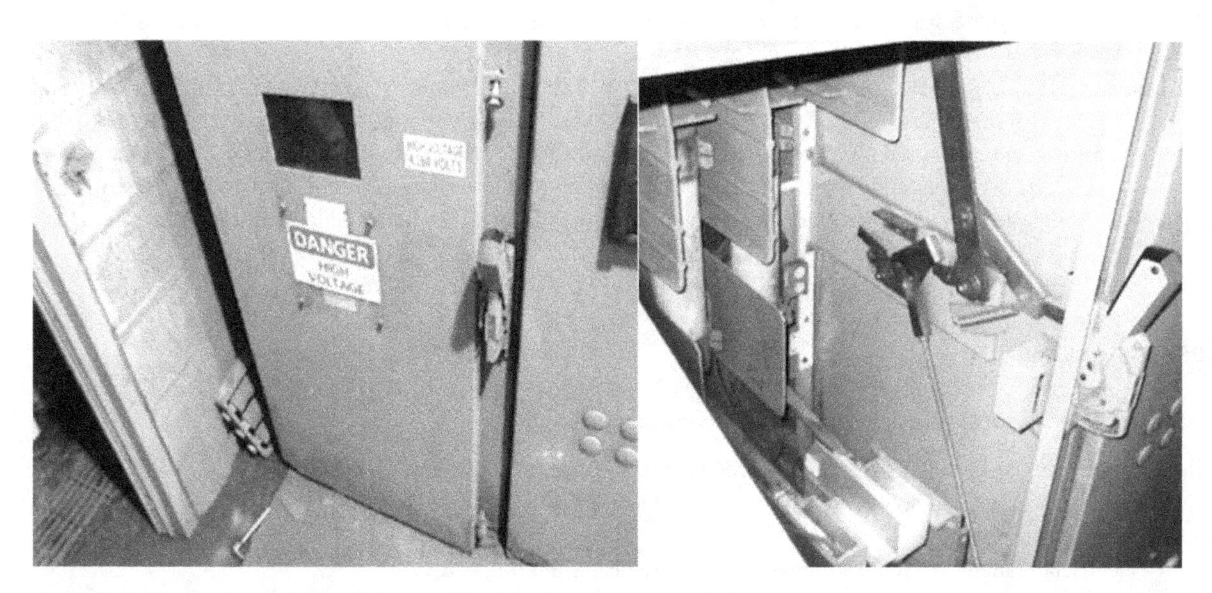

Figure 8.30.30 Location of Mine Safety and Health Administration (MSHA) reported electrocution fatality. Source: Photos courtesy of the Mine Safety and Health Administration, US Department of Labor.

- After performing the lock-out, always verify the equipment is deenergized by attempting to start the device using all starting locations, including the DCS system if so equipped.
- Control Hazardous Energy! Design and arrange MCCs so electrical equipment can be serviced without hazards. Install and maintain a main disconnecting means located at a readily accessible point capable of disconnecting all ungrounded conductors from the circuit to service the equipment safely.
- Install warning labels on the line side of terminals of circuit breakers and switches indicating that the terminal lugs remain energized when the circuit breaker or switch is open.
- Before performing troubleshooting or electrical work, develop a plan, communicate, and discuss the plan with qualified electricians to ensure the task can be completed without creating hazardous situations.
- Follow these steps BEFORE entering an electrical enclosure or performing electrical work:
  1. Locate the circuit breaker or load break switch away from the enclosure and open it to deenergize the incoming power cable(s) or conductors.
  2. Locate the visual disconnect away from the enclosure and open it to show that the incoming power cable(s) or conductors have been deenergized.
  3. Lock-out and tag-out the visual disconnect (follow the LOTO procedure).
  4. Ground the deenergized conductors.
- Wear properly rated and maintained electrical gloves when troubleshooting or testing energized circuits.
- Focus on the task at hand and ensure safe work practices to complete the service. A second qualified electrician should double-check to ensure you have followed all necessary safety precautions.
- Use properly rated electrical meters and non-contact voltage testers to ensure electrical circuits have been deenergized.

What Has Happened/What Can Happen

BP Texas City Refinery
Electrician Fatality – Knowingly Working an Electrical Circuit While Energized
5 June 2007

A BP Texas City Refinery, safety sharing report communicated and summarized this incident. It is also outlined here to help ensure that a similar incident will not happen again in the industry. Thanks to BP for sharing this incident.

An experienced and qualified contract electrician was fatally injured when working on a live single-feed 480-volt lighting circuit. The electrician was working with his co-worker, an experienced and qualified electrician who was the only eyewitness to the incident. The co-worker stated they disconnected the wires at the field junction box to remove the lighting fixture to a more convenient work location. During this Contractor's long history at the Texas City site, their stated practice has been to only work on these systems after energy isolation, and their past audits had not found deviations from this policy.

The Risk Assessment identified electrical shock as a known risk. The Authorization to Work (ATW) and the associated Risk Assessment issued for their work required energy isolation and verification. The contractor foreman stressed this at the morning safety meeting, attended by both contract craftsmen. Unfortunately, the contractor craftsmen accepted a known elevated level of personal risk by electing to work the electrical system while it was still energized. His co-worker did not stop the job even though he knew that the work was being performed in a manner that violated contractor and company policies.

Key Lessons Learned

The subsequent incident investigation determined the following causes:

- Violation by the individual – neither worker followed energy isolation procedures.
- Violation of group – worker, and co-worker were aware that applicable policies required that the lighting be deenergized and locked out before this work.
- Shortcuts – electrician failed to execute a known and required isolation procedure.
- Servicing of energized equipment – equipment was not locked out prior to servicing.
- Inadequate isolation of process or equipment – equipment was not locked out before servicing.

- Energized Electrical System – Breaker was not isolated with a personal lock.
- Poor judgment – both contract workers decided to work the job electrically live, knowing that removing the lighting fixture without locking it out violated energy isolation policies.
- A less experienced contractor craftsman, and a more experienced contractor craftsman, did not feel empowered to stop an unsafe work practice.
- Recommendations to prevent a recurrence of this incident included:
  - Implement the recommendations of a cross-functional team of contract and company representatives to enhance behavioral safety.
  - Incorporate the "Duty of Care" training into the Safety Council site-specific training.

Additional References

National Fire Protection Association (NFPA) NFPA-70 "The National Electrical Code", National Fire Protection Association (NFPA) NFPA-70E "Standard for Electrical Safety in the Workplace".

American Petroleum Institute API RP 500 and 505 "Hazardous Area Electrical Classification".

American Petroleum Institute API RP 54 (Section 9.14 Generators, Motors, and Lighting).

Relevant OSHA Standards (Electrical Safety – Not intended to be a complete list)

- 29 CFR 1910.147 "The Control of Hazardous Energy (Lock-out/Tag-out)."
- 29 CFR 1910.137 "Electrical Protective Devices."
- 29 CFR 1910.302–308, and 399 "Electrical Standards for General Industry."
- 29 CFR 1910.302–330 "Design Safety Standards."
- 29 CFR 1910.331-360 "Safety Related Work Practices."
- 29 CFR 1910.399 "Definitions."
- 29 CFR 1926.403 "General Requirements."
- 29 CFR 1926.404 "Wiring Design and Protection."
- 29 CFR 1926.405 "Wiring Methods, Components, and Equipment for General Use."
- 29 CFR 1926.416 "General Requirements."

Other OSHA Resources:

- Publication 3075 "Controlling Electrical Hazards"

OSHA Standard Interpretation: Protection of workers operating mechanical equipment near overhead power lines. https://www.osha.gov/laws-regs/standardinterpretations/1999-12-07-2

Facilities net:
Article: "Understand Arc Flash Codes, Standards, and Regulations". https://www.facilitiesnet.com/datacenters/article/Understand-Arc-Flash-Codes-Standards-and-Regulations--17741

EHS Daily Advisor:
Article: Electrical Safety Tips: Using NFPA 70E to Protect Workers, By Lee Marchessault 22 January 2020, *Electrical Safety, Personal Protective Equipment*, **Updated: 22 January 2020**. Electrical Safety Tips: Using NFPA 70E to Protect Workers – EHS Daily Advisor (blr.com).

CBSARCSAFE Article, Description of the 'Chicken Switch' and other remote electrical switching devices, Chicken Switch – CBS ArcSafe, 2017 Code Digest, Article 500 Hazardous (Classified) Locations, EATON'S CROUSE-HINDS BUSINESS 2017 Code Digest C. https://www.eaton.com/content/dam/eaton/products/low-voltage-power-distribution-controls-systems/crouse-hinds/crouse-hinds-codedigest2017.pdf

UK Health and Safety Executive, "Working safely near overhead electricity power lines." https://www.hse.gov.uk/pubns/ais8.pdf. "Avoiding danger from overhead power lines (Guidance Note GS6 (Fourth edition)." https://www.hse.gov.uk/pubns/gs6.pdf

Chapter 8.30. Electrical Safety Hazards for Operations Personnel

End of Chapter Quiz

1 What are the three types of electrical safety hazards that operations personnel must be aware of?

2 How much electrical current is required to stop your heart and take your life when the conditions are right?

3 How does a Ground Fault Circuit Interrupter (GFCI) device work to provide protection against accidental electrical contact?

4 A GFCI, when properly used, will stop the current flow at about ______ mA before the current can have devastating effects on the body.

5 What is an electrical "cheater" cord, and why are they not permitted in process areas unless a hot work permit is issued?

6 LOTO, in the context of this Chapter, refers to the isolation of __________ energy prior to work on circuits or other electrical equipment. LOTO is equally important for the isolation of other energy hazards, such when opening process equipment or where there is the potential for moving parts or other gravity, spring-loaded, thermal, or chemical hazards present.

7 How is work on electrical systems turned over to maintenance personnel to protect them from electrical shock hazards?

8 Why is it so important for the Operator to attempt a start of the equipment after completing the LOTO and before turning the equipment over to maintenance?

9 Once the equipment is turned over to the craftsmen before maintenance begins, the __________ should test for current in the isolated system and install their personal ________ on the breaker.

10 Use of 110 volts and higher electrical tools and equipment inside storage tanks and process vessels is a unique safety hazard because the vessel is a direct _________. Any incidental _________ between the power leads or other energized conductors and the vessel with people present can result in the potential for electrocution.

11 What is the hazard with the Motor Terminal Junction Box during a motor start?

12 If 110 volts tools or equipment must be used, each device must be _________ by a ground fault circuit interrupter (GFCI). The ________ must be physically located outside the vessel, and the power cables must be protected from damage, especially where they pass through the vessel manway(s) or other similar pinch points.

13 When you pick up an electrically powered tool, what are examples of things you should inspect before the tool is used?

14 The use of 110 volts-powered equipment inside process vessels should be viewed as an __________ to routine work practices, and requirements for each use should include a hazard assessment, including review by __________ specialists and approval by the appropriate site management representative(s).

8.31

Hazards of Chemical Incompatibility or Reactivity

This chapter aims to reduce or eliminate chemical-related storage risks, especially due to the inadvertent mixing of two or more incompatible chemicals. Chemicals are important in many petroleum refineries and petrochemical plants as additives and chemical treatments. The inherent hazards of chemicals can be reduced by minimizing the amount of chemicals on hand and by having a documented process for how required chemicals are properly stored and handled, including the physical separation of those chemicals incompatible with other chemicals.

Where Can We Find the Proper Storage Information for the Required Chemicals?

Chemical storage information can be located in Section 7 of the Safety Data Sheet (SDS). The SDS is required by the US Code of Federal Regulations (29 CFR) 1910.1200, and an SDS must be on hand for every hazardous chemical in the refinery or chemical plant. The following information can be found in Sections 7, 9, and 10 of the SDS:

- Is the chemical flammable?
- Is the chemical corrosive?
- Does the chemical need to be stored other than at ambient temperature?
- Is the chemical an oxidizer or reducer?
- Is the chemical light-sensitive?
- Does the chemical require any special handling procedures?

The UN/NA (United Nations/North America) Chemical Identification Numbers

The EPS Cameo response system and other chemical emergency response guides utilize the United Nations/North America (UN/NA) chemical identification numbers, which cross-reference to a detailed set of emergency response guides providing response information for emergency responders. UN/NA numbers are not required by OSHA on an MSDS, although many SDS sheets have them to simplify shipping requirements. DOT's 2020 Emergency Response Guide is indexed by "ID Numbers," which are the UN/NA numbers.

United Nations (UN) Numbers are four-digit numbers used worldwide in international commerce and transportation to identify hazardous chemicals or classes of hazardous materials. These numbers generally range between 0000 and 3,500 and are ideally preceded by the letters "UN" to avoid confusion with other number codes. North America (NA) numbers are issued by the United States Department of Transportation and are identical to UN numbers, except that some substances without a UN number may have an NA number. The additional NA numbers use the range NA 9000–NA 9279. For example, "UN1017" provides response information for Chlorine and cross-references to the Response Guide 124.

It is critical that receiving tanks for chemicals are labeled and properly identified. For example, the receiving lines and valves should be clearly labeled and identified with the chemical name and the UN/NA chemical identification numbers. This should be set up as a check for the tank truck driver or the person unloading a chemical railcar to ensure the correct chemical is being unloaded into the storage tank. The Unit Operator should do the final verification for any product being unloaded into a storage tank from a tank truck or railcar. The Operator should validate the cargo manifest with the signs and labels before unloading the product into the storage tank.

Process Operations Safety: The What, Why, and How Behind Safe Petrochemical Plant Operations, First Edition. M. Darryl Yoes.

Typical storage considerations may include the storage temperature, control of potential ignition sources, adequate ventilation, segregation from other chemicals, and proper labeling. Proper chemical segregation is required to prevent incompatible materials from inadvertently mixing and causing a violent reaction, a fire or explosion, or the evolution of toxic vapors or gases.

It is important to remember that acids should NOT be stored with bases when segregating chemicals, and oxidizers should not be stored with organic materials or reducing agents. Physical barriers with effective labeling or other identification and adequate distances between chemicals are effective for proper segregation.

Most of us know that acids and bases do not play well together, and mixing the two will result in a violent reaction, high temperatures, and rapid corrosion of most metals. When mixing acid and water, most of us remember the adage always to mix the acid into the water and never the other way around. Mixing water into the acid will rapidly boil the water and splash the acid onto people who may be nearby. An easy way to remember this informal rule is to remember the expression "Add the Acid." This should always be done working under a fume hood in a laboratory environment so the vapors will be safely vented away.

Remember that acids and caustics will readily corrode most metal cabinets when using storage cabinets to store relatively small volumes of chemicals. Consider the cabinet construction and its compatibility with the chemical to be stored. In the case of acids or caustics, a better choice may be nonmetallic or epoxy-painted cabinets. When storing acids and bases, always separate the two into separate cabinets or at least separate compartments to avoid inadvertently mixing the two. Some acids and bases may damage the epoxy-painted surfaces if a spill occurs. Also, perchloric acid should not be stored in a wooden cabinet.

There are cabinets explicitly designed to store flammable liquids, and it is essential to be aware of the maximum allowable container size and maximum quantities for storage in cabinets based on the category of the flammable.

Flammable liquids are categorized based on their flash point and boiling point. US OSHA defines a liquid to be flammable if the flashpoint is at or below 199.4°F (93 °C) and categorizes flammable liquids into four separate categories based on Table B.6.1 of 29 CFR 1910.1200, Appendix B.

Note: Table B.6.1 categorizes products with a flashpoint of less than 73.4°F (23 °C) as Category 1 and 2 flammable liquids (See Figure 8.31.1*).*

See Figure 8.31.1 for a copy of this table. Figures 8.31.2 and 8.31.3 also list the maximum amounts that can be stored in a single container or single storage cabinet, respectively.

Each storage cabinet shall contain no more than 60 gallons of Category 1, 2, or 3 liquids and not more than 120 gallons of Category 4 liquids, according to OSHA 29 CFR 1910.106(d3) (i).

In addition to the inadvertent mixing of acids and bases, many other chemicals may react when they come into contact with other noncompatible chemicals. The reactions may be mild or very violent, resulting in a severe exothermic reaction and may also result in an explosion and major fire. In some cases, the reaction may also produce toxic gases, which can be deadly to personnel.

An example of one of these reactions occurred at the Union Carbide plant in Bhopal, India, during 2–3 December of 1984 when water inadvertently entered a storage vessel containing Methyl Isocyanate liquid. Methyl Isocyanate violently reacts when contacting water, resulting in an exothermic reaction and generating massive amounts of highly Methyl Isocyanate vapor. At Bhopal, this vapor was released from an atmospheric vent into a densely populated area, affecting over 500,000

A flammable liquid shall be classified in one of four categories in accordance with Table B.6.1:
TABLE B.6.1 Criteria for Flammable Liquids

Category	Criteria
1	Flash point < 23 °C (73.4°F) and initial boiling point ≤ 35 °C (95°F).
2	Flash point < 23 °C (73.4°F) and initial boiling point > 35 °C (95°F).
3	Flash point ≥ 23 °C (73.4°F) and ≤ 60 °C (140°F).
4	Flash point > 60 °C (140°F) and ≤ 93 °C (199.4°F).

Figure 8.31.1 Table B.6.1 of OSHA 29 CFR 1910.1200, Appendix B. OSHA Classification of Flammable Liquids (text and annotation by OSHA).

Container type	Category 1	Category 2	Category 3	Category 4
Maximum allowable size of containers and portable tanks				
Glass or approved plastic	1 pint	1 quart	1 gallon	1 gallon
Metal (other than DOT drums)	1 gallon	5 gallons	5 gallons	5 gallons
Safety Cans	2 gallons	5 gallons	5 gallons	5 gallons
Metal Drums (DOT spec.)	60 gallons	60 gallons	60 gallons	60 gallons
Approved Portable Tanks	660 gallons	660 gallons	660 gallons	660 gallons

Figure 8.31.2 A chart listing the maximum volume of flammables that can be stored in a single container type.

Liquid class	Maximum storage capacity
Maximum storage quantities for cabinets	
Category 1	60 gallons
Category 2	60 gallons
Category 3	60 gallons
Category 4	120 gallons

Figure 8.31.3 A chart listing the maximum volume of flammables that can be stored in a single flammable storage cabinet.

people and resulting in thousands of deaths. This was one of the most severe process safety incidents that have ever occurred worldwide.

See Appendix K from the US National Institute of Health for the most common chemical incompatibilities (Chemical Segregation and Storage Table). The table provides information on some of the most common chemicals and their incompatible counterparts, those chemicals with which they will react. In this appendix, chemicals listed in the left-hand column should be stored and handled so they cannot contact the corresponding reactive chemicals listed in the right-hand column. This table contains some chemicals commonly found in laboratories, but it should NOT be considered a complete list.

Chemical incompatibilities for the specific chemicals you are using can usually be found in the "REACTIVITY" or "INCOMPATIBILITIES" section of the Safety Data Sheet. The Safety Data Sheet should always be consulted before mixing any chemical with another Chemical to ensure against inadvertently mixing two incompatible chemicals.

Wiley has published a "Rapid Guide to Chemical Incompatibilities" by Pohanish and Greene as a quick resource guide, which provides further information on the incompatibilities of hundreds of chemicals.

A much more comprehensive listing of reactive chemicals is available in Bretherick's Handbook of Reactive Chemical Hazards. (Urben, P.G., ed. – 6th ed. St. Louis, MO: Elsevier Science and Technology Books, 2000.) Bretherick's reference includes every chemical for which documented information on reactive hazards has been identified.

The following guidance is from the NIH (National Institute of Health concerning Chemical Incompatibility) (By Hazard Class). The NIH says that to help avoid incidents involving chemicals, we should do the following:

Separate acids from:

- Bases (possible violent exothermic reaction).
- Most metals (production of flammable hydrogen gas).
- Cyanides (forms toxic and flammable hydrogen cyanide gas).
- Sulfides (forms toxic and flammable hydrogen sulfide gas).
- Azides (may form explosive hydrazoic acid).
- Phosphides (may form toxic and flammable phosphene gas).
- Oxidizers (may form toxic and/or explosive compounds).

Separate oxidizers from:

- Acids (may form toxic and/or explosive compounds) (e.g., concentrated sulfuric acid mixed with chlorates or perchlorates forms explosive compounds).
- Organic materials (especially when mixed with flammables may ignite).
- Metals (may form explosive compounds).
- Reducing agents (e.g., boranes, hydrides, sodium hydrosulfite, etc.).
- Ammonia (anhydrous or aqueous).

It is important that we separate the chemicals that will react with water from water or moisture:

- In some cases, the moisture in the air will react with some chemicals. For example, moisture can react with metal hydrides and some metal dust that may be present in the workplace. Hydrogen gas can be generated from this reaction and can ignite readily. Other gases that may be toxic can also be formed from this kind of reaction. For additional information, please see the NIH website.

In October 2002, the Chemical Safety and Hazard Investigation Board (CSB) released a report titled "Improving Reactive Hazard Management." This report is available on the CSB website (CSB.Gov) and identifies and analyzes 167 serious reactive chemical incidents in the United States involving uncontrolled chemical reactivity from January 1980 to June 2001. Forty-eight of these incidents resulted in a total of 108 fatalities, averaging six reactive incidents a year and an average of five fatalities per year. Nearly 50 of the 167 incidents directly impacted the communities surrounding the sites, and over half of the total incidents involved chemicals not covered by existing OSHA or EPA process safety regulations. Approximately 60% of the total involved chemicals either are not rated by the NFPA or have "no special hazard" (NFPA "0"). Only 10% of the 167 incidents involved chemicals with an NFPA rating of "3" or "4."

Unfortunately, this trend appears to be continuing to this day. All of the incidents listed in the "What has Happened" section below occurred after this report was issued, plus others not discussed here. This indicates that continued vigilance, attention to detail, and perhaps additional regulations are needed in this important area.

What Has Happened/What Can Happen

First Chemical Corporation
Pascagoula, Mississippi
13 October 2002
Explosion and Fire due to the decomposition of mononitrotoluene inside a distillation column (three Injuries)

This incident was investigated by the US Chemical Safety and Hazard Investigation Board.

This incident occurred on 13 October 2002, when approximately 1,200 gallons of mononitrotoluene product (MNT), which had been left in a fractionator tower, suddenly exploded. The unit had been placed on total reflux on 7 September due to other operational problems and then totally shut down 22 days later (29 September). The 1,200 gallons remained inventoried in the column, and the manual steam reboiler valves remained closed.

When this explosion occurred, the top 35′ of the fractionator tower separated from the unit and was propelled into the surrounding area. The vessel's top head and about 30 feet of cylindrical shell were propelled off-site. The burning structured packing material inside the tower was also propelled off-site, resulting in secondary fires to surrounding equipment. A large section of the tower struck a storage tank approximately 500 feet away containing more than 2 million pounds of para-MNT, resulting in a tank fire. Several sections of debris also landed near crude oil tanks at a refinery across the highway from First Chemical Corporation.

This incident had the potential for very serious secondary impacts. For example, heavy shrapnel struck other equipment or landed near other process equipment and/or crude oil storage tanks at the adjacent refinery. One section weighing nearly 6 tons was hurled 1,100 feed and landed about 50 feet from a 250,000-barrel capacity crude oil tank. Several buildings were also significantly damaged, including the unit control room. Operators in the control room received injuries from flying glass and other debris.

As mentioned above, the entire plant was shut down on 29 September 2002. The steam boilers were then brought back online on 5 October. Following the restart of the boilers, the temperature at the fractionator bottoms steadily increased from 5 October until the explosion occurred on 13 October. During this period, 1,200 gallons of MNT remained in the tower and were exposed to the gradually increasing temperature. Also, during this period, there was no revalidation of tower isolation or indication that the tower temperature was being monitored.

The investigation by the CSB found that the steam to the tower reboilers likely leaked through the closed isolation valves heating the MNT in the tower. This was confirmed by examining the valve internals and the temperature data for the fractionator internals recorded by the plant computer system (DCS system).

The MNT chemistry played a significant role in this incident. The following characteristics are quoted directly from the mononitrotoluene SDS in the CAMEO Database of Hazardous Chemicals:

- "Normally stable but can become unstable at elevated temperatures and pressures."
- "Containers may explode when heated."
- "p-nitrotoluene and sulfuric acid explode at 80 °C (176°F) [Chem. Eng. News 27:2504]."
- "Many explosions have occurred in the distillation of this material owing to its heat sensitivity."
- "May become unstable in the presence of strong bases."

The CSB confirmed an exothermic reaction during their investigation of this incident. The CSB conducted a laboratory test where a small sample of MNT was slowly heated until it reached an exothermic reaction. This test confirmed that the exothermic reaction occurred very close to the actual temperature recorded by the First Chemical Corporation DCS computer system just before the explosion occurred. The CSB also found this data to be consistent with existing literature concerning the decomposition of MNT.

Another significant factor discovered by the CSB during their investigation was the metal loss to the upper section of the fractionator column involved in this incident. The tower had lost about 70% of its wall thickness in some locations, most likely due to corrosion under insulation (CUI) (see Chapter 8.13 for information on CUI). However, the contractor who conducted thermal stability testing for the CSB felt that the pressures that would have been developed during an

Figure 8.31.4 Photo of fractionator tower that exploded at First Chemical Corporation. This incident occurred when steam leaked through manual valves and heated mononitrotoluene (MNT) inside a distillation column. The MNT decomposed and exploded resulting in the vessel damage and projectiles which also damaged surrounding equipment, causing a secondary fire. Source: Photo courtesy of US Chemical Safety and Hazard Investigation Board. Remains of a 145-foot distillation column that exploded due to an uncontrolled chemical reaction.

uncontrolled exothermic reaction and decomposition event would have greatly exceeded the tower design pressure. It is believed that the tower would be expected to fail catastrophically under these conditions, regardless of the corrosion issue. See Figure 8.31.4 for a photo of the Fractionator that exploded at First Chemical Corporation (courtesy of the US Chemical Safety and Hazard Investigation Board).

Key Lessons Learned

The CSB report provided a detailed list of key lessons learned and recommendations to help ensure a reoccurrence does not occur. It is not intended to duplicate this list here; please refer to the CSB report for additional details.

Ensure safety reviews are conducted for all processes involving reactive chemicals using a Process Hazard Analysis (PHA) or equivalent process. These reviews should identifythe specific reactivity hazards and the key limits to prevent the excursion. This information should be specified in the procedures and the operator training and should be the basis for the layers of key protective systems.

Operators must be well-trained when operating equipment with a product or material known to be chemically reactive. Operators should be aware of the unit conditions or limits that must be in place to prevent the reaction from occurring and the actions to take should an exotherm occur. This is true for continuous flow processes and batch processes.

Ensure the design and use of critical instrumentation, alarms, and automatic trips (layers of protective systems). A process involving a potentially reactive product or process material should be designed with the critical instrumentation required to monitor the process and to bring the process to a safe park if an excursion occurs. This includes instrumentation such as temperature and/or pressure indicators, alarms, critical automatic shutdown trips, safety interlocks, or other safety protection layers to monitor the operation, identify a potential exotherm, and safely take the unit offline and isolate heat input or reduce pressure to stop the chemical reaction quickly. This is true for continuous flow processes as well as batch processes.

Ensure adequate overpressure protection for process equipment. During the Process Hazard Analysis (PHA), the overpressure protection should be confirmed. A potentially reactive chemical can rapidly create extreme pressures. Unless this is considered and addressed during equipment design (and in subsequent PHAs), the pressure can result in catastrophic failures of the process vessel. It is essential that either PRVs or an alternate method of stopping the reaction be specified to control the pressure to safe limits.

Ensure complete and detailed procedures are in place. Procedures should be detailed and cover all modes of operation, including on-stream operations and temporary off-stream periods, to ensure potential hazards are identified and adequately addressed. Procedures should be in checklist format with caution and warnings as appropriate. (Refer to Chapter 7, "Importance of Operating Procedures.")

Ensure complete energy isolation during periods of offline operations. This is especially true if the process equipment contains a product known to have reactive characteristics. This may include blocking and blinding, double block and bleed, or other similar practices to guard against leaking valves or valves inadvertently left in the open position.

What Has Happened/What Can Happen

Vacuum Truck Explosion and Fire
Santa Paula, CA
Santa Paula Waste Treatment Facility
18 November 2014
Vacuum Truck Explosion

This explosion resulted from the inadvertent mixing of two reactive chemicals inside the vacuum truck. The vehicle exploded at about 3:46 a.m. at the Santa Clara Wastewater Company in Santa Paula. This was initially reported as Organic Peroxides and Sulfuric Acid, but after several months, it was determined to be a mixture of several chemicals, including Peroxides and Sodium Chlorite.. Reports indicated that the vacuum truck operators were vacuuming waste material from several unlabeled but identical plastic totes when the incident occurred.

Following the initial explosion, the material formed a white crystalline reactive powder on the ground that continuously reacted, causing secondary fires. The fire department reported that the unstable and water-reactive chemical sparked spontaneous explosions, spread across a nearly 400-foot radius, and reacted with any incidental contact, such as by the responder's boots or vehicle tires when driving over the materials.

The incident evolved into a disaster when, later in the morning, additional materials began to burn and explode, resulting in a three-mile-long plume of toxic smoke. Due to the vapor plume, the incident resulted in a large-scale evacuation of residents in the surrounding areas and the closure of adjacent roadways, schools, etc.

Thirty-seven people were treated for chemical exposure, and a large-scale community evacuation and shelter-in-place was issued for the surrounding communities. Several people were hospitalized, including three firefighters. Officials also reported that the chemical sickened at least 12 hospital employees at the local hospital emergency room where the injured workers and responders were being treated. The three injured firefighters were reported to have remained off duty indefinitely due to their injuries.

As a result of this incident and the resulting investigation by the District Attorney, a grand jury indicted nine defendants for noncompliance and intentional malpractice of waste management regulations. This included Santa Clara Wastewater and its parent company, Green Compass. The indictments resulted in jail terms, suspended jail sentences, and significant fines or penalties for several defendants.

A good writeup was done by KTLA Channel 5 (Nexstar) in Los Angeles, which can be found on their website. The article also has some very good photographs, which can be viewed on their site. Unfortunately, the images could not be made available for this book.

Key Lessons Learned

The operating company may have been tolerating intentional noncompliance with local and federal regulations relative to waste recovery operations. Many regulations appear to have been violated in this case, such as failure to disclose the presence of hazardous materials at the site, nonlabeled waste containers, the apparent mixing of various chemicals, and apparent falsification of waste management records, among others (this is based on the number of grand jury indictments and subsequent guilty pleas by the majority of the nine defendants).

Detailed operating procedures and employee training must be in place at facilities of this type to ensure employees are aware of the chemical hazards involved and materials are properly labeled, including cautions and warnings relative to proper PPE and issues of noncompatibility between chemicals. This should be an ongoing process of audits and verifications to ensure compliance with local, state, and federal regulations, including verification of records and documentation such as employee training records, inventory of hazardous materials, inspection records for equipment and storage vessels, an inspection of emergency response equipment such as firewater systems, fire extinguishers, and gas detection.

Management must be involved daily to establish the expectations for all employees and ensure an ongoing culture of safe operations compliance and continuous improvement.

The potential for the handling and processing of reactive chemicals should always be a consideration during a PHA study. If determined to exist, the study should include considerations for the mitigation of the identified hazards.

What Has Happened/What Can Happen

Calpine
Los Moedanos Energy Center
Pittsburg, California
24 May 2007

During the delivery of the corrosion inhibitor, approximately 300 gallons of Nalco Trasar (phosphoric acid) was inadvertently unloaded into a storage tank containing about 378 gallons of 12.5% Sodium Hypochlorite solution. The mixing of the phosphoric acid and sodium hypochlorite resulted in a release of chlorine gas, exposing the Operator and two other employees to chlorine. The three employees were transported to the medical center for evaluation.

The incident resulted in the emergency response team isolating the plant entry and evacuating non-essential personnel to a safe location. The facility responded by flushing the neutralized contaminated tank contents to the cooling tower and ventilating the affected building.

Key Lessons Learned

The key learning from this incident is the realization that this could have been an incident with a different outcome had the chemicals been different or more reactive. This incident illustrates why the Operators should be trained to verify the chemical being unloaded (or loaded) by checking the bill of lading, the labeling on the truck or other conveyance, and the Material Safety Data Sheet before connecting to the receiving tank or compartment. The cargo contents should be verified against the labels or signage on the receiving tank to ensure the cargo being loaded is correct for the receiving tank.

Additionally, signage and color coding of offloading connections should be in place as an aid to the Operator to help prevent incidents of this type. The hose fittings should be incompatible with the different chemicals to help make it impossible to connect or mingle incompatible chemicals without fabricating adapters or the use of additional connections.

Additional References

Occupational Safety and Health Administration (OSHA), “29 CFR 1910.120 – Hazardous waste operations and emergency response.” https://www.osha.gov/laws-regs/regulations/standardnumber/1910/1910.120

Occupational Safety and Health Administration (OSHA), Safety and Health Topics/Chemical Reactivity Hazards. https://www.osha.gov/SLTC/reactivechemicals/standards.html

U.S. Chemical Safety and Hazard Investigation Board, “Hazard Investigation – Improving Reactive Hazard Management”, Report No. 2001-01-H; Issue Date: October 2002. https://www.csb.gov/improving-reactive-hazard-management/

U.S. Chemical Safety and Hazard Investigation Board, Safety Video: “Reactive Hazards; Dangers of Uncontrolled Chemical Reactions”. https://youtu.be/sRuz9bzBrtY

Bretherick’s Handbook of Reactive Chemical Hazards, Urben, P.G., ed., 6th ed. St. Louis, MO: Elsevier Science and Technology Books, 2000. (Includes every chemical for which documented information on reactive hazards has been identified).

“Rapid Guide to Chemical Incompatibilities”, Richard P. Pohanish, Stanley A. Greene, John Wiley & Sons, 29 January 1997.

US Chemical Safety and Hazard Investigation Board, “Investigation Report; First Chemical Corporation”, Pascagoula, Mississippi, October 13, 2002, Report No. 2003-01-I-MS October 2003, (Three Injured, Potential Off-site Consequences). https://www.csb.gov/assets/1/20/first_report.pdf?13794

Investigation Digest, Reactive Explosion at First Chemical Corp., Pascagoula, Mississippi, 13 October 2002. https://www.csb.gov/assets/1/20/first_digest.pdf?13795

EPA Chemical Compatibility Chart. https://orf.od.nih.gov/EnvironmentalProtection/WasteDisposal/Documents/chemical_waste_chemical_compatibility_chart.pdf

“Essential Practices for Managing Chemical Reactivity Hazards”, The Center for Chemical Process Safety, (CCPS, 2003).

Wikipedia, “Article on Santa Paula, California”, Including the Santa Clara Waste Water plant industrial disaster. https://en.wikipedia.org/wiki/Santa_Paula,_California

“On the wake of the Santa Paula explosion, many wastewater treatment plants do not comply with the 2012 NFPA standard 820.” Metropolitan Engineering Consulting & Forensics Services. https://sites.google.com/site/metropolitanenvironmental/on-the-wake-of-the-santa-paula-explosion-many-wastewater-treatment-plants-do-not-comply-with-the-2012-nfpa-standard-820

“NFPA 820 Standard for Fire Protection in Wastewater Treatment and Collection Facilities”, National Fire Protection Association. https://catalog.nfpa.org/Codes-and-Standards-C3322.aspx

“Major Accidents at Chemical/Refinery Plants in Contra Costa County”, Contra Costa County, California. https://cchealth.org/hazmat/accident-history.php

The Center for Chemical Process Safety (CCPS), “A Checklist for Inherently Safer Chemical Reaction Process Design and Operation”. http://www.aiche.org/ccps/safetyalert

DOT Emergency Response Guidebook 2020, ERG2020-WEB.pdf (dot.gov).

US National Institute of Health, Chemical Segregation and Storage Table (Appendix XX-K). https://ors.od.nih.gov/sr/dohs/Documents/chemical-segregation-table.pdf

Chapter 8.31. Hazards of Chemical Incompatibility and/or Reactivity

End of Chapter Quiz

1. Where can we find the proper storage information for the required chemicals?

2. How do we store chemicals in small quantities in storage lockers or cabinets?

3. What is very important about the chemical receiving lines or receiving tanks?

4. When two or more reactive chemicals mix, the reactions may be _____ or may be very _______, resulting in a severe ____________ reaction and may also result in explosion and/or major fire. In some cases, the reaction may also produce _________ gases which can be deadly to personnel.

5. United Nations (UN) Numbers are four-digit numbers used worldwide in international commerce and transportation to identify _________ ___________ or classes of hazardous materials. NA numbers (North America) are issued by the United States Department of Transportation and are identical to UN numbers, except that some substances without a UN number may have an NA number. For example, "UN1017" provides response information for Chlorine and cross-references to the Response Guide 124.

6. What is a requirement for the truck driver or the person accepting a shipment of chemicals into an onsite storage tank?

7. The Unit Operator should do the final ___________ for any product being unloaded into a storage tank from a tank truck or railcar.

   The Operator should validate the cargo manifest with the _______ and _______ before unloading the product into the storage tank.

8. What are several considerations for chemical storage? Choose from the following.
   a. Tank temperature.
   b. Control of ignition sources.
   c. Proper labeling.
   d. Proper segregation from other incompatible chemicals.
   e. None of the above.
   f. All the above.

9. What is the cardinal rule when it comes to storing acids and bases?

10. Where can we find a quick reference to incompatibility with other chemicals?

11. What safety process is recommended to be used to evaluate the safety hazards associated with a site that is handling or processing hazardous materials?

8.32

Hazards of Unintended Process Flow (The Mixing of Hydrocarbons and Air) in Fluid Catalytic Cracking Units (FCCU)

Fluid Catalytic Cracking Units (FCCU) are somewhat unique within the refining industry in that they rely on the maintenance of the pressure balance and internal catalyst barrier between the hydrocarbon-containing components of the process and the air-containing components of the process to prevent the inadvertent mixing of the two. The industry has experienced decades of very successful operations, essentially incident-free with this process. However, there have been two significant process safety incidents in the recent past where this balance was lost.

Both events occurred during abnormal operations, resulting in major explosions and severe equipment damage. In both cases, the FCC units were in end-of-run conditions, and slide valves that maintained the catalyst circulation between the oil side and air side of the units were severely worn.

FCCU operations involve a reactor in which hydrocarbons are exposed to high temperatures and catalysts to change the molecular structure of the feed. It is essential that the reactor is operated without the presence of air; otherwise, a catastrophic explosion can result. A typical FCC reactor operates at around 950 °F (510 °C).

After the reaction is complete, the catalyst is circulated into a regenerator vessel, where the carbon is burned from the catalyst at very high temperatures (typically 1,300–1,400 °F [704–760 °C]) in the presence of air. A controlled burn of carbon occurs in the regenerator in the presence of air, and the regenerated catalyst is circulated back into the reactor, where the reaction continues. No hydrocarbon must be allowed to enter the regenerator or other downstream equipment containing air; otherwise, a catastrophic explosion can occur. It is equally important that no air be allowed to enter the reactor or fractionator or mix with the hot hydrocarbons present in these vessels.

It is noteworthy that FCC units operate with the presence of pyrophoric materials (sulfur compounds) on the oil side of the unit and can also play a role when contacting air. However, we should understand the operating temperatures are well above the autoignition point for all hydrocarbons, and therefore, the pyrophorics are only a contributing factor when the unit is down and at ambient temperatures.

The theoretical separation of the "oil side" and the "air side" of the FCCU is shown in Figure 8.32.1. The pressure balance is maintained during normal operations, with the regenerator operating at a slightly higher pressure than the reactor, preventing an untended process flow or backflow into the regenerator. The regenerated catalyst flows from the regenerator, is mixed with the fresh feed, and is carried into the reactor by the fresh feed and steam.

The separation between the oil side and air side is also maintained by two slide valves highlighted in circles in Figure 8.32.1. The fluidized catalyst level above the reactor spent catalyst slide valve (valve 9 in Figure 8.32.1) helps to prevent hydrocarbons from entering the regenerator and the other equipment containing air downstream of the regenerator. The fluidized catalyst level above the regenerated catalyst slide valve (RCSV) (valve 8 in Figure 8.32.1) also serves as a barrier helping to prevent air from the regenerator from entering the reactor and other hydrocarbon-containing equipment downstream of the reactor (i.e., the fractionator). Should these catalyst levels be lost during the operation, or if catalyst circulation is lost, causing loss of the catalyst fluidization, air and hydrocarbons can mix. In this case, backflow can occur between the two major vessels, the reactor and regenerator, allowing air and hot hydrocarbons to come into contact with the process equipment. This can result in the potential for a devastating explosion.

The regen slide valve (valve 8) regulates the flow of the regenerated catalyst in a fluidized state to the catalyst riser and "feeds" the regenerated catalyst to the reactor. This valve maintains the pressure head in the standpipe from the regenerator and helps to protect the regenerator from a flow reversal. The spent catalyst slide valve (SCSV) (valve 9) controls the stripper catalyst level and regulates the flow of the spent catalyst, also in a fluidized state, to the regenerator. The spent slide valve helps to protect the reactor and main fractionator from a flow reversal. Flow reversal from either slide valve, for example, due to worn valves or failure of the control system, may result in the inadvertent mixing of hydrocarbons and air inside the process equipment. Normal catalyst flow is illustrated in Figure 8.32.2, while the reverse flow is illustrated in Figure 8.32.3.

Process Operations Safety: The What, Why, and How Behind Safe Petrochemical Plant Operations, First Edition. M. Darryl Yoes.

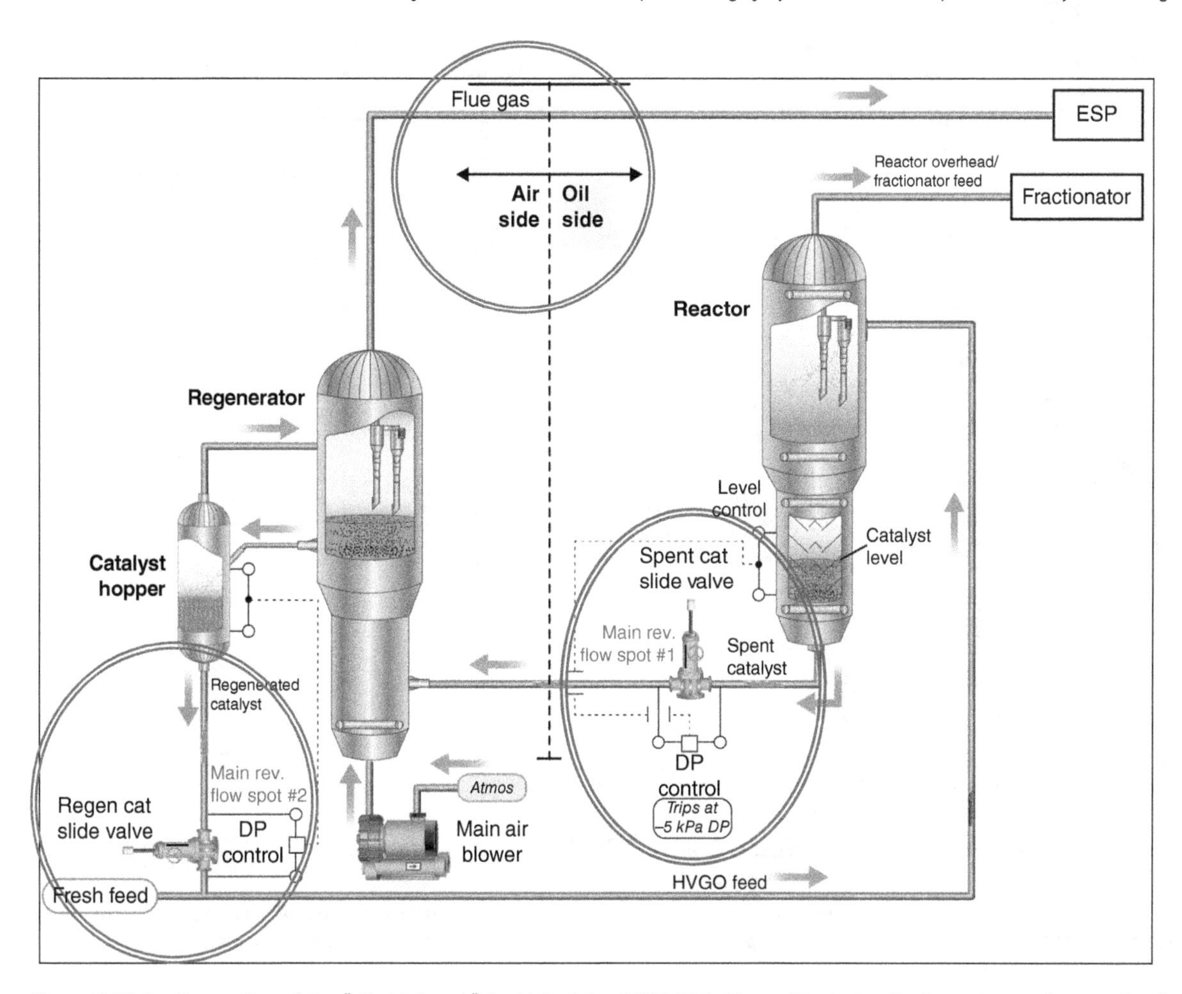

Figure 8.32.1 Separation of the "oil side" and "air side" of the FCCU. This Figure illustrates the importance of separation between the oil and catalyst circulation systems. This separation is controlled by maintaining the reactor/regenerator pressure differential and catalyst levels above the two slide valves (the spent catalyst and regenerated catalyst slide valves). These images are credit of Ecoscience Resource Group Visual Artists and were created to illustrate the hazards of FCC unintended reverse flow and are not representative of any company or specific incident. Note: This illustration is for a generic FCCU and does not represent either unit discussed in this section. Note for simplicity, the fractionator, the CO Boiler, the Electrostatic Precipitator (ESP), and other key FCC components are not shown here.

These catalyst levels are challenging to control during periods of abnormal operation, such as a brief downtime for maintenance work or during the initial periods of an FCC unit shutdown or startup. It has been demonstrated that catalyst can lose its fluidized state when catalyst circulation is lost, and air can pass through the catalyst. Also, when the unit is nearing the end-of-run conditions, the slide valves may be worn internally and allow the reverse flow to occur. Worn slide valves allow the catalyst to pass through the slide valve, even when the instrumentation shows the valve as completely closed. A worn SCSV can allow hot hydrocarbons from the reactor to enter the regenerator where the air is present, or a worn regenerated slide valve can allow air to pass from the regenerator into the reactor where hot hydrocarbons are present. This can result in the potential for an internal explosion.

Recognizing this hazard, some refiners have normally installed a second slide valve in its fully open position. When there are any indications of wear in the primary valve, they switch to this "spare" valve to continue the run until the next turnaround. With the spare valve in the fully open position, there is no differential pressure across the valve, and it is less subject to internal wear.

In the past, operators developed a dependence on the catalyst level above the slide to prevent reverse flow and provide separation between the air side and the oil side of the unit. However, we have experienced two significant events where this was attempted and resulted in the inadvertent mixing of hot hydrocarbons and air (see the abbreviated case studies below). In both cases, an explosion resulted. This has resulted in a change in strategy for maintaining separation between

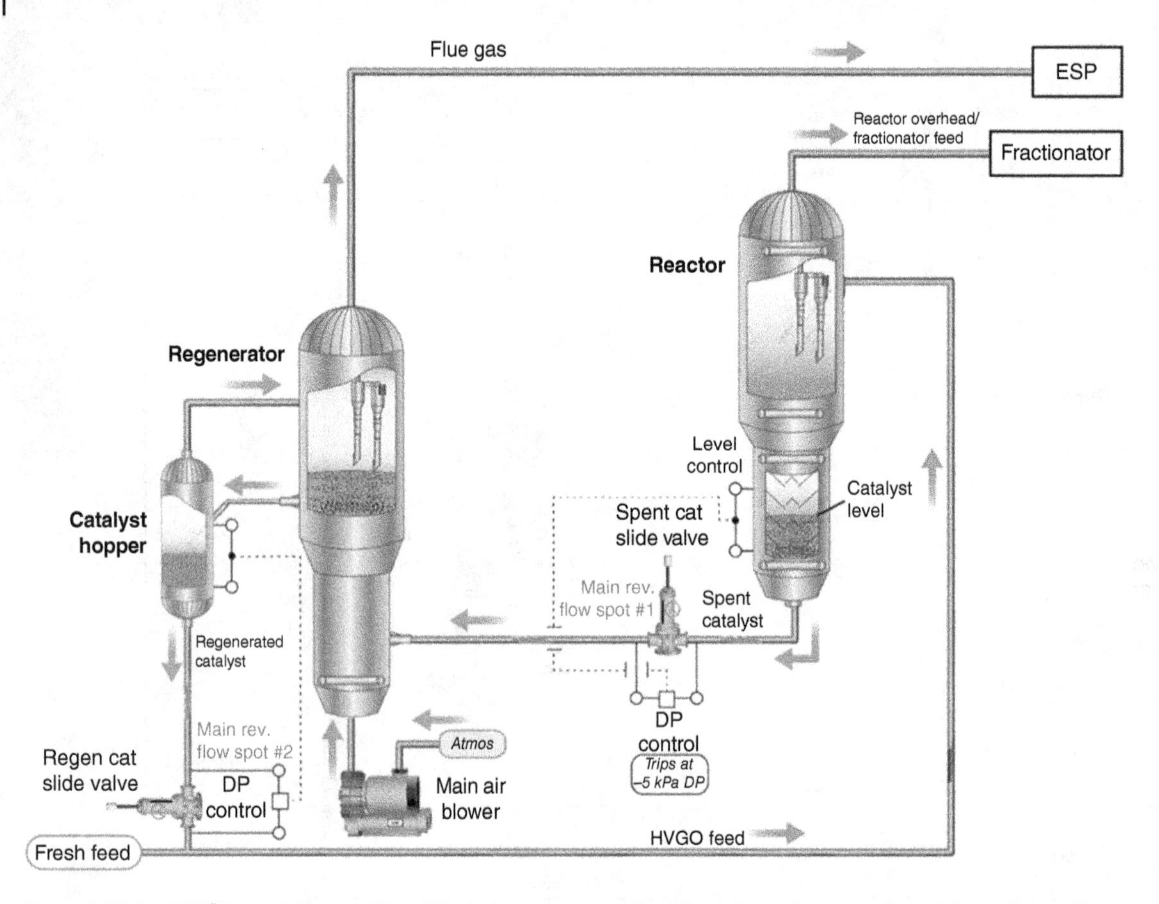

Figure 8.32.2 FCC "Normal Catalyst Flow." These images are credit of Ecoscience Resource Group Visual Artists. They were created to illustrate the hazards of FCC unintended reverse flow and are not representative of any company or specific incident.

the two sides of the unit. The current strategy is to do the following when catalyst circulation is lost to help prevent this from occurring:

- Always closely monitor the catalyst circulation rates, recognizing that catalyst fluidization depends on circulation and the injection of air on the regenerator side and hot vapors and steam on the reactor side of the unit. Without catalyst fluidization, air or vapors can pass through the catalyst.
- If catalyst circulation is lost for any reason, recognize that catalyst fluidization may be lost and, therefore, the catalyst is no longer dependable as a barrier.
- Increase the steam flow to the riser and stripper and shift the unit pressure balance so that the reactor is at a slightly higher operating pressure than the regenerator.
- Start the injection of noncondensable gas (nitrogen, fuel gas, or steam) into the fractionator overhead.
- Closely monitor the differential pressures between the reactor, regenerator, and the bottom of the cat fractionator. The differential pressure differences should be small, with the reactor pressure slightly above the regenerator and the fractionator bottom.
- Initiate steps to monitor O_2 levels in the fractionator overhead vessel (should not be above trace amounts of Oxygen).

In addition to the above, when shutting down an FCC unit following a long run, Operators should consider that the slide valves may worn. It is extremely important that the console operator closely monitors the delta pressure ("delta P") across the catalyst barrier above each of the two slide valves. The Console Operator should constantly be aware of the possibility of losing the barrier by leaking slide valves and creating the potential for backflow and mixing of hydrocarbons and air, which can lead to an internal explosion. This is equally true for both the spent catalyst and the regenerated catalyst slide valves.

Should the differential pressure across the slide valve (delta P) be lower than expected, the console operator should immediately verify that the steps outlined above have been implemented and increase the catalyst level above the slide valves to

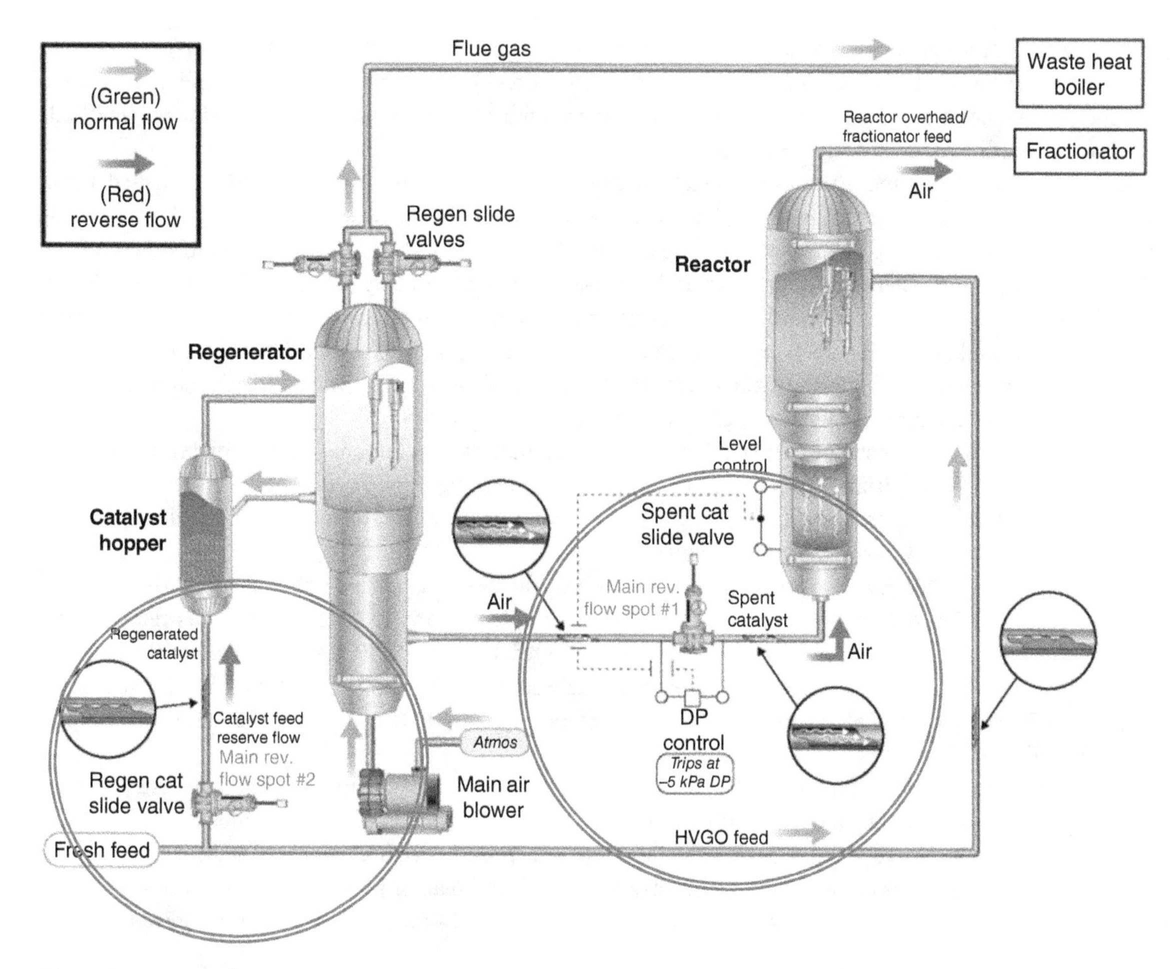

Figure 8.32.3 FCC "Reverse Catalyst Flow." Note that two different unintended reverse flow scenarios are illustrated. Source: Image credit of Ecoscience Resource Group Visual Artists. These images were created to illustrate the hazards of FCC unintended reverse flow and are not representative of any company or specific incident.

the extent possible. We are depending on the pressure instruments to read small deviations in differential pressure, typically less than 1 psi. It may be necessary to validate these values in the field.

This Figure indicates the normal catalyst flow between the FCC Reactor and Regenerator vessels, including the Spent Catalyst Slide Valve (SCSV) and the Regenerated Catalyst Slide Valve (RCSV). Note the catalyst levels above the two slide valves, preventing unintended reverse flow.

- The first scenario illustrated on the right side of the image is the loss of catalyst level above the Spent Cat Slide Valve and airflow reversal from the Regenerator into the Reactor and possibly the fractionator mixing with hot hydrocarbons.
- The second scenario is illustrated on the left side of the image with a loss of catalyst level above the RCSV, resulting in hot feed flowing into the Regenerator and mixing with air.
- Both of these scenarios can have catastrophic consequences.

What Has Happened/What Can Happen

ExxonMobil Refinery
Torrance, CA
18 February 2015
Explosion in Electrostatic Precipitator (ESP)

This incident was investigated by the US Chemical Safety and Hazard Investigation Board, and a detailed report and a great training video were released on this incident.

An explosion occurred in the FCCU electrostatic precipitator (ESP), an environmental control device. This incident occurred during a temporary FCC downtime for mechanical repairs to address high vibrations on an expander, a device to boost air pressure to the regenerator. During the unit downtime, ExxonMobil chose to use a "variance" using two methods for energy isolation, duplicating a similar task performed in 2012.

To isolate energy during the downtime, ExxonMobil established a catalyst level above the SCSV and established steam injection above the slide valve. The electrodes in the ESP were left energized during the downtime.

While the mechanics attempted to install a blind in the expander, they were hindered by excess steam flowing from the expander flange. The Operators then reduced the steam flow above the SCSV. Shortly afterward, the mechanic's H_2S monitors started alarming, indicating that H_2S and hydrocarbons had entered the expander, where no hydrocarbons were expected. The Operators started evacuating everyone away from the unit and rapidly increased the steam flow above the SCSV. However, it was too late; hydrocarbons had leaked through the SCSV and backflowed into the regenerator, expander, and ESP. The electrodes in the ESP ignited the hydrocarbons, resulting in an explosion.

The Chemical Safety and Hazard Investigation Board (CSB) concluded that the fractionator column pressure was operating higher than when the original variance was issued in 2012 due to a leaking naphtha exchanger. The naphtha is a flammable hydrocarbon that increased pressure in the fractionator and may have contributed to the flammable vapor leaking into the unit's airside during 2015, resulting in the explosion.

The CSB concluded that the SCSV was worn internally and allowed the catalyst barrier established above the SCSV to leak by the slide valve. This left only the steam purge to hold back the hydrocarbons, and when it was reduced, the hydrocarbons flowed from the fractionator into the reactor and from the reactor to the ESP, resulting in an explosion. The CSB also concluded that the electrodes in the ESP were the most likely source of ignition. This scenario is illustrated in Figure 8.32.4 from the Chemical Safety Board courtesy of the final report. Damage to the FCC Unit ESP and surrounding equipment is shown in Figure 8.32.5.

This incident also emphasized the hazards of hydrofluoric acid (HF), which is widely used in the refining industry. During the explosion, debris, namely large metal panels from the sides of the ESP, were propelled onto equipment in the adjacent HF alkylation unit. There was no release of HF; however, the potential was present because of this incident.

When the operators reduced the steam flow into the reactor riser, the reduced reactor pressure allowed hydrocarbons to backflow from the main column through the closed slide valve into the air side of the unit. The mixture flowed into the ESP and was ignited by the energized ESP.

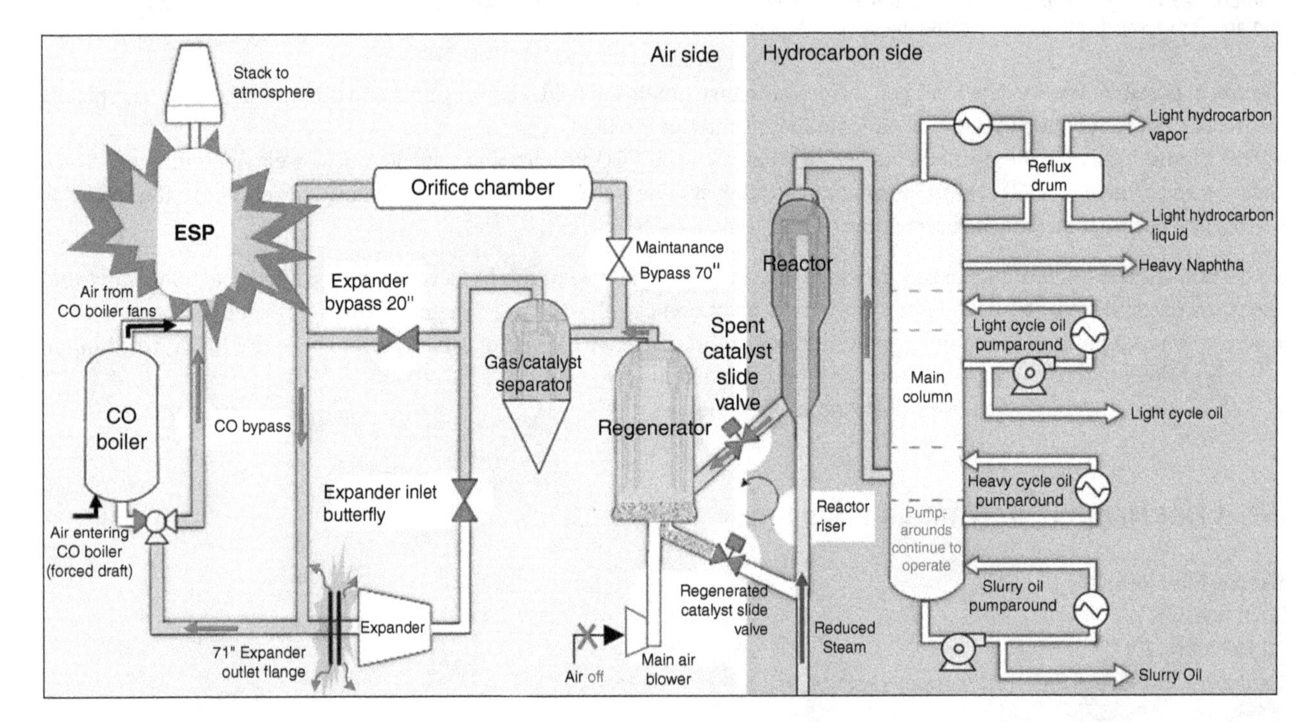

Figure 8.32.4 ExxonMobil Torrance FCC reverse catalyst flow. Source: Photo courtesy of the US Chemical Safety and Hazard Recognition Board. From the CSB Final Report "ExxonMobil Torrance Refinery Investigation Report". Text and annotation is by the CSB.

Figure 8.32.5 Damage from the ExxonMobil Torrance FCC ESP explosion. Source: Photo courtesy of the US Chemical Safety and Hazard Recognition Board. From the CSB Final Report "ExxonMobil Torrance Refinery Investigation Report."

Key Lessons Learned

This incident highlights the importance of effective equipment isolation during maintenance. Generally, this means, as a minimum, the use of a double block valve and bleed arrangement to ensure effective isolation. If the equipment is not designed for a double block and bleed isolation, then a detailed risk assessment should be conducted to ensure an acceptable alternate method is used.

When using an alternate isolation process, such as a catalyst level above a slide valve or providing a steam purge, there should be a process for ensuring that the isolation is maintained throughout the task. This means a way of measuring and ensuring the integrity of the isolation. For example, delta P instrumentation with continuous monitoring to ensure the catalyst barrier is maintained or continuous monitoring of reactor and regenerator pressures.

Please see the final report issued by the US Chemical Safety and Hazard Investigation Board for additional findings and recommendations.

What Has Happened/What Can Happen

Refinery explosion and subsequent fire; Husky Superior Refinery
Superior, Wisconsin
26 April 2018

This incident is under investigation by the US Chemical Safety and Hazard Investigation Board, and a report will be issued soon. The CSB has released a "Factual Update" and another great training video on this incident.

Thirty-six people sought medical attention, including eleven OSHA recordable injuries by refinery and contract workers. The incident also resulted in evacuations in the surrounding communities.

Although the Chemical Safety and Hazard Investigation Board's investigation is continuing, they have discovered some very close similarities between the Superior Refinery incident and the ESP explosion that occurred at the ExxonMobil Torrance Refinery in 2015. In both cases, the FCCU slide valves had eroded during the run, and in both cases, they lost the catalyst barrier due to leakage during abnormal operation.

In the Superior incident, the refinery was shutting down for a planned turnaround when air flowed unexpectedly from the air side of the FCC unit to the hydrocarbon side through the SCSV. This resulted in an internal explosion on the unit's hydrocarbon side, spreading explosion debris over a wide refinery area. Also, as was reported in the Torrance incident, the debris impacted equipment in the surrounding units. However, at Superior, the debris resulted in subsequent secondary fires and releases into the atmosphere.

The Superior Refinery was reported to have a separate control system to help prevent air from leaking from the SCSV into the hydrocarbon side of the unit. However, this system failed to function since the slide valve was eroded and could not provide an effective catalyst barrier. Shrapnel struck an asphalt tank, releasing hot asphalt into the surrounding area. The asphalt ignited a short time later, resulting in a large, extended fire that impacted other units and the surrounding community.

Another similarity between the Superior and Torrance incidents involved the hydrofluoric acid (HF) Alkylation unit. The HF Alkylation unit at Superior was also reportedly struck by shrapnel, resulting in a hydrocarbon release and subsequent fire and the release of a small amount of hydrofluoric acid into the atmosphere. As of this writing, no injury has been reported due to the acid release; however, this incident also adds to the concerns about using Hydrofluoric acid in the petrochemical industry.

Update:

The CSB Investigation Report on the Husky FCC incident was released on 23 December 2022 and included recommendations regarding the separation between the air and hydrocarbon sides of the FCC unit. Typically, when operating in a transient operations mode such as a shutdown, there is more than one preventative measure to prevent this from occurring. In addition to the catalyst barrier above the slide valve, at least two additional preventive measures should be in place.

The first of these is the provision of a steam purge between the reactor and the hydrocarbon side of the unit and the air side of the unit. The CSB reported that the reactor can be operated at a higher pressure by injecting steam to separate the reactor from the hydrocarbon-containing equipment. In this way, the reactor serves as an additional barrier to prevent the air from reaching the hydrocarbon side and preventing a potential explosion.

The second preventive measure that the CSB determined was not in place was establishing a purge of non-condensable gas to continually purge the unit's gas side, thereby preventing a build up of air that could mix internally and create a flammable or explosive mixture. The CSB said that this was not done.

See Figures 8.32.6 and 8.32.7 for images of the damage caused by the FCC internal explosion.

As was mentioned in the report above, some refineries have installed a "spare" slide valve, which is normally operated in the fully open position during the run to avoid wearing the valve internally by the high differential pressure. This spare valve can be commissioned before the unit goes into a transient operation to prevent what happened at Husky.

Key Lessons Learned

Following the Torrance ESP explosion, the US Chemical Safety and Hazard Investigation Board highlighted the importance of considering all modes of operations, including nonroutine operations and maintenance during unit standby. This is especially important when performing a unit process hazard analysis. This recommendation should also apply to unit

Figure 8.32.6 Image of the Primary absorber, sponge absorber, and stripper before and after the incident. The before image is in the small box on the lower right. From the CSB Final Report on the Husky FCC explosion. Source: Photo courtesy The US Chemical Safety Board, the Husky Superior Refinery, and Google Earth. The photo includes annotations by the CSB (text and annotation are by the CSB).

Figure 8.32.7 Image of the Primary and sponge absorbers after the incident with CSB annotations. Source: Photo courtesy The US Chemical Safety Board (text and annotation are by the CSB).

startup and shutdowns, such as the Superior incident, where hazards that are not present during routine unit operations may exist.

Both incidents also highlight the importance of developing an inspection and maintenance plan and strict adherence to the inspection and maintenance procedures. This is especially true for all safety-critical equipment. Catalyst slide valves perform an essential safety function on FCCU units in that they help create an additional barrier between the air and hydrocarbon sides of the unit. A detailed and rigorous inspection and maintenance plan should be in place for this and other similar safety-critical equipment.

FCC operators should ensure that the potential mixing of air with hydrocarbon-containing equipment is avoided by developing comprehensive procedures and ensuring compliance with those procedures, especially when undergoing transient operations such as startup, shutdown, and during unit emergencies. More than one way to separate the air and hydrocarbon-containing equipment should be considered.

Additional References

US Chemical Safety and Hazard Investigation Board, ExxonMobil Torrance Refinery, Torrance, California, 18 February 2015, "Electrostatic Precipitator Explosion". https://www.csb.gov/exxonmobil-refinery-explosion-/

US Chemical Safety and Hazard Investigation Board, Husky Energy Explosion and Fire, Superior, WI, 26 April 2018, "Factual Update". https://www.csb.gov/assets/1/6/husky_factual_update.pdf?16317

CIRCOR – TapcoEnpro, Article: "FCC Slide Valves". http://www.tapcoenpro.com/fccu-slide-valves/

US Chemical Safety and Hazard Investigation Board, FCC Unit Explosion and Asphalt Fire at Husky Superior Refinery, Superior, WI, Incident Date: 26 April 2018 | No. 2018-02-I-WI.

2020 American Fuel and Petrochemical Manufacturers, Presentation at AFPM Summit, "Safeguarding the FCCU During Transient Operations", 27 February 2020, Ziad Jawad (Phillips 66), CJ Farley, G.W. Aru LLC.

Chapter 8.32. Hazards of Unintended Reverse Flow – Fluid Catalytic Cracking Units (FCCU)

End of Chapter Quiz

1 Describe the two primary "sides" of the FCCU Unit.

2 What is the concern about shutting down an FCCU after a run of several years?

3 What is unique about how an FCCU functions compared to most other process units?

4 What should the Console Operator be aware of and constantly evaluating when shutting down an FCC following a long run?

5 What are two additional steps that should be taken by the Console Operator and Field Operators?

6 The industry has experienced decades of successful operations, essentially incident-free with this process. However, there have been two significant process safety incidents in the recent past where this balance was lost.

   Both events occurred during periods of ___________ operations, resulting in major explosions and severe equipment damage. In both cases, the _______ ________, the valves that maintain the pressure balance between the oil side and air side of the units, were severely worn.

7 What are the indications that this important barrier is being maintained or that the catalyst level is dropping and may be at risk of an unintended reverse flow?

8 How have some refiners decided to address the issue with worn slide valves on their FCC Units?

9 The Regenerator is the main process vessel on the air side of the FCC. Is it possible for hot hydrocarbon feed to enter the Regenerator? If so, how can this happen?

10 FCC operators should ensure that the potential mixing of air with hydrocarbon-containing equipment is avoided by developing comprehensive __________ and ensuring __________ with those procedures, especially when undergoing __________ operations such as startup, shutdown, and during unit emergencies.

More than ______ ways to separate the air and hydrocarbon-containing equipment should be considered.

8.33

Hazards of "Normalized Deviation"

"Normalized deviation," sometimes referred to as "normalization of deviance," is an expression for deviating or not following the rules or a procedure so often that the deviation becomes completely normal to the person not following the rule. For example, someone who develops a habit of not wearing a seat belt while driving the car to the corner grocery store may justify that they are just as safe, or even safer, by not following the law. The same thing can happen to someone who is not following a procedure. They can justify that it is okay to deviate. No one says anything to them, and nothing happens, reinforcing that the deviation was acceptable. After all, the task was done faster and with less effort by deviating from the established procedure. The next time they do the same job, they deviate again until the deviation becomes normal for that person or group. They continue with the deviation until one day, the situation is a little different, or the weather has changed, or something else has changed. A combination of the deviation plus the unexpected change results in an accident or injury or even may result in loss of life.

Normalized deviation can be by one person, or it can be by a group of people, and even by a whole organization. It is not unusual to find the deviation by a larger number of people since they tend to reinforce each other in the belief that the deviation is justified. By doing so, the individual or group has accepted the deviation to be the standard and has, therefore, lowered the organization's standard. This can be toward a wide range of business objectives such as production, environmental compliance, personal safety, process safety, or cost control. Every time they deviate and achieve their goal, they are reinforced that it is OK to do the same thing next time.

When I think of normalized deviation, I cannot help but think about Charlie Morecraft. Charlie was an Operator at the Exxon Bayway Refinery and was involved in a tragic manifold fire in an offsite area of the plant in 1980. He was burned over 45–50% of his body and spent about five years in the hospital going through multiple surgeries and skin grafts. Following this accident and after Charlie was back on his feet, he became an acclaimed motivational public speaker with a message of taking responsibility for your actions. In Charlie's presentations, he talks about his habit of not following the rules and procedures and not wearing the required personal protective equipment (PPE). He also talks about how his not turning off the engine in his pickup truck at the time of the accident and not wearing his PPE correctly contributed to the accident and the severity of his injuries.

I believe this is precisely what normalized deviation is; I believe that Charlie had convinced himself that his "shortcuts" were just as safe as following the procedure, and by not following the procedure, he could get the task done in a shorter time. I believe Charlie's story will positively affect anyone who hears it, and I strongly encourage everyone to get a copy of Charlie's video "Remember Charlie" and review it. The message is so strong that I believe those who hear it will want to share it with their peers and family. When I was the Safety, Health, and Environmental Manager at a major Gulf Coast refinery, I brought Charlie to our site and did multiple sessions with Charlie as the featured speaker. I tried as hard as possible to have everyone at the site attend one of Charlie's talks.

Lessons from NASA Shuttles Challenger and Columbia Disasters

The Challenger Space Shuttle Disaster

Many of us will remember, or we have certainly heard the story, that on 28 January 1986, the shuttle Challenger exploded 73 seconds into its launch, killing all seven crew members. The subsequent NASA investigation determined that a solid

Process Operations Safety: The What, Why, and How Behind Safe Petrochemical Plant Operations, First Edition. M. Darryl Yoes.

rocket booster (SRB) O-ring seal in one of the rocket booster joints failed, resulting in a fire that allowed flames to impinge on the external fuel tank. The impinging fire resulted in the rupture of a liquid hydrogen tank, which exploded, causing the rupture of a liquid oxygen tank. The resulting massive explosion destroyed the shuttle and resulted in the loss of the entire crew.

Before the Challenger disaster, NASA had flown 24 shuttle missions with four different vehicles without incident. However, the investigation revealed that the O-rings had malfunctioned (leaked) on seven missions, although without incident. In hindsight, it was evident that the NASA team had expected that the O-rings may leak and accepted this as "normal." NASA knew that the Morton Thiokol design of the O-ring seal was flawed nearly nine years before the loss of the Challenger but had not addressed the concern. On the morning of the tragedy, Florida's ambient temperature was frigid (31 °F or −6 °C), and the engineers cautioned about continuing with the mission. NASA management did not accept these warnings, and the mission deviations were accepted as normal, and the shuttle was launched. In hindsight, the investigation revealed that the O-ring seals leaked on all previous launches where the ambient temperature was below 65 °F (18 °C). See Figure 8.33.1 for images of the weather impacts on the morning of the Challenger launch.

Normalized deviation played a role in the Challenger incident. Had the flaws with the O-ring seals been addressed earlier, or had the mission been delayed due to the cold weather, the Challenger incident may not have occurred, and the seven astronauts' lives would not have been lost.

The following "Contributing Factors" were part of a report on the Challenger disaster published by NASA:

Normalization of deviance: The space shuttle's SRB problem began with the faulty design of its joint and increased as both NASA and contractor management first failed to recognize it as a problem, then failed to fix it, and finally treated it as an acceptable flight risk[1].

Organizational silence: The decision to launch Challenger was flawed. Those who made that decision were unaware of the recent history of problems concerning the O-rings and the joint and were unaware of the initial written recommendation of the contractor advising against the launch at temperatures below 53 °F and the continuing opposition of the engineers at Thiokol after management reversed its position.

- Launch day temperatures as low as 22°F at Kennedy Space Center.
- Thiokol engineers had concerns about launching due to the effect of low temperature on O-rings.
- NASA Program personnel pressured Thiokol to agree to the launch.

Figure 8.33.1 Image of the weather impacts on the morning of the Challenger launch. Source: Photo courtesy The National Aeronautics and Space Administration. From NASA report "Lessons from Challenger STS-51L: January 28, 1986." The Challenger exploded after 73 seconds of flight. All seven astronauts were killed when the cabin crashed into the ocean.

1 Boston College sociology professor Diane Vaughan, author of the book "The Challenger Launch Decision," referred to this as "The normalization of the technical deviation of the booster joints ... "

Silent safety organization: There were serious ongoing weaknesses in the shuttle Safety, Reliability, and Quality Assurance Program, which had failed to exercise control over the problem tracking systems, had not critiqued the engineering analysis advanced as an explanation of the SRM seal problem and did not provide the independent perspective required by senior NASA managers at Flight Readiness Reviews.

The Columbia Space Shuttle Disaster

Seventeen years after the loss of the Challenger, a similar series of circumstances surfaced in another NASA incident. On 1 February 2003, the space shuttle Columbia disintegrated while reentering Earth's atmosphere at 10,000 mph, scattering debris over 2,000 square miles of Texas, again with the loss of all seven crew members and destruction of the $4 billion spacecraft. As a result of this incident, NASA grounded the entire shuttle fleet for 2 1/2 years for investigation and modifications to the fleet.

This time, the investigation revealed that during the launch of shuttle Columbia on its 28th mission to space, a piece of foam insulation detached from the external tank, striking the shuttle's left wing. A similar loss of foam insulation had also occurred during previous shuttle launches, some with nearly catastrophic consequences. However, even with the prior incidents, the shuttle was allowed to continue to fly. Upon reentry to the earth's atmosphere, the failed insulation allowed the super-hot atmospheric gases to penetrate the heat shield, destroying the wing structure and disintegrating the shuttle.

Again, normalized deviation appears to play a role in the Challenger incident. Had the issue with the foam insulation been addressed before this mission, or had management postponed the mission until it was addressed, the Columbia incident may not have occurred. The Columbia Accident Investigation Board reported, "Cultural traits and organizational practices detrimental to safety were allowed to develop," "reliance on past success as a substitute for sound engineering practices," and "organizational barriers that prevented effective communication of critical safety information." I believe these are all symptoms of normalized deviation.

Please refer to Figure 8.33.2 for a memorial image of the Challenger crew and to Figure 8.33.3 for a memorial image of the Columbia crew. These men and women were truly pioneers in space, advancing our technology for the future. We must all learn from these tragic incidents and how they can be applied to our industry.

Did Normalized Deviation Play a Role at BP Texas City?

Based on the investigation completed by the US Chemical Safety and Hazard Investigation Board (CSB), it appears that normalized deviation played a significant role in the BP Texas City explosion and fire of 23 March 2005. BP Texas City had written procedures for establishing tower levels during start-up operations; however, the CSB found that Operators routinely deviated from these procedures. The CSB investigation reported, "In the majority of the 19 ISOM startups between April 2000 and March 2005; the tower was filled above the range of the level transmitter despite procedural instructions to fill the tower to a 50 percent reading on the transmitter. Operators frequently ran the valve sending liquid raffinate out of the unit to storage in "manual" instead of the "automatic" control mode as required by the procedure."

This is only one example of procedural deviations reported by the CSB in their report. From the report, it is easy to conclude that the Operators had deviated from the procedure so many times that the deviation had become "normal," or the way things are done, rather than following the procedures. This deviation is a dangerous practice because when things are not what is expected, or things are different on that day or at that time, the results can be catastrophic. At BP, the routine practice of running the level above the range of the level instrument on 23 March 2005, coincided with several other instrumentation and alarm failures and resulted in the overfill of the tower and the resulting catastrophic explosion and fire.

Unfortunately, other examples from the BP Texas City incident also exemplify normalized deviation. For example, the Operator on the ISOM unit left the post an hour before the end of the shift without making a turnover with their replacement, and turnover at the console position was marginal at best. During their investigation, the CSB found that there was "widespread tolerance of noncompliance with basic HSE rules"; in other words, at BP Texas City, not following the rules was accepted ("tolerated"). I believe this speaks volumes about the organizational culture (more on this later).

Frequently, operators may deviate from written procedures with the best intentions, thinking they are taking the safest path. However, they may not have the benefit of knowing the reason behind the written procedure, or they may not be fully aware of other conditions in the unit. Strict adherence to written procedures is the best course of action. If a concern exists, they should always discuss it with their supervisor and jointly decide on the best course of action. If the procedure is revised, it should be "redlined" with the changes, and a supervisor should review and approve the changes before proceeding.

More About Organizational Culture

The culture of the organization must come from the top. The site manager and their direct staff play a huge role in establishing the culture and expectations of all employees and contractors at the site. This includes the rules of conduct, both written and unwritten rules. I believe the nuclear industry has set an example for us to follow. Admittedly, federal and state regulations heavily influence the nuclear industry, but they appear to have this figured out. In addition to the strict regulations that govern nuclear power operations in the United States, the operators have formed the Institute of Nuclear Power Operations (INPO) as an industry oversight group with the mission to "promote the highest levels of safety and reliability and to promote excellence in the operation of commercial nuclear power plants."

The INPO works with the power plant operators by establishing performance objectives, criteria, and guidelines for the nuclear power industry, conducting regular detailed evaluations of nuclear power plants, and providing assistance to help them continually improve their performance. INPO also assists in reviewing any significant events at nuclear electric generating plants. Through information exchange and publications, they communicate lessons learned and best practices throughout the nuclear power industry.

Through INPO information exchange and publications, lessons learned and best practices are shared throughout the nuclear power industry. The following is from the INPO mission statement: INPO evaluation teams travel to nuclear electric generating facilities to observe operations, analyze processes, and observe plant activities with an intense focus on safety and reliability.

The nuclear industry is a regulated utility and is not subject to competitive pressures like the petroleum and petrochemical industries. However, I believe the petroleum industry can learn from the nuclear industry and seek ways to improve operations safety by working more closely together and sharing lessons learned from incidents where competition is not a factor. I believe this helps the employees understand the "why" behind the rules and procedures, which would go a long way toward helping avoid normalized deviation. The bottom line is that employees in the nuclear industry follow the rules and procedures to the letter, regardless of whether someone is looking on or not. That should be our objective, but this is not an easy goal.

This starts with management setting expectations that rules and procedures must be followed and communicating that to all employees. Then, recognize employees who appear to have gone over and above to complete a task while following the established procedures. We must also follow up with the employees involved when they are observed working outside of an established rule or procedure, clarifying our expectations. We must also recognize the employee(s) who come forward with suggestions to make a task easier or more efficient or help resolve an issue with a current rule or procedure. I make this sound simple, but changing the organizational culture is not simple; it requires a constant and consistent approach.

Strict Adherence to Written Procedures Can Help Prevent Normalized Deviation

Management should establish an organizational or corporate culture that ensures procedures are periodically reviewed to ensure they are maintained current. This organizational culture should also reinforce the importance of strict adherence to procedures and the expectation that concerns or issues with procedures can be discussed and addressed on an ongoing basis. Operators and others should feel free to bring up their concerns and to have the concerns and required procedure revisions discussed quickly. Remember that all changes to procedures should also go through the MOC process to ensure the revisions do not introduce hazards to the procedure or the process.

Periodic field verifications should reinforce this "culture" to ensure the procedures are being followed in the field and that Operators feel free to bring forward any issues or concerns. It helps to have the procedures, especially the critical ones, in a checklist format where the Operators sign off on each step of the procedure as it is implemented. The expectation is that these should also be field-verified during their use. Management expectations and strict adherence to written procedures can help prevent normalized deviation.

Key Lessons Learned

The following is from the same NASA report on the Challenger disaster:

- We cannot become complacent.
- We cannot be silent when we see something we feel is unsafe.

We must allow people to come forward with their concerns without fear of repercussion.

In Memorial

Figure 8.33.2 The Challenger crew. Source: Photo courtesy The National Aeronautics and Space Administration. From NASA report "NASA – STS-51L Mission Profile" 5 December 2005; Updated: 7 August 2017. STS-51L Crew (l-r): Payload Specialists Christa McAuliffe and Gregory B. Jarvis, Mission Specialist Judith A. Resnik, Commander Francis R. Scobee, Mission Specialist Ronald E. McNair, Pilot Michael J. Smith, Mission Specialist Ellison S. Onizuka.

In Memorial

Figure 8.33.3 The Columbia crew. Source: Photo courtesy The National Aeronautics and Space Administration. From NASA report "STS-107 Shuttle Press Kit." The STS-107 crew. Seated in front are astronauts Rick Husband, commander, and Willie McCool, pilot. Standing (from left) mission specialists Dave Brown, Laurel Clark, Kalpana Chawla, Mike Anderson (payload commander), and payload specialist Ilan Ramon, representing the Israeli Space Agency.

Additional References

Priceonomics, Nemil Dalal, "The Space Shuttle Challenger Explosion and the O-ring", Available from: https://priceonomics.com/the-space-shuttle-challenger-explosion-and-the-o/

National Aeronautics and Space Administration, "Lessons from Challenger", STS-51L: 28 January 1986.

The National Aeronautics and Space Administration. "NASA - STS-51L Mission Profile" 5 December 2005; Updated: 7 August 2017.

"Columbia Disaster: What Happened, What NASA Learned", Elizabeth Howell February 01, 2019 Spaceflight, Available from: https://www.space.com/19436-columbia-disaster.html

The National Aeronautics and Space Administration. NASA report "STS-107 Shuttle Press Kit".

Safety motivational video: "Remember Charlie.", Charlie Morecraft, Available from: https://www.safetyvideos.com/Charlie_Morecraft_s/51.htm

IRMI Expert Commentary, "Normalization of Performance Deviations", Available from: https://www.irmi.com/articles/expert-commentary/normalization-of-performance-deviations

HuffPost – Captain Alan Price, "Normalization of Deviation", Available from: https://www.huffpost.com/entry/normalization-of-deviatio_b_13059988

Institute of Nuclear Power Operations (INPO), Web Site Home Page, Available from: http://www.inpo.info/AboutUs.htm

Chapter 8.33. Hazards of "Normalized Deviation"

End of Chapter Quiz

1 Explain what is meant by the expression "normalized deviation."

2 Which group of managers are responsible for establishing the site's organizational culture?

3 Can you list three examples where Normalized Deviation resulted in catastrophic consequences?

4 They continue with the deviation until one day, the __________ is a little different, or the weather has changed, or something else has changed. A combination of the deviation plus the __________ change results in an accident or injury or even may result in loss of life.

5 How many things can you list that may reinforce that the deviation is acceptable? List as many as you can.

6 Is it unusual to find that the deviation is by a larger group of people, for example, several Console Operators doing the same deviation?

7 How can we guard against the occurrence of normalized deviation?

8 What role do written procedures play in helping avoid Normalized Deviation?

9 What were some of the critical learnings from the two space shuttle disasters? List as many as you can.

8.34

Criticality of Car Seal Valve Management Plan

Can these simple car seals and/or car-sealed valves save lives when incorporated into a safety management system and used properly in the field?

See Figure 8.34.1 for images of three types of safety car seals.

ASME International Boiler and Pressure Vessel Codes

The ASME International Boiler and Pressure Vessel Code requires each pressure vessel subjected to overpressure to be protected by a pressure relief device (i.e., Pressure Relief Valve [PRV], Rupture Disk). The standard also requires a clear and open relief path between the pressure vessel and the PRV and between the PRV and the relief location (the flare, a tank, or the atmosphere). This means that when block valves are installed on the inlets and outlets to PRVs or rupture disks, these block valves must be car sealed or locked in a fully open position to ensure the availability of the relief device. The PRV inlet and outlet block valves must also be maintained in the fully open position to ensure the availability of the relief valve for the PRV to provide the required overpressure protection.

The way we ensure that the valves are in the correct position and that they remain open is by applying the car seals to secure the block valves in the appropriate position (typically open) (CSO valves), but sometimes also used to ensure that valves are not opened inadvertently or car seal in the closed position (Car Sealed Closed valves or CSC).

CSO/CSC valves are also used in other safety-critical equipment to ensure they are available when needed. CSO/CSC Valves for Safety-Critical Equipment are critical to ensuring the availability of equipment designated as "safety or environmental critical" or "last line of defense." Operators are trained to respect car-sealed valves; therefore, it is not likely that an Operator will inadvertently close a CSO valve or a CSO valve by mistake.

Critical Valve Lineup Procedure

Each site should also develop a "lineup procedure" to help ensure that critical valves are being maintained in the proper position. This is especially true for valves associated with SHE critical equipment, including valves designated as CSO or CSC valves. Each site should also have a Safety Management Process for maintaining CSO/CSC valves, including a field verification or routine audit process and audit checklist to help ensure compliance with the valve lineup procedure.

SHE critical block valves shall normally be full-port, gate, ball, or plug-type valves. Block valves to and from PRVs shall normally be car-sealed open. All CSO and CSC block valves, including those associated with PR devices and flare headers, should be painted a distinctive color, and no other valves should be painted the same distinctive color, as this dilutes the effectiveness of the CSO/CSC valves.

A bleed valve shall normally be installed between the PR device and the inlet and outlet block valves to help facilitate the removal of the PRV from service for maintenance. CSO/CSC valves should also normally be installed with the valve stem oriented horizontally or vertically downward to help prevent the valve from inadvertently closing in the event the valve gate separates from the valve stem. Should the valve gate separate from the valve stem with the valve in the vertical orientation, the valve is likely to close by gravity, and the Operator would not have any indication that the critical valve is closed.

CSO/CSC valves should also be utilized for bypasses on safety-critical devices such as critical instruments and emergency trip devices such as furnace and boiler trips. These valves and all other CSO/CSC valves used in critical service should also be included on the site critical valve lineup procedure and audit checklist.

Process Operations Safety: The What, Why, and How Behind Safe Petrochemical Plant Operations, First Edition. M. Darryl Yoes.

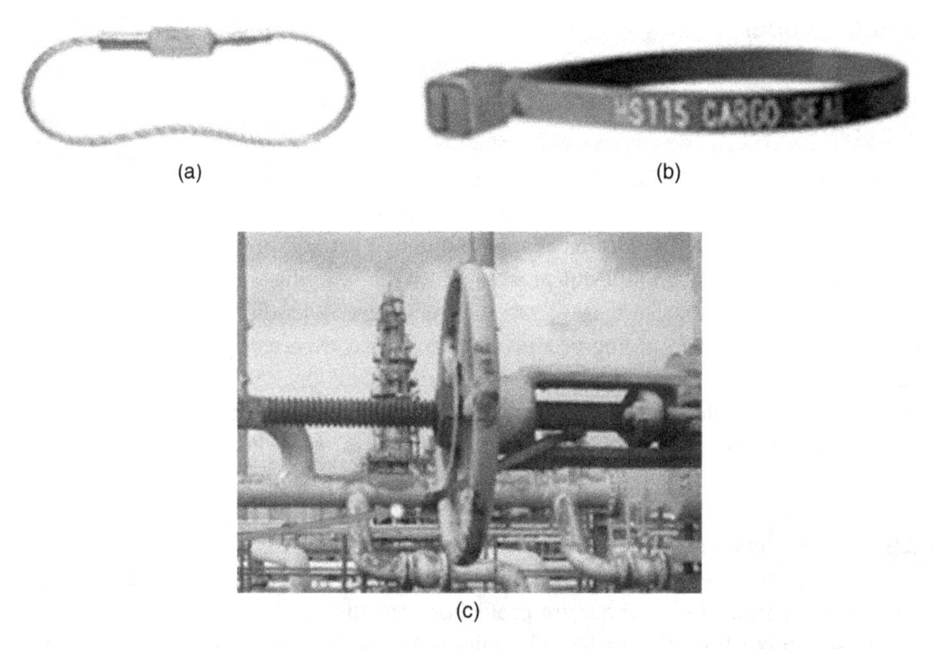

Figure 8.34.1 Car seals and car-sealed valves. Illustrated in these images are several types of car seals that are available in industry (all are not illustrated here). They may be fabricated from wire, cable, or plastic but what is important is that they must be broken when removed to show signs of potential tampering.

When defeating a critical instrument or PRV for maintenance, ensure approval of a critical device impairment or an approved "Control of Defeat" is in place for the device being defeated before breaking a Car Seal. Ensure the car seal has been reinstalled before the device is returned to service.

Operators should be trained to field verify that car seals are in place for CSO/CSC valves during field walkarounds, and any broken or missing car seals should be promptly replaced.

Please refer to Appendix I for additional technical guidance on the application of car seals (by PROtect/ERI Solutions, Inc.).

What Has Happened/What Can Happen

The explosion of a Liquid Nitrogen Storage Tank
Hokkaido Frozen Foods Distribution Facility
Japanese Food Processing Plant
28 August 1992

A double-walled liquid Nitrogen storage vessel ruptured catastrophically due to thermal expansion on 28 August 1992 at a food processing plant in Japan. The vessel maximum allowable working pressure (MAWP) for the Nitrogen storage tank was 135 psi (9.3 bar); the vessel ruptured at 996 psig (68.7 bar), or many times the MAWP. During the investigation, the team discovered that the block valves for the PRV were in a closed position. The isolation valves for a separate rupture disk, also designed to prevent overpressure, were also found to be in the closed position. With these two valves in the closed position, there was no overpressure protection to prevent the rupture from thermal expansion.

The explosion destroyed the food processing plant and damaged homes over 1,300 feet (400 meters) from the site. Major vessel sections were propelled over 1,100 feet (340 meters) from the site. Photographs of the area show the catastrophic damage to the facility and the surrounding community.

This incident emphasizes the importance of maintaining an effective Car Seal Valve Management Process and ensuring that the block valves around pressure relief valves and rupture disks are managed properly and not inadvertently isolated or closed.

What Has Happened/What Can Happen

New LNG Export Facility
Bontang, Indonesia – April 1983 (three Fatalities)

This incident occurred after the construction of a new LNG Export facility had essentially been completed. The new facility was undergoing preparation for the startup. Before startup, the team decided to install a new isolation valve between the low-pressure and high-pressure separator vessels, and although this change did not go through the formal management of change process, the new valve was installed. During the dry-out process and while purging the large vertical, main liquefaction column spiral-wound heat exchanger with warm natural gas, the exchanger violently ruptured due to rapid overpressure. As unit pressure was increased during the purging process, the exchanger experienced internal pressures greater than three times its design pressure before rupturing. The incident occurred before any LNG was introduced into the system. When the column was overpressured, it exploded, resulting in a major fire and propelling debris and coil sections about 165 feet (50 meters) from the column. Shrapnel from the column struck and killed three workers participating in the start-up activities.

During the investigation, the new valve was found to be installed between the low-pressure exchanger and low-pressure separator. However, the overpressure relief protection for the exchanger and the low-pressure separator was installed on the low-pressure separator. The team also found the new valve was left in the closed position. The location of the new valve effectively isolated the low-pressure exchanger from the overpressure protection, resulting in rapid overpressure of the low-pressure exchanger without overpressure protection. This incident highlights the use of an effective MOC process but also emphasizes the importance of a CSO/CSC valve lineup process. It also recognizes the hazards of last-minute changes without adequate review. See Figure 8.34.2 for a partial schematic of the new LNG unit.

The following was extracted from a report by Report by La'o Hamutuk listing incidents in the LNG industry. This is about the incident at the new LNG plant at Bontang, Indonesia, in April 1993: "A rupture in an LNG plant occurred as a result of overpressure of a heat exchanger caused by a closed valve on a blow-down line. The exchanger was designed to operate at 25.5 psig, and when the gas pressure reached 500 psig, the exchanger failed. This resulted in a hydrocarbon release and explosion."

What Has Happened/What Can Happen

Williams Olefins Plant
Geismar, LA – 13 June 2013 (two Fatalities)

The Williams Olefins case study is presented in more detail in Chapter 8.14, "Hazards of Thermal Expansion." However, this is another example of block valves being installed ahead of PRVs. Unfortunately, these valves were also left in the closed position, and the hazard was not recognized or addressed in the PSSR or the PHAs.

On the day of this incident, a Process Supervisor was troubleshooting a Depropanizer reboiler exchanger and started the hot water flow through the hot side (tube side) of the exchanger while the cold side (shell side) was isolated (blocked in). He did not realize that the exchanger was also blocked away from its overpressure protection on the shell side, and the exchanger ruptured violently after less than two minutes of flow on the hot side of the exchanger. The supervisor and another Operator working nearby were killed in this tragic incident. One hundred sixty-seven others on the unit were injured in the explosion.

A valve lineup process with field audits to verify the effectiveness of CSO and CSC valves may have made the difference in this case.

Goodyear Chemical Plant
11 June 2008 (one Fatality)

Unfortunately, another incident, like that at Williams, occurred at the Goodyear Chemical Plant on 11 June 2008. In this case, the Operator connected a steam hose to the exchanger on the tube side in preparation for returning the exchanger to service. She was unaware that the shell side was blocked in and the relief valve and rupture disk had been disabled. The exchanger exploded from internal pressure due to thermal expansion, instantly killing the operator. Chapter 8.14, "Hazards of Thermal Expansion," covers this incident in more detail.

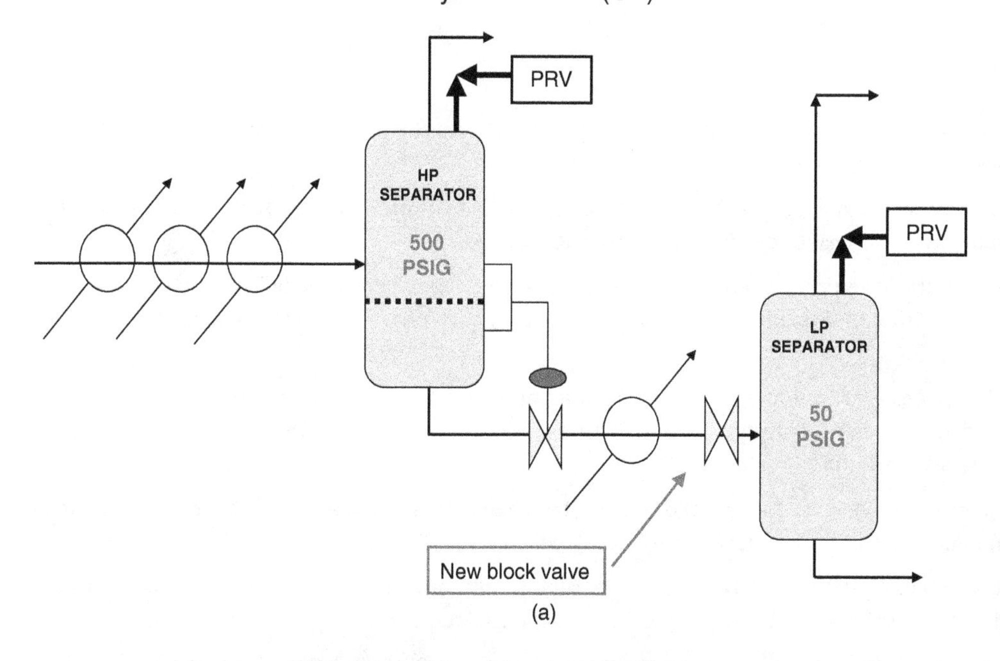

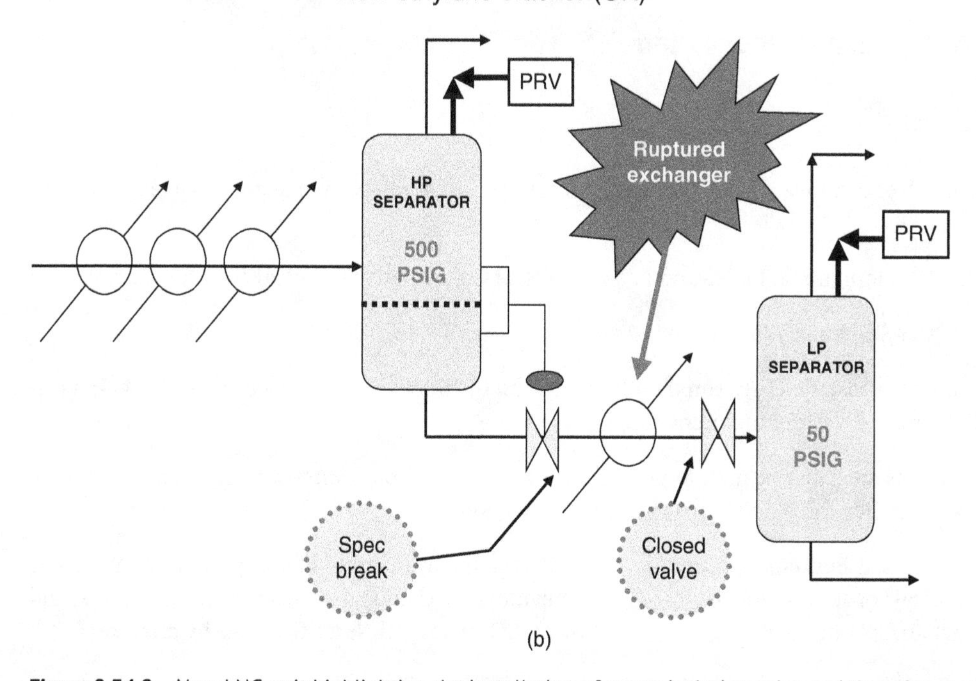

Figure 8.34.2 New LNG unit highlighting the installation of a new isolation valve and the subsequent failure of the low-pressure exchanger due to overpressure. Figure (a) illustrates a new isolation valve that was installed downstream of the specification break where the design pressure was lower. The location of the ruptured exchanger is illustrated in (b) which was caused by overpressure due to the location of the new isolation valve. These images were created to illustrate the hazards of failure to use or respect car-sealed valves.

All four of these incidents emphasize the importance and value of a strong CSO/CSC Safety Management and audit process and how this type of process can help prevent equipment accidental overpressure. In all four case studies, valves were inadvertently left in closed positions, resulting in equipment overpressure and damage. This also illustrates how this can also lead to loss of life.

Additional References

British Health and Safety Executive, Rupture of a Liquid Nitrogen Storage Tank, Japan, 28th August 1992. https://www.hse.gov.uk/comah/sragtech/caseliqnitro92.htm

Kobayash, Hideo; Tokyo Institute of Technology, Article: "Explosion of Liquified Nitrogen Storage Tank by Closing Shutoff Valve for Safety Valve". http://www.sozogaku.com/fkd/en/hfen/HB1011017.pdf

The Chemical Safety and Hazard Investigation Board, Williams Geismar Olefins Plant, Reboiler Rupture and Fire Geismar, Louisiana, 13 June 2013 (2 Fatalities, 167 Reported Injuries). https://www.csb.gov/williams-olefins-plant-explosion-and-fire-/

A Report by La'o Hamutuk, Timor-Leste Institute for Reconstruction Monitoring and Analysis, February 2008 (Three Fatalities), Appendix 4. History of accidents in the LNG industry, (Includes a brief summary of the Bontang, Indonesia incident). https://www.laohamutuk.org/Oil/LNG/app4.htm

The California Energy Commission, Liquefied Natural Gas Safety, (Includes a brief summary of the Bontang, Indonesia incident). http://laohamutuk.org/Oil/LNG/Refs/015CECLNGSafety.pdf

The Chemical Safety and Hazard Investigation Board, Heat exchanger rupture and ammonia release in Houston, Texas, 11 June 2008 (One Killed, Six Injured). https://www.csb.gov/goodyear-heat-exchanger-rupture/

LNG: Accidents and Malfunctions. https://grassrootsrendering.files.wordpress.com/2016/05/lngaccidentsmalfunctions-docx.pdf

Chapter 8.34. Car Seal Valve Management Plan

End of Chapter Quiz

1. Car seals and/or car-sealed valves can save _______ when incorporated into a safety management system and used properly in the field.
2. What is required by the ASME International Boiler and Pressure Vessel Code relative to a clear path for relief valves?
3. What is meant by the term "Car Seal Valve Management Plan"?
4. A _______ valve shall normally be installed between the PR device and the inlet and outlet block valves to help facilitate the removal of the PRV from service for maintenance.
5. PRV inlet and outlet block valves must also be maintained in the fully ________ position to ensure the availability of the relief valve for the PRV to provide the required overpressure protection.
6. CSO/CSC Valves for Safety-Critical Equipment are critical to ensuring the availability of equipment designated as "safety or environmental critical" or as "last line of ________." Operators are trained to respect car-sealed valves, and therefore, it is not likely that an Operator will ___________ close a CSO valve or close a CSO valve by mistake.
7. What should Operators inspect for regarding car-seal valves when making routine process rounds on their unit?
8. What is typically required for a field Operator to close a CSO valve or open a CSC valve?
9. What is the typical duration for the authorization for the defeat for the critical device or the changing the position of the CSO/CSC valve?
10. What are the typical requirements for SHE critical block valves, including car-sealed valves?

8.35

Small Piping Guidelines for Operations Safety

In some of the process safety incidents with catastrophic or near-catastrophic consequences, we find the cause of the initial loss of containment to have been a failure of small piping or small piping connections. For example, the major fire at the Formosa Plastics Corporation facility located at Point Comfort, Texas, on 6 October 2005, was due to a small piping failure that occurred when a forklift struck a three-quarter (3/4) inch bleeder valve. The resulting fire was estimated to reach about 800 feet high and burned for five days, causing significant damage to plant equipment. Several employees were severely burned in this incident, requiring hospitalization.

The bleeder valve was a threaded connection with no seal welds and no gussets or other bracing to provide additional support. Other factors played a role in this event. For example, not using a spotter to help the forklift driver ensure the forklift was clear of process equipment. Also, if this is a regular operation, bollards (a pipe embedded in concrete) should be in place as additional protection. This incident was investigated by the US Chemical Safety and Hazard Investigation Board (CSB), and a detailed report was issued, plus a great training video on this incident. There have been similar incidents involving small piping failures, unfortunately, with similar results.

Small Piping

Small-bore piping is typically considered pipe or valve size less than NPS 1-1/2 inch (40 millimeter) and is generally used in supporting systems or equipment connected to pumps, compressors, analyzers, and numerous other large process equipment. Most process units have a large amount of small-bore pipe, and Operators sometimes fail to realize the consequences that can result from the failure of small-bore piping systems and the large loss of containment that can occur should it fail. For example, at the Union Carbide polyethylene plant in Antwerp, Belgium, in 1975, a 1-inch diameter ethylene line failed and resulted in a devastating explosion, causing six fatalities and the destruction of most of the plant.

This chapter will discuss several types of failures involving small-bore piping or other small-bore equipment and the causes of these failures. Small-bore piping may be only threaded or may be threaded and seal welded at the threads. It may also be socket welded at the joints, between sections of pipe, or to valves or other fittings.

Small-bore piping is covered in ASME B31.3 "Process Piping Guide" and the corresponding code. ASME B31.1 clarifies that "the original and continued safe operation of a piping system depends on the competent application of codes and standards." The codes have different requirements for Category "D" services, such as water from hydrocarbon or other flammable or toxic products or services. This chapter will discuss the typical requirements for flammable or other toxic services. The requirements of small piping are discussed at a high level in this chapter, and therefore, the information provided should be used in conjunction with the appropriate standards and codes.

Piping Excess Vibration

One of the most frequent causes of failure of small-bore piping can be attributed to vibration. Small-bore piping is typically associated with equipment prone to vibration, such as rotating equipment like pumps and compressors. However, small piping is associated with many other equipment classified as vibrating services, such as pressure relief valves, mechanical mixers or agitators, process flow or pressure control valves, etc. When a failure occurs, such as a weld failure at a pipe junction where a small-bore pipe is connected in a branch connection to a larger pipe, the cause is quickly determined to be vibration. I think we need to look further to determine the actual root cause. Most often, the true root cause could be poor design or excess vibration of the equipment.

Process Operations Safety: The What, Why, and How Behind Safe Petrochemical Plant Operations, First Edition. M. Darryl Yoes.

Pipe design should include consideration for adequate pipe supports or braces. This is covered in more detail in the next section, but in most cases, small pipe connections should be gusseted or braced at intersections with other piping, anticipating that vibration or pipe stress will be present. Another consideration is the effect of temperature cycling, and exposing the piping to rapid changes in temperature results in a different type of stress. It is common to find cracks in piping in cyclic service, especially in welds.

To the extent possible, process designers should minimize connections of small-bore piping to larger piping when close to rotating equipment such as pumps and compressors. When these junctions are near the rotating element, the small connections can experience significant stress when the equipment is first started, especially in a compressor surge. Failures have occurred on small-bore piping close to rotating equipment and have resulted in significant loss of containment.

Some experts in vibration analysis can analyze the total system and help determine the true cause of the vibration and recommend solutions. While not recommending a particular firm, a couple has brief videos posted on YouTube that I believe are worth viewing if you are experiencing piping vibration. One good YouTube video by Wood, which illustrates the vibration of small-bore piping on typical process units, can be found online at: https://www.youtube.com/watch?v=ldG_fmN_ENs&t=21s. Another good YouTube video is by Beta Machinery Analysis (www.BetaMachinery.com) and can be found at https://www.youtube.com/watch?v=PadaiNRy9L8. Of course, a quick search may find others as well.

Always be aware of excess vibration in small-bore piping and report any excess vibration to your supervisor as soon as possible. Excess vibration is usually from rotating equipment such as pumps or compressors but can be from other reasons. A chattering relief valve or a control valve can present severe vibration and is most often an indication of a design issue with the relief valve or the control valve and requires prompt attention. Small piping has failed as a direct result of excess vibration and has the potential to result in a large unit fire should failure occur. If your site is experiencing excess vibration in small-bore piping or any piping, seek out the experts in the industry who specialize in piping and dynamic structural analysis and can determine the reason for the vibration and recommend strategies to address it.

Unsupported Small Piping

Some of these failures are like the small connection at Formosa Plastics, where they are not properly braced or supported. Lack of bracing or supports can lead to a failure of the pipe components, especially if the small connection is in vibrating service or if it is struck or impacted by motorized equipment. Small-bore connections, also sometimes referred to as branch connections, to the main process piping represent a potential serious vibration issue, especially on rotating and reciprocating machinery and associated process piping. In 2008, the Energy Institute reported that piping vibration and fatigue could account for up to 20% of hydrocarbon releases, many of which were the failure of small-bore connections. These vibration-induced failures can be a failure of the small pipe connection itself, or the vibration can result in cracking in the pipe or vessel to which the small pipe is connected.

Generally, small-bore piping is seal welded at the joints connecting with the main pipeline and through at least the first block valve. The connection to the main line should also be provided with gussets or braces welded from the main line out to the valve or a small connection. Braces are installed with two braces welded at 90° angles to one another. See the photo in Figure 8.35.1 for an example of a properly installed seal welded and gusseted small connection. Figure 8.35.2 Illustrates a pump case drain on an LPG pump that is not seal welded or braced (gusseted). This connection resulted in a significant vapor release when the pipe nipple failed (due to galvanic corrosion in the threads due to improper metallurgy – a carbon steel pipe nipple in a stainless-steel pump case). Think about this – LPG service, wrong metallurgy, no seal welds, and no gussets. Was this risk assessed?

The connection shown in Figure 8.35.3 was in the outlet line on a naphtha hydrotreater furnace operating more than 600°F (315°C) when a crack and leak were discovered in the pipe nipple where the connection was welded to the main line. The small connection was properly seal welded; however, there were no gussets or braces to guard against piping vibration. The pipe cracked, and a leak resulted from a fatigue failure due to excess vibration.

The unit was safely shutdown and repairs were made without incident. Following a stress analysis, bracing was added to reinforce the connection to protect against vibration. This is an excellent example of where engineering should be used to help prevent excess stress on small-bore piping, which can lead to failure and loss of containment.

A similar incident occurred at a chemical plant in the United States when a ¾″ bleeder connection nearly failed at a liquid propane control valve station. I was leading a Safe Operations Training class at the site shortly after this incident occurred when one of the process supervisors in the course brought in the failed bleeder to share with the class. The control valve station operates at 300 psi and is considered a vibrating service. The supervisor explained that one of the Operators discovered a leak at the bleeder, and they quickly isolated and depressured the control valve station. The supervisor explained that

Figure 8.35.1 A properly seal welded and gusseted small connection. Gussets are installed at 90 ° angle to each other and welded from the main line to the small connection.

Figure 8.35.2 An LPG pump case drain that is not seal welded or gusseted, and the pipe nipple is the wrong metallurgy; carbon steel is threaded into a stainless-steel pump case. This should be considered a recipe for trouble.

after the control valve was depressured and made safe, he determined the leak was from a crack in the pipe nipple leading to the bleeder valve. He said he was able to grasp the bleeder valve and break the pipe nipple completely off with his hand, indicating that the connection was approaching a total failure.

The leak was determined to be a crack in the 3/4″ pipe nipple leading to the bleeder valve and was consistent with a fatigue failure due to pipe vibration. The bleeder valve was normally used to depressure the control valve and was plugged at the time of the leak; no other pipe or equipment was connected to the bleeder at the time of the failure. A photo of the failed pipe nipple is available in Figure 8.35.4. The crack is visible in the photo, and the "old" crack is distinguishable by the discoloration from the "fresh" crack where the supervisor broke the connection. This clearly shows that this pipe nipple

Figure 8.35.3 Crack in small-bore piping on a naphtha hydrotreater furnace outlet line. A lack of bracing was determined to be the cause of this failure.

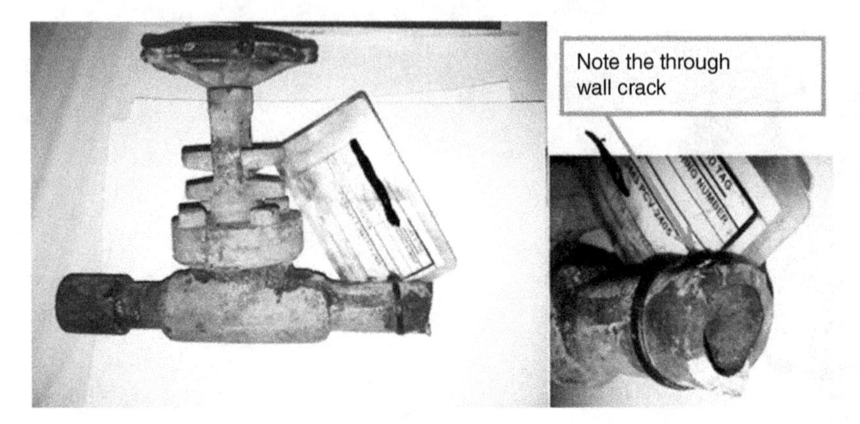

Figure 8.35.4 Photo of the ¾-inch bleeder valve connection with a fatigue crack in the pipe nipple. Note there is no gusset or brace to support the bleeder. Note that the packing gland bolts vibrated completely out of the valve.

was close to a total failure and could have quickly escalated to a significant incident. In discussions with the supervisor, I learned this was the second failure at the same control valve station in about three months.

The cause of this failure was pipe fatigue due to vibration at the control valve station. It was a socket weld pipe nipple, and the valve and the welds were still completely intact. The pipe nipple failed due to pipe fatigue since there were no gussets or supports to minimize the pipe fatigue and prevent failure. We can only speculate that the first failure was a similar type of failure. This incident helps reinforce the requirement for gussets on small-bore connections, especially in vibrating services such as control valve stations and rotating equipment such as pumps and compressors, relief valves, etc.

Another good example of small connections without adequate bracing is shown in Figure 8.35.5. This orifice flange connection is seal welded at the orifice flange but does not have adequate bracing or gussets to support against vibration. Note the heavy flanges, the leverage effect of the piping, and the resulting stress applied to the seal welds. These will fail after a period of time due to vibration and fatigue.

Figure 8.35.5 Photo of orifice flange connections without adequate bracing or gussets.

Excess "Moment Arm or Leverage"

Excess moment arm or pipe leverage results from an unsupported, long pipe connection. Moment arms are a process safety hazard because a small amount of force or "leverage" applied to the end of the piping can easily result in breaking the pipe or connection and loss of containment. A relatively small amount of force applied to one end of the moment arm is multiplied by the "moment or leverage," resulting in the failure. See Figure 8.35.6 for an example of a moment arm. Note in the photo that the pipe connection to the pump case is seal welded; however, there are no gussets, and the rest of the pipe fittings are not seal welded.

One can easily see how the pipe can be broken off, for example, if it is struck by a carry deck crane, forklift, or someone attempting to stand on the pipe. Moment arms should be avoided to the extent possible. All joints in the piping should be seal welded, and supports, or braces should be used to provide additional support for the pipe.

Small Piping "Christmas Trees"

Another significant issue involving small-bore piping is the issue with "Christmas trees," especially in flammable or toxic services. Christmas tree is frequently used to describe many small connections involving small-bore piping, valves, pressure gauges, and other similar pressure-containing equipment. A Christmas tree may be fabricated during initial construction but is often created by adding additional connections to existing equipment. Frequently, these additional connections are not seal welded, nor are they gusseted or supported other than the inherent support from the piping. See the examples provided in the photos in Figure 8.35.7a,b. Note that Figure 8.35.7a illustrates a large christmas tree on a vertical pump discharge. This one has other issues such as comingled metallurgy and lack of small piping seal welds or bracing. Figure 8.35.7b

Figure 8.35.6 Photo of a potential "moment arm" or pipe leverage.

(a) (b)

Figure 8.35.7 Christmas trees.

illustrates a christmas tree associated with small piping for an orifice flange connection. Both of these installations are subjec to significant vibration, and similar connections have failed resulting in signifiant loss of containment.

In the Christmas trees shown in the photos, the pipe attached to the main line is not seal welded or gusseted. The total weight of the pipe and all the additional equipment, including the pipe nipples, the valves, pressure gauges, and pressure transmitters, are placing stress on the weld connections. These two examples show that these are frequently installed with mixed metallurgy and without gussets, braces, or seal welding. In vibration service, these are stressed even more and have failed in service, resulting in the loss of containment.

(a)

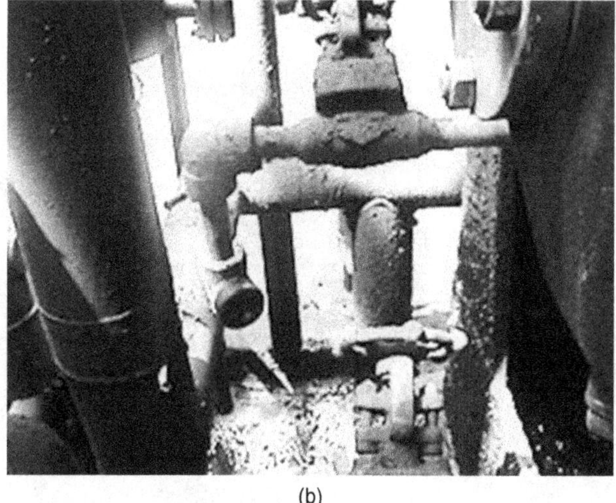

(b)

Figure 8.35.8 Photos of missing bar stock plugs. Missing pipe plugs are illustrated in Figures (a) and (b). Missing plugs in hydrocarbon or chemiocal service is a violation of US Environmental Protection Agency (EPA) standards and are subject to signifiant penalties.

Missing or Loose Pipe Plugs

All bleeders and vent valves or any other open-ended pipe should be properly fitted with a bar stock pipe plug to seal the pipe. Valves tend to leak, and open-ended bleeders can result in unexpected loss of containment and, therefore, should be sealed by a pipe plug when not in use. The plug should be the same metallurgy as the valve or coupling to prevent galvanic corrosion. Pipe plugs have vibrated out; therefore, the plug should always be installed with a wrench to ensure it is tight. Plugs fabricated from rolled bar stock and threaded are the most effective and should be the only plugs used in hydrocarbon or toxic service. See photos in Figure 8.35.8a,b, which are photos of missing bar stock plugs.

This is not just good practice but also the law in the United States and other countries. In the United States, the Environmental Protection Agency requires plugs to be installed in all open-ended piping. This is included in the Code of Federal Regulations; 40 CFR 63.167 – Standards: Open-ended valves or lines states "(a)(1) Each open-ended valve or line shall be equipped with a cap, blind flange, plug, or a second valve, except as provided in § 63.162(b) of this subpart and paragraphs (d) and (e) of this section. (2) The cap, blind flange, plug, or second valve shall seal the open end at all times except during operations requiring process fluid flow through the open-ended valve or line or during maintenance or repair."

What About Pipe Unions

Pipe unions are the source of frequent leaks or loss of containment, especially the threaded pipe boss on the union. I believe the use of unions should be minimized in equipment designs and piping layouts. However, sometimes, they are necessary to facilitate equipment maintenance. Due to the potential for leaks, unions should not normally be installed inside the first block valve off a main line or pressure vessel. If installed inside the first block valve, the union should be seal welded, and the union boss and the two threaded pipe connections should all be seal welded in this case. The issue is that the boss can loosen due to vibration, and the logic is that if the union is outside the main block valve, it can easily be isolated. See the photo in Figure 8.35.9 of a union in service. Notice that the union is leaking, creating a mess in the surrounding area.

Missing Check Valves

Check valves are used to prevent backflow, especially in temporary applications such as preparation for unit shutdown or startup. In these scenarios, we frequently connect utilities such as steam or nitrogen to the process piping or other process equipment. When doing so, we must prevent the inadvertent backflow of the process into the utility, and we do this by

Figure 8.35.9 Photo of a leaking pipe union. Note the leak and contamination of the surrounding area.

Figure 8.35.10 Photo of a missing check valve in a utility connection to process equipment (Nitrogen to purge a pump in preparation for maintenance).

installing a check valve (sometimes called an NRT – a nonreturn valve) at the point of connection to the process. If these are omitted, and the pressure on the process side is high enough, the process will backflow into the utility system, contaminating the utility. This can have adverse consequences, such as creating fire or explosion opportunities. This is covered in more detail in Chapter 8.24 ("Safe Use of Utilities and Connecting Utilities to Process Equipment"). See Figure 8.35.10 for a photo of a missing check valve.

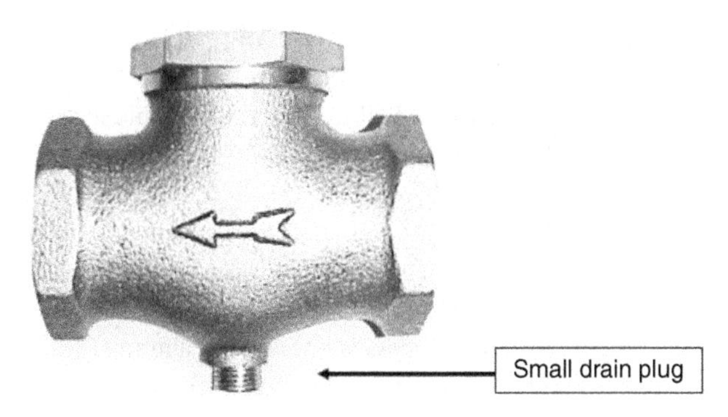

Figure 8.35.11 Photo of a small diameter check valve. Note the arrow indicating the direction of flow and the small diameter plug at the bottom of the valve.

Check Valve Failure

Check valves play an important role in process safety when installed correctly and properly inspected and maintained. Check valves are used to help ensure against the reverse flow in process lines, for example, on pump and compressor discharge piping, to prevent the process from backflowing into utility systems and a variety of other similar services. Some check valves are so important that they are classified as safety-critical equipment and are identified as such in the field. However, they must be installed and inspected periodically to ensure they function properly.

Check valves must be installed correctly, with the flow going in the right direction. A common mistake made by craftsmen when installing check valves is installing the check valve backward. This sounds all too basic, but it happens frequently. Of course, this will block process flow and is usually discovered quickly after unit startup. Most check valves have an arrow embossed or cast into the valve body to indicate flow direction. However, believe it or not, there have been cases where this arrow pointed in the wrong direction, making troubleshooting difficult. See Figure 8.35.11 for a photo of a check valve.

Another common issue is that some check valves have a small diameter hole in the side of the check valve, typically fitted with a pipe plug. This threaded hole may allow access to the check valve shaft or may be for a bleeder or vent. These pipe plugs have backed out in service, resulting in a loss of containment. If these small diameter holes are not used, they should be plugged and secured to prevent the plug from backing out. For example, seal welded or secured with a tie wire. Note that in Figure 8.35.11, the small diameter plug was installed in the check valve.

Do Not Mix Metallurgies

From a metallurgical perspective, we should treat small-bore piping no differently from other piping circuits. In other words, we should not mix metallurgies unless there is an engineering reason to do so, and only with review and approval from a metallurgist or materials engineer and with an approved Management of Change. Mixing metallurgies can result in galvanic corrosion and other forms of corrosion. For example, hot hydrocarbons containing sulfur compounds can result in aggressive sulfidation corrosion on carbon steel, whereas piping containing chromium is resistant to this type of corrosion. Therefore, the installation of carbon steel components in a chrome circuit in hot hydrocarbon service with sulfur compounds present can result in catastrophic failure and loss of containment. When assembling small-bore piping, the metallurgy of the new pipe and all piping components should be verified as correct, typically by the unit Inspector.

For example, the small-bore pipe nipple installed in the pump illustrated in Figure 8.35.2 was determined to be carbon steel, and the pump case was stainless steel. This resulted in galvanic corrosion in the threads of the pipe nipple, resulting in a small leak at the threads. The pump was in service as the reflux pump to a Deethanizer tower, a very light hydrocarbon service. The operator, thinking he was doing the right thing, attempted to tighten the leaking pipe nipple. While doing so, the pipe broke completely at the corroded threads, resulting in a very large light hydrocarbon release. Fortunately, there was no explosion or fire, but the potential was certainly present. This incident reinforces that metallurgies should not be mixed unless there is a good reason to do so.

Please refer to Chapter 8.17, "Metallurgy Matters; Importance of Positive Materials Identification (PMI)," for additional details on the hazards and effects of mixing metallurgies in piping and other process systems.

Be Aware of Temperature Embrittlement (Brittle Fracture)

The hazards of brittle fracture are covered in more detail in Chapter 8.11 ("Brittle Fracture Failure"); therefore, this section will be very brief. However, we need to be aware that exposing the steel to low temperatures, sometimes even for short periods, can have catastrophic consequences.

Carbon steel can fail catastrophically at temperatures of −20°F (−29 °C). Generally, process designers specify stainless steel for circuits that will operate at low temperatures when designing new equipment. This is to avoid the hazard of temperature embrittlement. While stainless steel is resistant to temperature embrittlement and brittle fracture, carbon steel is not. Should carbon steel inadvertently be substituted in place of the stainless, and the process is operated at a temperature below −20°F, even for short periods, the piping can fail catastrophically. This is true whether we are talking about large-bore piping circuits or small-bore piping. Small-bore piping is just as susceptible to brittle fracture as the large-bore.

I make this point here because occasionally, the Operators frequently assemble or fabricate temporary piping circuits with small-bore piping. When doing so, the Operators must be aware of the metallurgy concerns within their unit or area. Any metallurgy change requires the complete Management of Change process and the involvement of the site metallurgist or site materials engineer.

Excess Flow Valves on Gauge Glasses

Level sight glasses on process vessels typically contain glass lenses to facilitate verification of the level in the vessel by the Operator. Gauge glasses can fail, resulting in loss of containment of the fluids; flammable or toxic products can be lost to the atmosphere, resulting in explosion or fire. To help mitigate this hazard, gauge glasses in flammable service are fitted with excess flow valves designed to stop the loss of containment in the event of glass failure. Excess flow valves are effective when used in clean service and when operated in accordance with manufacturer guidelines. Dirty or fouling service will plug the narrow ports in the excess flow valve, making it challenging to read the level in the gauge glass.

A manufacturer's photo of an excess flow valve is included in Figure 8.35.12a and 8.35.13. Note that a ball check prevents loss of containment when the gauge glass fails. When failure occurs, the surge of liquid flowing against the ball pushes the ball against the seat, effectively stopping the liquid flow. These valves are effective when used in clean service and when operated correctly. For the valve to function, the valves, both gauge glass top and bottom valves, must be in the fully open position. Otherwise, a pin holds the ball off the seat, preventing the ball from seating.

The purpose of the pin is to allow the valve to be placed in service. When placing the valves in service, the valves must be opened only 1/4 to 1/2 turn until the level is established in the gauge glass. Once the level is established, the valves must be

Figure 8.35.12 Illustrates a cut-away view of a traditional Jerguson® Safety Ball Check Valve commonly used on the top and bottom of gauge glass level indicators, especially in toxic or flammable services. See the video references at the end of this section.
Source: Images and information Courtesy of Jerguson® Safety Ball Check Valve, Clark-Reliance Corporation, Strongsville, OH. Images are property of Jerguson®, a product line of Clark-Reliance® LLC.

Figure 8.35.13 Typical Process Vessel Equipped with Jerguson® Safety Ball Check Valves on the top and bottom of the gauge glass level indicator. Source: Images and information Courtesy of Jerguson® Safety Ball Check Valve, Clark-Reliance Corporation, Strongsville, OH. Images are property of Jerguson®, a product line of Clark-Reliance® LLC.

fully opened to function properly. Jerguson® provides videos describing the procedure to place the ball check excess flow valves in service (see References at the end of this chapter).

The newer Jerguson® 360 Series Metal Seated Safety Ball Check Valve has a different design, which includes additional features to increase safety (lockout/tagout, position visible) and makes commissioning a glass gauge easier, which allows for safer operation.

What About Integrally Reinforced or Extended Body Valves

While this should always be verified by your fixed equipment engineer, bracing is generally not required for integrally reinforced valves or pipe fittings, otherwise known as extended body valves. These valves and fittings are integrally reinforced or weld-reinforced and are designed to be used without external bracing or other supports. However, they must be welded so that the weld extends to the main body of the valve and the cavity between the pipe or nipple and the new extended body valve is filled in with weld metal. Some sites prefer the extended body valves for vents, drains, instrument connections, and similar small connections since this eliminates the external bracing or gussets that are otherwise required.

See Figure 8.35.14 for a photo of an extended body valve and Figure 8.35.15 for the specific welding requirements for extended body valves. Figure 8.35.16 shows two extended body valves that are properly welded and in service.

Figure 8.35.14 Photo or extended body or integrally reinforced block valve.

Figure 8.35.15 Specific welding requirement for extended body valves.

Figure 8.35.16 Extended body valves properly welded and in service (note that one of the valves in the photo is chained and locked in the open position due to ongoing maintenance work at the time this photo was taken).

What Has Happened/What Can Happen

At an Asian refinery, a large fire occurred on a Diesel Hydrotreater, which was quickly extinguished with minimal damage and no injuries. The cause of the loss of containment and fire was quickly determined to be the fatigue failure of a dead-ended 1-½ inch connection at the weld where the pipe ell was connected to the main line. The dead-ended connection included a valve and a flanged end that was blanked. This connection was subjected to continuous vibration of the piping, the extra weight of the valve, flange, and blank, and the imposed stresses on the weld. The connection was not equipped with gussets, bracing, or supports, contributing to the weld failure and resulting in the loss of containment and fire. See Figures 8.35.17 and 8.35.18 for photos of the failed pipe.

What Has Happened/What Can Happen

With over 50 years of service in the industry, I am rarely surprised at what we find in the field. At a South Louisiana refinery, the maintenance team received a request to check four ¾ inch bleeder valves on a 600-psi steam exchanger. The exchanger was out of service for maintenance.

When the maintenance team inspected the valves, they discovered that all four bleeders had been previously installed with the gates removed from the valves. During a prior downtime, the valve gates were removed to weld the socket-weld flanges and stress-relieve the piping.

Figure 8.35.17 Failed piping due to lack of gussets or other bracing.

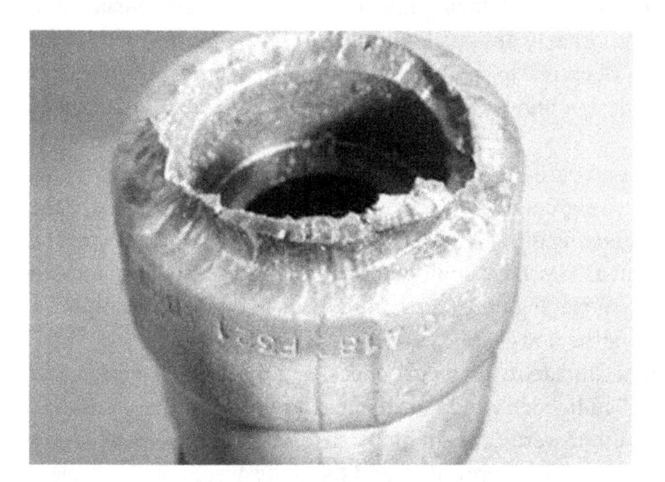

Figure 8.35.18 Closeup photo of the failed weld.

(a)

(b)

Figure 8.35.19 Four ¾ inch bleeder valves, previously installed without valve gates. Figure (a) illustrates the location of the four bleeder valves that were previously installed without valve internals. The internals are shown in Figure (b) and the internal gates are missing.

The welding and stress relieving were completed as planned, and the valves were returned to service without reinstalling the gates. Since someone could have been injured by removing a pipe plug, this was highlighted as a Near Miss. Had the plug been removed while the exchanger was in service, someone could have been seriously burned by the steam. See Figure 8.35.19a, b for photos of the four ¾ inch bleeder valves (valves installed with missing gates).

What Has Happened/What Can Happen

Shell Chemical Olefins Plant
Deer Park, Texas
22 June 1997

There have been failures of check valves in the industry, some with catastrophic results. For example, a Shell Chemical Olefins plant in Deer Park, Texas, experienced a devastating explosion and fire due to a 36-inch diameter check valve failure in the Pyrolysis fractionator overhead gas compressor 4th stage-discharge piping (between the compressor 4th and 5th stages). This check valve was a large pneumatic assist check valve in light hydrocarbon gas service. The failure resulted in a large flammable gas release and subsequent explosion and fire, which burned for ten hours. There were no fatalities in this incident, although several employees were injured. However, the damage to the olefins plant was extensive.

The US Environmental Protection Agency (EPA) and the US Occupational Safety and Health Administration (OSHA) investigated this incident, and a detailed report was issued. The investigation team found that the check valves installed in the Olefins plant process gas compression system were not appropriately designed and manufactured for the heavy-duty service they were subjected to. This resulted in the valves being susceptible to shaft blow-out during normal use. The report emphasizes that the valve failure was a combination of valve design and undetected damage to and eventual failure of critical internal valve components.

The valve failure occurred when a small dowel pin designed to secure the valve shaft to the internal valve components failed in service. This allowed the shaft and its associated counterweight weighing about 200 pounds to be blown out of the check valve body by the 300 psig internal pressure. This left an opening of 3.75 inches in diameter as the relief path for the highly flammable process gas to release into the surrounding area. The valve shaft and counterweight were found about 42 feet from the check valve under damaged piping, debris, and sludge. Figure 8.35.20, from the EPA report, is a simplified cross-sectional view of the check valve, illustrating the location of the dowel pins.

The investigation team also discovered and reported that other incidents involving check valves of similar design had also occurred at two other Shell-owned petrochemical plants in Saudi Arabia and Louisiana. However, these incidents had been handled as maintenance issues, and no root-cause investigations were done. In hindsight, there were common causes between these incidents and the Deer Park incident. In the prior cases, luck prevailed, and the shafts were not blown clear of the valve body.

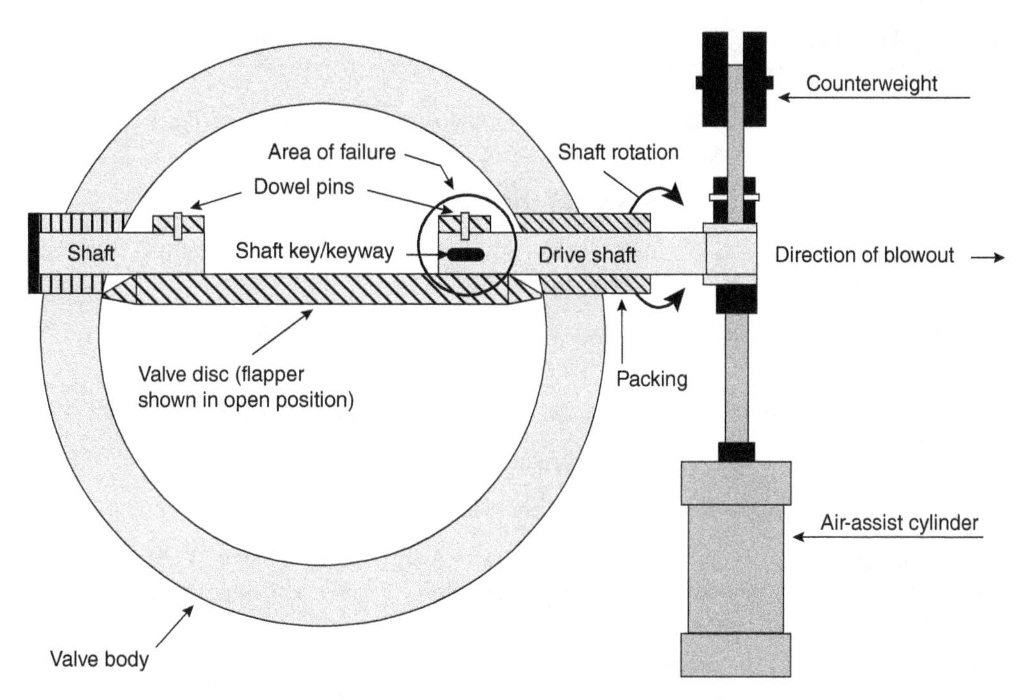

Figure 8.35.20 Simplified cross-sectional view of check valve, illustrating the location of dowel pins (Flow direction is on the page). Annotation by US EPA and OSHA Investigation Team. Source: Joint EPA/OSHA Investigation Report of the Shell Chemical Olefins Plant explosion, Deer Park, Texas, on 22 June 1997.

Key Lessons Learned

This incident shows us how a very small component, like dowel pins installed in check valves, can result in a catastrophic incident if not designed and installed properly. The post-incident analysis determined that the dowel pins used to retain the shaft in place were manufactured from carbon steel and used in a hydrogen-rich environment, leading to hydrogen embrittlement. This, plus additional stress from the check valve cycling open to close, rapidly weakened the dowel pins, resulting in their failure.

The report also stated that the check valve was subjected to additional stresses, such as valve "slamming" due to compressor surging (flow reversal) and intermittent compressor trips. This places additional stresses on the check valve and the internal components, including the dowel pins, and may have contributed to the failure. The report recommended changes to the operating procedures to minimize the potential for compressor surging or trips. For example, this included having the operator verify the position of the check valve before starting the compressor to avoid rapid action by the pneumatic check valve actuator.

The EPA/OSHA report also highlighted the value of good communication of incidents and the sharing of lessons learned between manufacturing sites. The report highlighted the value of sharing incidents with trade organizations, the trade organizations sharing with member companies, and individual companies sharing lessons learned with other companies. Had the lessons learned from the previous incidents in Saudi Arabia and Louisiana been shared with Deer Park, and had those lessons learned been acted on, it is possible that this incident would not have happened.

Additional References

API 570: "Piping Inspection Code: Inspection, Repair, Alteration, and Rerating of In-Service Piping Systems"
API 598: "Valve Inspection and Testing"
RP 574: "Inspection of Piping, Tubing, Valves and Fittings"
ASME B31 and B31.3: "Code for Pressure Piping"
ASME B16.34: "Valves - Flanged, Threaded, and Welding End"

The American Society of Mechanical Engineers (ASTM), Two Park Avenue, New York, NY, Continuing Education and Development, Inc., 9 Greyridge Farm Court Stony Point, NY 10980.

Process Piping Fundamentals, Codes, and Standards. https://www.cedengineering.com/userfiles/Process%20Piping%20Fundamentals,%20Codes%20and%20Standards%20%20-%20Module%201.pdf

Jerguson® Safety Ball Check Valve Commissioning video, Clark-Reliance Corporation, Strongsville, OH.

Jurguson® provided. as a courtesy, the following two videos that will provide additional information on Jerguson® Safety Ball Check Valves. https://www.youtube.com/watch?v=JuEgCdFTieg&t=13s

This video shows the standard commissioning procedure for glass gages with traditional Jerguson® Safety Ball Check Valves. https://www.youtube.com/watch?v=HeLZJj_imAs

This video details the Jurguson® 360 series Safety Ball Check Valve and includes the commissioning procedure.

Machinery Vibration Video, BETA Machinery Analysis. https://www.youtube.com/watch?v=B87gPJypUI0

Wood, Video; Examples of piping and small-bore vibration. https://www.youtube.com/watch?v=ldG_fmN_ENs

US Environmental Protection Agency (EPA), US Occupational Safety and Health Administration (OSHA), EPA/OSHA Joint Chemical Accident Investigation Report, Shell Chemical Company, Deer Park, Texas, 22 June 1997. https://archive.epa.gov/emergencies/docs/chem/web/pdf/shellrpt.pdf

U.S. Chemical Safety and Hazard Investigation Board, Final Investigation Report: Formosa Plastics Vinyl Chloride Explosion, April 23, 2004. https://www.csb.gov/formosa-plastics-vinyl-chloride-explosion/

U.S. Chemical Safety and Hazard Investigation Board, CSB Safety Training Video: Formosa Plastics Vinyl Chloride Explosion, 23 April 2004. https://www.youtube.com/watch?v=gDTqrRpa_ac

Chapter 8.35. Small Piping Guidelines for Operations Safety

End of Chapter Quiz

1 What is considered as 'small-bore piping' on most process units?

2 The cause of many process safety incidents with catastrophic or near-catastrophic consequences often starts as a loss of containment from the failure of _____ _______ or _______ _______ connections.

3 Do most Operators fully realize the consequences that can result from the failure of small-bore piping?

4 What are some examples of the source of excess vibration in small-bore piping and fittings? Name all you can.

5 Piping design should consider adequate support from ________ and _________.

6 Excess vibration is usually from rotating equipment such as pumps or compressors but can be from other reasons. A chattering relief valve or a control valve can present severe vibration and is most often an indication of a _______ issue with the relief valve or the control valve and requires prompt attention.

7 What examples of small-bore piping hazards have resulted or could result in major incidents? Name all you can.

8 Is bracing or gussets required for integrally reinforced valves or pipe fittings?

9 Thinking about small-bore piping, what is meant by the term "moment arm" or leverage?

10 What is the hazard with Christmas tree piping?

8.36

Overpressure Protection (All About Relief Valves and Rupture Disks)

Relief Valves come in a variety of types and sizes, but there is a common denominator. They are all our last line of defense and can save our lives. What is coming in this chapter? We will discuss some basic terminology associated with overpressure devices. We will briefly cover some basic code requirements for overpressure protection and discuss the types of safety relief devices and some basic design contingencies. We will also discuss safety valve chatter, what causes chatter, and what causes it. We'll also talk about what you should look for in the field regarding overpressure protection. Please refer to Figure 8.36.1 for a chart highlighting PRV characterization (characteristics of the system or vessel and characteristics of the PRV).

Some common PRV terminology includes:

- Operating pressure.
- Design pressure.
- Maximum Allowable Working Pressure (MAWP).
- Set pressure.
- Accumulation.
- Overpressure.
- Blowdown.

The PRV terminology will be described below using Figure 8.36.1 as a guide. This chart illustrates the relief valve characterized as a function of the pressure valve's maximum allowable working pressure (MAWP).

The following are examples of the various types of pressure relief valves.

- The direct acting type:
  - The oldest and most common type of pressure relief valve.
  - Kept closed by a spring or weight to opposing lifting force of the process pressure.
- Pilot-operated type:
  - Maintained closed-by-process pressure.
  - Also opens by process pressure.
- Buckling Pin or Rupture Pin Valves:
  - Kept closed by rupture pin (requires replacement after rupture).
- Pressure/Vacuum Vent Valves:
  - Protects against pressure and vacuum (primarily used on storage tanks).
- Rupture Discs:
  - Frangible disk (requires replacement after rupture).

See Figure 8.36.2 for images of relief valves and relief devices. We will cover these in greater detail later in this chapter.

Operating Pressure

As indicated on the chart, operating pressure is any pressure at a practical pressure level, providing the pressure remains within the prescribed operating range and below the MAWP.

Process Operations Safety: The What, Why, and How Behind Safe Petrochemical Plant Operations, First Edition. M. Darryl Yoes.

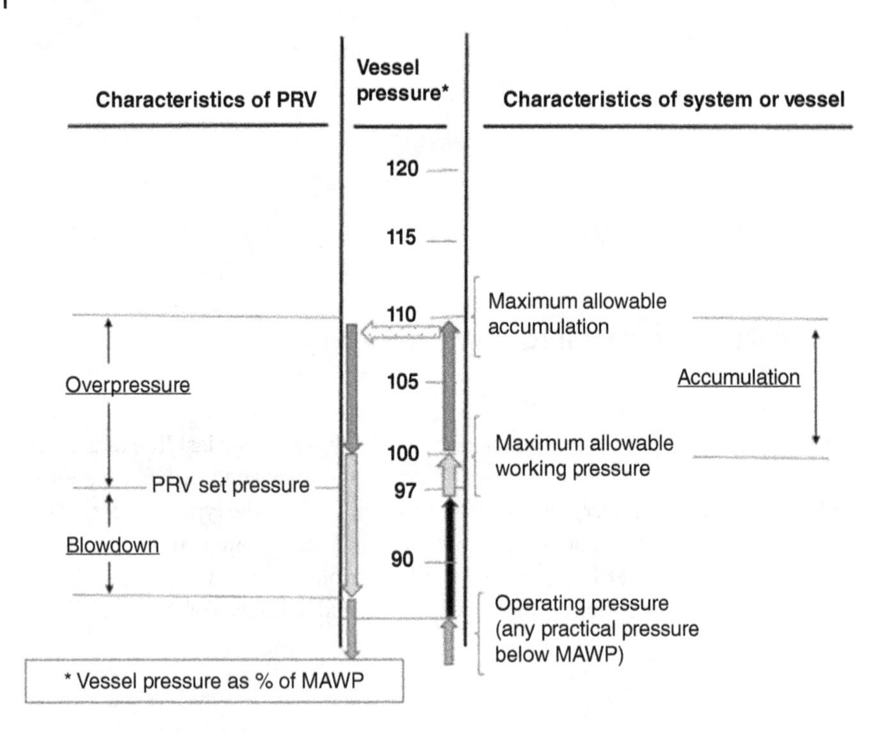

Figure 8.36.1 Pressure relief valve characterization chart.

Design Pressure vs. Maximum Allowable Working Pressure

There are many design codes for pressure vessels, including overpressure protection. We will not attempt to cover all here, but the most prominent are listed below:

- ASME Boiler and Pressure Vessel code (ASME Section VIII, Division 1)
- ASME B31.3/Process Refinery Piping
- ASME B16.5/Flanges and Flanged Fittings

According to ASME Section VIII Division 1, the definitions for Design Pressure and MAWP are as follows (VIII-1 App 3). (Note that these codes are subject to change. Always ensure your engineer has the latest versions of the design codes):

Design Pressure

"The pressure used in the design of a vessel component together with the coincident design metal temperature, for the purpose of determining the minimum permissible thickness of physical characteristics of the different zones of the vessel. When applicable, static head shall be added to the design pressure to determine the thickness of any specific zone of the vessel."

Maximum Allowable Working Pressure

"The maximum gauge pressure permissible at the top of a completed vessel in its normal operating position at the designated coincident temperature for that pressure. This pressure is the least of the values for the internal or external pressure to be determined by the rules of this Division for any of the pressure boundary parts, including the static head thereon, using nominal thicknesses exclusive of allowances for corrosion and considering the effects of any combination of loadings listed in UG-22 that are likely to occur at the designated coincident temperature. *It is the basis for the pressure setting of the pressure-relieving devices protecting the vessel.* The design pressure may be used in all cases in which calculations are not made to determine the value of the MAWP."

Note that the MAWP is the basis for the pressure setting of the pressure relief valves protecting the vessel. This is why we are using the MAWP as the reference point for describing the characteristics of the PRVs.

Figure 8.36.2 Images of various types of overpressure protection. The following types are covered in the figure: Figure (a) is an image of a conventional spring-loaded pressure relief valve (PRV), which by far is the most common type of overpressure protection. Figure (b) is an image of a pilot-operated PRV used typically when a conventional valve is chattering, or when rapid action is required. Figure (c) illustrates a balanced bellows PRV typically used when back pressure is present. Figure (d) shows a pressure/vacuum valve, which is typically used on an atmospheric storage tank. Figure (e) illustrates a buckling pin valve which may be used to replace a rupture disk, and allows the operator to replace the rupture pin. The final figure, Figure (f) is an image of a rupture disk, most frequently used for environmental purposes, or when rapid response to an overpressure event is required.

Critical Code Requirements

Embedded in these codes are the following requirements (Note: These were in place at the time of this writing. These are subject to change. Always ensure your engineer has the latest versions of the design codes).

- All pressure vessels subject to overpressure shall be protected by a pressure-relieving device. This excludes open-top tanks or tanks with floating roofs.
- Multiple vessels may be protected by a single relief device, provided there is a clear, unobstructed path to the device. This is generally done by ensuring any valves are fully opened and car sealed in the fully open position. More on this later.
- At least one pressure relief device must be set at or below the Maximum Allowable Working Pressure (MAWP).
- Relieving pressure shall not exceed MAWP (accumulation) by more than:
  - 3% for fired and unfired steam boilers.
  - 10% for vessels equipped with a single pressure relief device.
  - 16% for vessels equipped with multiple pressure relief devices.
  - 21% for fire contingency.

Relief Valve Set Pressure

The set pressure is the pressure set on the relief valve spring for when the relief valve begins to open under service conditions. As the pressure continues to rise, the relief valve continues to open. The set pressure is measured in pounds per square inch gauge (PSIG). The set pressure is about 10% above the operating pressure and never above the MAWP.

Relief Valve Accumulation

When the set pressure on a spring-loaded relief valve opens, the valve will begin to open. As the pressure continues to build, the valve continues to open. The pressure under the relief valve will reach a pressure higher than the set pressure on the relief valve spring. This pressure is referred to as accumulation. As stated above, depending on the relief valve service and what we are protecting against (the contingency), the design codes have limits on the amount of allowed accumulation. For example, if the only contingency we are protecting against is a fire, the accumulation allowed is 21% above the relief valve setting. Only 3% accumulation is allowed for a steam boiler. Except for steam boilers, an increase in accumulation is allowed for adding additional relief valves.

Overpressure

Overpressure is the difference between the relief valve set pressure and the maximum pressure reached during discharge. It is usually expressed as a percentage of set pressure.

Blowdown

Blowdown is the difference between the set pressure and the pressure when the valve reseats after a release. Following a release and when the vessel pressure has been brought back under control, the valve will begin closing. The valve **does not** close when the vessel pressure reaches the set pressure. The vessel pressure continues to drop until the valve reaches the blowdown pressure below the set pressure. This means that the Operators must reduce the vessel pressure below the set pressure for the valve to reseat. Blowdown is expressed as a percentage of the set pressure.

Design Contingencies

Numerous conditions in a refinery or petrochemical plant can lead to piping or vessel overpressure. For example, any of the following could result in overpressure: the inadvertent closure of a block valve, fire exposure, a check valve failure, thermal expansion in a heat exchanger, and the loss of a utility. These, and others like them, should be considered in the plant design and include provisions for overpressure protection. These are called design contingencies.

Double or Remote Contingencies

The failure of two systems or components that do not occur simultaneously or from a common cause should be considered a double or multiple contingency. Generally, these are only considered in the design if it is foreseen that a common cause could occur. For example, the failure of a control valve to fully open when the bypass valve is already fully open. Another example would be for an Operator to inadvertently close a valve that is car-sealed open.

Remote Contingency

A remote contingency is an abnormal process condition that could exceed the design conditions for pressure and temperature but has been deemed a very low probability. We typically design for these conditions only if the predicted pressures exceed the hydro test pressure (1.5 times design).

Contingency Examples

The following are examples of design contingencies (these would be covered in the base unit design and overpressure protection provided):

- Loss of containment and resulting fire.
- Utility failures include electricity, steam, instrument air.
- Electric power.
- Cooling water.
- Failure of a single valve.
- Operating errors.
- Thermal expansion (especially important in long lines).
- Runaway reaction in a reactor.

Double or Multiple Contingencies:

- The control valve failing open with the bypass valve fully open.
- Heat exchanger tube failure.

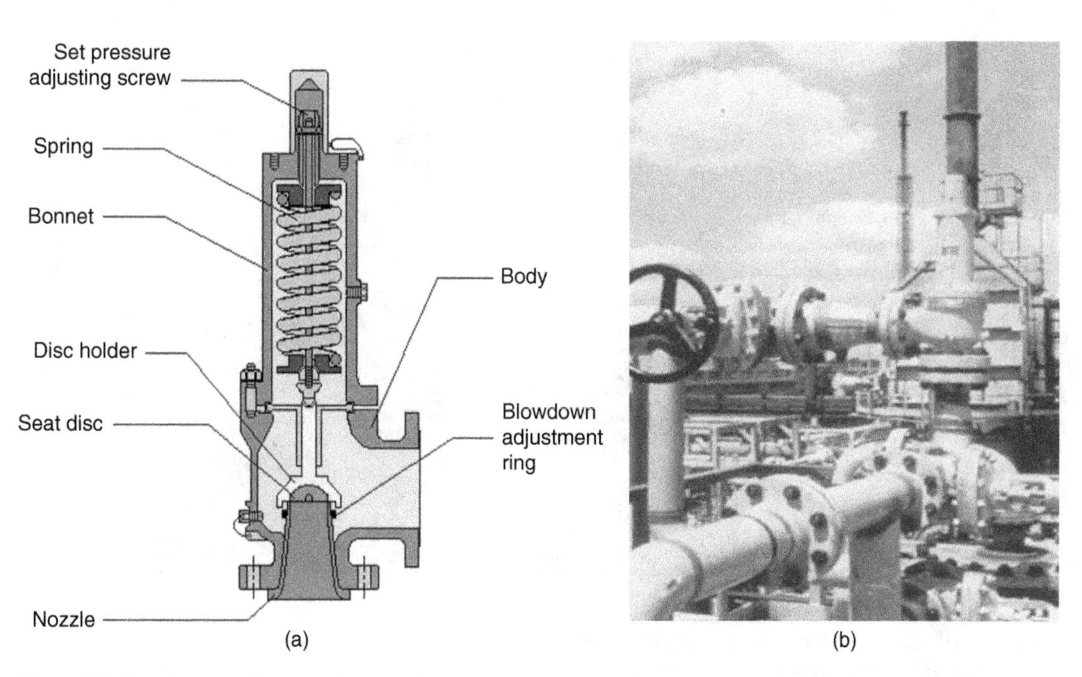

Figure 8.36.3 Images of conventional spring-loaded safety relief valve. Figure (a) illustrates the internals associated with a conventional spring-loaded relief valve; Figure (b) is an image of a conventional relief valve installed on a process unit.

- Inadvertent opening or closure of a car-sealed valve.
- The rupture disc installed upside down.

Figure 8.36.3 provides an overview of the typical conventional spring-loaded safety relief valve.

Conventional Spring-Loaded Relief Valve

The conventional spring-loaded relief valve opens when the process pressure on the seat disc overcomes the pressure from the opposing spring. When the process pressure overcomes the spring pressure, the valve opens, relieving the pressure by releasing liquid or vapor.

Advantages:

- The most reliable type is properly sized and operated.
- Versatile – can be used in many services.
- Very cost-effective as compared to other PRVs.

Disadvantages:

- Relieving pressure affected by back pressure.
- Susceptible to chatter if built-up back pressure is too high (chatter will be covered in more detail later).

Figure 8.36.4 provides an overview of the typical balanced bellows spring-loaded safety relief valve.

Balanced Bellows Spring-loaded Relief Valve

The balanced bellows spring-loaded safety relief valve appears similar to a conventional spring-loaded valve with one key exception. You will note that a bellows has been installed to encapsulate the seat disc, and the cavity above the seat disc is vented either to the atmosphere or a safe location. This is to prevent the back pressure from the relief valve from affecting the set pressure of the relief valve. Without the bellows, the back pressure is additive to the spring pressure and will affect the pressure at which the valve operates.

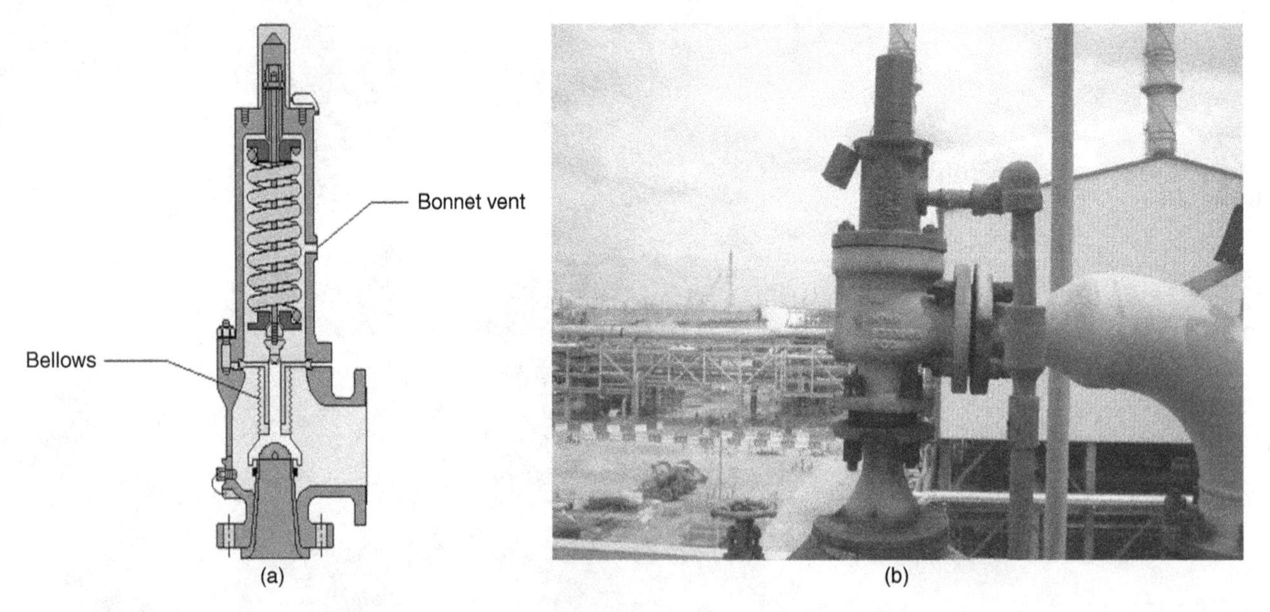

Figure 8.36.4 Images of balanced bellows spring-loaded safety relief valve. Figure (a) illustrates the internals associated with a balanced bellows spring-loaded relief valve. Please note the bellows installed immediately above the valve seat. Figure (b) is an image of a balanced bellows relief valve installed on a process unit. Note that the bellows is vented to a "safe location". More to come on this later.

Balanced bellows valves are typically specified where there is excessive fluctuation in superimposed back pressures or where the built-up back pressure exceeds 10% of the set pressure. They are also used in fouling or corrosive service since the bellows protect the spring from the process.

It is important to recognize that a balanced bellows PRV requires the bonnet vent to be always open to atmospheric pressure. If the vent is plugged or otherwise obstructed, the bellows are defeated, and the seat disc will be exposed to backpressure from the PRV outlet. Some sites identify the bellows valve in a special way to help the Operators identify the type of valve and to help ensure the vent is unobstructed. Note the paint scheme used in the photo on the right above. All bellows valves use this scheme at this site to help identify these valves. See Figure 8.36.5 for an image of a bellows from a balanced bellows spring-loaded safety relief valve.

Should this type of valve be subjected to chatter or internal corrosion, there is a possibility that the bellows could rupture or fail. The product will be released from the PRV outlet through the bellows and out the vent. This has resulted in a loss

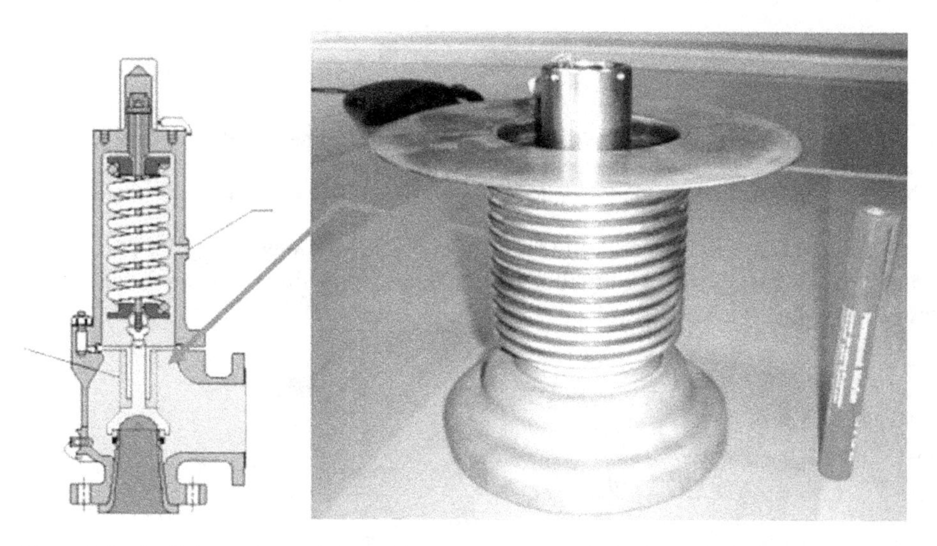

Figure 8.36.5 Bellows from a balanced bellows spring-loaded safety relief valve.

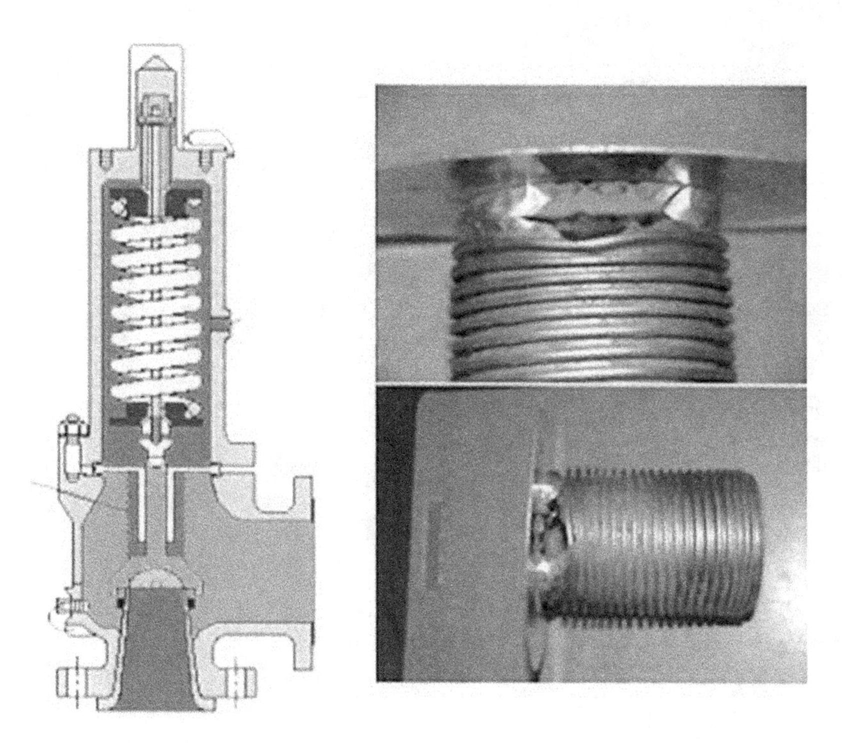

Figure 8.36.6 Failed bellows from a balanced bellows spring-loaded safety relief valve.

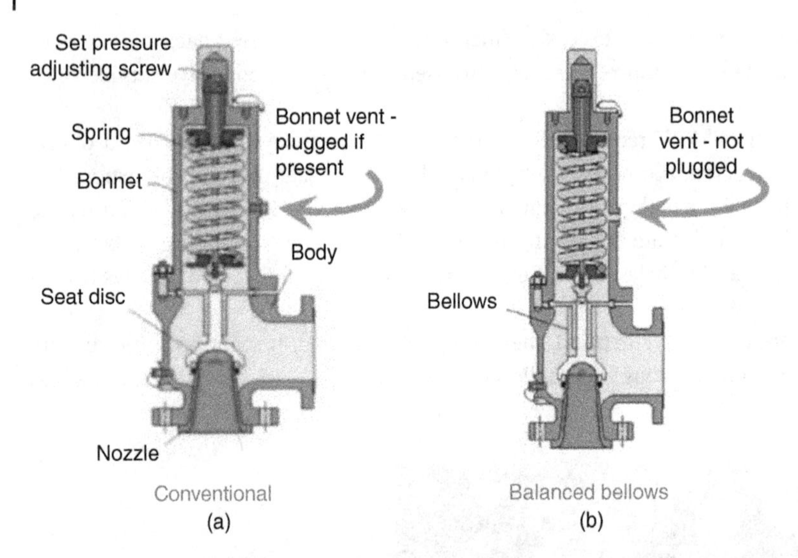

Figure 8.36.7 These images help illustrates when the vent should be plugged, and when they should be routed to the atmosphere without restriction. Please note that the vent plug is installed on the conventional spring loaded relief valve (a), while the vent on the valve with the bellows installed must have the vent unrestricted and vented to a safe location (as illustrated in b).

of hydrocarbons and/or H_2S containment to the atmosphere or to the disposition of the bellows vent piping. This is why some sites route the bellows vent high above the platform or to another safe location. See Figure 8.36.6 for an image of a failed bellows. Figure 8.36.7 illustrates when the vent should be plugged.

Advantages:

- Relieving pressure not affected by back pressure.
- It can handle higher built-up back pressure.
- Protects spring from corrosion.

Disadvantages:

- Bellows are susceptible to fatigue/rupture.
- May release flammables/toxics to the atmosphere.
- Requires a separate venting system.

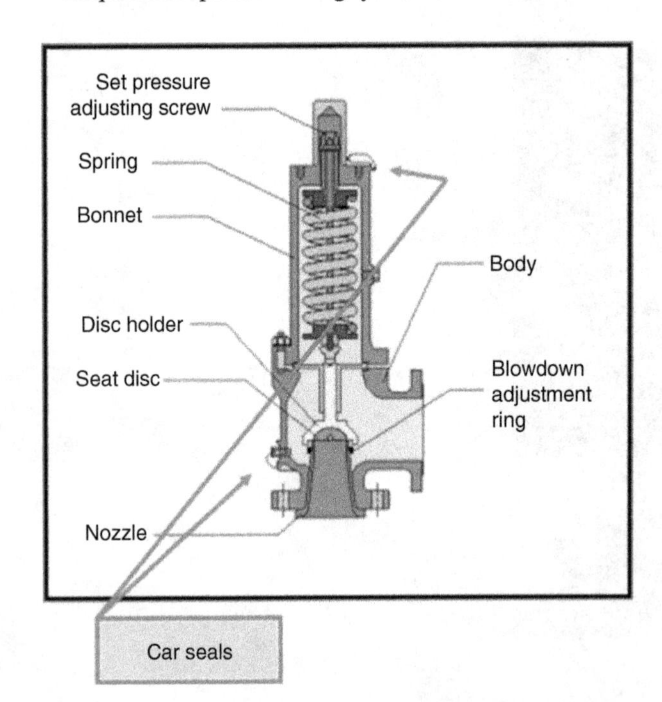

Figure 8.36.8 Set pressure and blowdown adjustments (shop adjustments only).

Valve Adjustments (Conventional and Balanced Bellows Valves)

Set Pressure (Initial opening pressure) is adjusted in the shop by rotating the adjustment screw above the spring. This screw is protected by a cap that should be car sealed and not removed in the field.

The blowdown Pressure (closing pressure) is also adjusted in the shop with a blowdown adjustment ring. This adjustment ring should also be car sealed and not removed or adjusted in the field.

Field verifications should be done periodically to ensure the car seals are in place and undisturbed. These adjustments are shown in Figure 8.36.8.

Piston-type Pilot-operated Relief Valve

The piston-type pilot-operated relief valve is not a new design but an innovative approach to overpressure protection. This valve uses process pressure to open it, but it also uses process pressure to keep it closed. So how does it work?

Note in Figure 8.36.9 that this type of pressure relief valve is fitted with a pilot valve that sits just above the relief valve. You will also note a pressure pickup from the relief valve inlet to the inlet of the pilot valve. This pressurizes the pilot valve, keeping it in the open position routing pressure to the top of the pilot-operated valve piston. This forces the relief valve to remain closed against process pressure. This applies the principle of area X pressure = force. The top of the piston is larger in diameter than the bottom. Thus, more downward force is applied with the same amount of pressure.

When the pressure on the inlet to the pilot rises above the setpoint of the pilot valve, it quickly closes, taking pressure off the main valve piston. When this happens, the main valve quickly moves from the fully closed position to fully open, releasing pressure on the drum or vessel. This is illustrated in Figure 8.36.10.

Function

- This type of valve uses process pressure to hold the valve seat in the closed position.
- When the pressure exceeds the set pressure of the pilot-operated valve, the pilot valve almost instantaneously releases the pressure on the top of the piston and the valve travels to the fully open position.
- Note the piston diameter is larger on the top than on the bottom. Pressure X Area = Force. The larger diameter results in higher force on the top of the piston, holding the valve in the closed position.

Advantages:

- Relieving pressure is not affected by back pressure.
- It can operate at up to 98% of set pressure.
- Less susceptible to chatter (some models) without significant simmering.

Figure 8.36.9 Image of a piston pilot-operated pressure relief valve.

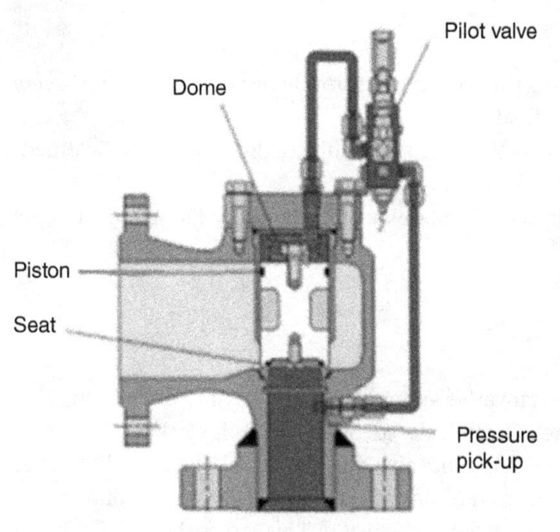

Figure 8.36.10 Illustration of the function of the pilot-operated pressure relief valve.

Disadvantages:

- The pilot valve is susceptible to plugging.
- Limited chemical and high-temperature use by "O-ring" seals.
- Vapor condensation and liquid accumulation above the piston may result in fouling or corrosion.
- Potential for backflow (can be corrected by the installation of check valves in the pickup lines).

Note: The inlet pressure sensing line may extend to the process vessel. If this valve is inadvertently closed or the line becomes clogged, it may result in the release of the pressure on the top of the piston. This will result in the PRV going to the fully open position. These valves should be car sealed in the fully open position and painted a distinctive color.

Buckling Pin Relief Valves – Also Called Rupture Pin Valves

Buckling Pin Valves are a new type of nonreclosing pressure relief device that utilizes a buckling pin instead of a spring or rupture disc to relieve pressure. Several features of the Buckling Pin valve (BPV) make it a candidate to replace rupture disks in selected services:

- The BPV is not subject to premature failure.
- It can be more accurate than rupture disks at low pressures (below 40 psig).
- The BPV can be reset in minutes without tools and without breaking flanges.
- BPV design can be balanced or unbalanced (similar to conventional or bellows relief valves). The balanced valve is less affected by back pressure.

See Figure 8.36.11 for an image of a typical buckling pin valve.

Figure 8.36.11 An image of a typical buckling pin valve.

Figure 8.36.12 Image of a typical pressure/vacuum valve.

Pressure/Vacuum Vent Valves

Pressure/vacuum Valves (PVV) are used on storage tanks and similar process vessels to protect against damage from overpressure and vacuum. Due to the potential for fouling, they should not normally be used in fouling services, such as asphalt or heavy waxy products. In some services, heat tracing may prevent fouling or plugging.

These valves utilize one or two either weighted or spring-loaded plates, typically one at the top of the valve that protects against overpressure and one to the side that protects against vacuum.

PVV Valves should be inspected periodically, typically at least annually, to ensure they are not fouled or plugged.

Never block or obstruct a PVV Valve without approved authorization for disablement, including mitigation (such as a Control of Defeat). Overpressure or collapse of the vessel can occur if the PVV valve is blocked or otherwise obstructed.

Figure 8.36.12 is an illustration of a typical pressure/vacuum valve.

Relief Valve Chatter

Relief valve chatter is caused by a PR Valve's rapid, alternating opening and closing. This can be a very loud and damaging scenario with a lot of vibration and may cause misalignment and valve seat damage and, if prolonged, can cause mechanical failure of the valve internals and associated piping.

I have seen situations where the chatter has resulted in the loosing of bolts and loss of containment and a resulting fire. Opening a flange also has the potential for vapor release, toxic release, and an explosion. Relief valve chatter may occur in either liquid or vapor services.

Caution – Relief valve chatter may cause violent vibration of the valve and indicates that a potentially serious design issue may exist. PRV chatter should be reported immediately.

Causes of relief valve chatter:

There are several causes of relief valve chatter. All are caused by or related to design issues.

The first and the most common is excessive inlet pressure drop in PRV inlet or outlet piping. Note in Figure 8.36.13 that a wedge plug-type valve is installed in the inlet piping to this relief valve. The restriction caused by the valve wedge will result in a pressure drop (delta pressure or dp) when the relief valve is flowing.

In this scenario, when the valve opens, the flow accelerates, and the pressure under the seat quickly drops. As soon as the flow increases, the dp across the wedge increases, and pressure under the seat decreases, causing the valve to close. The pressure under the seat rapidly increases as soon as the valve closes. This rapid flow and pressure fluctuation causes the relief valve to cycle rapidly between fully open and fully closed, slamming the seat disc against the valve seat each cycle. This results in the valve cycling rapidly between open and closed, causing the valve chatter. This cycling is so fast that I have heard this described as sounding like a submachine gun firing.

Other causes of relief valve chatter include excessive built-up back pressure, an oversized valve, and a relief valve handling widely varying rates. A PRV must flow at least 25% of its capacity to ensure the valve remains open; an undersized valve will most likely chatter. These are all design issues and should be addressed promptly.

Figure 8.36.13 A conventional spring-loaded relief valve with a wedge plug-type isolation valve was installed in the inlet. Note: A wedge plug valve has a tapered plug, which results in a pressure differential. This pressure differential can lead to valve chatter.

Potential Chatter Solutions

Avoid excessive pressure drop in PRV inlet and outlet piping by avoiding long piping runs in inlet and outlet piping and avoiding excessive piping bends or elbows. If the piping must have long runs or excessive bends, increase the piping diameter to reduce the differential pressure.

Other alternatives, if you are unable to replace the piping, include increasing the PRV blowdown to help keep the valve open longer or installing a smaller PRV (if the PRV valve is oversized). Another alternative is to install a different type of PRV (such as a pilot-operated valve), which is much more resistant to chatter.

Rupture Discs

A rupture disc is a thin diaphragm (generally a solid metal disc) designed to rupture (or burst) at a designated pressure. It is used as a weak element to protect vessels and piping against excessive pressure (positive or negative). Sometimes they may be used as the primary pressure relief device, especially in rapid pressure rise situations like runaway reactions when a pressure relief valve cannot respond quickly enough. See Figure 8.36.14 for an image of a rupture disc installed in a piping system. Rupture Disc must be installed correctly in the direction of flow. It is a good idea to verify this during field rounds. Figure 8.36.15 is an image of a typical data plate for a rupture disk (note this example is for an Asian refinery).

However, they can also be used in conjunction with a pressure relief valve to provide corrosion protection for the PRV (in corrosive services such as acid or caustic). They can also help prevent the loss of highly toxic or very expensive process chemicals or reduce fugitive emissions to meet environmental requirements. The rupture disc is normally placed directly

Figure 8.36.14 An image of a rupture disc installed in a piping system.

Figure 8.36.15 Example of a Rupture disc with data tab. The image in (a) illustrates the rupture disc data tab (an European version is displayed here), while the rupture disc is illustrated in (b).

below a conventional pressure relief valve in these cases. Note that the direction of flow is shown on the data plate. This is very important and should always be verified during installation.

When comparing a rupture disc to conventional pressure relief valves, they have some unique advantages but several disadvantages.

Rupture disc advantages:

- No simmering or leakage prior to bursting.
- More effective against overpressure from deflagrations or heat exchanger tube rupture.
- Less expensive to provide corrosion resistance.
- Less tendency to foul or plug.
- Provide both overpressure protection and depressuring.
- Reduce fugitive emissions when required by local authorities.

Rupture disc disadvantages:

- Do not reclose after relief.
- Requires full depressuring.
- Burst pressure cannot be tested.
- Require periodic replacement.
- Greater sensitivity to temperature issues and mechanical damage.

- Not as accurate as PRVs.
- Never "sure" because cannot "pop" test.

What Has Happened/What Can Happen

Ethylene Release and Fire
Kuraray America, Inc. EVAL Plant
Pasadena, Texas Incident Date: 19 May 2018
Investigated by the US Chemical Safety and Hazard Investigation Board

This incident occurred when the unit was undergoing a startup following a regularly scheduled maintenance turnaround. During the startup, a high pressure developed in the reactor, and the pressure relief valve activated, releasing Ethylene into the open area adjacent to the process vessel. Refer to Figure 8.36.16 for an image of the relief valve atmospheric vent location.

Plant workers were performing maintenance activities unrelated to the reactor startup, including welding. The CSB concluded that the most probable cause of the Ethylene ignition was the welding, as it was in the direct downstream area from the safety valve discharge. The fire was extinguished when the reactor pressure dropped below the PRV set pressure, and the valve closed.

Several workers caught in the fire attempted to escape, including some who jumped from the second or third story of the structure and ran to safety, some falling as they escaped. Others were in fall protection harnesses with their lanyards attached to the structure, delaying their escape and resulting in more severe injuries.

Two workers were life-flighted to a local hospital for treatment. Emergency responders transported as many as 19 other injured workers for medical treatment.

The US CSB concluded that the cause of the incident was Kuraray's long-standing emergency pressure-relief system design that discharged flammable ethylene vapor through horizontally aimed piping into the air near the workers' work. The CSB also concluded that had the pressure relief system discharged to a safe location, for example, near the top of the structure or to an enclosed unit blowdown system or flare, the discharge of flammable gas would not have harmed the workers. See Figure 8.36.16 for images of the Ethylene PRV atmospheric vent. Notice the orientation of the vent horizontally and directed into the adjacent open area. This is where the maintenance crew was welding at the time of the incident.

Figure 8.36.16 Image of the Kuraray America Ethylene PRV atmospheric vent. Source: Photo courtesy of the US Chemical Safety and Hazard Identification Board.

What Has Happened/What Can Happen

Williams Olefins Plant
Geismar, Louisiana
Incident date: 13 June 2013 (2 fatalities and 167 injured).

The Williams incident was investigated by the US Chemical Safety and Hazard Investigation Board. It is also reported in more detail in Chapter 8.14 of this book.

According to the CSB, three main factors in combination with one another contributed to the incident:

- The unexpected presence of liquid hydrocarbons in the reboiler in standby mode.
- Heat is introduced into this standby reboiler.
- The pressure relief system (PRV) was isolated from the reboiler, which was in standby mode.

The exchanger failed catastrophically due to internal thermal expansion.

Please refer to Chapter 8.14 for a quick review of this tragic incident involving the overpressure of a heat exchanger and two fatalities as a direct result of the lack of overpressure protection.

What Has Happened/What Can Happen

Goodyear Chemical Plant
Houston, Texas
11 June 2008 (One fatality, six injuries)

An incident almost identical to the one at Williams Olefins occurred at Goodyear Chemical Plant in Houston on 11 June 2008. In that incident, one person was killed, and six received injuries.

The exchanger at Goodyear was protected by a ruptured disk and a relief valve. The rupture disk was installed below and in line with the relief valve. However, both were blocked the day before the incident to allow maintenance personnel to replace the ruptured disk.

The following day, a Goodyear employee started the process of returning the exchanger to service, connected a steam hose, and started steaming the hot side of the exchanger (with the cold side blocked in and the relief valve and rupture disk still isolated).

This incident is also investigated by the CSB and is reported in Chapter 8.9.14 in more detail.

What Has Happened/What Can Happen

A Refinery in the Midwest US
April 2002

A fire occurred at the FCC, resulting in an unplanned FCC shutdown and two minor injuries.

This incident resulted from a release from the seven FCC main column relief valves and autoignition of the released vapors. The fire was of short duration. However, two Operators received noncritical injuries, and the unit was shut down for about twelve hours due to the incident.

Seven pilot-operating atmospheric relief valves protect this FCC main column from overpressure. The sensing lines from all seven pilot valves merge into one common column pressure sensing line and include an isolation valve located in the structure, remote from the pilot-operated pressure relief valves. This isolation valve has not designated a car seal valve nor car sealed on the day of the incident. The sensing line is also equipped with a Nitrogen purge to ensure it does not foul or plug.

On the day of the incident, some ongoing operational issues were associated with the FCC main column. The engineer was troubleshooting the tower to confirm the issues with the column and requested the Operators to conduct a pressure survey of the FCC main column. This is not a common task for the Operators; therefore, there were several involved and were attempting to locate the pressure monitoring locations. They decided to isolate the Nitrogen from the pressure sensing line and take the reading there. As soon as they closed the pressure-sensing header isolation valve, multiple relief valves opened to the atmosphere. This released a significant amount of hot hydrocarbon vapor directly on top of the main column to the atmosphere, where it promptly ignited from autoignition or from hot surfaces. Immediately after the release occurred, the Operator realized that the release occurred as soon as the pressure sensing line was closed, and the valve was reopened. This quickly resulted in the pilot-operated valves reclosing.

Due to the fire on the top of the main column, a signal was called to take the unit offline, and it was quickly shut down. It was offline for about 12 hours before being returned to service. The unit sustained minor fire damage on the top of the main column.

The investigation concluded that the root cause of this incident was a general lack of understanding of the operation of the remote-operated pilot pressure relief valves. More detail was needed in the FCC procedures and additional focus on the car seal valve program. The pilot sensing lines were added to the car seal valve program and were car sealed.

A Refinery in the United Kingdom

Vacuum Damage to a Cone Roof Storage Tank

This new internally lined cone roof storage tank was built at a UK refinery and lifted into place, complete with insulation. The tank was successfully hydrotested and blinded to be left empty over the Christmas Holidays. The site had recently adopted an updated blinding/isolation procedure, which required 100% blind all connections on enclosed pressure vessels. The tank was blinded according to this new procedure, including a slip blind below the pressure/vacuum vent.

Over the holidays, the weather changed when a cold front came through, dropping the temperature. The resulting change in the weather resulted in a vacuum forming in the tank resulting in shell distortion. See Figure 8.36.17 for images of the damaged storage tank.

In my career, I have seen several tanks where this has occurred. I was driving in to work early in my career, and my route took me by a large company-owned and operated tank farm (approximately 100 large storage tanks). As I passed by, I noticed one of the tanks had caved in, and I quickly notified the Refinery Superintendent. As it turns out, the story was almost the same. The tank had just been taken out of service for cleaning; all lines were blinded, including the pressure/vacuum vent; however, no manway plates had been removed. A change in atmospheric pressure resulted in the tank damage. Minimal changes in pressure, even atmospheric pressure, can affect these tanks due to the massive surface area involved.

Key Lessons Learned

- PRVs are the “last line of defense” and must be maintained and operable.
- Is there an effective process to ensure that CSO/CSC valves are properly managed? Is it being audited? See Chapter 8.34 for a discussion on the importance of a car seal valve management plan.
- Ensure all isolation valves associated with PRVs are maintained in the car seal program and are car sealed to prevent inadvertent opening or closing.
- Ask yourself these questions:
  - Do you know how many PRVs are overdue for inspection?
  - Are you doing inspection/analysis (“as received”) and adjusting inspection intervals based on inspection results?
- Ensure PRVs are evaluated during the PHA process (including safe routing of discharge).
- Recognize that tank Pressure/Vacuum vents are also critical and must be inspected periodically and never blinded or blocked away from the tank without authorization and a defeat plan.

Additional References

Industry Design Standards *(not a complete list):*

- National Board of Boiler and Pressure Vessel Inspectors, “National Board Inspection Code.”
- ASME Code Section I, Parts PG-67 thru PG-73, “Safety Valves and Pressure Relief Valves.”
- The American Society of Mechanical Engineers (ASME). Guidebook for the Design of ASME Section VIII Pressure Vessels, Fourth Edition

(a)

(b)

Figure 8.36.17 Images of a new cone roof storage tank damaged by vacuum due to defeating the pressure/vacuum vent. Figure (a) is a view of the atmospheric storage tank from a side view, while Figure (b) is a view looking directly up the side of the tank shell. The damage can be seen from either view.

- ASME Code Section VIII, Div. 1, Parts UG-125 thru UG-136, "Pressure Relief Devices."
- ASME Code Section VIII, Div. 1, Appendix M, "Installation and Operation."
- ASME B31.3/Process Refinery Piping
- ASME B16.5/Flanges and Flanged Fittings
- API Standard 510, "Pressure Vessel Inspection Code"

- API Recommended Practice 520, Parts I and II, "Sizing, Selection, and Installation of Pressure Relief Devices in Refineries."
- API Recommended Practice 521, "Guide for Pressure-Relieving and Depressurizing Systems."
- API Standard 526, "Flanged Steel Pressure Relief Valves."
- API Standard 527, "Seat Tightness of Pressure Relief Valves."

Most companies also have internal design standards and practices covering overpressure protection.

Chapter 8.36. Overpressure Protection (Relief Valves, Rupture Disks)

End of Chapter Quiz

1. Name the major types of pressure relief devices.
2. What are the seven most common terminologies used for pressure relief devices?
3. How would you describe operating pressure?
4. What is set pressure?
5. What is the key difference between design pressure and the maximum allowable working pressure?
6. Is the design pressure or the MAWP used for setting the relief valves on the vessel?
7. How does a conventional spring-loaded safety valve function?
8. What is the difference between a conventional spring-loaded valve and a balanced bellows relief valve?
9. After a conventional spring-loaded valve relieves, the seat will reclose when the pressure returns to the set pressure.
   - True
   - False
10. What is common between the buckling pin relief valve and the rupture disc?
11. What is critical about a pressure/vacuum vent?
12. All pressure vessels subject to __________ shall be protected by a __________ __________ __________.
13. Multiple vessels may be protected by a single relief device, provided there is a _______ and __________ path to the device.
14. Which of the following could result in overpressure?
    a. The inadvertent closure of a block valve.
    b. Fire exposure.
    c. Loss of feed.
    d. A check valve failure.
    e. Thermal expansion in a heat exchanger.
    f. The loss of a utility.
15. What effect would it have on the unit if a pressure sensing line to a pilot-operated relief valve was inadvertently closed?
16. What are three reasons why a rupture disc may be installed on your unit?

8.37

Special Hazards of Rotating Equipment

As I was wrapping up this book, I realized that I was so focused on hazards associated with the loss of containment that I had almost forgotten about some of the special or unique but serious hazards associated with rotating equipment. There are primarily three very specific hazards associated with rotating equipment that I felt should be covered in some detail. This is the (1) Reverse Rotation of pumps and drivers, (2) Steam Turbine Overspeed, and (3) Compressor Surge. I will cover these and include several other significant issues with rotating equipment in this chapter.

Fortunately, these three types of incidents do not happen with regularity. However, when they do occur, the results can be catastrophic. Operators should be aware of these hazards and alert for how to recognize when they may be occurring and how to respond. They should be alert for when and how they can help prevent their occurrence in the workplace.

Pump and Driver Reverse Rotation and Reverse Overspeed

Reverse rotation of pumps and drivers most commonly occurs when two or more pumps are operated in parallel. Reverse rotation can also result in catastrophic reverse overspeed and destruction of the driver, usually a motor, but it could be a steam turbine or a gas turbine. The most common cause of reverse rotation is caused by failure of the pump discharge check valve resulting in the reverse process flow through the idle pump from the operating pump, potentially causing an overspeed condition.

Pump reverse overspeed can damage the driver (motor, steam turbine, or gas turbine) and/or the pump. This may also result in the release of flying shrapnel with the potential for serious injury to personnel. Generally, a pump that has experienced reverse impeller rotation occurs due to internal failure of the pump discharge check valve.

A check valve is usually installed in the discharge line to prevent a reverse flow of liquid through a centrifugal pump. Submerged pumps may also be equipped with anti-reversing rachets in the motor to prevent reverse rotation. Reverse overspeed potential exists when a pump is operated in parallel with one or more other pumps or where the pump is subject to backflow from another pressurized source, such as a pressurized reactor, fractionator, or similar process vessel.

When a pump is idle (or has tripped), AND the discharge check valve fails, the process fluid flows backward through the idle pump, making the impeller behave like a turbine causing the pump to rotate in reverse and potentially overspeed.

There are two most common types of check valves used on pump discharge piping. A typical flapper-type check valve with an internal swing check plate is the most common. The check plate uses the flow and pressure from the downstream side to close the flapper, preventing reverse flow. Another type is the spring-loaded flapper check valve which uses the force of the opposing spring to keep the check valve in the closed position.

See Figure 8.37.1 for an illustration of a typical pump installation with a check valve installed in the pump discharge. Figure 8.37.2 shows the two most common check valve designs used in industry.

Typical check valve failure modes that can allow reverse flow include:

- Worn or broken hinge pin or flapper shaft (due to wear, the flapper can drop in the valve body).
- Loose or broken internal bolts or nuts.
- Seat failure (The threaded seat can back out and drop into the valve body, blocking the flapper in the open position).
- Trash or debris obstructing the flapper from closing.
- Broken or worn springs (flapper-type check valves).

Process Operations Safety: The What, Why, and How Behind Safe Petrochemical Plant Operations, First Edition. M. Darryl Yoes.

Figure 8.37.1 Illustration of a typical pump installation with a check valve installed in the pump discharge.

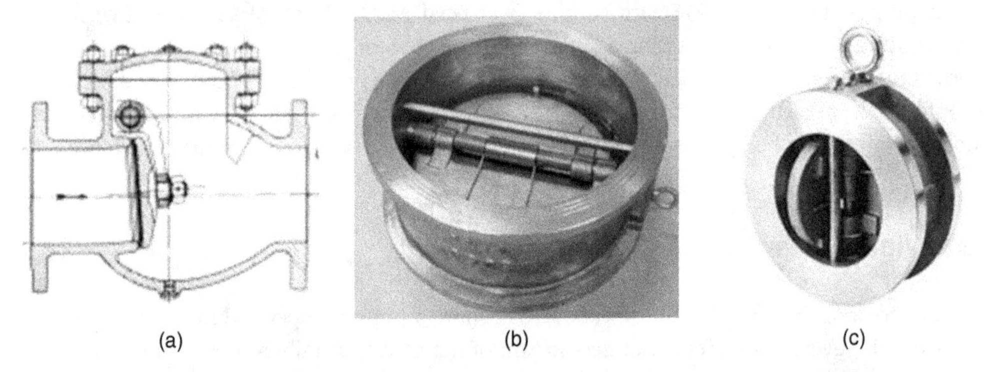

Figure 8.37.2 (a–c) Illustration of the two most common check valve designs. Figure (a) illustrates the flapper type check valve which opens to allow flow in only one direction, and is designed to close when the flow is reversed. Figures (b and c) shows the body and internals of a spring loaded in-line chek valve. This valve is assisted to close by spring tension to prevent reverse flow.

When starting a centrifugal pump:

Most pump manufacturers recommend starting centrifugal pumps against a closed or slightly open (throttled) discharge valve. This procedure prevents internal damage to the pump caused by cavitation; however, it also prevents backflow through the pump and reduces the potential for reverse overspeed if the discharge check valve fails internally.

Starting one pump in a parallel pump operation may result in reverse overspeed in one of the parallel pumps if the block valves are in the open position and the check valve has failed on the parallel pump.

It is a good idea to keep the discharge valves closed on pumps that are idle unless the pump is equipped with an auto-start or auto-kick-in device. When starting a pump, always ensure that the discharge valve on the pump being started and other pumps discharging into the same system are closed or throttled.

If you experience reverse rotation – immediately stop the reverse flow by stopping the pump that is in service. If possible, this should be done from a remote location to help avoid the line-of-fire with the plane of the impeller in reverse rotation. When the pump experiencing the reverse rotation slows, close the discharge valve to stop the reverse flow. In some cases, the pump supplying the energy is critical to the operation; stopping the pump may trip the unit. This procedure is critical to keeping the Operator out of the line-of-fire of the pump in reverse overspeed.

Always report any indication of reverse flow through a pump or indication of a driver (motor or turbine) turning in reverse rotation. In almost all cases, the issue is a failed check valve, and the pump should be removed from service until the check valve is repaired.

Some sites have installed panel indications of rotating equipment speed (remote reading tachometer) with alarms to detect and alert operators of potential reverse overspeed. The visual indication of pump speed is available from a remote location for these pumps. Also, in some cases, the pump discharge valves are motor-operated and can be closed remotely, keeping the Operator out of the line-of-fire.

I prefer to see a procedure readily available for the Operators to follow in the event of a reverse overspeed and used for periodic Operator training.

To recap the Operator's actions in the event of a reverse rotation or reverse overspeed:

- If a reverse rotation or overspeed is suspected, the pump rotation should be stopped by stopping the reverse flow.
  - If possible, trip or shut down the operating parallel pumps from a remote location.
  - If possible, close the pump discharge block valve or other valves in the pressure source to stop the reverse flow from a remote location.
  - Operators should be trained to remain out of the line-of-fire or to block flow from a remote location.
  - Centrifugal pumps should never be started against a reverse rotation or backflow condition.

Failure Analysis of Failed Check Valves

It is important to conduct a thorough failure analysis to determine the cause of the check valve failure. The root cause should be identified and corrected to prevent a reoccurrence. Consider similar actions for other check valves in parallel pumps of the same design.

The analysis should consider the design flow rate vs. actual flow conditions. The flow rate should be sufficient to maintain the check valve in the open position during normal operation. Low flow or erratic flow may cause the flapper to chatter, resulting in rapid wear of the internals and premature failure.

What Has Happened/What Can Happen

At this mid-west refining site, an Operator responding to a loss of process flow was severely injured by a piece of flying shrapnel from a motor driver, which failed while in service. As it turns out, the pump was in reverse rotation, driven by the process flow from the sister pump due to a failed check valve.

The FCC overhead pump was operating in parallel with other FCC Overhead pumps when the motor tripped due to exceeding maximum amps. When the Operator started another parallel pump to hold the OH drum level, the tripped pump went into reverse rotation. The pump discharge check valve failed, allowing reverse flow and reverse rotation when the flow reversed through the failed check valve.

Figure 8.37.3 Image of failed motor, damaged by reverse rotation.

The Operator did not realize that the pump was in reverse flow and had approached the pump while attempting to troubleshoot the cause of the loss of flow. Unfortunately, he was in the line-of-fire when the motor failed and was severely injured.

This incident illustrates that it may be difficult to tell if the pump is in reverse flow. The pump will be making all the normal bearing and fan noises and may be rotating so fast in reverse that the Operator may be unable to distinguish that it is rotating in reverse. If there is a loss of flow or discharge pressure, the possibility of pump reverse flow should always be considered. It may be possible to verify that the motor is running by checking the ammeter (if equipped), but in most cases, this will mean visiting the electrical substation. See Figure 8.37.3 for an image of the failed motor.

What Has Happened/What Can Happen

At another manufacturing site, a similar failure occurred; this time, a steam turbine driver was rotated in reverse by the reverse flow. The pump discharge check valve failed partially open, allowing backflow from other parallel pumps. The reverse overspeed drove the pump, gearbox, and steam turbine driver in reverse rotation to destruction. This incident resulted in shrapnel and metal debris being ejected over a wide area with potential personnel injury. When the pump is driving the turbine, the turbine overspeed trip, which shuts off the steam to the turbine, will have no effect. See Figure 8.37.4 for an image of the damaged steam turbine.

Figure 8.37.4 Damaged steam turbine due to reverse overspeed.

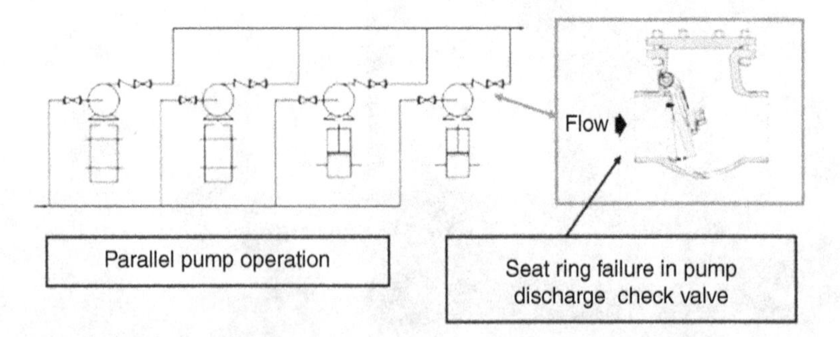

Figure 8.37.5 Pump configuration and illustration of the failed check valve.

In this incident, the pump discharge check valve failed when the internal seat ring unthreaded and dropped the seat ring into the valve body. The seat ring then held the flapper away from the seat, allowing the backflow. See Figure 8.37.5 for the pump configuration and an illustration of the failed discharge check valve.

Key Lessons Learned

- Ensure Field Supervisors and Operators are trained to recognize the potential for Reverse Overspeed and how to prevent it.
- Procedures should be developed and readily available to the Operators on the proper response to a reverse overspeed event. Procedures should be used periodically in operator training to help ensure readiness.
- Operators should also be trained in the correct response to Reverse Overspeed, especially the safe body position for stopping reverse flow (out of line-of-fire).
- Consider periodic inspection of discharge check valves for pump installations with multiple pumps operating in parallel (these have a higher potential for Reverse Overspeed). It is important to remember that Reverse Overspeed has occurred in pump installations with only two pumps operating in parallel.
- Ensure failure analysis is done for all failed check valves installed in pump discharge systems. The analysis should include verification of proper check valve sizing (for actual flow capacity).

References: (Note: The following has been provided as an additional reference only. Due to potential copyright restrictions, none of this material has been used in this book).

Power Magazine published a nice article titled "How to Prevent Circulating Water Flow Reversal." It is available at the following link.
https://www.powermag.com/how-to-prevent-circulating-water-flow-reversal/

Texas A&M also published "HAZARDS OF REVERSE PUMP ROTATION."
https://oaktrust.library.tamu.edu/bitstream/handle/1969.1/162591/Hazards%20of%20Reverse%20Pump%20Rotation.pdf?sequence=1

Centrifugal Pumps

Pumps come in all types and sizes depending on the design and required specifications for the application. The most common type of centrifugal pump on a typical process unit is the single stage overhung pump where the impeller is "overhung" outboard of the bearing housing. Other types include the double case split suction design, the multistage pump with the impeller designed between the bearings, the vertical in-line pump, and the deep well centrifugal pump, just to name the most common designs. Figure 8.37.6 illustrates some of the types of pumps you may experience on your unit.

There are many other types of pumps on the market and the following article has some great information on these. The article is titled "The Must Have Handbook for Centrifugal Pumps" by Crane Engineering and is available at this site: https://www.apprep.com/Site/images/pump_resources/centrifugal_pumps_handbook.pdf.

Pump Failure Modes

Pump failure modes include the following: bearing failure, mechanical seal failure, failure of the coupling, and pump failure at startup conditions.

A significant failure mode is the failure of the bearings or seal. These failures can quickly result in loss of containment of the product and a major fire or explosion. Seal and bearing technology have advanced significantly in the last decade or so, and these types of failures do not happen with regularity. However, when they do occur, the results can be devastating to the operation, with damage to equipment and potential injuries to personnel. Reverse overspeed is another significant cause of failure which is covered separately in the book.

The most important thing you, as an operator, can do is ensure that the lubricant is clean and that the lube levels are good. The bearings can fail quickly with inadequate lubrication, for example, with dirty lube oil or water in the lube oil. Failed bearings can lead to failure of the seal mechanism and loss of containment.

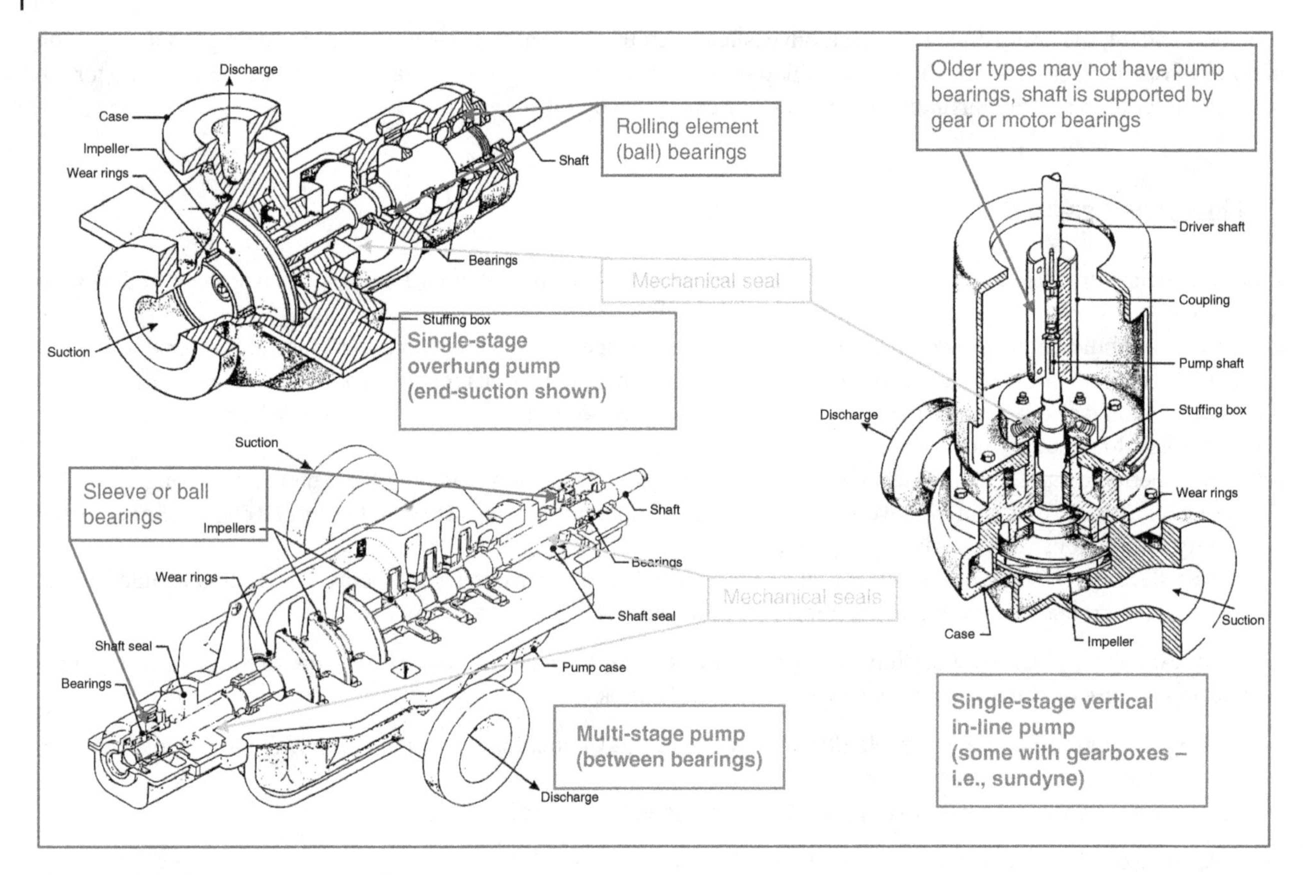

Figure 8.37.6 Several of the types of pumps you may experience on your unit.

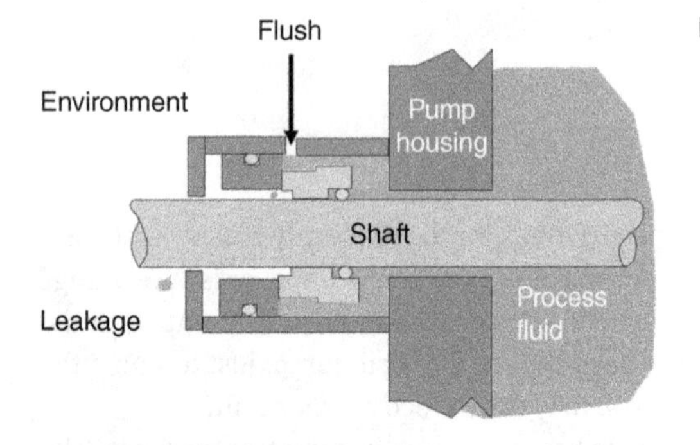

Figure 8.37.7 Basic seal design, including seal flush.

The Mechanical Seal

The mechanical seal provides a dynamic seal between the process fluid and the atmosphere. More information on mechanical seals, types, and seal designs can be found in API-682 (design, seal types, and application). Figure 8.37.7 illustrates the basics of seal design, including seal flush. The seal flush serves as a lubricant for the rotating seal face, removing heat and supplying clean product to the seal faces.

Figure 8.37.8 is an image of a cutaway mechanical seal showing the provision for seal flush. The mechanical seals have a stationary face and a rotating face, providing the seal that prevents the product from escaping into the atmosphere. You will note that there is some minor leakage around the seal. This is minimal, but all seals will have a very small amount of leakage, and this helps with lubrication and cooling of the seal faces. See Figure 8.37.8 for a cutaway image of a mechanical seal illustrating the seal flush and seal faces.

Pump Performance Curves

All centrifugal pumps are provided with a pump curve. This provides the best operating conditions for maximum pump reliability and pump life. The curve provides design conditions for pump discharge head (pressure) vs. pump flow rates.

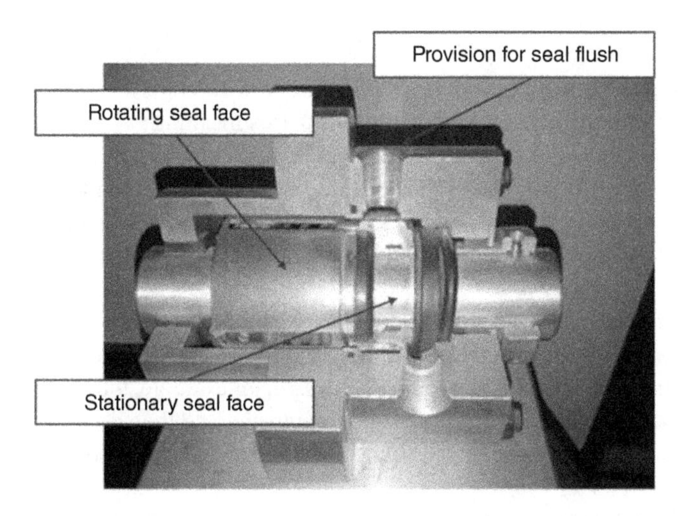

Figure 8.37.8 Cutaway of a mechanical seal illustrating the seal flush and seal faces.

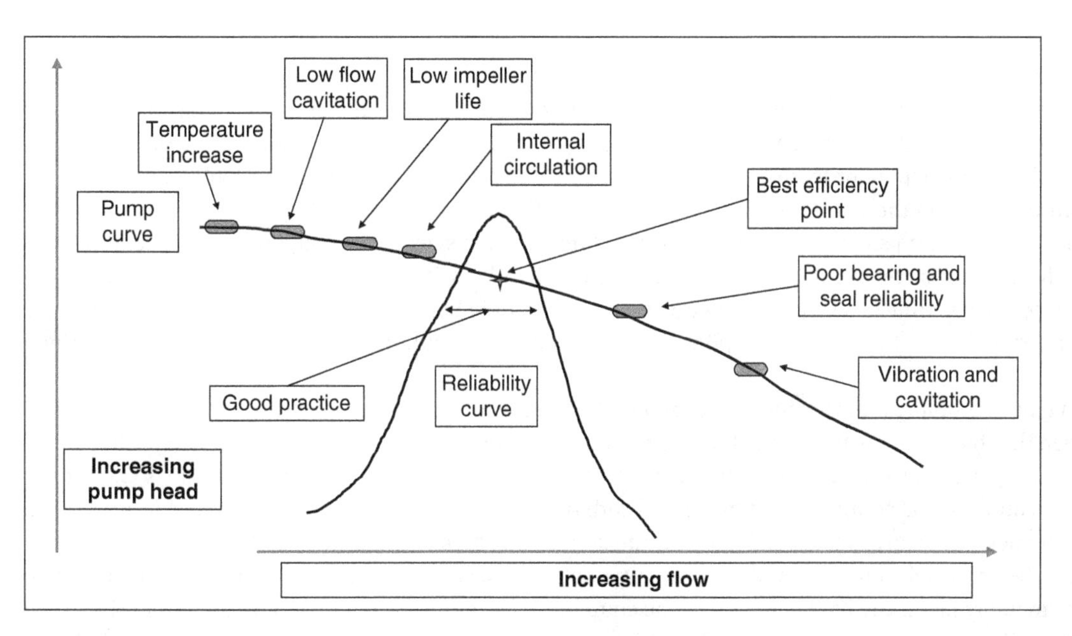

Figure 8.37.9 A simplified pump curve showing pump head vs. flow, including the pump reliability curve and BEP.

The example illustrated in Figure 8.37.9 highlights the best efficiency point (BEP), which is where the pump operates with minimum vibrations and cavitation and where the pump operating temperatures are optimum for a long and reliable service life. Generally, the best practice is to operate the pump in a range of about 15% above the BEP to about 10% below the BEP for the pump and its components to have a long life. Of course, this is only an example curve. Please refer to the rotating equipment specialists in your area, and they can provide curves specific to your equipment.

Centrifugal pumps operate best and more reliably when they are operating within the Equipment Reliability Operating Envelope or EROE. The EROE is a combination of pump head (pressure) and pump flow where pump vibrations and cavitation are at a minimum. Centrifugal pumps have what is referred to as the BEP or the midrange of the EROE. Operating outside the EROE will result in pump cavitation, higher vibrations, and a shortened seal life. Pumps should never be operated dry (with no liquid flow) as this generates extreme heat, and there is no flush to cool the seals.

Other pump curves are available with various parameters providing other pump performance data such as pump impeller size, operating speed (RPM), brake horsepower (the horsepower) required to operate the pump at a given capacity, shutoff head (the parameters of the pump were operating against a closed discharge valve), etc. These data are used by engineers to design not only the pump but the driver (motor or turbine) as well.

Dead Heading and Pump Cavitation

A pump is considered "deadheaded" when the discharge pressure is high and there is little or no flow through the pump. This can result in the pump recycling internally and results in extreme heat buildup in the pump. Deadheading also results in extreme side thrust against the bearings and can result in bearing failure, which leads to seal failure, which leads to loss of containment and fire or explosion. This can be resolved by the addition of a pump recycling system that automatically recycles fluid from the pump discharge, either back to the pump suction or back to the tank. The potential for deadheading should be considered during the initial pump design and a provision for recycling should be included in the design basis if warranted.

Cavitation is caused by either a low suction head, for example, a low level in the tank or vessel from which we are pumping, or by a restriction to the suction side of the pump. This is referred to as suction cavitation. Cavitation can also occur from high pressure on the pump discharge due to system back pressure or poor piping design, for example, too many pipe elbows or other restrictions. Cavitation occurs when vapor bubbles form around the pump impeller and then collapse. This is like mini explosions occurring around the impeller, and if not corrected, can cause severe damage or destroy the impeller.

Cavitation can also result in excessive pump vibration and can lead to premature bearing or seal failures. In severe cases, cavitation can also affect the pump efficiency, and the expected pumping rates cannot be achieved.

Coupling Hazards

The purpose of the Coupling is to transmit power from the driver to the pump while allowing for minor misalignment between the shafts. The pump should never be run without a properly fitted coupling guard in place. One of the hazards associated with a coupling is the Operator inadvertently contacting the coupling or having a small tool lanyard, wire, or other loose object getting caught in the coupling.

To prevent this, the coupling guard should cover the entire coupling, with the sides of the coupling guard extending well below the coupling. There should be no openings in the coupling larger than ½ inch to prevent someone from inadvertently getting their fingers into the coupling. The coupling guard should be securely bolted to the base plate. API Standard 671 provides additional information on the types of couplings used for centrifugal pumps, a couple of which are explained below.

The most common type of coupling is the flexible disc coupling. This design uses stainless steel shims to absorb any minor misalignment between the shaft from the driver and the pump shaft. The shim pack is designed with many very thin shims sandwiched together and placed between the two rotating elements. This shim pack rotates at either 1,800 or 3,600 rpm, depending on the application, and is continuously flexing to absorb the misalignment. The hazard with this design is that, eventually, these shim packs will start to fail. At this point, the shims start breaking apart, and small metal slivers, like small razor blades, can be propelled from the coupling. The coupling guard acts as a barrier. As the Operator, if you notice small shiny slivers of metal lying on the base plate under the coupling, this should be reported, and the pump should be removed from service until the coupling can be repaired. A shim pack coupling is illustrated in Figure 8.37.10, including a failed coupling.

The coupling in the right side of the image above failed while the pump was running in cooling water service. The coupling failed when the motor driver locked up while running at full speed. As you can see in the image, the shim pack sheared

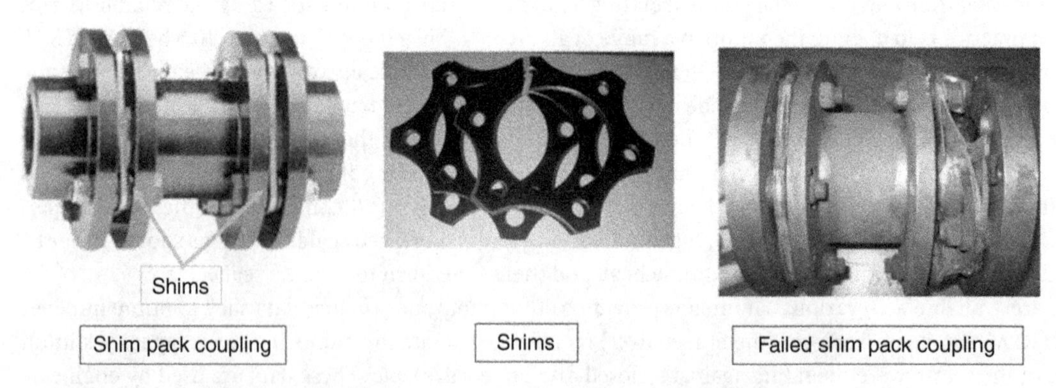

Figure 8.37.10 Illustration of a Shim Pack Coupling, including one that has failed.

due to the sudden extreme torque applied to the shaft. The coupling guard prevented the shims, which were ejected from harming personnel.

Lubrication, Lubrication, Lubrication

Continuous lubrication is essential for reliable pump operations. Most process pumps operate 24 hours a day and must be continuously lubricated with quality, clean lube oil. The lubricant should be dry (no water or moisture) and free of particulates or foreign matter of any kind. The lubricating system, either the permanently mounted oilers or an oil mist system, should be checked on each process round to ensure that the levels are correct and the pump bearings are being lubricated. Without lubrication, the bearings will quickly overheat and can fail catastrophically.

Hazards of Pump Startup

When you push the start button on a pump driver, the entire rotating assembly, including the motor, the pump, and the gearbox (if equipped), instantly accelerates from 0 rpm to either 1,800 or 3,600 rpm, depending on the design, in less than a second. When this happens, the motor is under a very high load and drawing six to eight times the amps than when the motor is running normally. The coupling receives a very high impact torque load. The bearing thrust loads instantly change from zero load to a massive thrust load within less than a second.

The point here is the Operator and anyone else near the equipment when it is being started must choose a place where they are not in the line-of-fire when the equipment is being started.

There are several key areas that should be considered for each startup:

- **The Motor Terminal Junction Box (MTJB) (This hazard is covered in more detail in Chapter 8.30) (Electrical Safety for Operations Personnel):**
  The motor terminal junction box (MTJB) contains the three leads (one for each phase) from the circuit breaker, typically located in the electrical substation, directly to the three leads to the motor. Each lead is typically connected by twisting the wires together or by bolting the leads together. In either case, the three connected leads are then securely taped.
  When the push button is pressed, this activates a relatively low voltage relay, sending the high current from the breaker directly to the motor. This very high current surge has resulted in an explosion in the MTJB, causing the cover plate to be ejected at high velocity. I am aware of one Operator who was severely injured from just such a failure and several other failures where only luck prevented injuries.
  The area around the MTJB should be avoided during a pump startup.
- **The area near the coupling:**
  This area should also be avoided by the Operator and all other personnel in the area during a pump startup. As stated earlier, the coupling received a very large thrust load during the pump startup and may fail. A failed coupling may eject shrapnel or even rotating shaft components and could injure personnel in the area.
- **The area near the pump bearings:**
  The bearings may also fail during a pump startup, and shrapnel could also be ejected. This area should be avoided by all personnel.

Please see Figure 8.37.11, which illustrates the area near a pump or other rotating equipment as a safer area for the Operator during equipment startup.

Starting a Pump Dry (No Fluid in the Pump)

Do not start a pump dry or operate the pump dry, as failure will occur quickly.

Centrifugal pumps should never be started dry. The pump case should be filled with liquid before attempting to start the pump. The liquid serves as lubrication to the seal faces and helps cool the seals and other pump components, such as the impeller, the shaft, and the bearings. The pump internals will quickly overheat and can easily be damaged if the pump is started dry or runs dry.

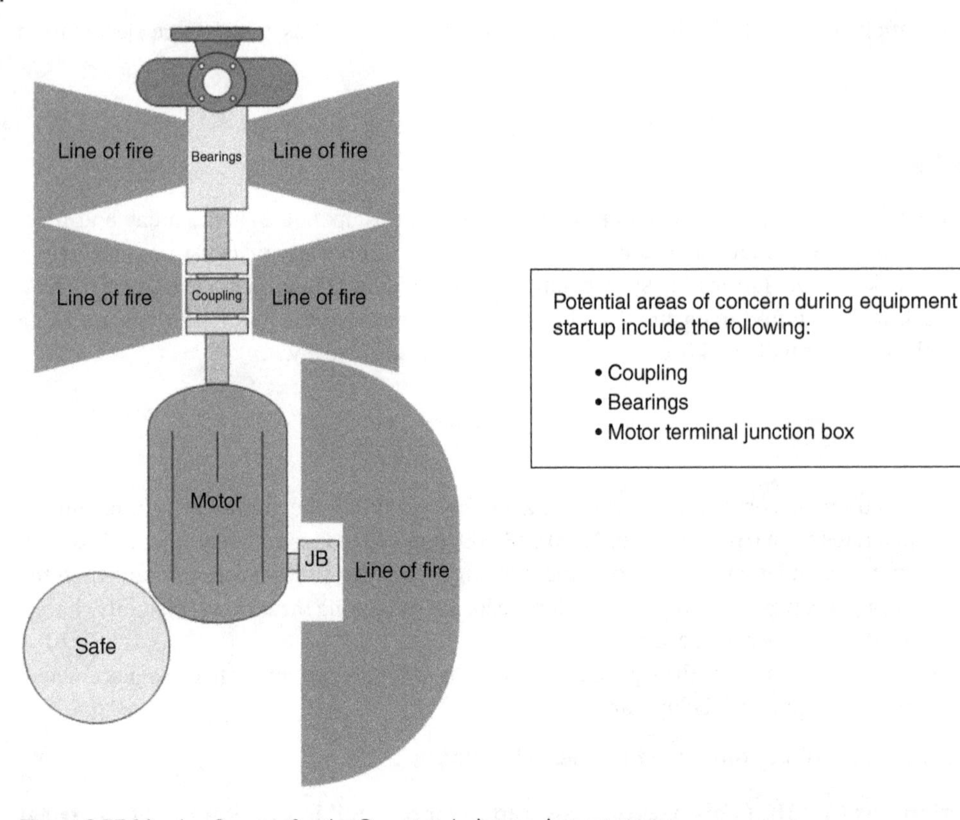

Figure 8.37.11 A safer area for the Operator during equipment startup.

Key Lessons Learned

- The Operator should always ensure that he and other personnel in the area are away from the line-of-fire from the MTJB and rotating elements before attempting to start the pump.
- Lubrication is essential for reliable pump operations. On each process round, verify the lubricant levels in pump oilers and oil mist systems and that the lubricant is water-free and clean of particulates and other debris.
- Be aware of slivers of metal that look like little razor blades lying on the pump base plate under the coupling. This may be an indication that the shim pack coupling is failing.
- Always attempt to operate centrifugal pumps within the Equipment Reliability Operating Envelope and as close to the BEP as possible.
- Be aware of pump deadheading (high discharge pressure, with low process flow).
- Be aware of pump cavitation. Cavitation results from collapsing vapor bubbles in the impeller, and it sounds like the pump is pumping rocks. This can severely damage the pump impeller. A pump that is cavitating typically has a low suction head due to a low tank level or a restriction in the suction lines, possibly a plugged strainer or partially closed valve.
- Ensure that pump couplings are equipped with a fully secured coupling guard that extends well below the coupling and there is no small opening in the guard.
- Always ensure the pump is full of liquid before attempting to start or return it to service. Starting a pump dry or operating it without liquid can result in severe internal damage.

References (Note: The following has been provided as additional reference only. Due to potential copyright restrictions, none of this material has been used in this book).

- API Standard 682: "Pumps - Shaft Sealing Systems for Centrifugal and Rotary Pumps."
- API Standard 610: "Centrifugal pumps for petroleum, petrochemical, and natural gas industries." API 610 also covers couplings.
- API 671: "Coupling and guard requirements."

- API 686: "Recommended practices for machinery installation and installation design" Also covers machinery alignment practices.
- "The Must Have Handbook for Centrifugal Pumps," by Crane Engineering. (https://www.apprep.com/Site/images/pump_resources/centrifugal_pumps_handbook.pdf)
- "Understanding Pump Curves," by MGNewell. (https://www.mgnewell.com/wp-content/uploads/2017/07/Understanding-Pump-Curves.pdf)
- "How to Read Pump Curves," by Zoeller Engineered Products. (https://www.zoellerpumps.com/content/general/how-to-read-pump-curves.pdf)
- "Tech Brief – Reading Centrifugal Pump Curves," by National Environmental Services Center. (https://actat.wvu.edu/files/d/92330560-f5d1-4597-8f57-b02b4b425e26/reading-cent-pump-curves.pdf)
- "SKF Couplings," by SKF Group. (https://cdn.skfmediahub.skf.com/api/public/0901d196806fd7be/pdf_preview_medium/0901d196806fd7be_pdf_preview_medium.pdf)
- "What is Pump Cavitation," by Crest Pumps (https://fluidhandlingpro.com/fluid-process-technology/pumps-pumping-systems/what-is-cavitation-in-centrifugal-pumps/#:~:text=Cavitation%20occurs%20in%20centrifugal%20pumps,and%20significant%20damage%20to%20both)
- "Mechanical Seals Technical Manual," by Fluiten Mechanical Seals. (https://www.fluiten.it/wp-content/uploads/2018/02/SEM001_ENG_L.pdf)
- "Common Centrifugal Pump Problems and Solutions," by C&B Equipment. (https://cbeuptime.com/signs-centrifugal-pump-needs-repair/)
- Process Safety Beacon, by American Institute of Chemical Engineers (AICHE). (https://www.aiche.org/sites/default/files/2002-07-Beacon-English_0.pdf)

What Has Happened/What Can Happen

Centrifugal pump failure; bearing failure, seal failure, broken pump shaft, thrown coupling.

Please refer to Figure 8.37.12a–c for images of a process pump failure. This is a fairly common cause of pump failures, which can be prevented with attention to detail and good pump design.

This pump was operated essentially deadheaded, with very little flow and high discharge pressure. We would describe this as "the pump was operating way back on the pump curve." The bearings failed first, resulting in the seal failure and a significant fire. During the event, the pump shaft failed and ejected the coupling and a portion of the pump shaft a long distance from the pump. Of course, this contributed to the major loss of containment and the significant fire that occurred. No one was injured, as there were no personnel in the area at the time of the failure.

The team redesigned the pumping system by adding a pump discharge recycle line and a control valve to recycle the pump discharge back to the suction automatically. This should prevent a reoccurrence of an event of this type by ensuring the pump is operating within the recommended reliability operating envelope. Figure 8.37.13 represents another significant pump fire as a result of the failure of the mechanical seal.

This image clearly shows the type of damage that can occur in the event of a pump seal failure. Fortunately, in this incident, there was no one nearby, which prevented injuries to personnel. As clearly outlined in the Process Safety Beacon and here, there are several things we can do to help prevent a repeat of an incident like this. For example, we can do the following:

- On process rounds, always be alert for equipment where liquids are leaking or seeping. Ensure these are properly reported for follow-up.
- Operate the pump within the Equipment Operating Envelope (the pump curve).
- Never start or operate a pump dry, without liquid in the pump.
- Never operate the pump deadheaded, with high discharge pressure and a low or no flow rate through the pump.
- Ensure the seal flush is operating and not plugged or restricted.
- If maintenance is completed on the pump, the maintenance supervisor should ensure the correct seal is installed following the seal manufacturer's guidance.
- Seal leaks do not heal by themselves. If a seal is identified as leaking, the pump should be removed from service.

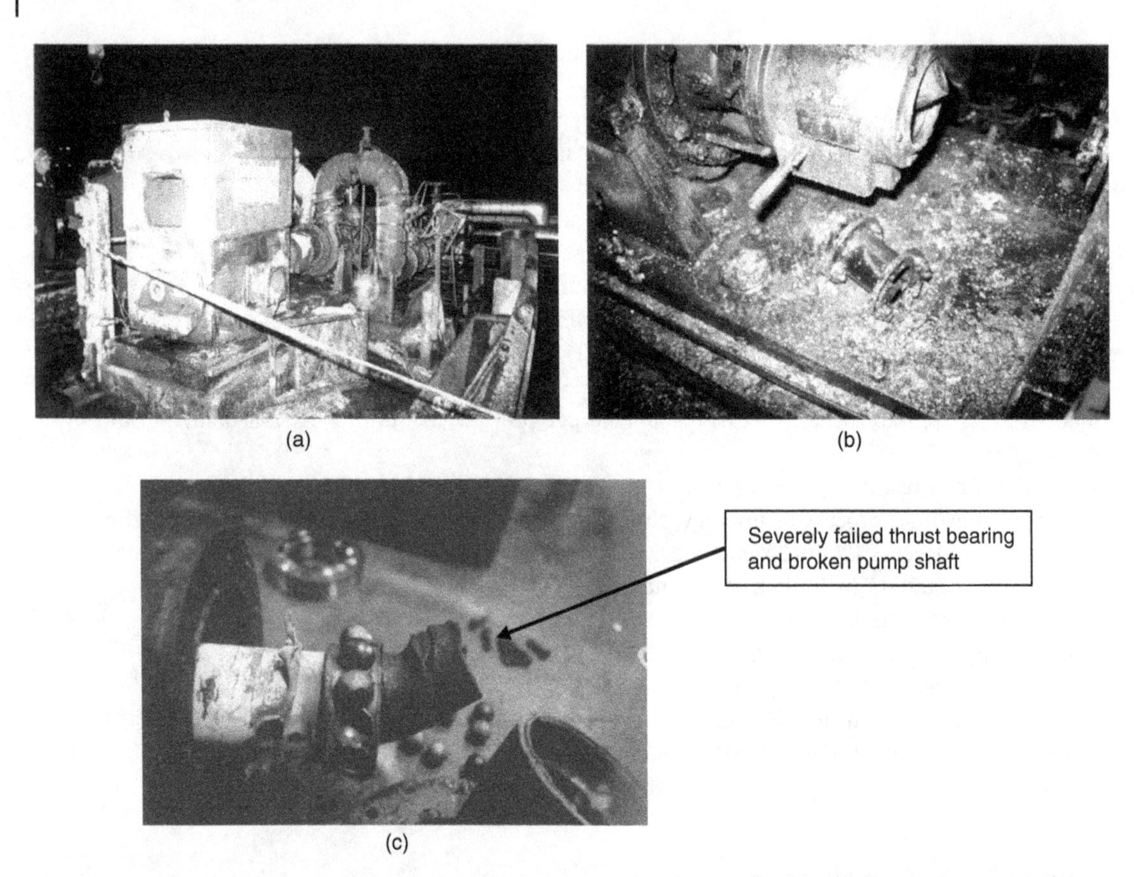

Figure 8.37.12 Images of a process feed pump failure that occurred some time ago. This image was in my files, and I am not sure where this occurred, although this does not matter because it can happen again at any one of our sites under the right circumstances. Figure (a) shows the location of the feed pump and some of the adjacent fire damage. The pump housing with the broken shaft is illustrated in Figure (b), and Figure (c) shows a close up of the broken shaft and the failed thrust bearing which led to this event.

Figure 8.37.13 Image of a process pump after a major refinery fire due to the failure of the pump seal. Source: Image was used in the Process Safety Beacon and is provided courtesy of the Beacon and American Institute of Chemical Engineers (AICHE).

Steam Turbine Overspeed

In a modern refinery or petrochemical plant, steam turbines are used for many applications, from general-purpose centrifugal pumps to super-large gas compressors. Steam turbines have several potentially significant safety issues in addition to turbine overspeed, such as overpressure, steam leaks, and handling hot condensate. However, these pale in comparison to the potential damage that can be done by one turbine overspeed incident. This discussion will focus on turbine overspeed and how to prevent it from occurring.

We will start with some reminders of some general safety precautions to remember and have in place where steam turbines are concerned. First, steam turbines are not designed to see full steam pressure. Most turbine cases are made from cast iron and, therefore, are very brittle and will fail if exposed to full steam pressure.

Many steam turbines may be equipped with "sentinel" warning devices to alert the Operator that the case is experiencing high pressure. These devices are for warning purposes only and are not designed to provide overpressure protection. A good practice is to car seal the turbine exhaust valves in the fully open position to prevent pressure buildup in the turbine case. When placing the turbine in service, the turbine exhaust valves must be opened before the steam inlet valves are opened again to prevent turbine rupture. It is also a good idea to have warning signs posted at each turbine to remind the Operator of this important procedure. Some newer turbines may be protected by relief valves to protect the turbine case from inadvertent overpressure. Some installations may have a Safety Interlock System designed to prevent steam turbine casing overpressure; these are most common on 150 psi (10 bar) to 600 psi (41 bar) turbines.

Now, let us focus more on steam turbine overspeed. The most common causes of overspeed include the loss of load, loss of suction, or a sudden change in the steam inlet or outlet pressures. Overspeed can also be caused by a malfunction of the turbine governor or a sudden failure of the coupling.

The governor controls the turbine speed, which is typically hydraulically controlled in most general-purpose steam turbines but may be electronically controlled in larger turbines. Turbine overspeed is detected and prevented by the Overspeed Trip device. When a turbine over speeds, the mechanical forces generated can exceed design conditions resulting in the failure of the turbine components, including the rotating assembly (shaft, turbine wheel, and blades) (also referred to as buckets).

In a mechanical overspeed device used on most general-purpose steam turbines, the overspeed trip works by centrifugal force. When the shaft speed goes overspeed, a spring-loaded "bolt" is ejected from inside the shaft and strikes a linkage,

Figure 8.37.14 An example of a general-purpose steam turbine.

tripping the turbine offline. If the overspeed trip is slow to respond or not working, the turbine will overspeed. The trip valve has the potential for sticking and may not close quickly enough. This is generally due to calcium buildup on internal components. The turbine governor linkage and stem must be free to move. Failure of an overspeed trip could allow for turbine overspeed and failure. See Figure 8.37.14 for an example of a general-purpose steam turbine.

Testing the Steam Turbine Overspeed Trip Device

There should be a documented and closely followed procedure for periodic preventative maintenance for each steam turbine, including testing the overspeed trip device. Preventive maintenance tasks are generally scheduled on an annual basis and only extended based on repeated good experiences with the machine. Images of the overspeed trip device used in most general-purpose steam turbines are illustrated in Figure 8.37.15a–c.

A Word of Caution

During the test of the overspeed trip, the coupling is removed, so there is NO load on the turbine. This means that extraordinary procedures must be in place and closely followed throughout the test. Remember, the turbine is intentionally placed in an overspeed condition to test the trip. This is the only place in a refinery or petrochemical plant

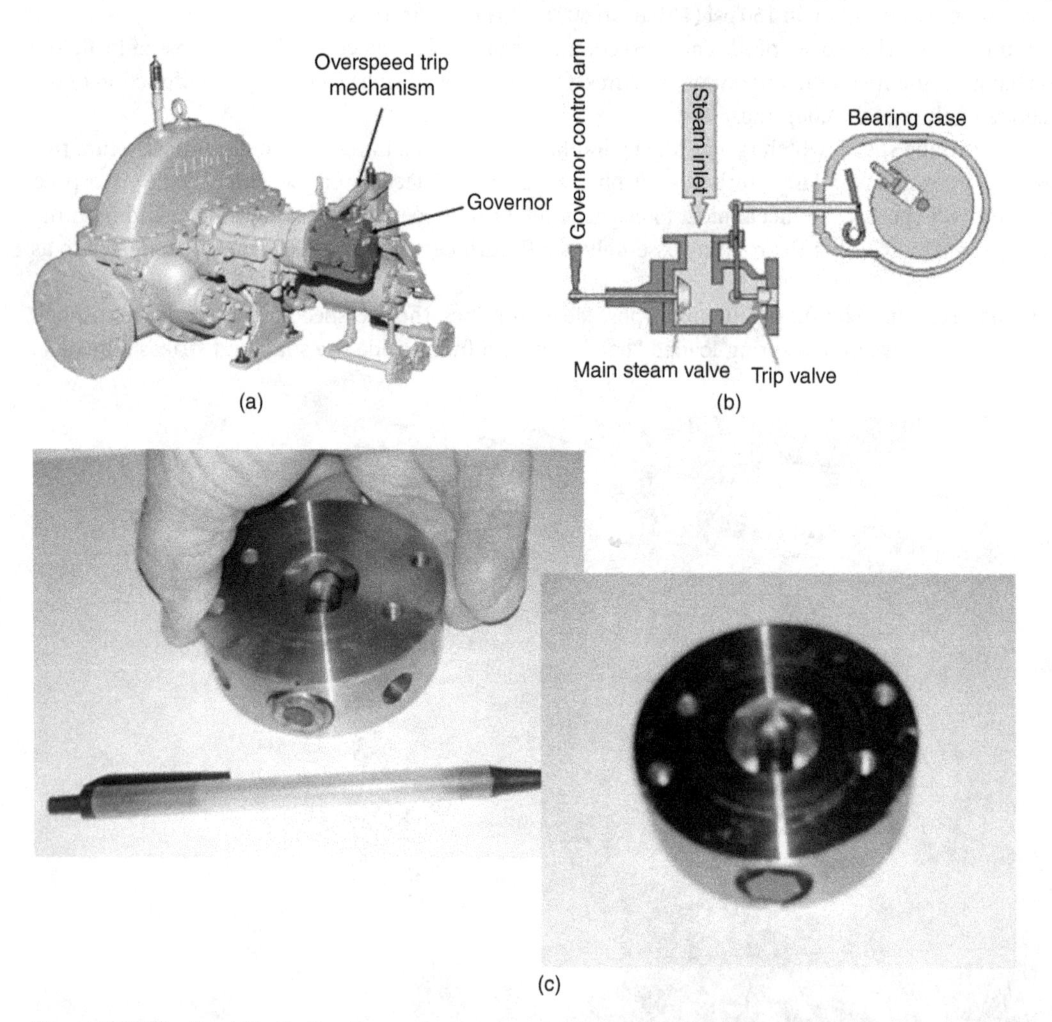

Figure 8.37.15 (a–c) Illustrates a general-purpose steam turbine and the elements for the governor and the overspeed trip device. Figure (a) is a general overview of a steam turbine illustrating the location of the speed control device (the governor) and the overspeed trip mechanism. The function of the mechanical overspeed trip device is shown in (b). In event of a turbine overspeed, this device functions by centrifugal force of the rotating shaft propelling a bolt which trips the steam valve closed causing the turbine to stop. Figure (c) shows the bolt assembly which is attached to the rotating shaft. This spring-loaded bolt rotates with the shaft and trips the turbine at a high rotating speed.

where I am aware that this is done. If the trip malfunctions, the tech must manually intervene and trip the turbine. Otherwise, the turbine can overspeed to destruction!

Example test procedure (Use your site procedure – the following is only for an example):

- Ensure that all personnel not involved in the overspeed trip test are removed from the vicinity.
- Perform routine preventative maintenance (adjust and lubricate all linkages, etc.).
- Test the trip linkages and the trip valve for proper seating (BEFORE the actual overspeed trip check).
- Close the main steam inlet valve and use the smaller bypass valve around the main steam inlet valve (if provided) to better control steam flow (small-bore bypass with a globe valve designed for throttling the steam during the overspeed trip test).
- Slowly increase the turbine to the specified operating speed, ensuring that the governor controls.
- Use a remote tachometer to check turbine speed and slowly override the governor to bring the turbine to an overspeed condition (usually 10–15% over the specified operating speed).
- Ensure that the turbine trips as specified and record the turbine RPM.
- Throughout the test, maintain position out of the line-of-fire as much as possible.
- Follow the procedure!!!

Alternative Steam Turbine Overspeed Test Process

There is an alternative process for testing the turbine overspeed test device without an online test. Mr. Jack Little, President of ILD, located in Baton Rouge, Louisiana, developed this process. This process uses a variable drive electric motor and a belt drive to drive the turbine to overspeed conditions. The turbine speed is closely controlled, and there is no chance it will go into an overspeed condition since no steam is involved. The Nuclear industry seems to be fully bought into this process and uses it in nuclear plants to test the overspeed trip devices on their largest steam turbines.

I met Jack, and I quickly felt this deserved a closer look and saw some real safety benefits in the process. As mentioned earlier in this chapter, I believe conducting turbine overspeed tests is one of the most hazardous things we do in a refinery or petrochemical plant. I brought into the technology and introduced Jack to the rotating equipment experts at my company.

Following their review, I received feedback that the engineers felt this technology needed to be a complete test of the turbine overspeed trip device. They felt that this was not a complete test since the turbine was not running on steam and was not operating at high temperatures, like during a conventional test. I guess the jury is still out on this, but it seems as though this could easily be overcome with a slight change in the procedure.

Starting a Steam Turbine (Use Your Site Procedure – This Is aOnly an Example)

A procedure should also be readily available for the Operator's use when starting a steam turbine. This is particularly important with less experienced Operators. Follow the procedure operating the valves in sequence with the steps in the procedure.

- Verify that steam lines and turbine case are free of condensate and that steam traps are functional. All steam traps should be hot.
- Check that all lubricants are dry (no moisture) and that the oil levels are correct.
- Ensure that the turbine steam outlet (exhaust) valve is fully open and is car sealed in the fully open position.
- Before opening the steam inlet valve, operate the overspeed trip device to ensure it is not stuck or binding (all linkage should be free).
- Stand away from the line-of-fire of the turbine wheel throughout the procedure.
- Slowly bring the turbine up to speed by slowly opening the steam inlet valve. Carefully observe the governor's control to ensure it is controlling.

Caution

No steam turbine shall be placed in service without a fully functional overspeed trip device!

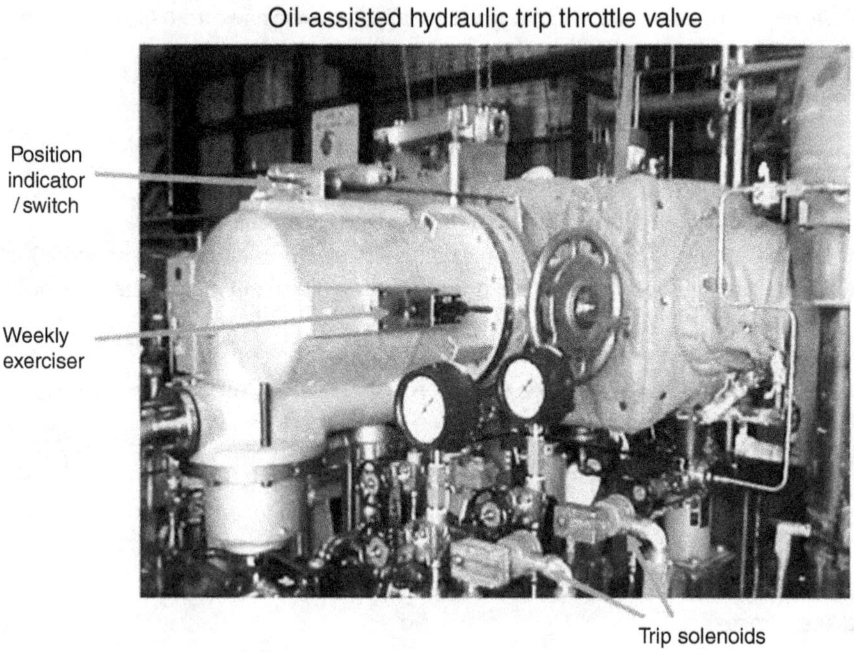

Oil-assisted hydraulic trip throttle valve

Position indicator / switch

Weekly exerciser

Trip solenoids

Figure 8.37.16 Image of a large steam turbine equipped with an electronically controlled, hydraulically assisted steam turbine trip and throttle valve. Note the trip electronic solenoids (arrow to the location of the trip solenoids).

Electronic Controls (Electronic Governor and Trip Controls)

Some newer steam turbines are equipped with electronic controls, and some older machines have been retrofitted with this newer, sophisticated electronic equipment. These operate much smoother and have been proven to be very reliable. The electronic systems allow smaller components, equipped with redundant controls and voting logic, helping prevent false trips. For example, it is common for these systems to be equipped with triple redundant speed sensors, which will trip only when two of the three sensors indicate that overspeed has occurred.

Figure 8.37.16 illustrates an electronically controlled, hydraulically assisted steam turbine trip and throttle valve installed on a large steam turbine driving a large centrifugal compressor.

What Has Happened/What Can Happen

To illustrate that issues with steam turbines have been around for a long time, the following article was found in a May 1917 New York Times article:

> **"Runaway Turbine Caused Wreck"**
> **"John A Topping, Chairman of the Republic Iron and Steel Company, said yesterday that the accident at the company's Youngstown plant on Sunday, which killed four men, was not a result of an explosion but was due to a runaway steam turbine which wrecked a section of the powerhouse."**

In the Safe Operations Training course, I frequently tell the attendees that testing the steam turbine overspeed trip mechanism is one of the most hazardous tasks in a refinery or petrochemical plant. This is because the test is not simulated. When we test the overspeed device, the turbine is intentionally put into an overspeed condition. If control is lost during the test, the turbine can fail catastrophically, which can happen in milliseconds. As you can see in the cases described below, this has happened, and the consequences can be catastrophic when it does.

In the Key Lessons Learned section, we will discuss the precautions and procedures that should always be followed to help prevent turbine overspeed.

What Has Happened/What Can Happen

Steam Turbine Overspeed and Failure during a test of the Overspeed Trip.
24 February 2001
One Fatality (Struck by flying objects when the turbine failed)

The National Institute for Occupational Safety and Health (NIOSH) and the Centers for Disease Control and Prevention (CDC) investigated and reported the following incident. The following is a summary of the report.

A 51-year-old journeyman Machinist was killed when he was struck by flying fragments projected from a steam turbine that failed catastrophically during a test of the overspeed trip device. The Machinist along with several other employees of three different companies, were standing on the platform observing the turbine test when the failure occurred.

Consistent with procedures for testing the overspeed trip system, three separate tests are required. The team had successfully completed the first test and was in the process of testing the turbine the second time. During this test, the overspeed mechanical steam shutoff valve failed to close, and the turbine immediately went into overspeed resulting in the failure of the turbine rotor. One hundred steam turbine buckets, part of the internal blades attached to the rotor, failed. The buckets penetrated the two-inch turbine housing, resulting in flying fragments. The victim was struck by these buckets, resulting in his death.

Investigation Key Observations

- A key observation in the investigation report is that the victim and six others were observing the test on the platform that supports the turbine. Also, six other workers were in the area but not on the platform. Procedures should ensure that the test is controlled remotely and that all personnel are away from the turbine during the overspeed test.
- There was no written, detailed procedure for the overspeed test that conformed with manufacturer guidelines and industry best practices (see References below).
- Operators were not adequately trained in the safe procedures for conducting the overspeed trip test.
- There was no clear backup system for immediately stopping the steam flow to the turbine if the overspeed trip device failed.
- There was confusion among the workers as to the correct speed of the turbine for conducting the overspeed trip test. The turbine manual stated the operating speed to be 3,370–4,100 rpm. The brass tag on the machine indicated that the turbine would trip at 4,500 rpm. The overspeed trip test speed and the expected trip point should be clearly documented in the test procedure.
- The RPM indicator mounted on the turbine near the emergency mechanical trip device was not accurate and did not function as designed. No one was positioned to physically trip the machine if it did not trip as designed. The turbine should not have been allowed to operate without a properly functioning turbine speed indicator.
- A person was stationed at the Compressor Control Corporation (CCC) computer speed panel located remotely from the platform and was using hand signals to communicate the turbine speed to another person who was operating the turbine speed electronically. Unfortunately, the hand signals being used were inconsistent, which could lead to additional confusion and delayed action to stop the turbine.
- No one was aware that the turbine overspeed test could have been completed from a CCC control unit computer had it been programmed to do so. The computer had not been programmed to conduct this test. Conducting the test from this computer would have ensured that no one was in the line-of-fire during the test.
- Barricades, warning signs and such should be in place to help ensure that all personnel are out of the area during the speed-control test (Figures 8.37.17 and 8.37.18).

Figure 8.37.17 Plant area illustrating the former location of Turbine 3 and the current location of Turbine 4. Turbine 3 is the turbine which failed due to overspeed. This also shows the location of the victim relative to the failed turbine. Source: Courtesy of Centers for Disease Control and Prevention/https://www.cdc.gov/niosh/face/stateface/mi/01mi011.html/last accessed Feburary 29 2024/Public Domain.

Figure 8.37.18 Turbine 3 shaft and turbine rings. Note the remaining exposed buckets. One hundred similar buckets were projected from the turbine in the overspeed incident. Source: Courtesy of Centers for Disease Control and Prevention/https://www.cdc.gov/niosh/face/stateface/mi/01mi011.html/last accessed Feburary 29 2024/Public Domain.

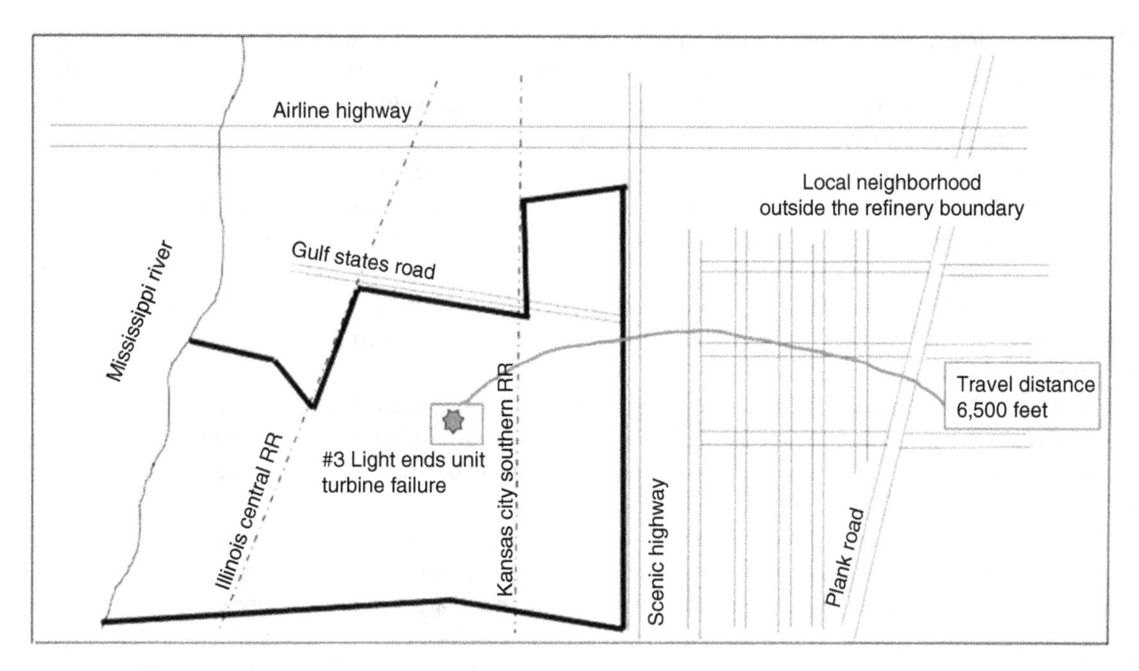

Figure 8.37.19 A catastrophic steam turbine failure. This sketch is not to scale and is an approximation only.

What Has Happened/What Can Happen

A catastrophic steam turbine failure.
A very long time ago (early 1960s)
No injuries, but significant equipment damage

This is an incident where I do not have any personal knowledge. However, I started at the refinery within a few years of the event. It was obvious that everyone knew about this incident, and it was discussed frequently for many years thereafter. The incident occurred during a steam turbine overspeed trip test on a large cooling water turbine on a refinery Light Ends Unit.

During the test, the turbine went into an overspeed, and the rotating element, the turbine wheel, came completely out of the turbine case, severed a small-bore pipe in the overhead pipe rack, and traveled a long distance before landing completely outside the refinery boundary. A large part of the rotating element landed approximately 6,500 feet (1,981 meters) or about 1.25 miles (about 2 kilometer) from where it started. The turbine wheel "flew" over local residences and landed in a backyard. See Figure 8.37.19 for an approximation of the turbine's flight path.

This incident clearly shows the extreme energy associated with an overspeeding turbine.

What Has Happened/What Can Happen

Escom Holding SOC Ltd.
Duvha Power Station 9 February 2011
Duvha Unit 4 Steam Turbine Overspeed
Steam Turbine Overspeed Incident during the Overspeed Trip Test

The following is a media statement by Duvha Power – 10 February 2011

> "Thursday, 10 February 2011: Unit 4 of Eskom's Duvha Power Station was damaged last night in the course of a routine test. No one was injured, but the unit is expected to take extensive time to repair."

"The event took place when the 600 MW unit at Duvha had been taken offload to perform a required turbine test. This is a statutory test that is carried out on every power station. In the execution of the test, the protection on the unit failed, causing severe mechanical damage and starting a fire…" The statutory test being performed was a test of the turbine overspeed trip device.

Later reports indicate that the primary cause was attributable to Operator error. However, it was also reported that changes had been made during 2004, and the operators may not have been appropriately trained. "The direct cause of the incident is attributed to operating error, in that the operator did not follow the set procedure while undertaking (a) test," CEO Brian Dames said at the release of Eskom's interim results in Johannesburg.

The damage to the turbine and generator was catastrophic, illustrating the energy available in these large pieces of rotating equipment. The projectiles from this failure damaged much of the surrounding buildings and other equipment.

The South African electrical power generation stations have some of the largest power generation equipment in the world. This super-large steam turbine is rated at 600 MW, which is enough electrical energy to power about 100,000 US homes. The machine operates at about 3,000 rpm, and the overspeed trip should function and automatically trip the machine at about 10% overspeed, or about 3,300 rpm. At the time of the overspeed test, it was reported that the turbine speed reached about 4,500 rpm before it failed catastrophically.

The building is one long structure with five almost identical turbines and generators. When the turbine failed, a large amount of shrapnel was ejected, damaging the turbine building and the massive building roof. Eskom reported that the turbine and generator damage was expected to take a year or more to complete before the turbine and generator could be returned to service.

Although this happened with this giant turbine, exactly the same thing can also occur with the more common smaller versions of the general-purpose steam turbines, which drive the smaller process pumps. It can also happen with larger-sized special-purpose turbines, such as those driving centrifugal compressors or other larger rotating equipment.

After an event like this one, we expect to see substantial changes in the overspeed test procedures, assignment of roles and responsibilities of the employees and contractors performing the test, and verification of the instrumentation and trip devices before attempting to conduct the test.

Most modern sites are designed to allow overspeed testing of the larger turbines from a process computer that is fully equipped to shut the steam flow to the turbine immediately in case of a false trip or overspeed. If testing with or without the computer, a designated person, located remotely if possible, with no other responsibilities, should be in direct contact with the turbine speed indicator and have complete control to stop the steam flow immediately if the turbine fails to trip after exceeding the trip speed. In the event of an overspeed, this action is required in milliseconds to prevent the turbine destruction and flying shrapnel. This is why I repeat here: a turbine overspeed test is one of the most hazardous tasks we do in our plants.

Overview

When I started writing this article for the chapter on Steam Turbines, I was planning to write about an overspeed incident at the Escom Duvha Power Station on 9 February 2011. This catastrophic incident destroyed a 600 MW steam-powered turbine and generator, quickly becoming a larger story. Escom is the largest electrical power producer in South Africa and operates a nuclear power station and fifteen coal-fired power stations. Escom is a state-owned conglomerate that produces and sells electrical power across the country, producing about 90% of the electricity consumed in South Africa. In addition to nuclear and coal-powered plants, they also produce electricity from diesel, water, and wind; however, the bulk of the energy is from nuclear and coal-fired plants.

As I attempted to secure the images of the turbine incident for this book, I researched Escom and other sites, looking for alternative images of the Unit 4 failure that I could obtain. In doing so, I realized how much Escom could benefit from a formalized Process Safety Management System and that this should be a much larger story. I will come back to this a little later.

Duvha Unit 4 Turbine Overspeed

The driver for this article is still the catastrophic failure of the Duvha Unit 4 steam turbine and generator that occurred on 9 February 2011. This incident happened while the steam turbine was undergoing a test of the overspeed trip device. Something went terribly wrong, and the turbine went into an uncontrolled overspeed and failed catastrophically. When the turbine failed, shrapnel was hurled in all directions, causing substantial damage to the turbine building, the building roof, and the infrastructure. As stated earlier, the turbine and generator were virtually destroyed.

It was reported that the turbine was designed to operate at about 3,000 rpm and should trip at about 10% over, or about 3,300 rpm. The turbine was reported to have three overspeed trip devices, and during the test, it was reported that all three of these devices failed to function, and the turbine failed. The last reported speed was 4,250 rpm, although it is believed that the turbine may have reached speeds even higher. I have seen many photographs of the damage, which are available online by searching for "Duvha V2." These and several additional images are also available on other websites. Of course, I attempted to get permission to use the images in this book, but to no avail. I encourage you to look at these because words cannot do enough to convey the amount of destruction that occurred here. Fortunately, and somewhat of a miracle, no one was killed or severely injured in this event.

What I gleaned from the images of the Unit 4 failure:

- There were large openings in the turbine case where steam buckets and other shrapnel exited the case.
- Two sections of large diameter shafts broke away as shrapnel. One was embedded into the concrete on the turbine deck, and the other penetrated through a steel floor deck plate.
- There were several large holes in the concrete and block wall sections where shrapnel penetrated. One large piece of shrapnel went through the concrete wall and severed a large steel I-beam that ran along the outer wall.
- There are gaping holes and many smaller ones in the turbine building roof where shrapnel exited the building. The roof for the turbine building is reported to be about 90 feet above the turbine deck.

The main message in this article is that this same kind of failure can happen on steam turbines at any of our petroleum refining or petrochemical sites if we do not have highly specialized procedures and if those procedures are not being followed to the letter. Also, remember, when we conduct a test of the turbine overspeed device, we are testing the overspeed by putting the turbine in an actual overspeed condition. This is not a simulated test like the testing of most of our safety-critical devices. If the overspeed trip fails to function, the turbine must be stopped immediately without delay by stopping the steam flow to the turbine. Someone must be stationed on this trip device, out of the line-of-fire, and with no other responsibilities other than to monitor the turbine speed and trip the turbine if it over speeds.

As usual, there are many things we can take away from a failure like this, and it is extremely important to do so. We must do all we can to ensure that this is not repeated, as lives are at risk when it does.

We will talk about the lessons learned from this and other steam turbine overspeed incidents in the "Key Lessons Learned" section of this chapter.

Eskom Observations

I mentioned earlier how I realized how much Escom could benefit from a formalized Process Safety Management System. I want to expand on that comment a little more here.

Eskom is in an enviable position; the company is owned and supported by the state and by earnings from its operations. Unfortunately, because of the unreliability of their operations and incidents like this one, Escom has become so accustomed to long periods of electrical loadshedding or rolling blackouts that this appears to have been normalized.

It appears that Escom accepts this loadshedding as a normal and accepted part of the business. They appear to operate with a large portion of their capacity, sometimes as much as 50% or more being offline, which is accepted as routine. They appear to accept reliability issues, operator errors, and poor planning as normal parts of the business. According to local media, they respond that they need additional generation capacity instead of addressing the issues.

It also appears that a considerable part of the reasons for the loadshedding is that the offline equipment is due to unplanned events that are process safety related. The steam turbine incident covered above is an example of this. According to local media, this 600 MW generator was out-of-service for nearly three years due to this incident.

There have been other similar significant incidents that fall into this same category. These incidents combined resulted in some of the poorest reliability and highest number of electrical loadshedding events for South African citizens and businesses the country has experienced. Unfortunately, these are continuing into 2024.

The following are examples of the kinds of incidents that have occurred. This is not intended to be a complete list:

- **8 January 2003:** Duvha Power Station – Unit 2 generator exploded while being returned to service after undergoing regular maintenance. The hydrogen gas-cooled turbogenerator exploded and caught fire in Unit 2 during recommissioning; the oxyhydrogen explosion destroyed the generator, leading to an extended outage.
- **30 March 2014:** Duvha Power Station – The steam boiler on Unit 3 experienced an over-pressurization event. The overpressure alarm sounded, and the boiler tripped on overpressure. Damage to the boiler was observed, and one person was treated for dust inhalation. The incident, caused by over-pressurization of a boiler furnace, resulted in the loss of 600 MW to the electrical grid. It was reported by Anton Eberhard in November 2021 to have never been repaired.
- **September 2020:** Medupi Power Station – A conveyor belt that feeds coal into the generation units failed, resulting in additional load-shedding.

- **August 2021:** Medupi Power Station – A large explosion and fire destroyed unit 4 after an operating error created a volatile mixture of hydrogen and oxygen. Unit 4 is expected to be offline for more than three years.
- **September 2021:** Kendal Power Station – A fire broke out at the power station. One unit was reported to be damaged by the incident.
- **October 2021:** Tutuka Power Station – Eskom, in their report, indicated that the power station had deteriorated to an unacceptable state, with large amounts of ash covering walkways and the instrumentation that is required for ongoing operations.
- **March 2022:** Koeberg Nuclear Power Station – Eskom nearly lost 920 MW of power when a worker closed a wrong valve. The workers reportedly closed a block valve to a relief valve on the operating Unit 1. He should have closed the same valve on the out-of-service unit (Unit 2). This was reported to be the second time that an incident like this occurred during the Unit 2 outage that was ongoing at the time. This could have been a very significant event.
- **13 June 2022:** Duvha Power Station – A fire was reported at Unit 2, but it was later reported as an oil spill. Massive black smoke was reported to be seen from the facility. The reported fire followed several incidents of sabotage at the facility.
- **September 2022:** Kusile Power Station – The station experienced a fire at a gas-fired air preheater on a unit that was not synchronized to the grid. The significant damage delayed bringing the new 800 MW generation unit at Kusile online for a year.
- **September 2022:** A fire occurred on the conveyor that moves coal from the Kriel mine to the Eskom Power Station.
- **September 2022:** Camden Power Station – A Technician opened the wrong valve and contaminated the plant's demineralized water supply. This required the power station to be taken offline.
- **October 2022:** Duvha Power Station – A fire resulted after the wrong oil was added to a generator unit at the Duvha power station. This delayed returning the unit to service.
- **10 October 2022:** Kusile Power Plant – A section of the Unit 1 flue gas duct failed, requiring the unit to be out-of-service for an extended period. As reported by Eskom: "The October 2022 failure of the chimney system at the Kusile Power Station, which removed more than 2,400 MW of capacity, is the major cause of the elevated stages of loadshedding. This, together with the planned extended outage of Unit 1 of the Koeberg Nuclear Power Station, are responsible for three stages of loadshedding."

Numerous other equipment outages and periods of loadshedding have reportedly occurred due to incidents of sabotage, operator error, lack of training, and poor planning. As recently as 2024, Eskom reported periods of significant loadshedding: On Monday, 29 January 2024, Eskom reported, "Over the past 24 hours, five generation units were taken offline for repairs. Additionally, the return to service of two generating units that were on planned maintenance was delayed due to an opportunity to perform preventative maintenance. These factors have contributed to the shortage of available capacity necessitating the implementation of Stage 3 loadshedding...."

Escom's Environmental Performance

In 2021, the Center for Research on Energy and Clean Air reported that Eskom's 15 coal-fired power plants were responsible for 1.6 million tons of Sulphur Dioxide (SO_2) per year. They reported that this makes Escom the highest producer of Sulfur Dioxide worldwide. It is amazing in 2024 that, we have one company with this level of environmental emissions. This speaks to the technology being used at the Escom facilities and the design and availability of the emission control devices.

Process Safety Management Systems

So, how would a comprehensive and formalized Process Safety Management System turn this performance around? I talk about this in length in Chapter 6, "An Introduction to Process Safety Management Systems," so I will summarize this here and refer you to Chapter 6 for the details.

Process Safety Management Systems (PSMS) are the guiding documents that transcribe the company's corporate Safety, Health, Environmental, and Security values and culture into documented and actionable management plans. The corporate values are documented and communicated to all employees, contractors, and anyone who works for the company. Some companies have also used a similar management system for addressing reliability issues.

The PSMS documents additional details and objectives and ensures the goals are translated into actionable plans and procedures aligned with the corporate values. The Objectives, Plans, and Procedures are communicated to all corporation members (or site, as the case may be) for implementation. Responsibilities are assigned to individuals, generally members of management with corresponding roles in the organization, or to subject matter experts for PSMS implementation. Those members Responsible are also Accountable for implementation and the ongoing verification and measurement to ensure complete execution and continuous improvement.

With a corporate PSMS, essentially every aspect of the company's business is touched and improved. The results are continually measured, and continuous improvement is expected and required. This touches all aspects of ongoing operations with management systems from Management Leadership, Training, Procedures, Equipment Design and Construction, Emergency Plans and Preparedness, Planning, Managing Change, Contract Resources, Environmental and Community Impact, and more.

In a typical PSMS, the above management systems are broken into a Main system and Subsystems with the following key elements and expectations for each system:

- Requirements and Action Plans (What is required by the system),
- Measurements (How are we performing versus our objectives),
- Reported to (Who do we report to, by name),
- Responsibility for follow-up (Who has responsibility, by name),
- Continuous improvement (Plans for getting better, our short and long-range performance objectives, and how we will get there).

I have seen how implementing a Process Safety Management System can turn a company around and make it a high-performing organization. The critical elements of a good PSMS are available in Chapter 6. I believe that if a process like this were implemented at the Eskom facilities, it could quickly become a high-performing organization, which would greatly benefit the citizens and businesses of South Africa. I am convinced that the loadshedding that appears to be normalized will become the exception.

I ask myself, who am I to be even attempting to give Escom guidance or a suggestion like this? However, if I were Escom, I would run toward a Process Safety Management System as fast as possible.

What Has Happened/What Can Happen

One More Example

Hydro Agri Trinidad's wholly-owned ammonia production plant
Savanotta near Point Lisas, Trinidad
October 1997 – three Fatalities

This site experienced a tragic incident during a routine steam turbine overspeed trip test. Three lives were lost (two died outright, and one died later due to his injuries).

"Hydro Agria, a unit of Norway's Norsk Hydro, had originally estimated that the approximately 240,000 – 250,000 tons per year capacity plant would be out of action until late October or early November. But it now estimates that the unit will be down at least to the end of November and possibly to the end of December."

"The accident, in which two men were killed and three were seriously injured, occurred when a steam turbine failed during a routine trip test. An official investigation has concluded that the incident was a result of a combination of errors occurring in a sequence that allowed the turbine to fail."

Key Lessons Learned

- An overspeeding or overpressurized turbine can destroy itself and cause extensive damage and serious injury.
- Overspeed trip mechanisms must function as designed. Enforce the requirement that a turbine without a fully functional overspeed trip cannot be placed in service.
- The following can cause a turbine to overspeed?
  - Loss of load; loss of suction.
  - Sudden change in steam inlet or outlet pressures.
  - Governor malfunction.
  - Coupling failure.

Figure 8.37.20 Illustration of a steam turbine "line-of-fire in event of overspeed and failure."

- Turbines must always have an unobstructed exhaust outlet. Always enforce the requirement to car seal the turbine exhaust valve in the open position.
- Take position out of the line-of-fire of the turbine wheel when starting the steam turbine. See Figure 8.37.20 for an illustration of a steam turbine "line-of-fire."
- Safety, sentinel, and hand valves must be in good operational condition.
- Sentinel Valves:
  - Turbine sentinel valves are designed to provide an overpressure warning only.
  - They are set near casing pressure limitations and not all have the same setting.
  - They do NOT provide adequate volumetric capacity for relief protection.
  - They are NOT installed on all steam turbines.
- If the overspeed trip is slow to respond or is not working, the turbine will overspeed.

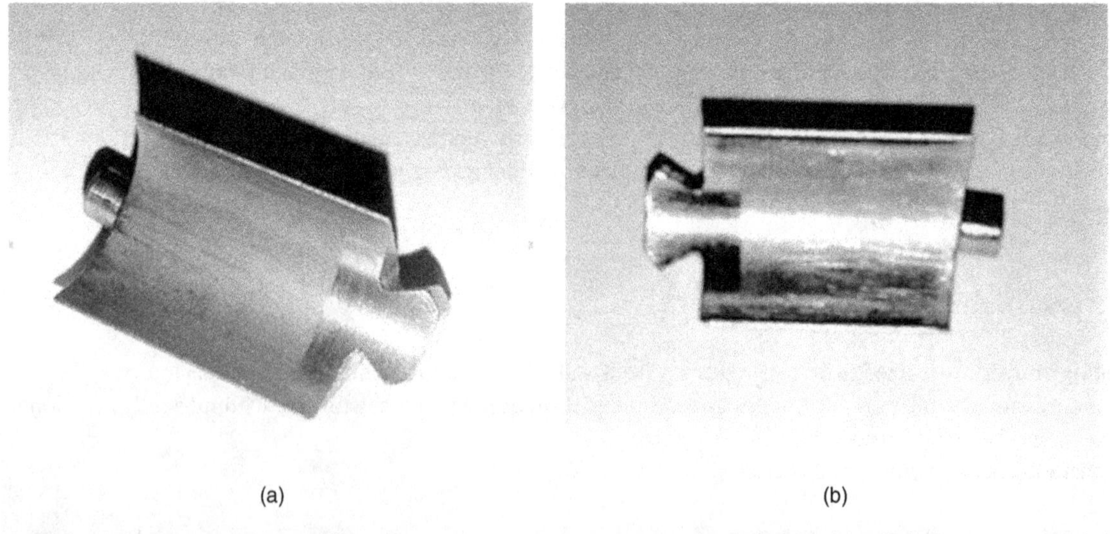

Figure 8.37.21 Examples of turbine components that can become shrapnel should a turbine fail (Steam turbine "buckets"). Two images are provided to illustrate the shape and general size of the general service steam turbine buckets. These are generally 1.5–3 inches wide, depending on the size of the steam turbine, and are a pressed fit into the steam turbine wheel.

- The steam trip valve has the potential for sticking and may not close quickly enough (calcium build up on internal components)
- A failure of the overspeed trip could allow for turbine overspeed and failure.

- Ensure that turbine overspeed trip devices are checked on schedule.
- During routine rounds, operating personnel check for unusual noise, vibration, hot bearings, and lubrication.
- Mechanical performs routine preventive maintenance tasks.
- Teamwork between process, mechanical, and technical is the key to safe, efficient, reliable turbine operation.
- Ensure that Operators are well-trained in turbine operations.
- When a turbine overspeeds and fails, many large and small projectiles are released, which can cause serious injuries and damage adjacent facilities. See Figure 8.37.21a and b for examples of components that can become shrapnel should a turbine fail.

Additional References

Reference: API 611 "General Purpose Steam Turbines for Petroleum, Chemical, and Gas Industry Services"

API Standard 612: "Petroleum, Petrochemical, and Natural Gas Industries – Steam Turbines – Special-purpose Applications"

U.S. Environmental Protection Agency Combined Heat and Power Partnership; Catalog of CHP Technologies, Section 4. Technology Characterization – Steam Turbines (https://www.epa.gov/sites/production/files/2015-07/documents/catalog_of_chp_technologies_section_4._technology_characterization_-_steam_turbines.pdf)

Accident Prevention Manual for Business and Industry, Engineering and Technology, 11th Edition, National Safety Council, Chicago, 1997.

American Society of Mechanical Engineers Performance Test Code, (PTC) Overspeed Trip Systems for Steam Turbine Generators, 20.2-1965 Reference 4.

MIOSHA Standards cited in the NIOSH/CDC report can be found at the Consumer and Industry Services, Bureau of Safety and Regulation Standards Division website at http://www.michigan.gov/lara/0,4601,7-154-61256_11407_15368-,00.html (Link Updated 3/27/2013) Follow the links to Workplace Safety & Health, then Standards & Legislation, to locate and download MIOSHA Standards. The Standards can also be obtained for a fee by writing to the following address: Department of Consumer and Industry Services, MIOSHA Standards Division, PO Box 39043, Lansing, MI. 48909-8143. MIOSHA's Telephone Number is (517) 322-1845.

OSHA Job Hazard Analysis can be found on the OSHA website at http://www.OSHA.gov/

Compressor Surge

What is a compressor surge, and how can it be recognized?

Compressors are large, complex machines requiring critical support equipment and trained personnel to ensure safe and reliable operations. See Figure 8.37.22 for a photograph of a typical refining or petrochemical plant centrifugal compressor.

Centrifugal compressors are large, complex machines and are the "heart" of many process units! See Figure 8.37.23 for an image of one of these machines in service. Figures 8.37.24 and 8.37.25 illustrate cutaway views of these complex machines.

Surge – is the point at which the centrifugal compressor cannot add enough energy to overcome the system resistance or backpressure. This is typically caused by flow restrictions on the discharge of the compressor, such as a throttled discharge valve. It can also occur due to restrictions on the inlet to the compressor.

This restriction results in a rapid flow reversal (i.e., surge) and can result in loud noise, high vibration, temperature increases, and rapid changes in axial thrust.

Surge can damage the rotor seals, bearings, compressor driver, and other equipment. It can result in a catastrophic failure of the machine if it is allowed to continue for a significant period.

A surge can also cause the machine to go into an overspeed condition before the control system can react. In the event of an overspeed, the overspeed trip mechanism' is designed to shut down the machine.

Figure 8.37.22 An image of a typical refining or petrochemical plant centrifugal compressor with a motor driver. This vertically split compressor design has the gas piping inlet and outlet lines mounted at the bottom of the compressor. They are not visible in this image. Source: The Board of Trustees of the Science Museum/https://collection.sciencemuseumgroup.org.uk/objects/co8415275/national-grid-partington-collection-gas-storage/last accessed January 18, 2024/CC BY 4.0.

Figure 8.37.23 A large, complex centrifugal compressor and gearbox. The driver, typically a motor or steam turbine, is out of view.

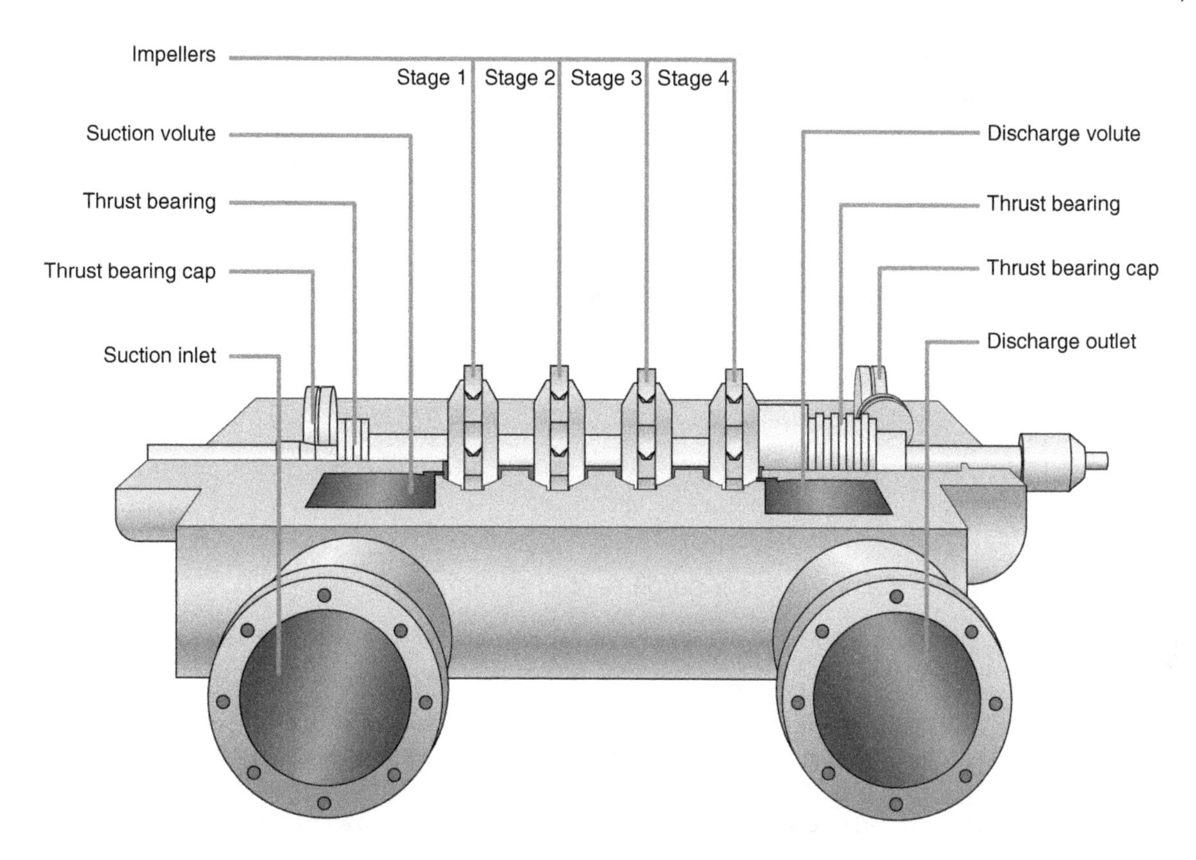

Figure 8.37.24 Artist rendition of a cutaway view of a centrifugal compressor. Source: Image courtesy of Ecoscience Resource Group Visual Artists.

Figure 8.37.25 Photo of an "opened" centrifugal compressor showing the impellers, compression stages, and flow volutes.

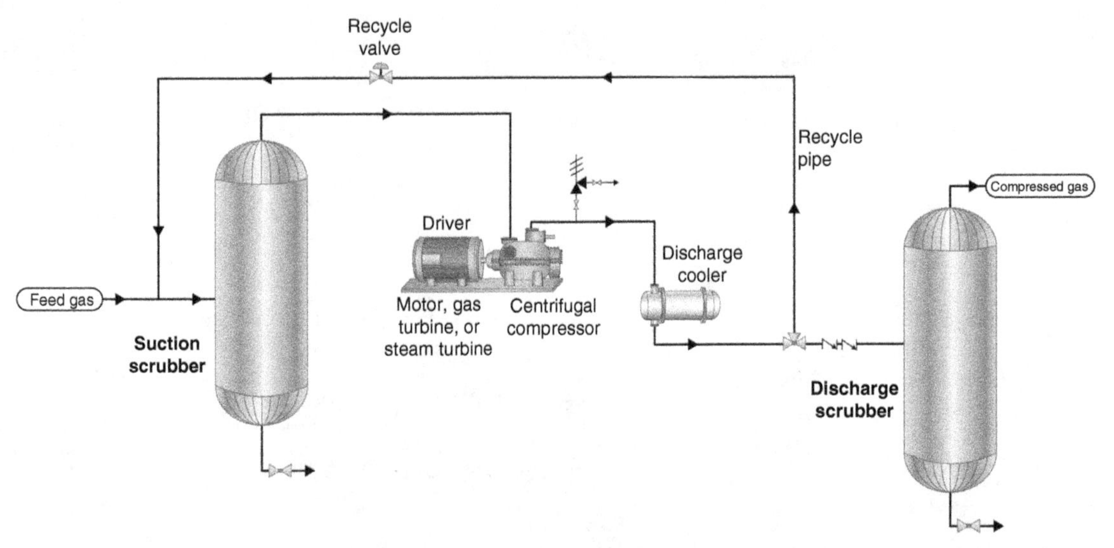

Figure 8.37.26 Centrifugal Compressor flow schematic highlighting the minimum flow recycle loop. Source: Image courtesy of Ecoscience Resource Group Visual Artist.

These problems of Surging are usually prevented by an anti-surge system and control valve, which, externally, recycles discharge gas back to the suction side to maintain a minimum flow rate to the machine. This is illustrated in the flow schematic in Figure 8.37.26.

Because compression causes the temperature to increase, the recycled gas is normally taken from the outlet side downstream of the aftercooler.

A Process Computer, such as the Distributed Control System (DCS), can be programmed to manipulate the compressor controls to help prevent compressor surge. It is always a good idea for the Process Operators to understand how this process works in the event manual control is required.

Mini Case Study of Compressor Surge (Courtesy of KnightHawk Engineering, Inc.):

This compressor failure occurred at a refining facility located along the Gulf Coast of the USA. This installation was a 6,000 HP two-stage process gas compressor in catalytic cracking gas service. It experienced a catastrophic failure as the unit was in the process of shutting down for a planned turnaround. A detailed failure analysis was performed on the compressor after the failure to help determine the cause and to recommend corrective actions to prevent another similar failure. The failure analysis concluded that all data strongly suggested the root cause of the failure was due to the surging of the compressor without adequate surge protection. The analysis also concluded that without surging, the failure would not have occurred. See Figures 8.37.27, 8.37.28a,b, and 8.37.29 for images associated with this incident.

Another serious example of compressor surge resulting in damage to the machine is highlighted in the McGraw Hill Compressor Handbook with a case study by Ted Gresh of a large centrifugal compressor damaged due to surge. The damaged machine is highlighted in Figure 8.37.30.

Hazard of Compressor Reverse Overspeed

A centrifugal compressor may reach a dangerous overspeed in reverse if the discharge nonreturn valve (check valve) fails to close. A good practice is to use either twin discharge check valves (as illustrated in Figure 8.37.26) or a power-operated trip valve in series with a nonreturn valve to guard against the potential for reverse flow.

Due to the "stored energy" in the pressurized outlet gas line, it is important to minimize the volume of gas between the compressor discharge and the nonreturn valve when the compressor is idle.

Reciprocating Compressors

Reciprocating compressors are a special breed of compressor, and I chose to cover these separately due to their uniqueness. These are typically referred to as "recips" in the industry and function as positive displacement machines. This means that they are designed to compress air or other gases and are not designed for liquids. In fact, reciprocating compressors have been severely damaged or destroyed when liquids enter the cylinders. I will talk more about this later.

Figure 8.37.27 Photo showing an overall view of the damaged centrifugal compressor. Source: Courtesy of KnightHawk Engineering, Inc.

Figure 8.37.28 Photo showing North and South views of the damage. These two images clearly the catastrophic damage that can occur with a compressor is operating in surge. Source: Courtesy of KnightHawk Engineering, Inc.

Figure 8.37.29 Photo showing Recreation of the damaged compressor as part of the failure analysis process. Source: Courtesy of KnightHawk Engineering, Inc.

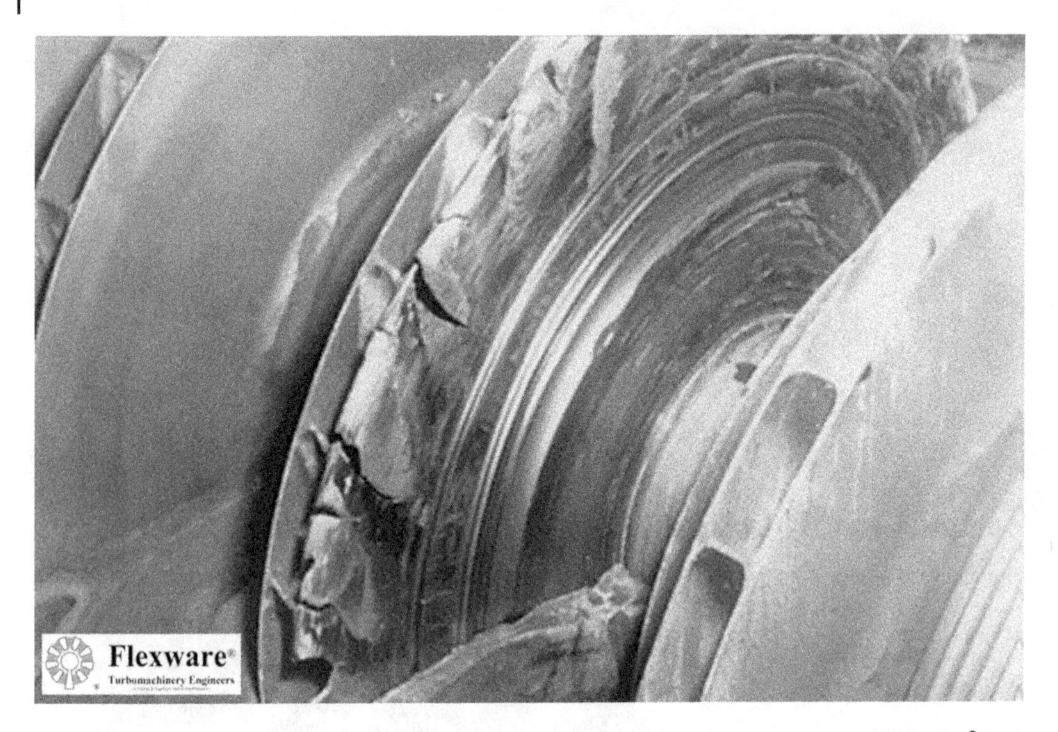

Figure 8.37.30 Centrifugal Compressor Damaged due to Surge. Source: Image courtesy of Flexware® Turbomachinery Engineers.

The "recip" compressor is designed with cylinders and reciprocating pistons driven by a rotating crankshaft to compress the gases. They can be a single stage, but most are multistage machines, meaning the pressures rise as the gases flow through the stages in series flow. Multistage machines can deliver very high pressure and are widely used by petroleum refineries and petrochemical plants for this purpose. A typical installation is a multistage reciprocating compressor in Hydrogen service on a gasoline or distillate hydro treater or hydrocracker. The reciprocating compressor delivers Hydrogen at pressures more than reactor pressure in order for the gas to enter the reactor. Pressures of 3,000 psi or higher are common for these machines. See Figure 8.37.31 for an illustration of the functionality of reciprocating compressors, and Figure 8.37.32 for an image of a modern multistage reciprocating compressor. Drivers are typically electric motors, typically with a gear reduction box, but in some rare cases steam turbines may also be used.

The way these machines work, the incoming gas enters the suction manifold for the compressor and flows through the intake valve to the compression cylinder. The reciprocating piston then compresses the gas before the discharge valve opens. Then when the discharge valve opens, the higher-pressure gas is discharged to either the intake of the next compressor stage, or to the outlet manifold. The compressor can be either single-or dual-acting design. In the single acting design, compression occurs in one direction only, while in the dual-acting design, compression occurs on both ends of the piston while stroking in either direction. Most modern designs are of dual-acting design since they are more efficient and take up about the same footprint.

There is a nice animation video on YouTube by ACI Services, Inc. that illustrates how the piston and valves function in a dual-acting reciprocating compressor. It can be found at the following link: https://www.youtube.com/watch?v=E6_jw841vKE.

In industrial applications, it is common to have the driver or gearbox located in the center of the compressor with two dual-acting compression cylinders on each side of the driver. This provides not one stage of compression, but multiple stages. The multiple stages can be piped into a single stage for one service, plus multiple stages for other services, all driven by one driver. This provides a lot of flexibility for designers.

In addition to Hydrogen services, reciprocating compressors are used in a wide range of services, ranging from light hydrocarbons, air, and other dry gases. It is essential that the gases are dry as these machines do not and cannot handle liquids.

Note the large motor driver to the right of the image with the rotating shaft between the two sets of reciprocating multistage pistons.

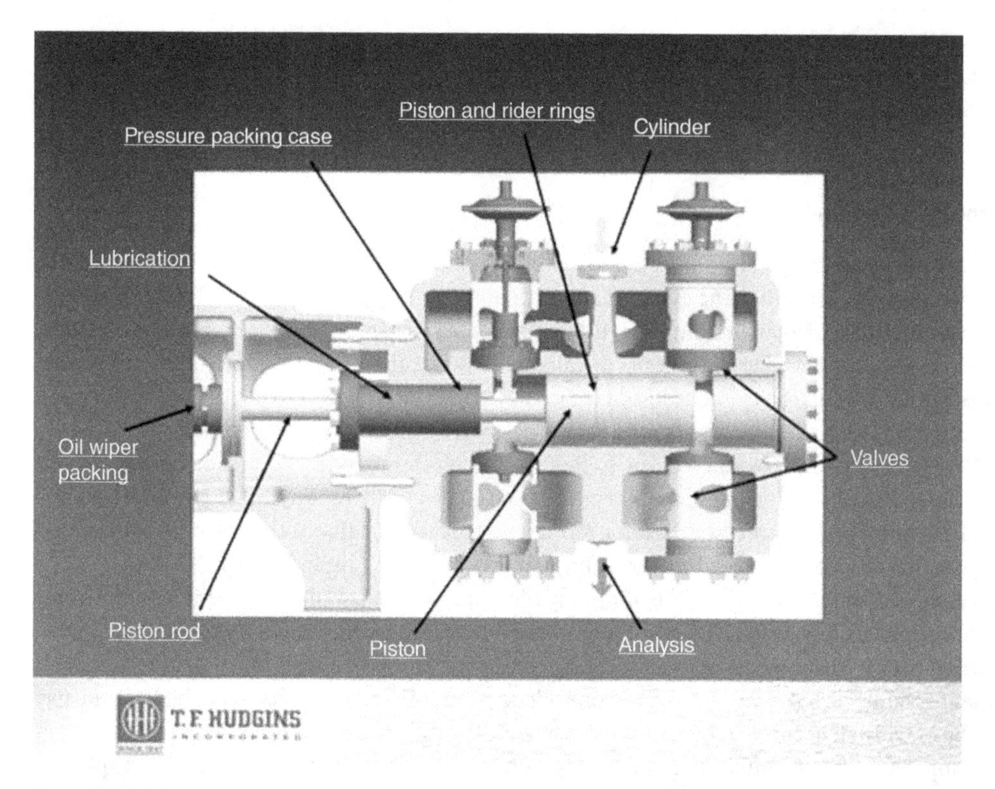

Figure 8.37.31 Illustration of a dual-acting Reciprocating Compressor. Source: This illustration provided courtesy of T. F. Hudgins company, an Allied Reliability Brand (Annotation by T. F Hudgins).

Figure 8.37.32 Photo of a large modern reciprocating compressor. Source: U.S. Department of Energy, National Energy Technology Laboratory Final Report "Advanced Reciprocating Compression Technology." Prepared by The Southwest Research Institute.

Typical Instrumentation Associated with Reciprocating Compressors

Most modern reciprocating compressors include the following instrumentation, most are cascaded to compressor shutdown systems to prevent damage to the machine.

- Low Lube Oil Pressure.
- Crankshaft Bearing Vibration.
- Frame Vibration.
- Rod Drop (indicating wear in the rider rings).
- Gas Discharge High Temperature.

- High Liquid Level in Suction Knockout Drums.
- Cylinder And Packing Lubricator Low Level
- Process gas pressure and temperature

Hazards of Liquids in Reciprocating Compressors

Reciprocating compressors are not designed to handle liquids. If you were to remove the cylinder head, for example during compressor maintenance, you would notice that when the cylinder is at the end of the compression stroke, fully extended in the cylinder, the piston is within thousandths of an inch from the cylinder head. And, of course, liquids are not compressible; therefore, should liquids be carried into the compressor there is a collision between the piston and the liquid at the end of the compression stroke. This has resulted in the cylinder heads being blown completely off the end of the cylinder and other major damage to the machine. In extreme cases, the compressor is destroyed by liquid in the cylinder.

An employee was killed in 1996 when a compressor failed due to liquid entering the cylinder while the compressor was running. This was reported by Federal OSHA in Incident summary Nr: 200810026. OSHA reported that employee #1 and coworkers were starting to load the number 2 compressor used for refinery vapors when a liquid slug was introduced into the cylinders. This caused a hydraulic overload of the cylinder, resulting in an explosion. Employee #1 was killed.

I have had a photograph of a damaged reciprocating compressor in my files for at least a couple of decades. I do not know where the image came from; unfortunately, I do not have permission to use it here. However, it clearly shows exactly what can happen when liquid is carried over into an operating reciprocating compressor. In this image, the machine's frame is broken, and the crosshead, the mechanism that connects the piston rods to the crankshaft, has been pushed completely out of the case and exposed to the outside environment. Of course, this machine is essentially destroyed by liquid carryover into the cylinders. When this happens, a loss of containment of the high-pressure gas can occur, resulting in an explosion or major fire. This exemplifies why it is important to have liquid knockout drums to remove the liquid before it goes to the compressor. It is also important to have high-level alarms and high-level trip devices set to take the compressor offline before it can be damaged by liquid carryover. These devices should be classified as safety-critical and be subjected to periodic testing to ensure their availability.

Figure 8.37.33 illustrates the key components in a reciprocating compressor that can be damaged by liquid carryover into the compressor.

Liquid in a reciprocating compressor has the potential to damage one or more of the following components:

1. Broken piston.
2. Broken cylinder head.
3. Broken piston rod.
4. Broken crankshaft.
5. Broken frame.

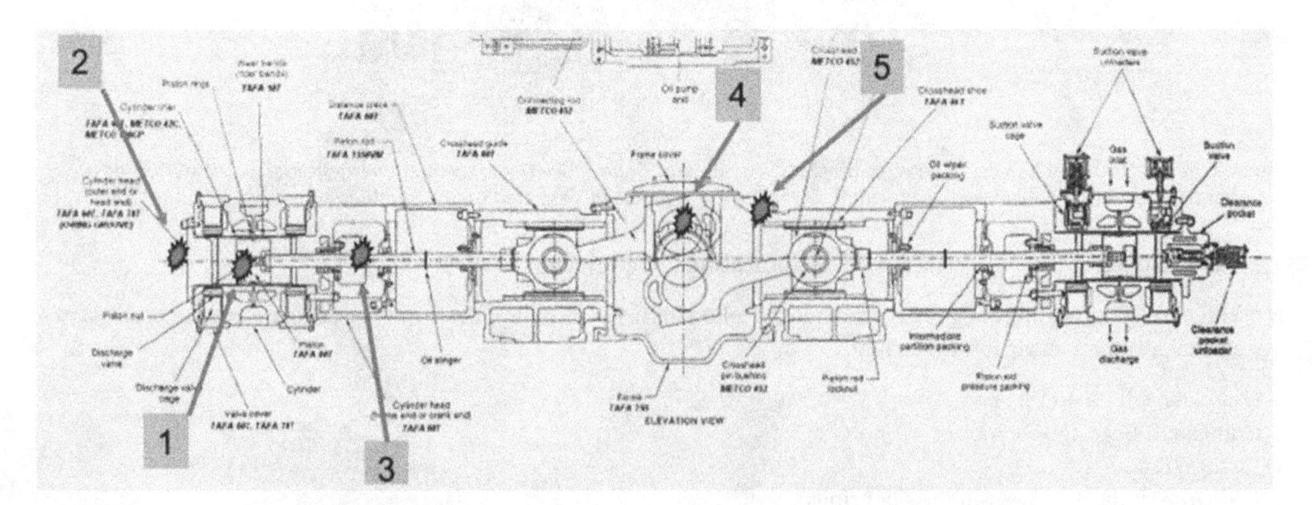

Figure 8.37.33 Reciprocating Compressor illustrating the potential damage due to liquid in the machine.

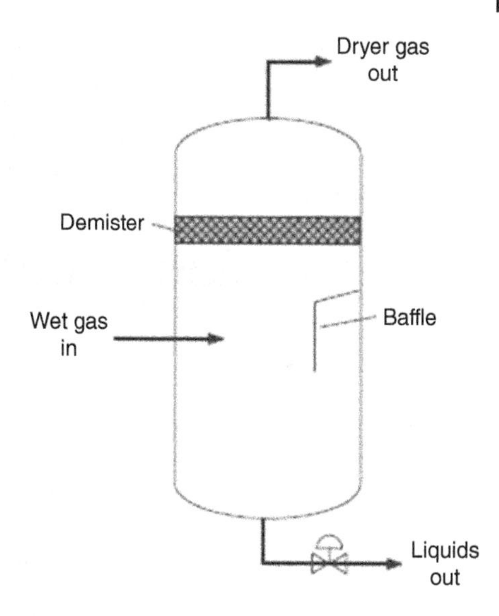

Figure 8.37.34 Typical refinery compressor liquid knockout drum. Source: Image courtesy of Ecoscience Resource Group Visual Artists.

Preventing Liquid Carryover into Reciprocating Compressors (Applies to Centrifugal and Reciprocating Compressors)

Frequent verification of liquid levels in compressor suction knockout drums is essential to ensure safe and reliable operation of these machines. Liquid must be periodically drained to prevent liquid build up and carryover into the compressor.

Liquid knockout drums shall also be equipped with redundant high-level alarms classified as Safety-Critical Devices, and all instrumentation, including these level indicators and high-level alarms and trips must be periodically checked and maintained to ensure they are working properly and to prevent failure. See Figure 8.37.34 for an example compressor suction knock out drum.

Other potential hazards associated with reciprocating compressors, all of which can result in damage to the machine:

- Failure of lubrication systems.
- Leaks or spills of lube oils (personnel safety and/or fire hazards).
- Coupling failure (motor or steam drivers).
- Startup and other transient operations.
- Motor terminal junction box (primarily a hazard to personnel during startup of motor-driven machines).
- Deadheading (running blocked in).

Additional References

McGraw Hill Compressor Handbook, Paul C. Hanlon Editor, Library of Congress Cataloging-in-Publication Data: Includes index ISBN 0-07-026005-2. http://www.irantpm.ir/wp-content/uploads/2014/12/Compressor-Handbook.pdf

Article on compressor surge, including how to prevent, by Wikipedia. https://en.wikipedia.org/wiki/Surge_in_compressors

Failure Analysis of damaged FCC compressor, by KnightHawk Engineering, Inc.. https://turbolab.tamu.edu/wp-content/uploads/2018/08/Case-Study-04.pdf

Understanding Centrifugal Compressor Surge and Control, by Anup Kumar Dey (Blog). https://whatispiping.com/understanding-centrifugal-compressor-surge-and-control/

Controlling Surge in Centrifugal Compressors, by Emerson (Jim Cahill and Mikhail Ilchenko – Bloggers). https://www.emersonautomationexperts.com/2019/control-safety-systems/controlling-surge-centrifugal-compressors/

What is compressor surge? by Turbomachinery staff and contributors. https://www.turbomachinerymag.com/view/what-is-compressor-surge

Article in Hydrocarbon Processing related to Reciprocating Compressors. https://www.hydrocarbonprocessing.com/news/2019/04/siemens-to-supply-reciprocating-compressors-for-steam-methane-reformer

U.S. Department of Energy, National Energy Technology Laboratory Final Report, "Advanced Reciprocating Compression Technology (ARCT)". https://www.osti.gov/biblio/876564. Prepared by The Southwest Research Institute.

Chapter 8.37. Special Hazards of Rotating Equipment

End of Chapter Quiz

1 What are the three special hazards associated with rotating equipment?

2 Explain the common cause of pump/driver reverse overspeed.

3 What are failure modes associated with pump discharge check valves? Name any three of the following.

4 What steps should you take if you suspect a pump is operating in reverse rotation?

5 Why is it important to take a position out of the line-of-fire when closing the pump discharge valve to stop the reverse rotation?

6 What should be available for training and use in the field in a reverse overspeed situation?

7 Following a reverse overspeed incident, what additional actions should be taken to help prevent a reoccurrence?

8 Steam turbine cases are designed to see full system steam pressure.
- True
- False

9 A steam turbine is subject to overspeed under which of the following conditions?

10 What is designed to prevent a steam turbine from overspeeding?

11 Briefly explain how a mechanical overspeed is designed to function, thereby preventing the turbine from overspeeding.

12 What are examples of checks that should be done before placing a steam turbine in service?

13 What is the first step in preparing to perform a test of the turbine overspeed trip device?

14 Briefly explain what is meant by surge in a centrifugal compressor.

15 How is compressor surge controlled or prevented?

16 What can result in reverse overspeed of a large centrifugal compressor?

17 What is the significant difference between a centrifugal compressor and a reciprocating compressor.

18 Please select the three significant hazards associated with the operation of a reciprocating compressor.
a. Liquid entrainment into the compressor.
b. Compressor surge.
c. Loss of lubrication.
d. Running the machine deadheaded.
e. All of the above.

8.38

Two Insidious Hazards Associated with Petrochemical Plants (Popcorn Polymer and Ethylene Decomposition)

As I was wrapping up this book, the catastrophic explosion occurred in Neches, Texas at the TPC Group Chemical Plant Butadiene Unit. This incident was investigated by the US Chemical Safety Board and their report identified the cause of the incident as "Popcorn Polymer." I cover this as a known hazard in Chapter 8.10 "Hazards of Piping Dead Legs." Dead-leg piping is identified as piping with little or no process flow. After this incident, I felt that Popcorn Polymer merited a separate chapter, along with another insidious hazard which is also typically limited to petrochemical plants that is known as ethylene decomposition. The two are unrelated other than that they can occur in petrochemical plants. Both will be covered in a little more detail here.

Popcorn Polymer

Popcorn Polymer forms by olefin polymerization which can occur when a conjugated diolefin is exposed to heat over a prolonged period, such as in a tower or piping system with little or no flow, for example, in a section of dead-ended piping or in a tray downcomer where there is limited flow. In areas of stagnant flow, the olefin is exposed to heat and time promoting the polymerization of these olefins. Popcorn polymer occurs predominantly in higher concentration 1,3-butadiene streams, although it has also been reported to occur in styrene and isoprene, although less frequently reported. The conjugated diolefin structure associated with butadiene makes it much more reactive and therefore more likely to produce popcorn polymer. Once this reaction is started, the polymerization reaction repeats, and will continue until it either reacts with a terminating radical (such as an inhibitor), or all monomer is consumed. Popcorn polymer is a polymeric resinous material with typically a white or light color and closely resembles popcorn in appearance. It is a porous type of material.

The exact cause of popcorn polymer is still the subject of study; however, it is thought that polymerization begins with a seed site of insoluble product and rapidly progresses from there at logarithmic rates. It has been reported that polymerization is limited to streams containing more than 70–80 wt% butadiene; however, cases have been reported in much lower concentrations. The growth formation is catalyzed by the popcorn itself, by corrosion by-products in the equipment such as rust scale, and even by water, particularly with small amounts of peroxides present. What is also known is that the polymerization reaction is extremely exothermic and can cause equipment damage and loss of containment. The rapidly expanding nature of the reaction results in extreme internal pressures and can rupture piping or vessels. Popcorn polymer has also been found to be pyrophoric and can result in the ignition of other hydrocarbons or combustibles nearby.

Please refer to Figure 8.38.1 for an image of popcorn polymer courtesy which was found in another unit, not from the TPC Group Chemical Unit discussed here.

The significant risk of popcorn polymer is that it is an extremely fouling material and can plug piping, fractionation columns, exchangers, and other process equipment. Even more hazardous, when polymerization occurs, the molecules expand dramatically and can rupture the piping or vessels that serve as containment. I have seen photos of an exchanger shell that was ruptured by the force of popcorn polymer. In the TPC Group explosion, the popcorn polymer formed in an idle section of piping that went to the spare pumps which had been out of service for an extended period. When it formed in the piping, it resulted in extreme internal pressure rupturing the fractionator bottoms line resulting in loss of containment and the explosion.

Once popcorn polymer is found, all the material must be removed to eliminate the seed sites to prevent a return of the phenonium. Inhibitors can be used to help control the polymerization rate and limit the growth of the polymer inside process equipment. Of these, the three most common are sodium nitrite, tertiary butyl catechol (TBC), and diethyl hydroxylamine (DEHA). Methyl ethyl mercaptan has also been reported as effective in mitigating popcorn polymer.

Process Operations Safety: The What, Why, and How Behind Safe Petrochemical Plant Operations, First Edition. M. Darryl Yoes.

Figure 8.38.1 Image of Popcorn Polymer discovered in a process Fractionation Column (not from the TPC Group Chemical Plant). Source: Image courtesy of the US Chemical Safety and Hazard Investigation Board and The International Institute of Synthetic Rubber Producers.

Units in butadiene, styrene, isoprene, or similar services should be mindful of this hazard and be aware of excess differential pressures in equipment in these services. Differential pressures that are trending higher may be an indication of internal fouling which may be caused by polymer forming in the fractionator or exchanger internals or in associated piping systems. It may be a good idea to periodically conduct differential pressure surveys on this equipment as a diagnostic method to help identify areas where polymerization may be occurring. Other nondestructive testing may also be useful in determining the presence of polymers in the equipment such as radiographs (X-rays).

So, what happened at TPC Group Chemical Plant on 27 November 2019? You will notice in Figure 8.10.6 that the plant was designed with two separate pumps for the fractionator bottoms system, a fractionator primary transfer pump, and a common spare transfer pump. The suction piping for these two pumps was configured as two separate runs of piping. This configuration means that when one of the pumps was in service, the other pump and its associated suction pipe were not in service and had no flow. This idle piping provides the perfect opportunity for polymerization to occur in the low-flow or no-flow regime. With this piping configuration, it was also practically impossible for the site to have known that fouling or plugging was occurring in the idle suction piping. Without some type of nondestructive testing, there would have been no indication that this was occurring.

Key Lessons Learned

- Popcorn Polymer is an insidious hazard for petrochemical plants, especially plants processing 1,3-butadiene.
- Each site processing butadiene should have practices in place to help prevent internal polymerization from occurring and to identify and control its growth.
- For example, these sites should identify and eliminate any areas exposed to the process and identified as dead legs or other areas where there is low or no flow.
- Consideration should be given to the development of nondestructive testing to help identify areas where polymers may be forming. For example, periodic radiograph scans and reviews.
- In areas where polymers have been found or where it has occurred in the past, consideration should be given to the use of inhibitors to help retard the continued polymerization.

What Has Happened/What Can Happen

Please refer to Chapter 8.10 of this book for information on the TPC Group Chemical Plant explosion and the causal factors (Dead-leg Piping Hazards).

Additional References

Digital Refining, January 2015 – Safe removal of polymeric deposits. https://www.digitalrefining.com/article/1001034/safe-removal-of-polymeric-deposits#:~:text=Popcorn%20polymer%20(also%20called%20%CE%B1,a%20small%20specific%20inner%20surface

US Chemical Safety and Hazard Investigation Board (US CSB), Final Report: TPC Group Chemical Plant Explosion – Port Neches, Texas, November 27, 2019. https://www.csb.gov/tpc-port-neches-explosions-and-fire/

For additional references please refer to the those listed in the CSB report.

Ethylene Decomposition (Decomp)

Under conditions of high pressure and temperature, ethylene can decompose into methane and carbon. This highly exothermic reaction, once initiated, is difficult to terminate. Under conditions of high pressure, typically greater than 1,000 psig (69 bar), and elevated temperature, typically over 600 °F (315 °C), ethylene can decompose with a significant release of energy in the form of heat. The resulting decomposition can be a low-level decomp resulting in very small carbon particles, and possibly the fouling of equipment such as instrumentation. Usually, these are not noticed until the equipment is opened for inspection or maintenance.

However, the decomp can also take the form of a thermal reaction and a rapid escalation of temperatures, resulting in equipment damage or even loss of containment. There have been several reports of thermal decomposition of ethylene in industry, which indicates the hazard of ethylene decomp is not just limited to only those facilities that are processing high-purity ethylene. Studies have shown that there are several factors which can increase the probability of ethylene decomposition, such as:

- The presence of small concentrations of oxygen or air in the system.
- The sudden compression of the ethylene pipeline, such as opening a valve from a system operating with significantly higher pressure. This results in the sudden compression of the ethylene and can trigger a decomp reaction.
- The exposure of ethylene-containing equipment to an adjacent equipment fire impinging on the ethylene pipeline or other equipment.
- The presence of a catalytic-like material such as corrosion by-products; rust scale such as iron oxides or similar materials.
- The flow rate of ethylene as opposed to the rate of flame propagation in the system.
- Other initiators such as a shock wave or impact.
- The physical size and configuration of the pipeline or equipment.
- The rate of heat dissipation affects the probability of a decomp occurring.

Facilities or operations where ethylene decomp can occur or has occurred:

- Hot-tapping an ethylene pipeline should normally be avoided. The pipe wall's high temperatures, along with the chaff from the tapping process, could trigger a decomp event. The temperatures can affect the ethylene, and the chaff may serve as a catalyst to help start the decomp event.
- Reciprocating compressors in ethylene service should normally be avoided if possible. The high temperatures of the compressor discharge have the potential of triggering an ethylene decomp. Also, hot spots, such as bad valves, can result in higher temperatures.
- Commission of ethylene pipelines requires detailed procedures which should be closely followed in the field. The use of high-pressure nitrogen to purge the pipeline can result in very high internal temperatures when pressurized from atmospheric pressure to pipeline pressure. These temperatures can exceed the decomp temperature of ethylene and may result in a decomp event.

The following is a generic procedure for returning an ethylene line to service and repressuring the line with ethylene. This is provided only as an example; use the procedure developed by your technical group specifically for your facility or system.

When purging with nitrogen for oxygen freeing, the procedure should ensure the nitrogen pressure is reduced to just above atmosphere pressure, and the pipeline is verified to be nondetectable for O_2 content, and to ensure no moisture is present. The pipeline should be at a very low dewpoint before admitting ethylene.

When admitting ethylene, the pipeline pressure should be limited to no more than around 50–100 psig for the first fill. After a waiting period of 10 minutes, slowly reduce the pressure to near atmospheric before slowly repeating this pressurization back to 50–100 psig. This low-pressure pressurization cycle should be repeated for a minimum of five times before beginning the line pressurization.

Following this, the pressure should be raised to 200 psig, followed by another waiting period. This cycle is then continued raising the pressure 100 psig each time and waiting for about 10 minutes between cycles. This should be continued until the line is at the ethylene system pressure.

The procedure should also ensure that other considerations as possible initiators include the elimination of corrosion by-products in the pipeline (rust scale), the possibility of external heat tracing failure, or electrical shorts in cathodic protection systems, and the potential for external fire that may impinge on the pipeline, etc.

- Direct-fired or electrical heaters should be avoided for ethylene service due to the potential for hot spots or electrical faults. Close controls must be in place for steam heaters to ensure continuous ethylene flow and close control of temperatures. Guidelines on How to Reduce the Hazards of a Sudden Decompression Decomp Event:
- Always ensure that technical is involved in developing/approving the procedure.
- If a small-bore bypass is available – use it to very slowly depressure the nitrogen high-pressure side into the lower ethylene pressure side. Very slowly depressure the nitrogen side using the small-bore bypass valve.
- Use the vent valve to depressure the nitrogen side to a safe location if one is available (e.g., to a flare, a blowdown system, or other process vessel).
- If the bypass or vent is not available, the Operator should pursue other avenues to safely depressure the nitrogen before opening the ethylene line to nitrogen pressure. Generally, a safer alternative is available.
- Generally, we want to avoid any scenario where we expose trapped ethylene to high-pressure nitrogen as this can result in ethylene decomposition.

Key Lessons Learned

- Ethylene decomposition is another insidious hazard for petrochemical plants, especially olefins plants handling large amounts of high-purity ethylene.
- Procedures should be detailed regarding commissioning ethylene pipelines and equipment, including precautions about avoiding rapid compression of ethylene-containing pipelines and equipment.
- Commissioning of ethylene pipelines and equipment requires patience of all concerned to ensure that procedures are fully complied with.
- Personnel should be aware of other potential initiators such as rust scale and potential external heating of the pipeline or equipment.

What Has Happened/What Can Happen

Ethylene Decomposition in a Pipeline

Dr. Trevor Kletz in "Lessons from Disaster" (IChE), has an example of an ethylene decomposition that occurred in a 12″ pipeline in ethylene service operating at 1,300 psig. In this case, a motor-operated valve in a side connection was in the closed position to facilitate the repair of an instrument. A motor-operated valve on the side connection was reopened and the ethylene rapidly pressurized air that was trapped between the valves resulting in a rapid temperature rise of about 800 °F (in about 3 seconds). Dr. Kletz reported that ethylene acted as a piston compressing the air and resulting in the temperature rise due to adiabatic compression. He reported that the resulting decomposition traveled about 290′ against the stream and the ethylene started decomposing. The rapid temperature rise resulted in failure of the ethylene line and loss of containment and fire.

The data suggests that the compression heating caused by one gas (in this case, the ethylene) of another gas can also happen to other gases as well. For example, the data suggests that this heating can happen with nitrogen, hydrogen, and possibly other gases as well.

It is interesting that Dr. Kletz says that this hazard can prevented by slowly pressurizing the static or dormant sections of piping.

Dr. Kletz provides a reference as McKay F.F. et al., Hydrocarbon Processing, November 1977, 56(11):487.

What Has Happened/What Can Happen

Ethylene Decomposition After a Compressor Startup

Another ethylene decomposition incident occurred at an olefins plant in Brazil when an ethylene compressor exploded during startup. The cause was determined to be the decomposition of ethylene in the presence of a diatomic gas, most likely air or nitrogen, at low pressure (reportedly around 140 psig).

Additional References

Britton, L. G., Taylor, D. A. and Wobser, D. C., "Thermal Stability of Ethylene at Elevated Pressures", Process Safety Progress – October 1986.

Melching, J. S., "Ethylene Decomposition, Process Safety Issues in Ethylene Plants", Ethylene Producers Conference; Houston, TX – April 2007.

McKay, F. F, Worrell, G. R., Thornton, B. C., and Lewis, H. L., "If an Ethylene Pipeline Ruptures", Hydrocarbon Processing – November 1977.

Chapter 8.38. Two Insidious Hazards of Petrochemical Plants (Popcorn Polymer and Ethylene Decomposition)

End of Chapter Quiz

1. What are the two insidious hazards associated with petrochemical plants?
   Answer: ______________________________

2. Which of the following are the unique hazards associated with Popcorn Polymer?
   a. Popcorn polymer can plug or foul equipment internals.
   b. Popcorn polymer may be pyrophoric.
   c. Popcorn polymer exerts extreme internal pressure and can damage equipment and result in loss of containment.
   d. Popcorn Polymer is a white flexible material that appears like plastic.

3. Which of the following chemicals are most susceptible to the formation of popcorn polymer?
   a. Ethylene
   b. Nylon
   c. 1,3-butadiene
   d. Propylene

4. In what areas of the plant is popcorn polymer most likely to form?
   Answer: ______________________________

5. Name one type of inhibitor that may be effective in mitigating popcorn polymer.
   Answer: ______________________________

6. What is the hazardous nature of ethylene decomposition?
   Answer: ______________________________

7. How can an ethylene compression explosion occur during the commissioning of an ethylene pipeline?
   Answer: ______________________________

8 Which part of the organization should be involved in developing the procedures for ethylene pipeline commissioning?
Answer: ____________________

9 What is the allowable amount of oxygen in an ethylene pipeline before admitting the ethylene?
a. No more than 5% O_2 is allowable.
b. 5–10% is allowable.
c. O_2 reading should be nondetectable before admitting ethylene to the pipeline.
d. There is no requirement for checking O_2 before admitting ethylene.

10 What are some other potential initiators for an ethylene decomposition incident? Rust scale from the inside of the pipeline.
a. Overheating due to faulty electrical heat tracing.
b. High discharge temperature on the ethylene reciprocating compressor.
c. Low temperature from the inhibitor injection system.

11 When commissioning an ethylene pipeline, the startup procedures are only a guide. What is more important is the experience level of the Operations Technicians who are performing the work.
- True
- False

12 When commissioning an ethylene pipeline the sudden exposure of ethylene to the high nitrogen pressure in the pipeline can result in an ethylene decomposition event and possibly an explosion.
- True
- False

9

Importance of Operations Emergency Response Drills and Exercises

For those who remember the successful water landing on the Hudson River on 5 January 2009, I believe all would agree that Captain Chelsea Sullenberger and his entire crew did an incredible job of safely bringing everyone home that fateful day. After striking a flock of birds and with zero thrust in both engines, Captain Sully, First Officer Captain Jeffrey Skiles, and the rest of their crew successfully water-landed an Airbus A-320 aircraft with 155 souls onboard.

It is also noteworthy that the operators and tenders for all the marine vessels that helped rescue passengers and crew from the aircraft's wings and the cold waters of the Hudson River also were well-trained. As a result, every person was saved by a successful water landing, and saved by the skills of the emergency response personnel.

This was no accident; these people knew what they were doing and carried out their mission well.

Our roles are no different; the emergency will happen when we least expect it to occur. It will be in the middle of the night and possibly during bad weather. One thing is sure, when it happens, we must be just as ready as Captain Sully and his crew and those boat crews on the Hudson River. We can affect as many people as Captain Sully and his crew, especially when a site incident can also impact many outside the facility. For example, on 27 November 2019, several major explosions occurred at the TPC Group Plant in Port Neches, Texas, evacuating more than 60,000 residents from the surrounding communities.

We must be well-trained, know our equipment, and know our emergency response procedures. Because when the incident happens, there will be no time to pull the P&IDs (drawings) or review the procedures. We must be ready to quickly respond and isolate the equipment, isolate or block in the leaking piping circuit, failed pump, or whatever has resulted in the loss of containment leading to an explosion or fire. And, of course, we must be able to do this remotely without putting ourselves or others in harm's way while isolating the equipment. We must know where the primary valves are and where the backup or remote valves are, and we must be ready to access those valves quickly.

For those who have not been through a major incident, please indulge me by hearing this out for a few minutes. First, every one of the large incidents I have been involved in appears to be absolute chaos and almost total confusion. Despite our best planning, our best drills, and exercises, it always seems like organized chaos. Most people are running away from the event, while members of the Operations group are running toward it to isolate the source. There are always a few others running in to help by turning valves; however, they need guidance on which valves to turn, so they need you or your attention. And again, they must be able to access the valves without putting themselves in harm's way. In most large incidents involving loss of containment, the other issue is the noise; the noise is just indescribably loud. You quickly realize that the radio on your hip is of absolutely no value. Most communications occur by standing next to the person you are attempting to communicate with and screaming into their ear. They are also screaming into yours. You quickly evolve to communication by hand signals.

Hopefully, you get the picture that you will be lost in this scenario without well-practiced emergency response drills and exercises. Of course, the reality is that you may go for years and possibly even an entire career without facing such an event. On the other hand, please just look at the long lists of incidents available just in this book, and I believe you will want to be prepared (see the partial listing just since BP Texas City in Chapter 8.1 of this book) ("Returning Equipment to Service").

There Are Basically Two Kinds of Drills

There are formal drills or exercises that are organized by the site leadership. There are various types of formal drills, for example, loss of containment or release drills, spill drills, etc. There are oil spill drills where we respond to a simulated spill onto the water, medical response drills, etc. These are the more formal drills scheduled and carried out by the site and stewarded to meet a leading indicator target. The formal drills may involve connecting the fire trucks and pumping water

Process Operations Safety: The What, Why, and How Behind Safe Petrochemical Plant Operations, First Edition. M. Darryl Yoes.

onto the equipment (wet drills or exercises). These are great, and I hugely support these types of drills. My take on these is that the more we practice, the better we should expect to perform during the event.

The Informal Drill

The other kind of drill is the informal variety; these are not generally scheduled, they are not stewarded or tracked in any way, but they are nonetheless just as important as formal drills. I am talking about training drills scheduled by the Unit or Area Operator, carried out by the Operator and their crew in the off-hours, weekends, or holidays when otherwise it is a relatively quiet period. If you are not currently doing these informal drills on your shift team, I believe you are missing out on a great learning opportunity.

Get your crew together when time is available on the night or weekend shifts. Discuss a very realistic scenario, not one that is super unrealistic. Select a scenario that could or has already happened, maybe on someone else's shift. Make sure your whole team is involved and discuss how your team would handle the incident if it happened during your shift. Talk in detail; what additional PPE will be required to respond? Where do we get the PPE? Who is going to do what? For example, who will turn on the fixed foam system or turret monitors in a significant pool fire incident? Who is going to stop the air fan coolers? Who is going to notify the emergency response team? Who is going to divert the feed from the unit, etc.?

Do not stop there! After a good discussion, the team should walk out to the field or unit as a group and identify and physically put their hands on the isolation valves. Do not turn valves, but physically locate the isolation valves on the unit and stop and think about where you are relative to where the fire would be if the incident were actual. Ensure that those valves would be accessible if this incident were real. Locate the emergency response equipment, the fixed foam system, the turret nozzles, the sphere water flood connections, etc. For example, talk through exactly how the foam system would be activated or connected to the water flood system. Try to picture what the scenario would look like if the incident were actual, and talk among yourselves about what you may do differently if this were real. Then, repeat this scenario several times until all team members are ready to respond, then move on to another equally important scenario. And if you do this discussion repeatedly, your team will be ready when the big one comes.

Ensure Each Response Drill or Exercise Is Critiqued

Critiques of each drill or emergency response exercise are very important. To the extent possible, ensure that the entire response group has an opportunity to participate in the critique and that each team member has a chance to discuss what went right and areas where improvement is needed. Documenting the areas identified for improvement and assigning follow-up responsibilities is also essential. These should be periodically tracked until the action items associated with the areas identified for improvement are completed and then tested in a subsequent drill or exercise.

Additional References

How to Plan for Workplace Emergencies and Evacuations U.S. Department of Labor Occupational Safety and Health Administration OSHA 3088 2001 (Revised) How to Plan for Workplace Emergencies and Evacuations (osha.gov)

BIC Magazine article "Practice makes perfect – emergency drills." Practice makes perfect – emergency drills – BIC Magazine

Principal Emergency Response and Preparedness Requirements and Guidance OSHA Occupational Safety and Health Administration OSHA.gov https://www.osha.gov/sites/default/files/publications/osha3122.pdf

Emergency Planning Guidance Notes – Refinery Emergency Planning Prepared on behalf of the CDNCAWE Major Hazards Management Group by the SpecialTask Force on Emergency Planning (MHISTF-1) W.A.G Bridgens (Chairman) rpt_88-6ocr-2004-01730-01-e.pdf (concawe.eu)

10

Ensuring Comprehensive Shift Turnover

By now, you are aware of the catastrophic process safety incidents where an ineffective or nonexistent shift turnover process played a significant role, either as a direct cause or as a major contributing factor. Certainly, this was the case in events such as the March 2005 BP Texas City Refinery explosion and fire (15 fatalities, 180 injuries), the July 1988 Piper Alpha oil platform explosion and devasting fire in the North Sea (167 fatalities and destruction of the entire platform), and the December 2005 Buncefield Terminal explosion and fire which occurred in Hemel Hempstead, Hertfordshire, England (20 tanks involved in the initial tank farm fire and most of the terminal was destroyed).

The following quote is directly from the report issued by the British Health and Safety Executive in their report on the Buncefield incident, "There is evidence to suggest that on the night of the incident, the supervisors were confused as to which pipeline was filling which tank." "This confusion arose because of deficiencies in the shift handover procedures and the overlapping screens on the Automatic Tank Gauging system."

These incidents, and others with similar results, indicate that we must improve the shift turnover process and establish expectations for a comprehensive shift turnover process by process shifting personnel. These improvements are typically made by each production or manufacturing side by including the shift turnover requirements in the site Integrity Management System (see Chapter 6 for more information on Process Safety Management Systems). The Integrity Management System will ensure that the process is subjected to the management system's continuous improvement and verification elements.

The basics of a successful and comprehensive shift turnover process include the following:

- The requirements of the shift turnover process are documented.
- The expected content of the shift turnover logs is documented and communicated to all affected personnel.
- The shift turnover is held at a "standardized" location where there is easy access to the shift log, the process computer, and other pertinent information needed for a successful shift turnover discussion.
- The required shift turnover verifications (audits) have been defined and are in place.
- The retention period for all shift turnover logs and associated documents has been defined.
- The expected benefit of a comprehensive shift turnover process is also communicated.

The shift turnover process should be documented in the Integrity Management System and spells out the general requirements for the process post personnel to follow. In other words, we do not make this up as we go; the process is documented to establish expectations that the shift turnover process is rigorous, with clear expectations for field compliance. It is a defined process to ensure accurate documentation and thorough communication between the incoming and outgoing process personnel responsible for the post assigned. The process sets clear expectations for a face-to-face discussion between the outgoing and incoming individual *before the outgoing person leaves their post*.

Each outgoing shift post shall develop a comprehensive shift log, which is the basis for the shift turnover process, and the log will be reviewed during the face-to-face meeting at shift turnover. The log should be "structured" to help ensure that all critical information is included during the shift turnover discussion. For example, the log should be structured with prompts for critical information such as any safety, health, or environmental incidents or near misses that may have occurred during the shift, any critical safety or environmental devices defeated or otherwise inoperable, and any critical alarms defeated or disabled, any mechanical work that is ongoing in the field, etc. This log can be a paper log or a computer-generated log.

I am a proponent of a computer log since the computer log is searchable, and entries can be easily retrieved for research or data analysis. Also, the computer log helps ensure that the log entries are legible. The log is expected to cover, at a minimum, the following critical information:

Process Operations Safety: The What, Why, and How Behind Safe Petrochemical Plant Operations, First Edition. M. Darryl Yoes.

- Information related to current plant operations, equipment, and changes during the shift (Safety, Health, and Environmental issues, any quality or reliability concerns).
- Status of abnormal and nonroutine operations and mechanical repairs.
- Status and contingency plan for equipment out of service or unavailable.
- Any outstanding issues with the potential to generate a process interruption or abnormal event and the plan to manage it.

The shift turnover should normally be held at a designated location with access to the shift log, the computer or console, and any other pertinent information or documents/records that may need to be accessed and reviewed as part of the shift turnover discussion. Also, there should be good access to the field operations whereby the participants can visit the field together to review and discuss some of the more complicated field issues where this is warranted. The outgoing and oncoming persons should visit the field to review complicated line-ups, the preparation of process equipment for opening or mechanical work, or where similar complex or difficult situations exist.

Both outgoing and incoming personnel should typically be dressed in standard work attire for the shift turnover discussion to accommodate the field visit and to ensure a readiness to respond should a major incident occur during the shift turnover discussion.

Immediately following the shift turnover discussion, the oncoming person should sign the shift log. This signature indicates that the shift turnover is complete, and the oncoming person has accepted responsibility for the shift and any ongoing activities.

Remember, shift logs and shift turnover entries may be considered legal documents following an unexpected personal or process safety incident. It is important that all entries in the shift logs are maintained in a professional and business-related manner.

The following is a list of typical shift team discussion items to be discussed at shift turnover (not a complete list and all items may not apply to all sites or areas):

- Any flawless operations items, including safety, health, and environmental incidents, near misses, and any issues with safety critical equipment.
- Any changes to the operations run plan.
- Alarms – especially any alarms that are inhibited or disabled.
- Any deviations to the operating limits or operating envelope and actions are taken to bring the parameters back within the normal range.
- Changes in circuit line-up and stream disposition.
- The status of fire protection equipment and other emergency response systems.
- Any environmental issues, for example, flaring or discharge to the sewer.
- Any emergency management of change documents generated during the shift.
- The status of safety or environmentally critical devices that are disabled or unavailable for any reason. Include the contingency plan for monitoring the process and protecting the equipment during the outage.
- Equipment preparation for maintenance - to be verified as prepared by the incoming shift (opening process equipment, lock-out/tag-out isolation, equipment that is drained and open).
- Any off-spec product streams with potential impact on the final product.
- Any inventory or tankage issues.
- Any shift maintenance activities or ongoing mechanical work (carried over from the prior shift).
- Staffing issues (i.e., unfilled posts, workers on extended hours, etc.).
- Any operating interfaces with upstream, downstream, sidestream, etc.

When done in this comprehensive way, the shift turnover results in no surprises later in the shift. After the shift turnover discussion is complete, and when the oncoming person is in the field making a process round, we do not want to find a process stream going to a disposition that we did not know about or mechanical work going on in the unit or process area that we have not heard about. A good process shift turnover will ensure that there are no surprises.

The Buncefield Terminal Incident

Let us look at The Buncefield Terminal incident located North of London in England, where poor shift turnover practices contributed to the outcome. In this incident, the terminal was essentially destroyed, has not been rebuilt, and probably will not be rebuilt due to the significant public outcry of this incident.

On the evening of 10 December 2005, tank number 912 at the Buncefield terminal was being filled with unleaded gasoline. The tank was equipped with two different level monitoring and control devices to help the Operators monitor the level in the tank during the fuel transfer operation and to automatically stop the transfer to prevent an accidental overfill in the event a high level is reached in the tank. The level monitoring device was a remote reading gauge to allow the Operators to monitor the level in the tank as the tank was filling. This device was linked to three different alarms: the "user" level, the "high" level, and the "high-high" level. The second device was an independent high-level switch (IHLS), which is designed to stop the flow to the tank in the event it is triggered by a high level in the tank.

The tank was accidentally overfilled, resulting in major fires in 20 storage tanks almost from the initial incident. This incident was investigated by the British Health and Safety Executive (HSE), the regulatory authority in the United Kingdom, and a detailed report was issued following their investigation. After the investigation, we know that the gauge on tank 912 "flatlined" or stopped moving at 3:20 a.m. Although the tank continued to fill, the level gauge was not changing. This meant that the user level was not changing, nor was the signal to the high-level or high-high-level alarm changing. When the level reached the independent high-level switch, it should have tripped and stopped the flow into the tank. As it turns out, this device was also disabled and failed to function. As a result, at 5:37 a.m., the gasoline level started spilling from the tank into the surrounding tank containment area. This continued unchecked until a large vapor cloud formed and reached an ignition source at around 6:01 a.m. Sunday, 11 December 2005. The massive explosion quickly engulfed the terminal, igniting 20 tanks. The HSE estimated that by the time the explosion occurred, over 66,000 gallons (250,000 liters) of petrol had escaped from the tank into the adjacent containment areas.

Fortunately, there were no fatalities, although the explosion and resulting fire injured 40 people. There is no doubt that the damage and human suffering would have been much more severe had the incident not occurred early on a Sunday morning when not many people were present. The fire burned for five days, nearly destroying the terminal and adjacent facilities. The primary failures reported by the HSE were the tank level gauge, the tank level monitoring system, and the independent high-level switch. However, the HSE report indicated several other significant contributing factors, including

Figure 10.1 An aerial view of the Buncefield Terminal before the fire on 11 December 2005. Contains public sector information published by the Health and Safety Executive and licensed under the https://www.nationalarchives.gov.uk/doc/open-government-licence/version/3/. Source: Reproduced by the kind permission of HSE. HSE would like to make it clear that it has not reviewed this product and does not endorse the business activity of Safety Consulting International, LLC.

Figure 10.2 An aerial photo of the Buncefield Terminal after the fire. Contains public sector information published by the Health and Safety Executive and licensed under the https://www.nationalarchives.gov.uk/doc/open-government-licence/version/3/. Source: Reproduced by the kind permission of HSE. HSE would like to make it clear that it has not reviewed this product and does not endorse the business activity of Safety Consulting International, LLC.

the shift turnover process. See Figures 10.1 and 10.2 for photos of the Buncefield Terminal before and after this devastating explosion and fire.

For example, quoting directly from the HSE report, "There is evidence to suggest that on the night of the incident, the supervisors were confused as to which pipeline was filling which tank. This confusion arose because of deficiencies in the shift handover procedures and the overlapping screens on the ATG system."

An effective shift turnover helps improve understanding and awareness of the status of operations across the team and helps ensure a seamless interface between the shifts. It provides continuity in operations, communicates changes, helps with problem resolution, and helps minimize abnormal and nonroutine operations. Shift turnover can also reduce duplication in effort, improve trust, and nurture teamwork as shift personnel learn to leverage each other's strengths. It helps to accelerate the transfer of knowledge and experience to newer operators and provides an accurate running record for updating shift personnel, especially those returning from extended absences.

An effective shift turnover process helps drive home the message; "We Are One Team."

Additional References

Human factors: Shift handover British Health and Safety executive – hse.gov.uk Human factors/ergonomics – Shift handover (hse.gov.uk)

BP America Refinery Explosion csb.gov BP America Refinery Explosion | CSB

Piper Alpha: The Disaster in Detail The Chemical Engineer July/August 2018 Issue 925/926 https://www.thechemicalengineer.com/features/piper-alpha-the-disaster-in-detail/

Buncefield: Why did it happen? The underlying causes of the explosion and fire at the Buncefield oil storage depot, Hemel Hempstead, Hertfordshire, on 11 December 2005 COMAH – the Competent Authority COMAH – Buncefield: Why did it happen? (hse.gov.uk)

https://journals.sagepub.com/doi/pdf/10.1177/0020294013479847 The Buncefield Incident – 7 Years on: A Review Measurement and Control 46(3): 76–82 © The Institute of Measurement and Control 2013 DOI: 10.1177/0020294013479847

Yokogawa America Shift Turnover Video Effective Shift Handover | Yokogawa America

11

The Importance of Leading and Lagging Indicators for Process Safety

BP experienced a catastrophic explosion and fire at its Texas City refinery on 23 March 2005, resulting in 15 fatalities and over 180 people injured, some very seriously. Following the explosion, the US Chemical Safety and Hazard Investigation Board conducted a comprehensive investigation resulting in a number of recommendations, including one for BP to appoint an independent panel to review BP's corporate safety culture and management oversight of its US refineries. This independent panel was chaired by James A. Baker III and became known as the "Baker Panel."

The Baker panel made several recommendations to help improve process safety at BP including a recommendation to implement a system of Leading and Lagging Indicators for process safety. The board's recommendation extended the initiative on Leading and Lagging Indicators beyond BP to the refining and chemical processing industries.

An excerpt from the Baker Panel's report is inserted below:

> "RECOMMENDATION #7 – LEADING AND LAGGING PERFORMANCE INDICATORS FOR PROCESS SAFETY
>
> BP should develop, implement, maintain, and periodically update an integrated set of leading and lagging performance indicators for more effective monitoring of the process safety performance of the U.S. refineries by BP's refining line management, executive management (including the Group Chief Executive), and Board of Directors. In addition, BP should work with the U.S. Chemical Safety and Hazard Investigation Board and with industry, labor organizations, other governmental agencies, and other organizations to develop a consensus set of leading and lagging indicators for process safety performance for use in the refining and chemical processing industries."

The US Chemical Safety and Hazard Investigation Board also issued a similar recommendation following their investigation of the BP incident.

An excerpt from the US Chemical Safety and Hazard Investigation Board Final Report is inserted below:

> American Petroleum Institute (API) and United Steelworkers International Union (USW) Work together to develop two new consensus American National Standards Institute (ANSI) standards.
> 2005-4-I-TX-R6 In the first standard, create performance indicators for process safety in the refinery and petrochemical industries. Ensure that the standard identifies leading and lagging indicators for nationwide public reporting as well as indicators for use at individual facilities. Include methods for the development and use of the performance indicators.

Leading vs. Lagging Indicators

Lagging indicators have been used by refineries and chemical plants for many years to help monitor safety performance. Traditionally these indicators have primarily trended performance of personnel safety by tracking the number of injuries, occupational injuries, and the respective OSHA indicators of Total Injury and Illness injury rates, e.g., the total recordable injury rate (TRIR) and lost time injury rate (LTIR). Prior to the BP incident, many sites also tracked lagging indicators for process safety performance by trending the data, such as the number of fires and environmental incidents (i.e., spills or releases to the environment). However, lagging indicators are hindsight since the incident has already occurred, and we are simply trending results, and for a large part, we were only tracking personnel safety performance.

The incident at BP reveals the unquestionable need (and benefit) of developing and tracking process safety performance and includes both leading and lagging performance tracking.

Process Operations Safety: The What, Why, and How Behind Safe Petrochemical Plant Operations, First Edition. M. Darryl Yoes.

Leading indicators help provide a more proactive view of process safety performance by evaluating the performance of the critical work processes and practices in place to prevent a process incident from occurring and/or mitigate the outcome should one occur. The challenge in defining and implementing leading indicators is that it typically takes a significant number of indicators to effectively measure the performance of the many safety management systems and work processes for process safety.

The Center for Chemical Process Safety has developed a comprehensive guide available online as a resource to the process industries in developing leading and lagging indicators. Further guidance for the development of leading and lagging indicators is available in API RP-754.

The API Recommended Practice for Process Safety Leading Indicators (API-754) focuses primarily on the loss of containment or potential loss of containment events and establishes a four-tier tracking/reporting stewardship process similar to the long-standing personnel safety triangle. The bottom tier (4) is primarily for reporting and tracking lower-level leading indicators, such as operating discipline and management system issues that could have resulted in a loss of containment. Tier 1, on the other hand, is for reporting and tracking the more severe type of process safety incidents, such as actual loss of containment resulting in greater consequences. A fact sheet, also developed by the API, is available at the following link: https://www.api.org/-/media/files/oil-and-natural-gas/refining/process%20safety/rp-754-fact-sheet.pdf. Another great reference for Leading Indicators is available from the American Institute of Chemical Engineers (AICHE) at this link: https://www.aiche.org/sites/default/files/docs/embedded-pdf/CCPS_ProcessSafety2011_2-24-web.pdf.

Each site should benefit significantly from implementing a comprehensive set of leading indicators to complement existing lagging indicators to help manage process safety. There is truth in the expression, *"We improve what we measure."* When functioning correctly, leading indicators should indicate where additional management emphasis or resources may be required. Both leading and lagging indicators are important in analyzing process safety performance.

The following are examples of leading indicators that may be used to analyze the effectiveness of process safety work practices:

Note: This list includes examples as consideration for potential leading indicators. It is not intended to be a complete list. API RP-754 includes additional examples of leading indicators.

- Demands on critical safety systems (e.g., pressure relief devices, equipment trips., furnace and compressor trips, electrical circuit breakers, electronic system trips, etc.)
- Overdue inspection of critical safety equipment, for example pressure relief devices, critical instruments or trip systems, fire or vapor release suppression equipment (e.g., fire protection or water deluge system checks, etc.).
- Small leaks of flammables or toxins (not exceeding regulatory limits). A loss of containment exceeding the regulatory limit is reported as a lagging indicator.
- All deviations to Operating Limits.
- Overdue fixed equipment inspections (e.g., piping, vessels, storage tanks, etc.).
- The number of outstanding high-priority safety recommendations (e.g., Process Hazard Analysis and other similar risk assessments).
- Near misses with causal factors linked to safety management systems.
- Failure of critical alarms or alarm system (e.g., inactive or suppressed alarms, alarm flood event, etc.).
- Failure of safety management systems (e.g., a change implemented without an approved Management of Change).
- Furnace flood events (fuel-rich scenario).
- Overdue operating procedures (e.g., procedures overdue for annual review).
- The Operator's refresher training is overdue.

Again, these are just examples and are intended to be "food for thought" as you develop the leading and lagging indicators for your site. Your indicators should be developed specifically for your site and include the parameters that are considered important to your site. If we measure what is important and implement corrective actions as needed, the overall performance of process safety can be significantly improved with the consistent application of leading and lagging indicators.

Additional References

American Petroleum Institute (API) Recommended Practice API RP 754 and RP 754 Fact Sheet Process Safety Performance Indicators for the Refining and Petrochemical Industries, Second Edition (June 2017)

US Chemical Safety and Hazard Investigation Board BP America Refinery Explosion Related Documents; The BP US Refineries Independent Safety Review Panel

US Chemical Safety and Hazard Investigation Board BP America Refinery Explosion Final Report: Recommendation to API and United Steelworkers Union (*"2005-4-I-TX-R6*).

American Institute of Chemical Engineers (AICHE) A comprehensive overview of Leading and Lagging Indicators with examples of both, and example threshold reporting. https://www.aiche.org/sites/default/files/docs/embedded-pdf/CCPS_ProcessSafety2011_2-24-web.pdf

Center for Chemical Process Safety A comprehensive guide that is available online as a resource to the process industries in developing leading and lagging indicators.

12

Process Safety Hazard Recognition and the Importance of Field Presence and the Value of Periodic Field Walk-arounds

Importance of Field Presence and Periodic Walk-arounds

Field presence provides an opportunity to observe compliance with process safety procedures and practices and the opportunity to coach or intervene to assure compliance with process safety rules. It also provides an excellent opportunity for field oversight for work permit compliance, JSA/JHA/JLA compliance, and interaction with operators, other employees, and contractors to discuss process safety.

While in the field, you will also be able to review the status of many critical process safety practices such as control of defeat, management of change, energy isolation, the status of car seal valves, compliance with operating procedures, etc. In addition, you will be able to observe and address unit/area housekeeping, the general state of repairs, etc.

Improved field presence means spending quality time in the field, interacting with the workforce, and intervening when required. You may be able to recognize and avoid "normalization of deviation" in your interactions. Increased field presence will allow you to recognize any potential degradation in process safety standards and help ensure they are properly addressed. You can participate in emergency response drills and exercises, follow-up critiques of drills, and in the overall improvement plan.

While in the field, you can be continuously alert for things that are not as they should be and strive to improve them. For example, small piping connections; many large explosions and fires have started small, directly from failed small piping connections. Become familiar with your company's standards for small connections, for example, when they should be seal welded and braced (gussets). Most companies require seal welding inside the first valve on vessels and large-diameter piping; the logic is that if a leak occurs inside the first valve, the isolation is exceedingly difficult or impossible.

Unsupported small piping is another significant issue that can lead to vibration and failure or connections without seal welding or bracing (gussets), "Christmas trees" and "moment" arms (piping with long unsupported extensions) are other examples of unsupported small connections that can easily be broken off due to either vibration or the effect of leverage. There are several photos of small piping connection-related concerns in Chapter 8.35 ("Small Piping Guidelines for Operations Safety"). We should continuously work to improve the small piping on our unit to help prevent the opportunity for loss of containment.

Other unsafe conditions that you should observe while in the field include the following (not intended to be a complete list; examples only):

- Unsealed sewer catch basins if hot work is going on in the area (can lead to a major sewer fire or explosion due to sparks entering the sewer). See Figures 12.1 and 12.2 for examples of inappropriately sealed or drainage partially restricted by unsealed sewers.
- Catch basins with the sewer covers left down after the hot work has been completed (can lead to a spill fire due to obstructed drainage).
- Proper water draw valves on spheres or all other light ends vessels (C3, C4, etc.) (double block valves; operated by opening the upstream valve fully and throttling on the downstream valve to prevent auto-refrigeration and freezing the valves in the open position).

Do not be surprised – If something does not look right, it is probably not right!

Do not be surprised by what you find while in the field. I like to say that if something does not look right, it is probably not right. An example of one that I am familiar with is where a leak occurred at a pressure gauge. When this was investigated, it was discovered that the gauge was threaded into the inside diameter (ID) of a standard pipe nipple. A pipe nipple has threads on the outside (the OD) and is neither designed nor furnished with threads on the ID. Also, the pipe nipple was not

Process Operations Safety: The What, Why, and How Behind Safe Petrochemical Plant Operations, First Edition. M. Darryl Yoes.

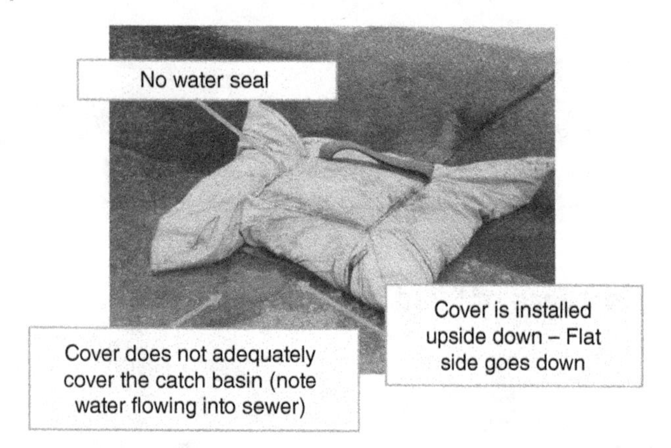

Figure 12.1 Illustrated unsealed sewers (but restricting drainage and may trap and accumulate a spill, potentially spreading fire throughout the area). Example of an Improperly Sealed Sewer.

Figure 12.2 Illustrated unsealed sewers (but restricting drainage and may trap and accumulate a spill, potentially spreading fire throughout the area). Are the Sewers Properly Sealed Before Starting Hot Work? Are the sewer covers picked up and properly stored after the hot work is completed?

(a)

(b)

Figure 12.3 (a) Illustrates a pressure gauge threaded directly into the ID of a pipe nipple. (b) Illustrates a pipe nipple-to-tee connection without seal welds or gussets (supporting braces).

seal welded to the pipe tee connection. It was reported that, at first glance, the pressure gauge had been installed correctly. Had someone been carefully observing the equipment during routine walk-arounds, it is possible this would have been noticed and corrected before the loss of containment occurred. See Figure 12.3 for a photo of the pressure gauge threaded into the pipe nipple.

Other examples are the status of equipment isolation for maintenance, including the lock-out-tag-out (LOTO) process, the status of electrical equipment, such as the location of the start/stop switch relative to the motor terminal junction box, or unauthorized "cheater cords" on the unit or in process areas. Other examples: are product sampling areas in H_2S areas properly identified with warning signs for proper PPE? Are there warning signs or relief valves to protect against thermal expansion on heat exchangers and double-seated block valves?

I believe you get the idea. There are many opportunities to make a difference in process safety by spending time in the field and conducting periodic field walk-arounds. Walk-arounds should have a purpose and be done with the knowledge of what things to look for while you are there. Walk-arounds conducted to improve process safety can identify and correct hazards before a large incident can occur.

The following photographs are provided as examples of the types of hazards discussed in this chapter (Figures 12.4–12.11).

Figure 12.4 Photo of warm-up line around the check valve on a pump discharge. All threaded connections with no seal welds or gussets were provided.

Figure 12.5 Photo of discarded LOTO chains and locks from previous LOTO task.

Figure 12.6 Photo of orifice piping connection with no seal welds or gussets.

Figure 12.7 Photo of sphere water draw piping with valves left in the open position (should be double-blocked).

Figure 12.8 Photos of sewer covers left in place after work was completed. Process sewers appear to be obstructed – potentially spreading fire in the event of loss of containment.

Figure 12.9 Image of a chain-operated valve wheel tethered to small bore piping.

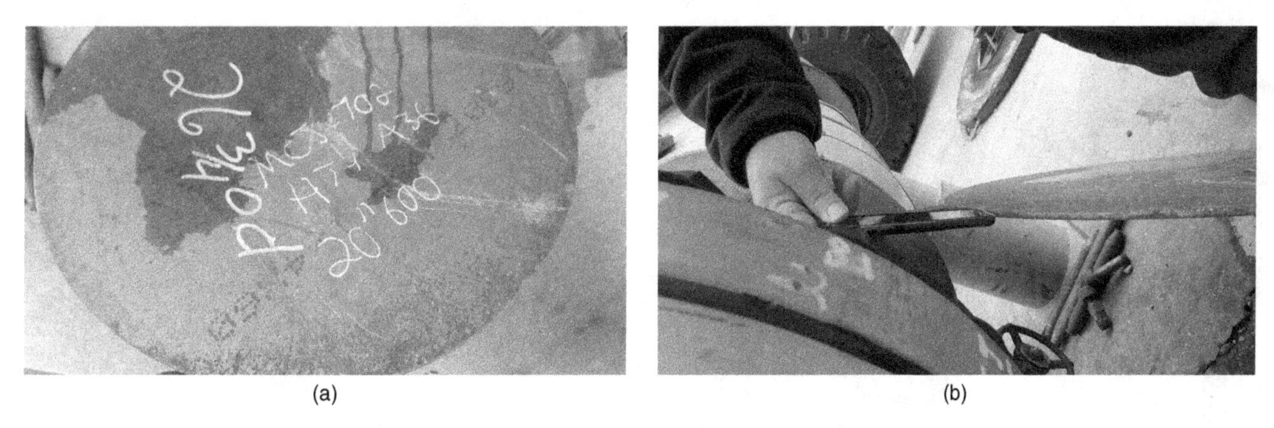

Figure 12.10 Photo of thin (non-rated) blind positioned for installation in 600 psig flange. Notice the thickness of this blind in the right image. Rated for 600 psig?

Figure 12.11 Photo of electrical "cheater cord" on process unit, a violation of the Electrical Hazardous Area Classification.

Additional References

What Is a Safety Walk-Around?, Posted by EHS Insight Resources, What Is a Safety Walk-Around? (ehsinsight.com.)

Process Safety: Walk The Line article 2016, Chemical Processing, Process Safety: Walk The Line | Chemical Processing.

Petroleum Refinery Process Safety Management National Emphasis Program, US Occupational Safety and Health Administration (osha.gov), Petroleum Refinery Process Safety Management National Emphasis Program | Occupational Safety and Health Administration (osha.gov).

OSHA PSM Regulation (Process Safety Management of Highly Hazardous Chemicals), US Occupational Safety and Health Administration (osha.gov), 1910.119 – Process safety management of highly hazardous chemicals. | Occupational Safety and Health Administration (osha.gov).

13

Process Safety Leadership

Have you ever thought about why your personal safety leadership is so valuable to your company and yourself? It is because you can make a huge difference in the site's safety performance and, most importantly, help to ensure that you and other workers can safely go home at the end of the shift or the end of the day. Your observations and, most importantly, your personal interventions can sometimes make all the difference between a safe return home or an unexpected trip to the hospital, and in my career, I have seen both. By now, you have seen many incidents and countless people (workers) who have lost their lives or were severely injured. The incentive is clearly in front of us to put forth an effort to reduce the number of incidents in our industry.

What is leadership? Do you have to be a supervisor or leader to have leadership?

Leadership as defined by Merriam-Webster:

- *The office of the position of a leader.*
- *The capacity to lead.*
- *The act or instance of leadership.*
- *Leaders (the party leadership).*

Of course, the obvious answer is NO – you can be a leader, regardless of whether you are a supervisor, especially in safety. Regardless of your company position or experience, you can influence others to cause them to work safely and follow the rules. Can you remember some of those you thought of as leaders in your earlier career? I am guessing many you considered to be "leaders" were not supervisors or managers.

This is particularly true where process safety or safe operations are involved. A process safety leader ensures that they actively participate in process safety practices and routinely communicate key process safety issues to ensure they receive proper attention and focus. They work to ensure that process safety considerations are included in significant business decisions and that major hazard risks are identified and managed. They also understand the importance of critical control measures that are communicated, championed, and enforced.

A true process safety leader also routinely practices intervention when noncompliance with process safety procedures or practices is observed. They also recognize that excellent housekeeping is practiced by his/her team and that it generally contributes to excellent SHE performance.

A process safety leader leads by example by being visible and active in safety activities and establishing high standards. They constantly communicate, teach, reward, and intervene constructively and consistently. They practice safe habits all the time (become unconsciously competent) and have the courage to speak up or intervene when process safety issues or noncompliance concerns are identified.

A process safety leader is also a coach and mentor to others, especially in the application of the principles and practices related to process safety. They always try to explain "the why" behind the principles/procedures, and they typically take the lead in safety meetings with peers and others. They must also be comfortable and willing to intervene when key principles or procedures are not followed. Process safety intervention is "lifesaving" work!

Demonstrate Commitment – Lead by Example

Some general characteristics of a leader in the process safety area:

- They are typically more Proactive vs. Reactive.
- They observe Safety, Health, and Environmental (SHE) rules and practices on and off the job.

Process Operations Safety: The What, Why, and How Behind Safe Petrochemical Plant Operations, First Edition. M. Darryl Yoes.

- They set, practice, and demand high standards – all the time, every time.
- They discuss some aspects of SHE daily with peers, supervisors, managers, and other employees.
- They are enthusiastic and sincere about SHE and have a positive attitude.
- They understand and work to manage process safety.
- They give priority to SHE matters.
- They communicate performance; correct deficiencies, including fair, consistent, and predictable accountability.
- They reward appropriately.

Process safety must be supported from all levels of the organization. Managers, supervisors, and employees must be actively involved in process safety. Personnel should be trained and competent to recognize process safety hazards and issues, recognize what is important, and ensure that hazards are addressed or raised to site management and then tracked to completion or resolution. It is also crucial that significant lessons learned from incidents or near misses are communicated and shared (both for internal and external incidents).

14

Process Task Management System

In a typical process unit, there are many safety and environmental tasks that the operating crew must carry out to assure ongoing safety and environmental compliance. A Process Task Management system is an effective way to ensure tasks are completed as scheduled and provides a user-friendly way to track the completion of each task.

The task management system can be paper-based or programmed into electronic data logging systems using a mobile handheld keypad and a code reader as the input device. The advantage of the data logger is that it provides an electronic record of when the field checks were made with a date/time stamp electronically tied to a bar code (1D or 2D). This helps ensure the process crew member is at the equipment when the check is done. The data recorder can also record the "as found" condition of the equipment at the time of the Operator visit or test and any actions that were taken by the process crew member. The electronic device also provides an easy sync with the computer for data recording, analysis, and archiving. Electronic tracking also helps ensure that tasks are done in the proper sequence, that the task was carried out as defined in the tracking system, and that the Operator complies with the critical device inspection schedule.

Intellitrac and PlantLog are only two examples of available electronic data logging systems. Each manufacturer has models designed explicitly for refining and petrochemical facilities and the type of data collection required in these facilities. Some of the electronic systems are designed to work with mobile phones and iPads; however, these devices should be verified as compatible with the electrical hazardous area classification (EHAC) before their use in the process areas (see Chapter 8.30 ("Electrical Safety Hazards for Operations Personnel") for additional information on the EHAC). While this is an efficient way to implement the task management process, a manual paper-based checklist can be very effective when properly implemented and enforced.

The list of process task items should be developed specific to the needs of the process and are typically used to ensure compliance with safety, health, and environmental regulations and company guidelines. A few of the typical task book items are listed below. Task management tracking should be developed specifically for each process unit or area.

Typical process task items requiring a tracking process:

- Verification of condition and status of critical safety and environmental equipment such as overpressure protection equipment (pressure relief valves, rupture discs, etc.), furnace protection devices, fugitive emissions checks, etc.
- Verification of car seals on safety or environmental critical valves (safety valve isolation valves and other car seal valves).
- External visual inspection checks, such as for storage tanks.
- Status of drains for storage tank dikes (open or closed).
- Verification of Fugitive Emission Devices.
- Condition of safety and fire equipment such as:
  - Breathing air packs.
  - Safety showers and eye wash stations.
  - Fire extinguishers.
  - Fire hydrants, fire hose reels, and fire monitors.
  - Firewater sprays or deluge systems.
  - Fixed firefighting foam systems.
  - Furnace header and coil injection steam.
- Rotating equipment (Pumps, Compressors, Turbines), periodic vibration, lubrication, seal oil, and other preventive maintenance inspections/checks.
- Product hose inspection, fire water hose inspection, or similar testing or inspection requirements.

Process Operations Safety: The What, Why, and How Behind Safe Petrochemical Plant Operations, First Edition. M. Darryl Yoes.

Unit: CHD-6
Equipment: **LHA-1A,** High-Level Trip for C-1 Compressor Liquid Knockout Drum (D-1)
Task Description: Verification of **LHA-1A**, C-1 Compressor Liquid Knockout Drum High-liquid Level trip instrument
Date: 1 July 2023
Time: 0845
Completed by: Joseph J. Smith (Unit Operator)
Ed T. McKnight (Instrument Technician)

Requirements:

1. An Instrument Technician and Unit Operator are required to carry out this field check.
2. The Console Operator and Unit Operator must be aware of and approve this equipment check prior to beginning work.
3. A Control of Defeat must be authorized, and an alternate protection plan must be fully implemented. The Alternate Protection Plan must be detailed on the Control of Defeat form.
4. The device (LHA-1A) must be on bypass prior to the test (to prevent the compressor from tripping during the test).

Test Procedure:

1. Verify with the Console Operator that LHA-1A (the C-1 compressor Liquid High-Level Trip instrument) is on bypass and that a mitigation plan is in place to prevent the compressor from tripping when the high level is simulated.
   Completed by: ______________ Date: ______________ Time: ________________

2. Break car seals and close the valves on both level taps from the Liquid Knock Out Drum (D-1) to LHA-1A, the High-Level Trip Instrument (two 3/4" valves located between the D-1 drum and LHA-1A High-Level Trip Instrument).
   Completed by: ______________ Date: ______________ Time: ________________

3. Connect the test water line (utility water) to the High-Level Instrument and raise the level float using the water. Verify that the console receives the High-Level Alarm and that the trip device has functioned as expected.
   Completed by: ______________ Date: ______________ Time: ________________

4. If the device functions as expected, go to step 5 in this procedure.
   In the event the instrument did not function as expected, record this below and in the equipment records system, including the actions taken to repair the device, including parts found that failed or were otherwise defective.
   ____
   ____
   ____
   ____

5. After receiving verification from the Console Operator that the instrument operated as expected, disconnect the water line and secure the instrument with caps and plugs.
   Completed by: ______________ Date: ______________ Time: ________________

1. Line up the instrument for service by opening the two 3/4" valves to D-1 and car seal the valves in the fully open position.
   Completed by: ______________ Date: ______________ Time: ________________

7. Notify the Console Operator and Unit Operator that the test has been completed that the instrument has been returned to service, and that the Control of Defeat can be closed.
   Completed by: ______________ Date: ______________ Time: ________________

8. Verify that the high-level trip device has been returned to full online service, that car seal valves are opened and car sealed and that the instrument has been returned to service by the Console.
   Completed by: ______________ Date: ______________ Time: ________________

Figure 14.1 Critical task management/task number: xxx (provided as example only).

The task frequency is established based on regulatory requirements and by local management and is input into a task schedule (calendar) for implementation by the Operators. Each task should have a checklist (manual copy or electronic) outlining the specific checks to be performed, the expected condition, and a place for the operator to record the actual condition found during the field check. The checklist should identify the person performing the check and provide a means for reporting any deficiencies found.

Like other Safety Management Systems, the Process Task Management System should include a self-assessment or verification process. An audit should be required on a periodic basis to ensure that the field checks are being completed and that identified deficiencies are being addressed. Audit results should be reported to site management, and follow-up to correct deficiencies found during the audit.

An example of a process task is attached in Figure 14.1. In this example, the task requires both a Process Operator and an Instrument Technician. However, another example could be only a Process Operator; for example, checking the status of a unit fire extinguisher or fire water deluge system.

Additional References

Intellitrac (tracking solutions). https://intellitrack.com/why-intellitrack/

Plantlog. https://plantlog.com/

15

Process Operator Rounds – Timely, Systematic, and Comprehensive

Comprehensive and meticulous process rounds by unit Operators are invaluable to sound and safe process operations. A good operator can detect an issue or problem long before it results in a loss of containment or a significant unit upset. Systematic and thorough process rounds can detect minor leaks, vibrating pumps or compressors, hot spots on furnace tubes, and process streams that are off-tested or routed to the wrong disposition long before these result in a major unit issue. Structured rounds provide the Operator with a preplanned route or path, highlighting the sequence of equipment checks to perform. It assists the Operator and functions as a guide to ensure no critical equipment is missed and is an aid to ensure the correct checks or field verifications are completed and documented.

If your site does not have a written process for ensuring detailed and comprehensive process Operator structured rounds, developing one would be worthwhile. If the process does not exist today, developing a paper system detailing the expectations for structured process rounds and getting the paper system up and running is a good idea. The paper system should provide information from the Operators based on their observations during the structured rounds and drive change in the organization. Once the system is in place, expectations are established, and the Operators are developing follow-up action items based on their "findings," the site can consider moving to an electronic data logging system. An electronic data logging type system will provide an opportunity to significantly expand the structured process rounds and facilitate the capture of many additional options for continued improvement.

The same type of electronic data logging system described in Chapter 14 ("Process Task Management System") can be used to help ensure complete and thorough Operator Rounds of process areas, typically at the beginning of the shift and again about midway through the shift. The data logger system can help ensure the round is "structured" to ensure that the equipment is visited in the correct sequence and that no equipment is missed during the field "rounds." Specific equipment checks can be easily programmed into the handheld datalogger for use by the process Operators. Using equipment or task bar codes can ensure that tasks and equipment checks are completed as scheduled, including data entry for specific tasks. Data generated in the handheld device can be easily downloaded into a process computer. The results are uploaded to a database where supervisors and subject matter experts quickly review the data (e.g., by rotating equipment specialists, furnace experts, steam boiler, cooling tower specialists, etc.).

For example, the system can be programmed to ensure that operations checks are completed for the following equipment (examples only; not intended to be a complete list):

- Rotating equipment (pumps, turbines, compressors).
  - Checks for excess vibration, seal leaks, lubrication, bearing temperatures, unusual noises or odors, coupling guards secured, etc.
- Furnaces and Heaters.
  - Verify feed pass balances, good burner patterns, tube condition (Tube Metal Temperatures and no visible hot spots, etc.),
  - Verify other furnace variables (excess O_2, draft, combustibles, CO, etc.).
- Unit rundown stream dispositions (product streams to correct tanks).
- Unit samples are caught, labeled, and sent to the laboratory as outlined on the sample schedule.
- Small leaks and drips are reported and documented as required.
- Verification of car-sealed valve positions and the condition of the car seals.
- Small piping condition (excess vibration, properly supported, seal welds and gussets or bracing, Christmas trees, moment arms, missing bar stock plugs, etc.).
- Verification of the mitigation plan for defeated safety-critical devices.
- Sewer covers (sealed for hot work, otherwise uncovered to ensure drainage).

Process Operations Safety: The What, Why, and How Behind Safe Petrochemical Plant Operations, First Edition. M. Darryl Yoes.

- Verify adequate mitigation plan for critical equipment that is out of service.
- Steam boiler operations (verify the following: water levels, proper blowdown rates, solids, chemical addition rates, etc.).
- Cooling Tower operations (number of fans in service, supply and return water temperatures, chemical addition rates, etc.).
- Verification of bar stock plugs in all vents and drains.
- Verification of check valve used in all utility connections to process equipment.
- Verification of storage tank levels and tank movements.
- Operators can also visit and observe mechanical work in progress, verifying compliance with equipment isolation, work permits, and PPE requirements.

Operators should conduct at least two comprehensive, structured rounds per shift on a scheduled basis and a third informal process round near the end of the shift as follows:

- A very detailed documented process round near the beginning of the shift, verifying valve lineups for stream disposition and tank lineups, proper operation of pumps and compressors, furnace operations, steam boiler, and cooling tower operations, and visits to mechanical tasks that may be in progress on their unit or process area of responsibility.
- A second comprehensive and documented process round near the mid-point of the shift.
- A third informal process round near the end of the shift and before completing the process turnover log entries.

A detailed analysis using a process like the Failure Modes and Effects Analysis (FMEA) should help determine the key elements of a thorough process round. This is an ideal opportunity for Operator involvement in developing the structured process round and for the Operators to have a role in determining their future. For example, it may be appropriate to have a small team of Operators develop a written, structured process round for the operators to follow. In many cases, the specific tasks may be developed from actual equipment failures or incidents that have occurred in the past, with the checks developed to help prevent a reoccurrence of the event.

However, the data developed from the structured process rounds is no better than the action taken after the data is collected. If the information is developed and then goes into a backlog or sits on a supervisor's desk with no action taken, the Operators will quickly lose faith in the process, and the data will stop flowing. The documented process for the structured process rounds should include a method for escalating issues identified by the Operators to their supervisor or to the business team in a timely manner. The data from the structured rounds should drive the development of work orders, and those work orders must be completed within a reasonable period to ensure the process's continuation.

The bottom line: a good Operator who performs thorough unit rounds and communicates their "findings" is an invaluable asset. They can help prevent significant unit upsets and even catastrophic events by "finding" the issues early, before they become a problem, and communicating the concerns to the proper personnel to address them promptly.

Additional References

Life Cycle Engineering Resource Library "How Operator Care Rounds Support a Culture of Continuous Improvement" https://www.lce.com/How-Operator-Care-Rounds-Support-a-Culture-of-Continuous-Improvement-1308.html

Wikipedia Failure Mode and Effects Analysis https://en.wikipedia.org/wiki/Failure_mode_and_effects_analysis

16

The Importance of Behavior-based Safety and Human Factors to Process Safety

What Is Behavior-based Safety?

Behavior-based Safety (BBS) is an ongoing process that helps inform management and employees of the overall safety of the workforce through peer-to-peer safety observations as a way to reinforce "good or safe" work behaviors and to identify and correct "at-risk" behaviors. BBS is an ongoing program to help focus workers' attention on their own and their peers' daily safety behaviors. The overall goal of the plant BBS program is to improve employee and contractor safety at the site.

Employees are typically trained as "observers" and to conduct on-site reviews of their peers' safety-related work behaviors. Observers are trained to note positive work behaviors and provide reinforcing feedback to the observed employee, such as wearing PPE or using fall protection when working at heights. The observers are also trained to identify the worker's at-risk behaviors and provide feedback on any identified areas for improvement. The observations or reviews are structured to be nonthreatening, with no discipline for observed deficiencies. The idea behind the peer-to-peer observation process is that the observation and follow-up discussions will result in more awareness of the observed employees of their at-risk behaviors, and over time, the employees will perform their tasks in a safer manner. BBS programs are based on a continuous feedback loop where employees and observers share input and feedback with each other, and as a result, the workplace becomes a safer place to work.

The BBS program observations are based on their operation's unique tasks and hazards. These are typically identified in advance by safety professionals and developed into a simplified checklist for the observers to use when conducting the jobsite behavior observations. For example, for unit Operators who are frequently turning valves, the checklist may consider the Operators' body positioning and adequate valve lubrication before undertaking the task. The BBS checklists often include the fundamentals of the observation (observation date and time, task location, behaviors observed, observer's name) and the number of safe and unsafe observations. The checklist also includes feedback or comments that were provided to the employee. The completed behavior-based checklist is then submitted to the safety professionals, who can analyze the data to identify trends or areas where additional site-wide or specific work area safety emphasis is needed.

Behavioral-based Safety is an employee-driven continuous safety improvement process that can deliver real safety improvements to the workplace.

Loss Prevention System+™ is not just a BBS process but a human and organizational performance process that improves the organization's systems, processes, as well as employees' risk management skills through a series of proactive tools and visible, measurable leadership activities. LPS+™ has been used by global Fortune 100 petrochemical companies across 100+ countries with industry-leading results in personal safety, process safety, reliability, etc.

The Loss Prevention System (LPS) website provides details on how the system is organized and details specific tools available to a facility to aid in the implementation and sustainment of the human and organizational performance process within an organization or site. LPS+™ focuses on improving organizational systems, processes, and individual risk management skills to prevent losses, including personal injury, safety-related incidents, process safety events, reliability issues, property damage, etc. When properly implemented, this process can enhance the organization's capacity to develop everyone's risk management skills and become better risk managers, which can lead to fewer process safety events involving loss of containment of a flammable or toxic.

Process Operations Safety: The What, Why, and How Behind Safe Petrochemical Plant Operations, First Edition. M. Darryl Yoes.

The Key LPS Tools

Loss Prevention Self-Assessment (LPSA)

An LPSA is a mental risk assessment used to identify and eliminate workplace practices and at-risk conditions that could lead to loss.

Job Loss Analysis (JLA)

JLA is a tool to identify, eliminate, and control higher-risk loss potential associated with a job or task before performing work.

Factors, Root Cause, Solutions (FRCS)

The FRCS is a flowchart that leads individuals through a range of possibilities to determine what causes or contributes to incidents.

Loss Prevention Observation (LPO)

LPOs are a standardized, systematic tool for observing a work process and determining if the process is being performed according to the site's standards. Flawless and at-risk behaviors and conditions are identified, root cause analysis is performed, and solutions are implemented to replicate successes and address items to prevent potential losses.

Near-Loss and Loss Learnings (NLL and LL)

NLLs and LLs are a systematic examination of near losses and losses for the purpose of eliminating risks and future losses. Similar to at-risk behaviors and conditions identified in LPO, a root cause analysis is performed, and solutions are implemented to prevent escalation of losses and recurrence.

Successful Task Learning

Flawless operational performance happens more regularly than at-risk items on an observation, near losses, or losses. Therefore, organizations should study and learn the root causes of successful task performance and implement recommendations to replicate future success. Nothing breeds success like success.

Stewardship Tool

Stewardship is a process for overseeing and monitoring the proper use and quality of tools to create a positive work environment and culture where LPS+™ is managed with the same importance as other major business activities and is a small investment of time for transformational change.

Unlike traditional safety programs that hold people accountable for safety statistics, LPS+ stewardship focuses on the quality aspects of the process for each tool over which employees have full control.

Companies that manage loss prevention activities instead of the "numbers" are the same companies that have directly reduced injury and loss measures. These organizations steward and manage LPS+™ activities just as they steward or manage other aspects of their business.

See the following website for more information on the Loss Prevention System https://www.lpscenter.net/services

Human Factors

Human factors (HF) identify issues regarding the safety behaviors of people and our skills and limitations regarding our safety behaviors. HF also considers the design of the hardware we use to do our job, the tools, equipment, machines, and tasks, and improves the interface between human behaviors and the tools and equipment. HF can significantly improve our individual and group behaviors for both personnel safety and process safety.

The Dirty Dozen (Use of the Dirty Dozen Courtesy of the US Federal Aviation Administration)

Thanks to the Aviation Glossary, available on the Federal Aviation Website https://www.faasafety.gov/files/gslac/library/documents/2012/Nov/71574/DirtyDozenWeb3.pdf.

I believe the principles in the Aviation Glossary apply equally to the refining and petrochemical sectors. The following is an overview of the "Dirty Dozen" with a few additional comments which were added for context:

Lack of Communication

Lack of Communication, failure to transmit, receive, or provide enough information to complete a task. Never assume anything. Only 30% of verbal communication is received and understood by either side in a conversation. Others usually remember the first and last part of what you say. Improve your communication.

- Say the most important things in the beginning and repeat them at the end.
- Use checklists.
- If the message is complex, you may want to present it in written form to ensure it is correctly received.

Complacency

Overconfidence from repeated experience performing a task. Avoid the tendency to see what you expect to see.

- Expect to find errors.
- Do not sign it if you did not do it.
- Use checklists.
- Learn from the mistakes of others.
- Always verify your work.
- There is a tendency to see what you expect, which can easily lead to an accident.
- The Aviation Glossary describes this as "self-satisfaction accompanied by a loss of awareness of the danger." Over time, an activity or task can become routine, especially if done often. The person can fail to see the hazard and miss an important signal or warning.

Lack of Knowledge

Lack of Knowledge Shortage of the training, information, and/or ability to perform successfully. Do not guess, know.

- Use current manuals.
- Ask when you do not know.
- Participate in training.

Aviation industry employees have a regulatory responsibility to ensure that their personnel have the required training.

This certainly also applies to our industries as well! We have a similar obligation in the refining and petrochemical sectors!

Distraction

Anything that draws your attention away from the task at hand. Distractions are the #1 cause of forgetting things, including what has or has not been done in a maintenance task. Get back in the groove after a distraction.

- Use checklists.
- Go back three steps when restarting the work.
- Only repair/work on what you are qualified for.
- Use up-to-date training manuals/guides.
- Request assistance from others knowledgeable/qualified in that field.

A distraction can be anything that draws your attention away from the task at hand. Psychologists report that this is the primary cause of forgetting things.

Like the Aviation industry, we are always thinking ahead. In the event, we are distracted before returning to the job, we may think we are further along than we are.

Lack of Teamwork

Failure to work together to complete a shared goal. Build solid teamwork.

- Discuss how a task should be done.
- Make sure everyone understands and agrees.
- Trust your teammates.

An effective team will:

- Maintain a clear mission.
- Maintain team expectations.
- Communicate to all team members.
- Maintain trust.
- Pitch in where needed.
- Accept critiques and opinions of coworkers.

Fatigue

Physical or mental exhaustion threatening work performance. Eliminate fatigue-related performance issues.

- Watch for symptoms of fatigue in yourself and others.
- Have others check your work.

The Aviation industry reports that studies have shown that being fatigued is like being under the influence of alcohol. We tend to underestimate the problem and overestimate our ability to cope. Studies have proven that after 17 hours of wakefulness, you are functioning as if you had an equivalent blood alcohol level of 0.05%, increasing to 0.1% after 24 hours! The more fatigued you are, the lower your IQ. It is also noteworthy that the more fatigued you are, the more easily you are distracted.

Lack of Resources

Not having enough people, equipment, documentation, time, parts, etc., to complete a task. Improve supply and support.

- Order parts before they are required.
- Have a plan for pooling or loaning parts.
- Maintain proper preventive maintenance.

Pressure

Real or perceived forces demand high-level job performance. Reduce the burden of physical or mental distress. Pressure can be brought on by our peers or may be self-induced.

- Communicate concerns.
- Ask for extra help.
- Put safety first.
- Make sure pressure is not self-induced.
- Notify leaders if more time is needed than what is allotted.

Frequently, people will bring pressure upon themselves. Self-induced pressures can occur when one accepts responsibility for tasks or responsibilities that belong to someone else. Not accepting responsibility when it belongs to someone else can help reduce the pressure. Not rushing through a repair relieves undue pressure and prevents mistakes.

Lack of Assertiveness

Failure to speak up or document concerns about instructions, orders, or the actions of others. Express your feelings, opinions, beliefs, and needs in a positive, productive manner.

- Express concerns but offer positive solutions.
- Resolve one issue before addressing another.
- Notify others when a risk/danger is noticed.
- Never compromise your standards if it risks safety, personnel, or equipment!

Assertiveness is not being aggressive. Express yourself in a positive manner and offer suggestions for resolving the issue. Seek others' opinions by asking open-ended questions or simply asking what they think.

Stress

A physical, chemical, or emotional factor that causes physical or mental tension. Manage stress before it affects your work.

- Take a rational approach to problem-solving.
- Take a short break when needed.
- Discuss the problem with someone who can help.
- Maintain a healthy eating, exercise, and sleep routine.

There are two types of stress: acute and chronic. Acute stress relates to the demands placed on the body because of current issues; for example, time constraints.

In the case of the refining or petrochemical sector, examples are attempting to oversee the loading of two or more vessels simultaneously or oversight of a reactor showing some signs of a potential exothermic reaction.

Chronic stress can accelerate the effects of acute stress. To handle acute stress, try to take short breaks and practice deep breathing. Dealing with chronic stress is more difficult and may require a change in lifestyle.

Lack of Awareness

Failure to recognize a situation and the potential consequences, understand what it is, and predict the possible results. See the whole picture.

- Make sure there are no conflicts with an existing repair or modifications.
- Fully understand the procedures needed to complete a task.
- Request a "second look" from a fellow coworker.

Ask yourself, "What am I missing?" Ask your team, "What have we forgotten?".

Norms

Expected, yet unwritten, rules of behavior. Help maintain a positive environment with your good attitude and work habits. Norms can be positive or negative and can be followed by most of the workforce.

- Existing norms do not make procedures right.
- Always follow proper safety procedures.
- Ensure others are following the same standards/procedures.
- Identify and eliminate negative norms.

A positive norm would be scanning the area where a task was completed before closing up.

In our case, a positive norm would be conducting a last-minute risk assessment before drawing water on an LPG sphere and asking, "What is the worst thing that could happen?" And then ensuring that the hazards have been mitigated before proceeding.

A negative norm would be pushing an aircraft into the hangar without assistance or a spotter.

Using our water draw example, this would be like proceeding with the water draw operation without thinking through the last-minute risk assessment and mitigating the hazards. For example, opening the valves in the incorrect sequence. Note: The correct sequence is to open the upstream valve fully and throttle on the downstream valve. Doing this in the wrong sequence can lead to auto-refrigeration and freezing of both valves in the open position, leading to a catastrophic loss of containment of LPG.

Application of BBS and Human Factors

As illustrated in the preceding paragraphs, the primary focus of BBS programs and HF programs is on the reduction of personnel injuries. These are the slip, trip, and fall type of injuries, the struck by injury, or the caught between types of injury. These programs have demonstrated that implementing these programs can dramatically reduce these injuries. The buy-in and support these programs have by the employees are a huge part of their success. These programs are indeed owned and stewarded by the employees, and once the employees are "bought in," the injury numbers will start to come down.

However, these programs can also be extremely helpful in reducing the number of loss of containment incidents that can lead to catastrophic explosions, major fires, or toxic release incidents. These types of incidents can affect far more people, as indicated in the case studies just in this book (and this book only reflects a very small percentage of the number of actual process safety incidents that have occurred).

For example, if an Operator is about to drain water from a spherical tank containing Propane the risk of a catastrophic loss of containment and a potential BLEVE is high if they get confused, open the downstream valve fully open and throttle on the upstream valve. Due to the pressure drop across the upstream valve, the LPG will vaporize and auto-refrigerate and can easily freeze the upstream and downstream valves in the open position and cannot be closed. This has led to a catastrophic loss of containment and the BLEVE of multiple spherical tanks with the loss of many lives.

Using the water draw as an example if the Operator were to take a couple of minutes before doing the task and conduct a Loss Prevention Self-Assessment (LPSA) by applying the LPS+™, it is quite possible that this type of incident could be avoided. Or if a trained BBS observer was observing the Operator, it is possible that the trained observer would have the opportunity to intervene in the process and, again, the incident could be avoided. Another factor is the application of Human Factors. If Human Factors were properly considered, including training for the Operator, including hazard recognition, and if the Operator had been following written procedures with visual cues to help the Operator, this may also have prevented the incident.

The point I am making here is that the BBS, Human Factors, and LPS+™ are all especially important tools to address personnel safety hazards and reduce the number of injuries. However, these same tools can go a long way toward addressing the process safety incidents, the very incidents that can lead to loss of containment of flammable or toxic material and can lead to massive explosions or major fires and can forever change the lives of many people.

Additional References

Aviation Glossary Defining the Language of Aviation https://www.faasafety.gov/files/gslac/library/documents/2012/Nov/71574/DirtyDozenWeb3.pdf

Wikipedia Article on Behavior-Based Safety https://en.m.wikipedia.org/wiki/Behavior-based_safety

Cambridge Center for Behavioral Studies https://behavior.org/help-centers/safety/

Center for Behavioral Safety http://cbsafety.com/

Loss Prevention System+™ An integrated process to maximize business performance https://lpscenter.net/

17

Use of Inherently Safer Design and Technologies

On 6 August 2012, a large loss of containment of a product from the Atmospheric four side stream draw on the Chevron, Richmond (CA) atmospheric crude unit resulted in a very large process unit fire. As a result of this incident, 15,000 people in the community reported for medical treatment, primarily due to breathing difficulties. Following an investigation by the US Chemical Safety and Hazard Investigation Board, the board recommended that the piping should have been replaced with piping of an "inherently safe design." Inherently safe design helps ensure that the pipe is fabricated of alloy steel that will not fail when exposed to corrosives or temperatures that would cause piping manufactured from carbon steel, as in this case, to fail. In the Chevron Richmond example, had the piping been fabricated with a relatively small amount of chrome, it would have been very resistant to the high-temperature sulfidation that resulted in this significant fire and, therefore, would be considered an inherently safe and corrosion-resistant design.

Another equally tragic incident occurred at the Tesoro Anacortes (WA) refinery on 2 April 2010, when a reactor feed/effluent heat exchanger on a naphtha hydrotreater failed catastrophically. In that incident, the exchanger was fabricated primarily from carbon steel and was exposed to hydrogen at high temperatures during routine operations. The exchanger failed due to high-temperature hydrogen attack resulting in the deaths of seven Operators. This is another example where using a corrosion-resistant alloy material would have been an inherently safer design and may have prevented this tragic incident.

The inherently safer design may be safer or more resilient metallurgy, as indicated in the Chevron and Tesoro incidents. Other strategies may be substituting a hazardous chemical with an inherently safer chemical, operating changes that reduce the operating temperature or pressure, using an inherently safer catalyst, or a safer procedure. New designs and modifications of existing equipment should further consider applying inherent safer design to reduce the risk of a process safety incident. For example, this is important when scheduling pipe or vessel replacement in systems where internal corrosion has been an issue. The replacement with piping or process vessels of inherently safer metallurgy and/or other materials of construction should be considered.

Dr. Trever Kletz has been credited with being the first person to identify the basis for inherently safe design. Following the catastrophic explosion at the Nypro chemical plant in Flixborough, England, on 1 June 1974, Dr. Kletz first coined the phrase "What you don't have, can't leak." He wrote this in a 1978 article for "Chemical and Industry" and emphasized the philosophy of minimizing the volume of hazardous materials stored and processed on-site.

Dr. Kletz also coined the expressions "intensification" and "attenuation" to emphasize the ideas of smaller facilities but with the same throughput, or "less is better" where hazardous materials are involved. At the Nypro plant, 28 people were killed in the explosion, 36 were injured on-site, and 53 others in the adjacent community. In several chapters, Dr. Kletz referred to the concept and benefits of inherent design in his book "Learning from Accidents."

The catastrophic incident at Bhopal (Madhya Pradesh, India) on 2 and 3 December 1984, is another harsh reminder of why inherent safe design is so important. This incident still stands as one of the worst process safety incidents ever recorded and resulted in the release of an estimated 42 tons (38,000 kilogram) of highly toxic methyl isocyanate (MIC) into the surrounding communities. More than 500,000 people were exposed to the toxic effects of the chemicals. The death toll estimates vary significantly, but estimates are that 8,000 people died within the first two weeks of the incident, and another 8,000 people have died as of 2006. However, unofficial estimates are tragically much higher.

MIC was an intermediate chemical used to manufacture pesticides and was produced and stored for later use. Other sites using MIC no longer hold the highly hazardous chemical on-site but have modified their process to immediately consume the highly hazardous material, minimizing storage on-site.

Wikipedia, in an article describing Inherent Safety, says, "an inherently safer design is one that avoids hazards instead of controlling them, particularly by reducing the amount of hazardous material and the number of hazardous operations in

Process Operations Safety: The What, Why, and How Behind Safe Petrochemical Plant Operations, First Edition. M. Darryl Yoes.

the plant"; and that "in the chemical and process industries, a process has inherent safety if it has a low level of danger even if things go wrong. Inherent safety contrasts with other processes where a high degree of hazard is controlled by protective systems".

The Center for Chemical Process Safety (CCPS) in the Final Report "Definition for Inherently Safer Technology in Production, Transportation, Storage, and Use" (developed for the Chemical Security Analysis Center Science & Technology Directorate, US Department of Homeland Security, Aberdeen Proving Ground) stated that "Inherently Safer Technology (IST), also known as Inherently Safer Design (ISD), permanently eliminates or reduces hazards to avoid or reduce the consequences of incidents."

"IST is a philosophy applied to the design and operation life cycle, including manufacture, transport, storage, use, and disposal. IST is an iterative process that considers such options, including eliminating a hazard, reducing a hazard, substituting a less-hazardous material, using less-hazardous process conditions, and designing a process to reduce the potential for, or consequences of, human error, equipment failure, or intentional harm. Safe design and operation options range from inherent to passive, active, and procedural risk management strategies. There is no clear boundary between IST and other strategies."

CCPS identifies four Inherently Safe Processes (ISP) strategies to consider when designing or modifying a process (CCPS 2008b). As adapted from that volume, one can:

- **Substitute:** Use materials, chemistry, and processes that are less hazardous.
- **Minimize:** Use the smallest quantity of hazardous materials feasible for the process, and reduce the size of equipment operating under hazardous conditions, such as high temperature or pressure.
- **Moderate:** Reduce hazards by dilution, refrigeration, and process alternatives that operate under less-hazardous conditions; reduce the potential impact of an accident by locating hazardous facilities remotely from people and other property; or
- **Simplify:** Eliminate unnecessary complexity and design "user-friendly" plants.

This certainly meets the vision set by Dr. Kletz of "What you don't have, can't leak."

Additional References

"Definition for Inherently Safer Technology in Production, Transportation, Storage, and Use" The Center for Chemical Process Safety Developed for Chemical Security Analysis Center Science & Technology Directorate CCPS 2008b Inherently Safer Chemical Processes, Bollinger et al., 1996 Center for Chemical Process Safety (CCPS), American Institurte of Chemical Engineers (AIChE)

US Department of Homeland Security Aberdeen Proving Ground, MD 21010-5424 https://www.aiche.org/sites/default/files/docs/embedded-pdf/ist_final_definition_report.pdf

"A Checklist for Inherently Safer Chemical Reaction Process Design and Operation" The Center for Chemical Process Safety (CCPS) http://www.aiche.org/ccps/safetyalert

"The Use and Storage of Methyl Isocyanate (MIC) at Bayer CropScience (2012)" Chapter: 4 The Concepts of Inherently Safer Processes and Assessment

Final Investigation Report Report NO. 2012-03-I-CA January 2015 Chevron Richmond Refinery Pipe Rupture and Fire US Chemical Safety and Hazard Investigation Board https://www.csb.gov/chevron-refinery-fire/

Investigation Report Report 2010-08-I-WA May 2014
Catastrophic Rupute of Heat Exchanger (Seven fatalities) US Chemical Safety and Hazard Recognition Board https://www.csb.gov/tesoro-refinery-fatal-explosion-and-fire/

Report on Flixborough (Nypro UK) Explosion, 1 June 1974 British Health and Safety Executive https://www.hse.gov.uk/comah/sragtech/caseflixboroug74.htm

The Bhopal disaster and its aftermath: a review Edward Broughton Environmental Health, Volume 4, Article number: 6 (2005) https://rdcu.be/bZktN

Textbooks by Dr. Trevor Kletz "Learning From Accidents". "What Went Wrong". "Still Going Wrong". "Plant Design for Safety". "Dispelling Chemical Engineering Myths". Trevor Kletz Gulf Publishing

The Concept of Inherently Safer Processes and Assessment National Academies of Sciences, Engineering, and Medicine. 2012. The Use and Storage of Methyl Isocyanate (MIC) at Bayer CropScience. Washington, DC The National Academies Press

Wikipedia Article on "Inherent Safety" https://en.wikipedia.org/wiki/Inherent_safety

Wikipedia Article on "Bhopal Disaster" https://en.wikipedia.org/wiki/Bhopal_disaster

18

Putting It All Together and Making a Difference

When one joins a company like Chevron, Microsoft, Humana, and many like these, they join with people from different backgrounds, ethnicities, religions, and beliefs. Under the best circumstances, they pool their backgrounds and become like one, joining together to be part of a large team or almost like a family. They still have what makes them human, their backgrounds and beliefs, but they function almost as one person with common goals and attitudes. What drives this, and how can it be achieved? Many psychologists and others who specialize in organizational behavior have attempted to answer this through the ages.

One thing that can help drive this is a tragedy or the loss of a loved one. When this occurs, people naturally come together to lift the people involved and support them during these difficult times. This is a natural and human response; regardless of their unique differences, we all want to contribute to help make things better for those affected. I have seen this happen during several tragic incidents at the refinery where lives were lost to unforeseen accidents. After the incident, many people from all walks of life stepped in to help and provide support for their teammates during this difficult time.

In this book, we have discussed and learned about many process safety incidents, many of them involving the loss of life, severe injuries, major damage to equipment and infrastructure, and extended downtimes. All these incidents were real, they happened, and they happened just as they were described in this book. Many of these people are not with us today due to what happened at their facilities. Husbands, wives, brothers, sisters, sons, and daughters are not here due to these accidents. Most got up one morning, got dressed to go to work, drove from home to the site, and did not go home that evening. This is hard for most of us to comprehend. Most of us naturally say to ourselves that cannot or will not ever happen to me.

Unfortunately, I am here to tell you that it can! Not only can it happen, but it also may happen if you are not following procedures or taking shortcuts. It can also happen if we are not paying attention to alarms, or the equipment is not being maintained as it should. It can happen if we are not paying attention to the many details of the job, details that can give us a "heads up" that something is not what it should be. It can happen if we are looking at the process and when we see what we expect to see instead of what is there. For example, if we have a reactor that is in thermal runaway, but we acknowledge the alarm without taking action to correct the temperature excursion.

I realize that this sounds kind of dramatic. I am attempting to get your attention so that an incident like those you have been reading about in this book does not happen to you or one of your teammates.

So, what if I told you that you, and your team members, will be involved in a major refinery incident next week or month? How would you behave? What would you do differently to help ensure that this will not and cannot happen? Let us briefly talk about several things that can make a difference and possibly prevent the kind of incident that we are discussing here.

I realize I have mentioned this before, but every time I discuss it with my peers and people I work with daily, I am reminded about Charlie Morecraft and his safety seminars. Charlie talks about what happened to him eloquently and with emotion as only he could having experienced this personally. He talks about how shortcuts and not wearing the personal protective equipment the way he should have contributed to the tragic incident where he was severely burned over about 50% of his body. He talks about how the incident impacted his family and changed not only his life but the lives of those he cared so much about. The best guidance I can give you is not to let something like this happen to you or your workmates.

While you are on the job, stay focused on what is important. Try to avoid prolonged discussions about nonwork-related activities or issues. Pay special attention to the details if there is a unique task. For example, if you are loading a railcar, blending into a tank, or maybe sulfiding a reactor, think through the hazards and ask yourself, "What could go wrong"? Develop a plan for monitoring the operation that includes indicators. Set milestones for checking the progress to ensure it is completely on track. Collaborate with your teammates, and do not be afraid to ask for help if you are unsure of what

Process Operations Safety: The What, Why, and How Behind Safe Petrochemical Plant Operations, First Edition. M. Darryl Yoes.

you are doing or what the next step should be. This is regardless of your level of experience. I have seen very experienced Operators make mistakes because "this is the way I have always done this."

Intervene if you see a teammate doing a task and not following the correct procedure. Always explain the correct way the task should be done, why this is important, and what could happen if they continue doing it the way they were.

If you see something that is not what it should be, take responsibility to correct it then and there. At least get the right person's attention to correct it as soon as possible. Just do not see it and then walk away as if it is not your responsibility. Think about how you would feel if you saw something unsafe, such as a broken guard rail, and you simply walked away. You failed to put up a barrier and did not report what you observed. And then, a week later, your best friend was killed by falling from the platform because of the broken guard rail. This would be hard to take, and you would probably never forgive yourself for failing to act responsibly.

Again, I realize that this sounds dramatic and may never happen. But how do you know? Remember, this could be you falling because your friend saw something that was not what it should be and failed to take action to get it barricaded or corrected. We are responsible for looking after others, just as we expect them to look out for us.

By working together, we can make a difference, and the benefit is that we all get to go home at the end of the shift or the end of the day.

Appendix A

End of Chapter Quiz with Answers

Chapter 1. Guiding Principles of Process Safety Management

End of Chapter Quiz

1 What are the six Guiding Principles of Process Safety Management?
Answer:
- Prevent Loss of Containment (of a Flammable or Toxic)
- Minimize Fuel and Air Mixtures
- Minimize Sources of Ignition
- Prevent Uncontrolled Exothermic Reactions
- Ensure Rigorous Field Verification and Audits
- Maintain a Strong Organizational Culture of Process Safety

2 What is meant by "Organizational Culture"?
Answer:
The set of values and behaviors contribute to an organization's unique social and psychological environment. They are the written and unwritten rules that drive the organization.

3 How do we ensure that the process safety management principles are continuously applied?
Answer:
Through a periodic (scheduled) and comprehensive process of Field Verification and Audits to ensure, these principles are consistently being applied.

4 How does a core safety management system degrade to a "formality"?
Answer:
By failing to conduct field checks and verifications and not having a continuous improvement mentality.

5 What was one of the contributing factors in the tragic Piper Alpha fire?
Answer:
The Permit to Work process was not enforced and became a "formality."

6 How can we prevent the reoccurrence of a major process safety incident?
Answer:
The Guiding Principles of Process Safety Management, when implemented through the Process Safety Management System and embedded into the company or site culture, will help prevent the reoccurrence of process safety incidents.

7 Guiding Principles of Process Safety are considered __________ ____ __________ to help prevent significant site incidents such as explosions or fire.
Answer:
"Layers of Protection"

Process Operations Safety: The What, Why, and How Behind Safe Petrochemical Plant Operations, First Edition. M. Darryl Yoes.

8 This chapter discusses nine practices that can help prevent the loss of a flammable or toxic containment. Can you name at least five?
Answer: List any five of the following.
- Ensuring safe design for processing or handling flammable or toxic materials, including Inherently Safer Technology where appropriate.
- Establishing and enforcing Safe Operating Limits.
- Ensuring Equipment Inspection and Follow-up.
- Maintaining Operating Procedures up-to-date and Procedure Compliance.
- Properly managed changes to equipment, procedures, chemicals, roles, responsibilities, etc.
- Ensure effective control of the Disablement of Critical Devices.
- Closely controlled response to "minor" drips/leaks to prevent the occurrence of a catastrophic release.
- Ensure the timely Operator Response to Alarms.
- Verify the proper process lineup of piping and valves to prevent vessel overfills.

9 Allowing fuel (hydrocarbons) to mix with air can result in a ___________ ___________ leading to a Vapor Cloud Explosion and an explosion in a confined space such as a tank or other process vessel.
Answer:
flammable mixture

10 Organizational Culture is the set of ________ ____ _________ that contribute to an organization's unique social and psychological environment. They are the written and unwritten rules that drive the organization.
Answer:
Values and behaviors

Chapter 2. Process Safety Background and Federal Regulations

End of Chapter Quiz

1 Name the guiding regulation for Process Safety Management.
Answer:
OSHA 29 CFR 1910.119 "Process Safety Management"

2 Which federal agency is charged with investigating significant process safety incidents?
Answer:
US Chemical Safety Board (CSB)

3 What is the most common cause of process safety events?
Answer:
Loss of containment of a flammable or toxic material.

4 Besides the employees of a petrochemical facility, who may be affected by a large process safety incident at a facility?
Answer:
Nearby residents and businesses.

5 What valuable training resources does the US Chemical Safety Board make available free of charge?
Answer:
Completed incident investigation reports, including the cause of failure, recommendations to prevent a reoccurrence, and training videos.

6 What other federal regulation is very similar to the Process Safety Management regulation?
Answer:
The Environmental Protection Agency regulation or the EPA Risk Management Plan regulation.

7 Process safety management focuses on preventing "______ __ _________" (releases) of flammable or toxic materials into the atmosphere or process vessels or spaces not intended to contain these materials.
Answer:
loss of containment

8 Good process safety management requires detailed knowledge of the _________ ___ __________ hazards associated with the operation of the plant.
Answer:
chemical and process

9 The Clean Air Act (CAA) of 1990 required __________ and __________ to issue regulations governing process safety.
Note: Select the correct answer:
a. EPA and NFPA
b. OSHA and EPA
c. DOT and USCSB
d. NTSB and OSHA
Answer:
b. OSHA and EPA

10 The United States Occupational Safety and Health Administration (OSHA) developed the Process Management Rule for facilities that handle or process ________ ________ chemicals. This rule was titled 29 CFR ______.____.
Answer:
highly hazardous
1910.119

11 Where can the quantity threshold for a specific toxic chemical be found?
Answer:
The OSHA standard lists the quantity threshold (29 CFR 1910.119).

12 How many different process safety elements are covered in the OSHA PSM rule?
Answer:
The PSM rule provides specific and detailed regulations for 14 process safety elements.

13 A significant step forward in addressing Process Safety occurred in the US in 1990 when the US Chemical Safety and Hazard Investigation Board was authorized by the Clean Air Act Amendments of 1990 to function as an _________ _________ agency for process safety incidents.
Answer:
independent investigation

14 The CSB website is a ready resource for outstanding process safety training materials with ready access to ________ _________ incident information.
Answer:
process safety

15 Both _________ ____ _______ investigations lead to new safety recommendations, which are the Board's principal tool for achieving positive Change.
Answer:
accident and hazard

Chapter 3. Properties of Hydrocarbons (Fire and Explosion Hazards)

End of Chapter Quiz

1 A hydrocarbon mixture in air must have the correct ratio of hydrocarbons to air for ignition to occur when exposed to a flame or other ignition source. What do we refer to this mixture?
Answer:
The flammable range.

2 Where can most of the hydrocarbon hazards be found?
Answer:
The Material Safety Data Sheet (MSDS, sometimes called the SDS).

3 Which hydrocarbon characteristic is referenced when considering the volatility of the product and especially when considering storage of the product?
Answer:
The products flash point.

4 What hazard may go undetected when gas testing a vessel for LEL?
Answer:
A gas test for LEL may not indicate the presence of trapped liquid hydrocarbons at a temperature below their Flash Point.

5 What property of Acetylene makes it particularly hazardous in welding operations, especially in or near confined spaces?
Answer:
The extremely wide flammable range makes Acetylene particularly hazardous, especially when used in confined spaces.

6 It is important to remember that when heated above this characteristic, a product will generate flammable vapors and ignite readily.
Answer:
The products flash point.

7 How is the autoignition temperature of a hydrocarbon defined?
Answer:
Autoignition temperature is when hydrocarbon vapors spontaneously ignite without an external ignition source.

8 Most hydrocarbons will spontaneously ignite at what temperature? Where can a more exact temperature be found?
Answer:
Most hydrocarbons will spontaneously ignite at around 480°F (249° C). The exact temperature can be found on the MSDS.

9 What is Vapor Density?
Answer:
Vapor Density is defined as the relative weight of a gas or vapor compared to air, with an arbitrary value of one. A gas with a vapor density of less than one will generally rise in the air. The gas will generally sink in air if the vapor density is greater than one.

10 What are the hazards of gases such as Propane, Butane, and Gasoline vapors which have a Vapor Density making them heavier than air?
Answer:
Vapors heavier than air, such as Propane, Butane, and Gasoline, have a vapor density greater than one and can accumulate in low places, making them more susceptible to explosions. Hydrogen Sulfide is also heavier than air and can collect in low places or at the bottom of vessels, increasing its deadly effects from the vapor concentration.

11 What is meant by the term Vapor Pressure?
Answer:
Vapor pressure indicates the pressure the vapor exerts on the atmosphere or the container, such as a floating roof tank or the gasoline tank in your vehicle.

12 What is the difference between Reid Vapor Pressure and True Vapor Pressure?
Answer:
Reid vapor pressure is the measured vapor pressure of the product at 100°F (37.7° C). The true vapor pressure is the pressure when tested at the actual temperature of the product.

13 What is the difference between deflagration and a detonation explosion?
Answer:
The difference between deflagration and detonation is the speed (velocity) of the flame front moving through the unreacted fuel.
A deflagration is a combustion event where the velocity of the flame front through the unreacted fuel is lower than the speed of sound. In a deflagration event, shock waves or overpressure does not typically occur.
A detonation is an event where the velocity of the flame front moving through the unburned fuel is greater than the speed of sound. The resulting shock waves or overpressure can result in catastrophic damage to people, plant, and equipment.

14 There must be _______, _______, _______, and the resulting chain reaction to start a fire. Note that items in parentheses below are not required for the answer.
Answer:
fuel, air (or oxygen), heat (source of ignition)

15 When gas testing for trapped liquids in a process system, the gas test for LEL may ______ indicate the presence of trapped liquid hydrocarbons if the liquid is at a temperature below the Flash Point.
Answer:
not

16 The boiling point is the temperature at which a liquid will ______ (measured in °F at sea level).
Answer:
boil

17 Increasing the pressure will _______ the boiling point; decreasing the pressure will _______ the boiling point, for example, in a vacuum.
Answer:
increase
decrease

18 How many of the properties of flammable liquids that can increase the risk of fire or explosion can you name?
Answer:
- Lower Explosive Limit (LEL).
- Wider Flammable Range.
- Lower Flash Point (Flammable.)
- Lower Autoignition Temperature.

- Lower Ignition Energy.
- Higher Vapor Pressure (RVP or TVP).

19 A BLEVE occurs when the liquid is heated above its atmospheric _________ _________ and rapidly released into the atmosphere (e.g., as the result of a _________ failure).
Answer:
boiling point
vessel

20 An explosion is a rapid transformation of _________ or _________ energy into mechanical energy and involves gas expanding rapidly.
Answer:
physical
chemical

Chapter 4. Vapor Cloud Explosions (VCE)

End of Chapter Quiz

1 What effects can increase the force and contribute to the destructive nature of a vapor cloud explosion?
Answer:
Confinement, such as the vapor cloud accumulating in a confined space or confinement because of unit congestion caused by pipe racks, structures, or process vessels in the explosion area.

2 How does confinement increase the force of a vapor cloud explosion?
Answer:
These "obstructions" create turbulence and contribute to a more thorough mixing of the hydrocarbon vapor and air, producing a much more uniform mixture of fuel and air. The rapidly expanding flame front and hot gases in the presence of the obstructions result in the high velocity of the hot gases. This high velocity results in "overpressure" or a pressure wave. The overpressure can result in catastrophic equipment damage and, in many cases, secondary loss of containment, contributing to additional fuel released into the fire area and a rapidly spreading event.

3 In a vapor cloud explosion, most personnel injuries occur outdoors or indoors?
Answer:
Most personnel injuries and fatalities generally occur indoors when people are struck by flying debris or by building collapse. Buildings can be severely damaged and may collapse if not designed for the overpressure and resulting pressure wave that can occur in an explosion.

4 What is the most effective way to prevent a vapor cloud explosion?
Answer:
The most effective way to prevent a vapor cloud explosion is to avoid the loss of containment from occurring.

5 How can a facility prevent loss of containment incidents which may lead to a vapor cloud explosion? What are several examples of management systems? Name three examples of a management system.
Answer:
The most effective way to prevent loss of containment incidents is to ensure that effective process safety management systems are in place and that these systems are routinely audited and continuously updated. Name any three of the following.
Examples of these Process Safety Management systems include the following:
- Management of Critical Procedures
- Dead Leg Piping identification and management

- Corrosion Under Insulation identification and management
- Injection and Mixing Point identification and procedures
- Critical Device Impairment procedures (Control of Defeat)
- Thermal Expansion identification and management
- LPG Water Draw facilities and procedures
- Winterization procedures

6 How can a site ensure that personnel can be protected in the event of a loss of containment incident?
Answer:
Ensure that an Overpressure Study has been completed and includes an evaluation of occupied buildings in or near process areas.
This study should include evaluating building design relative to potential overpressure should an explosion occur.

7 All loss of containment or near loss of containment events should be ________ and ___________ for potential lessons learned and include recommendations ______________ to help guard against a reoccurrence of the event. Even _______ flammable or toxic vapor release events should be reported and investigated.
Answer:
reported
investigated
implemented
minor

8 What is the purpose of the site overpressure study?
Answer:
The site overpressure study is a tool to help management understand the potential blast hazards associated with the hazardous materials at the site and to develop protection and prevention plans.

9 What are several causal factors that can result in a loss of containment and lead to Vapor Cloud Explosions? Name all you can.
Answer:
- Failure of Lock-out/Tag-out to isolate equipment
- Thermal expansion of trapped liquids
- Corrosion under insulation
- Corrosion or erosion caused by chemical injection or mixing points
- Failure of protection systems such as relief valves or shutdown trips
- Failure of dead-ended piping systems due to corrosion or freezing water
- Polymerization of chemicals resulting in rupture of process equipment
- Brittle fracture or other failures of LPG pressurized equipment
- Loss of containment while draining water from pressurized spherical tanks
- Inadvertently sending light hydrocarbons to atmospheric tankage

10 Name two references to learn more about Vapor Cloud Explosions.
Answer: Name any two of the following.
- US Chemical Safety Board website (csb.gov)
- American Petroleum Institute Standard API-752. "Management of Hazards Associated with Location of Process Plant Permanent Buildings."
- Guidelines for Vapor Cloud Explosion, Pressure Vessel Burst, BLEVE and Flash Fire Hazards, 2nd Edition (Center for Chemical Process Safety, Wiley)
- Any of the following investigation reports or other reports on a VCE event:
  - Phillips Pasadena VCE
  - Williams Olefins VCE
  - Killingholme, England VCE
  - Exxon Baton Rouge VCE

11 What causes extreme damage that can occur in a VCE?
a. Confinement
b. Fire
c. Overpressure
d. Velocity
Answer:
Overpressure

12 Why is it important to investigate all loss of containment incidents involving flammable or toxic releases, even the small ones that have no impact?
Answer:
A detailed investigation of the minor incidents with corrective actions implemented can prevent the more serious incidents, even those that could have had catastrophic consequences.

13 How do safety engineers and risk management specialists model the predicted overpressures from a potential VCE?
Answer:
Safety engineers and risk management specialists can model the overpressures from a potential vapor cloud explosion using specialized computer programs.

14 How can the risk management overpressure model help prevent incidents during unit turnarounds when many people are present on the unit?
Answer:
The model will help predict the building or structure design required that will best protect personnel assigned to the turnaround.

Chapter 5. Boiling Liquid Expanding Vapor Explosion (BLEVE)

End of Chapter Quiz

1 What is the mechanism or cause of a Boiling Liquid Expanding Vapor Explosion (BLEVE)?
Answer:
A Boiling Liquid Expanding Vapor Explosion (BLEVE) is a type of explosion that can occur when a pressurized vessel containing a liquid that is stored or heated to a temperature at or above its boiling point suddenly fails (a liquid above its boiling point if it were at atmospheric pressure).

2 What is by far the most common cause of a BLEVE occurring?
Answer:
The most common cause of BLEVE is a fire on or near the storage vessel resulting in flame impingement on the vessel causing failure of the vessel shell, and the resulting catastrophic loss of containment.

3 How long after the initial fire and vessel impingement must we prepare for the potential BLEVE?
Answer:
A BLEVE has been known to occur within minutes, hours, or even days after the initial fire.

4 What is meant by LPG at a temperature "above its boiling point"?
Answer:
A typical Propane rail car at ambient temperature would be about 150 psig, and the propane would be in a liquid state. However, the Propane wants to boil and expand. If the pressure were suddenly released to atmospheric pressure, for example, due to vessel failure, the Propane would rapidly boil and auto-refrigerate to its boiling point of −44°F (−42° C). The Propane would rapidly expand 370 to 1. Every gallon of Propane would create 370 gallons of highly flammable Propane vapor. This is the hazard of LPG and the underlying cause of BLEVE.

5 What is the hazard of a BLEVE?
Answer:
An explosion resulting from a BLEVE can be catastrophic and may result in numerous secondary fires or explosions due to the impact on adjacent facilities.
The explosion results in a significant pressure wave, and large sections of the tank or other containment vessel may become projectiles and "rocket" thousands of yards, causing further damage to adjacent equipment.

6 What are the warnings, and how do we know the vessel is about to BLEVE?
Answer:
The real hazard of a BLEVE is that it is instantaneous, with no warning when one does occur. BLEVE's have caught many responders off guard resulting in tragic consequences, including large loss of life.

7 What precautions do we take to help prevent or mitigate a potential BLEVE?
Answer:
- BLEVE emergency response preplanning is critical and should be periodically tested in drills and exercises.
- Consider equipment spacing during plant design.
- Emergency isolation equipment should be available to quickly isolate a leaking or failed system.
- Ensure effective LPG Water Draw procedures are in place.
- Ensure the Firewater Deluge Spray Systems are frequently tested and are well maintained.
- Ensure the capability to deliver an uninterrupted firewater supply in the event of flame impingement on a process or storage vessel.

8 How can a water draw operation on an LPG sphere result in a BLEVE of the vessel? Describe the proper procedure for draining water on an LPG sphere.
Answer:
Failure to follow procedures can result in freezing the water valves in the open position and a massive loss of containment of the LPG. The procedure should specify opening the upstream water draw valve fully and throttling the flow on the downstream valve.
When the water changes to LPG, the LPG will flash and auto-refrigerate, causing the valve to freeze. With the upstream valve in the fully open position, there is no pressure drop, therefore, no auto-refrigeration. The flow can be stopped quickly by closing the upstream valve. Also, the Operator must be in attendance monitoring the operation at all times when the water drain valves are open.

9 Can you name several sites where a BLEVE has occurred? Try to name at least three.
Answer: Not intended to be a complete list; you may know of others.
- Elf Refinery, Feyzin France
- Pemex Oil Company, Mexico City, Mexico
- Cosmo Oil Company, Chiba, Japan
- Union Oil Company, Romeoville, IL.
- Train derailment, multiple BLEVE's, Crescent City, IL.
- Texas City Refining, Texas City, TX.
- EniChem, Priolo, Sicily, Italy
- Train derailment, multiple Propane BLEVE's, Tuscany, Italy
- Williams Olefins, Geismar, LA.

10 When Propane flashes from a liquid to a vapor, as when a vessel initially BLEVEs, how much does it expand when it vaporizes?
a. By 140 times its initial volume.
b. By 560 times its initial volume.
c. By 270 times its initial volume.
d. By 4 times its initial volume.
Answer:
By 270 times its initial volume.

11 Is it possible for a vessel that is in a service other than hydrocarbon service to BLEVE?
Answer:
BLEVE is not limited to LPG or even to flammable materials. Vessels containing other liquids above the boiling point may also experience a BLEVE should the containment vessel be compromised. For example, a refinery steam drum can BLEVE should the vessel fail due to accidental damage, overpressure, or corrosion. This steam explosion and the resulting blast and release of hot water/steam can also be devastating.

12 Can metal debris be propelled long distances from a vessel during a BLEVE?
Answer:
Yes – Metal debris from the Phillips Pasadena BLEVE was found six miles from the site.

13 The BLEVE potential is considered during the initial vessel design. Name at least four special considerations incorporated into the LPG vessel design.
Answer: Name any four of the following.
- Firewater Deluge Spray System.
- Overpressure Protection is designed to avoid impingement (relief valve outlets directed away from the vessels).
- Fireproofing of Structural Supports.
- Firewater Emergency Flooding (to fill the sphere with firewater during a loss of containment or fire impingement).
- Drainage sloping away from the vessel to prevent the pooling of LPG spills under or near the vessels.
- The impounding of potential spills in the enclosed firewall (in some designs).
- Water and Sample facilities are designed to prevent the auto-refrigeration and potential freezing of LPG valves in the open position (double block valves, including spring-loaded to close).

Chapter 6. An Introduction to Process Safety Management Systems

End of Chapter Quiz

1 What is the purpose of the Process Safety Management System?
Answer:
Process Safety Management Systems (PSMS) are the guiding documents that transcribe the company's corporate Safety, Health, Environmental, and Security values and culture into documented and actionable management plans.

2 What are the five key written components of each Process Safety Management System?
Answer:
- The "What" (Defines the Scope/Objectives of the system)
- The "How" (Provides the Procedures/Practices to make it happen)
- The "Who" (Who is Responsible and Accountable for implementation?)
- The "How Do We Know?" (Are we getting the expected results?)
- The "How Can We Do Better" (Ensures continuous improvement)

3 Can you name the four key elements the OSHA example Safety Management system requires?
Answer:
The United States Occupational Safety and Health Administration (OSHA) prepared a sample Safety Management System named the "Safety and Health Program Management Guidelines."
The OSHA guideline outlines the following four elements as the guiding principles for the safety management system:
- Management Leadership and Employee Involvement
- Worksite Analysis
- Hazard Prevention and Control
- Training

4 Why is the OSHA sample Safety Management System unsuitable for a petroleum refining or petrochemical plant management system?
Answer:
Due to some of the chemicals used in the refining and petrochemical industry, it must meet the standard for Process Safety Management (29 CFR 1910.119), which is much more comprehensive.

5 Can you name the 14 key elements required by the OSHA Process Safety Management regulation (29 CFR 1910.119)?
Answer:
The OSHA Process Safety Management Regulation mentioned above is made up of fourteen elements which are:

1. Employee Participation	8. Mechanical Integrity
2. Process Safety Information	9. Hot Work Permit
3. Process Hazard Analysis	10. Management of Change
4. Operating Procedures	11. Incident Investigation
5. Training	12. Emergency Planning and Response
6. Contractors	13. Compliance Audits
7. Pre-startup Safety Review	14. Trade Secrets

6 What is the first Management Element in each of the referenced Management Systems?
Answer:
The element covering Management Leadership, Commitment, Employee Involvement, and Participation is in all management systems.

7 The Process Safety Management Systems (PSMS) are the guiding documents that transcribe which of the following values and culture into documented and actionable management plans? (Please select the correct answer):
a. Safety
b. Health
c. Environmental
d. Security
e. All of the above
Answer:
All of the above

8 The PSMS forms a Framework by which the corporate values are documented and communicated to all ________, __________, and _________ who works for the company.
Answer:
employees, contractors, and anyone

9 Match the components of the management system with the proper description.

A. The "What"	A. Who is Responsible and Accountable for implementation?
B. The "How"	B. Are we getting the expected results?
C. The "Who"	C. Defines the Scope/Objectives of the system.
D. The "How Do We Know?"	D. Ensures continuous improvement.
E. The "How Can We Do Better"	E. Provides the Procedures/Practices to make it happen

Answer:
A – C
B – E
C – A
D – B
E – D

10 The importance of Safety Management Systems is emphasized in what training video, where 167 lives were lost, and the critical work processes were characterized as "sloppy, ill-organized, and unsystematic."
Answer:
Piper Alpha (the oil platform fire in the North Sea that occurred in 1988)

11 PSMS sets the minimum standard/expectation for key ________ performance parameters. These are supported by management and reinforced by ongoing stewardship and management reviews to help ensure compliance and corrective actions when required.
Answer:
PSM (Process Safety Management)

12 The PSMS must be designed with a defined process for self-improvement, including periodic ________ to evaluate the PSMS program documentation and field compliance to meet the program's deliverables. Generally, these audits are conducted by peers with PSMS management system experience and experience in the ________ ________, ________, or ________ ________.
Answer:
audits
specific program, procedure, or assessed equipment.

13 The training required by specific organization members is detailed in the supporting documents to include training requirements for ________ Operator training, Operator ________ training, and training for ________ employees.
Answer:
initial
refresher
maintenance

14 The overall management system should also utilize ________ indicators to track compliance with the specific system requirements.
a. leading
b. lagging
c. leading and lagging
d. critical
Answer:
leading and lagging

Chapter 7. Importance of Safe Operating Procedures

End of Chapter Quiz

1 Which regulation mandates required procedures for covered processes at petrochemical facilities?
Answer:
The OSHA PSM Standard (29 CFR 1910.119) prescribes detailed requirements for operating procedures and ongoing owner verification for accuracy and relevance.

2 Why is it so important that this regulation be followed so closely?
Answer:
A significant number of catastrophic incidents, some with loss of life, have occurred due to deviation from established procedures and have contributed to significant process safety events, including those with the devastating loss of life, severe damage to site facilities, and resulting major impact on the community.

3 Operating procedures are critical to safe operations and must be maintained ________, and managers/supervisors should _________ procedure compliance in the field and on the console.
Answer:
current
enforce

4 The US Chemical Safety and Hazard Investigation Board found that Operators at BP Texas City (March 2005) were routinely __________ from written procedures during the startup of the Isomerization Unit.
Answer:
deviating

5 The OSHA PSM Standard recognizes the importance of procedures and requires PSM-covered sites to maintain detailed __________ operating procedures.
Answer:
written

6 What does the PSM Regulation say about the frequency of procedure reviews to ensure accuracy?
Answer:
The OSHA PSM Standard requires that "operating procedures be reviewed as often as necessary to ensure that they reflect current operating practice, including changes resulting from changes in process chemicals, technology, and equipment, and changes to facilities. The employer shall certify annually that these operating procedures are current and accurate."

7 Does the OSHA PSM Regulation require periodic training on the procedures?
Answer:
OSHA requires that "refresher training" shall be provided at least every three years and more often, if necessary, to each employee involved in operating a process to ensure that the employee understands and adheres to the current operating procedures.

8 Can you name a good practice for tracking the required Operator refresher training?
Answer:
A good practice is to track the Operator training as a process safety-leading indicator.

9 What is a good practice for the designation of procedures for higher-risk tasks that have already resulted in or could result in a catastrophic incident?
Answer:
A good practice is for procedures of this type to be designated so they receive special attention. For example, these procedures could be designated "SSHE Special" (Safety Security, Health, or Environmental) or a similar designation. Procedures so designated would allow special attention to their structure or layout during auditing and, most importantly, during their use by the Operators.

10 Can you name the four key signal words defined by ANSI and used in procedures to designate a degree or level of safety alerting and tie each to their associated colors?
Answer:
- Danger/Red
- Warning/Orange
- Caution/Yellow
- Notice/Black

11 Specific procedures require special emphasis due to the nature of risk and industry experience.
Name as many of these special emphasis procedures as you can.
Answer:
- Furnace and Heater operations, especially the following:

  - Purging of unburned fuel from firebox before lighting pilots and burners
  - Furnace flooding or fuel-rich scenarios
- Unit Operating Limits and Critical Alarms
- Air-Freeing Equipment Before Introducing Hydrocarbons (during startup or bringing equipment back into service following repairs or for new equipment)
- Start-up Procedures
- Procedures for prevention of brittle fracture failures
- Managing hydrocarbon thermal expansion hazards in process piping and vessels
- Identification and management of imported hazardous chemicals
- Pressure and Vacuum Hazards
- LPG Handling and Prevention of BLEVE (Boiling Liquid Expanding Vapor) – Includes procedures for draining water from LPG vessels (water drawing, decanting operations)
- LPG Emergency Response Procedures (include in Emergency Response Section of procedures)
- Unplugging process bleeders (process vents and drains)

12 What are the requirements laid out by the PSM regulation relative to process procedures for sites covered by the PSM regulation?
Name as many of the PSM requirements as you can.
Answer:
The OSHA PSM Standard recognizes the importance of procedures. The OSHA standard says, "The employer shall develop and implement written operating procedures that provide clear instructions for safely conducting activities involved in each covered process consistent with the process safety information and shall address at least the following elements."

- Steps for each operating phase,
- Initial startup,
- Normal operations,
- Temporary operations,
- Emergency shutdown, including the conditions under which emergency shutdown is required and the assignment of shutdown responsibility to qualified operators to ensure that emergency shutdown is executed in a safe and timely manner,
- Emergency Operations,
- Normal shutdown and startup following a turnaround or after an emergency shutdown,
- Operating limits, consequences of deviation, and steps required to correct or avoid deviation,
- Safety and health considerations,
- Properties of, and hazards presented by, the chemicals used in the process,
- Precautions necessary to prevent exposure, including engineering controls, administrative controls, and personal protective equipment,
- Control measures to be taken if physical contact or airborne exposure occurs,
- Quality control for raw materials and control of hazardous chemical inventory levels; and any special or unique hazards,
- Safety systems and their functions, and
- Operating procedures shall be readily accessible to employees who work in or maintain a process.

13 Emergency procedures should include all foreseeable scenarios, such as:
a. flammable or toxic vapor release
b. fire,
c. loss of plant utilities such as steam or instrument air,
d. major weather events such as a hurricane or severe freeze
e. none of the above
f. all the above
Answer:
All the above

14 Operators should be trained in emergency procedures and periodically practice procedure execution through __________ discussions and ___________/__________.
Answer:
tabletop
drills/exercises

15 How should all changes or modifications to procedures be approved and controlled?
a. Management of change process
b. Procedure modification and control documentation
c. Control of defeat process
d. Work permit process
Answer:
Management of change process

16 Which class of process equipment merits special attention to detail due to the nature of the hazards present?
Answer:
Furnaces and fired heaters

17 What special tools are available specifically for unplugging bleeders?
Answer:
The fully contained bleeder unplugging device.

Chapter 8.1. Returning Equipment to Service – Safe Startups and Other Transient Operations

End of Chapter Quiz

1 The start-up procedures and the start-up team must recognize and address many hazards. See how many you can list below – name as many hazards as possible.
Answer:
- Leak or release or other loss of containment.
- Air left inside process equipment.
- Water left inside process equipment.
- Overfilling process vessels.
- Loss of tower liquid level (tower bottoms level).
- Instrumentation failures.
- Backing hydrocarbons or other process chemicals into utility systems (through temporary hose connections).
- Inadvertently leaving a blind in a process line or system.
- Operator fatigue leading to operating errors.
- Inadequate information exchange at shift change or poor communication between shift and day organizations.

2 What procedure is completed during preparation for a startup to help ensure that a leak or major loss of containment does not occur?
Answer:
To prevent a leak of release during startup, we conduct a rigorous pressure test of all equipment, including all piping, pressure vessels, and flanges. This pressure test is generally called a "tightness test" and is done either with steam, nitrogen, or, in rare cases, Helium.

3 Explain how the start-up procedure should be used during the unit startup.
Answer:
The start-up procedure should be designed as a checklist with sign-off by the Operator for each step. The procedure should be signed for each step as the step is completed with the initial and date/time the step was completed.

4 Prior to unit startup, the Operators do a comprehensive unit walk down to ensure the unit is ready for startup. What are examples of what the Operators are looking for during this walk down?
Answer:
Operators should be trained to look for full thread engagement on all flange studs (no short bolting) and no missing studs. The Operators should verify that no vents or drains are left open and that all utility hoses are disconnected, rolled up, and safely stored off walkways and platforms. All scaffolds that interfere with the Operator's access to pumps and other process equipment should be dismantled or removed from the unit. All sewer covers should be removed and safely stored to ensure good drainage. And finally, and very importantly, all nonessential personnel should be off the unit and removed from the unit and surrounding areas before bringing in hydrocarbons.

5 How are instruments and other critical devices prepared and checked before startup?
Answer:
All instruments must be verified to be installed and connected. All instruments should be verified as functional from the control room, and control valves should be stroked to ensure they are fully functional before bringing in hydrocarbons. All devices classified as Safety Critical should receive full operational checks to verify the functionality before bringing hydrocarbons to the unit, including critical alarms such as low and high level, temperature, low or high flow, and pressure alarms.
For example, all unit trip devices, such as furnace and reactor trips, should be fully tested, from the initiating element to the final element, before start-up commences.
The start-up procedure should have a critical device checklist listing all the critical alarms and devices, and it should be followed to ensure none are missed.

6 Explain the process for air-freeing process equipment before bringing in hydrocarbons.
Answer:
The air-freeing process is detailed in the start-up procedure, and it must be followed in its entirety. Generally, the equipment is purged into the atmosphere using steam or nitrogen as the purge medium. Due to nitrogen's asphyxiation properties, everyone must be aware, and all vents must be well barricaded.
Pressure the equipment to a pressure less than the operating pressure, then quickly release the nitrogen into the atmosphere (sometimes referred to as concertina purging). Follow this process three times and test the vent at the top of the equipment with a recently calibrated oxygen analyzer. Continue the process until the target of less than 1% O2 (or the target outlined in the procedure) is achieved.
Ensure that all equipment is purged before allowing hydrocarbon into the unit. We use a checklist to help ensure that all equipment is air free and nothing is missed.

7 Are Operating Limits important during startup?
Answer:
Yes – Operating limits should be established for all key process parameters, and those limits should be enforced. Expectations should be in place for prompt reporting of all deviations to established limits, including an investigation to determine why the deviation occurred and corrective actions to help prevent a reoccurrence of the deviation.

8 What is the purpose of the Pre-startup Safety Review (PSSR)?
Answer:
The PSSR is a special safety review conducted on a new unit or modified equipment prior to the startup or commissioning of the equipment. The PSSR is intended to help ensure that the new or modified facilities meet the design criteria and operating intent and to detect any potential issue or hazard that may have been introduced during the equipment design and/or construction.
Simply stated, the PSSR is essentially the last audit to help ensure that the unit or equipment, the procedures, and the Operators are ready for startup.

9 Is there a process for tracking and accounting for blinds, and is this referenced during startup?
Answer:
Yes – The blinds are tracked using the equipment Blind Isolation List, sometimes referred to simply as the blind list, which is extremely important. Every blind installed in the unit must be accounted for on the blind isolation list. This is

to ensure that installed blinds are not forgotten and left in the unit, only to be discovered during startup or later when attempting to switch a stream to its final disposition.

Every blind installed must be listed on the isolation list; likewise, when a blind is removed from the unit, that blind must be signed off the blind isolation list. The list must always reflect the number of blinds currently installed and list each blind separately.

10 Do startups ever become routine and become "just another startup"?

Answer:

Every startup is unique, and every startup is different. For example, modifications may have been made to the unit since the last time the unit was started, or other interfacing process units or off-site facilities may have different situations. For example, it could be that a pump normally used for the startup is out of service or a different tank lineup is being used for this startup.

For this reason, it may be necessary to update the start-up procedure to address the conditions that will exist at the time; we will follow the MOC process when doing so. It is a good idea to get the start-up team together in advance, discuss the procedure that will be used, and discuss what is different from prior startups. Develop the start-up plan and then implement the plan.

11 What is the consequence of water left trapped in a fractionation column during startup?

Answer:

When introducing feed and increasing the unit's operating temperature, the water will flash violently inside the tower, resulting in a pressure surge inside the column. This can result in an unexpected pressure release from the relief valves to the atmosphere and can also result in internal damage to the column by upsetting the trays. This can result in an extended downtime to repair the internal damage.

12 Describe the potential consequences of losing the bottoms level in a fractionation column when the tower is operating.

Answer:

The column will blow through the bottoms to downstream equipment that may not be designed for the temperature, pressure, or type of materials. For example, loss of tower level can result in very light material being routed to a floating roof tank not designed to contain the lighter molecules. This can result in a vapor release and the sinking of a floating roof.

13 Why is a detailed shift log and shift turnover discussion important during a unit startup or similar unit transition?

Answer:

Unit operations change rapidly during a unit startup, and a detailed unit log and shift turnover discussion ensure the oncoming person fully understands the unit status and conditions.

14 List at least five topics that should be discussed during shift turnover during a unit startup.

Answer:

Note: Answer at least five of the following.

- Any changes made during the shift (i.e., instruments verified, levels established, blinds installed or removed, etc.).
- The status of safety or critical environmental devices that are disabled on unavailable for any reason. Include the contingency plan for monitoring the process and protecting the equipment during the outage.
- Any flawless operations items, including safety, health, and environmental incidents, near misses, or any issues with safety-critical equipment.
- For example, any environmental issues, flaring, or discharge to the sewer.
- Alarms – especially any alarms that are inhibited or disabled.
- The status of fire protection equipment and other emergency response systems.
- Any emergency management of change documents generated during the shift.
- Equipment preparation for maintenance – to be verified as prepared by the incoming shift (opening process equipment, lock-out/tag-out isolation, drained and opened equipment).
- Any interfaces with Mechanical, Technical, or Operations, and/or with upstream or downstream units, etc.

15 Why is it critical to maintain the unit blind list up-to-date?
Answer:
The blind list should reflect the blinds installed in or removed from the unit. During a turnaround, it is easy to miss logging a blind as it is installed or pulled. This resulted in tragic incidents later when toxic vapor was admitted to a vessel.

16 Operating limits should be ___________ for all key process parameters, and those limits should be ___________. Expectations should be in place for prompt ___________ of all deviations to established limits, including an investigation to determine why the deviation occurred and corrective actions to help prevent a reoccurrence of the deviation.
Answer:
established
enforced
reporting

17 At BP Texas City, the Operators had developed a habit of routinely ___________ from written procedures and filling the fractionator tower above the top tap on the level instrument.
Answer:
Deviating

18 What is meant by a "running" blind?
Answer:
A running blind is a blind that should be installed when the unit is running. It is important to capture these on the unit blind list when the blind is pulled for turnaround to ensure it is reinstalled before the unit is restarted.

Chapter 8.2. Safe Permit to Work, Breaking Containment, and Controlling Work

End of Chapter Quiz

1 What is a Work Permit?
Answer:
The work permit is an authorization by a competent, authorized person for a work group to proceed with the specified job under the stated conditions.
It is a "legal" binding document, and when the work permit is approved and issued to a work group, it is a "license" for work to start.
The key objective of the permit to work is to ensure that personnel are informed of hazards and the work is executed safely, protecting personnel safety, health, and the environment.

2 What is the expectation for compliance with the Permit to Work process in the facility?
Answer:
There should be clear expectations throughout the plant for full compliance (zero tolerance) with applying work permit processes and procedures.

3 Is this also true for work being done in the field?
Answer:
Fieldwork should fully comply with the work permit requirements, including the procedures for energy isolation, opening process equipment, and confined space entry.

4 Who is responsible for preparation of the equipment to be worked on?
Answer:
The Process organization, as the equipment owner, is accountable and responsible for ensuring that equipment or area handed over to the workgroup is safe to work on.

Each designated individual or position in the work permit system has specific roles and responsibilities with qualifications and training requirements specific to those roles.

5 What does the Permit to Work process do to help ensure the job is performed safely?
Answer:
The Work Permit process ensures that Jobs are planned, equipment is prepared, precautions are taken, and effective communications have occurred between the permit issuer, permit acceptor, and other impacted individuals/groups.

6 Is it common for a work site to have more than one work permit form? Please expand on your answer.
Answer:
Yes, it is common for a site to have several different types of work permits.
For example, work may be classified as hot work and require a "Hot Work Permit," hot work without an open flame may require a "Non-flame Hot Work Permit," cold work may require a "Cold Work Permit," entry into a confined space typically requires a "Confined Space Entry Permit," or opening process equipment, may require an "OPE" or "Opening Process Equipment Permit."
Each type of work permit is typically a different color to help identify the type of permit being authorized and has the specific precautions and types of gas testing required noted on the permit form for the type of work authorized.

7 Energy isolation is an important part of the work permit. What kind of energy is controlled?
Answer:
Hazardous energy may consist of process pressure, electrical, mechanical (such as springs), temperature, etc. Failure to recognize energy sources and adequately control hazardous energy has resulted in many process safety incidents, some with tragic outcomes.

8 When process equipment is involved, how is hazardous energy controlled?
Answer:
To achieve energy isolation, the operations group, typically Operators, will ensure the equipment is depressurized and drained; valves for isolation are identified, closed, and secured in the closed position. Most sites use a system of process isolation tags and locks (with a chain if required) to secure the valves in the isolated position. The objective should always be to achieve "zero energy," which should be confirmed with open (and unobstructed) process vents and drains.

9 What is this energy isolation process called?
Answer:
Lock-out Tag-out or LOTO

10 Describe the process for ensuring lock-out tag-out for electrical equipment.
Answer:
Electrical energy isolation is normally accomplished by opening the electrical breaker and installing the isolation tags and a process lock to secure the breaker in the open position. Before starting work, the personnel performing work must also install their individual locks on the opened breaker(s). An attempt to start the equipment at each pushbutton location should be made to ensure the electrical energy has been isolated BEFORE the work can begin.

11 A higher-level review and formal ______ __________ to identify mitigation will be required if zero energy cannot be achieved; in some cases, for example, when flammable or toxic materials are involved, this will require the equipment to be shut down to achieve proper isolation before work can be performed.
Answer:
risk assessment

12 What is the main purpose of gas testing before opening process equipment?
Answer:
When opening process equipment, it is important to verify that hazardous materials such as flammables or toxics have been properly purged from the equipment before work begins. For example, the gas test can help identify the need for

additional Personal Protective Equipment (PPE), such as self-contained breathing apparatus (SCBA), while installing blinds in the piping systems.

13 If a gas test instrument indicates 0% flammables, does this always indicate that the equipment is "hydrocarbon free"?
Answer:
No, it does not. A gas test for flammables will *not* detect combustibles trapped in piping or other process equipment due to the flash point of the combustibles (flash points above 100°F) (38° C).
This material will not be vaporizing at most ambient temperatures; therefore, vapor will not be present for the gas test apparatus to read. Therefore, a good gas test may not indicate that the equipment is hydrocarbon free.

14 When a task represents a potential for a more serious hazard, what can be done to identify and control hazards in addition to the traditional Permit to Work process?
Answer:
On jobs with a higher potential for hazards to be present or deemed to have a higher potential for injury, a Jobsite Hazard Analysis should be conducted before the work begins.
This process identifies the individual steps to be undertaken during the job, the hazards associated with each step, and a plan for mitigating or controlling the identified hazards.

15 Just before the permit is issued, what is the final process to help ensure good communication between the permit issuer and permit acceptor and that proper precautions are in place?
Answer:
Before work begins, there should be a joint field visit by the permit issuer and the person accepting the work permit to review the work to be done, the precautions and mitigation in place, and to verify equipment isolation for mechanical work.
The exact work to be done should be discussed in the field; for example, when opening process equipment, the visit should highlight which flanges to be opened and, if applicable, the proper sequence to be followed. The flanges or points where the equipment is to be opened should be clearly identified with tags, painting, or another method selected by the site.

16 How do we ensure that the work permit process is applied properly at our site and that it doesn't become a "formality"?
Answer:
Due to the importance of work permits, each site should have a work permit audit process to help ensure that work permits do not become a "formality" like what happened at Piper Alpha.
Work permits, both paper audits and fieldwork in progress, should be audited periodically to ensure full compliance with the site work permit standards and requirements.

17 If you are about to issue a permit to burn or weld on a process vessel, is it enough to have an acceptable gas test before issuing the permit?
If the answer is no, please explain your answer.
Answer:
The answer is no. A gas test simply confirms the lack of vapors from hydrocarbon or toxic vapors. The vessel may also contain diesel or gas oil with a flash point above the atmospheric temperature. In this case, the tank's liquid contents could still ignite. This must be confirmed with a visual observation or confirmation by openings (for example, bleeders) at the low points of the vessel.

18 Why should 'break-in' work be avoided to the extent possible?
Answer:
Break-in work is typically a rush-type job and not very well planned. This can introduce both safety and efficiency issues.

19 The work permit should specify the appropriate _____ _______equipment that should be readily available at the job site (for example, a pressurized firewater hose, properly rated fire extinguishers) (Class A, B, C, or D), etc.
Answer
fire protection

20 Opening Process Equipment (OPE) should always be considered "_____ ______" work and requires attention to detail and close monitoring of work activities during field execution.
Answer:
high hazard

Chapter 8.3. Managing Change

End of Chapter Quiz

1 Which US Regulation requires management of change?
Answer:
The US OSHA Process Safety Management Standard (29 CFR 1910.119) ("PSM") requires the employer to establish and implement written procedures to manage changes (except for "replacements in kind") to process chemicals, technology, equipment, and procedures; and changes to facilities that affect a covered process.

2 Why is the MOC process such an important part of process safety management?
Answer:
An effective Process Safety Management System focuses on proactively identifying risks, assessing them, and taking actions to control them. An effective process to identify changes, review the changes, and implement controls and approvals almost always includes Management of Change as a key component.

3 What kinds of changes require strict adherence to the MOC process?
Answer:
We generally think of physical changes to the equipment as requiring a "Management of Change" (MOC), and physical changes such as upgrading equipment metallurgy, replacement of a pump impeller with a larger one, and installing a new pipe connection are certainly changes covered under the OSHA Process Safety Management standard.
Other changes that are also straightforward MOC include revisions to operating procedures, introducing a new or different process chemical, and/or introducing a new processing technology.

4 Are there other changes that are not necessarily so easily recognized but still require compliance with MOC?
Answer:
Yes, many other changes may be more subtle, but never-the-less should be considered as requiring a MOC. Examples include changes in unit feed quality, changes in catalyst technology and/or activity, and program changes in the distributed control system (DCS) should all trigger the MOC process.

5 What is the Pre-start up Safety Review, and when is it required?
Answer:
After field construction is completed, the change should be subjected to a Pre-startup Safety Review (PSSR). The PSSR helps confirm that the field construction conforms with the original design and that procedures have been developed and are adequate and available to the Operators.
The PSSR also verifies that the Process Hazard Analysis (PHA) recommendations have been resolved or implemented and that workers have been trained in the new or modified process or equipment.

6 As defined by the MOC system, what are the main types of changes?
Answer:

- **In-Kind Change:** replacement that is identical to the original and satisfies all relevant requirements, standards, and specifications (does not require a MOC process).

- **Not-in-Kind Change:** is one in which there is any difference between the original and the replacement (requires the full MOC process).
- **Emergency Change:** an unplanned change needed immediately to reduce the risk of an SH&E incident or to respond to an unforeseen event or situation that affects continued efficient operations. Generally completed on-shift by a pre-designated and trained on-site emergency MOC contact leading the evaluation.

7 One thing that is certain about operating an asset-intensive operation like a petroleum refinery or petrochemical plant is ________. There is a saying, "Nothing is consistent but change."
Answer:
Change

8 At the Williams Olefins plant, if the MOC had been done as required by the standard, along with a rigorous Process Hazard Analysis (PHA) and a thorough Pre-startup Safety Review (PSSR), it is possible that the hazard would have been __________ and this incident would not have happened.
Answer:
Identified

9 Which of the following does NOT require a MOC to be developed?
a. Replacement of a pump impeller with a larger one.
b. Replacement of a seal on a floating roof tank with the same type of seal.
c. Modification of a turbine overspeed trip to make it more reliable.
d. Replacing the catalyst in a reactor with catalyst that is thought to be more active.
e. Updating a start-up procedure with better drawings.
f. Changing the unit feedstock to one with a higher yield.
Answer:
b. Replacement of a seal on a floating roof tank with the same type seal.

10 The MOC documentation should specify the __________ basis for the change and any ________ ____ ________ effects resulting from the change.
Answer:
Technical
safety and health

11 What is the purpose of the Process Hazard Analysis (PHA), and why is it done?
Answer:
The PHA ensures that the proposed change does not introduce hazards into the process or equipment. This review ensures that safety, health, and environmental risks arising from these changes remain acceptable.

12 Does "Approval to Progress" mean that the site has the approval to operate the equipment that the engineers have designed?
Answer:
No, it does not. "Approval to progress" means that the change is approved for work to begin in the field on the construction or modification. Approval to progress does not mean the change has been approved to operate or commission.

13 What is required by the MOC process to commission the equipment?
Answer:
The "approval to commission" is given once the PSSR has been completed, all follow-up items have been resolved, and personnel have been trained in the change. This means the change can be placed in service.

14 Describe what constitutes an emergency change and how this can be implemented.
Answer:
An Emergency Change is an unplanned change needed immediately to reduce the risk of an SH&E incident or respond to an unforeseen event or situation that affects continued efficient operations. They are generally completed on-shift by a pre-designated and trained on-site emergency MOC contact leading the evaluation.

15 The approved time period should be specified for the temporary change, and the temporary change should be re-reviewed and reapproved before the initially approved time period expires.
- True
- False

Answer:
True

16 The OSHA Process Safety Management standard (29 CFR 1910.119) requires that any change that may affect a process covered by that standard, except a "__________ ___ ________," requires detailed management of change (MOC) evaluation.
Answer:
"replacement in-kind,"

17 What other US regulatory standard has this same MOC requirement?
Answer:
The Environmental Protection Agency, under its Risk Management Plan rule.

Chapter 8.4. Ensuring Availability of Safety, Security, Health, and Environmental (SSHE) Critical Devices

End of Chapter Quiz

1 What does the term Safety, Security, Health, and Environmental Critical Devices mean?
Answer:
Safety equipment and safety systems are designed to protect personnel, the community, the environment, and equipment from process upsets, abnormal process conditions, and security intrusions. These devices are Safety, Security, Health, and Environmental (SSHE) critical devices or, most often, just Safety-Critical Devices (SCD).

2 Which is by far the most common of these devices?
Answer:
Protection against overpressure, like safety relief valves, is the most common type of safety-critical device. These are also known as relief or safety valves but may be other types of overpressure protection, such as rupture disks, rupture pin valves, or pressure vacuum valves.

3 What is the potential consequence if a SSHE Critical Device fails to function as expected?
Answer:
Safety Critical Equipment includes equipment, instruments, analyzers, or other devices, either fixed, mobile, or portable, whose failure to operate or function could result in one or more of the following:
- Serious personal injury to a plant worker.
- Loss of containment which is likely to result in a significant fire or explosion.
- Serious disruption to the off-site community.
- Loss of containment which is likely to have a serious environmental impact.
- Inability to mitigate or carry out emergency response to an adverse event.
- Unauthorized access to plant facilities or computer systems.

4 Can you name other examples of SSHE Critical Devices?
Name all that you can.
Answer:
The following are examples of devices that are generally defined as SSHE critical devices (this is not intended to be a complete list):

- Pressure/vacuum relief valves - Car-seal valves - Steam turbine overspeed trip devices - Fired equipment protective systems - Compressor protective systems - Process unit shutdown systems - Fixed and mobile firefighting facilities - Firewater pumps - Breathing apparatus sets - Safety showers - Flame/detonation arrestors	- High/low temperature and pressure protective systems - Storage tank high-level alarms - Level alarms/cut-outs on flare and blowdown drums - Protective systems for heaters on tanks - Hydrocarbon gas detectors - Smoke detectors, fire detectors, and fire suppression systems - Power-operated emergency isolation valves

5 What is the best practice for each manufacturing site relative to identifying and tracking safety-critical devices?
Answer:
Each manufacturing site should identify and develop a list of all SSHE critical devices at the site as a controlled document. The list should be periodically verified as current (up-to-date), with copies readily available to field and console Operators.
The individual SSHE devices should be clearly identifiable in the field and on the P&ID drawings. A common practice is to paint each SSHE device a distinctive color in the field, with no other devices painted the same distinctive color.

6 Is it important that the SSHE Critical Devices are periodically tested in the field to ensure full functionality? Why is this important?
Answer:
Yes, most SSHE devices lie dormant, although continuously monitoring the process against the trip target; if the process remains stable and on the safe side of the trip point, the device does not function. This means to ensure that the device is functional and will work when expected, it must be periodically tested.
The test must be scheduled and verify the device's full function, from the initiating element to the final element.

7 How is the test interval established for each Critical Device?
Answer:
The test interval is directly related to the expected service factor or device availability target. Therefore, the manufacturing site should develop the expected service factor or availability target for each SSHE device.
A typical availability target for most SSHE devices will fall between 98% and 99%; the higher the availability target, the more frequent the test to ensure the device will meet the expected availability.

8 How are the people and equipment protected when the critical device is out of service for testing or maintenance?
Answer:
Each site should have a well-documented and communicated process for ensuring the equipment is protected when one of these SSHE devices is out of service ("defeated"), that the defeat is controlled, has proper authorization, and is communicated to all involved with the process.
The process should also ensure that safety mitigation plans have been developed and implemented to ensure that alternate protection is in place while the critical device is unavailable.
It is also essential that the device is properly returned to full online service after the PM or maintenance is completed.

9 What are these systems or procedures called?
Answer:
These systems are typically referred to as Critical Device Impairment (CDI) or Control of Defeat (COD). There may also be other names for these systems.

10 Can you name the key features of a typical Critical Device Impairment plan or a Control of Defeat plan? Name all the features that you can.
Answer:
- The CID or COD process should be well-documented and communicated to all operations personnel.
- Each defeat should have proper authorization and be communicated to all involved with the process.
- The safety mitigation plans must be developed and implemented to provide alternate protection before the defeat and continue until the device is returned to full online service.
- The device should be properly returned to full online service as soon as possible after the PM, or maintenance is completed.
- A list of "defeated" safety-critical devices should be posted in clear view of the Unit Operator and the Console Operator. The Operators should be aware of the alternate mitigation in place.
- The device disablement authorization should be for a specific period, for example, one process shift. COD extensions for longer periods should require escalating levels of management approval. The longer the duration of the defeat, the higher the level of approval that is required. Defeat extensions should also require consideration for additional layers of protection.
- The CID/COD information must be communicated as a routine part of the shift turnover process at each shift turnover at the console and in the field.
- Another good practice is to require a "defeat tracking board" located in the control center in clear view of managers and supervisors, listing all active SSHE device defeats. This can help by providing emphasis on the number of defeats and duration of defeats.

11 In years past, these critical systems were simple pneumatic-controlled devices. In today's world, many of these systems are now ______________-______________ and are minicomputers with triple redundant technology.
Answer:
electronically-controlled

12 In addition to being triple redundant, these new electronic systems are designed with ________ __________ and onboard diagnostics. For example, they may require two sensors out of three for the device to act. This dramatically improves their reliability and eliminates false trips.
Answer:
voting logic

13 What is a good practice for the identification of SSHE critical devices?
Answer:
The SSHE devices should be identifiable in the field and on the P&ID drawings. A common practice is to paint each SSHE device a distinctive color in the field, with no other devices painted the same specific color.

14 The ______ ______ _______ management plan is a related process for protecting critical devices. This requires a protective car-seal placed on the isolation valves upstream or downstream of each critical device.
Answer:
car-seal valve

15 What is a good practice for keeping the site management and Operators aware of the number and types of critical devices that are bypassed or otherwise disabled?
Answer:
A list of "defeated" safety-critical devices should be posted in clear view of the Unit Operator and the Console Operator, and the Operators should be aware of the alternate mitigation in place. Another good practice is to require a "defeat tracking board" located in the control center in clear view of managers and supervisors, listing all active SSHE device defeats.

16 When do we most need the SSHE devices to be fully functional? Mark all that apply.
a. When the unit is in an upset condition.
b. During a unit startup or shutdown.

c. When a control variable is out of range (above or below limits).
d. When a relief valve is relieving.
e. During a utility failure.
Answer:
All the above

17 What is the only SSHE device where a disablement process like Control of Defeat does not apply? Why?
Answer:
A steam turbine overspeed trip device. A control of defeat does not apply to a steam turbine overspeed trip device. The turbine will be removed from service if the overspeed trip is not fully functional.
There is no mitigation available to protect the turbine from overspeed and the potential destruction of the turbine.

Chapter 8.5. Managing Critical Alarms

End of Chapter Quiz

1 What has changed that makes the Alarm Management System so very important?
Answer:
With the advent of computerized process control and today's sophisticated distributed control systems (DCS), it is now relatively easy to install alarms on just about all process variables. Therefore, it is essential to have a system to ensure that the new alarms are required, well thought out, and properly prioritized. Otherwise, the Operators can be overwhelmed by the number of alarms to the point where analysis of unit upsets, or emergencies is very difficult.

2 Why is Alarm Management one of the key focus areas?
Answer:
Alarm management is one of the key focus areas since alarm issues can lead to loss of containment events. Major refinery incidents have occurred, including loss of containment events attributable to alarm management issues.

3 What are the key elements of a site Alarm Philosophy document?
Answer:
The Alarm Management Philosophy should address the overall philosophy for the alarm systems used at the site and typically includes the following key information:

- The alarm definitions used at the site.
- The alarm priorities typically based on required operator response times (typically based on an alarm priority matrix).
- The alarm performance indicators.
- Alarm maintenance procedures.
- Alarm management of change procedures.
- The procedure for disabling alarms.
- Alarm interface design (interface to console and field operators).
- Alarm system training.
- A description of the Alarm Database (including the designated owner).

4 Alarm color and tone should be consistent with alarm ________ and expected Operator ________ time.
Answer:
priority
response

5 What are some of the issues that can be experienced with the alarm system that the Alarm Management System should address?
Name all the issues that you can.

Answer:
Alarm Flood:
An alarm flood is a condition where alarms are coming into the Operator faster than the Operator's ability to comprehend them. In the alarm flood condition, the Operator cannot easily distinguish between alarms needing immediate attention and those that are relatively lower priority.
Nuisance Alarms:
Another factor that often contributes to alarm issues is nuisance alarms. Alarms continuously in and out of the alarm state are distracting to the Operator and can lead to complacency.
Alarm Rate:
If the number of new alarms generated each hour is high, it can distract and overload the Operator.
Average Number of Alarms On:
Alarms in a continuous state of alarm can become a nuisance and quickly become a "a normalized deviation."
Chattering alarms:
Alarms constantly going in and out of alarm significantly add to alarm overload and may represent a mechanical or alarm setpoint issue.
Redundant alarms:
An alarm that warns the Operator of an event that is already being alarmed by a separate alarm. This is another form of a nuisance alarm.

6 Can these be utilized as Leading Indicators for process safety and improving the alarm management system?
Answer:
Yes, these can be used as Leading Indicators and tracked against continuous improvement targets to improve the overall alarm performance.

7 Can you name the key human factors issues that should be part of the alarm management system?
Answer:
- Alarm display should be in proximity to and visible to the Operator.
- Alarm priority should be established, which includes considerations such as alarm panel display, alarm indicator color, and the alarm auditable tone.
- Alarm color and tone should be consistent with alarm priority and expected Operator response time.
- Generally, alarms prioritized as "high priority" should not be DCS-based. These "hard-wired" alarms should also have a distinct sound and color to easily distinguish high-priority alarms from other lower-priority alarms.

8 What is meant by alarm rationalization?
Answer:
Rationalization is a thorough review of existing alarms and a process to eliminate or "rationalize" those found to be noncritical.

9 What is meant by alarm "enforcement"?
Answer:
The Alarm Management Enforcer (The Enforcer) is an option that can be built into the Alarm Management Database. The Enforcer is a software program that can automatically reset all alarms that have been disabled, inhibited, or changed from their original state back to their original state or value as described in the Alarm Management Database at the beginning of the following process shift.

10 Generally, alarms prioritized as "high priority" should not be DCS-based. Due to potential reliability issues, all high-priority alarms should bypass the ________ system and go directly to a "______-______" alarm, fully independent of the DCS system or subsystems. These alarms should also have a distinct sound and color to easily distinguish high-priority alarms from other lower-priority alarms.
Answer:
DCS
"hard-wired"

11 To whom should the routine Alarm Management Performance Tracking be routinely reported? Should this be an integral part of the site's Integrity Management Process?
Answer:
Site Management
Yes

12 All changes to the alarm database or individual alarms must be managed through the site's MOC process.
- True
- False

Answer:
True

13 If alarms are defeated or disabled for any reason, they should be subject to the control of defeat type process to ensure that the proper authority ____________ the disablement. The defeat process should also ensure that an alternate protection plan is in place that replaces the ______________ of the defeated device and that the alarm is properly ___________ ___ ___________ as soon as possible.
Answer:
Authorizes
functionality
returned to service

Chapter 8.6. Managing Safe Operating Limits or Operating Envelopes

End of Chapter Quiz

1 What are the OSHA Process Safety Regulation requirements respective to safe operating limits?
Answer:
The OSHA Process Safety Regulation (OSHA 29 CFR 1910.119) (d) requires each OSHA Process Safety Management covered process to define the safe upper and lower limits for such items as temperatures, pressures, flows, or compositions; and an evaluation of the consequences of deviations, including those affecting the safety and health of employees.

2 Generally, the safe operating limits are established by ______________ ______________ with involvement and concurrence by ______________.
Answer:
Generally, the safe operating limits are established by **technical support** with involvement and concurrence by **Operations**.

3 The normal operating limit must be within the ____________________.
Answer:
The normal operating limit must be within the **safe operating limit**.

4 Once the safe operating limits are established, what is the final step in the process?
Answer:
Once the safe operating limits are established, the final step is to evaluate and document the consequences of the deviations beyond that limit as required for OSHA PSM (for US facilities covered by OSHA PSM) (29 CFR 1910.119 (d).

5 Following development of the safe operating limits, what should the expectation be if there is an exceedance of these limits?
Answer:
Management should establish the expectation for the prompt reporting of all exceedances of operating limits. Each exceedance should be investigated to identify the cause and corrective actions developed to prevent a recurrence.

6 In the BP Texas City explosion, the Operators routinely ___________ from the safe upper limits of the fractionator level by running the tower level above the range of the level instrument. This eventually resulted in the overfill of the tower and loss of containment of hot light naphtha to the atmosphere resulting in a devastating explosion and fire.
Answer:
Deviating

7 Establishing a safe operating envelope or limit requires input from the key stakeholders in the ________, ________, and __________. The expectation is that all personnel will ensure full _____________ with the operating envelope once it is established.
Answer:
Process, technical, and mechanical
compliance

8 What are some examples of operating parameters for which operating envelopes should be developed?
Answer:
Flows, temperatures, levels, pressures, composition, and other key operating parameters such as PH and maximum or minimum velocities.

9 How should all deviations from the operating limits be reported and tracked?
Answer
As a process safety-leading indicator.

10 Follow-up for each deviation should be developed to understand what caused the deviation and ________________ should be developed to prevent a reoccurrence.
Answer:
Recommendations

Chapter 8.7. Maintaining Safe Product Rundown Control

End of Chapter Quiz

1 What is the vapor pressure limit for a rundown naphtha stream to product storage?
Answer:
A naphtha product with a true vapor pressure greater than 13 psia is considered LPG and must be handled/stored in a pressurized tank such as a sphere or horizontal storage drum. Sending a high vapor-pressure product to a floating roof tank can result in sinking the tank roof, a major vapor release, and a potential explosion/fire.

2 What are the potential consequences of inadequate cooling of rundown streams to storage tanks by the process units?
Answer:
A fixed roof tank is limited to a maximum temperature rundown of 200°F (93° C) due to the water bottoms. If the tank temperature exceeds the boiling point of water, it can result in a boilover of the tank contents. Inadequate cooling of the product stream may send hot products to a cold tank, resulting in a tank boilover and major tank fire.

3 What are the considerations for establishing the operating limits for temperature and vapor pressure for product rundown streams?
Answer:
Operating limits must be established for product rundown streams, and those limits must be enforced. The limits must consider the product involved and the design and limitation of the rundown tanks. Alarms should be set to identify escalating conditions BEFORE the limitation is reached.

4 What temperature limitations are in place to protect against the flashing of lower flash products and creating a hazardous vapor near fixed roof tanks?
Answer:
The temperature of products in fixed roof tanks must be limited to no more than 15°F (8° C) below the product's flash point. This is due to the tank being vented to the atmosphere – heating the product above the flash point will result in vaporization of the product, resulting in an explosive mixture around the base of the tank.

5 What are the typical guidelines for operation of fixed roof tanks in "hot service" such as asphalt and hot residuum (tanks that normally operate at temperatures above 200°F) (93° C)?
Answer:
Fixed roof tanks in "hot service" must be operated water-free, with no free water in or sent to the tank. The tank temperature is normally maintained above 265°F (130° C) and should not be operated at temperatures between 200°F (93° C) and 265°F (130° C).

6 The operating conditions, primarily the ___________ and _________ __________ of the product rundown, must be aligned with the receiving tank, or the results can be catastrophic.
Answer:
Temperature and vapor pressure

7 Loss of control of a product stream, for example, inadequate cooling of the product stream resulting in sending hot products to a cold tank, especially a tank with a _______ bottom which can result in a tank ______ ________ or froth over, or sending light materials to an atmospheric tank such as LPG to a _______ _______ roof tank which can result in sinking of the floating roof and a full surface tank fire.
Answer:
Water
boil over
floating roof

8 Why is it important to operate a fixed roof tank in hot service, for example, asphalt and residuum, at a temperature of 265°F (130° C) or above, and they should not be operated at temperatures between 200°F (93° C) and 265°F (130° C)?
Answer:
This ensures that trace amounts of water are quickly flashed off and that water does not accumulate in the tank.

9 Tank alarms designed to prevent exceeding operating limits should be classified as critical ________ devices and be subjected to periodic function ________ to ensure operability.
Answer:
Safety
testing

10 Running blinds should normally be installed on operating units to prevent the inadvertent sending of _______ streams to atmospheric tanks, _______ streams to cold tanks, ________ streams to hot tanks.
Answer:
Light
hot
wet

11 What types of operating conditions are more likely to lead to periods where refineries and petrochemical plants are most susceptible to deviations from operating limits?
Answer:
Periods of transient operations such as a unit startup, shutdown, unit upsets, running a plant test, etc.

Chapter 8.8. Storage Tank Safety

End of Chapter Quiz

1 Name the different types of storage tanks discussed in this chapter.
Answer:
- Fixed Roof or Cone Roof storage tanks.
- Floating Roof storage tanks (External floater).
- Floating Roof storage tanks (Internal floater).
- Spherical storage tanks (spheres).
- Refrigerated storage tanks (for LPG).

2 Do the safe operations guidelines for storage tank safety apply only to tank field personnel?
Answer:
No. Sending products to atmospheric tankage too hot, too light (product containing light hydrocarbons such as C_3, C_4's, or C_5's), or sending water to hot tanks has resulted in a significant loss of containment such as a vapor release and resulting fires and explosions. This is why tank safety is not just for off-site personnel. Unit personnel and good unit operations can have a significant positive impact on storage tank safety.

3 What type of products can be safely stored in external floating roof tanks?
Answer:
These tanks are designed to store products with a flash point less than 100°F (38° C), such as naphtha, gasoline, crude oils, and chemical products such as benzene or toluene. They are limited to products with a True Vapor Pressure (TVP) of less than 13 psia.

4 What is the primary reason that a cone roof tank would be converted to an internal floating roof?
Answer:
The primary reason a cone roof tank would be converted to an internal floating roof would be to change the tank service to a product with a lower flash point, for example, changing a product from a combustible such as diesel to a flammable like gasoline. The internal floating roof is less likely to have environmental emissions and less likely to leak rainwater into the product and accumulate water in the product.

5 What services are cone roof storage tanks more typically used for?
Answer:
Cone roof storage tanks are used for services where floating roofs are not required to suppress the vapors. They are typically used to store diesel, fuel oils, lubricants, waxes, and similar nonvolatile products.

6 What are some of the limitations or restrictions associated with cone roof tanks?
Answer:
They operate below 200°F (93° C) due to the likely presence of free water on the tank bottom and the risk of flashing water if operated at higher temperatures.
Cone roof tanks must also operate at temperatures 15°F (8° C) below the flash point of the material stored to prevent releasing the vapor to the atmosphere from the tank vents.
Due to the atmospheric vents, cone roof storage tanks cannot be used to store products at temperatures above the flash point of the product stored, even if the tank is inert or vapor enriched.

7 Can a cone roof tank be used in hot services such as asphalt or hot residuum?
Answer:
Yes. Cone roof or fixed roof tanks may occasionally be used in "hot Service" for products such as asphalt or other heavy, intermediate unit feed products such as hot Coker or hot FCC feed. To prevent free water accumulation on the tank bottom, these tanks should be operated at temperatures above 265°F, and precautions should be in place to prevent water from entering the tank.

8 What are spheres or horizontal storage vessels, sometimes referred to as bullets, used for?
Answer:
Pressurized storage tanks such as spheres or horizontal storage vessels ("bullets") are used for storage of liquefied petroleum gases (LPG) such as propane, butane, pentane, or light chemical products such as propylene or butadiene. Products with a vapor pressure of more than 13 psia require pressurized storage.

9 What is the primary hazard when working on or around the spheres or bullets?
Answer:
The most significant hazard is the loss of containment of the LPG. LPG is very volatile and ignites easily and can result in a pressure fire. A pressure fire on or near a sphere can result in a catastrophic explosion (a BLEVE or Boiling Liquid Expanding Vapor Explosion).

10 What are some of the safety features associated with spheres or pressurized horizontal storage vessels? Name all the safety features that you can.
Answer:
Pressurized storage tanks typically have a range of safety features in the event of exposure to fire or loss of containment.
The terrain under the sphere should slope away from the sphere to avoid low spots where a pool fire could impinge on the vessel.
A water deluge system that is designed to completely blanket the vessel with firewater in the event of fire exposure or another emergency.
The support legs are fireproofed to prevent failure during a fire event.
Most spheres and bullets also have firewater injection facilities to facilitate "water flooding" of the vessel in case of a leak or release near the bottom.
The firewall or containment dykes for LPG-containing storage tanks are typically equipped with firewall drain valves that are normally closed.

11 What are some of the hazards associated with storage tanks? Name as many as you can.
Answer:
Storage Tank Hazards include the following potential scenarios.
- Vapor release.
- Boil over or froth over.
- Sinking a floating roof.
- Explosion or fire due to hot work (welding, etc.).
- Tank exposed to vacuum or overpressure.
- Pyrophoric Materials (covered in Chapter 8.21, "Pyrophoric Ignition Hazards").

12 Can you give at least one example of a site that appears to have lost its "license to operate" due to a significant tank overfill and the resulting explosion and fire?
Answer:
For example, the Buncefield terminal has yet to be rebuilt and will most likely not be rebuilt due to the significant community impact of that incident.

13 To prevent overfill, what kind of device are tanks in low-flash service, such as naphtha and gasoline, equipped with?
Answer:
Tanks in volatile service, containing either flammable or toxic materials, should normally be equipped with independent high-level alarms to help prevent tank overfill.
The high-level alarms should be classified as a safety critical device and subjected to periodic testing to ensure their functionality.

14 What are two potential causes of a storage tank boilover and devastating fire?
Answer:
Sending hot product to a storage tank with a layer of water on the tank bottom results in flashing the water to steam.
A storage tank fire involving a tank that has a layer of water on the tank bottom.

15 What are the immediate response steps an Operator should take in the event of a tank overfill?
Answer:
- Immediately evacuate all personnel from the area.
- Stop the flow into the tank from a remote location by shutting down pumps and closing the remote valves. Do *not* approach the storage tank or drive into the area.
- Stop all traffic in the area from a remote location and ensure barricades are in place to prevent vehicles from entering the area.
- Notify emergency response personnel and advise of the overfill. Advise that no one drives into the area.
- Notify the Wastewater Treatment Unit of the overfill.
- Secure the area, prevent access, and support the emergency response personnel.

16 What are the potential consequences if the tower internal temperature on a Debutanizer is allowed to go low due to mechanical failure or Operator error?
Answer:
A Debutanizer tower produces Butane and lighter components on the overhead stream and light naphtha on the tower's bottoms stream. The naphtha is often routed to a floating roof tank.
If the temperature on the tower goes low, the overhead (LPG) drops into the bottoms and goes out with the light naphtha product.
When released into the storage tank, the LPG flashes and can result in a sunken tank roof and/or a catastrophic explosion at the tank or in the area around the tank.

17 For storage tanks in flammable service and during normal operations, tank procedure generally avoids pumping the level down to the point where the roof is "landed" on the roof support legs. What is the safety concern with landing on the roof?
Answer:
When the level is pumped down to where the roof is landed on the support legs, a vent is opened to break the vacuum under the roof. This vent allows air to enter the tank vapor space under the roof, and an explosive or flammable atmosphere may be formed.

18 Why do we always say that ALL work around or near storage tanks should be considered high-hazard work?
Answer:
This is primarily due to the hazardous nature of the products and the volume of material involved. Storage tanks can result in a very bad day if there is a large loss of containment.

Chapter 8.9. Injection and Mixing Point Failures

End of Chapter Quiz

1 From an Inspection perspective, what is the challenge with Injection Points or Mixing points?
Answer:
Injection point failures are particularly elusive due to the nature of the internal corrosion or erosion. The corrosion is very localized and, therefore, very difficult to locate.

2 What are some examples of types of failures that involved or could involve mixing points or injection points? Name all you can.
Answer:
- Hydrogen injection or mixing points on hydrodesulfurization or hydrocracking units, especially temperature mixing points.
- Acid neutralization mixing points for neutralization with caustic can result in aggressive internal corrosion at or downstream of the mix point.
- Water injection points for removing ammonium chloride salts from overhead condensers or coolers. Failures typically occur downstream of the first piping ell downstream of the water injection.

- Others include chemical injection points such as amines, inhibitors, acids, and/or caustics, basically any location where chemicals, or process additives, are introduced into a process stream.
- Other examples are desalter additives such as demulsifiers, neutralizers, and process antifoulants.

3 What additional steps may pipe designers consider when they realize that the concentration of caustic or chlorides may be higher than expected?
Answer:
During the design process, pipe designers may consider an upgrade of the pipe metallurgy that may be subjected to higher levels of caustic or chlorides to a metallurgy that is more resistant to stress corrosion cracking.

4 Which piping areas are most vulnerable to severe internal corrosion as a result of mixing points or injection points?
Answer:
A significant change in flow direction when it occurs immediately downstream of the injection point or mixing point. For example, a 90° bend in a pipe within a few feet of the injection point may experience severe internal erosion/corrosion and be subject to a significant piping failure.

5 How can procedures in a checklist format requiring Operator sign-off on each critical step in the procedure help ensure against operator error and/or equipment failure?
Answer:
Procedures in checklist format and requiring Operator sign-off for each critical step help ensure that critical steps are not omitted and are done in the correct sequence.

6 When it is necessary to start a new water wash to remove salting from an exchanger or condenser or start a new chemical injection to a unit, is it OK to tie the new connection directly into an existing bleeder?
Please explain your answer and the considerations related to this question.
Answer:
No. Tying a new water wash or chemical injection directly into a preexisting bleeder without a thorough analysis is generally not appropriate. The Management of Change process should be used, and an analysis of the hazards, including the nature of the chemical, its concentration, the potential for piping or vessel impingement, the potential for the creation of turbulence downstream, any enhanced piping inspection required, and similar considerations should all be reviewed.
The MOC documents should also include a full review with management and engineering approval before implementing the change.

7 Where can we find additional information on Mixing Point inspection procedures and failures involving Mixing Points or Injection Points?
Answer:
- American Petroleum Institute (API)
  - **API 570:** Piping Inspection Code: In-service Inspection, Rating, Repair, and Alteration of Piping Systems
  - **API RP 571:** Damage Mechanisms Affecting Fixed Equipment in the Refining Industry
- UK Health and Safety Executive Public Report
  - Public Report of the Fire and Explosion at the ConocoPhillips Humber Refinery (16 April 2001)
- Occupational Safety and Health Administration (OSHA) Inspection
  - Public Report of the explosion at the Shell Oil Company Refinery which occurred near New Orleans, LA (5 May 1988).

8 Temperature mixing points are less susceptible to piping failures than water or chemical injection points.
- True
- False

Answer:
False. Temperature mixing points generally fail from fatigue. However, they still occur, and the loss of containment can be catastrophic.

9 What management process is critical when installing a new chemical injection point on a process unit?
Answer:
Management of Change – It is very important to have a well-documented Management of Change to ensure the change is engineered and any potential hazards are identified and mitigated as an integral part of the design and change.

10 What are just a few issues with connecting to an existing bleeder to start a new chemical injection?
Answer:
- Pipe wall impingement.
- Chemical concentration.
- Internal turbulence.
- Lack of control or monitoring.

Chapter 8.10. Hazards of Piping Deadlegs

End of Chapter Quiz

1 How do we define the term "Dead Legs"?
Answer:
A dead leg is "a section of pipe where there is normally no flow, is filled with a stagnant process fluid, and is pressurized by the system pressure."
Dead Legs may also be referred to as "stagnant zones."

2 Severe internal corrosion can occur in dead leg piping as the result of the precipitation of ___________ and ___________such as iron sulfide scale when operating at a temperature and pressure below the dew point of water when water can condense.
Answer:
water
sediments

3 Name the most typical failure modes for piping with dead legs. Please name all you can.
Answer:
- Severe internal and uniform corrosion and pipe rupture, most typically in the bottom of the pipe, where sour water and solids can accumulate.
- Freezing of free water accumulated in the dead leg results in pipe rupture.
- Piping dead legs may rupture from an uncontrolled Butadiene chemical reaction resulting in piping failure due to the growth of "popcorn" polymer causing extreme internal pressure. This is more typical of a failure in Polyethylene chemical plants or similar equipment containing Butadiene.
- Severe external corrosion due to corrosion under insulation (CUI) on relatively long runs of cold dead leg piping.

4 Failures in piping dead legs have resulted in _______, _________, and _________ releases of flammable vapor and resulted in fires and explosions.
Answer:
sudden, unexpected, and catastrophic

5 How can dead legs be avoided or mitigated?
Answer:
One way to avoid dead leg piping is to ensure that the Management of Change process is used anytime Changes are made to piping systems. The MOC process will evaluate the piping design and should be used to help eliminate piping deadlegs. Another way dead legs can be avoided is to ensure that pipe dead legs are evaluated during each Process Hazard Analysis (PHA) or HAZOP study.
For mitigation, piping inspection is critical to evaluate the potential for internal corrosion and possible pipe failure. Another way is to ensure the dead legs are drained of all free water, especially when freezing temperatures are possible.

Dead leg piping can also be insulated for freeze protection; however, be aware of the potential for corrosion under insulation.

6 See how many examples of piping design or configurations you can name that may meet the definition of a dead leg.
Answer: (not a complete list, you may be able to name others).
- Piping "jumpovers" leaving a section of piping without flow.
- Piping geometry.
- Equipment bypasses (control valves, exchangers, relief valves, etc.).
- A dead-ended pipe extension used for piping support in the pipe rack.
- Some startup or shutdown lines.
- Uncontrolled change, which leaves sections of pipe abandoned in place.
- Piping installed for short-term use and then left in place.

7 Piping dead legs are particularly susceptible to this internal corrosion in _______ service, that is, services containing hydrogen sulfide or other sulfur compounds in a stream with small amounts of water.
Answer:
Sour

8 When is a major release likely to occur when a process pipe fails due to freezing water trapped in a dead-end piping section?
Answer:
Within a few hours after, the ambient temperature rises above freezing.

9 What is a major unit hazard in units processing high concentrations of Butadiene and Isoprene?
Answer:
Popcorn polymer forming in dead-ended sections of process piping. It can polymerize, swell, and rupture the piping resulting in catastrophic loss of containment, explosions, and fire.

10 What is an effective practice for handling control valve or exchanger bypasses to prevent free water accumulation in the dead leg piping?
Answer:
Develop a scheduled periodic process task in the task management system for the Operators to periodically crack the bypass open and allow the small amount of free water to be flushed out of the bypass. This prevents a large amount of water from accumulating in the bypass.

Chapter 8.11. Brittle Fracture Failures

End of Chapter Quiz

1 Brittle Fracture is a catastrophic type of failure that has resulted in devastating incidents. What types of vessels or equipment can be subjected to this kind of devastating failure?
Answer:
Brittle Fracture is a type of metallurgical failure that occurs to process vessels like drums, tanks, fractionation columns, and piping and occurs instantly and without warning.

2 What is the main cause of brittle fracture?
Answer:
The characteristic of the steel changes as the vessel temperature is reduced to below the vessel's Minimum Design Metal Temperature (MDMT). The steel changes from ductile to brittle. When the vessel is brittle and when a form of stress is applied, the vessel can fail suddenly and without warning due to a brittle fracture.

3 How is brittle fracture defined?
Answer:
The temperature where the metal properties change from ductile to brittle is characterized as the MDMT (also sometimes referred to as the "transition temperature").
The MDMT is a metallurgically determining minimum safe working temperature at the vessel's maximum design pressure. It is based on the type of steel used in the vessel, the wall thickness, and knowledge of the vessel's manufacture, including the type of welding, heat treatment, and material tests undertaken by the manufacturer.

4 The potential for Brittle Fracture should be considered during equipment ________ for scenarios where the vessel may experience cold temperatures due to climate (ambient temperature) or the flash vaporization of LPG such as Ethane or Propane.
Answer:
design

5 What are the three things that must be present for a vessel to pipe to experience a brittle fracture?
Answer:
Before a brittle fracture can occur, the following three elements must be present (See Figure 8.11.2) (Brittle Fracture Triangle):
- Low toughness. Generally occurs in carbon steel at a temperature below the MDMT.
- A flaw or defect in the steel or a weld.
- Sufficient residual or applied stress (above some minimum). Most often, this is internal pressure in process vessels, but it could be piping stress induced by thermal shock, misaligned pipe flanges, stress at welded joints, an external mechanical impact, etc.

6 How is the Maximum Allowable Working Pressure defined?
Answer:
The maximum pressure the vessel may experience to keep within code-allowable stress and is set by vessel design.

7 How is Critical Exposure Temperature defined?
Answer:
It is the lowest temperature the piping or vessel is expected to see during its lifetime. This can be either the ambient temperature or the temperature of the process, whichever is the lowest.

8 A new vessel should have a code stamp and data plate, plus a technical design datasheet that specifies the design parameters (including ________ and ________). New designs should ensure that the ________ is above the MDMT.
Answer:
MDMT and MAWP
CET

9 How is brittle fracture managed for older vessels?
Answer:
Operating procedures should be in place for older vessels to ensure the vessel temperature is at or above the MDMT before the vessel is pressurized.

10 Briefly describe post-weld heat treatment sometimes referred to as "stress relieving."
Answer:
Post-welding thermal heat treatment of the piping or vessel to refine the welding metallurgy and reduce residual stresses.

11 Name three operations tasks where the potential for brittle fracture should always be considered.
Answer:
- Hydrotesting pressure vessels.
- Transient operations such as startup or shutdown.
- Cold temperature excursions such as auto- refrigeration of LPG.

12 Ensure that ___________ _________ are established to prevent brittle fracture and that these are rigorously enforced. Deviations should be reported and analyzed, and corrective actions taken to help eliminate reoccurrences. Operate within pressure/temperature limits.
Answer:
Operating limits.

13 What about after the vessel exposure to cold temperatures; is it safe to return the equipment to service if the vessel has no apparent damage?
Answer:
Following a low-temperature excursion resulting in the vessel temperature experiencing temperatures below the MDMT, the vessel should be inspected to ensure fitness for return to service.

14 What are the key steps to take if a vessel is discovered to be operating at pressure and a temperature below the MDMT?
Answer:
- Recognize that a brittle fracture is possible.
- Minimize personnel exposure (evacuate personnel).
- Stabilize the unit operations – no sudden changes.
- Remotely drain or remove the remaining liquid if possible.
- Slowly reduce the vessel pressure to 25% of the MDMT or less.

15 What recommendations are from the National Board of Boiler and Pressure Vessel Inspectors following a pressure vessel exposure to temperature and pressure below the Minimum Design Temperature, a near miss involving a potential brittle fracture failure?
Answer:
Recommendations from The National Board of Boiler and Pressure Vessel Inspectors (http://www.nationalboard.org) includes the vessel being slowly warmed in a nonpressurized condition and that the vessel should be thoroughly inspected for cracks before returning to service.

16 Why must a Management of Change (MOC) review be completed after weld repairs on a certified API 510 pressure vessel?
Answer:
It is possible that the welding has altered the Minimum Design Metal Temperature, and the vessel could fail from brittle fracture if not detected and corrected.

17 What is the hazard associated with conducting a pneumatic pressure test on a process vessel or any large vessel?
Answer:
Pneumatic testing is particularly hazardous, regardless of the failure mechanism. A pneumatic pressure test should normally be the method of last resort for pressure testing process vessels. In the event of a brittle fracture, the consequences can be catastrophic due to the vessel failing in "shards" or large fragments of broken steel. These fragments can be propelled by the stored energy with great force for hundreds of feet.

18 Flash vaporization and auto-refrigeration of some light hydrocarbons such as Propane or Butane can result in temperatures as low as -____°F (-____° C) and, therefore, certainly have the potential to result in brittle fracture of vessels or piping.
Answer
−40°F (−40° C)

Chapter 8.12. Hazards of Furnaces or Heater Operations

End of Chapter Quiz

1 Why are furnace procedures deemed critical procedures and in checklist format for the Operator to sign-off each step as they are completed?

Answer:
Due to the critical nature of furnaces operating with fuel, air, and a source of ignition inside an enclosed structure and due to the history of furnace explosions and the number of people severely injured or worse on or near furnaces. Step-by-step procedures also help ensure that steps are not omitted or executed in the wrong sequence.

2 What are the correct actions to take if a procedure is outdated or incorrect?
Answer:
If a procedure is incorrect or outdated, the procedure should be updated by redlining the procedure to insert the corrections. The redlined procedure should be approved by plant supervision and have appropriate management of change documents. Then the revised procedure should be followed by signing each step as the step is executed.

3 What is one of the most important steps in lighting furnace pilots or burners from a cold startup or immediately following a furnace trip?
Answer:
Ensuring an effective air purge of the firebox and verifying that all flammable vapor has been removed before attempting to light the pilots or burners.

4 Please explain what is meant by furnace flooding or bogging.
Answer:
Furnace flooding is meant by a furnace that is operating fuel rich. It generally has too little air coming into the firebox resulting in incomplete fuel combustion. This is a dangerous situation as the furnace may explode.

5 When a furnace is operating in a flooded state, without Operator intervention, the fuel gas control valve will continue to _______, increasing _______ to the flooded furnace.
Answer:
open
fuel

6 The process flow through the tubes provides the _________ to keep the furnace tube at a safe operating ___________. Loss of flow results in the tube metal overheating to the point where the tube fails, resulting in the loss of _______________.
Answer:
cooling
temperature
containment

7 What are the indications or symptoms of a flooded or bogging furnace?
Answer:
A furnace that is flooded will exhibit many, if not all, of the following symptoms:

- Low O_2 (<0.5%) and/or high combustibles (>1,000 ppm) are the first indications of a potential firebox flooding situation.
- These indications may be followed by a rapid drop in coil outlet temperatures and/or firebox bridge-wall temperatures.
- Loud rhythmic "whoosh" sound or vibration.
- Pulsating flames.
- Local draft gauge swinging widely.
- Hazy appearance in firebox or smoke from stack (severe bogging).
- Smoking stack.

8 What steps can we take to correct a furnace flooding or bogging situation?
Answer:
(1) Immediately move people away from the furnace; do not send anyone to the furnace.
(2) Place the damper in manual to prevent the control from closing the damper.

(3) Break the cascade on the fuel gas control and manually reduce fuel to the firebox by 25%.
(4) Do not do anything to increase the air to the furnace (do not adjust the burners, do not adjust the damper, do not open the peep doors, etc.).
(5) Then be patient and wait for the firebox to start making heat again. See the procedure above for more details.

9 The safe response to a furnace operating in a flooded state is to quickly trip the furnace offline.
- True
- False

Answer:
False
In a flooding situation, do not trip the furnace. Tripping the firebox puts the firebox in a natural draft, pulling additional air into the fuel-rich firebox. Due to the hot firebox and a fuel/air mixture, the box may explode.

10 Why do vertical furnace tubes represent special hazards when doing hot work inside the furnace, for example, during turnarounds?
Answer:
Vertical furnace tubes can trap a significant amount of naphtha, resulting in a fire inside the furnace while mechanical work is underway. Vertical furnace tubes should be drilled in advance to ensure they are hydrocarbon.

11 What can happen if the flow is lost in a single pass of a multi-pass furnace?
Answer:
The tube without flow can coke up, resulting in total loss of flow, and can rupture due to overheating leading to loss of containment and fire.

12 Which of the following is typically protected by the furnace emergency shutdown system? What is the effect of disabling this critical system?
Answer:
The Emergency Shutdown Systems typically protect the furnace from low process flow, high bridge-wall pressures, low pilot or fuel gas pressures, loss of induced draft or forced draft fans, and high temperatures to the induced draft fan, indicating a potential fire in the induced draft fan.
If these critical devices are disabled during furnace operation, the furnace is not protected in the event of an abnormal event.

13 The furnace emergency shutdown system should not be _________ except for testing and maintenance.
Answer:
bypassed

14 Ensure the feed is circulating through the ______ before attempting to light the burners. The tubes can be severely damaged if the burners are placed in service before a _______ is established in the tubes. If the box is a multi-pass design, ensure feed is flowing through all the furnace _________.
Answer:
tubes
flow
passes

15 What can cause flame impingement from the furnace burners onto the adjacent tubes?
Answer:
There are several things that can result in flame impingement onto the furnace tubes. The more likely causes of flame impingement are burners that are out of adjustment or dirty burners. Occasionally, debris such as refractory on or closely adjacent to the burners may impact the burner patterns.

Chapter 8.13. Corrosion Under Insulation and/or Under Fireproofing

End of Chapter Quiz

1 What is the primary concern with corrosion under insulation or fireproofing?
Answer:
This type of corrosion is hidden from view due to the insulation or fireproofing and can cause failures in areas that are not normally of primary concern to an inspection program.

2 At what temperature range is corrosion under insulation an issue at our refining and chemical plant facilities?
Answer:
CUI is common in carbon steel and 300 series stainless steels operating in temperature ranges of about 25°F to about 350°F (−4° C to 175° C). CUI is most severe in carbon steels operating at about 200°F (93° C).

3 The most typical source of corrosion is due to water penetrating the insulation or fireproofing barrier. How does the water get into the insulation, and what are some potential sources of this water? Name as many as you can.
Answer: (This is not intended to be a complete list; you may know of others).
Water can be introduced by breaks in the insulation barrier, leaking steam tracing, internal system leaks, ineffective waterproofing, or improper maintenance of the insulation systems.
The water can be from:
- Rainwater.
- Cooling tower drift.
- Fire protection systems such as a deluge system.
- Sweating from temperature cycling or low-temperature operation such as liquefied petroleum gas (LPG) or refrigeration units.

4 All areas that are identified as potentially exposed to CUI should be __________ and risk ________ with pending actions tracked until they are completed.
Answer:
Prioritized
assessed

5 Insulation with ________ properties, such as mineral wool, should also be avoided as they will absorb moisture and hold the moisture against the pipe or vessel, resulting in aggressive corrosion.
Answer:
wicking

6 Why is particularly hazardous to attempt a repair of what appears to be a minor leak involving CUI?
Answer:
Caution should be used before attempting any on-stream repair of a leak involving CUI. Due to the potential for uniform corrosion, what initially appears to be a minor leak of a flammable, toxic, or hot hydrocarbon may result in unexpected and catastrophic loss of containment with the potential for explosion or fire.

7 What are some additional factors that can greatly increase the corrosion rate of CUI?
Answer:
CUI can be exacerbated by the chemical content of water, for example, by chlorides or acids, which may be present in the insulation or from minor process leaks. In cases where moisture becomes trapped on the metal surface by insulation, these corrosive constituents such as chlorides and sulfuric acid can concentrate to accelerate the corrosion rates, for example, chlorides from cooling tower drift.

8 Can you explain the current technology used to address the CUI issue?
Answer:
Current technology to address CUI includes the application of Thermal Spray Aluminum coating (TSA), an aluminum coating bonded onto the clean (sand and grit blasted) surface at elevated temperatures using an electric arc or flame spray. These thermal spray coatings provide corrosion protection by excluding moisture and acting as a barrier coating (like paints, polymers, and/or epoxies). But unlike a typical barrier coating, they also provide sacrificial anodic protection.

9 Currently, the only way to combat the challenges with CUI and CUF is with a rigorous _________ program, an effective pipe ________ regime, and an aggressive program to eliminate or minimize __________ intrusion into the pipe insulation.
Answer:
Inspection
coating
water

10 Why are special precautions required when responding to what appears to be a minor leak on piping with external insulation or fireproofing?
Answer:
Caution should be used before attempting any on-stream leak repair involving CUI. Due to the potential for uniform corrosion, what initially appears to be a minor leak of a flammable, toxic, or hot hydrocarbon may result in unexpected and catastrophic loss of containment with the potential for explosion or fire. In some cases, the unexpected loss of containment has occurred while attempting to remove the external insulation.

Chapter 8.14. Hazards of Thermal Expansion in Closed or "Trapped" Systems

End of Chapter Quiz

1 The facts are that with liquid-filled process piping or vessels with no ______ space, extreme ________ can be generated within those systems with exposure to very modest temperature increases.
Answer:
vapor
pressures

2 When returning equipment to service after an outage for repair or replacement the Operator is expected to _______ the system down to ensure all __________ are complete and the _____________ is ready to return to service.
Answer:
walk
repairs
equipment

3 List some examples of equipment that can be subjected to overpressure if a relief path is not made available.
Answer: (Not intended to be a complete list. You may know of others).
- Piping in intermittent service blocked in at both ends.
- Vessels are blocked when removed from service (or returning).
- Heat exchangers blocked while preparing for cleaning.
- Heat-traced piping or vessels (if the equipment is liquid-filled).
- Thermal relief valves were removed for servicing.
- Heat exchangers with the hot side flowing while the cold side is blocked.

4 If a process system is blocked in and a relief device is not available, what must be available and followed to prevent piping or vessel overpressure?
Answer:
Procedures (i.e., leaving a valve on one end of the pipe open to a system to prevent overpressure).

5 Anytime the thermal expansion relief valves are removed for inspection or maintenance, we must ensure that overpressure __________ is provided while the device is removed or otherwise disabled.
Answer:
protection

6 Another serious thermal expansion hazard exists with double-seated block valves. These valves may be equipped with external drains or thermal expansion relief valves. What is the hazard associated with these valves, and how is the hazard controlled?
Answer:
Double-seated isolation valves have internal cavities above and below the valve seats. These cavities can trap pressure and overpressure the valve body if fully open or closed. These cavities either should be drained or protected by thermal relief valves.

7 Extreme pressures can be generated in trapped piping or vessels, pipe or vessels that are liquid filled and blocked in with ___ ______________ protection.
Answer:
no overpressure

8 Procedures used to return equipment such as exchangers to service should be in __________ format, and each relevant step filled out __________ as the procedure is being followed.
Answer:
checklist
complete

9 Isolation valves under protective devices, such as pressure relief valves and rupture disks, should be designated as _____ _____ valves and should be painted a unique _______, and no other valves should be painted the same distinctive __________. They should be full port valves and subjected to periodic __________ to ensure the valves are fully open with car seals in place.
Answer:
"car seal"
valves
color
color
audits

10 What is the unique hazard associated with heat exchangers?
Answer:
Heat exchangers are notorious for creating thermal expansion. They have a hot side and a cold side, and the purpose of the exchanger is to exchange heat between the tube side and the shell side (between the hot side and the cold side). If the cold side is blocked in and the flow is started on the hot side, or flow continues on the hot side, the heat is transferred to the cold site resulting in extreme pressures from thermal expansion on the cold side (the side that is blocked in).

11 What can be done to remind the Operators of the extreme safety hazard associated with thermal expansion in exchangers and double-seated block valves?
Answer:
It is good to provide and post warning signs on exchangers and double-seated block valves to remind the Operators of the severity of the thermal expansion hazards associated with this equipment, especially if the exchangers and valves are not equipped with relief devices.
The signs should also be reminders of the proper sequence of valve operation when isolating exchangers from service and when returning exchangers to service, for example, after maintenance.

Chapter 8.15. Hazards of Pneumatic Testing (For Testing the Integrity of Process Equipment) (Hydrostatic Testing vs. Pneumatic Testing)

End of Chapter Quiz

1 What is the most significant difference between a hydrostatic pressure test using water as the test medium and a pneumatic pressure test using air or nitrogen as the test medium?
Answer:
The compressibility of the test medium. In conducting a hydrostatic test, the water or other fluid is noncompressible; therefore, the pressurization energy is directly applied to the equipment being tested.
In the pneumatic test, the gas must first be compressed to apply internal pressure to the equipment. Therefore, in the pneumatic test, a very large amount of energy is stored in the compressed gas. In the event of vessel failure during the test, the sudden release of the "stored energy" can be catastrophic.

2 What are some situations where a pneumatic test would be necessary to conduct the pressure test?
Answer:
There are a few occasions where a pneumatic test may be required. For example, where the foundation will not support the weight of water or where it is necessary to keep the equipment completely dry due to catalyst or water-reactive materials. Even in these cases, a detailed risk assessment with mitigation and precautions should be in place before conducting a pneumatic test. Ensure that only those directly involved in conducting the test are in the vicinity of the test while the test is underway.

3 If the pneumatic testing process is used, the process should receive higher level ________ and __________ review and approvals, including a ________ assessment to ensure all hazards are identified and mitigated before being carried out.
Answer:
safety
management
risk

4 Where are procedures readily available that cover Pneumatic Testing of Pressure Vessels and Piping, including detailed considerations and precautions for conducting the test?
Answer:
ASME code PCC-2: "Repair of Pressure Equipment and Piping." Section 6.2.

5 What are the key lessons learned from this module?
Answer:
- Use the hydrostatic testing process when it is possible to do so; avoid pneumatic testing if possible.
- If a pneumatic "strength" test is planned, involve the site management and safety personnel. Ensure that a detailed safety plan has been developed and that hazards have been identified and mitigated.
- Ensure the equipment is protected against potential overpressure while performing the test.
- Be sure the potential for brittle fracture has been considered, especially if using vaporized nitrogen to perform the pneumatic test.
- Barricade the area and keep nonessential personnel away while performing the pneumatic test. Only personnel necessary to perform the test should be present, and they should minimize time inside the barricaded areas to only that necessary to conduct and monitor the test.

Chapter 8.16. Hazards of Cold Wall-design Piping and Vessels (Refractory Lined Equipment)

End of Chapter Quiz

1 What is meant by the term "cold-wall design"?
Answer:
Cold designed equipment is in a service where the operating temperature exceeds the equipment design temperature. For example, pipe in an FCC catalyst system may operate above its design temperature. Internal refractory protects cold wall equipment for these harsh operating conditions, protecting the metallurgy from failure.

2 Cold wall-designed piping and vessels are lined with ___________ inside to protect against very high process temperatures and have no external insulation.
Answer:
refractory

3 What happens if the refractory starts to wear or fail?
Answer:
In the event of refractory failure, for example, spalling away from the pipe wall, hot spots will appear, which can be mitigated with steam lances using the MOC process, and eventually, the unit may require a shutdown for repairs if the failures continue.

4 How can hot spots or small refractory failures be identified?
Answer:
The best way is for Inspection to conduct periodic infrared surveys, scanning the piping and vessels and identifying local hot spots or areas of refractory concern. An alternative is to use temperature-sensitive paint designed to change color, indicating the location of the hot spots.

5 Depending upon the severity of the refractory damage, some hot spots can be controlled by applying __________, air lances, or spargers.
Answer:
steam

6 After the steam lances are installed on a hot spot, they should be regarded as ________-________and subject to controls such as Management of Change.
Answer:
safety-critical

7 What is one of the primary concerns with cold wall-design equipment?
Answer:
One of the most significant concerns is having someone place an insulating blanket or insulating material over the cold wall piping or vessel (the pipe/vessel that has the internal refractory). In this case, the piping or vessel very quickly overheats and can fail catastrophically.

8 We understand that external insulation should not be placed on cold wall-designed equipment, and when done, it is typically without an understanding of the hazards.
What are a couple of reasons why this may be done?
Answer: (Not intended to be a complete list. You may know of others).

An attempt to protect maintenance workers working near the very hot equipment.
An attempt to protect instrumentation of other equipment near the very hot equipment.

9 External insulation placed on cold wall-designed piping or equipment can result in severe damage to the internal refractory and rapid ____________ of the pipe or vessel steel, leading to a catastrophic failure or rupture.
Answer:
Overheating

10 To help ensure the refractory failures are not propagating, piping and vessels where the temporary steam application is controlling hot spots should be periodically _________ by the equipment ___________ using thermography or other similar inspection techniques.
Answer:
monitored
inspector

Chapter 8.17. Metallurgy Matters (The Importance of Positive Materials Identification – PMI)

End of Chapter Quiz

1 Many significant process safety __________ have started with metallurgy, specifically the wrong metallurgy for the service or the corrosion mechanism. Several of these have had severe consequences, including loss of life.
Answer:
fires, explosions, incidents, etc.

2 Using only your senses, is it possible to distinguish the difference between carbon steel pipe and a similar-sized section of chrome pipe?
Answer:
Of course, it is impossible to tell the difference between carbon steel and chrome using only your senses. They have the same appearance, they will both corrode or rust when left in the elements, and the welder cannot tell one from the other as they both have the same characteristics during the welding. This is exactly why PMI or Positive Materials Identification is so critical.

3 What are the differences between carbon steel and chrome that make proper identification of the materials used important?
Answer:
These are two different metallurgies with different properties and resistance to the operating conditions. The purpose of PMI is to determine not just the difference between these two materials but to determine specifically what material was installed in our piping circuit and to ensure that the correct material was installed. For example, chrome piping has good resistance to high-temperature sulfidation, whereas carbon steel has very little resistance and may fail after only a short run.

4 Why is Chrome piping used in piping circuits where the which operate above about 450°F (232° C) in the presence of Sulfur?
Answer:
Sulfidation corrosion – Under these conditions, carbon steel pipe will fail within a few years, whereas Chrome piping will last much longer (typically decades).

5 Carbon steel in Chrome piping circuits is our "Achilles' heel" regarding PMI. The issue is that it is impossible to ___________ the two metals apart using only our senses.
Answer:
distinguish

6 Why do we Conduct PMI?
Answer:
Unfortunately, mistakes do happen, material is mislabeled, and some alloys visually look the same or very similar. Sometimes the mill certificates or shipping papers may be incorrect, or materials are mixed up in shipment or material storage bins. We know from experience that the wrong metallurgy materials installed in the field can lead to unexpected failures.

7 The PMI process should be applied to _____ alloy piping and other alloy equipment that can "see" the process pressure. For example, stainless steel is produced in many different grades, and the corrosion resistance and other properties vary widely by grade. It is impossible to distinguish between the various grades of stainless and many other metals without PMI.
Answer:
all

8 When do we conduct PMI?
Answer:
PMI should always be conducted following the new installation of pressure-containing components in alloy circuits and/or replacement of alloy components during maintenance.
We also conduct PMI for quality control in the warehouse and when materials are delivered to the field, also for quality control.

9 How is PMI Conducted?
Answer:
PMI is conducted using completely portable X-ray Fluorescence Technology (XRF Spectrometry) known as X-ray fluorescence.

10 More recently, we have rediscovered that it is also essential to understand the type and composition of carbon steels. In 6 August 2012, Chevron Richmond refinery loss of containment and fire, it was rediscovered that corrosion is much more aggressive in Carbon Steels with low _________ content.
Select the correct answer.
a. chrome
b. moly
c. silicon
d. carbon steel
e. copper
Answer:
a. Silicon

Chapter 8.18. Hazardous of Nitrogen and Inert Entry into Nitrogen-filled Confined Spaces

End of Chapter Quiz

1 See how many different uses you can name for how nitrogen is used in refining or petrochemical plant operations. Name all you can.
Answer: This list is not intended to be all-inclusive. You may know of others.
- Air-freeing process equipment prior to startup.
- Air freeing for inert entry (e.g., for catalyst work).
- Atmospheric storage tank vapor space blanketing.
- Preparation of equipment for hot work.
- For clearing/pigging lines.

- For off-loading tank cars and railcars.
- Process and laboratory analyzer operations.
- Purge or quench for instruments.
- Specific welding operations.

2 Describe the characteristics of nitrogen. What are the hazards of a nitrogen atmosphere?
Answer:
Nitrogen is a colorless, odorless gas, and is nontoxic. It exists in the atmosphere at 79%.
The primary hazard of nitrogen is that it is an asphyxiant. It displaces oxygen from the environment, for example, a confined space like a tank or other process vessel.

3 Oxygen concentration in a normal atmosphere is 20.8%. We know that as we increase nitrogen concentration in a confined space, the oxygen concentration available for breathing drops. Below about what percentage of oxygen do we start feeling the negative effect of lack of oxygen?
Answer:
Below about 19% oxygen, we start feeling the effects of oxygen deprivation.

4 Are there other gases, other than nitrogen, that can also result in oxygen deprivation incidents at the plant or in the mill?
Answer:
Other gases have also been known to result in oxygen deprivation and loss of human life.
For example, incidents have occurred with Carbon Dioxide (CO_2) used in fire suppression systems and Argon used in welding applications. Helium, Methane, or natural gas have resulted in incidents involving oxygen deprivation or death.

5 Where are there good training resources for nitrogen safety?
Answer:
The US Chemical Safety Board (csb.gov) has several excellent nitrogen Safety bulletins, prepared PowerPoint presentations, and readymade videos for training personnel on the hazards of nitrogen safety.

6 Do you have to be inside a process vessel to be overcome with nitrogen?
Answer:
No. We have several examples of where people have been working outside and have been overcome by nitrogen being used nearby. When starting the use of nitrogen, always ensure that the area is appropriately barricaded and warning signs or barricades are in place.

7 What specialized response plan must be developed and available BEFORE an entry is approved for any space that meets the definition of a confined space?
Answer:
A personnel rescue plan

8 Statistically, about half of those who die in nitrogen-related incidents died while attempting to rescue someone else. How can this be prevented?
Answer:
Resist the human urge to rescue should you receive a call for help. First, call for help and make notifications of the emergency.
If you are trained, you should obtain and don a self-contained breathing apparatus (SCBA) before responding to the scene.

9 Who is allowed to carry out the duties associated with an inert entry job; for example, catalyst work inside a reactor prepared for inert entry?
Answer:
Working inside a process vessel under inert atmosphere conditions involves a specialized contractor company with highly trained catalyst support staff and specially designed equipment and personnel protective equipment (PPE) for their technicians.

10 When is the only time that nitrogen should be used to power pneumatic tools? Why?
Answer:
Nitrogen should be used to power pneumatic tools when the tools are used during an inert entry into a nitrogen-filled atmosphere. This is to prevent the allowing air into the inert space.
Other use of nitrogen to power pneumatic tools should not be allowed as the workers can be overcome by Nitrogen.

Chapter 8.19. Hazards of Hydrogen Sulfide (H_2S)

End of Chapter Quiz

1 Where does the Hydrogen Sulfide come from?
Answer:
Hydrogen Sulfide (formula H_2S) generally comes into a refinery either directly in the raw materials (crude oils and natural gas) or is generated from the sulfur compounds in the crude oils.

2 Hydrogen Sulfide is a dangerous and unique hazard because it is highly toxic and can _____ you in seconds at high concentrations. It is particularly toxic when inhaled and affects the central nervous system, interrupting breathing. H_2S is Immediately Dangerous to Life or Health (IDLH) at concentrations of ____ ppm.
Answer:
Kill
100

3 H_2S is__________, slightly heavier than air, and can explode if exposed to an ignition source when in higher concentrations.
Answer:
flammable

4 It can be detectable by odor in extremely low concentrations (as low as 0.47 ppb) and has a distinctive ______ _______ odor.
Answer:
rotten egg

5 What is the US OSHA eight-hour working limit for H_2S in ppm?
Answer:
The US OSHA 8-hour working limit for H_2S is 10 ppm.

6 Hydrogen Sulfide is a dangerous and unique hazard because it is highly toxic and can kill you in seconds at high concentrations. It is particularly toxic when inhaled and affects the central nervous system interrupting the ability to breathe. At what concentration is H_2S deadly?
Answer:
H_2S is considered Immediately Dangerous to Life or Health (IDLH) at concentrations of 100 ppm.

7 Since H_2S is heavier than air, it will most likely accumulate and concentrate in low-lying areas. What are some examples where higher concentrations of H_2S may be found?
Answer:
- Bottom of vessels
- Ditches or trenches
- Tank levees
- Dike walls
- Drains
- Sewers
- Containment areas

8 You are preparing the agenda for an upcoming job site safety briefing that will be done in a work area that has H_2S potential.
Please list as many of the topics that should be covered as you can.
Answer:
- Review the appropriate MSDS.
- Orientation of the job's specific rules and regulations.
- H2S hazards and where they can be found.
- Safety requirements such as:
- Flagging and warning signs.
- Flagging tags.
- Backup and standby personnel requirements.
- Evacuation procedures.
- Communication procedures.
- Gas monitoring requirements.
- Unit/area alarms.
- Emergency Assembly Area and Emergency Meeting Points.
- Emergency procedures in case of an H_2S leak.
- Eye wash stations and emergency shower locations.

9 Employees should practice the use of the self-contained ________ ________ and be knowledgeable with the fixed ________ monitors.
Answer:
breathing apparatus (SCBA)
H_2S

10 Employees should also be familiar with their personal ____ ________ and ensure they are calibrated, and bump tested at regular intervals.
Answer:
H_2S monitors

Chapter 8.20. Hazards of Product Sampling and Drawing Water from Process Equipment

End of Chapter Quiz

1 When collecting samples from process vessels, including storage tanks, ships, barges, tank cars or tank trucks, and similar conveyances, we have the "relaxation period." The relaxation period allows static electricity to dissipate before sampling or gauging. No sampling or gauging should occur during this relaxation period, and the tank or vessel should not be moving; that is, should not have production into or out of the vessel. How long is this relaxation period for small volumes and large volumes?
Answer:
For small volumes, less than 10,000 gallons (38 Cu meters), this relaxation period should be 5 minutes; for volumes larger than 10,000 gallons, which most of our samples will involve, the relaxation period is 30 minutes.

2 During this relaxation period, the vessel should be completely ______ with no movement in the tank. The mixer should be turned ______ if the vessel is equipped with a mixer.
Answer:
Still
off

3 What is the purpose of this relaxation period?
Answer:
To allow any buildup of static electricity to dissipate.

4 The approved sampling equipment for sampling ships, barges, and storage tanks will be ___________ material and an all-bonded design. It will have a bonding cable that should be _________ to the tank or conveyance before anything goes into the tank vapor space.
Answer:
conductive
attached

5 To prevent a static discharge, what type of rope should never be used to catch a petroleum sample?
Answer:
Never use nylon, polypropylene rope, or synthetic rope, as they generate static electricity.

6 What product characteristics and other information about the sample should you be aware of before attempting to collect a sample of the product? Answer all that you are aware of.
Answer: This is not intended to be a complete list. You may be aware of others.
Know the pressure and temperature of the sample.
- Know the physical and chemical properties.
- Catch the minimum volume for testing.
- Choose the proper container (depending on the product and temperature).
- Leave a vapor space for the expansion of liquids.

7 Describe the type of facilities required to collect sour water or foul amine samples.
Answer:
For routine sour water or foul amine samples, the sample point should be enclosed in a special vapor-tight box vented overhead. The sample valve should be physically located approximately 5 feet from the box and be equipped with a spring-loaded to-close valve.

8 Why are these special sample collection facilities required?
Answer:
The special sample collection facilities are required for the extremely high concentration of H2S in the collected sample.

9 In addition to the hydrocarbon vapor and potential H_2S hazards, what other significant process hazard must be managed when collecting samples?
Answer:
Static electricity is always a potential hazard that must be managed.

10 When collecting a sample of LPG like Butane or Propane, a special _________ is required to prevent the possibility of auto-refrigeration.
Answer:
procedure

11 What is the main hazard with collecting samples of sour water or foul MEA/DEA?
Answer:
Sampling sour water or foul Amines (MEA OR DEA) requires special precautions due to the relatively high concentration of hydrogen sulfide (H_2S). Operators have been overcome by the H_2S when precautions are not taken, or sampling procedures are not followed.

12 How many key points can you remember about taking a defensive posture while collecting product samples? List as many as you can.
Answer:
- Stand upwind, away from the vapors.
- Avoid inhaling vapors.
- Anticipate possible splashing of liquid.

- Anticipate possible increase in flow as viscosity lowers or any blockages clear.
- Be aware of dropping greasy, filled glass bottles.
- Use only special containers to carry multiple samples (sample carriers designed for this purpose).
- Insert plugs in bottles and other containers on a sturdy, level surface.
- Be aware of the locations of safety showers and eye wash fountains.
- Be aware of hot streams (asphalt, Coker feed, bitumen, and others).
- Never transport samples inside vehicles or allow anyone else to do so.

13 After a water draw operation is completed, we always leave the water lines in what condition before walking away?
Answer:
Always leave the water draw lines double-blocked when you have completed drawing water.

14 When draining water on an LPG sphere, why is it important to open the upstream water draw valve fully and control the operation by throttling the downstream water draw valve?
Answer:
To prevent the water draw valves from freezing in the open position by auto-refrigeration due to vaporization of the LPG. With the upstream valve taking no pressure drop, it does not auto-refrigerate and can still be closed to stop the flow.

15 What was the primary cause of the series of catastrophic BLEVEs at the Feyzin Refinery in France that resulted in 18 fatalities, 81 injuries, and catastrophic damage to the facility?
Answer:
The Operators were attempting to collect a sample of LPG from the spherical tank water draw. They did not follow the procedure correctly, resulting in auto-refrigeration at the water draw valves, freezing both valves in the open position. This resulted in a massive loss of containment of Propane to the atmosphere and the explosions/fires that followed.

Chapter 8.21. Pyrophoric Ignition Hazards, Including Fractionation Columns with Internal Packing

End of Chapter Quiz

1 What is the definition of a pyrophoric material?
Answer:
A pyrophoric material is a liquid or solid that, even in small quantities and without an external ignition source, can ignite after coming in contact with air.

2 Is this the real hazard of a pyrophoric, or is it something even more significant?
Answer:
The most significant hazard is that pyrophoric can ignite other flammable or combustible materials it comes into contact with, including hydrocarbons or metal components inside our equipment, like internal packing or demister screens.

3 Pyrophoric ignition is a concern in many petroleum refining or petrochemical processes, especially where _________ is present in low oxygen environments.
Answer:
sulfur

4 Pyrophorics are typically found in process equipment containing Hydrogen Sulfide in concentrations >1% (but can form in lower concentrations). Corrosion products such as iron scale or rust combine with the sulfur compounds creating _____ _______ ______ (FeS), which is pyrophoric.
Answer:
iron sulfide scale

5 What is another less-known hazard involving a pyrophoric fire in a confined space?
Answer:
A pyrophoric ignition in a confined space may also reduce the oxygen concentration below safe limits.

6 We should expect to find pyrophoric materials in process equipment and areas where sulfur or sulfur compounds like Hydrogen Sulfide (H_2S) will be present. For example, H_2S will go overhead, so not a surprise that pyrophoric materials will be found in ____________ systems like flare systems.
Answer:
overhead

7 Where can pyrophoric materials typically be found? Name all you can.
Answer: Not intended to be a complete list. You may be aware of others.
- Flare and blowdown system piping, drums, and associated equipment.
- Process vessels in sour service
- Gas compressors and compressor KO drums.
- Storage tanks in sour service
- Heat exchangers in sour service
- Process sewer systems
- API Separators
- Portable tanks, when used in sour service
- Marine tankers and barges (crude, sour naphtha's, and other sour service)

8 How is the simplest and most effective way to control pyrophorics?
Answer:
Where possible, an effective approach is to ensure that the equipment is maintained water wet when open to the atmosphere. Water flooding is generally not required; however, periodic water spray should be considered to ensure the pyrophoric materials do not dry.

9 What about fractionation column internal packing or demister screens?
Answer:
If hot work is to be done inside the tower and near the packed bed, it may be prudent to include removal of the packing in the work plan. Otherwise, plan to build internal barriers to protect the packing against accidental ignition by the hot work. It is important to maintain wetting starting immediately after opening the vessel to the atmosphere.

10 The bottom line is that pyrophoric materials can be __________ and require our attention to prevent uncontrolled spontaneous combustion inside or near process equipment.
Answer:
elusive

Chapter 8.22. Controlling Static Electricity

End of Chapter Quiz

1 Static electricity is frequently not the primary cause of an event but contributes as the ________ _________ to ignite the resulting vapor cloud following an accidental loss of containment.
Answer:
ignition source

2 Static electricity is generated when ___________ move in contact with other materials and is a common occurrence when _________ are being moved through nozzles, pipes, mixed, poured, pumped, filtered, or otherwise agitated.
Answer:
liquids
liquids

3 Other causes include the ________ of solids or an immiscible liquid through a liquid, the ejection of particles or droplets through a nozzle, and the _______ or spray of a liquid against a solid surface, for example, in a storage tank or railcar.
Answer:
settling
splash

4 Many hydrocarbons and other products will generate static electricity. Some hydrocarbon products will __________ the static electricity in the product. For those products, a spark can occur when an object, such as a gauging tape or sample device at a different electrical potential, is lowered into the product.
Answer:
accumulate

5 When a ________ charge occurs in the vapor space of a storage tank or a barge, and when the vapor space is in the __________ range, the ideal ratio of fuel and air, it can result in a catastrophic explosion and a significant fire.
Answer:
Static
flammable

6 What three reference documents help us understand and manage static electricity?
Answer:
- The National Fire Protection Association code 77 (NFPA-77) "Recommended Practice on Static Electricity."
- The American Petroleum Institute Recommended Practice 2003 (API 2003) "Protection Against Ignitions Arising out of Static, Lightning, and Stray Currents."
- The American Petroleum Institute Recommended Practice (API-545 Recommended Practice of Lightening Protection of Above Ground Storage Tanks for Flammable or Combustible Liquids).

7 The amount of static accumulation in liquids is directly related to what characteristic of the liquid?
Answer:
Conductivity – The amount of static accumulation in the liquid is directly related to the electrical conductivity of the liquid.
Conductive liquids typically do not represent significant static hazards; less conductive liquids are more susceptible to static hazards.

8 Name several products that are low conductivity and therefore suspectable to static hazards.
Answer: This is not intended to be a complete list, but examples only.
- Jet fuel
- Aviation fuel
- Gasoline
- Diesel
- Naphtha
- Kerosene
- Benzene
- Toluene
- Crude oils
- Solvents

9 What are the two most effective ways to control static electricity?
Answer:
Bonding and grounding are the two most effective ways to control static electricity.

10 Describe "switch loading" as it relates to static electricity.
Answer:
Switch loading is defined as loading a high-flash product such as kerosene, jet fuel, or diesel after loading a low-flash product such as gasoline or naphtha or loading onto a low-flash heel.

11 How long is the relaxation period, when the tank or vessel must be still, before sampling or gauging a tank or vessel involving a large quantity (greater than 10,000 gallons)?
Answer:
The tank or vessel must be still for 30 minutes before sampling or gauging to allow time for the static electricity to "relax" and dissipate through the liquid to ground.

12 The sampling and gauging equipment should be all conductive and should be ___________to the tank *before* it is placed in the vapor space.
Answer:
bonded

13 When gauging storage tanks, to shield any potential static buildup that may be in the tank, the procedure should ensure that gauging is conducted within the approved tank ______ __________ or ______ _______.
Answer:
"stilling well" or "sounding tube"

14 When initiating a load or transfer of a nonconductive petroleum product, the initial transfer should be at ______ ______ or______ ______until the fill line is fully covered (by 2 feet or 61 cm). This is normally accomplished by gravity flow with no pumps started until the fill line is well covered. If the transfer is to a ________ _________ tank, the transfer rate should be maintained at a low rate (by gravitation) until the roof is floating.
Answer:
low rates or low velocities
floating roof

15 Mists, aerosols, and dust clouds are also _________ accumulators, and static can also play a significant role in explosions involving accumulations or clouds of combustible dusts.
Answer:
static

Chapter 8.23. Hydrofluoric Acid Hazards (and How we Manage HF Safely in our Facilities)

End of Chapter Quiz

1 HF is a strong and highly __________acid with a strong, irritating odor, and it is highly ________, even at _______ ______ concentrations. HF is also a contact poison with the potential for deep and initially painless burns and can cause significant damage to the underlying tissues and bones.
Answer:
corrosive
toxic
very low

2 HF is also highly toxic to the respiratory system, and breathing HF can damage lung tissue and cause lung swelling and fluid accumulation (pulmonary edema).
- True
- False

Answer:
True

3 How is a small contact burn with HF treated?
Answer:
A small HF skin burn should be treated promptly with 2.5% calcium gluconate gel, a topical antidote for HF skin exposure.

The Calcium gluconate should be applied using disposable exam gloves. Even after applying calcium gluconate, it is essential that a medical evaluation be made.

4 Hydrofluoric Acid (HF) is typically used in some petroleum refineries as a __________ in the Alkylation process to produce a high-octane gasoline component known as Alkylate.
Answer:
catalyst

5 What are the two types of acid used as catalysts for alkylation units?
Answer:
Sulfuric acid and Hydrofluoric acid.

6 HF is highly _______ and _________ at room temperatures creating a large vapor cloud that can travel long distances. Therefore, HF must be handled with extreme care, and procedures must be in place and executed correctly in the event of loss of ____________.
Answer:
Volatile
vaporizes
containment

7 What are the three types of HF currently used in industry?
Answer:

- Anhydrous HF (at higher concentrations, anhydrous HF is a colorless gas or a fuming liquid).
- Modified HF (modified HF contains an additive to reduce the volatility and minimize vaporization in the event of a release to the atmosphere).
- Aqueous HF (HF in solution with water).

8 HF is very _______ to humans and has an IDLH of only 30 PPM. A ________ of HF can travel a long distance (miles). Due to the hazards of HF, the American Petroleum Institute (API) has issued a recommended practice specific to HF units which recommends _________ access to HF units.
Answer:
Toxic
cloud
restricted

9 What are the most important emergency response equipment associated with an HF Alkylation unit?
Answer:
The HF rapid de-inventory system and the water wall HF containment system

10 Why is it so important to immediately flood a release of HF with deluge amounts of water?
Answer:
Large amounts of water will mitigate an HF release as HF is readily soluble with copious amounts of water. An unmitigated release of HF can reach concentrations at or above the IDLH levels for miles downwind.

11 What are the OSHA exposure limits for working around HF vapors?
Answer:

- OSHA TWA (Time Weighted Average for 8 hours) is 3 ppm.
- OSHA STL (Short-term Exposure Limit for 15 minutes) is 6 ppm.
- OSHA IDLH (Immediately Dangerous to Life and Health) is 30 ppm.

12 Many refineries have implemented the ________ ___ ________ procedure which is documented and practiced at regular intervals. This procedure is designed to protect many people in the event of a loss of containment of HF.
Answer:
Shelter-in-place

13 In addition to the de-inventory and water wall water deluge systems, what other HF response equipment is readily available on most HF Alkylation units?
Answer: This is not intended to be a complete list. You may be aware of others.
- Remote reading video cameras.
- Strategically placed multiple remote reading HF detection devices.
- Calcium gluconate gel treatment kits.
- Fully encapsulated suits with supplied breathing air.
- activated water fog nozzles.
- sealed pumps with loss detection alarms.
- Modified HF used by some sites to reduce the vapor pressure of the HF.

Chapter 8.24. Safe Use of Utilities and Connecting Utilities to Process Equipment

End of Chapter Quiz

1 Many incidents and even fatalities have occurred by not following basic safety precautions when connecting utilities to process equipment. What are examples of how utilities are used in refining and petrochemical operations? Name all you can.
Answer: This is not intended to be a complete list. You may be aware of others.
We use utilities for many purposes, including the following:
- Preparing equipment for mechanical work.
- Preparing equipment for return to hydrocarbon service.
- Cleaning or purging process equipment.
- Powering pneumatic-powered tools and many other applications.

2 What are the "cardinal rules" of utilities and connecting utilities to process equipment?
Answer:
- Never use plant air as breathing air.
- Never connect potable water (drinking water) directly to process equipment.
- Ensure the hose being used is rated for the pressure and temperature and is compatible with the material to be used before it is connected to the equipment.
- Ensure that check valves are installed at the connection to the process and at the utility to prevent backflow from the process into the hose or the utility system.
- Ensure the utility stations are equipped with unique quick disconnects or fittings for each utility service (especially for plant air, nitrogen, service water, and steam).
- Instrument air should not be used in place of plant air.
- Do not use nitrogen as backup for Instrument air (due to the asphyxiation hazards of nitrogen).

3 How much does water weigh? What is the concern about the weight of water, especially when water-filling process equipment prior to startup or for hydrostatic testing?
Answer:
Water weighs 8.33 pounds per gallon or 62.428 pounds per cubic foot. Always ensure the equipment, including the foundation, is rated for the weight of the water.

4 If potable water use is required, a "_____ ______" should be used to allow a free fall of the water from the utility system into the water supply tank.
The physical "break" ensures that any __________ from the process cannot enter the potable water system causing contamination of the drinking water system.
Answer:
"break tank"
backflow

5 How many times will a water stream expand when introduced into a hot fractionation column operating above the boiling point of water?
Answer:
Water will expand 1,700 times when flashing from water to steam at atmospheric pressure. In a vacuum tower, water will expand about 4,200 times. In either case, the rapidly expanding water can do severe internal damage to the fractionation column.

6 How do we control what type of utility connections are allowed?
Answer:
The manufacturing site should have site standards to specify what utility connections are allowed. The site standard should specify the required color of each utility connection, the type of quick connect allowed for each utility, and the labeling for each utility connection. The standard should also specify the backflow prevention to prevent reverse flow into the utility system.

7 Another significant concern is the potential for _______ _______ during water fill operations. The water temperature should be above the minimum design metal temperature (MDMT) for the vessel before filling the vessel. In some cases, this may require _______ the water.
Answer:
brittle fracture
heating

8 What special precautions should be in place for purging equipment with nitrogen when preparing equipment for mechanical work?
Answer:
Special precautions should be in place for the venting of nitrogen during equipment purging operations.
Never vent nitrogen into a confined space or walkways or alleyways. Warning signs and barricades should be used to prevent against accidental asphyxiation.

9 The utility hose must be rated at a temperature and pressure _______ than the utility being used AND rated _______ than the temperature and pressure of the process line or vessel being connected (in the event of unexpected backflow).
Answer:
higher
higher

10 The hose rating should be provided by the hose _______ and should be clearly _______ or embossed on the hose.
Answer:
Vendor
stamped

11 Utility connects should be standardized with a _______ quick connection and utility station and hose _______ for each of the main utilities (_______, _______, _______, and _______). All site utility connections should meet this standard.
Answer:
unique
color
air, water, steam, and nitrogen

12 Utility air should not be used for which of the following:
a. Reformer regen air.
b. Sulfur unit combustor air.
c. Breathing air.
d. For powering maintenance impact guns.
Answer:
c. Breathing air

13 Why is extreme care required when working with or purging equipment with nitrogen?
Answer:
Nitrogen is an asphyxiant and should only be used with good air circulation.

Chapter 8.25. Hazards of Steam and Hot Condensate

End of Chapter Quiz

1 Steam and Hot Condensate are utilities, but they are also very _______ _________.
Answer:
hazardous material.

2 How many ways that steam is used in a refinery or petrochemical plant can you list? List as many as you can.
Answer: There are many ways that steam is used; this is not intended to be a complete list. You may know of others.
- Purging equipment before hot work.
- Air freeing equipment before a unit startup.
- Adding heat to equipment, for example, in a reboiler or exchanger.
- Steam coils in tankage to keep the liquid pumpable.
- To drive steam turbines for pumps and compressors.
- For stripping steam in fractionator columns.
- To ensure furnace velocity in furnace tubes (to prevent coking).
- To extinguish small fires (steam lances).

3 Personnel contact with Steam or Hot Condensate has resulted in serious burn injuries and even death. How Does Contact Typically Occur?
Name the most common types of burn injuries.
Answer:
- Contact burns from steam tracing.
- Leaking steam tracing or faulty steam traps may result in pools of scalding water ("hot holes").
- Open-ended vertical pipes connected to steam systems may result in scalding water and the "geyser effect".
- The potential release of scalding water when breaking containment of process equipment due to inadequate isolation and/or trapped steam or hot water ("stored energy") and,
- The release of scalding water from a steam turbine exhaust, safety valve outlet, or other similar steam vents.

4 Steam leaking from a high-pressure system, operating above about 400 psi, can cut through solid objects and can result in severe injury.
Can steam leaking from a high-pressure system like this be seen?
Answer:
No. Typically steam leaking from a high-pressure system cannot be seen. You must use all your senses to identify leaks from a high-pressure system.

5 What is a Whip Check, and why are they commonly used on steam hoses?
Answer:
A Whip Check is a safety device that, when properly installed, secures the steam hose and steam manifold or between two steam hoses or between the manifold and hose. The purpose of the Whip Check is to prevent personnel injury from a flailing hose in the event the connection fails.

6 Following steam out of a process vessel, rail car, or storage tank, what additional step is essential before allowing the vessel to cool?
Answer:
When the vessel cools, a strong vacuum will be formed, severely damaging the tank. Always ensure the vessel is vented to prevent tank damage from vacuum.

7 What is one of the most predominant issues with returning high-pressure steam lines to service?
Answer:
Returning steam lines to service requires incredible patience. The line must be heated very slowly, and as the condensate forms inside the pipe, it must be allowed to continuously drain to ensure the pipe does not accumulate free water inside the pipe.

8 The steam or hot condensate temperature is directly related to the system pressure. As the system pressure is increased, does the temperature rise or fall?
Answer:
The higher the system pressure, the hotter the steam or condensate

9 Explain the most common way a "hot hole" is created.
Answer:
A hot hole is usually created by a steam trap that has failed to close, releasing hot condensate and live steam through the tail line. If this occurs where the tail line impinges on the soil, it will gouge the soil leaving a depression or hole that contains water at its boiling point.

10 Explain the key elements of the procedure for commissioning a high-pressure steam line that has been out of service.
Answer:
The key elements of a procedure for commissioning a steam line:

- Ensure a detailed procedure has been developed for commissioning the steam line.
  - The procedure should list all the steam traps and water drains and followed as the procedure is implemented.
- Ensure all steam traps are functional before introducing steam
  - Verify list with the procedure.
- Ensure steam lines are drained of water before introducing steam
  - Verify list with the procedure.
- Introduce steam very slowly until line reaches operating temperature to avoid thermal shock or steam hammer.
- Verify that all drains and steam traps are open to remove air and free water while repressuring with steam
  - Verify with the procedure.
- Slowly repressure until the temperature is consistent with the saturated steam table (pressure vs. temperature).
- Close the drains after the line temperature is consistent with the saturated steam table (temperature vs. pressure).

11 How long of contact time is required for catastrophic burn injuries if a person comes into direct contact with live steam or hot condensate?
Answer:
A catastrophic burn injury can occur in under one second.

12 How can these painful injuries and suffering be prevented?
Answer:
When working on or near these systems, always wear expanded PPE such as heavy-duty rain suits, face shields, and goggles, heavy-duty gloves, among others.

13 What causes steam hammer in steam piping systems?
Answer:
A condensation-induced steam hammer is sometimes also referred to as condensation-induced water hammer or steam bubble collapse. This is a violent action caused by trapped liquid, usually steam condensate, in a steam line. When this liquid accumulates in a steam line, it causes some of the steam to condense, creating more liquid and causing the steam to give up heat. As the steam bubble collapses into liquid, it leaves a void, resulting in steam rushing in to fill the void. The steam flowing over the liquid creates a "wave" of liquid inside the pipe.
As this progresses, it gets much worse as the liquid fills the pipe on one end, effectively sealing off the pipe internally. As the steam bubbles collapse and the steam rushes in to fill the void, it results in a violent hammering and can destroy the pipe.

Chapter 8.26. Electrical Hazardous Area Classification System

End of Chapter Quiz

1. Where do the requirements for electrical area classification system in the United States come from?
   Answer:
   In industry, this is addressed by the Electrical Hazardous Area Classification System, a requirement of the National Electrical Code, or NFPA 70.

2. Can you explain an Electrical Hazardous Area Classification of Class 1, Division 2, Group "D"?
   Answer:
   - A Class 1 means that a flammable gas may be present,
   - A Division 2 means that flammable gas may only be present during abnormal operations or upsets, and
   - Group D means that the vapor may be gasoline or a similar component.

3. What is meant by intrinsically safe?
   Answer:
   An intrinsically safe device is one that, under normal or abnormal conditions, cannot release enough electrical or thermal energy to ignite vapors in the surrounding atmosphere, even in its most easily ignitable concentration.

4. How do you know if your device is intrinsically safe?
   Answer:
   An intrinsically safe device will have a code stamp attached, indicating that it is safe to use in hazardous locations.

5. What is meant by explosion poof?
   Answer:
   These devices contain arcing or sparking relays or other devices and are designed to contain an explosion of flammable vapors inside the enclosure. The enclosure cools the vapors before releasing them into the surrounding area, thereby preventing ignition of vapors outside the enclosure.

6. How many of the enclosure bolts can be left loose or missing, and the device still protects against an explosion outside the enclosure?
   Answer:
   Zero – all bolts must be installed, and all bolts must be tight. Should an explosion occur inside the enclosure, the hot vapors will be released through the flanges, thereby cooling the vapors to a temperature below the ignition point of vapors outside the enclosure.

7. What is meant by a Class I, Division 1 classified area?
   Answer:
   A Class I, Division 1 location is a location in which ignitable concentrations of flammable gases or vapors are likely to exist under normal operating conditions.

8. Devices classified ____________ _______ will have a stamp or seal of approval by one of the certification agencies (FM – Factory Mutual Research Corporation, UL – Underwriters Laboratories, or in Canada by the Canadian Standards Association, and by the CENELEC in the European Union).
   Answer:
   intrinsically safe

9. What types of common process equipment and devices are intrinsically safe?
   Answer:
   Process radios and flashlights are generally designed and certified to be intrinsically safe devices.

10 What are two additional good reference standards covering classification of electrical installations at petroleum refineries?
Answer:
API RP-500 "Recommended Practice for Classification of Locations for Electrical Installations at Petroleum Facilities"
API RP-505 "Recommended Practice for Classification of Locations for Electrical Installations at Petroleum Facilities Classified as Class I, Zone 0, Zone 1, and Zone 2"

Chapter 8.27. Hazards of Combustible Dusts

End of Chapter Quiz

1 How would you describe a combustible dust?
Answer:
Essentially any material that will burn in air as a solid form can become a combustible dust when reduced to a finely divided form such as granules or powders.

2 The typical fire triangle has three sides: fuel, air or oxygen, and Heat or an ignition source. What are the additional elements needed to form a combustible dust explosion?
Answer:
The combustible dust is the fuel, and the two additional elements are disbursement and confinement of the dust cloud.

3 Companies must develop and implement comprehensive ________ ________, ____________, and training programs.
Answer:
dust control, housekeeping

4 What resource is readily available to provide guidance and support to industry to help minimize the opportunity for dust explosions?
Answer:
The National Fire Protection Association (NFPA) Combustible Dust Standard (NFPA-652).

5 Any combustible dust that will burn can be explosive when dispersed into the air in fine particulates.
Please list as many materials that can result in combustible dust explosions as possible.
Answer:
- Coal
- Petroleum coke
- Grains (rice, corn, wheat, sugar, etc.)
- Metals (e.g., aluminum, magnesium, iron, etc.)
- Plastics
- Resins
- Wood

6 Most of these incidents are __________, with fatalities or serious injuries occurring in 71% of the events.
Answer:
catastrophic

7 Although there has been a significant amount of publicity, combustible dust hazards are still not __________ in most industries; several combustible dust explosions and fires have occurred at sites where the potential for a dust explosion was not recognized by those sites prior to the incident.
Answer:
recognized

8 The National Fire Protection Association (NFPA) has warned that amounts as small as 1/32nd of an inch in thickness over only about ___% of a room's surface can be a combustible dust hazard.
Answer:
5%

9 Where dust clouds are concerned, _________ electricity is a significant hazard since a dust cloud can accumulate a static charge and hold that _________ for as long as five hours.
Answer:
Static
charge

10 Companies with the potential for generating or accumulating combustible dust should comply with the NFPA Combustible Dust Standard (NFPA-652) requirements to help ensure against a devasting explosion.
Answer:
Combustible Dust Standard

Chapter 8.28. Hazards of Undetected or Uncontrolled Exothermic Reactions (Thermal Runaways)

End of Chapter Quiz

1 In the petroleum and petrochemical industries, exothermic reactions occur when hydrocarbons are exposed to _______, ________, and _______ catalyst in the presence of _________.
Answer:
heat, pressure, and catalyst
Hydrogen

2 However, ____________ exothermic reactions can occur if reaction temperatures are not closely controlled.
Answer:
uncontrolled

3 Please list the exothermic refining processes.
Please list all you can.
Answer:
- Hydrocracking
- Hydrotreating
- Sulfiding of Hydrotreating and Hydrocracking Catalyst
- Alkylation
- Isomerization
- Olefins Polymerization
- Claus Sulfur Plants

4 Exothermic reactions can also occur by the mixing of two or more _____________chemicals; for example, mixing acid and water will result in a violent exothermic reaction.
Answer:
Incompatible

5 Undetected or uncontrolled exothermic reactions, which can lead to temperature excursions, otherwise known as a "_________ __________," have occurred in hydrocracking and other processes, especially during startup and other transient operations.
Answer:
"thermal runaway"

6 Uncontrolled reactions can lead to temperatures exceeding the ________ limits for equipment, resulting in equipment failures, loss of containment, fires/explosions, and unit or catalyst damage.
Answer:
Design

7 An undetected or uncontrolled thermal runaway can create a catastrophic incident and the potential for the loss of ________.
Answer:
life

8 The console operator operating the Hydrocracker sees a steady rise in reactor temperatures and catalyst bed differential pressures. Should he troubleshoot to understand if there is a problem with the reactor thermocouples, or what other action should he take?
Answer:
Any indication of temperature excursion or high pressure in a reactor should be a sign of a potential temperature runaway, and actions should be taken immediately to stop the reaction. There is always time to troubleshoot later after the unit is returned to a safe state. This should be entrenched in their training and procedures.

9 How many of the potential causes for an exothermic reaction can you list? List all you can.
Answer:
This is not intended to be a complete list. You may know of others.
- Uneven flow and heat distribution in the catalyst bed.
- Internal reactor failures lead to catalyst migration and dead or "stagnant" zones.
- Incomplete sulfiding of catalyst.
- Raising reactor temperatures too quickly when using a fresh, highly reactive catalyst.
- Uncontrolled reaction temperatures or failure to respond quickly to a rapid increase in reaction temperatures.
- Feed temperature too high can increase reactor temperatures leading to uncontrolled exothermic reactions.
- Loss of recycle gas.
- Low recycle gas or oil flow rate.
- Inadequate reserve quench gas capacity or operating with the quench gas control valve wide open.
- Improper control, Operator inattention, or overreaction to some process change.
- Carry over the catalyst from the reactor to downstream vessels where H_2 may be present at high temperatures.

10 What are the critical factors needed to help guard against an uncontrolled exothermic reaction leading to an explosion or loss of containment event?
Answer:
- Operator training
- Inherently safe design
- Operating Limits; established and followed
- Enhanced process control
- Periodic inspections
- Manage design changes
- Reactor inspection following maintenance or prior to catalyst loading

11 What are the indications of an incipient exothermic reaction and a temperature runaway in a reactor catalyst bed?
Answer:
A rapid increase in the catalyst bed temperatures and/or an increase in reactor catalyst bed differential pressures may be an indication of a temperature runaway reaction.

Chapter 8.29. Response to "Minor" Leaks and "Temporary" Repairs

End of Chapter Quiz

1 This lesson highlights that sometimes "we know what we know" about a situation or event. In event of a small dripping leak for one of our process units, who should be in attendance and consulted before we touch the pipe or equipment?
Answer:
- The Fixed Equipment Engineer or the Unit Inspector
- Process management or supervision or the Shift Superintendent.
- Safety Department representative

2 What information do we need to know about the situation described above?
Answer:
- Is the corrosion mechanism known, is it localized or uniform?
- What about the product? Is it light or hot? Will it autoignite or create a vapor cloud?
- Is it toxic, is it sour (H_2S)?
- Do we have an isolation plan should something go wrong?
- Can this work be done safely; will the people be protected?

3 Can you explain the difference between localized corrosion and uniform corrosion?
Answer:
Localized corrosion means the corrosion will be very isolated, such as a pinhole leak. Whereas uniform corrosion occurs over a wide area of the pipe or vessel internals and occurs at about the same rate.

4 Is temperature sulfidation a localized or uniform corrosion mechanism?
Answer:
temperature sulfidation is a uniform corrosion mechanism.

5 In terms of these two corrosion mechanisms, which is subject to a catastrophic loss of containment?
Answer:
Uniform, by far, has the most potential for a catastrophic incident.

6 When responding to what appears to be a minor leak, possibly only dripping, what could go wrong if we proceed before getting the right people involved and the details mentioned above?
Answer:
Exactly what happened in this case study. The pipe was found to be paper thin and failed while the workers were attempting to remove the insulation.

7 Do we need to consciously make a go/no-go decision after carefully considering all the facts?
Answer:
After reviewing this mini case study, you be the judge. If you determine that the corrosion is uniform, that the piping has thinned considerably, and that the product is above the autoignition temperature. From my perspective, it is time to bring the unit down safely while you have the opportunity.

8 The US Chemical Safety and Hazard Identification Board (CSB) Final Report on the Chevron Refinery Fire contains a new "______ __________ _________" developed by Chevron, incorporating much of what is covered in this chapter.
Answer:
"Leak Response Protocol"

Chapter 8.30. Electrical Safety Hazards for Operations Personnel

End of Chapter Quiz

1 What are the three types of electrical safety hazards that operations personnel must be aware of?
Answer:
The three types of electrical hazards that operations personnel must be aware of are:
1. shock or electrocution by direct contact with a live circuit,
2. arc-flash/arc-blast, and
3. ignition of flammables or other process materials by electrical equipment or circuits.

2 How much electrical current is required to stop your heart and take your life when the conditions are right?
Answer:
Current is measured in amps (a) or milliamperes (ma). One milliamperes equals 1/1000th of an amp. It only takes 100 mA or .1 amp to stop the heart and take your life. 100 mA is about equal to that which it takes to burn a small night light in your bathroom.

3 How does a Ground Fault Circuit Interrupter (GFCI) device work to provide protection against accidental electrical contact?
Answer:
The GFCI device works by monitoring the current on the incoming hot power lead flowing through the device to the powered equipment and comparing the return current on the neutral lead. When the GFCI detects a loss of about 5 mA, it immediately opens the circuit.

4 A GFCI, when properly used, will stop the current flow at about ______ mA before the current can have devastating effects on the body.
Answer:
(5)

5 What is an electrical "cheater" cord, and why are they not permitted in process areas unless a hot work permit is issued?
Answer: A cheater cord is one that has had the proof end removed and replaced with a household-type electrical connection. This makes it much easier to connect standard (nonexplosion proof) equipment to the unit receptacle. With this modification, these cords have become an arcing/sparking device and may become an ignition source for flammable vapors if they are present. They should not be allowed on the unit without a hot work permit.

6 LOTO, in the context of this Chapter, refers to the isolation of __________ energy prior to work on circuits or other electrical equipment. LOTO is equally important for the isolation of other energy hazards, such when opening process equipment or where there is the potential for moving parts or other gravity, spring-loaded, thermal, or chemical hazards present.
Answer:
electrical

7 How is work on electrical systems turned over to maintenance personnel to protect them from electrical shock hazards?
Answer:
The first order of protection is to ensure that work on electrical systems is only done when the system is fully isolated and de-energized. This is done by electrical lock-out tag-out (LOTO). This is typically done by operations personnel opening the electrical breaker on the equipment scheduled for maintenance or other similar work and placing a lock and a tag on the breaker handle.
Once the breaker is open with the appropriate lock and tag, the operator should attempt to start or operate the protected equipment from each location or start switch.

Once the equipment is turned over to the craftsmen before maintenance begins, the mechanics should test for current in the isolated system and install their personal locks on the breaker. These locks remain in place during all periods their work is active.

8 Why is it so important for the Operator to attempt a start of the equipment after completing the LOTO and before turning the equipment over to maintenance?
Answer:
It is possible that the wrong breaker was isolated or that the disconnect internals have failed, leaving the equipment energized. Both have happened.

9 Once the equipment is turned over to the craftsmen before maintenance begins, the __________ should test for current in the isolated system and install their personal ________ on the breaker.
Answer:
mechanics (or craftsman)
locks

10 Use of 110 volts and higher electrical tools and equipment inside storage tanks and process vessels is a unique safety hazard because the vessel is a direct _________. Any incidental _________ between the power leads or other energized conductors and the vessel with people present can result in the potential for electrocution.
Answer:
ground
contact

11 What is the hazard with the Motor Terminal Junction Box during a motor start?
Answer:
The Motor Terminal Junction Box (MTJB) is mounted directly to the motor (normally on the side of the motor) and is the junction point for the three high-voltage motor leads (in a three-phase motor). During a motor start, these high-voltage connections are exposed to a current "in-rush or surge" of about six to seven times the normal motor running amps resulting in a high load on these motor lead junctions during a motor start.
MTJBs have exploded, injuring Operators by flying shrapnel. The Operator must stand clear of the MTJB and out of the line-or-fire in event of MTJB failure.

12 If 110 tools or equipment must be used, each device must be _________ by a ground fault circuit interrupter (GFCI). The ________ must be physically located outside the vessel, and the power cables must be protected from damage, especially where they pass through the vessel manway(s) or other similar pinch points.
Answer:
protected
GFCI

13 When you pick up an electrically powered tool, what are examples of things you should inspect before the tool is used?
Answer:
These are examples only and not intended to be a complete list. You may know of others.

- Is the cord damaged or frayed?
  - This includes inspection of extension cords.
- Is the ground attached?
- Is the tool being used in an electrically hazardous classified area location?
  - If this answer is yes, a hot work permit may be required.
- Are there any missing cover plates or electrical seals?
- Is the tool the correct voltage?
- If the tool has an integrated GFCI, is it functional?

14 The use of 110V powered equipment inside process vessels should be viewed as an __________ to routine work practices, and requirements for each use should include a hazard assessment, including review by __________ specialists and approval by the appropriate site management representative(s).
Answer:
exception
electrical

Chapter 8.31. Hazards of Chemical Incompatibility and/or Reactivity

End of Chapter Quiz

1 Where can we find the proper storage information for the required chemicals?
Answer:
Chemical storage information can be in Section 7 of the Safety Data Sheet (SDS). The SDS are required by the US Code of Federal Regulations (29 CFR) 1910.1200, and an SDS must be on hand for every hazardous chemical in the refinery or chemical plant.
The following information can be found in Sections 7, 9, and 10 of the SDS:
- Is the chemical a flammable?
- Is the chemical a corrosive?
- Does the chemical need to be stored other than at ambient temperature?
- Is the chemical an oxidizer or reducer?
- Is the chemical light sensitive?
- Does the chemical require any special handling procedures?

2 How do we store chemicals in small quantities in storage lockers or cabinets?
Answer:
It is important to remember that acids should NOT be stored with bases when segregating chemicals, and oxidizers should not be stored with organic materials or reducing agents.
Physical barriers with effective labeling or other identification and/or adequate distances between chemicals are effective for proper segregation.

3 What is very important about the chemical receiving lines or receiving tanks?
Answer:
It is very important that the chemical receiving lines, valves, and tanks are clearly labeled with the chemical name and the UN/NA chemical numbers to avoid unloading the wrong chemical into the storage tank.

4 When two or more reactive chemicals mix, the reactions may be _____ or may be very _______, resulting in a severe ____________ reaction and may also result in explosion and/or major fire. In some cases, the reaction may also produce _________ gases which can be deadly to personnel.
Answer:
mild
violent
exothermic
toxic

5 United Nations (UN) Numbers are four-digit numbers used worldwide in international commerce and transportation to identify _________ __________ or classes of hazardous materials. NA numbers (North America) are issued by the United States Department of Transportation and are identical to UN numbers, except that some substances without a UN number may have an NA number. For example, "UN1017" provides response information for Chlorine and cross-references to the Response Guide 124.
Answer:
hazardous chemicals

6 What is a requirement for the truck driver or the person accepting a shipment of chemicals into an on-site storage tank?
Answer:
To verify that the chemicals being received match what was ordered.

7 The Unit Operator should do the final ____________ for any product being unloaded into a storage tank from a tank truck or railcar.
The Operator should validate the cargo manifest with the ________ and ________ before unloading the product into the storage tank.
Answer:
verification
signs and labels

8 What are several considerations for chemical storage? Choose from the following.
a. Tank temperature.
b. Control of ignition sources.
c. Proper labeling.
d. Proper segregation from other incompatible chemicals.
e. None of the above.
f. All the above.
Answer:
f. All the above.

9 What is the cardinal rule when it comes to storing acids and bases?
Answer:
Never store acids and bases together or allow them to mix.

10 Where can we find a quick reference to incompatibility with other chemicals?
Answer:
The Material Safety Data Sheet (MSDS).

11 What safety process is recommended to be used to evaluate the safety hazards associated with a site that is handling or processing hazardous materials?
Answer:
The Process Hazardous Analysis (PHA) is sometimes called HAZOP or Hazard and Operability Review.

Chapter 8.32. Hazards of Unintended Process Flow (The Mixing of Hydrocarbons and Air) in Fluid Catalytic Cracking Units (FCCU)

End of Chapter Quiz

1 Describe the two primary "sides" of the FCCU Unit.
Answer:
The FCCU Unit has an "oil side" and an "air side" of the unit. The two sides are separated by catalyst level or barrier above the catalyst slide valves.
It is very important to maintain the two sides as separate parts of the unit; otherwise, if the air side mixes with the oil side, an explosion can occur inside the equipment.

2 What is the concern about shutting down an FCCU after a run of several years?
Answer:
The concern about shutting down an FCCU after a run of several years is the potential for worn slide catalyst slide valves. This could allow the slide valves to leak and lose the catalyst barrier above the slide valves.

Loss of the catalyst barrier can allow for the unintended reverse flow and mixing of the air and oil sides and an internal explosion.

3 What is unique about how an FCCU functions compared to most other process units?
Answer:
An FCCU unit relies on maintenance of the pressure balance between the hydrocarbon-containing components of the process and the air-containing components of the process to prevent the inadvertent mixing of the two. The regenerator operates at a slightly higher pressure when the unit is in service. However, when the unit is shutdown this pressure balance should be changed so that the reactor is at a slightly higher pressure.
Also, the FCC unit maintains a catalyst level above the spent catalyst slide valve to act as an additional barrier between the hydrocarbon side of the unit (the reactor) from the air side (the regenerator).

4 What should the Console Operator be aware of and constantly evaluating when shutting down an FCC following a long run?
Answer:
The Console Operator should constantly monitor the pressure balance between the Fractionator bottom, the Reactor, and the Regenerator with the reactor at a slightly higher pressure when the unit is down.
When shutting down an FCC unit following a long run, the console operator must closely monitor the differential pressure ("delta P") across the catalyst barrier above each of the two slide valves.
The Console Operator should be constantly aware of the possibility of losing the catalyst barrier by leaking slide valves and creating the possibility of backflow and the potential mixing of hydrocarbons and air, which can lead to an internal explosion.

5 What are two additional steps that should be taken by the Console Operator and Field Operators?
Answer:
The Operators should increase the steam flow to the riser and stripper and shift the unit pressure balance so that the reactor is at a slightly higher operating pressure than the regenerator.
Start the injection of noncondensable gas into the fractionator overhead (nitrogen, fuel gas, or steam).

6 The industry has experienced decades of successful operations, essentially incident-free with this process. However, there have been two significant process safety incidents in the recent past where this balance was lost.
Both events occurred during periods of ___________ operations, resulting in major explosions and severe equipment damage. In both cases, the _______ _________, the valves that maintain the pressure balance between the oil side and air side of the units, were severely worn.
Answer:
Abnormal (or transient)
slide valves

7 What are the indications that this important barrier is being maintained or that the catalyst level is dropping and may be at risk of a reverse flow?
Answer:
The Console Operator can closely monitor the delta pressure ("delta P") across the catalyst barrier above the two slide valves. A low differential pressure across the slide valve (delta P) is an indication that the catalyst level or barrier is dropping. The console operator should take steps to increase the catalyst level above the slide valves.

8 How have some refiners decided to address the issue with worn slide valves on their FCC Units?
Answer:
Recognizing this hazard, some refiners have installed a second slide valve in series which is normally in its fully open position, where is not controlling and therefore experiencing no wear.
When there are any indications of wear in the primary valve, they switch to this "spare" valve to continue the run until the next turnaround.

9 The Regenerator is the main process vessel on the air side of the FCC. Is it possible for hot hydrocarbon feed to enter the Regenerator? If so, how can this happen?
Answer:
Yes
The principle is the same on the Regenerator side of the unit. Loss of the catalyst barrier above the regenerator slide valve or loss of the pressure balance can allow fresh feed to flow into the Regenerator, where it can mix with the air at extremely high temperatures.

10 FCC operators should ensure that the potential mixing of air with hydrocarbon-containing equipment is avoided by developing comprehensive __________ and ensuring ___________ with those procedures, especially when undergoing __________ operations such as startup, shutdown, and during unit emergencies.
Answer:
Procedures
compliance
transient

11 More than ______ way to separate the air and hydrocarbon-containing equipment should be considered.
Answer:
one

Chapter 8.33. Hazards of "Normalized Deviation"

End of Chapter Quiz

1 Explain what is meant by the expression "normalized deviation."
Answer:
"Normalized deviation," sometimes referred to as "normalization of deviance" is an expression for deviating or not following the rules or procedure so often that the deviation becomes completely normal to the person who is deviating. The person or group no longer recognizes that they are deviating from a rule or procedure until it results in a significant incident.

2 Which group of managers are responsible for establishing the site's organizational culture?
Answer:
The culture of the organization must come from the top. The site manager and his or her direct staff play a huge role in establishing the culture and expectations of all employees and contractors at the site. This starts with management setting expectations that rules and procedures are to be followed and communicating that to all employees across the board.

3 Can you list three examples where Normalized Deviation resulted in catastrophic consequences?
Answer:
- The Space Shuttle Challenger disaster (O-ring seal in one of the rocket booster joints failed, resulting in an explosion of the liquid Hydrogen tank).
- The Space Shuttle Columbia disaster (foam insulation on the external fuel tank failed, striking the shuttle's left wing).
- The BP Texas City Refinery explosion (failure to follow the procedures and overfilling the Isom Fractionator tower).

4 They continue with the deviation until one day, the __________ is a little different, or the weather has changed, or something else has changed. A combination of the deviation plus the __________ change results in an accident or injury or even may result in loss of life.
Answer:
Situation
unexpected

5 How many things can you list that may reinforce that the deviation is acceptable?
List as many as you can.
Answer:
This is not intended to be a complete list; you may know of others.
a. They think their process is just as safe or even safer.
b. We have done this before.
c. No one says anything to them, so it must be ok.
d. Nothing happens, so it must be ok.
e. After all, the task was done faster and with less effort.

6 Is it unusual to find that the deviation is by a larger group of people, for example, several Console Operators doing the same deviation?
Answer:
Not at all. It is not unusual to find the deviation by a larger number of people since they tend to reinforce each other in the belief that the deviation is justified.

7 How can we guard against the occurrence of normalized deviation?
Answer:
Management can establish an organizational culture that ensures procedures are periodically reviewed to ensure they are maintained current and reinforce the importance of strict adherence to procedures. Concerns or issues with procedures should be discussed and addressed on an ongoing basis.

8 What role do written procedures play in helping avoid Normalized Deviation?
Answer:
Procedures can play a huge role in avoiding Normalized Deviation.
Managers should ensure that procedures are periodically reviewed to maintain current, and the organizational culture should reinforce the importance of strict adherence to procedures.
Operators and others should be able to openly express any issues or concerns with procedures and promptly address them.

9 What were some of the critical learnings from the two space shuttle disasters?
List as many as you can.
Answer: This is not intended to be a complete list. You may know of others.
- Normalization of Deviance: Faulty design of the rocket booster, failure to recognize the problem, failed to fix it after they were aware, and finally treated it as an acceptable flight risk.
- Organizational Silence: Decision makers unaware of the recent history of O-ring joint problems, unaware of the initial written recommendation against the launch by the contractor, and the continuing opposition of the engineers reversed its position.
- Silent Safety Organization: Failed to oversee the problem tracking systems, had not critiqued the engineering analysis for the rocket booster seal issues, and did not provide the independent perspective required by senior NASA managers.
- NASA Reports Indicated Key Lessons were:
  - We cannot become complacent.
  - We cannot be silent when we see something we feel is unsafe.
  - We must allow people to come forward with their concerns without fear of repercussion.

Chapter 8.34. Car-seal Valve Management Plan

End of Chapter Quiz

1 Car seals and/or car-sealed valves can save _______ when incorporated into a safety management system and used properly in the field.
Answer:
lives

2 What is required by the ASME International Boiler and Pressure Vessel Code relative to a clear path for relief valves?
Answer:
The ASME International Boiler and Pressure Vessel Code requires each pressure vessel to be protected by a pressure relief device (i.e., Pressure Relief Valve [PRV], Rupture Disk).
The standard also requires a clear and open relief path between the pressure vessel and the pressure relief valve (PRV) and between the PRV and the relief location (the flare, a tank, or the atmosphere, etc.).
This means that when block valves are installed on the inlets and outlets to PRV's or rupture disks, these block valves must be car sealed or locked in the fully open position to ensure the availability of the relief device.

3 What is meant by the term "Car-seal Valve Management Plan"?
Answer:
Each site should develop a site "lineup procedure" to help ensure that critical valves are maintained in the proper position. This is especially true for valves associated with SHE critical equipment, including valves designated as either car-sealed Open (CSO) or Car-sealed Closed (CSC) valves.
Each site should also have a Safety Management Process for maintenance of CSO/CSC valves, including a field verification or routine audit process and audit checklist to help ensure compliance with the valve lineup procedure.
Operators are expected to be trained to respect car-sealed valves. Therefore, it is not likely that an Operator will inadvertently close a CSO valve or close a CSO valve by mistake.

4 A ________ valve shall normally be installed between the PR device and the inlet and outlet block valves to help facilitate the removal of the PRV from service for maintenance.
Answer:
Bleed

5 PRV inlet and outlet block valves must also be maintained in the fully ________ position to ensure the availability of the relief valve for the PRV to provide the required overpressure protection.
Answer:
Open

6 CSO/CSC Valves for Safety-Critical Equipment are critical to ensuring the availability of equipment designated as "safety or environmental critical" or as "last line of ________." Operators are trained to respect car-sealed valves, and therefore, it is not likely that an Operator will ____________ close a CSO valve or close a CSO valve by mistake.
Answer:
defense
inadvertently or accidentally

7 What should Operators inspect for regarding car-seal valves when making routine process rounds on their unit?
Answer:
Operators should be trained to field verify that car seals are in place for CSO/CSC valves during field walkarounds, and any broken or missing car seals should be promptly replaced.

8 What is typically required for a field Operator to close a CSO valve or open a CSC valve?
Answer:
The position of a CSO/CSC valve cannot be changed without an approved authorization from management. This includes a defeat plan with mitigation for the defeated device.

9 What is the typical duration for the authorization for the defeat for the critical device or the changing the position of the CSO/CSC valve?
Answer:
The authorization must be in place before the device is defeated and remain in place throughout the defeat.

10 What are the typical requirements for SHE critical block valves, including car-sealed valves?
Answer:
- SHE critical block valves shall normally be full port, gate, ball, or plug type valve.

- Block valves to and from PRV's shall be Car-sealed Open.
- All CSO and CSC block valves, including those associated with PRV devices and flare headers, should be painted a distinctive color, and no other valves should be painted the same distinctive color.
- A bleed valve shall normally be installed between the PR device and the inlet and outlet block valves to help facilitate removal of the PRV from service for maintenance.
- CSO/CSC valves should also normally be installed with the valve stem oriented horizontally or vertically downward to help prevent the valve from inadvertently closing in the event the valve gate separates from the valve stem.

Chapter 8.35. Small Piping Guidelines for Operations Safety

End of Chapter Quiz

1 What is considered as 'small-bore piping' on most process units?
Answer:
Small-bore piping is typically considered pipe or valve size less than NPS 1-1/2 inch (40 millimeter) and is generally used in supporting systems or equipment connected to pumps, compressors, analyzers, and numerous other large process equipment.
Most process units have a large amount of small-bore pipe and fittings.

2 The cause of many process safety incidents with catastrophic or near-catastrophic consequences often starts as a loss of containment from the failure of ______ _______ or _______ ________ connections.
Answer:
small piping
small piping

3 Do most Operators fully realize the consequences that can result from the failure of small-bore piping?
Answer:
Operators sometimes fail to realize the catastrophic consequences that can result from failure of small-bore piping systems and the large loss of containment that can occur should it fail.

4 What are some examples of the source of excess vibration in small-bore piping and fittings?
Name all you can.
Answer:
Not intended to be a complete list. You may be aware of others.
- Rotating equipment like pumps and compressors.
- Pressure relief valves.
- Mechanical mixers or agitators.
- Process flow or pressure control valves.

5 Piping design should consider adequate support from ________ and _________.
Answer:
support and braces

6 Excess vibration is usually from rotating equipment such as pumps or compressors but can be from other reasons. A chattering relief valve or a control valve can present severe vibration and is most often an indication of a ________ issue with the relief valve or the control valve and requires prompt attention.
Answer:
design

7 What examples of small-bore piping hazards have resulted or could result in major incidents?
Name all you can.
Answer: This is not intended to be a complete list. You may be aware of others.

- Small-bore piping excess vibration.
- Unsupported Small Piping, lack of gussets & braces,
- Missing or failed check valve (check valves not installed per procedure).
- Failed or leaking union (union installed at wrong location – inside the first isolation valve).
- Piping small-bore "Christmas trees" accepted as the norm.
- Piping moment arm (allows leverage to break the piping)
- Failure of dead-ended small-bore pipe.
- Failure to install or operate excess flow valves properly.
- Failure of small-bore piping due to wrong metallurgy (Positive Materials Identification issue) (PMI).
- Missing bar-stock plugs in vents and drains.

8 Is bracing or gussets required for integrally reinforced valves or pipe fittings?
Answer:
While this should always be verified by your fixed equipment engineer, bracing is generally not required for integrally reinforced valves or pipe fittings, otherwise known as extended body valves.
These valves and fittings are integrally reinforced, or weld reinforced and are designed to be used without external bracing or other supports.

9 Thinking about small-bore piping, what is meant by the term "moment arm" or leverage?
Answer:
Excess moment arm or pipe leverage results from an unsupported, long pipe connection. Moment arms are a process safety hazard because a small amount of force or "leverage" applied to the end of the piping can easily result in breaking the pipe or connection and loss of containment.

10 What is the hazard with Christmas tree piping?
Answer:
Christmas tree is frequently used to describe many small connections involving small-bore piping, valves, pressure gauges, and other similar pressure-containing equipment. A Christmas tree may be fabricated during initial construction but is more often created by adding additional connections to existing equipment.
The full weight of the pipe and all the additional equipment, including the pipe nipples, the valves, pressure gauges, pressure transmitters, etc., is placing stress on the weld connections. Frequently, these are installed with mixed metallurgy and without gussets or braces, or seal welding.
In vibration service, these are stressed even more and have failed, resulting in loss of containment.

Chapter 8.36. Overpressure Protection (Relief Valves, Rupture Disks)

End of Chapter Quiz

1 Name the major types of pressure relief devices.
Answer:
- Conventional spring-loaded pressure relief device.
- Balanced bellows spring-loaded pressure relief device.
- Piston pilot-operated pressure relief device.
- Buckling pin or rupture pin pressure relief device.
- Pressure/vacuum vent valve.
- Rupture disc.

2 What are the seven most common terminologies used for pressure relief devices?
Answer:
- Operating pressure.
- Design pressure.

- Maximum Allowable Working Pressure (MAWP).
- Set pressure.
- Accumulation.
- Overpressure.
- Blowdown.

3 How would you describe operating pressure?
Answer:
Operating Pressure is any pressure at a practical pressure providing the pressure remains within the prescribed operating range and below the Maximum Allowable Working Pressure.

4 What is set pressure?
Answer:
The set pressure is the pressure set on the relief valve spring for when the relief valve begins to open under service conditions.

5 What is the key difference between design pressure and the maximum allowable working pressure?
Answer:
The design pressure is used in the design of a vessel component together with the coincident design metal temperature. The MAWP is the maximum gauge pressure permissible at the top of a completed vessel in its normal operating position at the designated coincident temperature for that pressure.

6 Is the design pressure or the MAWP used for setting the relief valves on the vessel?
Answer:
The MAWP is the basis for the pressure setting of the pressure relief valves protecting the vessel.

7 How does a conventional spring-loaded safety valve function?
Answer:
The process pressure overcomes the spring pressure causing the valve to open and relieve the vessel pressure.

8 What is the difference between a conventional spring-loaded valve and a balanced bellows relief valve?
Answer:
The balanced bellows relief valve is equipped with a bellows that protects the valve seat from back pressure from the valve outlet.

9 After a conventional spring-loaded valve relieves, the seat will reclose when the pressure returns to the set pressure.
- True.
- False.

Answer:
False, The valve closes when the pressure reaches the blowdown pressure.

10 What is common between the buckling pin relief valve and the rupture disc?
Answer:
They both require replacement of the ruptured element after they function.

11 What is critical about a pressure/vacuum vent?
Answer:
They should never be defeated without authorization and a defeat plan.

12 All pressure vessels subject to __________ shall be protected by a __________ __________ __________.
Answer:
Overpressure
pressure relieving device

13 Multiple vessels may be protected by a single relief device, provided there is a _______ and ___________ path to the device.
Answer:
clear
unobstructed

14 Which of the following could result in overpressure?
a. The inadvertent closure of a block valve.
b. Fire exposure.
c. Loss of feed.
d. A check valve failure.
e. Thermal expansion in a heat exchanger.
f. The loss of a utility.
Answer:
a, b, d, e, f

15 What effect would it have on the unit if a pressure sensing line to a pilot-operated relief valve was inadvertently closed?
Answer:
The pilot-operated relief valve would immediately go to the fully open position.

16 What are three reasons why a rupture disc may be installed on your unit?
Answer:
To ensure rapid release, Environmental (fugitive emissions) and to protect the relief valve from corrosive chemicals.

Chapter 8.37. Special Hazards of Rotating Equipment

End of Chapter Quiz

1 What are the three special hazards associated with rotating equipment?
Answer:
Steam turbine overspeed
Pump/driver reverse overspeed
Centrifugal compressor surge.

2 Explain the common cause of pump/driver reverse overspeed.
Answer:
A pump running in parallel trips and the discharge check valve fails, allowing reverse flow.

3 What are failure modes associated with pump discharge check valves? Name any three of the following.
Answer:
- Worn or broken hinge pin or flapper shaft (due to wear, the flapper can drop in the valve body).
- Loose or broken internal bolts or nuts.
- Seat failure (The threaded seat can back out and drop into the valve body blocking the flapper in the open position).
- Trash or debris obstructing the flapper from closing.
- Broken or worn springs (flapper-type check valves).

4 What steps should you take if you suspect a pump is operating in reverse rotation?
Answer:
Immediately stop the reverse flow by stopping the pump that is in service from a remote location if possible. Avoid the line-of-fire with the plane of the impeller in reverse rotation. When the pump experiencing the reverse rotation slows, close the discharge valve to stop the reverse flow.

5 Why is it important to take a position out of the line-of-fire when closing the pump discharge valve to stop the reverse rotation?
Answer:
The pump or driver that is in reverse rotation can fail catastrophically.

6 What should be available for training and use in the field in a reverse overspeed situation?
Answer:
Written procedures to follow to stop the reverse rotation.

7 Following a reverse overspeed incident, what additional actions should be taken to help prevent a reoccurrence?
Answer:
Ensure a failure analysis is completed on the check valve with corrective actions implemented.

8 Steam turbine cases are designed to see full system steam pressure.
- True
- False

Answer:
False

9 A steam turbine is subject to overspeed under which of the following conditions?
Answer:
- A Sudden loss of load.
- Loss of suction.
- A sudden change in the steam inlet or outlet pressures.
- A malfunction of the turbine governor.
- A sudden failure of the coupling.

10 What is designed to prevent a steam turbine from overspeeding?
Answer:
The turbine governor and overspeed trip device.

11 Briefly explain how a mechanical overspeed is designed to function, thereby preventing the turbine from overspeeding.
Answer:
A spring-loaded bolt is installed in the rotating shaft. Centrifugal force moves the bolt to contact with linkage, which trips the steam valve to the turbine.

12 What are examples of checks that should be done before placing a steam turbine in service?
Answer:
- Verify that steam lines and turbine case are free of condensate and that steam traps are functional. All steam traps should be hot.
- Check that all lubricants are dry (no moisture) and that the oil levels are correct.
- Ensure that the turbine steam outlet (exhaust) valve is fully open and is car sealed in the open position.
- Before opening the steam inlet valve, operate the overspeed trip device to ensure it is not stuck or binding (all linkage should be free).
- Stand away from the line-of-fire of the turbine wheel throughout the procedure.
- Slowly bring the turbine up to speed by slowly opening the steam inlet valve.
- Carefully observe the governor's control to ensure it is controlling.

13 What is the first step in preparing to perform a test of the turbine overspeed trip device?
Answer:
Ensure that all personnel not involved in the test are out of the vicinity.

14 Briefly explain what is meant by surge in a centrifugal compressor.
Answer:
Surge is the point at which the centrifugal compressor cannot add enough energy to overcome the system resistance or backpressure. This is typically caused by flow restrictions on the discharge of the compressor, such as a throttled discharge valve. It can also occur due to restrictions on the inlet to the compressor.

15 How is compressor surge controlled or prevented?
Answer:
By the design and installation of a recycle control valve or system to allow flow back to the compressor suction in event of unexpected back pressure.

16 What can result in reverse overspeed of a large centrifugal compressor?
Answer:
Same as a centrifugal pump – higher pressure on the discharge side and failure of the discharge check valve.

17 What is the significant difference between a centrifugal compressor and a reciprocating compressor?
Answer:
The reciprocating compressor is a positive displacement machine, but the centrifugal is not.

18 Please select the three significant hazards associated with the operation of a reciprocating compressor.
a. Liquid entrainment into the compressor.
b. Compressor surge.
c. Loss of lubrication.
d. Running the machine deadheaded.
e. All of the above.
Answer:
a, c, d.

Chapter 8.38. Two Insidious Hazards of Petrochemical Plants (Popcorn Polymer and Ethylene Decomposition)

End of Chapter Quiz

1 What are the two insidious hazards associated with petrochemical plants?
Answer:
Popcorn Polymer and Ethylene Decomposition.

2 Which of the following are the unique hazards associated with Popcorn Polymer?
a. Popcorn polymer can plug or foul equipment internals.
b. Popcorn polymer may be pyrophoric.
c. Popcorn polymer exerts extreme internal pressure and can damage equipment and result in loss of containment.
d. Popcorn Polymer is a white flexible material that appears like plastic.
Answer:
a, b, c.

3 Which of the following chemicals are most susceptible to the formation of popcorn polymer?
a. Ethylene
b. Nylon
c. 1,3-butadiene
d. Propylene
Answer:
c

4 In what areas of the plant is popcorn polymer most likely to form?

Answer:

In piping and equipment containing butadiene or other similar products in areas that are stagnant or with little or no flow.

5 Name one type of inhibitor that may be effective in mitigating popcorn polymer.

Answer:

Any one of the following: sodium nitrite, tertiary butyl catechol (TBC), diethyl hydroxylamine (DEHA), and Methyl ethyl mercaptan.

6 What is the hazardous nature of Ethylene Decomposition?

Answer:

Ethylene decomposition results in very high temperatures and can lead to loss of containment and explosion.

7 How can an Ethylene Compression explosion occur during the commissioning of an ethylene pipeline?

Answer:

By inadvertently opening valves to other equipment containing nitrogen under high pressure causing rapid pressurization of the ethylene.

8 Which part of the organization should be involved in developing the procedures for ethylene pipeline commissioning?

Answer:

The Technical Group should be involved in the development of procedures for commissioning ethylene pipelines.

9 What is the allowable amount of oxygen in an ethylene pipeline before admitting the ethylene?

a. No more than 5% O_2 is allowable.
b. 5–10% is allowable.
c. O_2 reading should be nondetectable before admitting ethylene to the pipeline.
d. There is no requirement for checking O_2 before admitting ethylene.

Answer:

c

10 What are some other potential initiators for an ethylene decomposition incident?

a. Rust scale from the inside of the pipeline.
b. Overheating due to faulty electrical heat tracing.
c. High discharge temperature on the ethylene reciprocating compressor.
d. Low temperature from the inhibitor injection system.

Answer:

a, b, c

11 When commissioning an ethylene pipeline, the start-up procedures are only a guide. What is more important is the experience level of the Operations Technicians who are performing the work.

- True
- False

Answer:

False; The experience levels of the Technicians are very important but the procedures must be followed.

12 When commissioning an ethylene pipeline the sudden exposure of ethylene to the high nitrogen pressure in the pipeline can result in an ethylene decomposition event and possibly an explosion.

- True
- False

Answer:

True

Appendix B

Worldwide Process Safety Incidents and Numbers of Lives Lost

Contains process safety incidents - Personnel safety incidents, like slips, trips, falls, struck by, caught between, etc., are omitted. Contains excerpts from federal investigation reports (CSB.gov, OSHA.gov, EPA.gov). Not intended to be a complete list. Some are from media reports and cannot be verified.

Year:	Location:	Incident description:
1879	Point Breeze (South Philadelphia), PA.	Atlantic Refining Company – Lightning struck the facility, at the time, a major petroleum blending and shipping company, virtually destroying the complex. The ensuing fire destroyed over 25,000 cases of petroleum products at the Schuylkill River Dock. Also destroyed in the devastating fire were five foreign ships, while others were towed out of harm's way before they were reached by the fire. Almost every structure at the site was destroyed, including the tin shop where the metal oil containers were made and most of the refining equipment, offices, and warehouses.
1921	Ludwigshafen-Oppau, Germany	BASF – An ammonium Nitrate Explosion (561 Fatalities, About 2,000 injured) (casualty number updated from a 2018 news article).
1957	Windscale, Cumberland, UK	The Windscale fire was the worst nuclear incident in the history of Great Britain. The incident occurred in Windscale, Cumberland (now Sellafield, Cumbria), releasing substantial amount of radioactive material into the surrounding area.
1947	Texas City, TX.	Ship Explosion "Grand Camps" (Ammonium Nitrate fertilizer explosion) (576 Fatalities, Injured?)
1948	Ludwigshafen am Rhein, Germany	BASF – A tank car with 30 tons of dimethyl ether suddenly ruptured, resulting in this catastrophic explosion. The investigation revealed that the railcar may have been overfilled and was exposed to 30°C (86°F) temperature. The tank capacity may have been incorrectly calculated, and the car failed at a weak weld seam. About 7,000 buildings within a radius of 500 meters (over 1,600 feet) were destroyed. (207 dead and 3818 injured).
1951	Baton Rouge, LA.	Standard Oil Refinery – Major explosion and fire at the Naphtha Treating Plant. Two employees killed and ten injured (2).
1956	Dumas, TX.	Shamrock Oil Tank Farm (also known as McKee) – A pentane sphere released vapors from the relief valve which quickly flashed from a flame on an adjacent asphalt tank. The resulting explosion ignited several other tanks containing light materials. Fifteen men burned to death almost instantly when a hot wall of fire shot across the ground when the first of four tanks exploded and burned. Four others died later of horrible burns. Some 31 other persons were burned by the blast that shot an orange fireball thousands of feet and seared everything within a quarter mile radius. This was obviously a BLEVE, although not understood at the time. A memorial was erected at the Texas state capitol with their names. A second memorial is near the Moore Country courthouse. (19).
1963	Plaquemine, LA.	Dow Chemical Plant, Major explosion thought to be caused by sight glass failure. The Ethylene and Propylene plants were destroyed.
1965	Lake Charles, LA.	PCI – Cold brittle fracture when Methane dumped into CS pipe, Control room was destroyed.
1967	Cornwall, England	The supertanker Torrey Canyou shipwrecked off the West coast of Cornwall, England, resulting in a large oil spill and an environmental disaster. 123K tons of oil spilled, impacting shorelines in England and France – considered the first major oil spill.

(Continued)

Process Operations Safety: The What, Why, and How Behind Safe Petrochemical Plant Operations, First Edition. M. Darryl Yoes.

Year:	Location:	Incident description:
1966	Feyzin, France	Elf Refinery, Five Propane sphere BLEVE's, multiple tank fires. Incident occurred while Operators were attempting to collect samples from LPG sphere water draw. Water draw valves froze due to auto-refrigeration resulting a in large loss of containment, massive fire, and multiple BLEVEs. Damage reported in Vienna, 1475 homes damaged. (18 Fatalities, 88 Injured)
1969	Texas City, TX.	Union Carbide: A heat-triggered decomposition of Vinyl-acetylene and Ethyl-acetylenes (Five towers were toppled and/or heavily damaged)
1969	Santa Barbara, CA.	Union Oil – The company started a fifth oil well on the offshore oil platform (Platform A) which was located a few miles off the coast. The oil well blew out on 28 January releasing oil and gas into the sea. The sea floor quickly cracked in five places releasing more crude oil. This was followed by a second release on another well less than a month later. The crude oil accumulated all along the California coastline making this the largest oil spill at the time.
1970	Crescent City, IL.	Train derailment and multiple BLEVE's (0)
1970	Bayway, NJ.	Exxon Refinery – Catastrophic explosion of New H-Oil Unit. H-Oil unit was destroyed and severe damage to the refinery and surrounding area. Design issues with reactor internals led to the explosion (design) (0)
1971	Beaumont, TX	Mobil Refinery – BLEVE of process vessel while draining water (1)
1971	Castellon, Spain	Four workers were killed when a refinery compressor exploded (4)
1971	Darvaza; Turkmenistan	The Darvaza gas crater, sometimes also referred to as the Gates of Hell, is a burning natural gas field that has collapsed into a cavern. It is unknown how the crater was formed or how it was ignited. It has been burning for nearly 50 years and has become somewhat of a tourist attraction.
1971	Woodbine, GA.	Thiokol Chemical Plant – A series of explosions erupted due to the ignition of a pyrotechnic compound, an igniter for military flares, at this flare assembly plant. (29 Fatalities, "many injured")
1972	San Pedro, CA.	General American Transportation Co – A major fire hit the tank farm when 23 storage tanks were destroyed. Several firemen escaped death when a tank, containing more than 30,000 gallons of highly volatile acetates, exploded and was launched 300 feet into the air.
1973	Japan	Idemitsu Petrochemical; Uncontrolled exothermic reaction. The outlet piping ruptured due to overheating. (Explosion/Fire). (1)
1973	Staten Island, NY	Texas Eastern Transmission Corporation (TETCO) – An LNG tank exploded, ripping the concrete dome from the 108′ tank. The dome was detached from the anchors landing on workers about 100′ below. This incident remains the worst safety industrial incident for the Staten Island borough. 40 Fatalities, Injured?
1974	Flixborough, UK	Nypro Cyclohexane Plant – A temporary replacement was done on a section of interconnecting reactor piping, between two reactors in series, without engineering support. Due to lack of support structure and piping misalignment, a bellows joint failed catastrophically, resulting in massive loss of containment and explosion. The entire plant was destroyed, 1,821 homes, 167 shops or factories damaged (29 Fatalities, 89 Seriously injured, hundreds more affected).
1974	Plaquemine, LA.	Dow Chemical; Failure of a 24″ bellows expansion joint in a Propylene compressor system. The incident occurred during an electrical power failure causing loss of cooling water and quench water pumps (major Fire). (0)
1975	Netherlands	Dutch State Mines Steam Cracker – Brittle Fracture (14)
1975	Antwerp, Belgium	Union Carbide – Ethylene small piping failure and resulting explosion and fire (6)
1975	Beek, Netherlands	Dutch State Mines Ethylene Plant – The Ethylene Unit was in startup mode when a large vapor release occurred in the vicinity of the depropanizer overhead drum. The release quickly ignited causing a massive vapor cloud explosion causing major damage to the entire unit, including additional explosions in the adjacent storage areas. 14 people were killed and over 100 were injured, including several who were off-site. The belief is that a relief valve opened on the depropanizer drum, and the auto-refrigeration led to a brittle fracture failure of the 2″ PRV inlet line. (14 Killed, >100 Injured).
1976	Seveso, Italy	Industrie Chimiche Meda Società (Dioxin release) – A reactor explosion resulted in loss of containment of dioxin, a highly toxic substance. This major contamination incident resulted in sweeping regulatory changes in Europe and Great Britain. (0 fatalities, 1,800 people were affected)

Year:	Location:	Incident description:
1977	Qatar	Qatar NGL Processing Plant – A 260,000-barrel tank storage tank failed to release 236K bbls of propane at −44°C. The release quickly swept over the processing area and ignited. An adjacent tank containing butane was also destroyed. The explosion killed six workers (6).
1978	Texas City, TX	Texas City Refining, VCE & multiple BLEVE's (7) explosions ripped through the Texas City Refining Co. plant in Texas City, killing five people and injuring 10 others. The blasts rocked the area for 35 minutes and sent flames shooting 200 feet high
1978	English Channel (NW coast of France)	The Amoco Cadiz supertanker sank in the English Channel, spilling 1.6M barrels of crude. Currently, the largest oil spill in history.
1979	Bayway, NJ.	Exxon Refinery, Vapor cloud explosion/fire – dead-leg pipe failure (0)
1979	Gulf of Mexico	Pemex (Petróleos Mexicanos) was drilling a 3 kilometer (1.9 mi) deep oil well when the drilling rig Sedco 135 (Ixtoc) lost circulation of the drilling mud. This resulted in a blowout on the semi-submersible drilling platform resulting in a fire and the sinking of the rig. Over the following 10 months, 3.3M barrels were spilled into the Gulf, making this one of the most significant spills in history up to this time.
1979	Harrisburg, Pennsylvania	The Three-Mile Island nuclear power plant partial reactor meltdown. Mechanical failures in the nonnuclear secondary system and a stuck relief valve released radioactive material to the atmosphere. Operators initially failed to recognize the loss of cooling. There were reports of radiation causing cancer in Pennsylvania and other areas in the NE US.
1980	Honolulu, HI.	Chevron Fuels Terminal; Gasoline tank overfilled, explosion and fire (2)
1980	Lake Peigneur, LA	A Texico drilling rig drilled into a salt mine, collapsing the lake into the salt mine. Lake Peigneur was changed from a freshwater lake to a saltwater lake. The drilling rig, several smaller boats, and the mine were lost.
1980	North Sea	Alexander L. Kielland – The oil platform overturned (capsized) into the sea during a violent storm, making it the worst industrial accident in Norwegian history and a significant change for safety in the Norwegian oil and gas industry. (123 Fatalities)
1981	Port Jerome, France	Exxon Refinery, brittle fracture of reactor vessel while in stand-by (0)
1981	Fawley, England	Exxon Fawley Refinery – fire during maintenance (1)
1982	Venezuela	Tacoa Power Station; Heavy fuel oil tank was inadvertently heated above the flash point of the fuel during a routine tank transfer resulting in boil over. Homes, vehicles, and most of power station were destroyed (160)
1982	Salang, Afghanistan	A tanker truck exploded while in a highway tunnel. It is thought that a collision between an army truck and a fuel truck inside the tunnel resulted in the hydrocarbon release. The explosion and fire were thought to be an attack, and the exits in both directions were closed. The tunnel ventilation system was not working, and levels of carbon monoxide in the air increased drastically. Most casualties were due to burns and carbon monoxide poisoning. (Approx 3,000 people were killed in this incident, most were Soviet soldiers traveling to Kabul) (3,000).
1982	Grand Banks – Newfoundland, Canada	Ocean Ranger mobile drilling platform (Ocean Drilling and Exploration Company [ODECO] working for Mobil Oil): While drilling an exploratory well off the coast of Canada, the platform was struck by a rogue wave, followed by waves 65 feet and above. The platform was reported to be listing followed by a second report that they were abandoning the platform. All 84 people on board (46 Mobil employees and 38 contract employees) died in this tragic incident, from drowning or hypothermia. Other vessels located nearby were not equipped for sea rescue, especially in inclement weather.
1984	Romeoville, IL.	UNOCAL Refinery (Vessel Failure – BLEVE, Wet H_2S) An Operator conducting his normal duties noticed a 6″ crack at a circumferential weld on an amine absorber column, spraying propane to the atmosphere. As he was attempting to isolate the column the crack continued to grow to about 2′. The crew started evacuating the area, and about the same time, the emergency response team started arriving on-site. At this time the column failed catastrophically and rocketed from its foundation creating a massive fireball. OSHA conducted an extensive investigation and determined the non-post heat-treated vessel failed due to H_2S corrosion cracking. Other damage included the unsats gas plant, the FCC, and the Alkylation Unit. A large process vessel on the Alkylation Unit BLEVEed during the event. OSHA released an information alert to all of the industry about the hazards of H_2S corrosion which kicked off an intense inspection effort industry-wide. (17 Fatalities).

(Continued)

Year:	Location:	Incident description:
1984	Paraguaná Peninsula, Falcón, Venezuela	Amuay Refinery Fire – A massive explosion and fire occurred at this refinery due to the failure of a weld in a section 8-inch-diameter line carrying hot oil from the high-pressure separator to the low-pressure stripper in a refinery hydrotreater. The resulting spray of hot oil (700 psi and 650°F) reached the hydrogen units where it ignited. A 16-inch gas line ruptured due to the flame impingement and became huge expansion of the intense fire. Additional piping ruptured, continuing to add fuel to the ever-expanding fire. The entire refinery was quickly shut down, but the damage was catastrophic. Several units were essentially destroyed. The line that failed experienced high vibration and is thought to have failed due to fatigue and hydrogen embrittlement.
1984	Mexico City, Mexico	PEMEX Propane Fire, LPG storage terminal massive explosions multiple BLEVE's of LPG spheres and horizontal storage vessels (approximately 650 Fatalities, most in the surrounding community, 4,248 Injured, 31,000 left homeless).
1984	Bhopal, India	Union Carbide – Highly toxic MIC Release due to an uncontrolled thermal runaway reaction (2,153 Fatalities; However, some estimates are as high as 25,000 fatalities and 100,000 permanently disabled. The region is still experiencing health issues as a result of the release).
1984	Fort McMurray, Canada	Syncrude – Piping failure due to metallurgy (PMI). A 10-inch diameter pipe in the Fluid Coker slurry recycle oil line failed releasing liquids at or close to their autoignition temperature and the vapor cloud ignited almost immediately. The fire impinged on other piping resulting in the failure of six or seven additional lines. Post-incident metallurgical examination revealed that a 1.8-inch-long piece of carbon steel pipe had inadvertently been inserted into the slurry recycle line which was constructed of 5-chrome and failed due to high-temperature sulfidation corrosion. The unit suffered heavy damage because of the fire. About 2,700 barrels of hydrocarbon liquids were released from process equipment during the fire. Some pumps could not be shut down and contributed to the loss of product. A 900-psig steam line providing high-pressure steam to the turbine drivers of the compressors ruptured during the fire which played havoc on the firefighting efforts. (0)
1984	Bhopal, India	Union Carbide (MIC Release) (2153) (Note: 16,000 deaths reported as of 2006)
1984	Cubatao, Brazil	Petrobas Oil pipeline explosion and fire – A shantytown about 30 miles southeast of Sao Paulo had about 9,000 people living in makeshift homes, most built on stilts above swampland owned by Petrobas, a state-run oil company. Petrobas gasoline pipelines were next to the slum. When workers opened the wrong pipeline valve, highly flammable gasoline was released into the ditches resulting in an explosion and fireball. (508 Fatalities, many Injured).
1984	Campos Basin near Rio de Janeiro, Brazil	Petrobras' Enchova drilling platform – A well blowout led to an explosion and fire on the platform. The workers evacuated the platform; however, 42 workers died during the evacuation. Thirty-six died of these died when a lifeboat lifting mechanism failed (42).
1985	Priolo, Sicily, Italy	EniChem; A major fire originated at the base of the Deethanizer tower about the same time the pressure-relief system activated. Major fire, including multiple BLEVE's toppling several towers and spreading to storage tanks and plant infrastructure. (Brittle Fracture?)
1985	Wesseling, West Germany	Rhenische Olefinwerke Olefin Unit; 29 Injured, Pipe rupture during cold weather operations and freezing condensate in the piping. The fire burned for nine days, and the Olefins unit was shut down for over eight months (Explosion/Fire) (0)
1986	Thessaloniki, Greece	EKO, the state-owned refining and petrochemical company, Tank farm fire – seven tanks involved (0)
1986	Pripyat Ukraine, Soviet Union (Chernobyl)	The Chernobyl Nuclear Power Plant meltdown released about 5% of the reactor core (highly radioactive material) into many parts of Europe. This was due to poor plant design and a severe lack of training for the plant operators and running a plant test at the time. 335,000 people were evacuated from surrounding areas, and many areas have been permanently off-limits due to radioactive contamination. The number of fatalities is unclear, but it has been reported that there were 47 direct deaths, with many also dying in the surrounding areas. One New York Academy of Sciences report reported that nearly a million people have died due to this nuclear power incident and the radiation released.
1987	Grangemouth, Scotland	BP Refinery – A major fire occurred on the flare piping while maintenance workers were attempting to replace a leaking isolation valve. It was recognized that the isolation valve was passing, and a decision was to wait until the FCC and the flare system were down to replace the valve. When four workers were removing the spacer, liquid drained from the piping and flashed. Unfortunately, two of the workers were engulfed in the fire and were killed. (two Fatalities).

Year:	Location:	Incident description:
1987	Grangemouth, Scotland	BP Refinery – A massive explosion occurred on Hydrocracker Unit when the Low-Pressure Separator vessel failed. During a unit startup following a routine downtime, the unit tripped, and the Operators assumed this was a spurious trip and began preparing to do another restart. It was while the unit was still in stand-by mode when the explosion occurred. During the stand-by, the air-operated valve on the High-Pressure Separator was opened and closed several times. This allowed the level in the High-Pressure Separator to go low. Several years before this incident, the low-level trip on the HP separator had been disconnected. The Operators did not trust the main level control reading and referred to a chart recorder for a backup level reading. When this level was lost, the high pressure from the reactor blew through to the Low-Pressure Separator and overpressurized the vessel resulting in the explosion. There was an offset on the chart recorder which led the Operators to assume that the level in the HP separator was normal.
1987	Pampa, TX.	Hoechst Celanese Corp- A boiler explosion killed three workers and injured several others (3).
1987	Saint-Herblain, France	Explosion and fire in petroleum liquids depot. Depot was heavily damaged, with several tanks destroyed. (two Fatalities, eight Injuries, including five seriously burned)
1987	Texas City, TX.	Marathon Oil Co. – A release of hydrofluoric acid (HF) resulted in the evacuation of 4,000 residents. More than 1,000 residents were treated for eye and respiratory problems after a pipeline ruptured releasing the cloud of HF into residential neighborhoods. A crane carrying a large section of a convection furnace dropped its load onto an anhydrous hydrogen fluoride tank within the HF alkylation unit. Two pipelines were broken from the top of the tank resulting in the release of HF into the community.
1988	New Orleans, LA. (Norco)	Shell Norco explosion and fire (VCE) – Pipe corrosion downstream of an injection point was believed to have caused the failure of an 8″ overhead vapor line from the Depropanizer column on the FCC. Seven shell workers were killed during the explosion, and six others were hospitalized. Forty-eight residents and Shell workers were injured in the explosion. Approximately 4,500 people were evacuated because of the explosion and debris was found about five miles from the refinery. (7 Fatalities, 48 Injured).
1988	North Sea	Piper Alpha Oil Platform – A massive explosion and fire occurred on the oil platform resulting in the destruction of the rig. The incident occurred during maintenance activities to overhaul an LPG condensate pump and its relief valve. The sister pump failed overnight, and the overhauled pump was returned to service. Unfortunately, the piping to the relief valve was not adequately secured, resulting in a large release of light material, which quickly ignited. There are lots of additional lessons learned from this incident. Those who survived were rescued after jumping over 100 feet into the freezing waters of the North Atlantic. (167 Fatalities)
1988	Campos Basin near Rio de Janeiro, Brazil	Petrobras' Enchova drilling platform – The second major incident on this platform in four years occurred when the well suffered a gas blowout while being converted from oil to gas. A major fire occurred when sparks from a drill pipe that was pushed from the well struck an adjacent structural member igniting the released gas. In this incident, the crew was able to safely evacuate the platform. The fire burned for just over a month.
1988	Singapore, Singapore	Pulau Merlimau refinery – A massive fire occurred in the naphtha storage area of this facility. The fire originated while workers were attempting to free a jammed floating roof on one of the naphtha storage tanks. While they were working, naphtha spilled from the tank and subsequently ignited. The initial fire spread ultimately to three other naphtha tanks making this one of the worst industrial fires to occur in Singapore at that time. As a footnote, this refinery had experienced another fire in 1984 which damaged the crude unit.
1988	Henderson, NV.	Pepcon – Welding results in an explosion of ammonium perchlorate tanks. 2 Fatalities, 372 Injured.
1989	Morris, IL.	Quantum – Brittle fracture of the Deethanizer overhead exchanger results in large vapor release, explosion, and fire. (2)
1989	Pasadena, TX.	Phillips 66 Polyethylene Plant, Loss of containment during a maintenance operation (VCE) The explosion of ethylene and isobutane registered a 3.5 on the Richter scale about 20 miles from the site. (23 Fatalities, 314 Injured)
1989	Prince William Sound, Alaska	Exxon Valdez (Oil spill) The oil tanker was bound for Long Beach, CA, when it ran aground on Bligh Reef, spilling about 250K barrels of crude into the eco-sensitive Prince William Sound, AK. At the time, this was considered to be the most devastating man-made environmental disaster to date. (0)

(Continued)

Year:	Location:	Incident description:
1989	Baton Rouge, LA.	Exxon Refinery – Thermal expansion of off-site piping (VCE). Piping had been isolated during a hard freeze overnight and failed the following afternoon due to being overpressurized due to thermal expansion (2)
1989	Richmond, CA.	Chevron Refinery – A massive explosion and fire occurred on the Hydrocracker Unit. The Hydrocracker was in shutdown mode preparing for a turnaround when a 2″ exchanger bypass line carrying Hydrogen gas at 3,000 psi failed. The resulting fire impinged on the reactor skirt weakening the vessel. The reactor failed catastrophically and fell onto other unit equipment. Four Chevron workers were engulfed in the ensuing fire and three were critically burned, five other employees suffered less serious injuries.
1990	Cincinnati, OH.	BASF Chemical Plant (2)
1990	Channel View, TX.	ARCO Refinery – A chemical wastewater tank exploded, destroying the area about the size of a city block. No one inside the area survived. (17 Fatalities)
1990	Nagothane, India	Indian Petrochemicals (IPCL): Explosion/Fire when the ethane gas feed line failed outside the battery limits. The site installed block valves to accommodate the on-line exchanger cleaning. This included welding on piping that operates below −40°F. At the time of the explosion, personnel were responding to a leak on one of these exchanger inlet lines. (Brittle Fracture?) (32)
1990	Stanlow, U.K.	Shell Stanlow – An explosion occurred in the Fluoroaromatics Unit when the halogen exchange reactor ruptured. This was the result of a high pressure generated from an uncontrolled exothermic reaction. The explosion hurled multiple missiles very long distances from the facility and the plant was heavily damaged in the explosion. Six employees were injured, and one died from his injuries (one Fatality, six Injured).
1991	Lake Charles, LA.	Citgo Petroleum – Pressure vessel explosion during startup. (5).
1991	Sterlington, LA.	Angus Chemical Company (IMC Fertilizer Company) (IMCF) explosion. A fire occurred in the area of a waste gas vent compressor (RJ-291) in the nitroparaffins (NP) plant. A few moments after the fire started, a series of explosions destroyed a large section of the NP plant, sending shrapnel north and east of the plant. Large debris weighing up to 150 pounds was hurled almost a mile away. Employees #1 through #8 of IMCF were killed, and 42 were injured. In addition, approximately 70 residents of the town were injured and numerous businesses and residences were severely damaged. (8 Fatalities, 112 Injured, including 70 residents)
1991	Charleston, South Carolina	Albright & Wilson Chemical Plant (9)
1991	Seadrift, TX.	Union Carbide Chemical Plant – The ethylene oxide refining column was destroyed in the explosion and the ethylene glycol unit was substantially damaged, as was the co-generation unit. A pipe rack near the ethylene oxide storage area was struck by a large piece of shrapnel releasing various hydrocarbons into the fire area. All production at the facility was shut down. One person was killed and 26 were injured in the explosion(1).
1992	La Mede, France	La Mede Total Refinery – Explosion (VCE)/fire in dead-leg piping. The gas plant, the FCC unit, and their control building were totally destroyed in the explosion. Two processing units under construction were also damaged. The local community was also impacted by the force of the explosion including homes and businesses with damaged roofs and broken windows. Windows were broken as far as six miles from the refinery.
1992	Guadalajara, Mexico	A massive gasoline explosion in the city municipal sewer system. The cause was found to be galvanic corrosion between the Pemex steel gasoline pipeline and the zinc-coated municipal water pipe. >200 Fatalities, although many believe that this fire resulted in at least 1,000 deaths – >500 were missing, 500 were injured, and over 15,000 were left homeless.
1992	Los Angeles, CA.	Carson Refinery – A loss of containment and explosion occurred at this refinery resulting from the rupture of the outside radius of a 6-inch-diameter carbon steel 90° elbow. The failure released a hydrocarbon-hydrogen mixture which quickly ignited. A post-incident inspection revealed nearly full design thickness a short distance away from the failure. Therefore, the analysis confirmed the failure was the result of the thinning of the carbon steel elbow due to long-term erosion/corrosion.

Year:	Location:	Incident description:
1992	Sodegaura, Japan	Fuji Sekiyu Corporation – 10 people died, and seven people were injured when a cover plate failed on a heat exchanger releasing hot Hydrogen. The Hydrogen quickly flashed resulting in the explosion and fire. In this incident, the exchanger channel cover and lock ring of this heat exchanger were hurled into an adjacent plant about 650 feet from the refinery. Both components were five feet in diameter and weighed 4,000 and 2,000 lb, respectively. (10).
1993	Baton Rouge, LA.	Exxon Refinery – A loss of containment occurred when a piping ell failed on a Delayed Coker hot feed pump discharge. The resulting release quickly ignited engulfing the four-drum coke structure. The Delayed Coker unit was totally destroyed and three workers were killed in this tragic incident. The failed pipe ell was found to be constructed of carbon steel while all the remaining piping components in this circuit were 5% Chrome. The original unit was built in 1963 and this circuit was specified to be 5% Chrome. (3)
1994	Baton Rouge, LA.	Exxon Chemical – Sponge tower fire. Oversized control valve resulted in pipe vibration and flange failure. (0)
1994	Fawley, England	ExxonMobil Refinery, Reactor failure (uncontrolled exothermic reaction) (0)
1994	Milford Haven, Wales	Texaco Pembroke, Lightning strike and failure of flare piping due to liquid loading (0)
1995	Romeoville, IL	Pennzoil Products Co., A refinery worker was fatally burned while working on a refinery heater (1).
1995	Rouseville, PA.	Pennzoil Products Co., Five workers were badly burned when an explosion occurred at the refinery. Three were killed instantly, and two others died in the hospital (5).
1995	Corpus Christi, TX.	Coastal Refining and Marketing, A refinery worker was fatally burned in a Coker Unit release and resulting fire (1).
1995	Nipomo, CA.	Unocal Refinery, Workers were inside a storage tank, cleaning the tank when H_2S and sulfur dioxide exploded (5).
1995	Kawasaki, Japan	Exxon – H_2S release during maintenance operations (OPE) (4)
1996	Channelview, TX.	Lyondell; Ethylene 12″ product pipeline flange leak and fire results in plant shutdown damage to electrical system (0).
1996	Anacortes, WA.	Texaco Refinery, Heat exchanger exploded, and an employee was struck by cover plate (1).
1996	Borger, TX.	Phillips Petroleum Refinery, Reactor exploded, separating the top from the shell. A worker was killed when he was reportedly standing next to the reactor when it exploded (1).
1996	Bell Chase, LA.	BP Refinery, Contractor was overcome while working inside a tank to remove solids and was asphyxiated; he was wearing a chemical cartridge respirator (1)
1996	Sunray, TX.	Diamond Shamrock Refining, one employee was killed when a compressor failed due to liquid ingestion to the cylinder resulting in fire and explosion (one Fatality, two injured).
1996	Chiapas State, Mexico	Mexico's state petroleum company Petróleos Mexicanos – A significant amount of Mexico's gas processing is out of service due to several large explosions that severely damaged two gas plants. Although still under investigation, it appears that maintenance crews may have accidentally caused a release of natural gas. Six workers were killed and more than 40 injured in the explosion (6).
1997	Martinez, CA.	Tosco Avon Refinery – Uncontrolled hydrocracker exothermic reaction, explosion. Operators were troubleshooting the indications of erratic reactor temperatures when the explosion occurred (1 Fatality and 46 Injured).
1997	Deer Park, TX.	Shell Chemical; Major fire and damage to Olefins plant due to vapor release, explosion/fire. The release was due to an internal failure of the cracked gas compressor discharge check valve (a 36″ ID Clow model GMZ, weighing over three tons). The CV design includes a split shaft with two half shafts attached to the internal flapper with dowel pins. The attachment failed, allowing one of the half shafts to be propelled out of the valve body, providing a leak path for the hydrocarbons to the atmosphere. The fire burned for ten days, severely damaging the olefins unit.
1997	Bintulu Bintulu, Indonesia	Shell Gas to Liquids Plant – The explosion occurred in the air separation unit (ASU) which supplied oxygen for the production of the synthesis gas feedstock.
1997	Visakhapatnam, India	LPG Piping surge (>60)

(Continued)

Year:	Location:	Incident description:
1998	Texas City, TX.	Valero Energy Refinery, an Employee on top of a sour water tank was overcome by H_2S and died (1).
1998	Norco, LA.	Shell Refinery – injection point failure, explosion/fire (7)
1998	Anacortes, Washington	Equilon Enterprises Refinery, Coke drum hot water release while attempting to restart the unit after a power failure (6)
1998	Old Ocean, TX.	ConocoPhillips Refinery, Employee was exposed to sulfuric acid and died from inhalation and acid burns (1)
1998	Convent, LA.	Motiva Enterprises Refinery, Employee was overcome by H_2S and died while preparing to launch a pipeline scraper (1).
1998	Longford, Australia	Exxon Longford Gas Plant – Brittle Fracture (2)
1998	Carson, CA.	Arco Refinery, Fatality due to an explosion while welding on an in-service lime slurry tank (1).
1998	Pitkin, LA.	Sonat Exploration Co. – A catastrophic vessel failure and fire occurred near the Temple 22-1 Common Point Separation Facility. Four workers near the vessel were killed, and the facility sustained significant damage. The vessel lacked a pressure-relief system and ruptured due to overpressurization during startup, releasing flammable material which ignited. Also, workers at the facility were not provided with written operating procedures addressing the alignment of valves during purging operations. The vessel was rated for 0 PSIG and may have been exposed to as much as 800 PSIG (four Fatalities).
1999	Corpus Christi, TX.	Citgo Refinery, two workers were badly burned while attempting to restart a steam boiler (one Fatality, one injury)
1999	Martinez, CA.	Tosco Avon Refinery – Large refinery fire occurred during Crude Unit overhead naphtha piping maintenance. Workers were attempting to replace piping attached to a 150-foot-tall fractionator tower while the processing unit was in operation. During removal of the piping, naphtha was released onto the hot fractionator and ignited. The flames engulfed five workers located at different heights on the tower. Four men were killed, and one sustained serious injuries (four Fatalities, one critically injured).
1999	Richmond, CA.	Chevron Refinery – The bonnet on a block failed in the high-pressure section of the Hydrocracker resulting in a large vapor release, explosion, and fire. The explosion collapsed a large section of an adjacent pipe rack and overhead mounted fin-fan cooler. There was other significant damage to the refinery. A shelter-in-place was ordered for the adjacent community.
1999	Smackover, AR.	Cross Oil Company, three workers were killed in a flashfire while working on a naphtha tank valve when it exploded. (3)
1999	Sriracha, Thailand	Thai Oil Refinery – Gasoline tank overfill, explosion, and fire – 4 tanks burned (8)
1999	Brewton, AL.	Desoto Oil and gas Company, one worker died due to H_2S exposure.
1999	Westlake, LA.	Conoco Refinery, A refinery flash fire resulted in critical burns to one employee who later died (1)
1999	Hahnville, LA.	Union Carbide Corp. Nitrogen Asphyxiation Incident – One worker was killed, and another was seriously injured when they were asphyxiated by nitrogen while inside a temporary enclosure. The workers had erected over the end of a large open gas pipe. The workers were conducting a black light inspection and were unaware that the pipe was being purged with nitrogen, creating an oxygen-deficient atmosphere (one Fatality, one Injured).
1999	Bellingham, WA.	Olympic Pipeline Co. – A gasoline pipeline failure spilled gasoline into Whatcom Creek, which flowed through the Whatcom Park area, a local public park. The spill was eventually ignited, resulting in a major fire throughout the park. Three people in the park were killed (two 10-year-old boys playing in the stream and an 18-year-old who was salmon fishing were killed). The park area required extensive refurbishment. (three Fatalities).
1999	Tupras Korfez, Turkey	Korfez Refinery – An earthquake measuring 7.4 on the Richter scale collapsed a 312-foot-tall concrete chimney which fell onto a crude unit at this refinery. This resulted in a large fire on the crude unit as well as several storage tanks. on one of the crude units, setting off fires at this 226,000 bbl-per-day refinery. Firefighting efforts were limited to water drop by aircraft due to broken water mains. There were no reported injuries.
2000	Mina Al-Ahmadi, Kuwait	Mina Al-Ahmadi Kuwait Refinery – failure of LPG piping due to external corrosion, explosion/fire. Massive damage to refinery and LNG facility including 2 control rooms destroyed (5 Fatalities, 49 Injured).

Year:	Location:	Incident description:
2000	Corpus Christi, TX.	Valero Energy Corp, while wrapping up a sulfur plant turnaround, two workers were overcome by H_2S exposure. One died, and one recovered in the hospital (one Fatality, one injury)
2000	Baton Rouge, LA.	ExxonMobil Chemical; Internal pyrophoric fire in fractionating column
2000	South Philadelphia, PA.	Sunoco Refinery – A fire occurred at the refinery following an equipment failure in the crude unit. One person was taken to a local hospital, and another was treated at the scene. A thick, black cloud was released into the area due to the fire.
2000	Jurong, Singapore	ExxonMobil Chemical – Steam boiler explosion occurred during restart following a boiler trip (2)
2000	Carlsbad, NM.	El Paso Natural Gas Co. – A 30-inch natural gas pipeline ruptured near the Pecos River and ignited resulting in a large fire. Twelve civilians who were camping under a bridge that supported the pipeline were killed in the ensuing fire. Other impacts included destruction of their three vehicles and fire damage to two nearby gas pipeline suspension bridges. The NTSB reported that the rupture was caused by severe internal corrosion that had been undetected (12 Fatalities).
2000	Grangemouth, U.K.	BP Refinery – A medium-pressure steam main ruptured resulting in a very significant release of steam, noise, and vibration. The loss of steam also affected the steam balance in the refinery until it could be isolated. Damage to refinery infrastructure (fencing) and debris was released. A member of the public was injured while walking his dog.
2000	Grangemouth, U.K.	BP Refinery – During the restart of the FCC unit following an electrical power failure, a loss of containment and fire occurred in the FCC light ends section. The release occurred in an unsupported reducing tee branch pipe in the transfer line between the Debutanizer and Rerun columns due to metal fatigue. Firewater runoff containing unburned hydrocarbons was released to the River Forth. There were no injuries reported.
2000	Pasadena, TX.	Phillips 66 Houston Chemical Complex (HCC) – An explosion and fire occurred at the chemical complex in Pasadena producing huge plumes of black smoke that spread over the Houston Ship Channel and into residential areas. This explosion in the K-Resin unit involved a plastic made with butadiene. The tank that exploded was out of service for cleaning and there were no gauges (temperature or pressure) that would have alerted the workers of a potential problem. The explosion killed one worker and injured 71 others – 32 employees and 39 contractors were taken to local hospitals for burns and other injuries. The search for the missing person took five hours due to the amount of rubble (1 Fatality, 71 Injuries).
2001	Lake Charles, LA.	Citgo refinery – Internal vessel explosion during startup Operating Procedures (air freeing) (0)
2001	Texas City, TX.	BP Refinery, a worker overcome by Nitrogen while working near the top manway fell into the reactor and later died. (1)
2001	Torrance, CA.	ExxonMobil Refinery – A catalyst worker working inside a hydrocracker reactor under inert entry fell after being tangled in his hose line. He later died at the hospital (1).
2001	Delaware City, DE.	Sulfuric acid tank explosion during maintenance (1)
2001	Humber, England	ConocoPhillips Refinery – injection point failure and explosion (0)
2001	Corpus Christi, TX.	Citgo Refinery – A welder working in a confined space became overcome by Argon and died (1).
2001	St James, LA.	Chevron Phillips Chemical – Styrene fractionation column toppled due to internal pyrophoric fire. No injuries reported, although the large column was destroyed.
2001	Pasadena, TX.	A Pasadena, Texas refinery reported a release of hydrofluoric acid (HF) occurred at the refinery. The leak happened reportedly due to an Operator checking the operation of a valve. The plant was evacuated, and six workers were sent to the local hospital for observation. They were working several feet away and were exposed to the HF and were having problems breathing. They were later released.
2001	Delaware, DE.	Motiva Enterprises Refinery, A contractor was killed and eight others injured while repairing a walkway above a spent acid tank. Vapors released from the tank exploded causing the tank to release its contents. Other tanks in the area also released their contents. The acid spill reached the Delaware River where aquatic life was affected. (one Fatality and eight injured).

(Continued)

Year:	Location:	Incident description:
2001	Dutch Antilles	Aruba Oil Refinery – During a Visbreaker maintenance operation on a pump strainer, a block valve failed to hold on a Visbreaker unit, and a spill of hot oil occurred. The hot oil autoignited resulting in the fire destroying the Visbreaker unit and damaging nearby equipment, including the firefighting equipment.
2001	Mithapur, near Jamnagar, India	Tata Chemical Company – A fire was reported to be in the Boilerhouse and spread from there.
2001	Aruba, Caribbean	El Paso Corporation – The 280K B/D refinery has been shut down due to a fire in the process areas. There were three injuries reported (0).
2001	Lemont, Il.	Citgo Refinery – Internal Pyrophoric fire in Crude Atmospheric column and tower collapse. A pool fire developed in the crude unit due to failure of a pipe elbow which failed due to incorrect metallurgy. The refinery shut down due to the fire. Three days later, the crude column collapsed due to an internal fire caused by air ingress from the previously ruptured piping which reacted with pyrophoric material in the column. The crude unit was down for nearly a year for replacement of the column. No injuries were reported.
2001	Bristol, TN.	Necessary Oil Company reclamation plant, a worker died due to burns from an explosion and fire when oil was released from a ruptured overflow line.
2001	Corpus Christi, TX.	Citgo Petroleum, Contractor overcome by Argon gas and died while welding in a confined space (1)
2001	Toulouse, France	AZF Fertilizer Company – An explosion of ammonium nitrate explosion destroyed the facility and structurally damaged was extensive to the surrounding homes and businesses. (30 Fatalities, About 2,500 Injured)
2001	Los Angeles, CA.	Tosco Carson Refinery – A large fire occurred in the Coker unit due to a release in the refinery piping. Smoke was reportedly visible 90 miles from the site. No injuries were reported. The delayed coker was shut down for about two months as a result of the fire damage.
2001	New Orleans, LA.	Orion Refinery; 270′ (82 meters) gasoline tank fire – Tropical Storm Allison resulted in a sunken roof and full surface fire.
2001	Bottineau, N.D.	Dome Pipeline Co. – A pipeline carrying gasoline just west of the city ruptured and burst into flames. It was estimated that approximately 1.1 million gallons of gasoline burned before the pipeline could be shut down. Dome Pipeline Company was dissolved in 1998.
2001	Lemont, IL.	Citgo Refinery – A crude unit fractionation column collapsed due to an internal packing fire due to pyrophoric ignition. The incident occurred while firefighters were in the process of battling a fire in the unit. The crude distillation unit was shut down for 12 months. The initial fire was caused by the incorrect metallurgy in one piping elbow, which failed resulting in loss of containment and a pool fire. (0).
2001	Roncador Oil Field – Off Brazilian coast	Petrobras' P-36 oil platform – At the time, this was the largest semi-submersible oil drilling platform in the world. The platform suffered 2 back-to-back explosions resulting in the deaths of 11 workers. 175 were on board at the time of the incident. The platform started listing after the explosions and sank within five days (11).
2001	Augusta, GA.	BP Amoco Polymers, Loss of containment & fire during maintenance operations (OPE). Three people were killed when they opened a process vessel containing hot plastic, unaware that it was pressurized. The partially unbolted cover blew off the vessel, expelling hot plastic. A fire occurred after nearby tubing failed releasing hot fluid. (3)
2001	Three Rivers, TX.	Ultramar Diamond Shamrock – A significant fire occurred on the Alkylation Unit at this 100K B/D refinery causing the refinery to be shutdown. Three people were injured.
2001	St James, LA.	Chevron Phillips chemical plant – tower fire and collapse due to fire in packed bed during maintenance operations
2001	Wood River, IL.	Tosco Corporation Refinery – A fire occurred in the #2 Crude Unit.
2002	Medford, OK.	Koch Hydrocarbon Refinery, an explosion occurred while workers were preparing a Peco filter for catalyst replacement (one Fatality, two injuries)
2002	Meraux, LA.	Murphy Oil Refinery, one worker burned to death while removing a muffler from a triangle gate valve. (1)

Year:	Location:	Incident description:
2002	Festus, MI.	DPC Enterprises, A chlorine transfer hose ruptured during a rail car unloading operation at the chlorine repackaging facility. The ruptured hose released 48,000 pounds of chlorine, causing three workers and 63 residents to seek medical treatment. The US Chemical Safety and Hazard Investigation Board (CSB) determined that the ruptured hose was constructed of stainless-steel braid rather than Hastelloy C, a metal alloy (CSB, 2002) (0 Fatalities, 66 Injured).
2002	Samir Port of Mohammedia, Morocco	Port of Mohammedia Refinery – Flooding waters from heavy rains floated wastewater containing hydrocarbons contacted hot equipment resulting in a large surface fire. Several storage tanks also exploded and burned. The refinery was heavily damaged, and two employees were killed and three others missing. Some of the production units were restored within a couple of weeks but damaged units were expected to be out of service for nearly a year (two fatalities, three missing).
2002	Paulsboro, NJ.	Valero Energy Company, Nitrogen exposure inside a shed that was being purged (1)
2002	Norco, LA.	Orion Refining, one worker asphyxiated inside a tank (1)
2002	Pascagoula, MS.	ChevronTexaco Refinery, A worker died from burns sustained in a fire at the refinery. One other was injured in the refinery fire (1).
2002	Pascagoula, MS.	First Chemical Corporation – A violent explosion occurred in a chemical distillation tower sending heavy debris over a wide area; the flying debris, fortunately, missed hitting nearby storage vessels containing ammonia, chlorine, sulfuric acid, and other hazardous materials. The force of the explosion blew off the upper 35 feet of the tower and sent tons of debris flying up to a mile away. Three workers in the control room were injured by shattered glass. One nitrotoluene storage tank at the site was punctured by explosion debris, igniting a fire that burned for several hours. The CSB determined the cause was the inadvertent heating of mononitrotoluene (MNT) and an uncontrolled chemical reaction resulting in the explosion (zero Fatalities, three Injured).
2002	Pennington, AL.	Georgia-Pacific Naheola mill – Highly toxic hydrogen sulfide gas leaked from a sewer manway near several people working, exposing them to the deadly gas. Two contractors from Burkes Construction, Inc., were killed. Eight others were injured, including seven employees of Burkes Construction and one employee of Davison Transport, Inc. Choctaw County paramedics who transported the victims to hospitals reported symptoms of hydrogen sulfide exposure (two Fatalities, eight Injured).
2002	Friendswood, TX.	Third Coast Industries – A fire of relatively small magnitude destroyed the entire facility, which blended and packaged motor oils, hydraulic oils, and engine and other lubricants. Firefighters arrived at the scene within minutes but had insufficient means to fight the fire, which burned for more than 24 hours. The fire consumed 1.2 million gallons of combustible and flammable liquids and destroyed the site. One hundred nearby residents were evacuated, a local school was closed, and significant environmental cleanup was necessary due to fumes and runoff. The facility did not conduct an adequate fire protection analysis to ensure the implementation of fire protection measures. The origin of the fire was not determined (zero Fatalities, zero Injured)
2002	Garyville, LA.	Marathon Petroleum Refinery, Contractor fatality while doing an inert entry for catalyst work inside a reactor. He was engulfed by the catalyst; his face shield broke, allowing air to mix with the pyrophoric catalyst. He received severe burns from the fire and died. (1)
2002	Cincinnati, OH.	Environmental Enterprises, a hazardous waste treatment company – A maintenance employee was overcome when he inhaled hydrogen sulfide gas from a waste processing vessel. No other injuries or damage were reported, and the injured worker was subsequently pulled to safety.
2003	Kingston, NC.	Pharmaceutical plastics plant – Combustible dust explosion. The explosion and fire destroyed the West Pharmaceutical Services plant in Kinston, North Carolina, causing six deaths, dozens of injuries, and hundreds of job losses. The facility produced rubber stoppers and other products for medical use. The fuel for the explosion was a fine plastic powder, which accumulated above a suspended ceiling over a manufacturing area at the plant and ignited (6 Fatalities, 38 Injured).
2003	Glendale, AZ.	DPC Enterprises, the chlorine repackaging facility, experienced a chlorine release injuring 14 people, including 10 police officers. The release occurred when the chlorine scrubber's absorbent chemicals were exhausted and ineffective. (0 Fatalities, 14 injured).

(Continued)

Year:	Location:	Incident description:
2003	Torrance, CA.	ExxonMobil Refinery, a worker was electrocuted while installing an air conditioner unit (1)
2003	Ponca City, OK.	Conoco Phillips Refinery – While removing a pump inside a gas plant, a vapor was released, and an explosion occurred. From the OSHA report "At approximately 11:00 a.m. on July 21, 2003, Employee #1 and coworkers were removing a vertical pump and motor inside a gas plant. For some reason, a hydrocarbon was released inside the plant, causing an explosion and fire. Employee #1 suffered second- and third-degree burns to his upper body. He was hospitalized and treated for his injuries. Employee #1 died ten days later." Contractors were evacuated from the refinery (one Fatality, five Injured).
2003	Baytown, TX.	ExxonMobil Refinery – Worker burned when a filter exploded and died (1)
2003	Nagoya, Japan	Exxon Nagoya Distribution Terminal Gasoline tank explosion/fire during maintenance (6)
2003	Puertollano, Spain	REPSOL Refinery; Unstabilized naphtha was sent from the debutanizer to a floating roof tank during FCCU startup, vapor cloud from the tank ignited by the contractor's vehicle, explosion caused large damage to APS and resulted in 6 other tank fires (9)
2003	Glenpool, OK.	ConocoPhillips Co. – An 80,000 Barrel storage tank suddenly exploded while being filled with diesel and contained about 7,300 barrels of diesel when it exploded. The tank previously contained gasoline. The fire burned for 21 hours and damaged two other large storage tanks. Residents in the vicinity were evacuated and local schools were closed for two days. The NTSB concluded the cause was ignition of a flammable fuel-air mixture by static electricity. Filling with a cargo of high flash point over a heel of low flash point product is known as switch loading.
2003	Rosharon, TX.	BLSR Operating Ltd. an oilfield waste disposal facility. A major fire happened while two vacuum trucks were delivering flammable gas condensate waste for disposal. Two employees and one truck driver were fatally burned. Four other workers suffered serious burns (three Fatalities, four Serious burn injuries). CSB investigation completed and determined the most probable cause to be the diesel engines, although multiple other potential ignition sources were present at the facility.
2003	Staten Island, NY.	ExxonMobil Fuel Terminal – A barge exploded while offloading gasoline at the terminal. A large black smoke cloud hung over most of New York City for most of the day. Two barge crewmen were killed, and a terminal employee was critically burned (2).
2003	CSB investigations into three combustible dust incidents.	The CSB launched investigations of three major industrial explosions involving combustible powders. In North Carolina, Kentucky, and Indiana, these explosions cost 14 lives and caused numerous injuries and substantial property losses. The Board responded by launching a nationwide study to determine the scope of the problem and recommend new safety measures for facilities that handle combustible powders. The final report is available on www.CSB.gov.
2003	Corbin, KY.	CTA Acoustics manufacturing plant, A combustible dust explosion killed seven workers and severely damaged the facility. The facility produced fiberglass insulation for the automotive industry. CSB investigators discovered the explosion was fueled by resin dust accumulated in a production area, likely ignited by flames from a malfunctioning oven. (7 Fatalities, 37 injured)
2003	Louisville, KY.	Williamson food additive plant, A process vessel, was overpressurized and failed catastrophically. One worker was killed, and the vessel failure caused extensive damage to the facility and released aqueous ammonia. The CSB reported that the process was controlled manually and that the feed tank most likely failed because of overheating the caramel color liquid, which generated excessive pressure (1).
2003	CSB releases two presentations on Nitrogen Safety (A PowerPoint Training and a Nitrogen Safety Bulletin)	Message from the CSB: Every year people are killed by breathing "air" that contains too little oxygen. Because 78 percent of the air we breathe is nitrogen gas, many people assume that nitrogen is not harmful. However, nitrogen is safe to breathe only when mixed with the appropriate amount of oxygen. These two gases cannot be detected by the sense of smell. A nitrogen-enriched environment, which depletes oxygen, can be detected only with special instruments. If the concentration of nitrogen is too high (and oxygen too low), the body becomes oxygen-deprived, and asphyxiation occurs.
2004	Skikda, Algeria	LNG Plant; loss of containment – Boiler initiated Vapor Cloud Explosion (VCE) (27 Fatalities, More than 80 injured)

Year:	Location:	Incident description:
2004	Coffeyville, KS.	CVR Refining – While drilling a pilot hole in preparation for decoking the coke drum on the Delayed Coker, the water came in contact with hot coke. The eruption of steam fatally burned the worker (1).
2004	Texas City, TX.	BP Refinery, three workers burned working on condensate pump (two Fatalities, one critical injury)
2004	Corpus Christi, TX.	Valero Energy, two workers were exposed to excessive heat while working inside a reactor. One worker died due to heat stress (one Fatality, one injury).
2004	Gallup, NM.	Giant Industries Ciniza refinery – Four workers were seriously injured when highly flammable gasoline components were released and ignited. The release occurred as maintenance workers were removing a malfunctioning pump from the refinery's hydrofluoric acid (HF) alkylation unit. Unknown to personnel, a shut-off valve connecting the pump to a distillation column was apparently in the open position, leading to the release and subsequent explosions (Fatalities, four seriously injured).
2004	Dalton, GA	MFG Chemical Company – A chemical reactor overheated at the MFG Chemical manufacturing plant, releasing toxic allyl alcohol vapor. The resulting cloud sent 154 people to a local hospital and forced the evacuation of nearby residents. Vegetation and aquatic life near the plant were killed. (0 Fatalities, 154 Injured).
2004	Illiopolis, IL.	Formosa Plastics, PVC release during maintenance operations (OPE). Five workers were fatally injured, and two others were seriously injured when an explosion occurred in a polyvinyl chloride (PVC) production unit. The explosion followed a release of highly flammable vinyl chloride, which ignited. The explosion forced a community evacuation and ignited fires that burned for several days at the plant. (two fatalities, two Injured).
2004	Houston, TX.	Marcus Oil and Chemical polyethylene wax facility – A storage tank failed catastrophically, resulting in a blast, which was felt up to 20 miles from the plant site. Secondary large fires occurred that burned for several hours. Several buildings near the facility were damaged by the blast. Three firefighters were injured during the emergency response. The CSB concluded that the tank was pressurized with a mixture of Nitrogen and air – air was used by the Operators to compensate for an undersized Nitrogen generator. The tank failed due to a non-standard weld at a prior tank alteration (zero Fatalities, two Injured).
2004	New Jersey, US	Singapore Flag Vessel Bow Mariner – The US Coast Guard reported the Singapore-flagged tanker sank off the New Jersey coast. A cargo of about 3.6 million gallons of ethyl alcohol and 55,000 gallons of diesel was spilled.
2005	Offshore oilfield West coast of Mumbai, in the Gulf of Cambay.	Mumbai High North Oil and Gas Platform, operated by Oil and Natural Gas Corporation Ltd. (ONGC) – During a monsoon storm, a supply vessel struck the platform risers while performing a rescue mission from the platform. The ruptured export gas lift risers released a large quantity of gas at the platform deck level which quickly ignited engulfing the platform. 22 workers were killed, and 362 others were rescued over next fifteen hours. Six divers had to be left behind during the rescue but fortunately were rescued later. The high-value platform was a total loss (22).
2005	Texas City, TX.	BP Refinery, Isom Unit explosion after fractionation column overfilled during unit restart startup (VCE). Many victims were in or around work trailers near an atmospheric vent stack. The explosions occurred when a distillation tower flooded with hydrocarbons and was overpressurized, causing a geyser-like release from the atmospheric blowdown vent stack. (15 Fatalities, Over 180 injured)
2005	Bakersfield, CA.	Kern Oil and Refining – Explosion while workers were cleaning pumps (one Fatality, two injuries).
2005	Fort McMurray, Alberta, Canada	Suncor Oil Sands – Ruptured cycle line resulted in a large fire in the Suncor refinery and the evacuation of 250 people. The fire burned for 9 hours. (0).
2005	Cherry Point, WA.	BP Refinery – A worker was found deceased in a pool of water while he had been hydroblasting inside the Coker unit. BP reported that his death was due to natural causes.
2005	Hertfordshire, England	Buncefield Terminal; Gasoline tank overfilled, resulting in vapor cloud explosion – Initial fire involved 20 tanks. The terminal was destroyed, and there are no plans to rebuild (0).
2005	Delaware City, DE.	Valero Refinery – Reactor Inert Entry exposure to Nitrogen. Two contract employees were overcome and fatally injured by nitrogen as they performed maintenance work near a 24-inch opening on the top of a reactor. One of the workers died attempting rescue. (2)

(Continued)

Year:	Location:	Incident description:
2005	Port Comfort, TX.	Formosa Plastics – small piping failure when struck (VCE and major fire). A forklift towing a trailer collided with a line containing highly flammable liquid propylene, causing a release and a vapor cloud explosion. Sixteen workers were injured, the processing unit was heavily damaged, and a nearby school was evacuated (0 Fatalities, 16 Injured).
2005	Perth Amboy, NJ.	Acetylene Service Company, A gas explosion killed three workers at the plant. The blast originated in a wooden shed near six large storage tanks that received liquid waste from the plant's acetylene-generating system. (3) The CSB investigation is complete
2006	Torrance, California	ExxonMobil Refinery – Loss of containment, fire (piping metallurgy – PMI) (0)
2006	Daytona Beach, FL.	Bethune Point wastewater plant, An explosion occurred on a methanol Storage tank while workers were using a cutting torch above the tank to remove a steel canopy. The torch ignited ethanol vapors resulting in the explosion and ensuing fire. Two municipal workers died, and another was seriously injured, and the explosion resulted in the release of the total tank contents (about 3,000 gallons of methanol) (two Fatalities, one Injured). The CSB investigation is complete.
2006	Jackson, MS.	Partridge-Raleigh Oilfield – Three contractors died, and one contractor suffered serious injuries in an explosion and fire at the oilfield site. The contractors were standing on top of a series of four oil production tanks preparing to weld piping to the tanks. Tanks one, three, and four were empty. However, tank two contained a low level of hydrocarbon. All four tanks were interconnected by a vent pipe, and the contractors were installing overflow piping between tanks three and four. Almost as soon as the welding began, the tanks exploded, resulting in the 3 fatalities and serious injuries. The CSB investigated and made several hot work recommendations (three Fatalities, one serious injury).
2006	Mazeikiu Nafta, Lithuania	Mazeikiu Nafta refinery, owned by Poland's PKN Orlen PKNA.WA – Piping Failure, Fire, and Collapse of Vacuum Tower (Carbon Steel Piping Installed in Chrome Service/Lack of Fireproofing – Pipestill vacuum tower collapsed onto the feed train exchangers).
2006	Danvers, MA.	CAI/Arnel ink and paint manufacturing facility – A powerful explosion destroyed the ink and paint facility and damaged many nearby homes and businesses beyond repair. Ten local residents were injured, and some were hospitalized. The plant was not occupied at the time. The CSB investigation is complete. (0 Fatalities, 10 injured).
2006	Apex, NC.	EQ Hazardous Waste Plant – Explosions and fire at a hazardous waste facility forced the evacuation of approximately 16,000 nearby residents. The CSB reported that the incident likely began in the oxidizer section of the EQ North Carolina waste facility, where chemicals such as pool chlorination tablets were stored. The fire was allowed to burn out, and the facility was destroyed (0).
2006	Morganton, NC.	Synthron, LLC – A runaway chemical reaction and subsequent vapor cloud explosion and fires killed one worker and injured 14 (2 seriously). The explosion destroyed the facility and damaged structures in the nearby community. The CSB found that the reactor lacked basic safeguards to prevent, detect, and mitigate runaway reactions and that essential safety management practices were not in place (1 Fatality, 14 Injured, 2 seriously).
2006	CSB Issued Safety Bulletin on Positive Materials Identification (PMI)	Following the metallurgy-related failure on the BP Hydrocracker Unit in July of 2005 and the major fire that resulted, the CSB issued a Safety Bulletin on the importance of Positive Materials Identification to ensure the appropriate metallurgy is installed in critical systems.
2007	Singapore, Singapore	ExxonMobil Refinery – Loss of containment and fire during Maintenance operations (three Fatalities, one Critical burn injury)
2007	Robinson, IL.	Marathon Refinery – H_2S exposure Fatality (1)
2007	Garyville, LA.	Marathon Refinery – A truck driver accidentally drove his truck into a retention pond and drowned (1).
2007	Billings, Mt.	ExxonMobil Billings Refinery – mixing point failure, explosion/fire (0)
2007	Campeche Sound, Mexico	Pemex Oil Rig (Kab 121), Explosion and fire, (21) workers.
2007	Billings, MT.	ExxonMobil Refinery – A large fire occurred at the refinery when a piping connection failed on the hot hydrogen line to the Hydrocracker reactor. The failure occurred in a mixing tee connection where two different temperature hydrogen streams joined. No injuries were reported.

Year:	Location:	Incident description:
2007	McKee, TX.	Valero Refinery, Loss of containment and fire (piping dead-leg failure – trapped water froze, rupturing the pipe) A liquid propane release from cracked control station piping resulted in a massive fire in the propane Deasphalting (PDA) unit at Valero's McKee Refinery near Sunray, Texas, injuring three employees and a contractor. The fire caused extensive equipment damage and resulted in the evacuation and total shutdown of the McKee Refinery. A refinery pipe rack collapsed creating an even larger fire and additional damage. The refinery remained shut down for two months. The following are key findings of the Chemical Safety Board's (CSB) investigation: The propane release was likely caused by the freeze-related failure of high-pressure piping at a control station that had not been in service for approximately 15 years. The control station was not isolated or freeze-protected but left connected to the process, forming a dead leg. (zero Fatalities, four Injured).
2007	St Paul Park, MN.	Marathon Petroleum Refinery – One Operator was badly burned inside a tank that exploded; OSHA report available (1).
2007	Westlake, LA.	ConocoPhillips Refinery – Contractor fatality from H_2S exposure and possibly Methyl Mercaptan, OSHA report available (1).
2007	Texas City, TX.	BP Refinery – An electrician was electrocuted while working to prepare an idle hydrotreater unit for a restart. Electrocution (1).
2007	Valley Center, KS.	Barton Solvents, Storage tank explosions, and fire erupted during the offload of solvents into the storage tanks. The incident led to the evacuation of thousands of residents and resulted in projectile damage off-site and extensive damage to the facility. The cause was determined to be ignition by static discharge by the CSB. The CSB investigation is complete.
2007	Georgetown, CO.	Xcel Energy Company – five people were killed and three others injured when a fire erupted 1,000 feet underground in a tunnel at a hydroelectric power plant located approximately 45 miles west of Denver. The fatally injured workers were trapped deep underground during an operation to coat the inside of the tunnel with epoxy using highly flammable solvents. The tunnel is several thousand feet long and connects two reservoirs with electricity-generating turbines (five Fatalities, three Injured).
2007	Pascagoula, MS.	Chevron Refinery – A major fire broke out in the crude unit at this refinery. The fire was extinguished after about six hours. The remaining crude unit and the rest of the refinery continued in operation. No injuries were reported.
2008	Tyler, TX.	Delek Refinery – Sat Gas piping failed, resulting in explosion and fire. From OSHA: "The pipe rupture and fire occurred at the Saturated Gas Unit when a corroded pipe carrying naphtha (a flammable mixture of hydrocarbons) ruptured and the naphtha ignited.... During the fire, naphtha, and its constituents, including but not limited to butane. Carbon monoxide, ethane, ethylene, hydrogen, isobutene, isopentane, methane, pentane, propane, and toluene were released into the atmosphere." (two fatalities, three injuries). (2).
2008	Big Springs, TX.	Alon Refinery – A significant propylene release occurred when a pump failed while restarting a propylene splitter tower. The release quickly ignited resulting in a major fire. Five people were injured in the incident, including a passerby. All other personnel were evacuated from the site. OSHA investigated the incident and issued Alon penalties for OSHA PSM shortcomings.
2008	Institute, WV.	Bayer CropScience – A chemical runaway reaction occurred inside a 4,500-gallon pressure vessel known as a residue treater, causing the vessel to explode violently in the methomyl unit. Highly flammable solvent sprayed and immediately ignited, causing an intense fire that burned for more than four hours. Two workers were killed, and eight were treated for potential chemical exposure (two Fatalities, eight Injured). The CSB investigation is complete.
2008	Texas City, TX	BP Refinery – One worker was killed when a cover plate blew off a pressurized water filter system (1).
2008	Port Wentworth, GA.	Imperial Sugar, A massive combustible dust explosion occurred at the Imperial Sugar refinery mill northwest of Savannah, causing 14 deaths and injuring 38 others, including 14 with serious and life-threatening burns. The explosion was fueled by massive accumulations of combustible sugar dust throughout the packaging building. (9 Fatalities, 38 Injured)
2008	Sløvåg, Norway	Vest Tank Farm – While treating Coker light naphtha with hydrochloric acid to remove the sulfur compounds the mix tank suddenly exploded. The resulting fire quickly spread to two other tanks in the tank farm destroying all three tanks, the nearby office building, and three lorries parked nearby. There were no injuries. The cause was determined to be self-ignition in a charcoal filter being used to filter the vented tank vapors.

(Continued)

Year:	Location:	Incident description:
2008	Houston, TX.	Goodyear Heat Exchanger Rupture due to thermal expansion. During a maintenance operation on a heat exchanger, one worker was killed and approximately six others were injured. At the time of the failure, the Operator was attempting to steam through the tube side with the shell side full of ammonia and isolated from the rupture disk (one Fatality, six injured)
2008	Chesapeake, VA.	Allied Terminals, Inc., A 2-million-gallon liquid fertilizer tank, catastrophically failed, resulting in loss of containment (liquid fertilizer) and seriously injuring two workers. The adjoining neighborhood was evacuated due to the fertilizer spill. The CSB investigation is complete (zero Fatalities and two injuries).
2009	Tuscany, Italy	Train derailment and multiple Propane BLEVE's (32)
2009	Torrance, CA.	ExxonMobil Refinery – Operator critically burned from hot water while attempting to open a coke drum on a Delayed Coker unit (1).
2009	Texas City, TX.	Valero Refinery, A boiler exploded while lighting pilots after the boiler tripped (one Fatality, one injured) (1).
2009	Catano, Puerto Rico (Baymon)	Caribbean Refining Company; Gasoline tank overfilled while being filled from a ship. The resulting VCE and ignition of other nearby tanks – 17 tanks involved. The massive fire and explosion sent huge flames and smoke plumes into the air, and the resulting pressure wave damaged surrounding buildings and impacted moving vehicles (0). The CSB investigation is complete.
2009	Martinez, CA.	Shell Refinery – A plant worker died when he fell into a water tank. He drowned when he was unable to escape (1).
2009	Corpus Christi, TX.	Citgo, Alkylation Unit fire and Hydrofluoric Acid release (CSB investigation complete) (0)
2009	Jaipur, India	IOC Product Depot; Tank overfilled, explosion and fire. 12 Tanks reported involved with multiple fatalities (12) (unconfirmed)
2009	Garner, NC.	ConAgra Foods – An explosion fatally injured 4 workers and injured dozens of others. Three workers were crushed to death when a large section of the building collapsed. The explosion critically burned 4 others and sent a total of 71 people to the hospital. This incident occurred when a new natural gas line was being purged into an indoor room where several non-Div II electrical devices were present. The CSB investigation is complete including recommendations to prevent a reoccurrence. (4 Fatalities, 71 Injured).
2009	Garner, AR.	Teppco Gasoline Terminal: Gasoline tank explosion during maintenance (3)
2009	Woods Cross, UT.	Silver Eagle Refinery – This refinery experienced two significant Process Safety incidents about 10 months apart. In January 2009, two refinery operators and two contractors suffered serious burns resulting from a flash fire at the Silver Eagle Refinery in Woods Cross, Utah. The accident occurred when a large flammable vapor cloud was released from an atmospheric storage tank containing an estimated 440,000 gallons of light naphtha. The vapor cloud found an ignition source, and the flash fire spread up to 230 feet west of the tank farm. During November 2009, a second accident occurred on the Mobil Distillate Dewaxing Unit when the 10′ Reactor outlet pipe, which was severely corroded, failed catastrophically, resulting in an explosion and a powerful blast wave. Nearby homes were damaged by the overpressure.
2009	South Philadelphia, PA.	Sonoco Refinery – A release of deadly hydrogen fluoride (HF) was released from the refinery causing 13 workers to be sent to the hospital. OSHA reported that the probable cause was partly due to a design change a few years earlier. The heat exchanger tubes had been replaced with carbon steel, replacing the much more expensive nickel-alloy tubes.
2009	Billings, MT.	Conoco Phillips Refinery – A fire occurred in a storage tank containing heavy fuel/pitch, a feedstock for the Coker unit. Reports indicate that the fire occurred when the tank level fell below the internal heating coils causing the contents to be overheated. No injuries were reported.
2010	Belle, WV.	DuPont Chemical, Release of Phosgene (CSB investigation complete) (1)
2010	Chalmette, LA.	Chalmette Refining – H_2S Fatality – Contractor died due to H_2S exposure as he was leaving the worksite (1).
2010	Artesia, NM.	HollyFrontier Refinery – Vapors ignited from vapors causing explosion (two Fatalities, two injured)

Year:	Location:	Incident description:
2010	Buffalo, NY.	Dupont Company, Fatal Hot Work Incident (one Fatality, one serious injury)
2010	Anacortes, WA.	Tesoro Refinery – An exchanger in the Reformer Unit failed catastrophically during a period as the Operators were commissioning a parallel bank. The exchanger explosion was the result of HTHA. (seven Fatalities).
2010	Russia	Zabaikalsky Refining Company – Refinery explosion during open flame work in refinery pumping station (5)
2010	Buffalo, NY.	E.I. duPont de Nemours and Co. Inc., Yerkes chemical plant – A contract welder was killed and a foreman injured in a hot work explosion while repairing the agitator support atop an atmospheric storage tank containing flammable vinyl fluoride. The explosion blew most of the top from the tank. The CSB investigated and issued a report and a hot work training video (one Fatality, one injured).
2010	Gulf of Mexico	BP Deepwater Horizon Macondo Oil Rig Blowout Explosion and Fire. A sudden explosion and fire occurred on the oil rig resulting in the deaths of 11 workers and causing a massive, ongoing oil spill into the Gulf of Mexico. The rig was located approximately 50 miles southeast of Venice, Louisiana, and had a 126-member crew onboard. During temporary well-abandonment activities on the drilling rig, control of the well was lost, resulting in a blowout—the uncontrolled release of oil and gas (hydrocarbons) from the well resulting in the explosion, fire, and a massive oil spill (11 Fatalities, 17 Injured).
2010	Middletown, CT.	Kleen Energy Power Plant construction site – natural gas explosion (6)
2010	CSB Issues Safety Bulletin on Hot Work on or Near Storage Tanks	In February of 2010, the CSB issued a Safety Bulleting titled "Seven Key Lessons to Prevent Worker Deaths During Hot Work in and Around Tanks – Effective Hazard Assessment and Use of Combustible Gas Monitoring Will Save Lives." This follows several tragic incidents involving hot work on or near hydrocarbon storage tanks and is aimed at reducing the reoccurrence of such events.
2010	Fort McKay, AB Canada	Canadian Natural Resources Limited (CNRL) Refinery – A major fire on the top deck of a Delayed Coker occurred when the Operators inadvertently opened the top head of an in-service coke drum. Significant damage to coke structure. No fatalities but five personnel injuries, one with 3rd degree burns. The Operators had bypassed the interlock in order to open the top head. (zero Fatalities, five Injured, one critically).
2010	Gulf coast (offshore Louisiana)	Mariner Energy – The Vermillion Oil Rig 380 exploded while working off the Louisiana coast only about 200 miles from the Deepwater Horizon where 11 workers were killed. In this case, all 13 crew members were safely rescued by a supply ship. None were seriously injured.
2011	Pembroke, Wales	Chevron tank explosion, which occurred during tank cleaning operation (4)
2011	Kuwait	Ahmadi refinery – Four workers were killed, and two others injured Saturday in a gas unit blast at an oil refinery in Kuwait, as the Kuwait National Petroleum Corporation (KNPC) reported. The explosion was caused by a gas leak during routine maintenance work at the KNPC Oil refinery. n (four Fatalities, two Injured)
2011	Fukushima, Japan	The Fukushima nuclear accident in Japan is considered the largest nuclear disaster since the Chernobyl incident in 1986. The incident resulted from loss of electrical power caused by the Japanese earthquake and tsunami. 154,000 people were evacuated from the surrounding area.
2011	Chiba, Japan	Cosmo Refinery – Earthquake and multiple LPG tank farm sphere BLEVE's and Fire. Numerous spheres were destroyed, some catapulted into the Sea of Japan (0)
2011	Sendai, Japan	Nippon Oil & Energy, Sendai Refinery – The refinery was hit with the largest earthquake in the country's history, followed by a devasting tsunami. The large refinery fire began from product shipping facility. All plant workers had been evacuated and no ability to extinguish the fire (no electricity or firewater available). The off-site areas and the FCC were damaged in the resulting fire.
2011	Singapore, Singapore	Shell Refinery, Pulau Bukom Refinery – A major fire occurred at the gasoline blend header (referred to as the pumphouse). The blend header was being prepared for maintenance and the product was being drained from the piping systems. Reportedly, the product was being removed from two different locations, one with a vacuum truck and by draining into a plastic drum and one with a portable pump. The cause was determined to be ignition of vapors by static electricity. No injuries were reported although the refinery units were shut down during the fire. All units were returned to service a few weeks later (0).

(Continued)

Year:	Location:	Incident description:
2011	Buncefield, England	Gasoline tank overfilled while being filled from the pipeline -The terminal was destroyed, and there are no plans to rebuild. The initial fire involved 20 petroleum storage tanks. (0)
2011	Norco, LA.	Valero Energy – H_2S Fatality – A Koch Specialty contractor was working on a fractionation column and was overcome by H_2S. He reportedly fell about 100 feet. (1)
2011	Gallatin, TN.	Hoeganaes Corporation, Fatal Flash Fires – Three combustible dust incidents over a six-month period occurred at the Hoeganaes facility in Gallatin, TN, resulting in fatal injuries to five workers. The facility produces powdered iron and is about twenty miles outside Nashville. (five Fatalities)
2011	Mont Belvieu, TX.	Enterprise Products Company – A loss of containment resulted in an explosion and large fire at the Enterprise Products Pipeline, leaving one worker missing. His body was found the following day after the fire was extinguished, and the area was safe to enter. Enterprise handles natural gas and natural gas liquids (LPG) in pipelines and underground caverns. (one Fatality, zero Injuries).
2011	Louisville, KY.	Carbide Industries facility, Two workers were killed and two others injured as a result of a fire and explosion that occurred at the facility that produces calcium carbide products (two Fatalities, two Injured)
2011	Tula, North Mexico	Pemex Refinery – An explosion and fire occurred while the company was running a trial of a visbreaker, a Pemex spokesman said. The Visbreaker recently underwent maintenance work, and Pemex workers were testing it on Saturday to restart the unit. (two Fatalities, one serious injury)
2012	Memphis, TN.	Valero Energy Corp. – The explosion and resulting fire occurred while maintenance work was performed on a refinery flare system. (two Fatalities, one critical injury)
2021	Qatar Offshore LNG Facility	Condensate spill and fire while attempting to replace offshore condensate loading hoses. Fire onboard and around tugboat involved in this operation. Seven fatalities and five others injured (seven Fatalities).
2012	Memphis, TN.	Valero Energy Corp – According to Memphis Fire Department, two workers and two firefighters were taken to the MED's burn unit after a sight glass ruptured. Cook said the victims were exposed to a mixture of propane and hydrofluoric acid. The two firefighters were transported in non-critical condition. One worker died due to exposure to Hydrofluoric acid. (one Fatality, three Injured)
2012	Cherry Point, WA	BP Refinery – A loss of containment occurred at a section of dead-leg piping resulting in a significant fire (0 Fatalities, one Injured).
2012	Egypt	An explosion occurred at this Egyptian refinery during a maintenance operation. (5 Fatalities, 22 Injuries)
2012	Map Ta Phut,Thailand	Synthetic rubber plant owned by Bangkok Synthetics – A large explosion occurred at this plant which resulted in the deaths of 12 people and injured more than 140 others. The plant was shut down for turnaround when the initial explosion occurred in a benzene tank that had been cleaned (12 Fatalities, 140 others injured).
2012	Malaysia	Bunga Alpinia Tanker terminal explosion – A Malaysian tanker exploded as it was being loaded with methanol at the PETRONAS Chemicals Methanol Sdn Bhd terminal in Labuan at the time of the incident. Five crew members onboard the ship were killed in the ensuing fire. The incident occurred during heavy rain with lightning in the area. The vessel was reported as 80% destroyed (5).
2012	Punto Fijo, Venezuela	Amuay Refinery/Falcón State Refinery/Paraguaná Refinery Complex – A propane leak and catastrophic Explosion and fire (48 Fatalities, Approx 200 Injured)
2012	Bangkok, Thailand	Bangchak Petroleum – Apparently a loss of containment resulted in a massive fire, sending a thick column of smoke into the air that could be seen across the Thai capital. The large fire did not cause any injuries; although, the government closed the 120,000 barrels-a-day refinery for at least 30 days. There were no reported injuries in the blast and subsequent fire, which prompted the government to close the 120,000 barrels-a-day refinery for at least 30 days.

Year:	Location:	Incident description:
2012	Richmond, CA.	Chevron Refinery piping failure during maintenance attempting to install a clamp on the crude unit atmos four sidestream line while in service. 15,000 residents reported to the hospital. CSB investigation was complete, and a great CSB video issued. (zero Fatalities)
2012	Reynosa, Tamaulipas Mexico	Pemex gas plant explosion (33)
2012	Jubail, Saudi Arabia	Environment Development Limited Company (EDCO) – Explosion at the Industrial waste treatment plant (six Fatalities)
2012	Wynnewood, OK.	CVR Refining – Two workers were killed when a steam boiler exploded (2).
2013	Beaumont, TX.	ExxonMobil Refinery – A major fire occurred during maintenance hot work operations on a hydrotreater heat exchanger (2 Fatalities, 10 Injured).
2013	West, TX.	West Fertilizer Company Chemical Explosion and Fire. The violent detonation fatally injured 12 emergency responders and three members of the public. Local hospitals treated more than 260 injured victims, many of whom required hospital admission. The blast destroyed the WFC facility and caused widespread damage to over 150 off-site buildings. The WFC explosion is one of the most destructive incidents ever investigated by the US Chemical Safety and Hazard Investigation Board (CSB) as measured by the loss of life among emergency responders and civilians; the many injuries sustained by people both inside and outside the facility fence line; and the extensive damage to residences, schools, and other structures. Following the explosion, WFC filed for bankruptcy. The explosion happened at about 7:51 p.m. central daylight time (CDT), approximately 20 minutes after the first signs of a fire were reported to the local 911 emergency response dispatch center. Several local volunteer fire departments responded to the facility, which had a stockpile of 40–60 tons (80,000–120,000 pounds) fertilizer-grade ammonium nitrate (FGAN), not counting additional FGAN not yet offloaded from a railcar. (15 Fatalities, more than 260 Injured).
2013	Detroit, MI.	Marathon Refinery – The fire at the refinery resulted in the evacuation of 3,000 residents.
2013	Lac-Megantic, Quebec, Canada	A railroad derailment resulted in a devastating fire and 47 deaths in the community. About thirty buildings were destroyed in the blaze; 36 of the remaining 39 buildings had to be destroyed due to contamination. This was one of the worst rail accidents in Canadian history. The investigation determined that the train was left unattended on a 1.2% grade without enough brakes set. (47 fatalities)
2013	Mishazi, China	Jilin Baoyuanfeng poultry slaughterhouse – An ammonia release and fire resulted in 120 Fatalities and at least 60 injuries. The reason for the high number of fatalities is the building was locked, and the worker's escape was blocked. (120 Fatalities, 60 Injured).
2013	Lamezia Terme, Italy	An explosion at an oil storage occurred while undergoing maintenance and claimed three people's lives. The blast occurred at a facility near Lamezia Terme owned by fuel producer Ilsap Biopro, but the reason behind it has not yet been disclosed. Two men died immediately at the scene, while a third, who suffered 90% burns, passed a day later.
2013	Mexico City, Mexico	Pemex Oil Headquarters office explosion/fire (37)
2013	Geismar, LA.	Williams Olefins – Explosion (BLEVE) and fire in a propylene fractionator reboiler due to thermal expansion of trapped liquids. The incident occurred during nonroutine operational activities that introduced heat to a heat exchanger called a "reboiler," which was offline, creating an overpressure event while the vessel was isolated from its pressure-relief device. The introduced heat increased the temperature of the liquid propane mixture confined within the reboiler shell, resulting in a dramatic pressure rise within the vessel due to liquid thermal expansion. The reboiler shell catastrophically ruptured, causing a boiling liquid expanding vapor explosion (BLEVE) and fire (2 Fatalities, 167 reported injuries).
2013	Visakh, India	HPCL's Visakh Refinery – cooling tower explosion and fire. Occurred due to how work nearby. (27 Fatalities).
2013	Pascagoula, MS.	Chevron Refinery furnace explosion during relight following a furnace trip. OSHA reportedly cited Chevron for failing to provide adequate procedures and training for furnace purging and relighting following a trip (1).
2013	Antwerp, Belgium	Total Refinery; Valve bonnet bolts failed (stress corrosion) (SCC) on 1,000psi steam system during on-stream repairs (2)

(Continued)

Year:	Location:	Incident description:
2013	Ellesmere Port, Cheshire, England	Shell Stanlow Refinery – This major fire is believed to have started in a furnace. Firemen fought the fire for about six hours and local communities were issued a warning due to the heavy smoke from the fire.
2013	Ardmore, OK.	Valero Energy, Confined Space (1) Reported to be health-related.
2013	Port Arthur, TX.	Chevron Phillips Chemical – A fire at this site, which occurred during a period when the unit was undergoing a planned turnaround, resulted in 8 injured. (zero Fatalities, eight Injured)
2013	Jeddah, Saudi Arabia	Jeddah Refinery – Exposure to H_2S resulted in one fatality. (one Fatality, zero Injured).
2013	Shanghai, China	According to local authorities, a liquid ammonia leak from a refrigeration unit at a cold storage facility killed 15 people and injured 26 others. (15 Fatalities, 26 Injured).
2014	Borger, TX.	Chevron Phillips Refinery – A refinery fire resulted in injuries to six workers, of which two were airlifted to the burn unit at Galveston, Texas. (6 injuries, 2 seriously).
2014	Cumberland, WV.	AL Solutions – A Fatal Combustible Dust Explosion. An explosion ripped through the New Cumberland A.L. Solutions titanium plant in West Virginia on December 9, 2010, fatally injuring three workers. The workers were processing titanium powder, which is highly flammable, at the time of the explosion. (3 Fatalities, 1 Injured)
2014	Martinez, CA.	Tesoro Refinery – A Sulfuric Acid Spill burned two workers in the refinery's alkylation unit, who were transported to the nearest hospital burn unit by life flight. The incident occurred when the operators opened a block valve to return an acid sampling system back to service. Shortly after this block valve had been fully opened, the tubing directly downstream of the valve came apart, spraying two operators with acid (zero Fatalities, two Injured).
2014	Rugao, Jiangsu, China	Shuangma Chemical plant fire (5)
2014	Achinsk, Russia	Rosneft's Refinery explosion in a distillation column. No other details available – two engineers sentenced to prison as a result of this incident. (8 Fatalities, 30 Injured).
2014	La Porte, TX.	DuPont Chemical Plant – A release of approximately 24,000 pounds of highly toxic Methyl Mercaptan from an insecticide production unit killed three Operators and a Shift Supervisor. The release occurred into a manufacturing building, where they died from a combination of asphyxia and acute inhalation exposure to the chemical. The CSB concluded that the release occurred due to flawed engineering design and the lack of adequate safeguards. Contributing to the severity of the incident were numerous safety management system deficiencies (four Fatalities)
2015	Gibbstown, NJ.	PBF Energy – Malfunctioning air compressor (1)
2015	Torrance, CA.	ExxonMobil Refinery, Explosion in FCC Electrical Precipitator due to FCC Gas carryover from Fractionator/Reactor to Regenerator (0)
2015	Tiamin, China	Port District Warehouse explosion – incompatible mixed chemicals (120 Fatalities, 700 Injured)
2015	Convent, LA.	Motiva Refinery – Major fire on H-Oil Unit (0)
2015	South Philadelphia, PA.	Philadelphia Energy Solutions, Crude Distillation Unit fire (0)
2015	Bay of Campeche, Mexico	Pemex Oil Platform (Abkatun-Permanente platform) – A fire onboard the platform killed four workers. 300 workers were evacuated from the platform (4). The news article highlighted that the Pemex company has had several deadly accidents in recent years, which included an explosion at their company headquarters that killed 37 people.
2015	Off the Brazilian coast	An oil production and storage ship, the FPSO Cidade de Sao Mateus, owned by the Norwegian firm BW Offshore, and leased by Petrobras exploded off of the Brazilian coast. The incident killed 5 workers and injured more than 25 others.
2015	Lima, OH.	Husky Refinery Explosion – A fire that started after an explosion on Saturday at Husky Energy Inc's 155,000-barrel-per-day (bpd) crude oil refinery in Lima, Ohio, has caused extensive damage to a unit at the plant, a fire official said Sunday. According to local media reports, no injuries were reported in the blast, which was heard across the city and shattered nearby windows. Husky said all personnel were accounted for at the refinery. Originally reported as a Crude unit explosion and fire.

Year:	Location:	Incident description:
2015	Gunashli oilfield (Caspian Sea, Baku, Azerbaijan)	A storm damaged a high-pressure subsea gas pipeline resulting in a major fire. The fire quickly spread to adjacent oil wells. Ten of the sixty people at the site were killed in the incident, and another twenty were reported to be missing. Nine were hospitalized for their injuries (10 killed, 20 missing, 9 hospitalized).
2015	Delaware City, DE.	Delaware City Refining Co., An operator on the Kellogg Alkylation Unit, suffered second-degree burns to his face and neck while performing de-inventorying activities on a vessel in preparation for removing a pipe spool from a connected process. This incident follows two other incidents at the same facility, which occurred on August 21 and August 28, 2015. The CSB issued a Safety Bulletin, "Key Lessons for Preventing Incidents When Preparing Process Equipment for Maintenance."
2015	Dalton, GA.	MFG Chemical Company – One worker died due to a chemical release when a reactor was overpressurized due to an uncontrolled chemical reaction. A similar incident occurred at the same site in 2004, where 154 people reported to hospitals, and five were hospitalized. (one Fatality, one Injured). OSHA cites MFG Chemical Inc. for repeated safety hazards after 2 workers were injured. "An MFG Chemical Inc. worker died after hazardous chemical vapors released from an overpressurized reactor burned his respiratory system. A second employee was treated at a hospital and released. A July 2014 inspection by the US Department of Labor's Occupational Safety and Health Administration resulted in citations for MFG for 17 safety and health violations. OSHA initiated the inspection after a media referral alleged that a chemical release at the manufacturing facility had occurred." "MFG continues to violate OSHA standards, exposing workers to serious hazards associated with process safety management," said Christi Griffin, director of OSHA's Atlanta-West Area Office. "Allowing repeated violations demonstrates the company's lack of commitment to worker safety and health." MFG was inspected by OSHA previously in 2012 and received 19 serious citations related to process safety management standards.
2015	Brazil	Petrobras Refinery – Three workers were severely burned due to an explosion while working inside a process vessel. (Fatalities 0 – 3 severely injured)
2015	Gulf of Mexico	Pemex Oil Platform explosion and fire (0)
2016	Bayport, TX.	PeroxyChem – Chemical cleaning tank exploded (1)
2016	Nederland, TX.	Sunoco Logistics Partners – This terminal facility experienced a flash fire during hot work, injuring seven workers, three critically injured. The fire and explosion caused overpressure within the pipe segment, causing CARBER isolation tools that had been installed as part of the project and residual crude oil in the pipe to be ejected from the ends of the piping, resulting in impact and burn injuries to the seven workers (zero Fatalities, seven Injured).
2016	Ludwigshafen am Rhein, Germany	BASF – Two men of the plant fire brigade and a sailor were killed during a maintenance operation. According to media reports, investigators believe that an employee of an outside company triggered the explosion when he cut into the wrong pipeline with a cutting disc. (3 Fatalities, 8 Seriously injured, 22 others were also injured).
2016	Texas City, TX.	Marathon Refinery, Fire during maintenance turnaround (0)
2016	Coatzacoalcos, Mexico	Pemex Refinery explosion and fire (3 Fatalities, 136 Injured).
2016	Veracruz, Mexico	Pemex Veracruz Refinery Explosion & Fire (32) (Petroquimica Mexicana de Vinilo facility near the city of Coatzacoalcos)
2016	Moss Point, MS.	Enterprise Pascagoula Gas Plant Explosion and Fire (0)
2016	Geismar, LA.	Williams Olefins, Exchanger explosion due to overpressure from thermal expansion. The CSB reported the incident occurred during nonroutine operational activities that introduced heat to a type of heat exchanger called a "reboiler," which was offline, creating an overpressure event while the vessel was isolated from its pressure-relief device. The introduced heat increased the temperature of the liquid propane mixture confined within the reboiler shell, resulting in a dramatic pressure rise within the vessel due to liquid thermal expansion. The reboiler shell catastrophically ruptured, causing a boiling liquid expanding vapor explosion (BLEVE) and fire. (2 Fatalities, 167 Injured).
2016	Gazipur, Bangladesh	Multifabs, LTD – Boiler Explosion (23)

(*Continued*)

Year:	Location:	Incident description:
2016	Baton Rouge, LA.	ExxonMobil Isobutane release and Fire while preparing equipment for maintenance seriously injured four workers in the sulfuric acid alkylation unit. During the removal of an inoperable gearbox on a plug valve, the operator performing this activity removed critical bolts securing the pressure-retaining component of the valve known as the top cap. When the operator then attempted to open the plug valve with a pipe wrench, the valve came apart and released isobutane into the unit, forming a flammable vapor cloud. The isobutane reached an ignition source within 30 seconds of the release, causing a fire and severely burning four workers who were unable to exit the vapor cloud before it ignited (zero Fatalities, four serious burn injuries).
2016	Bay of Campeche, Mexico	Pemex Oil Platform (Abkatun-Permanente platform), Fire (3) (see also fire on same platform during 2015)
2016	Convent, LA.	Motiva Refinery, Loss of containment and major fire on the H-Oil Unit (0)
2016	Haifa, Israel	Carmel Olefins Refinery, Major Gasoline Release, and Gasoline Tank Fire (0)
2016	Cantonment, FL.	AirGas Inc., A nitrous oxide trailer truck exploded at the manufacturing facility, killing the only Airgas employee present and heavily damaging the facility. The CSB investigated and concluded this was a thermal decomposition of nitrous oxide-initiated explosion.
2016	Pascagoula, MS.	Enterprise Products Pascagoula Gas Plant – A major loss of containment (LOC) resulted in the release of methane, ethane, propane, and several other hydrocarbons, which ignited, initiating a series of fires and explosions, which ultimately shut down the site for almost six months. The CSB found that the probable cause was the failure of a brazed aluminum heat exchanger (BAHX) due to thermal fatigue. The CSB report provided recommendations to help ensure equipment integrity. There were no injuries, although primarily due to the incident in the late evening when few people were present.
2017	St Louis, MO.	Loy Lange Box Co. – Explosion of a steam pressure vessel due to corrosion (4)
2017	Delaware City, DE.	Delaware City Refining Company, Flash Fire while draining equipment for maintenance (0)
2017	Salina Cruz, Mexico	Pemex Refinery, Crude spill, major fire (during tropical storm & heavy rains) (1)
2017	Calabar, Nigeria	Linc Oil Terminal – Gasoline Explosion & Fire (11)
2017	Shippingport, PA.	First Energy Power Plant – Confined space entry results in 5 H_2S exposures (2)
2017	Kenner, LA.	Clovelly Oil Well Platform (Lake Pontchartrain) – Explosion and Fire (seven burn injuries – one missing)
2017	Ferndale, WA.	Phillips Refinery – The refinery experienced a leak of hydrofluoric acid (HF) that sent seven workers to the hospital. The incident in question reportedly occurred on the Alkylation Unit when a contractor disconnected an enclosed rod-out tool from an open drain valve.
2017	Gestoci, Abidjan	Ivory Coast Refinery – A significant fire erupted on the Hydrocracker unit resulting in extensive damage to the main reactor. The reactor required replacement to return the Hydrocracker to full service. (zero Fatalities, no injuries were reported).
2017	Pernis, The Netherlands	Shell Refinery, Major electrical system failure results in furnace explosion and total refinery shutdown. Furnace explosion and fire during restart. (0)
2017	Abu Dhabi	Abu Dhabi National Oil Company – Electrical failure and refinery shutdown. Major Fire occurred in a recently completed $10B expansion (0)
2017	Crosby, TX.	Arkema Chemical Plant, Flooding from Hurricane Harvey disabled the refrigeration system at the facility, which manufactures organic peroxides. The following day people within a 1.5-mile radius were evacuated due to concerns about the spontaneous combustion of the peroxides. The trailers slowly increased in temperature until the peroxides spontaneously combusted. Later the remaining trailers were ignited by responders. The CSB investigation is complete.
2017	Paradis, LA.	Phillipa 66 – A pipeline exploded while six workers were cleaning the pipeline. Two workers were taken to the hospital with injuries, including a contract worker who was later flown to a burn unit in Baton Rouge. One other worker was missing for several days until his body was recovered from the site. The pipeline was still burning at the time the body was recovered (one Fatality, plus several others injured. Sixty homes were evacuated during the fire) (1).
2017	Fort McMurray, AB Canada	Syncrude Refinery – A large fire occurred on an interconnecting pipe between two hydrotreaters at this oil sands refinery. The fire took two days to extinguish and caused significant refinery damage. The cause was freezing water in the piping dead-leg resulting in pipe rupture. (zero Fatalities, one serious burn injury).

Year:	Location:	Incident description:
2017	Cambria, Wisconsin,	Didion Milling (Didion) facility – An explosion resulted in 5 worker deaths and an additional 14 workers' injured. CSB Investigation. CSB Investigation Update: Shortly before the explosion(s) at Didion, workers saw or smelled smoke on the first floor of one of the mill buildings. In trying to find its source, workers focused on a piece of equipment called a gap mill. While inspecting the equipment, workers witnessed a filter connected to an air intake line for the mill blow-off, resulting in corn dust filling the air and flames shooting from the air intake line, followed by one or more explosions. The CSB investigation is ongoing.
2018	Pittsburg, OK.	Pryor Trust Gas Well – A blowout and rig fire occurred at Pryor Trust 0718 gas well number 1H-9, located in Pittsburg County, Oklahoma. The fire killed five workers inside the driller's cabin on the rig floor. They died from thermal burn injuries and smoke and soot inhalation. The blowout occurred about three-and-a-half hours after removing the drill pipe ("tripping") from the well. The CSB concluded the cause of the blowout and rig fire was the failure of both the primary barrier—hydrostatic pressure produced by drilling mud—and the secondary barrier—human detection of influx and activation of the blowout preventer—which were intended to be in place to prevent a blowout. (five Fatalities)
2018	Bijnor, India	Mohit petrochemical factory – The explosion of a methane gas tank, apparently used for welding, while repairing a boiler resulted in six fatalities and two hospitalized. One other person is reported as missing. (six Fatalities, two Injured, one missing).
2018	Vohburg, Germany	Bayernoil Refinery complex, Major refinery fire results in evacuation of 1,800 people from their homes (0).
2018	Superior, WI.	Husky Refinery, Major FCC Reactor Explosion, and Fire quickly spread to other parts of the refinery. A major refinery fire occurred when debris from the FCC reactor shell punctured a hot asphalt tank. The cause was a reverse flow on the FCC Reactor (0 Fatalities, 36 sought medical attention, 11 sustained OSHA recordable injuries).
2018	Pasadena, TX.	Kuraray America, Inc. – This incident occurred during a chemical reactor system startup following a scheduled turnaround. During startup, a reactor high-pressure condition activated the emergency pressure-relief system, discharging flammable ethylene vapor through horizontally aimed piping into the area where contractors were working performing tasks not essential to the reactor startup, including welding. The resulting fire caused injuries to 23 workers (0 Fatalities, 23 Injuries).
2018	Belle Chasse, LA.	Phillips 66 Refinery – A refinery worker was killed when he accidentally fell into the refinery cooling tower. The refinery was temporarily shut down to facilitate the recovery. OSHA is investigating the accident (1).
2018	Cheshire, England	Stanlow Oil Refinery – Major fire resulted in the evacuation of the refinery.
2018	Ingolstadt, Germany	Bayernoil Refinery – Major refinery explosion and fire, ten employees were injured, 600 firefighters were called to the scene, and nearly 2,000 evacuated from the surrounding area.
2019	Deer Park, TX.	Intercontinental Terminal Company (ITC) Tank Terminal Fire (0) Multiple tanks burned. CSB Investigation Update: The fire originated in the vicinity of Tank 80-8, an 80,000-barrel aboveground atmospheric storage tank that held naphtha, a flammable liquid typically used as a feedstock or blend stock for the production of gasoline. ITC was unable to isolate or stop the release of naphtha product from the tank, and the fire continued to burn, intensify, and progressively involved additional tanks in the tank farm. The fire was extinguished on the morning of March 20, 2019. The CSB Investigation is ongoing. (zero Fatalities, zero Injuries)
2019	Crosby, TX.	KMCO, LLC – CSB Investigation Update: A vapor cloud of isobutylene formed at the KMCO facility after a 3-inch gray iron (a type of cast iron) y-strainer, a piping component, failed. The resulting vapor cloud found an ignition source and ignited, causing an explosion killing one worker and seriously burning two others. At least 30 others were injured in the explosion. A shelter-in-place was issued to community members within one mile of the facility. The CSB Investigation is continuing. (1 Fatality, 30 Injuries).
2019	Sarnia, Canada	Imperial Oil – Collapsed Fractionator due to internal pyrophoric fire during a planned turnaround (0)

(*Continued*)

Year:	Location:	Incident description:
2019	Jiangsu, China	Tianjiayi Chemical Company – A massive explosion and fire occurred in March 2019, for which the cause has not yet been determined. It was reported that there was a long-term practice of illegal storage of hazardous waste, resulting in spontaneous combustion and explosion. One worker reported that the initial explosion was from a fire in a natural gas tanker that spread to the benzene storage tank, although this is unconfirmed. The explosion resulted in 78 fatalities, at least 94 severely injured, 32 of whom were critically injured. Around 640 people required hospital treatment and were taken to 16 hospitals. The blast was believed to have caused a tremor equivalent to a magnitude-2.2 earthquake. Thousands were evacuated from surrounding areas. (78 Fatalities, 94 seriously injured), and a total of 640 requiring hospital treatment.
2019	Philadelphia, PA.	Philadelphia Energy Solutions Refinery, Massive explosion – refinery announced the permanent refinery closing following the incident. On the morning of the incident, a pipe elbow in the HF Alkylation Unit failed, releasing primarily light hydrocarbons with about 2.5% HF which quickly ignited, resulting in a series of catastrophic explosions and a major fire. The control room operator then activated the Rapid Acid De-inventory (RAD) system, a safety system designed to quickly route HF to a separate drum in the event of a loss of containment incident. The CSB made several recommendations in their report to help prevent a reoccurrence and address the HF concerns (zero Fatalities, six injured).
2019	Baytown, TX.	ExxonMobil Baytown Olefins Plant, Fire on Propylene Fractionator (0 Fatalities, 66 reportedly injured)
2019	Central KY.	Enbridge Gas Pipeline Explosion and Fire (0)
2019	Istanbul, Turkey	Chemical Plant Explosion & a massive fire – one storage tank was launched in the explosion. (0).
2019	Rouen, France	Lubrizol Chemical Plant Fire in a warehousing facility resulting in large amount of dense black smoke into the community (0)
2019	Port of Ulsan, South Korea	S/S Stolt Groenland Explosion (0), heated cargo tanks in nearby compartments likely caused dangerous warming of a temperature-sensitive cargo, causing the explosion.
2019	Rodeo, CA.	NuStar Energy Company, Two Ethanol Tanks burned and were destroyed (0)
2019	Port Neches, TX.	TPC Chemical Plant – After highly flammable butadiene was released from a pipe failure, a series of explosions occurred. The explosions caused a process column to propel through the air, landing inside the facility. Other columns collapsed, falling on other process equipment, causing extensive damage and fires that burned for over a month. The butadiene unit was destroyed. More than 60,000 Residents were Evacuated from the surrounding areas. The CSB determined the cause to be internal piping pressure due to the formation of "popcorn polymer." The materials "exponentially" expanded until the pressure was too great, and the pipe ruptured. (zero Fatalities, three Injured).
2019	Yima, China	Henan Coal Gas – A massive explosion rocked a gas plant in central China with a death toll that has risen to 15. 15 others have been seriously injured, according to state media. There are many others reported to be "lightly injured." The blast shattered windows and doors of buildings in a 2-mile radius (3-kilometer radius). According to local media, the explosion occurred in the air separation unit of Henan Coal's factory facility. (15 fatalities, 15 seriously injured).
2019	Waukegan, IL.	AB Specialty Silicones, A massive explosion, and fire occurred at the facility, killing four workers and causing extensive damage to nearby businesses (4). CSB investigation is complete
2019	Carson, CA.	Phillips 66 Refinery – Media reports two separate fires at this refinery about 47 days apart. The first was reported on 15 March as a massive fire involving three pumps, the second on 2 May with smoke billowing from the facility. No injuries were reported in either incident.
2019	Odessa, TX.	Aghorn Operating Inc., Pumper A at the Waterflood station was fatally injured from his exposure to the released H_2S. Subsequently, the spouse of Pumper A gained access to the waterflood station and searched for Pumper A, was also overcome by H_2S, and died from the exposure. The CSB investigation is complete, and CSB released a training video on this incident.
2020	Meaux, LA.	Valero Energy Refinery – An explosion with unknown causes damaged the hydrocracking unit and injured one worker, who was quickly transported to the hospital. (zero Fatalities, one Injured).

Year:	Location:	Incident description:
2020	Puertollano, Spain	Repsol Refinery – Reports indicate that a large fire has occurred at the refinery. The fire occurred after a crude oil tank exploded following a lightning strike. A large plume of heavy black smoke can be seen miles away from the refinery.
2020	Baton Rouge, LA.	ExxonMobil Refinery – A large fire occurred in a pipe rack, and most of the refinery and chemical plants shut down. The cause was determined to be pyrophoric material in a crude unit rundown line (0).
2020	Carson, CA.	Marathon Refinery, a Major fire, initially reported as an explosion, later identified as a cooling tower fire (0).
2020	Cape Town, South Africa	Astron Energy Refinery – A Reformer furnace exploded, resulting in a large fire. The Milnerton plant, restarting after undergoing extended maintenance, was shut down after the fire. The explosion resulted in two fatalities; seven other people were injured, two of whom remain in hospital where they are receiving treatment for their injuries. (two Fatalities, seven injured).
2020	Durban, South Africa	Engen Refinery (Majority owned by Malaysia's Petronas) – Seven people were injured after an explosion rocked South Africa's second-largest crude oil refinery in Durban. (zero fatalities, seven Injured).
2020	Johor, Malaysia	Malaysia Petronas and Saudi Aramco Diesel Hydrotreater Explosion and Fire. This was the second fire in less than a year at the $27 billion Pengerang Integrated Complex (PIC) in Malaysia's southern state of Johor. (five Fatalities, one Injured).
2020	Cairo, Egypt	Crude Oil Pipeline release and major fire (0 Fatalities, 17 injured)
2020	Gujarat, India	Yashashvi Rasayan Chemicals – A massive blast in a storage tank and resulting fire at the Yashashvi Rasayan Private chemical plant in Dahej Industrial Estate killed 10 workers. Six workers were killed on the spot, and another four later succumbed to burn injuries after they were hospitalized. 77 other injured workers were sent to nearby hospitals. (10 Fatalities, 77 injured)
2020	Beirut, Lebanon	A warehouse storing ammonium nitrate exploded – at least (35 Fatalities and at least 5,000 injured) (many left homeless).
2020	Pasadena, TX.	Chevron Refinery. A fire at the Chevron Pasadena Refinery sent thick smoke billowing into the air (0).
2020	Canton, NC.	Evergreen Packaging Mill – Two contract workers were fatally injured when a fire occurred during a scheduled maintenance event on a processing unit. Two different companies were working on interconnected equipment, one using flammable material to apply fiberglass resin and an electrical heat gun to help cure the resin. This resulted in an internal fire in the interconnected equipment. Two of the contractors safely escaped the confined space, and the other two were badly burned and died of their injuries (2). The CSB issued a report with additional recommendations.
2020	Belle, WV.	Optima Belle LLC – While dehydrating a chlorinated isocyanate compound inside a double cone dryer, the dryer exploded, killing one Belle employee and injuring two others. The explosion significantly damaged the dryer unit, and projectiles struck a methanol pipeline at a co-located facility, causing a fire. Another large fragment landed approximately 1,000 feet off-site on a major highway damaging two vehicles and injuring one of the drivers. The CSB Investigation is ongoing. (one Fatality, three Injuries, two seriously injured from a fall while attempting to escape).
2020	Seosan, South Korea	Lotte Chemical Co., Naphtha cracker fire injured at least 56 people.
2020	Charleston, TN.	Wacker Polysilicon North America LLC – CSB Investigation Update: A graphite heat exchanger cracked, releasing hydrogen chloride (HCl) during maintenance activities. The release caused chemical burns to one contract worker. Another contract worker was injured fatally, and two others were injured seriously when they fell from an elevated structure while attempting to escape the release. The CSB Investigation is ongoing. (one Fatality).

(Continued)

Year:	Location:	Incident description:
2020	Houston, TX.	Watson Grinding and Manufacturing Co – An explosion fatally injured two employees and injured two other employees, a third individual, a nearby resident, died a week later, reportedly from injuries caused by the explosion. The explosion also injured other residents and damaged hundreds of nearby structures, including homes and businesses. The explosion was fueled by propylene that had accidentally released and accumulated inside an enclosed workshop. Watson Grinding used propylene as fuel to apply protective coatings to metal parts. The CSB Investigation is ongoing. (three Fatalities, two Injured + several local residents)
2020	Burleson County, TX.	Chesapeake Energy, Wendland 1H Well Fatal Explosion – A gas well explosion fatally injured three contractors. The CSB Investigation is ongoing. (three Fatalities, one seriously injured)
2020	Lake Charles, LA	Hurricane Laura heavily damaged the roof of this facility, allowing rainwater to contact with trichloroisocyanuric acid (TCCA), initiating a chemical reaction and subsequent decomposition. The reaction released Chlorine vapors into the adjacent community. A portion of the nearby Interstate 10 was closed for over 28 hours, and the Calcasieu Parish Office of Homeland Security and Emergency Preparedness issued a shelter-in-place order due to the release of hazardous gases. This incident was investigated by the US Chemical Safety Board.
2021	Newell, West VA.	Ergon Refinery reported an explosion followed by a major fire. Homes were evacuated one mile in all directions (1).
2021	Southern Iran	A major fire was reported at the Kangan Petro Refining Co. (KPRC) located along the Persian Gulf in southern Iran. Mysterious fires sank Iran's largest warship and burned a big Tehran oil refinery — seemingly unconnected blazes.
2021	Livorno, Italy	Eni Refinery – According to emergency services, an explosion and fire at the refinery has been brought under control. The refinery was shut down for maintenance at the time. The energy company said that the incident occurred on one of the furnaces, which resulted in smoke but no damage to the plant or to the surrounding areas.
2021	Columbus, OH.	Yenkin-Majestic OPC Polymers resin plant – Explosion and fire killed one and injured eight others. CSB Update: During the addition of a solvent to a Kettle, a worker realized that the stirring mechanism had not been running for over an hour. When he restarted the agitator, the product inside the kettle began to vaporize quickly. This increased pressure inside the kettle, which continued to rise until a mixture of resin liquid and the flammable solvent vapor was released from the Kettle manway into the enclosed room where the kettle was located. The flammable vapor quickly spread throughout different areas of the building, including the furnace room, and ignited. The explosion also ignited other flammable material at the plant, resulting in a large fire. Two employees were rescued from collapsed parts of the plant, one employee was fatally injured, and eight others were transported to area hospitals for treatment. The CSB Investigation is continuing (one Fatality, eight Serious injuries).
2021	West Java, Indonesia	Balongan Oil Refinery – An "apocalyptic oil refinery explosion" occurred after a lightning strike at the Balongan oil refinery. 4 of the refinery's 72 tanks caught fire. This incident resulted in 6 injuries and 1,000 residents evacuated and was the third fire in recent years. Other incidents occurred in 2007 and 2019.
2021	St. Croix, US VI	Limetree Bay Oil Refinery – The refinery experienced a fire in May 2021 and was shut down by the US Environmental Protection Agency in May 2021 after a series of chemical releases into the environment sickened neighboring residents. An August 2022 fire within the petroleum coke conveyor loading system burned for two weeks, prompting the inspection by the EPA. The EPA regional administrator said that Inspectors found corrosion on process valves, flanges, pipes, nuts, bolts, and pressure-relief devices, adding that gaskets were in poor condition. The EPA said the owners also failed to provide hazard assessments and other documentation for the facility.
2021	Rockton, IL.	Chemtool Inc. plant near Rockton experienced a large explosion and massive fire. 80 Fire Units responding. An estimated 1,000 residents who lived within a 1-mile (1.6-kilometer) radius of the plant were asked to evacuate and anyone within 3 miles (4.8 kilometers) to wear masks due to the threat posed by airborne impurities.
2021	LaPorte, TX.	LyondellBasell – A release of 100,000 pounds of Acetic resulted in two deaths and 30 people being hospitalized for treatment (2 Fatalities, 30 injuries). The CSB Investigation is continuing.
2021	El Dorado, AR.	Lion Oil/Delek Refinery – A release and large unit fire occurred during exchanger cleaning – six were admitted to burn unit (zero Fatalities, six injuries).

Year:	Location:	Incident description:
2021	Gainesville, GA.	Foundation Food Group ("FFG") facility – Liquid nitrogen was released from a freezer located in Plant Four, resulting in the fatal injuries of six employees and the serious injury of three employees and one emergency responder. The liquid nitrogen quickly vaporized, expanded, and accumulated inside a partially enclosed room within a building. Nitrogen, including liquid Nitrogen, is a colorless, odorless, non-flammable, non-toxic gas abundant in the air. High concentrations of nitrogen gas in an enclosed area can create an oxygen-deficient atmosphere. Atmospheres containing less than 19.5 percent oxygen can lead to asphyxia (low oxygen), brain damage, and death. The CSB Investigation is ongoing. (six Fatalities, four Seriously Injured).
2021	Baytown, TX.	ExxonMobil Refinery – Firefighters extinguished a large fire at a Houston-area oil refinery that broke out early Thursday, injuring four people. A large explosion was reported at an Exxon/Mobile refinery in Baytown, Texas, after what local sheriff's deputies called "a major industrial accident." Four people were injured, but everyone else on-site has been accounted for. Harris County Sheriff Ed Gonzalez said that three of the injured were taken to hospitals by helicopter while the fourth was taken by ambulance. (zero Fatalities, four Injured)
2021	Ahmada, Kuwait	Mina al-Ahmadi Refinery – A fire broke out in the atmospheric residue desulphurization (ARDS) unit resulting in several minor injuries and smoke inhalation cases among workers, according to the. According to local reports and social media posts, an explosion was heard in the area. The fire was then brought under full control, the national company announced later in the day. The cause was not reported. (zero Fatalities, "several"? Injured)
2021	Tehran, Iran	State-owned Tondgooyan Petrochemical Co. – A major fire occurred at the site caused by a leak in two waste tanks at the facility. Authorities initially suggested the flames affected the refinery's liquified petroleum gas pipeline. (0 Fatalities, At least 11 people were injured).
2021	Oregon, OH.	Toledo Refining Company – A fire occurred on a gasoline processing unit at the 170,000 bpd refinery. The company released a statement noting that "there have been no injuries, and everyone is safe and accounted for at the site. We have notified appropriate public officials and regulatory agency representatives and at this time, are unaware of any community impact." Two processing units were shut down, but the rest of the refinery was operated as normal Tuesday.
2021	Lima, OH	Cenovus Refinery – Four people were injured in a fire at the refinery this morning.
2022	Westlake, LA.	Westlake Chemical South Plant – The site experienced an explosion of an Ethylene Dichloride storage tank, used during the production of plastics and types of vinyl. Due to the nature of the chemical (a known EPA-listed carcinogen), a shelter-in-place was ordered for local businesses and residents. Six plant workers were injured in this event, five of whom were transported to the hospital for treatment. (one Fatalities, six Injured).
2022	Ingleside, TX.	Oxychem Chemical Plant – An equipment failure resulted in a significant chlorine release and fire. Media reports indicated that the failure was related to a valve rupture when the equipment was placed in service. The chemical release required Local businesses and residents to shelter-in-place.
2022	Jinshan District, Shanghai, China	Sinopec Shanghai Petrochemical Company – A massive fire occurred at the Sinopec facilities on 18 June 2022. One transport driver was killed, and one employee was injured. Reportedly there was a widespread shutdown of the Sinopec facilities as a result of the fire. (one Fatality, one injured).
2022	Benicia, CA.	Valero Refinery – Worker died while inspecting a confined space, determined to be asphyxiation by Argon (1).
2022	Billings, MT.	ExxonMobil Refinery – A large explosion rocked an ExxonMobil oil refinery in Montana on Saturday night, sparking a massive blaze at the site. The following day the local media reported that the crude distillation unit (CDU) was reported to be shut down due to the fire. (zero Fatalities, zero Injured).
2022	Oregon, OH.	BP-Husky refinery – An accidental release of flammable chemicals ignited, creating a fire that fatally injured two workers (two brothers) and resulted in substantial property damages at the refinery. The CSB and OSHA investigations are ongoing and focused on the Fuel Mix Drum for the Crude 1 Unit. (two Fatalities, zero Injured).
2022	Dalinpu, Taiwan	Chinese State-Owned Oil Refinery – Major refinery fire on an HDS Unit. Residents are concerned and demonstrating due to high air and soil pollution levels. Local officials have "promised" the relocation of 20,000 local residents.

(Continued)

Year:	Location:	Incident description:
2022	Corpus Christi, TX.	Valero Energy Company – A fire occurred at the Valero East facility near Corpus Christi, and a loud boom was heard around 6:00 a.m. The fire was contained on the property, and outside assistance was stated as not needed. (zero Fatalities, zero Injured).
2022	Abuja (Owerri), Nigeria	More than 100 people were killed and "many injured, some catastrophic" in an explosion at an illegal oil refining depot in Nigeria's Rivers (Southern Nigeria). Women and children were reported to be among the victims. Over fifty unidentified and unclaimed bodies were buried in three mass graves, all burned beyond recognition. The fire outbreak occurred at an illegal bunkering site. Unemployment and poverty in the oil-producing Niger Delta have made illegal crude refining an attractive business but with deadly consequences. Crude oil is tapped from a web of pipelines owned by major oil companies and refined into products in makeshift tanks. The hazardous process has led to many fatal accidents and has polluted a region already blighted by oil spills in farmland, creeks, and lagoons.
2022	Plaquemine, LA.	Olin Dow Chemical – A fire occurred, and a significant release of chlorine occurred, which triggered a shelter-in-place for local businesses and residents. Initial reports indicate that a compressor failure occurred, resulting in the fire and release of the toxic chemical.
2022	Garyville, LA.	Marathon Petroleum Company – an explosion occurred in the 110,000-BPD gas oil hydrocracker, which makes diesel. Marathon was overhauling multiple units at the refinery at the time of the fire. It was unclear if work was being done on the hydrocracker. Five contract workers sustained injuries, four of which were treated on-site, and one is currently being evaluated at a local healthcare facility as a precaution. The number of injuries reported was later reduced to 3. (zero Fatalities, three Injured).
2022	Meraux, LA.	Valero Energy Company – A leak and fire on a processing unit lands two employees and six contractors in the hospital for burn treatment. (zero Fatalities, eight treated at local hospital, two being treated for severe burns).
2022	Ulsan, So Korea	S-Oil Onsan Industrial Complex – Firefighters received a report of an explosion at the refinery before 9 p.m. local time on Thursday. The explosion took place during the processing of crude oil into petroleum. According to local news, at least eight people were injured in the explosion and a fire at the 580,000-bpd refinery. (zero Fatalities, eight Injured).
2022	Litvinov, Czech Republic	A hydrocracker at the 108,000 b/d refinery in the has been closed following an explosion and fire, the latest in a spate of unplanned shutdowns to hit Europe's downstream sector in recent weeks. Czech firm Orlen Unipetrol — a subsidiary of Poland's PKN Orlen, said the hydrocracker remained shut today and that excess fuel is being burned in a controlled manner. Besides the hydrocracker, the refinery continues to operate, as does the integrated chemicals complex.
2022	Valdez, AK	Petro Star Inc. (PSI) – A fire occurred at the Valdez refinery truck loading rack. The fire was contained and has been fully extinguished. There were no reported injuries to employees or responders. The refinery was not damaged in the fire. (zero Fatalities, zero Injured).
2022	Angarsk (Siberia), Russia	A fire caused by an explosion at an oil refinery in Angarsk, Russia, has killed two people and injured five others. The governor of the Irkutsk region, Igor Kobzev, said the fire did not affect operations at the facility. (two Fatalities, five Injured)
2022	Ahmada, Kuwait	Mina al-Ahmadi oil refinery – A fire in January at the LPG Unit killed 2 contractors and injured 10 others; the second fire to have happened at the country's largest refinery in the last three months after a number of people were injured in October. The cause was not reported. (4 Fatalities, 10 Injured – 3 critically).
2022	Neuquen Province, Argentina	New American Oil (NAO) Company – An explosion at the refinery killed three workers in Plaza Huincul. (3 Fatalities, Injuries ?)
2022	Quintana, Island (near Freeport), TX.	Freeport LNG – A large loss of containment of LNG was released from off-site piping and quickly ignited, resulting in an explosion and fire. The investigation revealed that a relief valve protecting the piping had previously been removed for maintenance/inspection and had not been fully returned to service. Subsequently, a block valve on the protected pipe was inadvertently closed, isolating the system. This went undiscovered for several days, and the piping failed due to overpressure from thermal expansion. (zero Fatalities, zero Injured) (See 1989 Exxon Baton Rouge Refinery for a similar incident)

Year:	Location:	Incident description:
2022	Calahorra, Spain	Iniciativas Bioenergeticas SL – Two workers were killed on Thursday in an explosion at a biodiesel plant. A tank containing biodiesel feedstock exploded, the regional government said. The tank is one of four located next to one another, two containing biodiesel and two with crude oil. According to the government, the deflagration in one of the tanks caused cracks and leaks in others. (two Fatalities)
2022	Matanzas, Cuba	Lightning struck an oil storage tank resulting in a large fire that quickly spread to a second tank. Firefighters from Mexico and Venezuela were responding. Seventeen firefighters were missing, and one body had been located. It is unclear if he was one of the missing. (? Fatalities)
2022	El Segundo, CA.	Chevron Refinery – The city's fire department said in a statement issued earlier that crews responded to the blaze inside the refinery at approximately 6:15 p.m. local time. According to the El Segundo Fire Department, the fire department from the nearby city of Manhattan Beach and the Chevron Fire Department also responded.
2022	Westlake, LA	A release and fire occurred at the Sasol chemical plant on the Ziegler alcohol unit. The fire was quickly contained, and no injuries of outside impact were reported.
2022	Kuala Lumpur, Malaysia	An explosion and fire occurred at the Pengerang refinery-petrochemical complex, a joint venture that operates with Saudi Aramco in the Malaysian state of Johor. No injuries were reported, and the incident Petronas reported no immediate threat to the surrounding communities.
2023	Panjin, Liaoning province, China	Panjin Haoye Chemical Company – Authorities said that the death toll from a petrochemical plant explosion in northeastern China has risen to twelve, with one person still missing. More than 30 people were injured when the refinery caught fire and erupted in a large blast at around 1:30 p.m. local time on 15 January, according to China's state broadcaster CCTV, which cited local officials. One resident said he heard the explosion and felt its vibrations 25 miles away. Media reports the "widespread" shutdown of the Sinopec facilities due to the explosion and fire. (12 Fatalities, 1 Missing, more than 30 Injured)
2023	Borger, TX.	Phillips 66 Refinery – Loss of containment during maintenance activities and significant refinery fire. According to local news reports, seven workers were injured- four victims were flown in critical condition directly from the site to University Medical Center in Lubbock. Three others were treated for second-degree burns at GPCH. (zero Fatalities, seven Injured) (four critically)
2023	Geismar, LA.	Honeywell – The Honeywell complex near the Ascension/Iberville Parish line had an apparent explosion and leak of toxic hydrogen fluoride and chlorine Monday night. An undetermined amount of Hydrogen Fluoride and chlorine gas was released. Hydrogen fluoride becomes hydrofluoric acid when mixed with water. Local roads were closed, and a shelter-in-place order was issued. A Honeywell employee was killed in October 2021 after contracting hydrofluoric acid.
2023	Central province of Samutsongkram, Thailand	Mae Klong River – An explosion on an oil tanker in Thailand has taken the lives of eight people. Ten people were on board at the time. Authorities think the explosion might have been caused by a spark from repair work igniting the ship's oil tank. (eight Fatalities).
2023	East Palestine, Ohio	on 3 February, a train derailed in East Palestine, Ohio, a village of about 4,700 residents about 50 miles northwest of Pittsburgh. About 50 of the train's 150 cars ran off the tracks. The Norfolk Southern train was carrying several toxic chemicals, with vinyl chloride being of most concern to investigators. A huge fire erupted from the derailment, sending thick billowing smoke into the sky and over the town. Residents on both sides of the Ohio-Pennsylvania border were ordered to evacuate,
2023	Hidalgo, Mexico	Pemex Refinery – Following a vehicle collision inside the Pemex refinery, two people were reported as killed in the accident. (two Fatalities)
2023	Veracruz, Mexico	Pemex Oil Company – Three fires occurred at different facilities in Mexico and the United States operated by the Mexican oil company. Five people were missing, and eight others were injured in the event. Five people could not be accounted for at a storage facility following a fire at the facility. A separate fire at the company's Minatitlan refinery, also in Veracruz, was under control after injuring five people. The company's Deer Park, Texas refinery also reported a fire on the same day. (five Missing, eight Injured)

(*Continued*)

Year:	Location:	Incident description:
2023	Tuapse city, Russia	An explosion and fire were reported at this Russian refinery overnight. According to local media reports, a fire was reported in an outbuilding at the oil refinery at 02:15 followed by a loud explosion that blew out windows in the adjacent community.
2023	Jakarta, Indonesia	Pertamina, the state-run oil company, reported that nineteen people were killed following a large fire at the Plumpang fuel storage station, and three remain missing. It was reported that a technical problem resulted in excess pressure during a fuel receipt at the depot from the Balongan Refinery. (19 Fatalities, 3 missing)
2023	Deer Park, TX.	A major fire erupted at the Shell Deer Park facility on the Olefins Unit in Deer Park, TX. On 5 May. Nine workers were sent to the hospital for precautionary medical evaluations and were released. The huge plume of smoke was visible from miles from the site.
2023	Deer Park, Galveston Bay, Corpus Christi, Texas	Media reported three refinery fires in East Texas in the past few weeks leaving one dead and a dozen injured. I this period, dangerous fires have occurred at the Marathon, Pemex-Shell, and Valero refineries in Deer Park, Galveston Bay, and Corpus Christi. According to the media report, these incidents follow a year-long sting of refining incidents in this oil-rich area.
2023	Lemont, IL	Seneca Petroleum plant. One person was killed and one injured in an asphalt tank explosion and fire.
2023	Plaquemine, LA	Dow Chemical Company. A large explosion and fire occurred at this facility near the Mississippi River banks. Six separate explosions were detected at the site. 350 households were instructed to shelter-in-place until all clear was given.
2023	Lake Charles, LA	Calcasieu Refining Company. A lightning strike reportedly struck a light naphtha tank resulting in a large tank fire. The local community was sheltered in place for a three-mile radius around the facility.
2023	Garyville, LA	Marathon Oil Refinery. A release of light naphtha from a storage tank quickly erupts into a major inferno with two large naphtha storage tanks fully involved. All residents within a two-mile radius of the refinery were issued mandatory evacuation orders from the local responders. The following day Marathon announced that they were shutting the 565K B/D refinery down. No injuries have been reported at this time.
2023	Port Arthur, TX.	BASF Total Petrochemical Company. A fire inside the pyrolysis gasoline unit at the refinery resulted in the collapse of the extractive dissolution tower. The refinery was evacuated and shut down because of the incident. All personnel were accounted for.

It is notable to see the number of incidents and the number of fatalities/injuries that have occurred since the BP Texas City explosion in 2005.

This list provides some insight. Details for those incidents that have occurred in the US are available on the OSHA and CSB websites.

Abbreviations used in this document:

VCE = Vapor cloud explosion
BLEVE = Boiling liquid expanding vapor explosion
OPE = Opening Process Equipment
HTHA = High-Temperature Hydrogen Attack
PMI = Positive Materials Identification
CCTV = Closed Circuit Television
GPCH = Local Hospital near Borger, TX.
CSB = US Chemical Safety and Hazardous Investigation Board
OSHA = Occupational Safety and Health Administration
EPA = Environmental Protection Agency
CDU = Crude Distillation Unit
HDS = Hydrodesulfurization Unit (known as a hydrotreater)
KNPC = Kuwait National Petroleum Company

Appendix C

Auto-Refrigeration, When Bad Things Happen to Good Pressure Vessels

Auto-Refrigeration, When Bad Things Happen to Good Pressure Vessels (Reprinted with permission of the National Board of Boiler and Pressure Vessel Inspectors [NBBI]).

The document can be found here:
https://www.nationalboard.org/index.aspx?pageID=164&ID=249

Appendix D

OSHA Hazards of Combustible Dust Poster

Credit: Occupational Health and Safety Administration
The document is available here:
https://www.osha.gov/sites/default/files/publications/combustibledustposter.pdf

Process Operations Safety: The What, Why, and How Behind Safe Petrochemical Plant Operations, First Edition. M. Darryl Yoes.

Appendix E

HF Water Mitigation – Techniques and Design Considerations

Reprinted with permission of BakerRisk

(Risk Consultants and Engineering Services)

Appendix E. BakerRisk "HF Water Mitigation, Techniques, and Design Considerations" BakerRisk Best Practices – August 2021

The document is available here:

https://www.bakerrisk.com/wp-content/uploads/BakerRisk_Best_Practices-HF_Water_Mitigation-FNLv3-WebRes_Spreads-1.pdf

Process Operations Safety: The What, Why, and How Behind Safe Petrochemical Plant Operations, First Edition. M. Darryl Yoes.

Appendix F

Authorization for Disablement (Example Only) Safety, Security, Health, and/or Environmental Critical Devices

- This form must be filled out in its entirety and approved BEFORE the device is defeated.
- This completed form becomes a legal record for the duration of the disablement and for 10 days following the device return to service.

Unit/Process Area: ______________________ Device/System Defeated: ______________________________________
Date/Time Defeated: ____________________ Date Device Expected Returned to Service: _______________________
Why Is This Device Defeated?____
Describe the Alternate Protection Plan (APP): ____

____

____

Initial Defeat (Defeat less than one day or two twelve-hour shifts):

- All affected parties notified of the defeat and the APP: YES: _______by: _______________
- Authorization Form Posted at Console and Operator Station: YES: _______ by: _______________
- If this defeat extends past the first two twelve-hour shifts the Extended Defeat portion must be completed and signed by each oncoming shift.

Approved by Unit or Process Area FLS. FLS Approval: ______________________________________

Extended Defeat (Defeat more than two twelve-hour shifts and up to fourteen twelve-hour shifts).

- Reason for Extended Defeat? ____
- All affected persons informed of Extended Defeat and APP: Yes: ______ By: ___________________________

Approved by Unit SLS. SLS Approval: ____

Shift no.	Console operator signature	Shift no.	Console operator signature

Process Operations Safety: The What, Why, and How Behind Safe Petrochemical Plant Operations, First Edition. M. Darryl Yoes.

Longer-Term Defeat (Defeat longer than fourteen twelve-hour shifts)

- Longer-Term Defeat Plan must be approved by the site Operations Manager.
  - Longer-Term Defeats must have the equivalent of a site risk assessment with the potential for additional mitigation to support the Longer-Term Defeat.
  - Each oncoming shift Console Operator must notify affected members of the defeat and the APP.

Longer-Term Defeat Approval ______________________________ **Date: /Time:** ________________
Operations Manager

Appendix G

Petroleum Refining and Petrochemical Plant Process Operations Safety

Review of OSHA Database for Burns due to Steam and Condensate
(US Refining Industry – OSHA SIC 2911, NAICS 324110 (Petroleum Refining)
REFERENCE INFORMATION
US Refining Industry
Severe Steam and Hot Water Burn Injuries February 2002 – February 2012
26 hospitalized with severe burns 4 fatalities
Resource: Information extracted from the OSHA injury database

Examples of steam and hot water incidents
From OSHA database (2002–2012)

Date	Company	Incident	Comments
12/4/2010	ConocoPhillips (Los Angeles, CA)	Employee doused with hot water while working near the bottom of a coke drum. Employee **hospitalized with 1, 2, and 3° burns.**	The employee was sprayed with hot water while he was repositioning the chute between the coke drum and crusher. He was unaware that the chute was plugged with coke. The coke plug broke free releasing the hot water.
3/13/2010	ConocoPhillips (CA)	Employee sustained 2° and 3° burns; **hospitalized for 30 days**	Employee fell into an unguarded trench (Hot Hole) containing hot condensate while conducting a steam trap survey.
6/4/2009	Hovensa (St Croix, VI)	Three employees sprayed from hot condensate – **all three were hospitalized** with burn injuries	Employees were sprayed with hot condensate from blow-down piping while installing sight glass.
4/11/2009	ExxonMobil (Torrance, CA)	Employee splashed with hot water from bottom of coke drum – **Died four days later from the burn injuries**	Hot water release occurred during coke drum unheading operation.
7/12/2008	Motiva	Three Contractors sprayed with steam and 275° water – all **three were hospitalized** with burn injuries	Steam and hot water were released from 600# steam line while the contractors were installing blinds in the line.
7/3/2008	P2s LLC Belle Chase, LA	Three employees breaking a flange on 600# steam line were sprayed with hot condensate. All **three were hospitalized** due to burn injuries.	Employees used a grinder and torque wrench to remove the corroded bolts. As most bolts were removed the flange opened spraying all three workers with boiling water.
4/5/2008	ConocoPhillips (CA)	Employee sprayed with **steam** and **molten sulfur sustaining 1st and 2nd° burns.** Employee was transported to hospital for treatment.	Three employees were clearing a suspected sulfur plug in exchanger outlet piping to the sulfur pit using a steam hose. When the steam was turned on the sulfur and steam were released from sulfur pit spraying the injured employee.

(Continued)

Process Operations Safety: The What, Why, and How Behind Safe Petrochemical Plant Operations, First Edition. M. Darryl Yoes.

Examples of steam and hot water incidents
From OSHA database (2002–2012)

Date	Company	Incident	Comments
5/9/2007	Big West of CA	Employee drenched in hot water, steam, and oil while using a steam hose to steam waste oil drum **Hospitalized with burn injuries**	Two employees steaming a drum top down for opening. Outlet line plugged and while attempting to unplug line remove top plate from strainer –plug suddenly released at open strainer
10/27/2006	ConocoPhillips (CA)	Employee sprayed with hot material from bottom of coke drum – **hospitalized** with burn injuries	Hot water release occurred during coke drum unheading operation
6/22/2006	Kern Oil and Refining Co. (Bakersfield, CA)	Employee sprayed with steam, hot water, and crude oil – **hospitalized** with 2nd and 3rd° burn injuries	Employee was steaming out blockage from a crude oil regulator when line plugged. While attempting to remove plug with wire the release occurred suddenly enveloping the employee
5/16/2006	ConocoPhillips (CA)	Employee splashed with hot water from open-top coke drum. – **hospitalized** with 2nd and 3rd° burn injuries	While working to clear a clogged coke drum 150 psi steam was introduced near the bottom of the coke drum. Hot water, steam, and coke were released from the top of the open coke drum slashing onto the employee.
9/2/2004	BP Products North America (Texas City, TX)	Three employees were sprayed with hot condensate while opening process equipment. All three sustained serious burn injuries resulting in **two fatalities** and **one serious burn injury**	Separate industry sharing indicated employees were removing 10″ check valve on discharge of boiler water feed pump. Trapped hot water between the check valve and downstream block valve was released spraying employees with hot water.
4/17/2004	Coffeyville Resources Refining and Marketing LLC	Employee sprayed with steam and hot water. Transported to hospital but **died from the severe burn injuries**	Employee was cutting the pilot hole on a coke drum, struck a "hot spot" in the drum and was sprayed with steam and hot water.
12/8/2002	Valero Refinery (CA)	Employee was sprayed with hot condensate while steaming out a debutanizer tower and related piping. Transported to **hospital with severe burns.**	Employee completed the steaming process and was attempting to confirm that the vessel was empty. When he probed into an open flange with a piece of wire to clear debris the flange suddenly released steam and hot condensate.
8/10/2002	Shell Oil Products US	Two employees were in the process of restarting a turbine-driven pump and were sprayed with hot condensate from the turbine exhaust. **Both received burns; one was treated at the hospital.**	Two Operators and other co-workers were restarting the flare knockout pump turbine, standing 3′ from the turbine exhaust when the turbine started.
5/31/2002	Phillips Refinery	Trainee was sprayed with hot water while reconnecting the 12″ coke drum inlet line. He **sustained 1st, 2nd**, and **3rd ° burns** from the hot water.	Trainee was being observed by a coworker while connecting inlet line. As bolts were loosened a puff of steam appeared followed by a surge of hot water spraying the trainee.
5/22/2002	Tesoro Refining and Marketing Co.	A Maintenance foreman was sprayed with 160°F water sustaining 2nd and 3rd° burns. Employee was **hospitalized** for treatment.	The employee was working on a scaffold 6′ high attempting to move a ¾″ steam line. As he moved the pipe, it failed spraying him with the hot water. He jumped from the scaffold in an attempt to escape the spraying hot water.

Appendix Ha

The Hazards of Nitrogen Asphyxiation

Credit: US Chemical and Hazard Investigation Board
This document is available at the following:
https://www.csb.gov/assets/1/20/nitrogen_asphyxiation_bulletin_training_presentation.pdf?15062

Process Operations Safety: The What, Why, and How Behind Safe Petrochemical Plant Operations, First Edition. M. Darryl Yoes.

Appendix Hb

Safety Bulletin – Hazards of Nitrogen Asphyxiation

Credit: US Chemical Safety and Hazard Investigation Board
This document is available at the following:
https://www.csb.gov/assets/1/20/sb-nitrogen-6-11-03.pdf?13883

Process Operations Safety: The What, Why, and How Behind Safe Petrochemical Plant Operations, First Edition. M. Darryl Yoes.

Appendix I

Technical Guidance – Car Seals

Reprinted with permission of ERI Solutions/PROtect, LLC.

TECHNICAL GUIDANCE

Car Seals

May 19, 2016

Process Operations Safety: The What, Why, and How Behind Safe Petrochemical Plant Operations, First Edition. M. Darryl Yoes.

Definition

A metal or plastic cable used to fix a valve in the open position - Car Seal Open (CSO), or closed position - Car Seal Closed (CSC)

Applicability

Car seals work much like a plastic cable tie. For valves, you pass the cable around the wheel or lever, then around the body of the valve or nearby steel work, push the end of the cable into the seal body. The cable is now locked in place and cannot be removed. The valve is now 'sealed'. To restore operation of the valve, the seal has to be cut.

Car seals are a relatively low cost, consumable item and are available in different cable lengths and colors. The seals can be uniquely marked with serial numbers for documentation purposes. See below of an example of a car seal, a car seal being utilized and an excerpt of a Piping and Instrumentation Diagram (P&ID) where car seals are identified.

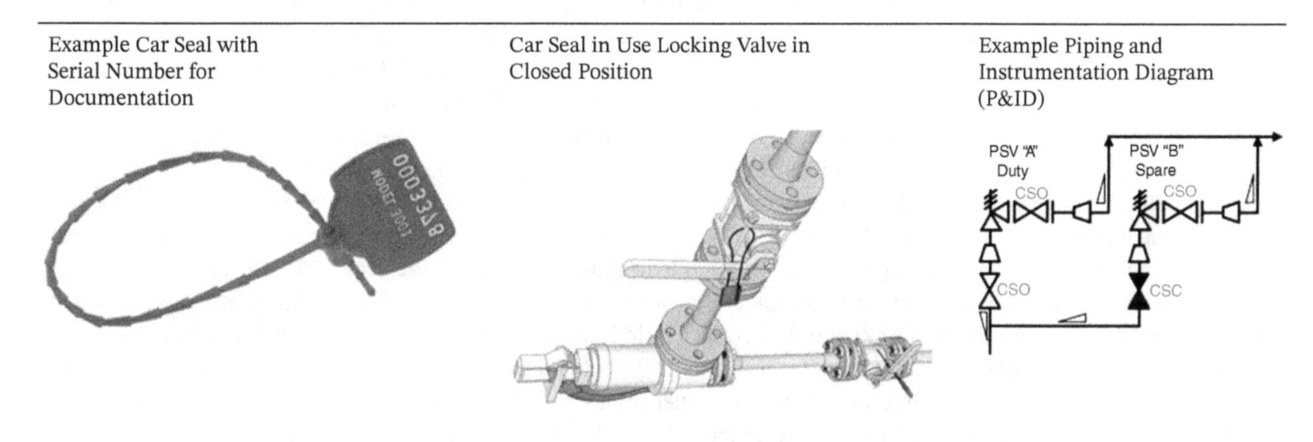

Example Car Seal with Serial Number for Documentation | Car Seal in Use Locking Valve in Closed Position | Example Piping and Instrumentation Diagram (P&ID)

Car seals are generally used to seal a rail car, truck, cargo container or other openings to prevent/indicate tampering are also used in industry processes. A car seal is typically placed on a valve but it can be placed on any type of physical equipment to prevent unauthorized operation.

A common installation of a car seal is on an isolation valve under a Process Safety Valve (PSV) or Pressure Relief Valve (PRV) that must be inspected frequently without shutting down the entire process each inspection. It is important to note that in the aforementioned installation, redundant Process Safety Valves (PSVs) or specific protocols are in place to ensure safety in the event of an over pressure situation while the isolation valve is in use.

Car seals may also be used for valves that periodically or in special instances are open or closed but must not remain open or closed during normal operations. These valves are noted on Piping and Instrumentation Diagrams (P&IDs) as Car Seal Open (CSO) for a valve that must remain open, and Car Seal Closed (CSC) for valve that must remain closed.

Case Study

In 2008, a tube and shell heat exchanger catastrophically failed when workers accidentally closed two different valves at two different times between the vessel and its relief valve. On June 10, 2008, operators closed an isolation valve between the heat exchanger shell (ammonia cooling side) and a relief valve to replace a burst rupture disk under the relief valve that provided over-pressure protection. Maintenance workers replaced the rupture disk on that day; however, the closed isolation valve was not reopened. On the morning of June 11, an operator closed a block valve isolating the ammonia pressure control valve from the heat exchanger. The operator then connected a steam line to the process line to clean the piping. The steam flowed through the heat exchanger tubes, heated the liquid ammonia in the exchanger shell, and increased the pressure in the shell. The closed isolation and block valves prevented the increasing ammonia pressure from safely venting through either the ammonia pressure control valve or the rupture disk and relief valve. The pressure in the heat exchanger shell continued climbing until it violently ruptured at about 7:30 a.m.

Regulatory Requirements/Best Management Practices (BMP)

- A procedure or Car Seal Program is simply a valve management program that contains the following: o A list of safety critical valves
  - Their safe position (Could be either open or closed)
  - A description of their location (e.g., serial numbers), and
  - Some type of seal that will prevent the valve from being deviated from its safe position. This seal can be a plastic tie wrap, chain, or some type of metal banding. Some facilities even using color coding for their seals (e.g., green is for open, and red is for closed valves). The physical seal should have suitable mechanical strength to prevent unauthorized valve operation.
- The program should also include updating your Piping and Instrumentation Diagrams (P&IDs) to reflect which valves are in this program. Most Piping and Instrumentation Diagrams (P&IDs) will use the designations Car Seal Open (CSO), and Car Seal Closed (CSC) to show the valves in the Car Seal Program.
- The program (an administrative control) is only as good as the workers who use it and rely on it. Some facilities build the Car Seal deviation procedure into their Management of Change procedure, which means anytime a valve in the program needs to be deviated from its safe position, a temporary Management of Change (MOC) will be completed and this temporary Management of Change (MOC) will always require a Pre-Start Up Safety Review (PSSR). The Pre-Start Up Safety Review (PSSR) is not required by the Occupational Safety and Health Administration (OSHA) however it is a best management practice to ensure this valve is returned to the safe position before returning the Highly Hazardous Chemical (HHC) and/or Extremely Hazardous Substance (EHS) to that part of the process.
- Some facilities will have a very basic board and tagging program where each sealed valve will be tagged with a weather resistant tag that is in two parts. The tab on the end of the tag is the permission tab. This is the one that is removed and is signed by the appropriate personnel granting permission to deviate the valve. On the back of the tab there is a matching number that also appears on the part of the tag that is attached to the valve. The removed tab is hung on a Valve Deviation Board in plain sight in the control room and the action should be entered into the shift log book. Workers then go back to the valve and with the signatures and the tag hung on the board and the action communicated to all workers on shift (can be via radio announcement, but then all workers must acknowledge they received the transmission) they will now deviate the valve. The program will also require an entry into either a paper log book that is used to communicate shift change data or an electronic method, but it is critical that any and all valve deviations be communicated to all workers, including contractors whose work will involve any deviated valve.
- Some facilities will even work valve deviations into their safe work permitting process to provide another layer of communication.
- The next critical path is returning the valve(s) back to its safe position. This should be done as soon as possible. Some facilities will even have a policy that no car sealed valve can be deviated longer than a single shift. Putting the valve back into its safe position will entail a new tag and a new seal, as well as another entry into the log book and communication to the workers on shift about the new status of the valve.
- The final step may be the most critical of all. The program needs to be audited to ensure its integrity. This auditing begins as part of a Pre-Start Up Safety Review to act as another layer of assurance that the valves have been placed back into their safe position before the process is started up. Some facilities will also include these critical valves in their Start-Up Standard Operating Procedures (SOPs) to ensure they are inspected before start-up. At some point, the facility will need to do a full audit of the program and this will include a physical inspection of each valve to ensure it is in its proper position, tag is in place, and seal integrity intact.
- Locks are sometimes used in the process for similar restrictions. If locks are used, a procedure should be in place for proper controls and authorized key access, even if it is not noted on the Piping and Instrumentation Diagram (P&ID). The process locks should differ from locks placed on sample ports to prevent access to the contents of the process and locks used in the Lockout/Tag out program.

Summary

There are many different variations of Car Seal Programs, some very basic and some very advanced, but they all serve a critical role in process safety. The first place to consider using a car seal program would be on all valves that are between a pressure vessel and its relief protection or even downstream of relief valves. If your facility falls under Occupational Safety and Health Administration's (OSHA's) Process Safety Management (PSM) standard and the Environmental Protection Agency's (EPA's) Risk Management Plan (RMP) rule, you could be cited for not having some form of a car-seal program if you have situations where this program would be a necessity (e.g. ASME Section VIII UG-135(d). Occupational Safety and Health Administration (OSHA) has recently included car-seal programs in their Process Safety Management (PSM) audits. It made its first official appearance in the Refinery Process Safety Management (PSM) Compliance Directive, and it is making its way into the Process Safety Management (PSM) Covered Chemical Facilities Compliance Directive (CPL) as well.

Appendix J

Example Blind List and Simplified Blind Sketch

Rev.1

Example Blind list

System No.12-05 (E-14 Exchanger Train)

Equipment: Lube **E-14 Clean exchanger and check ferrules**

Extraction; E-14A

Item	Blind No.	Description	Type	Size	Installation		Removal		Note
				In PSI	Date	Name	Date	Name	
1	032	CWS INLET to E-14A	H	10"X150					
2	033	CWR OUTLET from E-14A	H	10"X150					
3	034	OVERHEAD T-2 to E-14A *	H	24"X150					**Common with ABC Project**
4	035	CONDENSATE to D-36	H	4"X150					
5	036	VAPOR to J-36 A/B	H	12"X150					

*** Blind No.034 is common blind with system 12-08 and ABC Project**

Process Operations Safety: The What, Why, and How Behind Safe Petrochemical Plant Operations, First Edition. M. Darryl Yoes.

Example of Simplified Flow Plan for Blind System

Lube Extraction System No.12-05 (E-14 Exchanger Train)

E-14 Clean exchanger and check ferrules

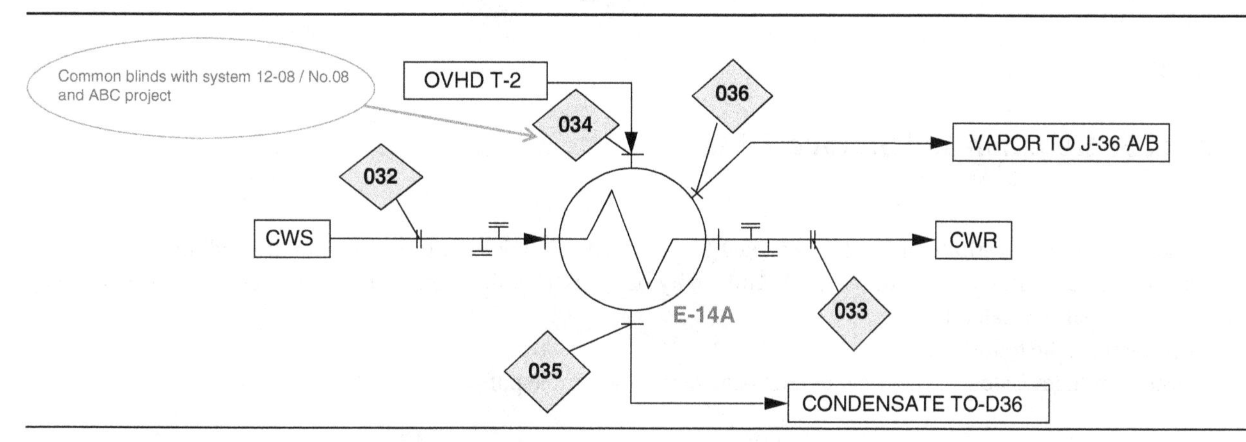

Appendix K

Chemical Segregation and Storage Table

Credit: The US National Institute of Health (Adapted from Prudent Practices in the Laboratory; Handling and Disposal of Chemicals, National Research Council, 1995, University of Texas/Health Science at Houston and Boston University Environmental Health & Safety)

The document can be found here:
https://ors.od.nih.gov/sr/dohs/Documents/chemical-segregation-table.pdf

Process Operations Safety: The What, Why, and How Behind Safe Petrochemical Plant Operations, First Edition. M. Darryl Yoes.

Appendix L

Process Safety Field Walkabouts (Example)

The following are examples of the types of process safety concerns that I look for when walking on a process unit. This is only a guide intended to help reinforce and improve process safety. This list is not in any order and is not intended to be a complete list:

- Pay attention to what people are doing in the field – feel free to discuss their activities; ask them to explain what they are doing ("what's going on today?").
  - Are observed practices in line with site procedures?
  - Ask what they talked about in their pre-job briefing.
- Housekeeping – general state of unit/equipment (no clutter, good access to critical equipment, personnel walkways and platforms clear, temporary hoses rolled up and stored off platforms and out of walkways, no hydrocarbon spills, etc.).
  - Observe for accumulation of trash and debris on scaffolds and upper unit structures or tank work platforms.
- Is "Control of Defeat" properly executed for safety critical devices (SCD's)?
  - Does the listed mitigation replace the functionality of the defeated device?
  - Are SCD's properly identified in the field?
  - Are SCD systems properly equipped with bypass for on-stream testing?
- Are Operating Procedures current, in checklist format, and being used as a checklist by the Operators?
  - Do the procedures cover brittle fracture requirements (including the Minimum Design Metal Temperature (MDMT) and/or Minimum Safe Operating Temperature (MSOT) for process vessels)?
  - Are Operators receiving refresher training on the procedures periodically (every 3 years in the US)?
- Cold wall equipment (piping and vessels) identified in the field (with labels and signs)?
  - Are Operators aware of concerns with the application of external insulation on cold wall equipment?
  - Are they aware we should have a MOC when external cooling is used?
- Relief valves:
  - Proper orientation of isolation valves (vertical)?
  - Bellows vents unobstructed and routed to a safe location?
  - Are isolation valves identified and car seals in place?
  - Are isolation valves full port and fully opened?
- Are bar stock plugs installed in vents/drains for hydrocarbon and chemical systems? Are they the same metallurgy as the vent or drain?
- Sample points (are they closed-loop systems)?
  - Esp. for H_2S or other toxic streams?
  - Double blocked or plugged?
- Static Electricity?
  - Operators aware of basic static electricity procedures?
  - No draining of hydrocarbons into "kiddie pools"? Are any plastic containers lying about in the field?
  - Look at tank gauging tapes. Are they manufactured tape with bonding wire/clip attached?
  - Grounding/bonding cables attached to tank trucks/tank cars, etc.?
- Are occupied temporary trailers and other buildings located adjacent to process areas?
  - Evidence of MOC for occupied temporary buildings?
  - Are non-occupied buildings secured to prevent unauthorized occupancy?
- Are there chain-operated valves without a supplementary retention device?

Process Operations Safety: The What, Why, and How Behind Safe Petrochemical Plant Operations, First Edition. M. Darryl Yoes.

- Utility connections to process equipment:
  - Are utility stations well identified, with specific coupling for each utility, equipped with check valves and vents or drains?
  - Are all utility connections to process equipment protected by a check valve?
  - Are utility hoses color-coded and stored appropriately when not in use?
- Small Piping Standards:
  - Are small piping connections to process equipment properly seal welded and gusseted; esp. vibrating services (pumps, compressors, PRV's, control valve stations, etc.)?
  - Is unsupported small-bore piping in place (esp. temporary connections that remain after the work was completed)?
  - Any "Christmas trees" and/or small piping "moment arms"?
  - Are pipe unions located inside the first block valve?
  - Are bar stock plugs properly installed in vents/drains for systems containing hydrocarbons, chemicals, or toxic materials?
  - Are temporary connections still in place after work was completed?
- Are sewers properly sealed for hot work (not just covered)?
  - And are the covers removed when work is complete to ensure proper drainage?
- Have dead legs and other stagnant areas been identified and is there an inspection program in place for dead legs?
  - Are dead legs protected from freezing?
- Have mixing points been identified, and is there an inspection program to monitor internal corrosion or erosion (water, chemical, temperature mixing points)?
- Are HP gas cylinders properly identified and stored (chained in place and not sitting on concrete where the bottom can corrode)?
- Is flash protection equipment readily available in electrical substations (for electricians?)
  - Are flash protection boundaries established near high-voltage equipment?
- Is mechanical work in compliance with Work Permit and Energy Isolation Requirements?
  - Are proper isolation and locks in place to positively isolate energy (electrical, pressure, mechanical, etc.)?
  - Do the workers know where/how the energy is isolated?
  - Discussion item: what was discussed in the joint job-site visit?
- LPG Storage:
  - Are spheres and other LPG vessels properly equipped with two water draw valves? Are both valves closed except when in use?
  - Are the water draws and other connections at the bottom of the vessel protected from freezing?
  - Are Operators familiar with proper water draw procedures for LPG vessels (open the upstream valve fully and throttle on the downstream valve, and never leave the water draw unattended during operation)?
  - Are there indications of defeating spring-loaded valves in place (unusual wire, rope, chain, bolts, nuts, etc., where there should be none)?
  - Are connections in place for internal flooding LPG storage vessels with fire water? Are Operators knowledgeable of water deluge and water flooding procedures?
- Piping/Flanges:
  - Are piping flange bolts fully engaged (no threads showing in the nut)?
  - Any obvious dissimilar metals?
  - If temporary pipe clamps are observed, are they engineered?
  - Evidence of spring hangers or pipe shoes out of normal tolerance (spring hanger pegged or pipe shoe off support)?
  - Are instrument thermowells seal welded?
- Are pump and compressor coupling guards in place and secured?
- Are any steam traps blowing through or creating hot condensate "hot holes" in the area?
- Steam turbines:
  - Can the craftsman test the turbine overspeed trip device while out of the line-of-fire of the turbine wheel?
  - Is there a way to restrict steam to the turbine during the overspeed turbine test to prevent an accidental runaway?
  - Are the steam turbine exhaust vents routed to a safe location?
  - Is the turbine exhaust valve properly identified and car sealed open?
- Are analyzer shelters controlled against unauthorized access (potential asphyxiation hazard – typically locked)?
- Are vessel gauge glasses fitted with excess flow valves for clean services (flammable or toxic)?

  - Are excess flow valves in the correct (run) position (fully open)?
  - Are Operators familiar with procedures for excess flow valves?
- If temporary pipe clamps are observed, are they engineered? Is there a Management of Change document for the clamp?
- Are any fire hydrants or turrets blocked in for maintenance? If yes, is there control of defeat and mitigation in place?
- Electrical:
  - Are switches for motor start/stop located out of the line-of-fire of the motor terminal junction box?
  - Are Operators aware of the safe position when starting pumps and compressors (out of line-of-fire of motor terminal junction box)?
  - Are electrical seals and conduits fully closed (no missing or loose covers, plugs, or bolts missing or loose in electrical junction boxes?)
  - Are flash protection suits readily available in substations for electricians' use?
  - Are flash protection boundaries established near high-voltage equipment?
  - Are there any electrical "extension cords" or "cheater cords" that violate the electrical classification in the area?
- Are any confined space entry tasks underway during the walkabout?
  - Verify current permit and proper gas test?
  - Verify dedicated "hole watch" is in place and is in communication with workers in confined space?
  - Verify hole watch has communication with rescue party?

Appendix M

OSHA® FactSheet

Internal Combustion Engines as Ignition Sources

Internal combustion engines present an ignition hazard when used in facilities processing flammable liquids and gases. If flammable vapors or gases are released in these facilities, an internal combustion engine could ignite the flammable materials with catastrophic consequences. Investigations by OSHA and the US Chemical Safety Board (CSB) document a history of fires and explosions at workplaces (oilfields, refineries, chemical plants, and other facilities) where an internal combustion engine was identified as or suspected to be the source of ignition.[1]

Understanding the Hazard

Internal combustion engines, whether fueled by gasoline, diesel, propane, natural gas, or other fuels, can act as ignition sources. Examples include:

- Stationary engines such as compressors, generators, and pumps.
- Mobile equipment or transports such as vans, trucks, forklifts, cranes, well servicing equipment, drilling rigs, excavators, portable generators, and welding trucks.
- Contractor vehicles and motorized equipment.
- Emergency response vehicles such as fire engines and ambulances.
- Vehicle-mounted engines on vacuum trucks, tanker trucks, and waste haulers.
- Small portable engines such as mowers, blowers, generators, compressors, welders, and pumps. This includes hand tools unrelated to a process, such as chain saws, brought in by contractors.

Internal combustion engines require a specific fuel-to-air ratio to work properly. Air enters the engine through the intake that leads to the combustion chambers (cylinders). If employers allow internal combustion engines in areas where flammable vapors or gases exist, then the vapors and gases can enter the cylinders of the engine along with the air. Additional flammable material in the cylinders provides an external fuel source and increases the fuel-to-air ratio in the engine. Changes in the fuel-to-air ratio create ignition hazards by:

- **Elevating engine operating temperatures.** Increasing the fuel-to-air ratio causes an increase in the energy output which results in increased surface and exhaust temperatures.

1 From Bureau of Labor Statistics and OSHA Integrated Management Information System (IMIS) databases, and CSB reports 2003–2010.

Process Operations Safety: The What, Why, and How Behind Safe Petrochemical Plant Operations, First Edition. M. Darryl Yoes.

An explosion at a refinery site killed 15 and injured nearly 200; an idling diesel pickup truck was the most likely ignition point.
Source: US Chemical Safety Board

Increasing the fuel-to-air ratio also causes pre-ignition within internal combustion engines. Pre-ignition occurs when a fuel-rich mixture in the cylinder ignites before the spark plug fires. Pre-ignition creates damaging pressure surges and higher engine surface and exhaust system temperatures.

If the temperature of the surface of the engine in contact with the fuel/air mixture reaches the autoignition temperature of that mixture, a fire or explosion will occur.[2]

- **Creating sparks.** Fuel-rich conditions in an engine can result in incomplete combustion.

When uncombusted fuel from the cylinders enters the exhaust system, it can ignite due to the hot surface, discharging sparks and flames (backfire). These can ignite flammable vapors and gases in the surrounding area.

- **Causing overspeed and runaway engines.** Overspeed occurs when flammable vapors and gases in the intake air cause engines to run faster than designed. This increases the wear and tear on the engine, causing overheating and risking autoignition. If allowed to continue, overspeed can result in mechanical failure causing the engine to blow apart, igniting flammable materials in the area, and causing a flash fire or explosion.

Three workers were killed and four injured in a fire resulting from a runaway diesel engine.
Source: US Chemical Safety Board

2 The autoignition temperature of a chemical is the lowest temperature at which an air mixture of the chemical will ignite without a spark or flame.

For a gasoline engine, overspeed is stopped by turning off the ignition switch, which shuts down the ignition source (spark plugs) in the cylinders. This is not the case for diesel engines. Diesel engines do not use spark plugs; turning off the engine ignition switch does not shut down the ignition source. Stopping the fuel supply is also ineffective because the fuel is present in the intake air. The only way to prevent mechanical failure and possible explosion is to cut off the intake air supply, using systems such as automatic engine overspeed shutdown devices.

Preventing Engines from Becoming Ignition Sources

Workplace Evaluation

- Identify areas where flammable liquids or gases are used or stored.
- Evaluate where internal combustion engines are located.
- Assess contractor use of internal combustion engines in flammable material areas.

Whenever possible, do not install permanently mounted internal combustion engines in areas where flammable vapors or gases could be present.

The OSHA General Industry and Construction standards contained in Subparts S and K (29 CFR 1910.307 and 29 CFR 1926.407) of the Code of Federal Regulations (CFR) define hazardous (classified) locations as areas with:

- Flammable gases or vapors (Class I)
- Combustible dust (Class II)
- Easily ignitable fibers (Class III)

In Class I locations, an installation must be classified as using the division classification system complying with paragraphs (c), (d), (e), and (f), or using the zone classification system specified in paragraph (g), of 29 CFR 1910.307 or 29 CFR 1926.407.

If employers cannot remove internal combustion engines from areas processing flammable materials, then the following preventive measures should be used.

These measures include administrative procedures for the safe use of portable or mobile equipment with internal combustion engines.

Control Measures to Reduce Risk

- Ensure that materials and equipment are stored and used in accordance with OSHA standards such as:
  - 29 CFR 1910.106 or 29 CFR 1926.152, *Flammable liquids;*
  - 29 CFR 1910.107, *Spray finishing using flammable and combustible materials;*
  - 29 CFR 1910.119 or 29 CFR 1926.64, *Process safety management of highly hazardous chemicals;*
  - 29 CFR 1910.178, *Powered industrial trucks.*
- Ensure that worksite safety programs and safe work permit systems:
  - Address internal combustion engines as ignition sources;
  - Evaluate and establish acceptable areas, boundaries, and entry routes for mobile internal combustion engines based on applicable standards and recognized and generally accepted good engineering practices;

An idling diesel pickup truck believed to have ignited a vapor cloud from a nearby process area.
Source: US Chemical Safety Board

  - Mark and enforce acceptable traffic routes through hazardous (classified) locations;
  - Account for special procedures, which might include the use of portable gas monitors, or emergency evacuation routes for vehicles.
- Use other preventive measures such as:
  - Installing automatic overspeed shutdown devices on permanently mounted engines.
  - Installing intake flame arrestors and exhaust system spark arrest systems on permanently mounted engines.
  - Installing flammable gas and vapor detectors in processing areas.
  - Installing shutdown systems (positive air shut-off for diesel or ignition kill for gasoline), intake flame arrestor, exhaust system spark arrest, or other appropriate protective systems[3] for mobile internal combustion engines.
  - Using a safe work permit system to control mobile combustion engine access into areas that could contain flammable vapors and gases.
  - Using a safe work permit system to control the use of open flames and spark-producing operations and equipment (e.g., welding, grinding, brazing, etc.)

Training

Provide training to workers and contractors on hazards in areas that contain flammable vapors and gases. The training should include instruction on:

- Hazards of internal combustion engines as ignition sources and the specific worksite areas that are subject to these hazards.

3 Forklifts require specific unit ratings to be used. See 29 CFR 1910.178.

One of two vacuum truck diesel engines believed to be the ignition source for a fatal fire at an open-air collection pit.
Source: US Chemical Safety Board

- Areas, boundaries, and acceptable routes for mobile engines, and applicable facility safety rules. Instruct workers to avoid driving in areas

The Process Safety Management (PSM) standard

(29 CFR 1910.119) applies to processes in facilities that have flammable liquids or gases on site in one location in a quantity of 10,000 pounds or more. For covered processes, the PSM standard addresses controlling ignition sources primarily by requiring employers to:

- Compile written process safety information for working with hazardous chemicals, including identifying areas where flammable materials are or may be present, 29 CFR 1910.119(d).
- Complete a process hazard analysis that addresses the hazards of the process, including control of ignition sources such as internal combustion engines, 29 CFR 1910.119(e).
- Develop and implement safe work practices, including control of vehicle access to process areas, 29 CFR 1910.119(f)(4).
- Develop and implement written procedures (f) and training for operators (g), maintenance personnel (j), and contractors (h), 29 CFR 1910.119(f), (g), (h), and (j).
- Control hot work, 29 CFR 1910.119(k).
  where flammable vapors and gases may be present, to stay on acceptable routes, and to follow site procedures for safe access in areas where flammable materials are being processed.
- Signs and hazards of flammable vapor and gas clouds, and associated precautions.
- Signs and hazards of internal combustion engine overspeed and runaway.
- Emergency procedures for flammable materials emergencies, including response to engine overspeeds and runaways.

Standards and Resources

In addition to following
applicable OSHA standards, employers should also refer to the American Petroleum Institute (API), the National Fire Protection Association (NFPA), and other applicable industry and consensus standards, which provide helpful guidance.

1. OSHA Standard 29 CFR 1910.106, *Flammable liquids:* 29 CFR 1910.106(e)(6)(i), (f)(6), and (h)(7)
   - (i)(A), *Sources of Ignition.*
2. 29 CFR 1910.107, Spray Finishing Using Flammable and Combustible Materials.

3. 29 CFR 1910.178(c)(2)(i), (ii), "Power-operated industrial trucks shall not be used in atmospheres containing hazardous concentration of acetylene, butadiene, ethylene oxide, and hydrogen ... metal dusts ... " For other listed chemicals, only certain unit ratings are allowed. See the standard for more details.
4. 29 CFR 1910.119, Process Safety Management of Highly Hazardous Chemicals.
5. OSHA – Petroleum Refinery Process Safety Management National Emphasis Program (CPL 03-00-010, dated 8/18/2009). *Page A-55,*
   Motorized Equipment. "Does the employer have a safe work practice that it implements for motorized equipment to enter operating units and adjacent roadways?"
6. Mine Safety and Health Administration (MSHA) Standard 30 CFR 7.98 – *Diesel Engine Technical Requirements.* Requires explosion-proof design and a safety shutdown system, as well as intake flame arrest and an exhaust system spark arrest system. See also MSHA Standard 30 CFR 36.23 – *Engine Intake System.*
7. Department of Interior, Materials Management Services (MMS) 30 CFR 250.510, 250.610, and 250.803(b)(5)(ii). Requires diesel engine automatic shutdown devices/systems.
8. Coast Guard Standard 46 CFR 58.10-5. Requires gasoline engine intake system backfire flame control.
9. American Petroleum Institute (API) Recommended Practice (RP) 54, *Recommended Practice for Occupational Safety for Oil and Gas Well Drilling and Servicing Operations.*
   - 6.1.15 "On land locations, vehicles not involved in the immediate rig operations should be located a minimum distance of 100 feet from the wellbore ...";
   - 9.14.2 "Rig generators on land locations should be located at least 100 ft (30.5 m) from the wellhead upwind considering the prevailing wind direction to isolate a possible source of ignition...";
   - 9.15.3 "Spark arrestors or equivalent equipment shall be provided on all internal combustion engine exhausts located within 100 ft (30.5 m) of the wellbore...";
   - 14.2.3 "Gasoline engines and other possible sources of ignition should be located at least 100 ft (30.5 m) from the wellbore during snubbing operations...";
   - 15.2.2 "Any engine within 100 ft (30.5 m) of the well (within 35 ft of the well for offshore) should not be operated during the drill stem testing operations without having a heat and spark arresting system for the exhaust...";
   - 18.3.1 "Where terrain permits, compressors should be located at least 100 ft (30.5 m) from the wellbore...".
10. API RP 7C-11F – *Recommended Practices for Installation, Maintenance, and Operation of Internal Combustion Engines.*
11. API RP 505 – *Recommended Practice for Classification of Locations for Electrical Installations of Petroleum Facilities Classified as Class I, Zone 0, Zone 1, or Zone 2.*
12. API RP 2001 – *Fire Protection in Refineries.*
13. API RP 2216 – *Ignition Risk of Hydrocarbon*
    Liquids and Vapors by Hot Surfaces in the Open Air.
14. API RP 2219 – *Safe Operations of Vacuum Trucks in Petroleum Service.*
15. National Fire Protection Association (NFPA) 37 – *Installation and Use of Stationary Combustion Engines and Gas Turbines.*
16. NFPA 70 – *National Electrical Code.*
17. NFPA 497 – *Recommended Practice for the Classification of Flammable Liquids, Gases, or Vapors and of Hazardous (Classified) Locations for Electrical Installations in Chemical Process Areas.*
18. NFPA 505 – *Fire Safety Standard for Powered Industrial Trucks Including Type Designations, Areas of Use, Conversions, Maintenance, and Operations.*
19. NFPA FPH – *Fire Protection Handbook.*
20. Diesel Engine Manufacturers Association (DEMA) – *Standard Practices for Low and Medium Speed Stationary Diesel and Gas Engines.*
21. California OSHA Standard, Subchapter 14, Article 35 – *Petroleum Safety Orders – Drilling and Production.*
22. California OSHA Standard, Subchapter 15, Article 21 – *Refining, Transportation, and Handling – Gas Conveyors and Engines.*
23. Chemical Safety Board (CSB) Investigation Report – *BLSR Operating Ltd. Vapor Cloud Deflagration and Fire, 2003.* Also, see other CSB reports.
24. Center for Chemical Process Safety (CCPS)

Process Safety Beacon, October 2009.

How Can OSHA Help?

OSHA has compliance assistance specialists throughout the nation who can provide information to employers and workers about OSHA standards, short educational programs on specific hazards or OSHA rights and responsibilities, and information on additional compliance assistance resources.

Contact your local OSHA office for more information.

OSHA's On-site Consultation Program offers free and confidential advice to small and medium-sized businesses with fewer than 250 employees at a site (and no more than 500 employees nationwide) to help identify and correct hazards at your worksite. On-site consultation services are separate from enforcement and do not result in penalties or citations. To locate the On-site Consultation office nearest you, visit OSHA's website or call 1-800-321- OSHA (6742).

Worker Rights

Workers have the right to:

- Working conditions that do not pose a risk of serious harm.
- Receive information and training (in a language and vocabulary they understand) about workplace hazards, methods to prevent them, and the OSHA standards that apply to their workplace.
- Review records of work-related injuries and illnesses.
- Get copies of test results that find and measure hazards.
- File a complaint asking OSHA to inspect their workplace if they believe there is a serious hazard or that their employer is not following OSHA's rules. OSHA will keep all identities confidential.
- Exercise their rights under the law without retaliation or discrimination.

  For more information, see OSHA's web page for workers.

Disclaimer

This Fact Sheet is not a standard or regulation, and it creates no new legal obligations. It contains recommendations as well as descriptions of mandatory safety and health standards [and other regulatory requirements]. The recommendations are advisory in nature, informational in content, and are intended to assist employers in providing a safe and healthful workplace. The *Occupational Safety and Health Act* requires employers to comply with safety and health standards and regulations promulgated by OSHA or by a state with an OSHA-approved state plan. In addition, the Act's General Duty Clause, Section 5(a)(1), requires employers to provide their workers with a workplace free from recognized hazards likely to cause death or serious physical harm.

This is one in a series of informational fact sheets highlighting OSHA programs, policies, or standards. It does not impose any new compliance requirements. For a comprehensive list of compliance requirements of OSHA standards or regulations, refer to Title 29 of the Code of Federal Regulations. This information will be made available to sensory-impaired individuals upon request. The voice phone is (202) 693–1999; teletypewriter (TTY) number: (877) 889–5627.

Appendix N

The History of Diesel Engine Runaway Accidents and Related Regulations

1984 – API issued Recommended Practice 2001 (4.2.10) for refineries stating "Consideration should be given to the use of spark arrestors on exhaust pipes and rapid shutoff valves on air inlets to diesel engine maintenance vehicles."

1985 – Saskatchewan Oil & Gas Conservation Regulations require operators to install "adequate air intake shutoff valves" for engines within 75 ft (23 m) from the well. Other Canadian provinces issue similar requirements for "positive air shutoff devices." In Edmonton, Canada, Roda Deaco and Rigsaver manual and electrical swing gate valves are developed and installed to meet these Regulations.

May 1989 – US Federal Minerals Management Service (MMS) regulations require automatic shutoff devices to be fitted to unattended diesel engines on offshore installations in the Gulf of Mexico.

September 1996 – Triodyne Inc. consultants published a Safety Bulletin "The Runaway Diesel–External Fuel Ingestion."

August 1999 – American Petroleum Institute (Drilling) Recommended Practice 54 9.15 states: "Emergency shutdown devices that will close off the combustion air should be installed on all diesel engines. Spark arrestors or equivalent equipment shall be provided on all internal combustion engine exhausts located within 100 ft (30.5 m) of the wellbore."

January 2003 – BLSR, Ltd., Rosharon, Texas, has a fatal accident where hydrocarbon vapors were drawn into the intakes of two idling vacuum trucks, causing runaway and explosion. Two workers are killed and three more are injured. The Chemical Safety Board investigation of the BLSR accident reported "We have located government records of other incidents in Texas and elsewhere where diesel engines revved up just before deadly fires and explosions due to the presence of flammable petroleum vapors."

March 2005 – BP Texas City refinery (near Houston) has a fatal accident where a vapor cloud is ignited by an idling diesel pickup truck, reported to over-rev and backfire. The workers fled, unable to switch off the racing diesel engine. The flashback ignited the flammable cloud. Much of the refinery is damaged by the blast. Fifteen people died and 170 are injured. The burned Ford truck is shown at right.

Process Operations Safety: The What, Why, and How Behind Safe Petrochemical Plant Operations, First Edition. M. Darryl Yoes.

March 2007 – The Final Investigation Report (2.5.13) stated “This truck was parked about 25 ft from the blowdown drum, and several eye witnesses reported seeing or hearing the truck's engine over-revving when the vapor cloud reached it. Two witnesses saw the truck catch fire, followed shortly by the vapor cloud explosion.”

2007 – BP a draft of new internal standards with many safety requirements including an “air intake system shall have approved device to automatically stop engine if overspeed above governed speed occurs.”

June 2007 – OSHA Directive CPL 03-00-004. The Compliance Guidance explains “Motorized equipment, if not properly controlled, can be a potential ignition source…”

OSHA asks: “Does the employer have a safe work practice which it implements for motorized equipment to enter operating units and adjacent roadways?” If the answer is “no,” possible violations include 119(f)(4) if the employer either: (1) did not develop a safe work practice to control fire or explosion hazards when motorized equipment enter or travel (includes parking with the equipment running) on adjacent roadways to operating units that contain flammable or combustible materials, or (2) developed but failed to implement the safe work practice.

October 2007 – Explosion of leaking propane tanker truck in Tacoma, Washington, killed one and injured four.

January 2008 – Greely, Colorado. Three workers are killed and 17 injured when a truck sparks an oil tank.

August 2008 – Bristow, Oklahoma. One killed and three injured when a truck ignites a gas well.

September 2008 – Chickasha, Oklahoma. One killed when a truck ignites a gas pocket.

October 2008 – Calumet Refinery, Shreveport, Louisiana. The fire Chief reports that refinery tank vapors are ignited when a pickup truck ingests vapors from a nearby parked vacuum truck. The runaway soon led to an explosion. Two firefighters were slightly injured during the firefighting operations.

April 2010 – Offshore Rig Explosion at Macondo Well, Gulf of Mexico – BP Deepwater Horizon. ELeven killed. Diesel engine overspeed in the engine/generator room is considered one contributing factor and identified as the source of ignition of the escaping gas. The three engine intakes were at least 70 ft from the gas leak on the rig.

August 19, 2010 – Ackerly, Texas. One death and one serious burn injury results when a leak is ignited. The local Sheriff reported a fireball from the gas being ignited by the running vehicle.

July 2011 – North Dakota. A truck driver is seriously burned when his truck has diesel runaway and ignites the fumes of the oily salt water cargo transferring into a flow back tank. The driver sued the employer as they had not followed the safety policies of the site oil field operator requiring a positive inlet air shutoff.

August 2012 – Explosion at Chevron Refinery, Richmond, CA, USA – Don Holmstrom, investigator for US Chemical Safety Board overseeing the investigation noted, “… that a possible source of ignition was the idling rig's diesel engine” (Picture of the burned fire truck at right).

2012 – OSHA publishe Fact Sheet 3589 “Internal Combustion Engines as Ignition Sources.” This highlights Evaluation and Preventive Measures.

September 2012 – Cal OSHA adopt new safety standards for California oil and gas drilling and production facilities. These require approved automatic air intake shutoff valves for diesel engines within 50 ft of the well bore. The standard also extends to chemical production areas including refinery-type plants.

November 2012 – Wyoming OSHA introduce safety standards for oil and gas drilling and production facilities. It states "Emergency shut-down device(s) that will close off the combustion air shall be properly installed and identified on all diesel engines that are an integral part of the drilling rig or are operated as a stationary or mobile engine of a drilling rig within the radius of the rig anchors or within seventy five feet of the well bore, whichever is greater."

2014 – OSHA reports three workers killed and two hospitalized following an explosion during Frac flow-back operations. The source of ignition was the runaway of a diesel-powered light tower parked 12 ft from the trucks.

2016/2017 – Transport Canada and Canadian CSA introduce regulations requiring automatic shutoff valves on all trucks carrying hazardous cargo Class 2 and Class 3 including gasoline.

> *"as of 1 Jan 2016, diesel engines on highway tanks and portable tanks containing Dangerous Goods of primary Class 3, or subsidiary Class 3, and being used during loading or off-loading shall be equipped with an automatic engine air intake shut off device that will prevent engine runaway in case of exposure to flammable vapors. The device shall activate automatically if engine runaway is detected and remain activated until manually reset."*

Appendix O

Potential Startup Issues – M. Darry Yoes's "Worry List"

This list is based on actual events which have occurred during startup.

- **Failure to adequately purge any remaining fuel from the furnace firebox prior to lighting pilots or burners** or failure to purge furnace when relighting burners following a furnace trip (Inadequate procedures or failure to follow procedures).
  - The operator was killed attempting to restart a process unit furnace after the unit tripped due to a power interruption. US OSHA report indicated that proper furnace purging procedures were not performed during the attempted restart (November 2003). ***ACTUAL EVENT*** *Multiple other examples are available.*
  - Never attempt to light a pilot or burner without first purging the firebox and never light a burner from another burner or from a hot wall **(CRITICAL).**
- **Operating a furnace in a flooded (fuel-rich) state** (inadequate procedures or failure to follow procedures).
  - The gasoline hydrotreater furnace flooded and exploded during an instrument check of the O_2 analyzer when computer console supervisor (CSS) failed to place the instrument on bypass (May 2017). ***ACTUAL EVENT*** *Multiple other examples are available.*
  - Ensure field and console operators are familiar with indications of furnace flooding and appropriate response actions **(CRITICAL).**
  - Ensure the burner air registers are closed if the burner is turned off.
- **Failure to follow general furnace operating procedures** (inadequate procedures or failure to follow procedures).
  - The fired steam boiler exploded due to an unauthorized bypass method used to start the boiler. The two fuel gas shutdown valves were in the bypass mode with the control valve in the open position allowing the uncontrolled gas flow into the boiler. Two operators were killed in the explosion (December 2000). ***ACTUAL EVENT***
  - Recognize the high frequency of emergency events (fires and explosions) associated with furnaces and fired boilers. Pay special attention to furnace operations, furnace procedure compliance, and operator training on furnace operations.
  - The emergency shutdown device (ESD) system should never be bypassed except for testing and maintenance and then only with mitigation procedures and authorization for bypass. Ensure an effective control of defeat process is in place and is enforced **(CRITICAL).**
  - Ensure that all critical instruments and shutdown systems are checked before startup.
  - Enforce frequent operator rounds to monitor furnace operations and critical parameters.
  - Frequent tabletop drills and "what-if discussions."
  - Closely monitor all critical flows/temperatures during startup/shut-down situations.
  - If all burners and pilots are extinguished, repurge the firebox before relighting any pilots or burners.
  - Operating above the maximum allowable Tube Metal Thermocouples (TMTs) will greatly reduce tube life.
  - Timely problem-solving is needed when a sudden change in TMT occurs.
  - Never ignore alarms indicating loss of pass flow. Prompt significant reduction in firing should accompany the loss of pass flow **(CRITICAL)**.
  - Be aware of the potential for fires in air preheaters.
  - Install pilot and fuel gas blinds promptly after a furnace shuts down.
  - Furnace entry is governed by the same rules as other confined spaces.
  - Vertical tube furnaces require special testing near tubes needing hot work.
  - Special precautions are needed for nonroutine jobs on furnace structures.

Process Operations Safety: The What, Why, and How Behind Safe Petrochemical Plant Operations, First Edition. M. Darryl Yoes.

- **Failure to control the tower/vessel liquid level,** allowing the tower to overfill, or blow through hot/light hydrocarbons from tower bottoms into downstream equipment/tanks (failure to verify instrumentation ready for startup)
  - Stabilized naphtha was sent from debutanizer bottoms to a floating roof tank during fluid catalytic cracking unit (FCCU) startup. The explosion caused major damage to the atmospheric crude unit and resulted in six additional storage tank fires. Nine fatalities, seven storage tanks were destroyed, and major damage to the crude unit (August 2003). ***ACTUAL EVENT****. Other examples are available.*
  - Overfilling of process vessel leading to loss of containment/explosion (inadequate procedures or failure to follow procedures, failure to verify instrumentation ready for startup, and failure of shift turnover) (March 2005) (15 fatalities). ***ACTUAL EVENT****. Other examples are available.*
  - For towers with hot or light materials, **ensure tower bottoms level controls *AND* the low/high-level alarms are verified** as fully functional before startup **(CRITICAL).**
  - Failure to **identify safe upper and lower operating limits** or failure to enforce compliance with operating limits (inadequate procedures) **(CRITICAL).**
- **Allowing hydrocarbons to enter a process vessel that is not air free** (improper air-freeing procedure or inadequate separation of hydrocarbons and air)
  - Internal detonation in the Unicracker hydrogen supply filter system as the system was being returned to service following a routine filter change. The company safety manager reported, "We had a vessel in which there was an accidental mixing of air and hydrogen." ***ACTUAL EVENT*** (September 2001)
  - Ensure **checklist to verify unit and equipment are adequately purged of air before startup** or bringing equipment into service **(CRITICAL).**
- **Failure of instrumentation or unit shutdown systems** during startup or operating with instrumentation bypassed (i.e., furnace trip system, tower bottoms level control, tank, or process vessel high-level alarm)
  - Inadequate procedure or failure to ensure level, flow, pressure, and temperature controllers are working correctly and readiness for startup.
  - Failure to verify instrument tubing has been connected and is verified as fully functional in the field and console.
- **Starting up with critical alarms or shutdown systems bypassed** or out of service (inadequate procedures or failure to follow procedures)
  - Failure to require control and mitigation of instrumentation bypass (inadequate control of defeat).
- **Failure to conduct adequate shift turnover** during startup activity (inadequate procedure/policy or failure to follow procedure)
- **Failure to follow LPG vessel water draw procedures** resulting in potential auto-refrigeration and freeze-up of the water draw valves. Potential large vapor release, explosion, fire, and resulting LPG vessel BLEVE (inadequate procedure or failure to follow the procedure)
  - Operators are attempting to collect propane samples from the bottom of the sphere water draw valve. The valves froze open due to auto-refrigeration of propane; operators were unable to close the valves. The vapor cloud ignited after 35 minutes, and 5 spheres BLEVE'd. Eighteen fatalities, including 12 firefighters, and 81 injured (January 1966). ***ACTUAL EVENT***
  - Procedures should specify "Always open the upstream water draw valve fully and throttle on the downstream valve. Never throttle with the upstream valve or with both valves" **(CRITICAL).**
- **Release of flammable hydrocarbons** to the atmosphere (failure to conduct rigorous equipment tightness test, vents or drains left open, and inadequate bolting or gaskets)
- **Free water left in process equipment** and later contacts hot hydrocarbons or flashes during the warm-up process (inadequate procedures or procedures not followed)
  - Major explosion and fire while starting up the FCCU. US OSHA determined the cause to be a failure to adequately drain water from the slurry settling drum during startup. The drum was overpressured and failed catastrophically when hot slurry contacted the accumulated water in the vessel. The resulting explosion and fire resulted in six operator fatalities (March 1991).
  - The procedure should specify a cold flush before startup, including draining water from low points and dead legs, flushing the control valve bypasses, and changing over pumps during cold flush – run spare pumps **(CRITICAL).**
- **Thermal expansion of trapped liquids** in process lines, valves, exchangers, vessels, or other process equipment where there are no thermal relief valves or pressure relief.

  - Overpressure and rupture and separation of a 6″ LPG pipeline, 20+ miles long (32 km) resulted in a catastrophic refinery explosion and multiple tank and piping fires. The incident resulted in two fatalities and extended refinery and chemical plant downtime (December 1989). ***ACTUAL EVENT***
  - Ensure potential overpressure of piping and other equipment is considered, and overpressure protection is provided, either by thermal relief valves or by procedures such as leaving process lines vented to other equipment (**CRITICAL**).
- **Brittle fracture of process equipment** during startup (inadequate procedures or failure to follow procedures).
  - Slugs of liquid LPG resulted in a process upset at a natural gas processing unit and the loss of lean oil in the LPG absorber tower. LPG flashing resulted in auto-refrigeration and very cold temperatures and brittle fracture of an exchanger. A major vapor release and fire resulted (September 1998). ***ACTUAL EVENT***
  - Brittle fracture occurs when vessels are operated while pressurized and at temperatures below the vessel's minimum design metal temperature (MDMT). For example, when LPG flashes inside a tower or vessel, it results in auto-refrigeration of the vessel or surrounding process equipment such as exchangers.
  - Ensure minimum and maximum operating limits or operating envelopes are in place and rigorously followed, especially during startup and other abnormal operations **(CRITICAL).**
- **Failure of alarms or failure to recognize and respond to alarms** (failure to verify alarm readiness for startup or failure to provide an effective alarm management policy or procedure). The lack of an effective alarm management procedure results in:
  - A major loss of containment occurred, resulting in an explosion and a fire and resulting in injury to 26 people, and caused major refinery damage. The incident followed an electrical storm and a refinery upset. It was reported that a critical alarm was missed when the two operators had to recognize, acknowledge, and correctly act on 275 alarms in the final 11 minutes before the explosion (July 1994). ***ACTUAL EVENT***
    - A "**Refinery Alarm Management Policy and Procedure**" should be in place and should address the following alarm management issues **(CRITICAL).**
    - **Alarm flooding:** "TOO MANY ALARMS" in a short period of time, distraction from normal duties, and stress for the operators. Important alarms may get missed inadvertently, potentially resulting in an incident.
    - Standing alarms are always there. A standing alarm hinders the operator from spotting the new emergency alarm; a standing alarm means no alarm reset; if the situation worsens, the operator will not be alarmed.
    - **Nuisance alarms:** a crying wolf is an alarm activated by an event other than the one intended.
    - **Chattering:** alarm going in and out of alarm state at very high frequency.
    - **Redundant:** they are all telling the same event, nothing new.
    - **Message, status, alert:** no action is required; eventually, operators lose trust in the alarm system.
    - **Disabled alarms:** hidden killers, poor alarm inhibition work process, poor shift handover.
    - Improper prioritized alarms.
- **Failure to ensure the unit is ready for** startup following new construction or modifications (failure to conduct the management of change, the prestart-up safety review [PSSR], the process hazard analysis [PHA], or failure to respond to PSSR results).
  - A major fire occurred at the Williams Olefins Plant due to the catastrophic failure of an exchanger due to thermal expansion. The exchanger had been accidentally blocked in due to valves that had been installed 12 years prior to the incident. The valves were not identified as a hazard in the MOC, PHA, or PSSR. The fire resulted in 2 fatalities and 167 injuries and the Olefins plant remained down for 18 months (13 June 2013). ***ACTUAL EVENT***
  - Ensure the key process safety management systems (MOC, PHA, and PSSR) are effectively used, and the results are documented and used to drive compliance and change in the organization. The Control of Defeat **(COD)** or an equivalent process should also be in place to control the defeat or disablement of safety-critical devices **(CRITICAL).**
- **Nonessential people** not removed from the unit during startup or other transient operations (inadequate procedures or failure to follow procedures).
  - Twenty-three lives were lost due to the explosion that occurred during the startup of the BP Isom Unit. These people were not involved in the unit startup (March 2005). ***ACTUAL EVENT***
  - Ensure all nonessential people are removed from the unit during startup procedures or other abnormal operations **(CRITICAL).**
- Ensure the refinery is conditioned to **question the fitness of any stream that is going to atmospheric storage tanks** (both the sender and the receiver).
  - Procedures should prevent water from entering storage tanks that are in "hot" service (this could result in a tank boilover).

- Procedures should also require the sender to routinely check and verify the fitness of the stream being sent to storage tanks (prevent light or hot streams from going to atmospheric tankage).
- Failure to designate **Safety Critical procedures** or to enforce safety critical procedures (inadequate procedures or failure to follow procedures).
  - The following recommendation was made following the explosion that occurred at a Lake Charles, LA, refinery when a small amount of free water was left in a process vessel, which later flashed and over-pressured the vessel, resulting in an explosion (six operators died in this incident) ***ACTUAL EVENT***
  - **US OSHA Recommendation:** "A feasible and useful method among others of correcting this hazard is to provide and strictly enforce the use of a detailed written startup procedure with signed and dated logs for specific task(s)" **(CRITICAL).**
- Making **modifications to the startup procedure** while the startup is in progress without the use of the management of the change process (inadequate procedure or failure to follow procedure).
- Individual steps in **procedures not in the correct sequence** or checklist format (inadequate procedures).
- **Blinds left in piping/vessels** or failure to adequately blind process vessels (no blind control procedure or failure to follow procedure).
- Ensure that fire water pumps and fire water systems are fully functional before startup, including all fire water deluge systems. **A petroleum terminal fire water system failed during a** major fire (December 2002). ***ACTUAL EVENT***
- Failure to define or enforce **fatigue prevention guidelines** (inadequate policy or failure to enforce policy) (major explosion during unit startup) (March 2005). ***ACTUAL EVENT***
- A flammable release from **temporary hoses or other temporary connections** that were made to support startup (inadequate procedure or failure to follow procedure)
  - When temporary hoses are used to support the startup, ensure **we use the proper hose for the service and temperature**, and only the proper hose.
  - **Disconnect and remove all temporary drain lines and hoses before startup** and do not leave hoses in structure or pipe racks (roll up and store properly) **(CRITICAL).**
- **Specific and detailed procedures should be in place for the commissioning of steam or condensate piping** to ensure that the steam is slowly introduced to the system, and condensate is continuously drained from the piping until the steam system has reached normal operating pressure. This is to prevent steam and liquid "hammer" from resulting in damage to the piping and valves.
- During startup, ensure **rigorous control of systems containing high concentrations of hydrogen sulfide**. Systems such as acid gas can contain as much as 98% H_2S, and a release can result in toxic concentrations for a considerable distance downwind.
- A significant release of H_2S resulted in the death of an employee and his spouse who attempted to locate him when he failed to return home (Gas Plant Water Flood Station, Odessa, Texas) October 2019, ***ACTUAL EVENT***
- Be aware of **last-minute changes, especially during startup**. Avoid pressure to take shortcuts or to not follow procedures or work practices such as MOC, PHA, and PSSR.
  - During the final stages of construction of a new Olefins plant, the operating crew requested the installation of an additional block valve between the high-pressure and low-pressure separator drums. This valve was installed without MOC, PHA, or PSSR and was left in the closed position. This resulted in the rupture of an exchanger in the low-pressure H_2 system and an explosion. UK chemical plant ***ACTUAL EVENT***

Other Issues Not Related to Unit or Refinery Startup:

- Reactor thermal temperature excursion and runaway
- Corrosion under insulation
- Proper metallurgy (positive metals identification or PMI)
- Pipe corrosion (high-temperature sulfidation, naphthenic acid, acidic corrosion, wet H_2S cracking, H_2/H_2S corrosion, etc.)
- Stress corrosion cracking
- High-temperature hydrogen attack (HTHA)
- Control of static electricity
- Combustible dust explosion
- Hazards of steam and hot condensate
- Electrical safety and work near overhead power lines

- Pneumatic pressure testing of process equipment
- Hot work around or near process sewers
- Pyrophoric materials in towers/vessels (especially packing and demister screens)
- Steam turbine overspeed testing hazards
- Storage tank high-level alarms and high-level shutdown trips
- Vacuum truck operations
- Pump and driver reverse overspeed (failed discharge check valve)
- Nitrogen asphyxiation hazards – especially during inert entry
- Hydrofluoric acid exposure hazards – toxic limits can be reached at a considerable distance downwind of a release
- Legionnaires disease – inadequately treated cooling towers may promote the growth of potentially infectious bacteria, including *Legionella pneumophila* (Legionnaire's disease)
- **Permit to work:** especially energy isolation, opening process equipment, and confined space entry
- Ethylene decomposition – ethylene can decompose exothermically/explosively in the absence of air, potentially resulting in equipment failure due to overheating and rapid pressure rise
- The potential catastrophic rupture of piping or vessels with "cold wall design". Cold wall design equipment is piping or vessels equipped with refractory lining on the equipment internals. The most significant hazard is the placement of external insulation on the pipe or vessel exterior, resulting in a rupture and loss of containment and a potential fire or explosion.

A Discussion Guide Used for an Actual Pre-startup Safety Meeting:

- Pace of activities will be controlled – no one is to be seen running or even walking too fast – remove the pressure of schedule from the workforce. Our success will be determined by the absence of any incidents, not by speed.
- EVERYONE is authorized and expected to stop any work or startup process if there is an unsafe condition or perceived unsafe condition.
- Safety policies and procedures are non-negotiable and must be complied with – regardless of any schedule delay!
- Supervision is highly visible and continuously involved in the startup process.
- ALL shifts will start with a quality safety meeting.
- Supervision is actively involved in shift relief, assuring all critical information is transferred.
- Any deviation from startup procedures or subcritical procedure requires notification and approval from the shift super.
- DO NOT proceed with questionable instruments or conditions – call out support or get help.
- Alarm, operating envelope summaries must be reviewed by the unit engineers with the and BTL as a second set of eyes. DO NOT overlook and account for "startup."
- Shift super – field verification, FLS and CSS, and field operator alignment – procedure use being followed to the letter, be on site for critical startups – check pace of work, safety awareness, task overload, distraction (help folks avoid tunnel vision).
- The control center will be in full startup mode. Distractions to the console are to be minimized, and no unnecessary personnel are allowed. The focus shall be on procedure review and walkthroughs with the field, instrument readings being field verified, sight glasses verified as clear and readable, and thorough shift handover and documentation.
- **Control Center:** communication between units is critical; operators are expected to provide frequent updates to the consoles on startup activities and potential impacts on others.
- A written signed-off procedure is required for all startup activities.
- All procedures are walked through before execution – all team members must understand their responsibilities.
- Procedures will be signed off as the steps are completed – not at the end of the shift.

Index

Process Operations Safety: The What, Why, and How Behind Safe Petrochemical Plant Operations, First Edition. M. Darryl Yoes.

C

e

f

g

h

m

n

q

r

t

www.ingramcontent.com/pod-product-compliance
Lightning Source LLC
Chambersburg PA
CBHW080849050125
19875CB00005B/36

9 781394 271931